New Trends in Lithium Niobate: From Bulk to Nanocrystals

New Trends in Lithium Niobate: From Bulk to Nanocrystals

Editors

Gábor Corradi
László Kovács

MDPI • Basel • Beijing • Wuhan • Barcelona • Belgrade • Manchester • Tokyo • Cluj • Tianjin

Editors

Gábor Corradi
Crystal Physics
Wigner Research Centre
for Physics
Budapest
Hungary

László Kovács
Crystal Physics
Wigner Research Centre
for Physics
Budapest
Hungary

Editorial Office
MDPI
St. Alban-Anlage 66
4052 Basel, Switzerland

This is a reprint of articles from the Special Issue published online in the open access journal *Crystals* (ISSN 2073-4352) (available at: www.mdpi.com/journal/crystals/special_issues/Lithium_Niobate_Nanocrystals).

For citation purposes, cite each article independently as indicated on the article page online and as indicated below:

LastName, A.A.; LastName, B.B.; LastName, C.C. Article Title. *Journal Name* **Year**, *Volume Number*, Page Range.

ISBN 978-3-0365-3340-7 (Hbk)
ISBN 978-3-0365-3339-1 (PDF)

Cover image courtesy of László Kovács

Contents

crystals

Editorial

New Trends in Lithium Niobate: From Bulk to Nanocrystals

László Kovács * and Gábor Corradi

Wigner Research Centre for Physics, Institute for Solid State Physics and Optics, Konkoly-Thege M. út 29-33, H-1121 Budapest, Hungary; corradi.gabor@wigner.hu
* Correspondence: kovacs.laszlo@wigner.hu

Abstract: The recent Special Issue on lithium niobate ($LiNbO_3$) is dedicated to Prof. Schirmer and his topics and contains nineteen papers, out of which seven review various aspects of intrinsic and extrinsic defects in single crystals, thin films, and powdered phases; six present brand-new results of basic research, including two papers on $Li(Nb,Ta)O_3$ mixed crystals; and the remaining six are related to various optical and/or thin film applications.

Keywords: lithium niobate; bulk crystals; thin films; nanocrystals

Citation: Kovács, L.; Corradi, G. New Trends in Lithium Niobate: From Bulk to Nanocrystals. *Crystals* **2021**, *11*, 1356. https://doi.org/10.3390/cryst11111356

Received: 5 November 2021
Accepted: 5 November 2021
Published: 8 November 2021

Publisher's Note: MDPI stays neutral with regard to jurisdictional claims in published maps and institutional affiliations.

This Special Issue was originally planned to celebrate the forthcoming 85th anniversary of Prof. Ortwin F. Schirmer, but his unexpected passing in 2020 made the occasion painfully more personal. A considerable number of his colleagues, students, and observers and followers of his fundamental lifework responded to the call, which resulted in a clarifying new synopsis of his topics.

1. Reviews on Defects in $LiNbO_3$

A general review on lithium niobate ($LiNbO_3$, LN) single crystals, as well as powders, which have been given a new emphasis, is presented in two parts by Sánchez-Dena et al. [1,2]. They discuss the crystal structure, the methods for the determination of chemical composition, the defect structures induced by the incorporation of hydroxyl ions, and the so-called optical damage resistant ions (e.g., Mg^{2+}), also discussing the origins of ferroelectricity, together with its possible association with ferromagnetism in $LiNbO_3$ doped with paramagnetic *3d* cations.

Extrinsic and intrinsic paramagnetic point defects in LN and lithium tantalate ($LiTaO_3$, LT) crystals have been reviewed by Grachev and Malovichko [3] based on electron paramagnetic resonance (EPR) and electron nuclear double resonance (ENDOR) studies. They focus on transition metal and rare-earth ions and give detailed information about the structures of impurity defects in LN crystals with various Li/Nb composition ratios (charge state, point symmetry, hyperfine interactions with neighboring nuclei, with corollaries for incorporation sites and charge compensation mechanisms).

A similar detailed review on defects in LN and LT crystals studied by nuclear methods has been presented by Kling and Marques [4]. They focus on ion beam methods under channeling conditions for the direct determination of the lattice site of dopants and intrinsic defects. Results of perturbed angular correlation measurements probing the local environment of dopants in the host lattice are also included in the analysis, yielding independent and complementary information.

The editors also initiated the clarification of the highly disputed issue of defects in the anionic sublattice of LN, but this only resulted in a repetition of earlier unconfirmed or refuted interpretations based on a superficial comparison disregarding the deep constitutional differences between LN and $SrTiO_3$ such as the extreme instability of the Li positions compared to the rigid Sr sublattice [5]. This forced the editors to present a review of their own on defect generation and annealing mechanisms in LN [6]. The oxygen sublattice of the LN bulk was demonstrated to play an essentially passive role, with oxygen loss and

Li_2O segregation mostly taking place in external or internal surface layers of a thickness of a few nanometers. Processes during thermo- and mechano-chemical treatments and irradiations of various types in LN were instead shown to be governed by as-grown and freshly generated Nb_{Li} antisite defects as traps for small polarons and bipolarons, while highly mobile lithium vacancies, also acting as hole traps, were shown to provide flexible charge compensation required for the stability of the defects just formed. The close relationship between $LiNbO_3$ and the Li battery materials $LiNb_3O_8$ and Li_3NbO_4 has also been pointed out.

A concise gap-filling review on the epitaxial growth of $LiNbO_3$ thin films has been published by Zivasatienraj et al. [7]. This review demonstrates that the highest crystalline quality obtainable by halide-based molecular beam epitaxy is comparable to that of bulk $LiNbO_3$ crystals while admitting that the slow growing rate presently limits most practical applications requiring substantial thicknesses. Research studies using the competing LNOI technology (LN On Insulator) based on ion-sliced LN films bonded on insulator substrates are addressed in the final part of this editorial.

2. Basic Research on Small Polarons and Optical, Electronic, and Acoustic Properties

Small polarons are the main topic of three recent research papers [8–10] published in this Special Issue, while the other papers are also at least indirectly concerned with transfers of polaronic charges or Li ions. Messerschmidt et al. [8] studied the laser-pulse-induced transient absorption of Nb^{4+}-type free polarons in heavily Mg-doped LN crystals. Their decay, attributed to polaron hopping leading to recombination with O^--type hole polarons, was found to be strongly temperature dependent between 45 K and 225 K. The Arrhenius-type behavior of the decay rate at high temperature and the non-Arrhenius one at low temperature could be coherently described by earlier theories of one of the co-authors (D. Emin), yielding good agreement for reasonable values of the hopping activation energy and the material's characteristic phonon frequency.

Using the same theoretical framework, Vittadello et al. [9] carried out Monte Carlo simulations of similar small-polaron-hopping processes for Fe-doped LN crystals, where, in addition to deep $Fe_{Li}{}^{3+}$ traps, shallow traps, also represented by $Nb_{Li}{}^{5+}$ antisite defects on Li sites, have to be taken into account. Decay regimes for fast direct trapping on Fe^{3+} for polarons generated in the vicinity of the dopant and slow hopping-governed trapping for others could be discerned, with the transition between both regimes depending on the concentrations of the traps and the temperature.

Schmidt et al. [10] further developed their first-principles calculations on intrinsic free and trapped-electron small-polaron and bipolaron states based on density-functional theory. Improved calculation of the electron–electron and electron–hole interactions allowed reliable comparisons with optical absorption spectra and other properties of various selected defect species. Special attention was given to symmetry-breaking distortions that further lower the total energy.

Luminescence of $LiNbO_3$ crystals single-doped with various rare-earth ions (Sm^{3+}, Dy^{3+} or Tb^{3+}) and $LiTaO_3$ doped only with Tb^{3+} was studied in a wide temperature range by Lisiecki et al. [11]. Following an initial temperature-independent stage, the luminescence lifetimes showed a steep decrease with increasing temperature with the onset at about 700, 600, and 150 K for Sm^{3+}, Dy^{3+}, and Tb^{3+} ions in LN, respectively, which was interpreted by a phenomenological temperature-dependent charge-transfer model. It was concluded that LN:Sm^{3+} is suitable as an optical sensor between 500–750 K, while LN:Dy^{3+} offers the highest sensitivity between 300–400 K.

Electrical conductivity and acoustic loss of single crystalline $Li(Nb,Ta)O_3$ solid solutions (LNT) were studied as a function of temperature and compared to those in LN and LT crystals by Suhak et al. [12]. The dominant transport mechanism in LNT in the range of 400–700 °C was identified as lithium-ion migration via lithium vacancies, as found in LN. It was also shown that the acoustic loss in LNT strongly increases at elevated temperatures, originating from a conductivity-related relaxation mechanism. LNT bulk acoustic resonators were found to exhibit significantly lower loss as compared to LN crystals.

Nanocrystalline $LiNb_{1-x}Ta_xO_3$ samples with x between 0 and 1 were synthesized by high-energy ball milling and subsequent high-temperature annealing by Vasylechko et al. [13]. The milling and annealing parameters were optimized to obtain single-phase LNT nanopowders checked by X-ray diffraction. In the Raman spectra of non-optimal milling and annealing runs, bands typical for the $Li(Nb,Ta)_3O_8$, Nb_2O_5, and/or Ta_2O_5 parasitic phases were observed. The Arrhenius-plots of the electrical conductivity measurements performed up to 820 °C revealed activation energies between 0.86 and 1.09 eV.

3. Applications of $LiNbO_3$

Among the contributions dealing with nonlinear optical applications of LN and LT in this Special Issue, papers using both traditional bulk crystals [14–16] and thin films on LNOI platform [17–19] can be found. All are concerned with internal electric fields resulting in nonlinear response, another topic of Prof. Schirmer.

In the paper by Zhao et al. [14], a laser-damage-resistant MgO-doped LN crystal is used as a nonlinear optical mirror to realize a high-power wavelength-tunable mode-locked picosecond $Yb:CaGdAlO_4$ laser. The tunable range was between 1039 and 1062 nm, the maximum output power reached at 1049 nm was 1.46 W, and the output pulse duration was about 8 ps, with a repetition rate and bandwidth of 115.5 MHz and 1.7 nm, respectively.

Due to its high second-order nonlinear-optical coefficients, high transparency, and large Pockels coefficients, LN is an excellent material for whispering gallery resonators (WGR), allowing for optical parametric oscillation and frequency comb generation. Minet et al. [15] studied the average strength of the electric field along the z-direction inside the region of the optical mode for different configurations and geometries of the LN WGR using the finite element method. Their simulations may be useful for the optimal design of similar resonators in future applications.

The generation of entangled photon pairs based on spontaneous parametric down-conversion was investigated theoretically and numerically by Kim et al. [16] in undoped and Mg-doped LN for non-poled and periodically poled cases. The spectral positions of the generated photon pairs fall into the mid-infrared range and are expected to find important applications for free-space quantum communications, spectroscopy, and high-sensitivity metrology.

Taking into account the enormous potential of LN thin films for integrated quantum photonics produced by the LNOI technology and comparable only to silicon-based photonics, progress in this field is of utmost interest. The contribution of Gainutdinov and Volk [17] addresses the problem of understanding and influencing the unique properties of nanodomains written by AFM-tip voltages in LN films formed on insulator structures where the electrical conductivity is essentially due to polaronic charges trapped by nanodomain walls.

Various nanometer-to-micrometer-scale imaging techniques required for optimizing performance parameters of periodically poled thin-film devices on LNOI platforms were compared by Reitzig et al. [18]. Piezoresponse force microscopy (PFM), second-harmonic generation (SHG), and Raman spectroscopy (RS) were found to monitor differently relevant sample properties. For standard imaging, SHG was found to be best-suited, in particular when investigating the domain poling process in x-cut thin films.

Modeling of a thin film device on lithium tantalate has been carried out by Yao et al. [19] by using a full-vectorial finite difference method. Design, simulation, analysis, and optimization of optical microring resonators on a lithium tantalate on insulator (LTOI) platform operating near 1.5 μm has been presented.

Conflicts of Interest: The authors declare no conflict of interest.

References

1. Sánchez-Dena, O.; Fierro-Ruiz, C.D.; Villalobos-Mendoza, S.D.; Carrillo Flores, D.M.; Elizalde-Galindo, J.T.; Farías, R. Lithium Niobate Single Crystals and Powders Reviewed—Part I. *Crystals* **2020**, *10*, 973. [CrossRef]
2. Sánchez-Dena, O.; Villalobos-Mendoza, S.D.; Farías, R.; Fierro-Ruiz, C.D. Lithium Niobate Single Crystals and Powders Reviewed—Part II. *Crystals* **2020**, *10*, 990. [CrossRef]
3. Grachev, V.G.; Malovichko, G.I. Structures of Impurity Defects in Lithium Niobate and Tantalate Derived from Electron Paramagnetic and Electron Nuclear Double Resonance Data. *Crystals* **2021**, *11*, 339. [CrossRef]
4. Kling, A.; Marques, J.G. Unveiling the Defect Structure of Lithium Niobate with Nuclear Methods. *Crystals* **2021**, *11*, 501. [CrossRef]
5. Crespillo, M.L.; Graham, J.T.; Agulló-López, F.; Zhang, Y.; Weber, W.J. Real-Time Identification of Oxygen Vacancy Centers in $LiNbO_3$ and $SrTiO_3$ during Irradiation with High Energy Particles. *Crystals* **2021**, *11*, 315. [CrossRef]
6. Corradi, G.; Kovács, L. 'Horror Vacui' in the Oxygen Sublattice of Lithium Niobate Made Affordable by Cationic Flexibility. *Crystals* **2021**, *11*, 764. [CrossRef]
7. Zivasatienraj, B.; Tellekamp, M.B.; Doolittle, W.A. Epitaxy of $LiNbO_3$: Historical Challenges and Recent Success. *Crystals* **2021**, *11*, 397. [CrossRef]
8. Messerschmidt, S.; Krampf, A.; Vittadello, L.; Imlau, M.; Nörenberg, T.; Eng, L.M.; Emin, D. Small-Polaron Hopping and Low-Temperature (45–225 K) Photo-Induced Transient Absorption in Magnesium-Doped Lithium Niobate. *Crystals* **2020**, *10*, 809. [CrossRef]
9. Vittadello, L.; Guilbert, L.; Fedorenko, S.; Bazzan, M. Polaron Trapping and Migration in Iron-Doped Lithium Niobate. *Crystals* **2021**, *11*, 302. [CrossRef]
10. Schmidt, F.; Kozub, A.L.; Gerstmann, U.; Schmidt, W.G.; Schindlmayr, A. Electron Polarons in Lithium Niobate: Charge Localization, Lattice Deformation, and Optical Response. *Crystals* **2021**, *11*, 542. [CrossRef]
11. Lisiecki, R.; Macalik, B.; Kowalski, R.; Komar, J.; Ryba-Romanowski, W. Effect of Temperature on Luminescence of $LiNbO_3$ Crystals Single-Doped with Sm^{3+}, Tb^{3+}, or Dy^{3+} Ions. *Crystals* **2020**, *10*, 1034. [CrossRef]
12. Suhak, Y.; Roshchupkin, D.; Redkin, B.; Kabir, A.; Jerliu, B.; Ganschow, S.; Fritze, H. Correlation of Electrical Properties and Acoustic Loss in Single Crystalline Lithium Niobate-Tantalate Solid Solutions at Elevated Temperatures. *Crystals* **2021**, *11*, 398. [CrossRef]
13. Vasylechko, L.; Sydorchuk, V.; Lakhnik, A.; Suhak, Y.; Wlodarczyk, D.; Hurskyy, S.; Yakhnevych, U.; Zhydachevskyy, Y.; Sugak, D.; Syvorotka, I.I.; et al. Investigations of $LiNb_{1-x}Ta_xO_3$ Nanopowders Obtained with Mechanochemical Method. *Crystals* **2021**, *11*, 755. [CrossRef]
14. Zhao, L.; Tong, L.; Cai, F.; Yuan, Y.; Cai, Y. Wavelength-Tunable Nonlinear Mirror Mode-Locked Laser Based on MgO-Doped Lithium Niobate. *Crystals* **2020**, *10*, 861. [CrossRef]
15. Minet, Y.; Zappe, H.; Breunig, I.; Buse, K. Electro-Optic Control of Lithium Niobate Bulk Whispering Gallery Resonators: Analysis of the Distribution of Externally Applied Electric Fields. *Crystals* **2021**, *11*, 298. [CrossRef]
16. Kim, I.; Lee, D.; Lee, K.J. Study of Type II SPDC in Lithium Niobate for High Spectral Purity Photon Pair Generation. *Crystals* **2021**, *11*, 406. [CrossRef]
17. Gainutdinov, R.; Volk, T. Effects of the Domain Wall Conductivity on the Domain Formation under AFM-Tip Voltages in Ion-Sliced $LiNbO_3$ Films. *Crystals* **2020**, *10*, 1160. [CrossRef]
18. Reitzig, S.; Rüsing, M.; Zhao, J.; Kirbus, B.; Mookherjea, S.; Eng, L.M. "Seeing Is Believing"—In-Depth Analysis by Co-Imaging of Periodically-Poled X-Cut Lithium Niobate Thin Films. *Crystals* **2021**, *11*, 288. [CrossRef]
19. Yao, S.; Han, H.; Jiang, S.; Xiang, B.; Chai, G.; Ruan, S. Design, Simulation, and Analysis of Optical Microring Resonators in Lithium Tantalate on Insulator. *Crystals* **2021**, *11*, 480. [CrossRef]

crystals

MDPI

Review

Lithium Niobate Single Crystals and Powders Reviewed—Part I

Oswaldo Sánchez-Dena [1,*], Cesar David Fierro-Ruiz [2], Sergio David Villalobos-Mendoza [1], Diana María Carrillo Flores [1], José Trinidad Elizalde-Galindo [1] and Rurik Farías [1,*]

1 Instituto de Ingeniería y Tecnología, Universidad Autónoma de Ciudad Juárez, Av. Del Charro 450 Norte, Ciudad Juárez 32310, Mexico; sergio.d.v.m@hotmail.com (S.D.V.-M.); diana.carrillo@uacj.mx (D.M.C.F.); jose.elizalde@uacj.mx (J.T.E.-G.)

2 Departamento de Mecatrónica y Energías Renovables, Universidad Tecnológica de Ciudad Juárez, Avenida Universidad Tecnológica 3051, Colonia Lote Bravo II, Ciudad Juárez 32695, Mexico; cesar_fierro@utcj.edu.mx

* Correspondence: ossdena@gmail.com (O.S.-D.); rurik.farias@uacj.mx (R.F.); Tel.: +52-656-304-8486 (R.F.)

Received: 1 September 2020; Accepted: 20 October 2020; Published: 27 October 2020

check for updates

Abstract: A review of lithium niobate single crystals and polycrystals in the form of powders has been prepared. Both the classical and recent literature on this topic are revisited. It is composed of two parts with sections. The current part discusses the earliest developments in this field. It treats in detail the basic concepts, the crystal structure, some of the established indirect methods to determine the chemical composition, and the main mechanisms that lead to the manifestation of ferroelectricity. Emphasis has been put on the powdered version of this material: methods of synthesis, the accurate determination of its chemical composition, and its role in new and potential applications are discussed. Historical remarks can be found scattered throughout this contribution. Particularly, an old conception of the crystal structure thought as a derivative structure from one of higher symmetry by generalized distortion is here revived.

Keywords: lithium niobate; lithium tantalate; crystal structure; chemical composition; ferroelectrics; second harmonic generation; lead-free piezoelectrics

1. Introduction

Aside from being an important ferroelectric material, lithium niobate ($LiNbO_3$, LN) is linked to a gamut of pronounced physical properties. Practically little less than a myriad of technologically significant usages unfold from it in single crystal form. The keenest advocates of applications may easily list its versatility: "acoustic wave transducers, acoustic delay lines, acoustic filters, optical amplitude modulators, optical phase modulators, second-harmonic generation, Q-switches, beam deflectors, phase conjugators, dielectric waveguides, memory elements, holographic data processing devices", among others (the cornerstone paper by Weis and Gaylord is herein quoted) [1]. Although it is an entirely synthetic material, back in the late 1970s, LN was important for the development of surface acoustic wave (SAW) devices; it took second place only to quartz in the market of single-crystalline piezoelectrics [2]. Today, nearly 70% of the radio-frequency filters based on SAW, are fabricated on LN single crystals [3,4]. As for the future, a glimpse of the role of this material to achieve practical integrated on-chip micro-photonic devices might be grasped as it is already considered the cornerstone in photonics, just as silicon has been for electronics [5,6]. Indeed, for little more than a decade, LN has been dubbed as 'the silicon of photonics', a statement which seems to be more precise nowadays given the breathtaking results recently reported by Zhang et al. [7]. Such implications have been confirmed and updated by the same group [8]. It is "the workhorse material in optical communication applications", it has been elsewhere stated [9]. Hence, credit is here given to Pang et al., having stated that LN is one of the most favorite multifunctional crystals [10]: other important applications (potential and already achieved) are to be succinctly described all over the review.

The main goal of this review is to expose in a convenient and ordered way the primary elements necessary to explicitly state/understand the scientific paradigm shared by the community advocated to investigate LN [11]. In short, and on a first level, it relies upon solid state physics, chemistry, and crystallography: the crystalline state of matter is very stable, ruled simply in terms of electric charge stabilization and bonding considerations, and the ascribed physical properties to a given crystal are deeply related to its chemical composition and crystal structure. Secondly: on regular circumstances of growth, LN is a congruent melt crystal whose chemical composition at room temperature (and for a broad range of higher temperatures) is nothing more than a macrostate of its crystal structure, the latter being characterized by the presence of intrinsic defect clusters composed of Li vacancies and Nb_{Li} antisites. However, solid solutions can adopt a continuous range of chemical compositions. Upon different mechanisms, a physical instance close enough to the stoichiometric point can be reached, among which, doping with various kinds of elements (filling the voids) is aesthetic.

Some terms and concepts are to be revisited in accomplishing this goal, such as how the crystal structure is introduced in recent literature (Section 3). In this respect, we reintroduce the idea once developed by famous crystallographer Helen Megaw on conceiving the structure of ferroelectrics as variants of the family of distorted perovskites [12–14]. To our judgment (and best of our knowledge), the best introduction to the LN crystal structure has been made by Räuber [2], who acknowledged the work by Megaw, so that it builds it up, step by step, starting from a reparameterization trick of the atomic positions introduced by the latter. Also, some of the indirect methods to determine the chemical composition accurately (CC) of LN powders (LNPws) are discussed; important references are highlighted for the case of single crystals (Section 4). On the other hand, ferroelectricity is often associated with LN: it is both historically and practically relevant. In this review, a succinct recount is also given on how the main ideas to explain the mechanisms behind ferroelectricity have changed over time (Section 5). In this 2020, we celebrate that a century ago, Ph. D. student Joseph Valasek gestated the field of ferroelectricity by presenting his results on the Rochelle salt at the American Physical Society meeting [15,16].

Further insight could be obtained by adopting the ideas by Megaw. For example, the slight differences between the stoichiometric (ST) composition and near-ST (nST) compositions could be studied on mathematical grounds (theoretically or by simulation) in terms of a continuous set of distorted structures and local symmetry variations parametrized by a kind-of-ordering parameter of small change. It would hardly be explained in terms of the CC itself: as it is the case of the congruent composition, which is disputed to fall in between 48.38 and 48.60 mol % of the Li content [17], the ST point is probably yet undetermined as well (Section 4). This subject is discussed along with the basic principle of adjusting the stoichiometry by elemental doping. Lately, the mainstream idea of reverting the nominal off-stoichiometry of LN by an effective elimination of site and antisite point defects through doping with metal ions has permeated increasingly. It implies nothing else than the controlled interchange of intrinsic and extrinsic defects, and although it seems quite natural or intuitive, such conception is revisited in Part II of this paper. The potential use of nanocrystalline LN powders (LNPws) in a wide range of applications has been recently highlighted [18,19]. These will be revisited and further extended within the present contribution. Remarkably, the role of LNPws towards effectively reducing the amount of lead in commercial piezoelectrics based on complex ternary and four-component systems is treated, among others (Section 4).

2. Basic Notions and Early Developments

In the words of Volk and Wöhlecke, authors of the most comprehensive monograph written up to date on LN, this type of crystals are "colorless, chemically stable and insoluble in water and organic solvents, and have high melting points" [20]. By colorless, it is meant a high transparency window of the order of <0.1 cm^{-1} for radiation between 400 and 5000 nm [21]. This, combined with the fact LN crystals are non-centrosymmetric and have a high nonlinear coefficient, makes them suitable materials for the second harmonic generation and optical parametric generation in this range of the electromagnetic spectrum. High melting points along with a high Curie temperature (T_C~1200 °C) translates into the exhibition of a spontaneous polarization (ferroelectricity) in a broad range of temperatures. In its classic text, Räuber states, on the other hand, that LN "is a fairly useless material from a chemical point of view, its only application being as a starting material for crystal growth" [2].

The $LiNbO_3$ phase was first described by Zhachariasen in 1928 [22], who is considered one of the giants in crystal structure analysis, alongside Pauling and Belov [23]. The first-ever synthesis of LN can be traced back to 1937 [24,25]: small crystals having the shape of small prisms were obtained, although this information could not be corroborated in the present bibliographical survey. Remeika is credited with having grown large-single crystals for the first time a little before 1949 and, in this year, the same and Matthias published the first report on the ferroelectric behavior of such crystals [26,27]. In 1965, Ballman—and Fedulov et al., this is not often mentioned in recent literature—managed to grow larger crystals using the Czochralski method [27–29]. Since then, a parallel growth in scientific and engineering activities took place, remarkably at the Bell Laboratories, and ever since this pioneering work (published in 1966, five papers), details on the structure of this material are known to a high precision [30–34]. Particularly, [32–34] constitute a paramount contribution to the whole scientific community involved in the study of LN: they form the basis of all further discussions on the LN structure.

Solid solutions (SSs) of LN are characterized by showing a compositional change, parametrized by the ratio R = {Li}/{Nb} (the brackets denoting concentration in mol %). Single crystals have an intrinsic nature to deviate from the stoichiometric (ST) point R = 1. It can be readily seen in Figure 1, showing the binary phase diagram that describes the Li_2O—Nb_2O_5 system. Another important point—or perhaps the most important—is the congruent (CG) one R = 0.944, for which the highest uniformity of properties is attained. This point is so-called CG because, as shown in Figure 1, within the LN phase, an inflection on the liquidus–solidus curve takes place at this point, implying preservation of the CC in the process of passing from the Li_2O:Nb_2O_5 melt to the growing crystal. Said differently, CG means $R_{melt(liq)} = R_{crystal(sol)} = 0.944$. For off-congruent melts, the as-grown crystals are compositionally non-uniform, particularly along the growth axis, due to slight variations in the melt and crystal compositions in the growth process [20]. Different two-component phases delimit the LN phase, homogeneous mixtures of LN, and a SS of a secondary phase: methaniobates Li_3NbO_4 and $LiNb_3O_8$. Because of the preference of single or pure phase compounds over those with a mixture of phases, added to the fact that sub-CG compositions of LN present poor physical properties, these have been elsewhere referred to as 'parasitical' [35]. Interestingly, in the presence of Li_3NbO_4 ($LiNb_3O_8$), the existent SS of single-phase LN is of fixed ST (most defective) composition: in these cases, the formation of the parasitical phases compensates the excess (deficiency) of Li. Figure 1 is redrawn from the publications by Volk and Wöhlecke [20], and Hatano et al. [36].

In general, under regular growth circumstances, high-quality ST crystals are not obtained even for a $R_{melt(liq)} \geq 1$. In fact, nST crystals were available until 1992, year in which three independent methods for this purpose were reported, entailing a modification to the growth method combined with Li enrichment in the melt (double crucible Czohralski method) [37], lowering the crystallization temperature by use of melts containing 6 wt % K_2O [38], and diffusion of Li from a powder richer in this element than the crystal (this process takes several days and requires high-temperature annealing treatments, c.a. to 1100 °C) [39]. Five years later, the K_2O-based flux method was revisited by Polgár et al. [40], and from a conceptual refinement, it has been after that addressed as high temperature top seeded solution growth (HTTSSG). In the early 2000s, this method became popular; Polgár et al. (2002) demonstrated that the stoichiometry of the crystal could be further improved by using a solution or flux with starting compositions [K_2O]/[$LiNbO_3$]~0.16–0.195 and R = 1 [41]. Today, the synthesis of

LN single-large crystals with ST composition still is a state-of-the-art matter: a piece of an ST crystal costs nearly 12 times more than an equivalent (cut and dimensions) CG one [19]. LN crystals with a strict stoichiometry are important because a perfect lattice translates into a lower density of local field distortions and the lessening of anharmonic crystal interactions (phonon coupling) prerequisites for any kind of resonance with a small linewidth [20]. The unsolved problem of supply and demand of sLN crystals has already been identified, and the development of a growth technique to fabricate them is urgently needed, without being unpractical or expensive [17]. Recently, the HTTSSG method has been reviewed, the earliest developments in this field and its dynamic progress leading to new generation technologies of crystal growth are also discussed [42].

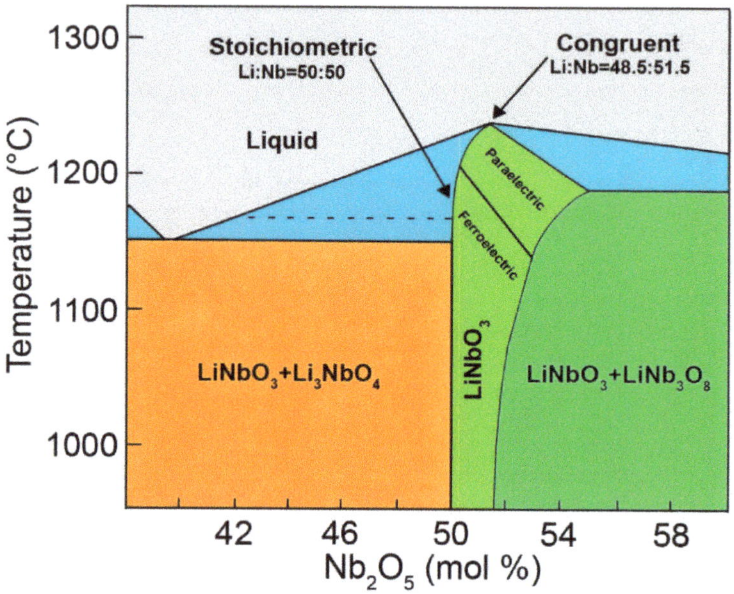

Figure 1. Schematic phase diagram of the Li_2O—Nb_2O_5 pseudobinary system in the vicinity of $LiNbO_3$.

On the other hand, from our paradigm (explicitly stated above, see the Introduction), it follows that the exacerbation/detriment of a physical property attributed to LN can be tuned by proper control of point defects. Considering only intrinsic defects, most of the measured coefficients follow a tendency to favor nST compositions with respect to compositions with a higher population of defects, as shown in Table 1; as it is discussed in Part II of this review, sLN crystals are particularly important for applications involving interaction with high energy laser light. In the case of the Curie temperature T_C—the temperature at which the paraelectric-ferroelectric phase transition turns out in most perovskite-type materials—its sensitivity to the CC can be readily seen in the phase diagram (Figure 1), not only for two fixed points but rather continuously on the whole range of the LN single-phase. To determine the CC (or the Li content within the crystal c_{Li}) using an empirical linear equation having T_C as the independent variable is nowadays a standard method [43].

Table 1. Optical and non-optical coefficients for LN single crystals with congruent (CG) and near-stochiometric (nST) compositions—as given by Volk and Wöhlecke in Ref. [20].

Property	CG	nST
OPTICAL		
Transparency region (nm)	320–5000	300–5000
Refractive index @ 633 nm (n_o, n_e)	(2.286, 2.203)	(2.288, 2.190)
Electro-optical coefficient @ 633 nm		
r^T_{33}, r^T_{31}, r^T_{22} (pm V^{-1})	32, 10, 6.8	38, 10.4, —
r^S_{33}, r^S_{31}, r^S_{22} (pm V^{-1})	31, 8.6, 3.4	—, —, 4.5
Nonlinear optical coefficient @ 1060 nm		
d_{33}, d_{31}, d_{22} (pm V^{-1})	34, 6, 3	42, 5, 2.5
NON-OPTICAL		
Crystal structure	Trigonal	
Space and point group (RT, hex)	R3c, 3m	
Lattice constant a_H (pm)	515.0	514.7
Lattice constant c_H (pm)	1386.4	1385.6
Melting point (°C)	1255	<1200
Curie temperature (°C)	1140	1206
Density (g cm^{-1})	4.647	4.635
Thermal expansion @ 300 K		
α_a 10^{-6} (K^{-1})	14.1	14.1
α_c 10^{-6} (K^{-1})	4.1	6.0
Specific heat at RT (kJ kg^{-1} K^{-1})	0.628	0.651
Thermal conductivity at RT (W m^{-1} K^{-1})	3.92	5.97
Spontaneous polarization (μC cm^{-2})	71	62
Dielectric constant		
ε^T_{11}, ε^T_{33}, $\varepsilon^S_{11}\varepsilon^S_{33}$	84, 30, 44, 29	54, 42, 42, 41

Because it is a light element, the Li concentration within SSs of LN can be directly estimated only by chemical methods: chromatography [44], atomic absorption analysis [45], and inductively coupled plasma atomic emission spectroscopy (ICP-AES) [46]. Apart from being destructive, these methods also consume large amounts of material and offer low precision (no better than 0.2 mol %) [3,47]. Likewise, the use of calorimetric methods for the determination of T_C has the main drawback of reaching T values very close to the melting point of the material (see Figure 1). In 1993, Schlarb et al. [48] and Malovichko et al. [49] independently revived the practical importance of other indirect methods to describe the CC of single crystals. These are shown in Table 2, among others. Notice that [48,49] were published just one year after nST single crystals were available, their novel content can be outlined in three main aspects: (1) introducing a new chemistry, the ternary system K_2O—Li_2O—Nb_2O_5, "responsible for the unexpected growth nST LN single crystals" [49]; (2) characterization covered almost over the whole compositional range; and (3) information on the crystal composition instead of the melt composition. We consider these contributions are nowadays seminal because they reviewed and informed our community about the number of different methods to describe the CC of LN single crystals. Such information was back then scattered throughout several publications. The review was extended—perhaps improved—three years later [47]. Also, it might be more precise to credit Vartanyan (1985) for point 1 above [42,50].

Table 2. Some indirect optical and non-optical methods for the determination of the chemical composition of LN single crystals.

Method	Measured Parameter	Equation; Accuracy (mol %)	References
	OPTICAL		
Fundamental UV optical absorption	fundamental absorption edge	nonlinear; 0.02	[51,52]
Polarized Raman spectroscopy	linewidth of Raman modes	linear; 0.05	[47–49]
Unpolarized infrared spectroscopy	intensity ratio I_{3480}/I_{3465} of peaks located at the wavenumbers in subscripts	linear; 0.01	[53,54]
Sellmeier equation	refractive index (extraordinary)	nonlinear; —	[55]
Dispersion of birefringence	refractive index (ordinary and extraordinary)	linear; <0.01	[47–49]
Phase matching T for second harmonic generation (SHG)	phase matching T for SHG	linear; <0.01	[39,56–59]
spontaneous noncolinear frequency doubling	cone angle	nonlinear; —	[60]
Holographic scattering (photorefractive effect)	reading and writing angles with respect to the normal of the crystal surface	—; —	[61]
	NON-OPTICAL		
Melt composition	Li_2O content of the melt	nonlinear; 0.3	[46,47,62]
Differential thermal analysis	Curie temperature	nonlinear/linear; 0.1	[39]/[43]
X-ray and neutron diffraction + structure refinement	cell volume	linear; 0.3	[45,63]
Density measurements	Density	—; —	[45,64–66]
Nuclear magnetic resonance (NMR) and electron paramagnetic resonance (EPR)	linewidth of NMR and EPR signals	linear for Fe doping concentrations smaller than 0.01 mol %; —	[49,67,68]
Velocity of surface acoustic waves (SAW)	velocity of SAW	—; 0.01	[69,70]

The inexorable tendency of LN to crystallize with the CG composition has been introduced above from a phenomenological point of view. To understand the why and how of the formation of the type of intrinsic defects or lattice imperfections involved, a formal discussion of a proper defect model must be given (Part II). In this respect, the description of the crystalline structure should precede. It will soon be treated in the next section. Not without addressing, however, additional comments of historical relevance.

Nowadays, a broad consensus exists in our community regarding the LN crystal structure as a 'pseudoilmenite' [20]. Thus, a side has been taken on the once important debate on whether it could be formally considered ilmenite or if it would better be conceived as a highly distorted perovskite [14]. As already said, in 1928, Zachariasen described it for the first time in history [22]. Reference [22] could not be consulted to write the present manuscript. However, by reading the first report on the ferroelectric behavior of LN (and isostructural $LiTaO_3$) by Matthias and Remeika (1949), it can be inferred that Zachariasen originally described the crystal structure of LN as being isostructural to that of ilmenite [$FeTiO_3$]. By putting the original mineral composition inside square brackets, the whole family of ilmenite structures is meant, according to the convention introduced by Muller and Roy [71]. Matthias and Remeika had already pointed out that something did not fit entirely by assuming that LN crystals adopt [$FeTiO_3$], stating that whereas the latter is a centrosymmetric structure, at the same time, "the existence of a spontaneous polarization [...] indicates the absence of a center of symmetry" [26]. If LN manifested ferroelectric behavior, then why would it adopt a centrosymmetric structure?

By 1952, the ilmenite-type picture persisted, a year in which a second article was written by Schweinler [72], anecdotally under the same title as that of Matthias and Remeika: "Ferroelectricity in the Ilmenite Structure" (Schweinler cites the work of the other!). Although the title is invariant to references [26,72], apart from the content, they differ by a trifle: whereas the most recent one indeed labeled LN as $LiNbO_3$, the one that precedes did it as '$LiCbO_3$'. Historically, niobium (in Greek mythology, the daughter of Tantalus) was originally named 'columbium' in honor of America, the continent whence the mineral arose. In 1950 (149 years after its discovery), the International Union of Pure and Applied Chemistry (IUPAC) adopted 'niobium' as the official name [73]. In 1954, Megaw put perovskite into the scene, refuting kinship between LN and ilmenite. In her own words: "it is misleading to classify the $LiNbO_3$ structure as a member of the ilmenite family" [12]. Her explanation was based on the results presented in the doctoral thesis by Bailey [74], which is acknowledged to be the first extensive study on the LN crystal structure (at room T, RT). Bailey was the first to show that the LN crystal structure at RT is not identical to ilmenite due to a different stacking sequence of the cations. Nevertheless, according to Räuber, the ideas by Megaw were widely accepted in the 1970s [2].

3. Crystal Structure

3.1. Modern/Practical Viewpoint

When the crystal structure of lithium niobate ($LiNbO_3$, LN) is nowadays addressed in a theoretical framework, little or nothing new can be added. This subject has been extensively studied, formulated, and re-formulated over time. All angles and edges have been covered. One thus has no other option than to re-tell the story as close as possible to an original and recent source because, after decades of work and a myriad of written reports, the story can hardly be better told. This being said, the present section thus saves ink and paper by being limited in describing the crystal structure of LN succinctly. The minimum of references needed to acquire comprehensive knowledge on this subject is provided.

We believe that the story of the LN crystal structure has been better told in the classic text by Räuber (1978) because the concept, which is a complex one, is therein built up step by step, details on pertinent symmetry issues are given along with a series of intermediate visual aids that contribute to a better understanding of the actual array of atoms [2]. Conversely, most consulted references limit themselves to treat it very succinctly by a text 1–2 pages long and a lone figure. See, for example, [1,75] which also excel in this subject and, because they are more recent, they may provide more reliable experimental data than [2]: atomic positions, lattice dimensions, phase transition temperatures, and compositional variations. Weis and Gaylord (1985) provide a detailed discussion on the symmetry elements of the point group pertaining to the ferroelectric phase (trigonal, space group $R3c$, point group $3m$), sense of the c-axis, and cleavage plane [1]. They also describe the two most common choices of axes or unit cell abstractions (ferroelectric phase): hexagonal and rhombohedral. These two choices are convenient for crystallographic purposes, the latter being the most popular to describe the LN crystal structure. Both are merged into Figure 2; redrawn from the publication by Sanna and Schmidt [75,76]. For most physical applications, the tensor properties are neither described in terms of the hexagonal system nor the rhombohedral. The cartesian system is rather used, denoted as 'orthohexagonal' in [2]. Its conventional definition in terms of the hexagonal system is given in [1]. Hence, while the structure description given by Räuber stands out from the others, perhaps the best approximation to it relies on properly combining and adapting the information found in [1,2,76]. Complementary information or a different way to introduce the concept may be consulted from [17,77]. Besides, Räuber considered the description by Megaw [12–14], namely by assuming that the LN crystal structure results from a set of large distortions of the atomic arrangement within a reference structure of higher symmetry, the ideal cubic perovskite. In this case, a fourth unit cell abstraction is introduced, the 'pseudocubic' setting [2].

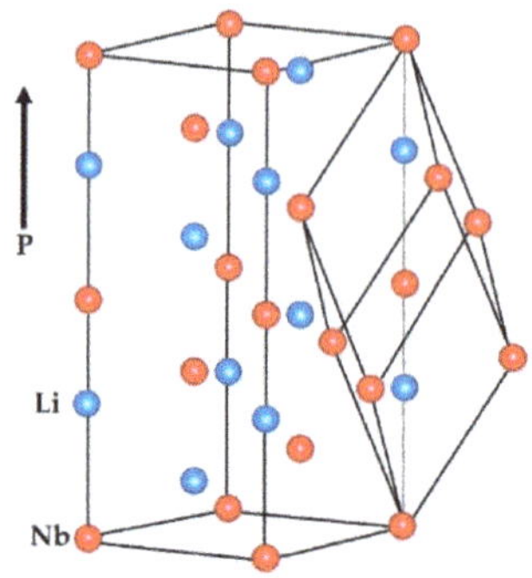

Figure 2. Conventional rhombohedral and hexagonal unit cells in $LiNbO_3$, having two formula weights (10 atoms) and six formula weights (30 atoms), respectively. The oxygen atoms are omitted.

Figure 3 (redrawn from the publication by Gopalan et al. [77]) shows a scheme of the LN crystal structure (stoichiometric composition) stable for $T < T_C \sim 1210$ °C (ferroelectric); as before mentioned, the Curie point moves according to the chemical composition, roughly from 1100 to 1200 °C. The basic net is formed from six equidistant plane layers of oxygen per unit distance, stacked in the direction of the polar axis *c* [2]. The essential feature to be noticed from the low-*T* structure (non-centrosymmetric, space group *R3c*, point group *3m*) is the partial filling of the octahedral interstitials in *c*-row: one-third filled by Li ions, one-third by Nb ions, and one-third are empty. This situation can be schematically depicted as –Li—Nb—□—Li—Nb—, where □ denotes a vacant octahedral site, also referred to as the structural vacancy. The Li octahedron is larger than the Nb one. Such a distorted octahedral environment is addressed to a transition from the nonpolar or paraelectric phase (for $T > T_C$, centrosymmetric, space group $R\bar{3}c$, point group $\bar{3}m$) to the paraelectric phase as *T* decreases. Small displacements of the Li and Nb cations are involved with respect to the oxygen layers and along the c-axis, measured to be 45 and 25 pm, respectively [34,75]. The displacement of oxygens is neglected in first order (~6 pm), and, thus, the oxygen framework is assumed to be fixed [34]. In the paraelectric phase, the Li cations are localized within the oxygen planes, whereas the Nb cations are in the center of the oxygen octahedra, that is, in between the planes.

It is important to realize, however, that such a scheme is an idealized one. Remind that, under regular circumstances of growth, LN single crystals are off-stoichiometric. Instead, the congruent (CG) composition prevails, characterized by an excess of Nb. It can be understood in terms of crystal chemistry: during crystallization, Li vacancies form in the unit cell with ease, since in the crystal the Li–O bond is significantly weaker than the Nb–O one [17]. Thus, it is natural to reason that Li deficiency implies an Nb surplus, or better said, "a decreasing Li content is accompanied by increasing content of the heavier Nb" [75]. However, how do these tendencies simultaneously conciliate as to have a stable LN solid solution? The occurrence of a stacking fault envisioned as the relatively excess Nb partially occupying Li vacancies, has been proposed and termed Nb antisite (Nb_{Li}) [64,78]. It is schematically shown in Figure 4, redrawn from the publication by Volk and Wöhlecke [75] The existence of this type of antisite defects has been repeatedly proven by detailed structure studies [45,63,65,79].

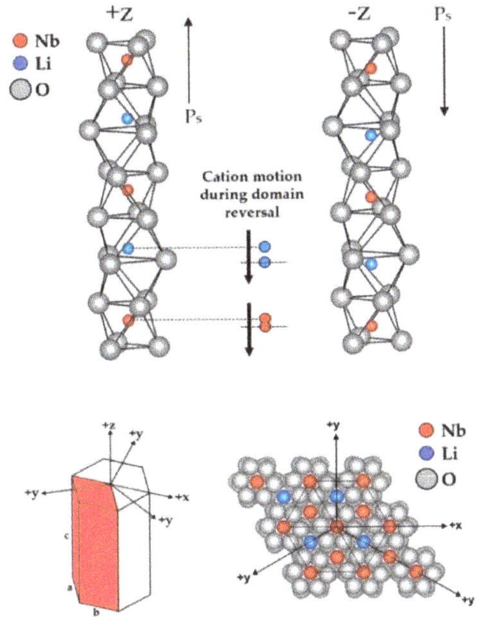

Figure 3. Crystal structure of ferroelectric $LiNbO_3$ under two cation displacement possibilities with respect to each equidistant oxygen octahedra. **Upper left:** Z+ orientation of cation displacement. **Upper right:** Z- orientation of cation displacement. **Bottom left:** conventional choice of axes for the 'orthohexagonal' setting (*XYZ*) with respect to the hexagonal setting (*a*, *c*). Notice that whereas in the hexagonal setting, all axes have an equal length, they all are different in the cartesian one. **Bottom right:** projection of the atomic arrangement along the *c*-axis.

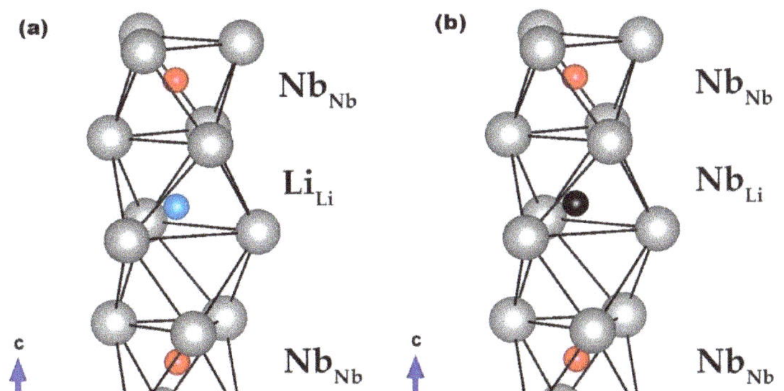

Figure 4. Comparison between the ideal and nonideal $LiNbO_3$ crystal structure. (**a**) Free from point defects describing the stoichiometric composition. (**b**) Under the most likely situation in which Nb antisites are present.

3.2. Earlier Thoughts: Relation to Basic Structures of Higher Symmetry

Following the short discussion presented by the end of Section 2, Megaw referred to the structures of ferroelectrics (that of LN per se) as pseudosymmetric structures of polar symmetry that result as a competing effect between isotropic (ionic character) and directed (covalent character) binding.

She assumed that the LN crystal structure for the paraelectric phase at high T was the ideal cubic perovskite—back then, the experimental information was only available for temperatures far below from the Curie temperature of LN. Then, she proposed the continuous transition to the ferroelectric phase structure under the hypothesis of a monotonical increment of the covalent character as T decreases, as far as the overall binding characteristics are still mainly ionic [12]. Indeed, 12 years later, Abrahams et al. (1966) concluded (also supported on the results by Peterson [80], cited in Ref. [28]) that LN "is not primarily an ionic crystal, but one in which directed, largely covalent, bonds play a determining role" [32]. Remarkably, the prediction by Megaw (the covalency role in ferroelectrics) reverberates until today because it agrees to the current most widely accepted theoretical framework regarding the innermost cause of ferroelectricity, that is, the stabilization of structural distortion through second-order Jahn Teller effects (see further discussion in Section 5). Moreover, it is interesting to notice that (apparently) back in 1954, Megaw was not aware of the work by Buerger (1947), who in the spirit of explaining the 'genesis of twin crystals' and superstructures, developed a formalism to describe some of the relations pertaining to a crystal structure called the basic structure, as contrasted to derivative structures derived from it by (mathematical) generalization [81,82]. The derivative structures would always present lower symmetry with respect to the basic structure. According to Bärnighausen, who refined and further extended the ideas of Buerger into the so-called Bärnighausen trees [83,84], Megaw (1973) herself introduced the terms arystotype and hettotype as synonyms of the basic and derivative structures, respectively [85]. These terms are nowadays widely accepted and used.

The structural family of perovskites is perhaps the best available example of distorted hettotypes. Most perovskite compounds are distorted and do not crystallize in the ideal cubic structure as $SrTiO_3$, and other few compounds do at RT [84,86–88]. It is believed that of all the structures adopting this crystal structure, only about a 10% adopt the ideal cubic one [88,89]. Even the original mineral composition $CaTiO_3$ is known to be slightly distorted [86,88]. Thus, $SrTiO_3$ is naturally regarded as the high-symmetry reference, the aristotype prototype within this crystal structure family. It also happens to be the stable phase at a high T (paraelectric) for most ferroelectrics based on perovskite oxides [90]. Before introducing the main aspects of the conceptualization conceived by Megaw, first, the documented quantitative analysis of how the perovskite crystal structure relates to some of its derivative structures by distortion is discussed. An introduction to the details of this crystal structure can be consulted elsewhere [84,86,87,91,92].

Three main types of distortion or framework perturbation in ABO_3 compounds adopting a perovskite crystal structure have been identified: (1) tilting or puckering of the octahedral framework, (2) B-cation displacements within these octahedra, and (3) tilting of the BO_6 octahedra relative to one another as practically rigid corner-linked units [88,92]. An example of how the ilmenite structure can be derived in successive steps from the ideal perovskite structure by octahedral tilting has been described by Johnsson and Lemmens [86]. The distortions may occur separately or in combination, just as one is hardly addressed to a single effect responsible for the distortion, these being mainly: size effects, deviations from the ideal composition, and the Jahn–Teller effect [86]. Focusing on the size effects, these can be quantitatively described in terms of the Goldsmith tolerance factor (GTF), which "reflects the structural distortion, force constants of binding, rotation and tilt of the octahedrons" [87]. Naturally, one may ask: how far from the ideal packing can the framework of BO_6 octahedra be perturbed by substituting cations of different ionic radii, and still being 'tolerated' by the perovskite structure? The GTF estimates the degree of distortion, or better said, "the structural frustration of the cubic perovskite structure regarded as an ionic solid" [91]. It is defined as the ratio [84,86,89,91,92] (in the discussion that follows, these references have been indistinctly used)

$$t = \frac{r_{AO}}{\sqrt{2}r_{BO}} = \frac{r_A + r_O}{\sqrt{2}(r_B + r_O)}, \tag{1}$$

where the r terms are bond lengths calculated as sums of ionic radii and $a = \sqrt{2}r_{AO} = 2r_{BO}$ is the geometrical definition of the a-axis within the ideal cubic cell. In using this equation, the 'effective' ionic radii are typically used, after Shannon and Prewitt [93]. Both cations can be accommodated in the ideal cubic perovskite $SrTiO_3$ if $t = 1$, thus having the optimal cation-anion bond lengths, in which case $(r_A, r_B, r_O) = (1.44, 0.605, 1.40)$ Å. Perovskites still could form for deviations from unity, typically in the range t: 0.8–1.1. In such cases, the degree of distortion from cubic symmetry increases, whereas the thermodynamic stability decreases with increments of the absolute value of $(1 - t)^2$ [92,94]. A tolerance factor less than unity indicates that the A cation is small compared to the site bounded by the oxygen octahedra. Under these circumstances, the A cation cannot be effectively bond to as many as 12 oxygen atoms, a situation that typically leads to a structural distortion by means of cooperative tilting of the octahedra. It is the most commonly occurring type of distortion. The prototypical cases include the mineral perovskite $CaTiO_3$ and the compound $GdFeO_3$; for both $t = 0.81$, as $(r_A, r_B) = (1.107, 0.78)$ Å. The arrangement of atoms is still considered to be of cubic symmetry in the range of $0.89 < t < 1$. The ilmenite structure is stable for values of t less than 0.8. According to Kubo et al., either ilmenite or corundum are adopted by compounds for which $r_A \sim r_B \approx 0.7$ Å. Ilmenite is favored in the cases when the A and B cations are largely different in terms of electric charge and electronegativity [95]. By further lowering the value of t, a strongly distorted structure would be favored down to the extreme of having a very low coordination number of 6 for both cations, as in the LN structure.

A review on phase transitions between these three polymorphs (LN, ilmenite, and perovskite) on titanates and related systems reveals a trend of ilmenites transforming into perovskite [$CaTiO_3$] at high pressures, which then quench to the LN structure on pressure release [92]. By applying pressure, the LN structure transforms reversibly to the perovskite one. Interestingly, at RT for both phase transitions [$FeTiO_3$] (ilmenite) → [$CaTiO_3$] and [$LiNbO_3$] → [$CaTiO_3$], the same amount of pressure is needed (around 16 GPa) [92,96–98]. Thus, the type and extent of distortion are sensitive to the (T, P) pair, and in the account of the same discussion, the LN phase can only be obtained, in the sense of a hettotype, as a quench product of perovskite at a high P, previously obtained by exerting pressure on an ilmenite structure. The GTF for the initial ilmenite must be restricted to be less than 0.78 in general, or in the case of titanates, the restriction translates into divalent cation radii less than 0.8 Å [92]. Interestingly, the use of Equation (1) with r_A (Li^+, 6-fold coordination) = 0.74 Å, r_B (Nb^{5+}, 6-fold coordination) = 0.64 Å, r_O = 1.40 Å (after Shannon and Prewitt [93]), gives $t = 0.74$ for the LN structure. However, it should be acknowledged that the GTF is only a rough estimate given that perovskites are not completely ionic compounds. Still, as expected, it is highly correlated to the electrophysical parameters featured by important perovskites. It has been a useful figure in guiding the synthesis of intelligent materials which "have been the heart and soul of several multibillion-dollar industries", such as polycrystalline ceramics based on $BaTiO_3$ [87,99].

The conceptualization conceived by Megaw relied on a profound analysis of the crystallographic data obtained for LN (Bailey, 1952 [74]) and ilmenite (Barth and Posnjak, 1934 [100]). Disregarding details, she concluded that whereas the transition paraelectric-ferroelectric in LN is continuous and reversible, implying however large atomic displacements of the order of 200 pm for Li and 70 pm for oxygen, the ilmenite structure cannot be attained by any distortion of a perovskite structure given that such a transition would involve a serious rearrangement of the structure and is irreversible [12]. Instead of considering a kinship between the LN and ilmenite structures, it was instead proposed that the former be closely related to the rhombohedral variant of $BaTiO_3$, which possibly "represents the early stages of distortion which $LiNbO_3$ shows fully developed" [12]. Megaw herself found an "obvious difficulty" in her explanation due to the large magnitude of the displacements involved. In 1957 they changed to 100 pm for Li and 60 pm for O in a book she wrote, where she also acknowledges that such displacements might not be physically possible [13]. These values conflicted with those reported by Abrahams et al. in a series of three papers, which together are considered the most extensive and detailed structural analysis of LN [32–34]. Reference [34] is a study of the atomic arrangement in polycrystalline LN as a function of T (RT to 1200 °C). In it, and accounting also for the results in a single crystal at RT by X-ray and neutron diffraction [32,33], it is shown that in this range of temperatures, the atomic arrangement is essentially unchanged: whereas a fixed oxygen framework exists, given the non-significant displacement of 6 pm within the oxygen layers, the Nb and Li cations exhibit small displacements along a common direction of magnitude 25 and 45 pm, respectively [34].

Besides this fact, the model proposed by Megaw was refuted, given the absence of evidence for a change in the crystal system between the ferroelectric and paraelectric phases. The comparison of the diffractograms obtained below and above T_C did not show appreciable changes in the positions of the lines, only did the intensities change in a non-reproducible way [34]. The large discrepancy to the calculated displacements by Megaw relied on the original ambiguity in determining the Nb position relative to those of Li and O, as originally done by Bailey [32,74]. Also, on the large error of about 100 pm existing in the determination by Bailey on the atomic coordinate of Li [34]. Together, these two aspects constitute the main criticism from Abrahams et al. to the work of Megaw (1954). Because of the ambiguity and inaccuracy present in the work of Bailey, Shiozaki, and Mitsui (1963) did neutron diffraction on polycrystalline LN to improve the results [101]. However, these also conflict with those obtained by Abrahams et al. (1966) [33]. Hence, it can be said that this researching group favored kinship of the LN crystal structure to that of ilmenite instead of perovskite, and probably due to it, they erroneously addressed the space group for the nonpolar paraelectric phase as $R\overline{3}$ instead of $R\overline{3}c$ [14,34]. That $R\overline{3}c$ is the proper space group for the paraelectric phase was shown in 1968 by Niizeki et al. [102]. In the case of the ferroelectric phase, the space group assigned by Bailey was confirmed, $R3c$, with lattice constants a = b = 5.148 $\dot{A}$, c = 13.863 $\dot{A}$, and atomic positions in the hexagonal reference [34]

$$\begin{gathered}\text{Nb: } (0, 0, 0) \\ \text{Li: } (0, 0, 0.2829) \pm (0, 0, 0.0023) \\ \text{O: } (0.0492, 0.3446, 0.0647) \pm (0.0003, 0.0005, 0.0004)\end{gathered}$$

In $R3c$, the z-position of the origin can be arbitrarily chosen. Clearly, in the ferroelectric phase, neither Nb nor Li is aligned to the oxygen lattice. Hence, Abrahams et al. chose the Nb site as the origin since it is the principal X-ray and neutron scatterer [32]. In 1968 Megaw, based on all the results by Abrahams et al. [32–34], found an improved argument to relate the LN crystal structure to perovskite by a reparameterization trick [14]. The origin in z was proposed to be reassigned, so that it lies midway between two oxygen layers, in the vicinity of an Nb site. A high-symmetry structural reference is obtained in this way, an idealized structure with hexagonally-closed packed oxygen atoms. Upon successive approximations in terms of small quantities expressing displacement parameters, the actual Nb–O framework, as determined by Abrahams et al. [14], can be obtained. Another set of successive approximations or displacement parameters would lead to the perovskite framework. The reparameterization gives the following coordinates for the actual structure in terms of the ideal [14].

$$\begin{aligned}\text{Nb}: &\ (0,\ 0,\ w) \\ \text{O}: &\ \left(u,\ \tfrac{1}{3}+v,\ \tfrac{1}{12}\right) \\ \text{Li}: &\ \left(0,\ 0,\ \tfrac{1}{3}+w'\right)\end{aligned}$$

With $u = 0.0492$, $v = 0.0113$, $w = 0.0816$, $w' = -0.01318$. The value of the axial ratio is $\frac{c}{a} = 2\sqrt{2}\left(1+\frac{1}{2}\beta\right)$, with $\beta = -0.0936$. The idealized structure is then described when all the parameters are equal to zero. It can be regarded as the zeroth-order approximation to the actual Nb–O framework, a centrosymmetric structure with axis ratio $\frac{c}{a} = 2\sqrt{2}$. In the first-order approximation $u \neq 0$ while the rest of the parameters remain equal to zero. A higher symmetry structure is described where the axis ratio tends to decrease from its ideal value. The perovskite framework is described by this first-order approximation when $u = \frac{1}{6}$. Insertion of Li into the octahedral interstices causes symmetry reduction, thus allowing parameters u, v, w to be different from zero, that is, the second-order approximation, describing the actual LN crystal structure. Moreover, regarding the effect of temperature, unexpectedly, as T increases, the actual structure (ferroelectric phase) would not tend towards the idealized structure (the zeroth-order approximation), but rather towards the first-order model with increasing u, while v and w would tend to zero. Again, by stating this, Megaw avoids saying that the ferroelectric LN crystal structure can be derived from the ideal cubic perovskite. In her own words: "it is not suggested that a change of framework configuration from one extreme to the other can take place in any actual material" [14]. That the ferroelectric phase follows from a small distortion of a high symmetry reference phase has demonstrated to be a powerful principle [91]. As already said, the detailed discussion of the LN crystal structure, as presented by Räuber [2], assumes the reparameterization trick herein discussed.

Although this conceptualization might not necessarily be correct, the LN crystal structure might be grasped more naturally by understanding its pros and cons. For a deeper insight, previous knowledge on the corundum, ilmenite, and perovskite structures would be advantageous for a proper comparison: these can be consulted from references [92,96,103], [92,96,100], and [84,86,87,91,92], respectively. The book by Lima-di-Faria provides useful descriptive charts in pages [23]: 99, 100, and 103, respectively.

4. Powders

4.1. Accurate Description of the Chemical Composition

Until recently, the lack of an accurate (indirect) method to determine the CC of LN powders (LNPws) was a reality. We demonstrated that none of four conventional indirect characterization techniques could be used for this purpose as they are nominally used for the case of single crystals. These are (for the case of single crystals, further details may be consulted in the given references):

- X-ray diffraction + structure refinement [45,47,63]. Abrahams and Marsh did earlier measurements by use of a Bond diffractometer [65,104]. Some of the available program packages for Rietveld refinement, commercial and public, and an introduction into this subject can be consulted in [104–112].
- Polarized raman spectroscopy [47–49,113,114]. The group theory elements in this field and the assignment of phonon modes in LN are discussed elsewhere [115–119], having a common root in the seminal paper by Schaufele and Weber (1966) [120]. The scattering geometry in polarized Raman experiments is described after Porto and Krishnan (1967) [121]. The high sensitiveness of the functional form (and intensity) of the recorded Raman spectra in LN to the experimental configuration might be observed from the educational video 'Convenient Application of Polarized Raman Spectroscopy', provided by the HORIBA Raman Academy [122]. The resolution of Raman bands or fitting techniques is critical for achieving great accuracy in determining the CC by this method [123,124]. Regarding this method of CC characterization, the works by Scott and Burns (1972) [125] and Balanevskaya et al. (1983) [126], are considered pioneers (both done on LNPws).
- Fundamental absorption edge [47,51,52]. One of the earliest reports on this subject was written by Redfield and Burke (1974) [127]. For practical reasons, a direct transition can be assumed regarding the intrinsic nature of the bandgap (no phonons involved for momentum conservation) [128], in which case the fundamental band gap is proportional to the square of the absorption coefficient α^2 [129].
- Differential thermal analysis [43–45,47]. Measurement of the Curie temperature (T_C) is one of the earliest calibration methods for determining the CC in LN single crystals [62]. Regarding this type of transition, as previously stated, a change in the crystal structure occurs in which the symmetry of the system decreases. The symmetry-breaking relation between the high-symmetry paraelectric structure and the ferroelectric one is consistent with a second-order transition, described by the Landau order-disorder theory [130,131]. A finite discontinuity in the heat capacity of the system having this transition has been addressed as a direct thermodynamic consequence.

Research was conducted in our group, where these characterization techniques were used (modified when needed) to obtain a set of equations that appropriately describe the CC of LNPws. These results have been published in [19]. Within the following lines, we shortly describe the main aspects of such publication. The current necessity for the correct characterization of LNPws in this context is also further discussed. We believe such a missing point in the literature for so long is that, historically, LNPws have been overlooked for research in favor of single crystals and thin films mainly because of their poor performance in most applications. For example, within the context of nonlinear optics, powders have been relegated compared to single crystals due to surface inhomogeneities and strong scattering effects, although they are tacitly considered easier and far less expensive to synthesize. Besides, the urgent demand for large single crystals of high quality to match the microfabrication process requirement in modern applications has already been outlined [17].

Our recent contribution arose from noticing that at least one of the equations given in [47–49] describing the CC of LN single crystals by using polarized Raman spectroscopy—is not accurate for the case of powders. Also, contrary to what could be expected, the approach of using an empirical equation formulated for single crystals to describe powders does not even work for T_C measurements (see equations in [43–45]). Since temperature is a scalar quantity (light propagates and interacts with matter in a vector-like form), it would be permissible to expect a single description of the CC of LN in terms of T_C, regardless of the solid solution being a single crystal or a powder. However, the differences in surface and grain size/boundary effects between single crystals on polycrystalline powders cannot be neglected. Before our work, the systematic measurement of lower T_C values (about 10 °C) for LNPws compared to equivalent single crystals, had been addressed [47,132]. As for the case of optical absorption, there is no point in using an equation that describes single crystals [52], since the terms 'refractive index' and 'absorption coefficient' lack sense when related to powders; a powdered sample can be conceived as a material of infinite thickness from the optical point of view. Instead of conventional optical absorption measurements, the UV–vis diffuse reflectance technique is usually employed while relying on the Kubelka–Munk approximation to describe the measured functional form or reemission function, a proxy the actual absorption spectrum [133–135]. The experimental details to make this approximation valid can be consulted in [133]. Thus, recently we gave a first step towards the accurate description of the CC of LNPws by developing a facile method based mainly on imposing X-ray diffraction as a seed characterization technique. Raman spectroscopy (non-polarized), UV–vis diffuse reflectance, and DTA enrich the work, representing various alternatives for the independent determination of this fundamental quantity in LNPws. An empirical equation that describes it in terms of a corresponding experimental parameter is given for each of these four characterization techniques (valid for LNPws with crystallite and particle dimensions close to 160 nm and 2.6 μm, respectively) [19]

$$\langle c_{Nb} \rangle = (8.6207 V_{cell} - 2692.5216) \text{ mol \%} \pm 0.5 \text{ mol \%}, \tag{2}$$

X-ray diffraction (XRD)

$$\langle c_{Nb} \rangle_L = \left(256.4103 * \left(\frac{\Gamma_L}{2x_c}\right) + 43.5385\right) \text{mol \%} \pm 0.4 \text{ mol \%}, \tag{3}$$

$$\langle c_{Nb} \rangle_G = \left(588.2353 * \left(\frac{\Gamma_G}{2x_c}\right) + 42.7059\right) \text{mol \%} \pm 0.5 \text{ mol \%}, \tag{4}$$

(non-polarized) Raman spectroscopy (RS)

$$\langle c_{Nb} \rangle = \left(3.9078 * E_g + 34.6229\right) \text{mol \%} \pm 0.4 \text{ mol \%}, \tag{5}$$

UV–vis diffuse reflectance (DR)

$$\langle c_{Nb} \rangle = (-0.0515 T_C + 110.8505) \text{ mol \%} \pm 0.4 \text{ mol \%}, \tag{6}$$

Differential thermal analysis (DTA)

where the brackets in the Nb concentration stand for an average of this quantity in the sampled crystallites. $V_{cell} = \sqrt{3}a^2c/2$ stands for the cell volume in (angstrom)^3 units, calculated by a standard structure refinement method. In Equations (3) and (4), Γ_i stands for the FWHM in cm^{-1} of the Raman band around 876 cm^{-1}, resolved by linear fitting either using a Lorentzian (i:=L) or a Gaussian (i:=G) line shape, x_c denotes the center of this Raman band. Normalization of the full Raman spectra precedes the linear fitting and, regardless of the line shape, enlargement around this band is suggested, extending as much as possible (precise determination of the baseline). Application of a single or double-peak fitting, rather than performing a multi-peak fitting of the full Raman spectra, is also convenient. In Equation (5), E_g stands for the fundamental band gap in eV units, being determined by assuming a direct interband transition, that is, the obtained reemission function is elevated to the power of 2. In Equation (6), T_C is the Curie temperature measured in Celsius degrees. Also, notice that in contrast to pioneering works (on single crystals, [47–49]), the equations are given in terms of the averaged Nb content in the crystallites $\langle c_{Nb} \rangle$, rather than $\langle c_{Li} \rangle$. By putting these equations in terms of $\langle c_{Li} \rangle$, a more straightforward comparison could have been attained to data available in the literature. However, it was decided to do it in terms of $\langle c_{Nb} \rangle$ because of a simpler interpretation and association with a phase diagram describing LN, like that given in Figure 1. It has been noticed that most of the phase diagrams existent in the literature to describe LN are presented in terms of Nb_2O_5 mol %. Because in the synthesis of the studied powders in [19], Li_2CO_3 was used as the other precursor, and also relative not too high temperatures were used in the calcination process (850 °C), a one-to-one correspondence exists between the Nb and Li contents in Equations (2)–(6). The equivalent relations in terms of $\langle c_{Li} \rangle$ can be consulted in Appendix B in [19]. We now wish to underline the main aspects of the method devised to obtain such equations.

The calibration of the cited methods is based on the quantification of pure and secondary phase percentages by XRD, followed by Rietveld structure refinement. Secondly, relying on the LN phase diagram (Figure 1), the chemical composition of the studied samples is inferred, and thereafter labeled in terms of the Nb content in the crystallites. Lastly, having done this, any of the four mentioned characterization techniques could be used to relate such labeling with their corresponding experimental parameter. In the case of a user who wants to determine the CC of LNPws only, she/he would only need to perform the last step and use any of Equations (2)–(6). On the other hand, in wanting to describe other powders apart from LNPws, the whole method (three main steps described above) might be further applied inasmuch as akin materials are investigated, lithium tantalate ($LiTaO_3$) powders for example.

The validity of this methodology is proven self-consistently with the determination of the CC of several samples, where the content of Nb and Li is varied in a controlled way. However, the main shortcoming of this investigation is the large uncertainty associated with Equations (2)–(6). Rigorously, they should not be used for a practical composition determination and, instead, it only could be stated with more confidence that, by using these equations, the composition of LN powder would be closer to the stoichiometric or congruent point, or rather in an intermediate one. It will soon be solved, noticing that both the resolution and the associated uncertainties of this methodology can be significantly improved by analyzing larger powder quantities. The major contribution to uncertainty emerges from the determination of the boundaries of the pure ferroelectric LN phase (details given in [19]). The associated uncertainty to Equations (2)–(6) can be significantly reduced if a larger number of samples are synthesized in this range, which can be more easily achieved if larger quantities of powder are prepared. For example, it is expected that by synthesizing approximately 10 g of powder, around 40 points would be available for analysis, assuming an increasing step of 0.1% in the Nb precursor mass. As a result, the overall uncertainties would decrease by about 50–80% (noticing that the linear fitting uncertainty would also be reduced significantly). Conclusively, apart from providing four distinct alternatives to describe the CC of LNPws accurately, the innovative character of our recently published work is the self-consistency character of the whole method. No other direct method is needed to confirm the CC of the powders since the determination of the pure ferroelectric LN phase boundaries by XRD analysis suffices for this purpose. The four involved characterization techniques are standard, accessible nowadays to large scientific communities in developing countries.

4.2. Role in New Potential Applications

Nonlinear optical powders such as LNPws have served in the past as survey materials. In 1968, Kurtz and Perry developed a method to predict the second-order nonlinear susceptibilities of materials unavailable under large single crystals by performing second harmonic generation (SHG) experiments on the powdered version of the material [136]. Thus, SHG experiments on powders are conceived initially as secondary within this method, which is by far the most popular approach to powders in relation to SHG. However, recent experiments have proposed a new paradigm to quantitatively study SHG efficiency with notable implications for pharmaceutical materials [137]. However, another feature that makes the SHG that arises from powders attractive nowadays is the possible tuning of the SHG intensity from LNPws, ascribed to proper control of the CC and grain size of the powders [138]. This feature has already been outlined in our recent publication, among others [19]. Here, we further extend details on such recent developments on LNPws (for which the necessity of a proper CC description of the LNPws is implied). In recent years:

- (2020) Strong and weak light scattering effects can be present simultaneously in random media (random in terms of its refractive index), such as powders. Here, accounting for optical nonlinear powders, photons can undergo multiple scattering in the sense that they undergo one SHG scattering event and single or multiple linear scattering events at the fundamental and harmonic frequencies. In contrast, ballistic photons undergo a single SHG scattering event and no linear scattering event. Depth profiling with polarization resolution of the SHG intensity from a powder stack of sLN microparticles/nanocrystals has been done in back-scattering or retro-reflection configuration. The results, supported by modeling, show that competition between multiple scattered and ballistic photons contributions holds during the focus longitudinal translation, which can be disentangled in observing polarization distortions occurring as the beam focus moves from air into powder [139]. The contribution from the multiple scattered photons dominates at all depths, whereas, at the maximum of the intensity depth profiles, the contribution of ballistic photons is at its maximum and enhanced due to collection efficiency. This work paves the way for a close and quantitative investigation of the SHG response from nonlinear optical powders and, combined with the derived conclusions in [138], the experimental configuration might also play a role in the discrimination of information: transmission experiments for the evaluation of averaged properties such as SHG efficiency (micrometer size scale of the powder) and retro-reflection experiments for overall scattering properties and local crystalline properties (nanometer size scale). The latter is a general idea somehow (not explicitly) postulated in a revision of the Kurtz and Perry method done by Aramburu et al. [140].
- (2018) Fe-doped LNPws show, after a post-thermal treatment in a controlled reducing atmosphere, a rather strong ferromagnetic response at room temperature for a doping concentration of the order of 1 mol % [141]. This may be considered a first report of the manifestation of ferromagnetism in nanocrystalline LNPws within the regime of very low doping concentrations. Post-thermal treatment in a controlled atmosphere is a key point for inducing this behavior, which could be explained as the recombination of unpaired electrons from the donor sites (Fe impurities) to the acceptor sites (oxygen vacancies) in the surfaces of the material. A statement that needs to be further scrutinized since the opinion on the existence of this type of vacancies has changed over time; lately they are neglected, at least with respect to single crystals and within the volume (see discussion in Part II). Bulk diffusion of oxygen has been exluded from the explanation of the results recently presented by Kocksor et al. [18], where LN nanocrystals were prepared by ball-milling the crucible residues of a Czochralski grown congruent crystal. Anywise, neither the enhanced factor of surface effects nor the depletion of oxygen at the surfaces can be overruled in LNPws prepared by a mechanochemical-calcination route.
- (2017) The same method of synthesis has been used to prepare LN micropowders whose SHG intensity is maximal at a certain λ_{max}, in terms of the fundamental excitation wavelength.

Possible tuning of λ_{max} could be ascribed to the control of the composition and grain size of the powder [138]. Also, there is the possibility of obtaining major technical benefits by exploiting the SHG from disordered materials such as LNPws, given that neither a critical adjustment of the orientation/temperature in the material (phase-matching condition) nor the accurate engineering of a microstructure (quasi-phase matching condition), are substantially needed [142].

- (2017) A novel fabrication process based on the powder-in-tube method to realize polarization-maintaining optical fibers has been demonstrated. It relies on the principle of "inducing an anisotropy of the refractive index in the core region by internal stress", in which the use of powdered material with a thermal expansion coefficient (TEC) higher than that of silica (fiber core) is the key point [143]. Glass complex systems based on SiO_2-Al_2O_3-La_2O_3, with a TEC around 10 times larger ($5.32–6.46 \times 10^{-6}$ K^{-1}) compared to silica (0.54×10^{-6} K^{-1}), has been used. Single crystalline LN has a TEC along the *c*-axis of the same order as the complex systems just mentioned ($\sim 5 \times 10^{-6}$ K^{-1}) and almost three times higher for either of the other axes (14.1×10^{-6} K^{-1}); see Table 1. In the case of LNPws, a TEC value between these two is expected, and thus they are, in principle, good candidates for the fabrication of polarization maintaining optical fibers.
- (2013) Cementation materials based on LN have been proposed as potential materials for an effective formation of eco-friendly end products through artificial photosynthesis; this is considered important for the global warming reduction problem [144]. Despite its wider band gap (3.8–4.1 eV) compared to that of TiO_2 (3.2 eV), LN is considered a better artificial photocatalyst due to its strong remnant polarization (70 $\mu c(cm)^{-2}$) [145]. Paraphrasing Nath et al.: "the photocatalyst $LiNbO_3$ can be used as a construction material to emit oxygen using atmospheric CO_2 and water reliably and inexpensively. The reduction of the global warming problem through the use of this photocatalyst would have a remarkably positive impact on the environment in the near future" [144]. Regarding LNPws, we emphasize that powders would not only be easier to implement than single crystals into cement-based materials but also they would enhance surface effects, perhaps improving this way the lifetime of the carriers (photo-generated electrons and holes) involved in artificial photosynthesis [145]. The importance of LNPws on these crucial matters must not be overlooked.
- (2012) LN and LT powders of several particle sizes at the microscale and with averaged crystallite size between 30 and 300 nm, have proved to be antimicrobial agents in aqueous solutions based on cyclical thermal excitation [146]. The powders were directly obtained from the crushing of single crystals by high-energy ball milling and various solution percussion routes. The powders obtained by the latter method show smaller particle and crystallite sizes, showing improved disinfection properties due to the increased direct surface contact with the bacterium.

On the other hand, perhaps an even more fascinating feature of LNPws is their role in the field of lead-free piezoelectric materials. This research field is nowadays vast and very active due to the environmental laws and regulations imposed by the government in 2003, the so-called RoHS directive, which may be considered as "the most disruptive event in the history of electronic manufacturing" [147]. It was approved in 2006 to minimize health and safety risks, dictating that new generation electrical and electronic devices are to be manufactured without any of the following compounds: lead, mercury, cadmium, hexavalent chromium, polybrominated biphenyls, or polybrominated diphenyl ethers [147]. The truth is that neither the RoHS directive has been fully applied nor research on lead-based materials has been stopped so far. This because lead zirconate ($Pb(Zr_{rx}Ti_{1-x})O3$), PZT), and lead titanate ($PbTiO_3$, PT) perovskites by far feature the best performances on these grounds (sensors, actuators, and other electronic components) at reasonably low production costs [148]. The existence of a morphotropic phase boundary (MPB) is acknowledged to be the driving force. It is a region of a structural phase transition, accompanied by extremes of electrophysical parameters, particularly, high dielectric and piezoelectric parameters: the anomalous increase in these properties shown by MPB compositions is the result of the enhanced polarizability that arises from the coupling of two equivalent energy states, that is, the coexistence of two structural phases [149,150]. The MPB is nearly temperature independent, and its existence is attributed to high polymorphism (apart from the high piezoelectric characteristics) in which PZT outstands. Nowadays, piezoelectric PZT-based ceramics dominate a $1 billion market of actuators, sensors, and transducers that merge into the realization of novel applications such as medical ultrasound, high precision accelerometers, sonar systems, fish finders, thermal sensors, and precise positioners, among others (in the future it may also earn a position in the development of the Internet of Things) [151].

Researching interests on lead-free materials is far from dropping, and, today, most experts share the common belief that a possible way to replace lead-based materials is the synthesis of solid solutions of alkali metal niobates (AMN) [148,150,152]. LN belongs to this category of materials: recently, LNPws have been mixed with alike powders in the sense of forming a novel green compact to be sintered into a high-performance ceramic. Complex ternary and four-component systems involving LNPws have been studied: $NaNbO_3$–$LiNbO_3$–$PbTiO_3$ and $(1-x)(Na, Li)NbO_3$–$xPb(Ti, Zr)O_3$, respectively. The latter has shown a considerable increase of electrophysical parameters [152]. Although AMN systems are still inferior to systems mainly based on PZT, the four-component transition already features good figures of merit for applications in microwave devices (for a low content of x) and low-frequency electromechanical converters (for $0.70 < x < 0.85$) [152]. The influence of the CC of LNPws on the performance of these complex ceramics is evident and, thus, the urgent need to properly characterize it. However, in our first aim to furnish an accurate description (discussed above), we have missed one important point, that of the influence of the quality and macroproperties of the precursors on the product. Reznichenko et al. have pointed out the possible influence of the quality of niobium pentoxide (Nb_2O_5) on the resulted four-component system ceramics [152]. Almost 30 years ago, this question was settled regarding the CC of LN single crystals (see Section 4.4).

4.3. Methods of Synthesis

Conventionally, LNPws are prepared by solid state-reactions between the precursors at a temperature above 1000 °C. Although this method is simple, it may lead to the composition deviation from stoichiometry because of lithium evaporation at higher temperatures [153–155]. Concerning the work done so far in our group, the synthesis of the studied powdered materials practically summarizes into the successive combination of the following process: grinding (high-energy ball milling) and thermal annealing or calcination. In short, the precursors lithium carbonate (Li_2CO_3) and niobium pentoxide (Nb_2O_5), are mixed in solid phase (room temperature) and ground, and then subjected to a thermal annealing process (calcination). The precursors are bought with high purity, and the technique used in the grinding process is closely related to that of mechanochemical synthesis.

Strictly speaking, mechanochemical synthesis denotes a grinding process in which chemical reactions are necessarily involved: the simultaneous grinding of more than one precursor (chemically distinct). With the breaking of certain chemical bonds, the formation of new ones is favored. When no chemical reactions are involved, as in the investigations so far being held in our group, the precursors are ground only, and it is convenient to denote this process as merely grinding. The concepts and involved variables are, however, very similar for both processes. Detailed reviews are available after Suryanarayana (2001) and Kong et al. (2008) [156,157]. Recently, the available knowledge regarding this synthesis route connected with LNPws has been extended [18]. On the other hand, to state that no chemical reactions occur in the grinding process might not be accurate to some extent. Apart from contamination issues depending on the chemical and physical characteristics of the used vial and grinding medium [18,158,159], in the preparation of LNPws by high-energy ball milling, the product usually shows a greyish coloration (before calcination), also observed for LN single crystals after chemical reduction (near the surface) [160]. Another common finding is the remaining non-reacted and amorphous milled powder [161,162]. The preparation of pure LNPws by a mechanochemical method assisted by subsequent calcination at relatively high temperatures has been achieved and reported elsewhere [163,164]. The main mechanisms in the high-energy ball milling process are now discussed.

In the process, the particles that constitute the milled powder are exposed to an iterative combination of the following mechanisms: flattening, cold welding, fracture, and recombination. Because there is high mobility regarding the balls, there exist mainly three types of collisions: ball-ball, ball-powder, and ball-walls (the internal walls of the vial). However, ball-powder-ball type collisions also exist; in some instances, a given fraction of the powder is being pressed by two or more balls. Approximately a thousand particles with a total (aggregated) mass of 0.2 mg are confined in this type of collision [156]. At the initial stage, the impact force deforms the particles inelastically. As a result, work hardening and fracture take place at the surface level. New surfaces are formed, allowing the particles to weld. Thus, an effective increment of the averaged size of the particles follows. The size distribution widens in cases for which the new particles can be as three times larger than the original ones [156]. Then, deformation keeps its pace, and the particles harden due to mechanical fatigue or fragmentation of their layers or fragile sheets. The formation of this type of fragments might continue as far as no significant agglomeration forces exist. It translates into the reverse effect: an effective decrease in the average size of the particles. At this stage, the tendency of the particles to fragment dominates their tendency to weld [156]. Because the ball impacts continue, the structure of the particles keeps being refined and refined, no matter the size of the particles remains constant (it repeatedly increases and decreases due to the mechanism competition above described). Consequently, the inter-layer spacing decreases, whereas the number of layers per particle increases.

Now, so far, only the refinement mechanism of the particles has been discussed. The chemical reactions between the precursors have not been considered. Assuming that the precursors are oxides, then the particle refinement (and fragmentation) results in defect formation and an effective reduction of the diffusion distances. The interaction between the precursors is thus enhanced. Moreover, given the formation of new surfaces and interfaces, a substantial increment in the reactivity of the precursors is achieved [157]. This is what is understood by the 'chemical activation' of the precursors. If the process keeps going on, the chemical activation will also continue, and, in some instances, the necessary chemical reactions for the formation of a new phase start to appear. If the process still goes on further, it is possible that other processes also occur, such as nucleation, particle growth, and crystalline phase formation. The formation of an amorphous phase can even occur, which is not other than the extreme case of defect formation [157]. The localized heat transfer right at the impact zones (ball-powder-ball collisions) might contribute to this process. This hypothesis is justified by the fact that even when the temperature inside the vial hardly reaches 100 °C, the in situ temperature at each impact event can be large enough to activate the solid state reactions (~>800 °C) [165,166].

A Chinese group has extensively discussed other non-conventional methods to produce LNPws of high quality at the State Key Laboratory of fine Chemicals, Dalian University of Technology. This group has mainly focused on soft-chemistry methods such as sol-gel and hydrothermal processes [132], combustion method with urea as fuel [154], and wet chemical synthesis [155]. Within such references, Liu et al. have provided information on other synthesis methods for this purpose. Some are: sol–gel [167–169], Pechini method [170], metal alkoxides [171], hydrothermal process [172], and the peroxide route [168]. The synthesis of other ferroelectric powders by similar methods can also be consulted in the literature: chemical coprecipitation [173–175], sol–gel process [176–178], hydrothermal process [179–182], combustion [183], and molten salt [184,185].

Back to LN, the wet chemistry methods are based on the adoption of a kind of an α-carboxylic acid (citric acid) to coordinate with hydrated niobium acid ($Nb_2O_5 \cdot nH_2O$, often called as niobic acid), which is more reactive than the starting precursor based on niobium oxide, as described by Liu et al. [155]. In contrast, the mechanochemical-route is a dry environment method where, as mentioned, the reactivity of the Nb_2O_5 precursor is carried out through the high-energy ball milling process. While wet chemistry-based synthesis methods effectively tend to achieve the product at lower calcination temperatures, by employing high-energy ball milling the use of any kind of acid (organic or inorganic) is avoided. It is advantageous since the use of acids usually implies regulations to the exposure of hazardous or toxic fumes and vapors (ventilation issues), thus following strict security protocols.

4.4. Are the Stoichiometric and Congruent Compositions Univocally Determined?

In 1993 Kuz'minov and Osiko pointed out that the 'disturbance' of stoichiometry in LN crystals is primarily due to the lack of oxygen in Nb_2O_5, used as a raw material initial melt (Czochralski method) [186]. Accordingly, depending on its production technology, niobium pentoxide can adopt a wide range of compositions Nb_2O_{5-x}. Thus, once ST LN single crystals were believed to be only obtained for x = 0 at the synthesis stage. Although this is a severe statement, and even less likely to be correct because LN single crystals are hardly obtained without entailing a modification to the Czochralski method or an assisting process (as described in Section 2), the systematic study of the role of the CC of the precursor on that of the product would give interesting results. Among others, the documented indetermination of the CG point could be solved, since it is nowadays disputed to be in the range 48.38 mol % to 48.60 mol % of Li content [17]. Such an indetermination has been recently exposed for isostructural $LiTaO_3$ (LT) [187]: different defect concentrations were found on two CG LT single crystals synthesized by the same method but by distinct manufacturers. This type of analysis has not yet been conducted, according to the consulted literature. To answer these questions is a fundamental science problem, research of the kind of normal science concerning our scientific community. In principle, it can be solved without much effort or resources if treated at the level of synthesis and characterization of polycrystalline samples in the form of powders.

Research is here proposed where a mechanochemical-calcination route synthesizes LNPws. Before synthesizing several samples, the Nb_2O_5 precursor is to be processed to the extent of controlling the number of oxygen defects and characterized. The role of high-energy ball-milling is of interest, but no more than the role of annealing treatments under a controlled atmosphere in which at least both extremes must be studied: (1) passivated precursor powder (optimized conditions under an oxidizing treatment) and (2) highly-defective precursor powder (optimized conditions under a reduction treatment). The processed precursors would then be mixed with Li_2CO_3 and high-energy ball-milled for chemical activation; the milling atmosphere may or may not be controlled; an inert atmosphere would prevent oxygen defect formation. Calcination under an inert atmosphere would follow for the obtention of the products. Characterization by standard techniques such as XRD and Raman spectroscopy are recommended at all stages. For the case of the products, research like that described at the beginning of this section (and reported in [19]) could be conducted. However, in this case, the use of a direct method for determining the CC of the resultant LNPws is also suggested. The influence of the quality of the Nb_2O_5 precursor on that of the product should be traceable this way.

5. Ferroelectricity behind the Curtain

The origin of ferroelectric behavior in LN relies on the non-centrosymmetric arrangement of the constituent ions and their corresponding electrons (at low temperature) [188,189]. So far, this accounts only for the so-called 'spontaneous' polarization but not for its switching. The first concept stands for the existence of at least "two discrete stable or metastable states of different nonzero electric polarization in zero applied electric field", whereas switching describes the possibility to switch between these states with an applied electric field $\vec{E}$, which "changes the relative energy of the states through the coupling of the field to the polarization $-\vec{E}\cdot\vec{P}$" [91]. Some non-centrosymmetric polar structures, the wurtzite structure as an example, do not exhibit ferroelectric behavior since its intrinsic polarization cannot be switched at known experimental conditions [188]. Ferroelectricity is sensitive to *T*, the same as other related properties like piezoelectricity, pyroelectricity, and SHG. With increasing *T*, most ferroelectrics exhibit a "phase transition from the ferroelectric state, with multiple symmetry-related variants, to a nonpolar paraelectric phase, with a single variant" [91]. The measured Curie temperatures range for a gamut of ferroelectric materials from 1 K to 1000 K; for very high Ts, the probability of the material melting down before the transition occurs increases.

As previously stated, the symmetry-breaking relation between the high-symmetry paraelectric structure and the ferroelectric one is consistent with a second-order transition, described by the Landau order–disorder theory [130,131]. Apart from a finite discontinuity in the heat capacity of the system, this analysis also implies that the dielectric susceptibility diverges at the transition, and this can be linked to the vanishing frequency of a polar phonon. It is the central idea of the 'soft-mode' theory of ferroelectrics [189]. According to Rabe et al., during the 1960s and 1970s, the observation of the T-dependence of polar phonons was a "key ingredient in the great progress made in understanding the physics of ferroelectricity" [91]. Phonon spectroscopy is still a benchmark concerning the characterization of this type of transition, both via neutron scattering and optical spectroscopy [190,191]. Despite the success of this theoretical framework, no consensus exists today regarding the origin of ferroelectricity or ionic off-centering in known ferroelectrics [188,189].

Prototypical displacive transitions, characterized by a zone-centre vibrational mode—the soft mode—"with vanishing frequency at the phase transition and an eigenvector similar to the displacements observed in the ferroelectric phases", were long considered to be behind this type of phase transitions [189,192]. Displacive type transitions might be pictured as the multitudinous physical shifting of either the A or B cations (or both) off-center relative to the oxygen anions and, because of the dipole moment created by this shift, the emerging of a spontaneous polarization [188]. Nevertheless, in this respect, the so-called order-disorder type transitions have become popular lately. The difference between these types of transitions resides on whether a collective displacement of the cation sublattice is involved (displacive) or, instead, several local displacements of individual cations (order-disorder) [189]. In LN and LT, the latter type of displacements are thought to be induced by the second-order Jahn–Teller (SOJT) effect of the NbO_6 and TaO_6 octahedral units, respectively [189].

These ideas can be traced back to 1992 when Cohen made a paramount contribution regarding perovskite oxides [192]: he realized that the "great sensitivity of ferroelectrics to chemistry, defects, electrical boundary conditions, and pressure arises from a delicate balance between long-range Coulomb forces (which favor the ferroelectric state) and short-range repulsions (which favor the nonpolar state)." Most importantly, he concluded that "for ferroelectric perovskites in general, hybridization between the B cation and O is essential to weaken the short-range repulsions and allow the ferroelectric transition. Most ferroelectric oxide perovskites have B cations, whose lowest unoccupied states are d-states (Ti^{4+}, Nb^{5+}, Zr^{4+}, etcetera). It allows for d-hybridization with the O that softens the B–O repulsion and allows the ferroelectricity instability at low pressures" [192]. Hence, generally speaking (not only perovskites), the existence or absence of ferroelectric behavior relies on a balance between these short-range repulsions and additional bonding considerations, which act to stabilize the distortions that enable the transition to the polar state [188]. The 'additional bonding considerations' were pointed out by Megaw in 1954 [12]. The discussion in the next paragraph has been mostly adapted/reprinted from the content available in [188].

In terms of coordination chemistry, the stabilization of distorted structures is recognized to have its origin on chemical binding perturbations, denoted as SOJT effects [193–195], or sometimes pseudo-Jahn–Teller effects [196]. On the other hand, in plain physical terms, such effects can be seen by writing down a perturbative expansion of the energy of the electronic ground state $E(Q)$, as a function of the coordinate of the distortion Q [197]

$$E(Q) = E(0) + \left\langle 0 \left| \left(\frac{\delta H}{\delta Q} \right)_0 \right| 0 \right\rangle Q + \frac{1}{2} \left(\left\langle 0 \left| \left(\frac{\delta^2 H}{\delta Q^2} \right)_0 \right| 0 \right\rangle - 2 \sum_n{}' \frac{\left| \left\langle 0 \left| \left(\frac{\delta H}{\delta Q} \right)_0 \right| n \right\rangle \right|^2}{E_n - E(0)} \right) Q^2 + \cdots, \quad (7)$$

where $E(0)$ is the energy of the undistorted ground state, and the E_ns the energies describing the excited states. The term that is linear in Q is the first-order Jahn–Teller contribution, which is zero except for degenerate states. It is responsible for the characteristic tetragonal distortions in d^1 and d^4 perovskites, for example. Of the second-order terms, the first is always positive, whereas the second is always negative (provided that it is different from zero). If the second term is larger than the first, then distortion will cause an energy reduction. An increase in energy on distortion is described by the first second-order term, under the assumption of the absence of redistribution of electrons. Thus, it is dominated by the Coulomb repulsions between electron clouds and is the smallest for closed-shell ions that lack spatially extended valence electrons. On the other hand, two criteria must be satisfied for the second term to be large: (1) the energy denominator, $E_n - E(0)$, must be small, therefore, there must be excited states available of energy not largely separated from the ground state; and (2) the matrix element $\left\langle 0 \middle| \left(\frac{\delta H}{\delta Q}\right)_0 \middle| n \right\rangle$ must be nonzero; this occurs if the product of the symmetry representations for the ground and excited state and the distortion is totally symmetric. Thus, for a non-centrosymmetric distortion, if the ground state is centrosymmetric, then the lowest excited state must be non-centrosymmetric. Note that this term represents the mixing of the ground state with the excited states as a result of the distortion perturbation, and as such, it is associated with the formation of new chemical bonds in the low-symmetry configuration [198]. A non-centrosymmetric distortion then results if the lowering in energy associated with the mixing of term two is larger than the repulsion opposing the ion shift, described by term one [199].

In short, nowadays, ferroelectricity is often associated with materials having d^0 transition metal ions (V^{5+}, Nb^{5+}, Ta^{5+}, Ti^{4+}, Zr^{4+}) and cations with a lone pair s^2 electrons (Pb^{2+} and Bi^{3+}) [89,200]. In terms of crystal chemistry, these are SOJT active cations and, by breaking the centrosymmetric structure, have the potential of energy gain [89,189]. The d^0 ions tend towards the more covalent character in their bonding, leading to asymmetric coordination geometries that favor the development of a spontaneous polarization [200]. In LN and LT, the Nb and Ta ions have 5^+ formal charges with d^0 configuration, leading to cation displacement along the trigonal *c*-axis [189].

6. Conclusions

The scientific paradigm concerning lithium niobate (LiNbO3, LN) research has been explicitly stated and partially justified by presenting a conceptual framework with a convenient order. The concepts have been introduced having in mind both kinds of audience, the experienced researcher and the young contributors to this community. The present review is supported by the revision of several specialized references on this topic. We apologize in advance for the non-deliberated omission of any relevant piece of literature. The justification is to be completed with the content presented in Part II.

In the current part, special attention was put on the subject of the crystal structure. In this respect, we add that, rigorously, a correct statement would be that LN belongs to no other family of crystal structures, but it forms its own. Perhaps this may be implicitly stated when referred to as a 'pseudoilmenite' in recent literature. In other words, LN belongs to the family of pseudoilmenites, probably acknowledging the fact that LN is not a mineral that naturally forms on the surface of the Earth; it might be just probable that LN, or at least a close relative, may exist deep beneath the sea level (accounting for polymorphism and the extreme conditions in temperature and pressure) [201,202]. Then, the question arises: why not refer to it as the family of 'pseudoperovskites', or simply as a member of the larger family of highly distorted perovskites? Would this be somehow of advantage for future investigations?

Besides the exposition of various indirect methods to determine the chemical composition of LN powders (LNPws), the role of this polycrystalline version of the material in complex piezoelectric ceramics related modern electronic components and the innermost cause of ferroelectricity have also been discussed. The reader may have noticed that these two functionalities—the former concerning powders only—have had their corresponding progress out of this field based on the perovskite framework. Perovskites are currently shaping our modern life with computers, smartphones, and medical ultrasound, and seem to keep doing it with the advent of efficient solar cells and the

Internet of Things [151,203,204]. Unprecedented profit might be obtained by relaxing our conceptions about our favorite material and exhorting its kinship to perovskites. In this respect, it is interesting that in a contribution in which its fresh ink still can be smelled, LN and isostructural $LiTaO_3$ are classified as perovskite ferroelectrics [16]. Finally, the possibility of the stoichiometric and congruent compositions yet not being univocally determined, regardless of the solid solution of LN (single crystal or polycrystal), has been signaled. A strategy that would not convey much effort nor expense has been designed to answer this conundrum.

Author Contributions: Conceptualization, O.S.-D.; validation, D.M.C.F. and J.T.E.-G.; investigation, O.S.-D., S.D.V.-M. and C.D.F.-R.; writing—original draft preparation, O.S.-D.; writing—review and editing, O.S.-D. and R.F.; visualization, S.D.V.-M. and D.M.C.F.; supervision, R.F. and J.T.E.-G.; project administration, R.F.; funding acquisition, R.F. All authors have read and agreed to the published version of the manuscript.

Funding: The publishing fee was covered by Universidad Autónoma de Ciudad Juárez.

Acknowledgments: O.S.-D. thanks the post-doctoral grant PRODEP Oficio no. 511-6/2020-2054.

Conflicts of Interest: The authors declare no conflict of interest.

References

1. Weis, R.S.; Gaylord, T.K. Lithium niobate: Summary of physical properties and crystal structure. *Appl. Phys. A* **1985**, *37*, 191–203. [CrossRef]
2. Räuber, A. Chemistry and Physics of Lithium Niobate. In *Current Topics in Materials Science*, 1st ed.; Kaldis, E., Ed.; North-Holland: Amsterdam, The Netherlands, 1978; Volume 1, pp. 481–601.
3. Bartasyte, A.; Margueron, S.; Baron, T.; Oliveri, S.; Boulet, P. Toward High-Quality Epitaxial $LiNbO_3$ and $LiTaO_3$ Thin Films for Acoustic and Optical Applications. *Adv. Mater. Interfaces* **2017**, *4*, 1600998. [CrossRef]
4. Streque, J.; Aubert, T.; Kokanyan, N.; Bartoli, F.; Taguett, A.; Polewczyk, V.; Kokanyan, E.; Hage-Ali, S.; Boulet, P.; Elmazria, O. Stoichiometric Lithium Niobate Crystals: Towards Identifiable Wireless Surface Acoustic Wave Sensors Operable up to 600 °C. *IEEE Sensors Lett.* **2019**, *3*, 2501204. [CrossRef]
5. Harvard John, A. Paulson School of Engineering and Applied Sciences, "Now Entering, Lithium Niobate Valley: Researchers Demonstrate High-Quality Optical Microstructures Using Lithium Niobate" by Leah Burrows. Available online: https://www.seas.harvard.edu/news/2017/12/now-entering-lithium-niobate-valley (accessed on 31 July 2020).
6. Kösters, M.; Sturman, B.; Werheit, P.; Haertle, D.; Buse, K. Optical cleaning of congruent lithium niobate crystals. *Nat. Photon* **2009**, *3*, 510–513. [CrossRef]
7. Zhang, M.; Wang, C.; Cheng, R.; Shams-Ansari, A.; Lončar, M. Monolithic ultra-high-Q lithium niobate microring resonator. *Optica* **2017**, *4*, 1536–1537. [CrossRef]
8. Desiatov, B.; Shams-Ansari, A.; Zhang, M.; Wang, C.; Loncar, M. Ultra-low-loss integrated visible photonics using thin-film lithium niobate. *Optics* **2019**, *6*, 380–384. [CrossRef]
9. Osborne, I.S. An active platform for integrated optics. *Science* **2019**, *364*, 448. [CrossRef]
10. Pang, C.; Li, R.; Li, Z.; Dong, N.; Cheng, C.; Nie, W.; Böttger, R.; Zhou, S.; Wang, J.; Chen, F. Lithium Niobate Crystal with Embedded Au Nanoparticles: A New Saturable Absorber for Efficient Mode-Locking of Ultrafast Laser Pulses at 1 μm. *Adv. Opt. Mater.* **2018**, *6*, 180035. [CrossRef]
11. Kuhn, T.S. *The Structure of Scientific Revolutions*, 1st ed.; University of Chicago Press: Chicago, IL, USA, 1962; p. 172.
12. Megaw, H.D. Ferroelectricity and crystal structure. II. *Acta Crystallogr.* **1954**, *7*, 187–194. [CrossRef]
13. Megaw, H.D. *Ferroelectricity in Crystals*, 1st ed.; Methuen: London, UK, 1957.
14. Megaw, H.D. A note on the structure of lithium niobate, $LiNbO_3$. *Acta Crystallogr. Sect. A Cryst. Phys. Diffr. Theor. Gen. Crystallogr.* **1968**, *A24*, 583–588. [CrossRef]
15. Valasek, J. Piezo-Electric and Allied Phenomena in Rochelle Salt. *Phys. Rev.* **1921**, *17*, 475–481. [CrossRef]

16. Brennecka, G.; Sherbondy, R.; Schwartz, R.; Ihlefeld, J. Ferroelectricity—A revolutionary century of discovery. *Am. Cer. Soc. Bull.* **2020**, *99*, 24–30.
17. Kang, X.; Liang, L.; Song, W.; Wang, F.; Sang, Y.; Liu, H. Formation mechanism and elimination methods for anti-site defects in $LiNbO_3$/$LiTaO_3$ crystals. *Cryst. Eng. Comm.* **2016**, *18*, 8136–8146. [CrossRef]
18. Kocsor, L.; Péter, L.; Corradi, G.; Kis, Z.; Gubicza, J.; Kovács, L. Mechanochemical Reactions of Lithium Niobate Induced by High-Energy Ball-Milling. *Crystals* **2019**, *9*, 334. [CrossRef]
19. Sánchez-Dena, O.; Villagómez, C.J.; Fierro-Ruíz, C.D.; Padilla-Robles, A.S.; Farías, R.; Vigueras-Santiago, E.; Hernández-López, S.; Reyes-Esqueda, J.-A. Determination of the Chemical Composition of Lithium Niobate Powders. *Crystals* **2019**, *9*, 340. [CrossRef]
20. Volk, T.; Wöhlecke, M. Introduction. In *Lithium Niobate. Defects, Photorefraction and Ferroelectric Switching*, 1st ed.; Springer Series in Materials Science 115; Hull, R., Osgood, R.M., Jr., Parisi, J., Warlimont, H., Eds.; Springer: Berlin/Heidelberg, Germany, 2009; Volume 115, pp. 1–8.
21. Cudney, R.S.; Ríos, L.A.; Orozco-Arellanes, M.J.; Alonso, F.; Fonseca, J. Fabricación de niobato de litio periódicamente polarizado para óptica no lineal. *Rev. Mex. Física* **2002**, *48*, 548–555.
22. Zachariasen, F.W.H. Standard x-ray diffraction powder patterns. *[Kl.] 1 Mat. Naturv-Idensk. Kl* **1928**, *4*, 1–8.
23. Lima-de-Faria, J. *Structural Mineralogy. An Introduction*, 1st ed.; Springer Science+Business Media: Dordrecht, UK, 1994.
24. Sumets, M.P.; Dybov, V.A.; Ievlev, V.M. $LiNbO_3$ films: Potential application, synthesis techniques, structure, properties. *Inorg. Mater.* **2017**, *53*, 1361–1377. [CrossRef]
25. Crystals: Special Issue "Recent Progress in Lithium Niobate", Special Issue Information (Guest Editors: Robert A. Jackson and Zsuzsanna Szaller). Available online: https://www.mdpi.com/journal/crystals/special_issues/Lithium_Niobate (accessed on 31 July 2020).
26. Matthias, B.T.; Remeika, J.P. Ferroelectricity in the Ilmenite Structure. *Phys. Rev.* **1949**, *76*, 1886–1887. [CrossRef]
27. Nassau, K. Early History of Lithium Niobate: Personal Reminiscences. In Proceedings of the Guided Wave Optoelectronic Materials I, Los Angeles Technical Symposium, Los Angeles, CA, USA, 26 September 1984; Holman, R.L., Smyth, D.M., Eds.; SPIE Digital Library; Proc. SPIE 0460. pp. 2–5.
28. Ballman, A.A. Growth of Piezoelectric and Ferroelectric Materials by the Czochralski Technique. *J. Am. Ceram. Soc.* **1965**, *48*, 112–113. [CrossRef]
29. Fedulov, S.A.; Shapiro, Z.I.; Ladyzhenskii, P.B. Apparatus for Growth of Single Crystal, Single Domain $LiNbO_3$. *Kristals* **1965**, *10*, 268–273.
30. Nassau, K.; Levinstein, H.; LoIacono, G. Ferroelectric lithium niobate. 1. Growth, domain structure, dislocations and etching. *J. Phys. Chem. Solids* **1966**, *27*, 839–888. [CrossRef]
31. Nassau, K.; Levinstein, H.; LoIacono, G. Ferroelectric lithium niobate. 2. Preparation of single domain crystals. *J. Phys. Chem. Solids* **1966**, *27*, 989–996. [CrossRef]
32. Abrahams, S.C.; Reddy, J.M.; Bernstein, J.L. Ferroelectric Lithium Niobate. 3. Single crystal X-ray diffraction study at 24 °C. *J. Phys. Chem. Solids* **1966**, *27*, 971–1012.
33. Abrahams, S.C.; Hamilton, W.C.; Reddy, J.M. Ferroelectric Lithium Niobate. 4. Single crystal neutron diffraction study at 24 °C. *J. Phys. Chem. Solids* **1966**, *27*, 1013–1018. [CrossRef]
34. Abrahams, S.C.; Levinstein, H.J.; Reddy, J.M. Ferroelectric Lithium Niobate. 5. Polycrystal X-ray diffraction study between 24 °C and 1200 °C. *J. Phys. Chem. Solids* **1966**, *27*, 1019–1026.
35. Bartasyte, A.; Plausinaitiene, V.; Abrutis, A.; Stanionyte, S.; Margueron, S.; Boulet, P.; Kobata, T.; Uesu, Y.; Gleize, J. Identification of $LiNbO_3$, $LiNb_3O_8$ and Li_3NbO_4 phases in thin films synthesized with different deposition techniques by means of XRD and Raman spectroscopy. *J. Phys. Condens. Matter* **2013**, *25*, 205901. [CrossRef]
36. Hatano, H.; Kitamura, K.; Liu, Y. Growth and Photorefractive Properties of Stoichiometric $LiNbO_3$ and $LiTaO_3$. In *Photorefractive Materials and Their Applications 2*; Springer: New York, NY, USA, 2007; Volume 114, pp. 127–164.

37. Kitamura, K.; Yamamoto, J.; Iyi, N.; Kirnura, S.; Hayashi, T. Stoichiometric $LiNbO_3$ single crystal growth by double crucible Czochralski method using automatic powder supply system. *J. Cryst. Growth* **1992**, *116*, 327–332. [CrossRef]
38. Malovichko, G.I.; Grachev, V.G.; Yurchenko, L.P.; Proshko, V.Y.; Kokanyan, E.; Gabrielyan, V.T. Improvement of $LiNbO_3$ Microstructure by Crystal Growth with Potassium. *Phys. Status Solidi (a)* **1992**, *133*, K29–K32. [CrossRef]
39. Bordui, P.F.; Norwood, R.G.; Jundt, D.H.; Fejer, M.M. Preparation and characterization of off-congruent lithium niobate crystals. *J. Appl. Phys.* **1992**, *71*, 875–879. [CrossRef]
40. Polgár, K.; Péter, Á.; Kovács, L.; Corradi, G.; Szaller, Z. Growth of stoichiometric $LiNbO_3$ single crystals by top seeded solution growth method. *J. Cryst. Growth* **1997**, *177*, 211–216. [CrossRef]
41. Polgár, K.; Péter, Á; Földvári, I. Crystal growth and stoichiometry of $LiNbO_3$ prepared by the flux method. *Opt. Mater.* **2002**, *19*, 7–11. [CrossRef]
42. Lengyel, K.; Péter, Á.; Kovács, L.; Corradi, G.; Pálfalvi, L.; Hebling, J.; Unferdorben, M.; Dravecz, G.; Hajdara, I.; Szaller, Z.; et al. Growth, defect structure, and THz application of stoichiometric lithium niobate. *Appl. Phys. Rev.* **2015**, *2*, 040601. [CrossRef]
43. O'Bryan, H.M.; Gallagher, P.K.; Brandle, C.D. Congruent Composition and Li-Rich Phase Boundary of $LiNbO_3$. *J. Am. Ceram. Soc.* **1985**, *68*, 493–496. [CrossRef]
44. Grabmaier, B.; Otto, F. Growth and investigation of MgO-doped $LiNbO_3$. *J. Cryst. Growth* **1986**, *79*, 682–688. [CrossRef]
45. Iyi, N.; Kitamura, K.; Izumi, F.; Yamamoto, J.; Hayashi, T.; Asano, H.; Kimura, S. Comparative of defect structures in lithium niobate with different compositions. *J. Solid State Chem.* **1992**, *101*, 340–352. [CrossRef]
46. Furukawa, Y.; Sato, M.; Kitamura, K.; Yajima, Y.; Minakata, M. Optical damage resistance and crystal quality of $LiNbO_3$ single crystals with various [Li]/[Nb] ratios. *J. Appl. Phys.* **1992**, *72*, 3250–3254. [CrossRef]
47. Wöhlecke, M.; Corradi, G.; Betzler, K. Optical methods to characterise the composition and homogeneity of lithium niobate single crystals. *Appl. Phys. B* **1996**, *63*, 323–330. [CrossRef]
48. Schlarb, U.; Klauer, S.; Wesselmann, M.; Betzler, K. Determination of the Li/Nb ratio in lithium niobate by means of birefringence and Raman measurements. *Appl. Phys. A* **1993**, *56*, 311–315. [CrossRef]
49. Malovichko, G.I.; Grachev, V.G.; Kokanyan, E.; Schirmer, O.F.; Betzler, K.; Gather, B.; Jermann, F.; Klauer, S.; Schlarb, U. Characterization of stoichiometric $LiNbO_3$ grown from melts containing K_2O. *Appl. Phys. A* **1993**, *56*, 103–108. [CrossRef]
50. Vartanyan, E.S. Lecture on the All-Union Conference on Crystal Growth. Tsachkadzor-Aghveran: Armenian SSR. 1985. Available online: https://aip.scitation.org/doi/abs/10.1063/1.4929917?crawler=true&mimetype=application%2Fpdf&journalCode=are (accessed on 18 October 2020).
51. Földvári, I.; Polgár, K.; Voszka, R.; Balasanyan, R.N. A simple method to determine the real composition of $LiNbO_3$ crystals. *Cryst. Res. Technol.* **1984**, *19*, 1659–1661. [CrossRef]
52. Kovacs, L.; Ruschhaupt, G.; Polgár, K.; Corradi, G.; Wöhlecke, M. Composition dependence of the ultraviolet absorption edge in lithium niobate. *Appl. Phys. Lett.* **1997**, *70*, 2801–2803. [CrossRef]
53. Dravecz, G.; Kovács, L. Determination of the crystal composition from the OH- vibrational spectrum in lithium niobate. *Appl. Phys. A* **2007**, *88*, 305–307. [CrossRef]
54. Dravecz, G.; Kovács, L.; Péter, Á.; Polgár, K.; Bourson, P. Raman and infrared spectroscopy characterization of $LiNbO_3$ crystals grown from alkali metal oxide solvents. *Phys. Status Solidi* **2007**, *4*, 1313–1316.
55. Schlarb, U.; Betzler, K. Refractive indices of lithium niobate as a function of temperature, wavelength, and composition: A generalized fit. *Phys. Rev. B* **1993**, *48*, 15613–15620. [CrossRef] [PubMed]
56. Byer, R.L.; Young, J.F.; Feigelson, R.S. Growth of High-Quality $LiNbO_3$ Crystals from the Congruent Melt. *J. Appl. Phys.* **1970**, *41*, 2320. [CrossRef]
57. Luh, Y.; Fejer, M.; Byer, R.; Feigelson, R. Stoichiometric $LiNbO_3$ single-crystal fibers for nonlinear optical applications. *J. Cryst. Growth* **1987**, *85*, 264–269. [CrossRef]
58. Jundt, D.; Fejer, M.; Byer, R. Optical properties of lithium-rich lithium niobate fabricated by vapor transport equilibration. *IEEE J. Quantum Electron.* **1990**, *26*, 135–138. [CrossRef]

59. Schmidt, N.; Betzler, K.; Grabmaier, B.C. Composition dependence of the second-harmonic phase-matching temperature in $LiNbO_3$. *Appl. Phys. Lett.* **1991**, *58*, 34–35. [CrossRef]
60. Reichert, A.; Betzler, K. Characterization of electrooptic crystals by non-collinear frequency doubling. *Ferroelectrics* **1992**, *126*, 9–14. [CrossRef]
61. Arizmendi, L. Simple holographic method for determination of Li/Nb ratio and homogeneity of $LiNbO_3$ crystals. *J. Appl. Phys.* **1988**, *64*, 4654–4656. [CrossRef]
62. Carruthers, J.R.; Peterson, G.E.; Grasso, M.; Bridenbaugh, P.M. Nonstoichiometry and Crystal Growth of Lithium Niobate. *J. Appl. Phys.* **1971**, *42*, 1846. [CrossRef]
63. Zotov, N.; Boysen, H.; Frey, F.; Metzger, T.; Born, E. Cation substitution models of congruent $LiNbO_3$ investigated by X-ray and neutron powder diffraction. *J. Phys. Chem. Solids* **1994**, *55*, 145–152. [CrossRef]
64. Lerner, P.; Legras, C.; Dumas, J.P. STOECHIOMÉTIE DES MONOCRISTAUX DE MÉTANIOBATE DE LITHIUM. *J. Cryst. Growth* **1968**, *3*, 231–235. [CrossRef]
65. Abrahams, S.C.; Marsh, P. Defect structure dependence on composition in lithium niobate. *Acta Crystallogr. Sect. B Struct. Sci.* **1986**, *B42*, 61–68. [CrossRef]
66. Kovács, L.; Polgár, K. Density Measurements on $LiNbO_3$ Crystals Confirming Nb Substitution for Li. *Cryst. Res. Technol.* **1986**, *21*, K101–K104. [CrossRef]
67. Peterson, G.; Carruthers, J. 93Nb NMR as a sensitive and accurate probe of stoichiometry in $LiNbO_3$ crystals. *J. Solid State Chem.* **1969**, *1*, 98–99. [CrossRef]
68. Malovichko, G.; Grachev, V.; Schirmer, O. The effect of iron ions on the defect structure of lithium niobate crystals grown from K_2O containing melts. *Solid State Commun.* **1994**, *89*, 195–198. [CrossRef]
69. Yamada, K.; Takemura, H.; Inoue, Y.; Omi, T.; Matsumura, S. Effect of Li/Nb Ratio on the SAW Velocity of 128°Y–X $LiNbO_3$Wafers. *Jpn. J. Appl. Phys.* **1987**, *26*, 219. [CrossRef]
70. Kushibiki, J.; Takahashi, H.; Kobayashi, T.; Chubachi, N. Characterization of $LiNbO_3$ crystals by line-focus-beam acoustic microscopy. *Appl. Phys. Lett.* **1991**, *58*, 2622–2624. [CrossRef]
71. Muller, O.; Roy, R. *The Major Ternary Structural Families*; Springer: Berlin/Heidelberg, Germany; New York, NY, USA, 1974.
72. Schweinler, H.C. Ferroelectricity in the Ilmenite Structure. *Phys. Rev.* **1952**, *87*, 5–11. [CrossRef]
73. Alves, A.R.; Coutinho, A.D.R. The Evolution of the Niobium Production in Brazil. *Mater. Res.* **2015**, *18*, 106–112. [CrossRef]
74. Bailey, P. Measurement of the Proton Structure Function F_2 at Very Low Q_2 at HERA Doctoral Thesis. Bristol: 1952. Available online: https://www.sciencedirect.com/science/article/abs/pii/S0370269300007930 (accessed on 31 July 2020).
75. Volk, T.; Wöhlecke, M. Point defects in $LiNbO_3$. In *Lithium Niobate. Defects, Photorefraction and Ferroelectric Switching*, 1st ed.; Hull, R., Osgood, R.M., Jr., Parisi, J., Warlimont, H., Eds.; Springer Series in Materials Science 115; Springer: Berlin/Heidelberg, Germany, 2009; Volume 115, pp. 9–50.
76. Sanna, S.; Schmidt, W.G. Lithium niobate X-cut, Y-cut, and Z-cut surfaces from ab initio theory. *Phys. Rev. B* **2010**, *81*, 214116. [CrossRef]
77. Gopalan, V.; Dierolf, V.; Scrymgeour, D.A. Defect–Domain Wall Interactions in Trigonal Ferroelectrics. *Annu. Rev. Mater. Res.* **2007**, *37*, 449–489. [CrossRef]
78. Peterson, G.E.; Carnevale, A. 93Nb NMR Linewidths in Nonstoichiometric Lithium Niobate. *J. Chem. Phys.* **1972**, *56*, 4848–4851. [CrossRef]
79. Wilkinson, A.P.; Cheetham, A.K.; Jarman, R.H. The defect structure of congruently melting lithium niobate. *J. Appl. Phys.* **1998**, *74*, 3080–3083. [CrossRef]
80. Peterson, G.E. Private Communication. 1965. Available online: https://aip.scitation.org/doi/abs/10.1063/1.1840478 (accessed on 18 October 2020).
81. Buerger, M.J. Derivative Crystal Structures. *J. Chem. Phys.* **1947**, *15*, 1–16. [CrossRef]
82. Buerger, M.J. The genesis of twin crystals. *Am. Mineral.* **1945**, *30*, 469–482.
83. Bärnighausen, H. Group-Subgroup Relations between Space Groups: A Useful Tool in Crystal Chemistry. *Match Commun. Math. Chem.* **1980**, *9*, 139–175.

84. Benz, K.-W.; Neumann, W. *Introduction to Crystal Growth and Characterization*, 1st ed.; Wiley-VCH Verlag GmbH & Co.: Weinheim, Germany, 2014.
85. Megaw, H.D.; Templeton, D.H. Crystal Structures: A Working Approach. *Phys. Today* **1974**, *27*, 53. [CrossRef]
86. Johnsson, M.; Lemmens, P. Crystallography and Chemistry of Perovskites. In *Handbook of Magnetism and Advanced Magnetic Materials*; Kronmüller, H., Parkin, S., Coey, M., Inoue, A., Kronmüller, H., Eds.; Wiley: Hoboken, NJ, USA, 2007; Volume 4, pp. 1–9. Available online: https://onlinelibrary.wiley.com/doi/abs/10.1002/9780470022184.hmm411 (accessed on 18 October 2020).
87. Bhalla, A.; Guo, R.; Roy, R. The perovskite structure—A review of its role in ceramic science and technology. *Mater. Res. Innov.* **2000**, *4*, 3–26. [CrossRef]
88. Howard, C.J.; Stokes, H.T. Structures and phase transitions in perovskites—A group-theoretical approach. *Acta Cryst.* **2005**, *A61*, 93–111. [CrossRef] [PubMed]
89. Woodward, P.M.; Mizoguchi, H.; Kim, Y.-I.; Stoltzfus, M.W. Chapter 6. The Electronic Structure of Metal Oxides. In *Metal Oxides: Chemistry and Applications*, 1st ed.; Fierro, J.L.G., Ed.; CRC Press Taylor & Francis Group: Boca Raton, FL, USA, 2006; pp. 133–193.
90. Lufaso, M.W.; Woodward, P.M. Prediction of the crystal structures of perovskites using the software program SPuDS. *Acta Crystallogr. Sect. B Struct. Sci.* **2001**, *B57*, 725–738. [CrossRef] [PubMed]
91. Rabe, K.M.; Dawber, M.; Lichtensteiger, C.; Ahn, C.H.; Triscone, J.-M. Modern Physics of Ferroelectrics: Essential Background. In *Physics of Ferroelectrics: A Modern Perspective*, 1st ed.; Rabe, K.M., Ahn, C.H., Triscone, J.-M., Eds.; Springer: Berlin/Heidelberg, Germany, 2007; pp. 1–30.
92. Navrotsky, A. Energetics and Crystal Chemical Systematics among Ilmenite, Lithium Niobate, and Perovskite Structures. *Chem. Mater.* **1998**, *10*, 2787–2793. [CrossRef]
93. Shannon, R.D.; Prewitt, C.T. Effective Ionic Radii in Oxides and Flourides. *Acta Cryst.* **1969**, *B25*, 925–946. [CrossRef]
94. Takayama-Muromachi, E.; Navrotsky, A. Energetics of compounds ($A^{2+}B^{4+}O_3$) with the perovskite structure. *J. Solid State Chem.* **1988**, *72*, 244–256. [CrossRef]
95. Kubo, A.; Giorgi, G.; Yamashita, K. $MgTaO_2N$ Photocatalysts: Perovskite versus Ilmenite Structure. A Theoretical Investigation. *J. Phys. Chem. C* **2017**, *121*, 27813–27821. [CrossRef]
96. Wilson, N.C.; Muscat, J.; Mkhonto, D.; Ngoepe, P.E.; Harrison, N.M. Structure and properties of ilmenite from first principles. *Phys. Rev. B* **2005**, *71*, 075202. [CrossRef]
97. Leinenweber, K.; Utsumi, W.; Tsuchida, Y.; Yagi, T.; Kurita, K. Unquenchable high-pressure perovskite polymorphs of $MnSnO_3$ and $FeTiO_3$. *Phys. Chem. Miner.* **1991**, *18*, 244–250. [CrossRef]
98. Mehta, A.; Leinenweber, K.; Navrotsky, A.; Akaogi, M. Calorimetric study of high pressure polymorphism in $FeTiO_3$: Stability of the perovskite phase. *Phys. Chem. Miner.* **1994**, *21*, 207–212. [CrossRef]
99. Haertling, G.H. Ferroelectric Ceramics: History and Technology. *J. Am. Ceram. Soc.* **1999**, *82*, 797–818. [CrossRef]
100. Barth, T.F.W.; Posnjak, E. The Crystal Structure of Ilmenite. *Z. Kristallographie Cryst. Mater.* **1934**, *88*, 265–270. [CrossRef]
101. Shiozaki, Y.; Mitsui, T. Powder neutron diffraction study of $LiNbO_3$. *J. Phys. Chem. Solids* **1963**, *24*, 1057–1061. [CrossRef]
102. Niizeki, N.; Yamada, T.; Toyoda, H. Growth Ridges, Etched Hillocks, and Crystal Structure of Lithium Niobate. *J. Appl. Phys.* **1967**, *6*, 318–327. [CrossRef]
103. Heinrich, V.E.; Cox, P.A. *The Surface Science of Metal Oxides*, 1st ed.; Cambridge University Press: Cambridge, UK, 1994.
104. Bond, W.L. Precision lattice constant determination. *Acta Crystallogr.* **1960**, *13*, 814–818. [CrossRef]
105. XRD Software-DIFRAC.SUITE, TOPAS Software. Available online: https://www.bruker.com/products/x-ray-diffraction-and-elemental-analysis/x-ray-diffraction/xrd-software/topas.html (accessed on 31 July 2020).
106. BGMN Home Page. Site Map. Available online: http://www.bgmn.de/ (accessed on 31 July 2020).
107. Malvern Panalytical, HighScore Plus. Available online: https://www.malvernpanalytical.com/en/products/category/software/x-ray-diffraction-software/highscore-with-plus-option?creative=338893424388&keyword=x%20pert%20highscore%20plus&matchtype=e&network=g&device=c&gclid=EAIaIQobChMIwK708OSg5QIVDtvACh1vLAubEAAYASAAEgJf3PD_BwE (accessed on 31 July 2020).

108. FullProf Suite. Available online: https://www.ill.eu/sites/fullprof/ (accessed on 31 July 2020).
109. GSAS-II. Available online: https://subversion.xray.aps.anl.gov/trac/pyGSAS (accessed on 31 July 2020).
110. Young, R.A. *The Rietveld Method*, 1st ed.; Oxford University Press: New York, NY, USA, 1993.
111. Pecharsky, V.K.; Zavalij, P.Y. *Fundamentals of Powder Diffraction and Structural Characterization of Materials*, 2nd ed.; Springer: New York, NY, USA, 2009.
112. Dreele, R.B. Rietveld Refinement. In *Powder Diffraction: Theory and Practice*, 1st ed.; Dinnebier, R.E., Billinge, S.J.L., Eds.; The Royal Society of Chemistry: Cambridge, UK, 2008; pp. 266–281.
113. Bish, D.L.; Howard, S.A. Quantitative phase analysis using the Rietveld method. *J. Appl. Crystallogr.* **1988**, *21*, 86–91. [CrossRef]
114. Zhang, Y.; Guilbert, L.; Bourson, P.; Polgar, K.; Fontana, M.D. Characterization of short-range heterogeneities in sub-congruent lithium niobate by micro-Raman spectroscopy. *J. Phys. Condens. Matter* **2006**, *18*, 957–963. [CrossRef]
115. Fontana, M.D.; Bourson, P. Microstructure and defects probed by Raman spectroscopy in lithium niobate crystals and devices. *Appl. Phys. Rev.* **2015**, *2*, 040602. [CrossRef]
116. Sanna, S.; Neufeld, S.; Rüsing, M.; Berth, G.; Zrenner, A.; Schmidt, W.G. Raman scattering efficiency in $LiTaO_3$ and $LiNbO_3$ crystals. *Phys. Rev. B* **2015**, *91*, 224302. [CrossRef]
117. Pezzotti, G. Raman spectroscopy of piezoelectrics. *J. Appl. Phys.* **2013**, *113*, 211301. [CrossRef]
118. Hermet, P.; Veithen, M.; Ghosez, P. First-principles calculations of the nonlinear optical susceptibilities and Raman scattering spectra of lithium niobate. *J. Phys. Condens. Matter* **2007**, *19*, 456202. [CrossRef]
119. Margueron, S.; Bartasyte, A.; Glazer, A.M.; Simon, E.; Hlinka, J.; Gregora, I.; Gleize, J. Resolved E-symmetry zone-centre phonons in $LiTaO_3$ and $LiNbO_3$. *J. Appl. Phys.* **2012**, *111*, 104105. [CrossRef]
120. Schaufele, R.F.; Weber, M.J. Raman Scattering by Lithium Niobate. *Phys. Rev.* **1966**, *152*, 705–708. [CrossRef]
121. Porto, S.P.S.; Krishnan, R.S. Raman Effect of Corundum. *J. Chem. Phys.* **1967**, *47*, 1009–1012. [CrossRef]
122. Horiba Scientific: Webinar: Applied Polarized Raman Spectroscopy Webinar (by David Tuschel). Available online: http://www.horiba.com/us/en/scientific/products/raman-spectroscopy/raman-academy/webinars/applied-polarized-raman-spectroscopy-webinar/ (accessed on 31 July 2020).
123. Repelin, Y.; Husson, E.; Bennani, F.; Proust, C. Raman spectroscopy of lithium niobate and lithium tantalate. Force field calculations. *J. Phys. Chem. Solids* **1999**, *60*, 819–825. [CrossRef]
124. Thermo Fisher Scientific, Application Note: Curve Fitting in Raman and IR Spectroscopy. Available online: https://www.thermofisher.com/search/results?query=Curve%20Fitting%20in%20Raman&focusarea=Search%20All (accessed on 31 July 2020).
125. Scott, B.A.; Burns, G. Determination of Stoichiometry Variations in $LiNbO_3$ and $LiTaO_3$ by Raman Powder Spectroscopy. *J. Am. Ceram. Soc.* **1972**, *55*, 225–230. [CrossRef]
126. Pyatigorskaya, L.I.; Shapiro, Z.I.; Margolin, L.N.; Bovina, E.A. Determination of the composition of $LiNbO_3$ specimens by Raman spectroscopy. *J. Appl. Spectrosc.* **1983**, *38*, 491–493. [CrossRef]
127. Redfield, D. Optical absorption edge of $LiNbO_3$. *J. Appl. Phys.* **1974**, *45*, 4566–4571. [CrossRef]
128. Thierfelder, C.; Sanna, S.; Schindlmayr, A.; Schmidt, W.G. Do we know the band gap of lithium niobate? *Phys. Status Solidi* **2010**, *7*, 362–365. [CrossRef]
129. Fox, M. *Optical Properties of Solids*, 1st ed.; Oxford University Press: New York, NY, USA, 2001.
130. Jona, F.; Shirane, G. *Ferroelectric Crystals*; Dover: New York, NY, USA, 1993.
131. Chandra, P.; Littlewood, P.B. A Landau Primer for Ferroelectrics. In *Physics of Ferroelectrics: A Modern Perspective*, 1st ed.; Rabe, K.M., Ahn, C.H., Triscone, J.-M., Eds.; Springer: Berlin/Heidelberg, Germany, 2007; pp. 69–115.
132. Liu, M.; Xue, D.; Li, K. Soft-chemistry synthesis of $LiNbO_3$ crystallites. *J. Alloy. Compd.* **2008**, *449*, 28–31. [CrossRef]
133. Torrent, J.; Barrón, V. Diffuse Reflectance Spectroscopy. In *Methods of Soil Analysis Part 5—Mineralogical Methods*, 1st ed.; Ulery, A.L., Drees, R., Eds.; Soil Society of America: Dane County, WI, USA, 2008; pp. 367–385.
134. Kubelka, P. New Contributions to the Optics of Intensely Light-Scattering Materials Part I. *J. Opt. Soc. Am.* **1948**, *38*, 448–457. [CrossRef] [PubMed]

135. Kubelka, P. New Contributions to the Optics of Intensely Light-Scattering Materials Part II: Nonhomogeneous Layers. *J. Opt. Soc. Am.* **1954**, *44*, 330–335. [CrossRef]
136. Kurtz, S.K.; Perry, T.T. A Powder Technique for the Evaluation of Nonlinear Optical Materials. *J. Appl. Phys.* **1968**, *39*, 3798–3813. [CrossRef]
137. Chowdhury, A.U.; Zhang, S.; Simpson, G.J. Powders Analysis by Second Harmonic Generation Microscopy. *Anal. Chem.* **2016**, *88*, 3853–3863. [CrossRef]
138. Dena, O.S.; García-Ramírez, E.V.; Fierro-Ruiz, C.; Vigueras-Santiago, E.; Farías, R.; Reyes-Esqueda, A. Effect of size and composition on the second harmonic generation from lithium niobate powders at different excitation wavelengths. *Mater. Res. Express* **2017**, *4*, 035022. [CrossRef]
139. Sánchez-Dena, O.; Behel, Z.; Salmon, E.; Benichou, E.; Reyes-Esqueda, J.-A.; Brevet, P.-F.; Jonin, C. Polarization-resolved second harmonic generation from $LiNbO_3$ powders. *Opt. Mater.* **2020**, *107*, 110169. [CrossRef]
140. Aramburu, I.; Ortega, J.; Folcia, C.L.; Etxebarria, J. Second harmonic generation by micropowders: A revison of the Kurtz-Perry method and its practical application. *Appl. Phys. B* **2004**, *116*, 211–233. [CrossRef]
141. Fierro-Ruíz, C.D.; Sánchez-Dena, O.; Cabral-Larquier, E.M.; Galindo, J.T.E.; Farías, R. Structural and Magnetic Behavior of Oxidized and Reduced Fe Doped $LiNbO_3$ Powders. *Crystals* **2018**, *8*, 108. [CrossRef]
142. Skipetrov, S.E. Disorder is the new order. *Nat. Cell Biol.* **2004**, *432*, 285–286. [CrossRef]
143. Kudinova, M.; Humbert, G.; Auguste, J.-L.; Delaizir, G. Multimaterial polarization maintaining optical fibers fabricated with the powder-in-tube technology. *Opt. Mater. Express* **2017**, *7*, 3780–3790. [CrossRef]
144. Nath, R.K.; Zain, M.F.M.; Kadhum, A.A.H. Artificial Photosynthesis using $LiNbO_3$ as Photocatalyst for Sustainable and Environmental Friendly Construction and Reduction of Global Warming: A Review. *Catal. Rev.* **2013**, *56*, 175–186. [CrossRef]
145. Yang, W.-C.; Rodriguez, B.J.; Gruverman, A.; Nemanich, R.J. Polarization-dependent electron affinity of $LiNbO_3$ surfaces. *Appl. Phys. Lett.* **2004**, *85*, 2316–2318. [CrossRef]
146. Gutmann, E.; Benke, A.; Gerth, K.; Böttcher, H.; Mehner, E.; Klein, C.; Krause-Buchholz, U.; Bergmann, U.; Pompe, W.; Meyer, D.C. Pyroelectrocatalytic Disinfection Using the Pyroelectric Effect of Nano- and Microcrystalline $LiNbO_3$ and $LiTaO_3$ Particles. *J. Phys. Chem. C* **2012**, *116*, 5383–5393. [CrossRef]
147. Bretos, I.; Lourdes-Calzada, M. Chapter 5. Approaches Towards the Minimisation of Toxicity in Chemical Solution Deposition Processes of Lead-Based Ferroelectric Thin Films. In *Multifunctional Polycrystalline Ferroelectric Materials: Processing and Properties*, 1st ed.; Springer Series in Materials Science 140; Hull, R., Osgood, R.M., Jr., Parisi, J., Warlimont, H., Pardo, L., Ricote, J., Eds.; Springer: Berlin/Heidelberg, Germany, 2011; Volume 140, pp. 145–216.
148. Galassi, C. Chapter 1. Advances in Processing of Bulk Ferroelectric Materials. In *Multifunctional Polycrystalline Ferroelectric Materials: Processing and Properties*, 1st ed.; Springer Series in Materials Science 140; Hull, R., Osgood, R.M., Jr., Parisi, J., Warlimont, H., Pardo, L., Ricote, J., Eds.; Springer: Berlin/Heidelberg, Germany, 2011; Volume 140, pp. 145–216.
149. Saito, Y.; Takao, H.; Tani, T.; Nonoyama, T.; Takatori, K.; Homma, T.; Nagaya, T.; Nakamura, M. Lead-free piezoceramics. *Nature* **2004**, *432*, 84–87. [CrossRef]
150. Shrout, T.R.; Zhang, S.J. Lead-free piezoelectric ceramics: Alternatives for PZT? *J. Electroceramics* **2007**, *19*, 113–126. [CrossRef]
151. Troiler-McKinstry, S. Impact of ferroelectricity. *Am. Cer. Soc. Bull.* **2020**, *99*, 22–23.
152. Reznichenko, L.A.; Verbenko, I.A.; Shilkina, L.A.; Pavlenko, A.V.; Dudkina, S.I.; Andryushina, I.N.; Andryushin, K.P.; Abubakarov, A.G.; Krasnyakova, T.V. Chapter 1. Binary, Ternary and Four-Component Systems Based on Sodium Niobate: Phase Diagrams of States, the Role of the Number of Components and Defectiveness in the Formation of the Properties. In *Springer Proceedings in Physics 207: Advanced Materials: Proceedings of the International Conference on "Physics and Mechanics of New Materials and Their Applications", PHENMA 2017*, 1st ed.; Parinov, I.A., Chang, S.H., Gupta, V.K., Eds.; Springer: Cham, Switzerland, 2018; Volume 207, pp. 3–23. Available online: https://link.springer.com/chapter/10.1007/978-3-319-78919-4_1 (accessed on 31 July 2020).

153. Kalinnikov, V.T.; Gromov, O.G.; Kunshina, G.B.; Kuz'Min, A.P.; Lokshin, E.P.; Ivanenko, V.I. Preparation of $LiTaO_3$, $LiNbO_3$, and $NaNbO_3$ from Peroxide Solutions. *Inorg. Mater.* **2004**, *40*, 411–414. [CrossRef]
154. Liu, M.; Xue, D. An efficient approach for the direct synthesis of lithium niobate powders. *Solid State Ionics* **2006**, *177*, 275–280. [CrossRef]
155. Liu, M.; Xue, D.; Luo, C. Wet chemical synthesis of pure $LiNbO_3$ powders from simple niobium oxide Nb_2O_5. *J. Alloy. Compd.* **2006**, *426*, 118–122. [CrossRef]
156. Suryanarayana, C.; An, I.-S. Mechanical alloying and milling. *Prog. Mater. Sci.* **2001**, *46*, 1–184. [CrossRef]
157. Kong, L.B.; Chang, T.S.; Ma, J.; Boey, F. Progress in synthesis of ferroelectric ceramic materials via high-energy mechanochemical technique. *Prog. Mater. Sci.* **2008**, *53*, 207–322. [CrossRef]
158. Chadwick, A.V.; Pooley, M.J.; Savin, S.L.P. Lithium ion transport and microstructure in nanocrystalline lithium niobate. *Phys. Status Solidi* **2005**, *2*, 302–305. [CrossRef]
159. Heitjans, P.; Masoud, M.; Feldhoff, A.; Wilkening, M. NMR and impedance studies of nanocrystalline and amorphous ion conductors: Lithium niobate as a model system. *Faraday Discuss.* **2007**, *134*, 67–82. [CrossRef] [PubMed]
160. Sugak, D.-Y.; Syvorotka, I.I.; Buryy, O.A.; Yakhnevych, U.V.; Solskii, I.M.; Martynyuk, N.V.; Suhak, Y.; Suchocki, A.; Zhydachevskii, Y.; Jakiela, R.; et al. Spatial distribution of optical coloration in single crystalline $LiNbO_3$ after high-temperature H2/air treatments. *Opt. Mater.* **2017**, *70*, 106–115. [CrossRef]
161. De Figueiredo, R.S.; Messai, A.; Hernandes, A.C.; Sombra, A.S.B. Piezoelectric lithium niobate obtained by mechanical alloying. *J. Mater. Sci. Lett.* **1998**, *17*, 449–451. [CrossRef]
162. Pooley, M.J.; Chadwick, A.V. The Synthesis and Characterisation of Nanocrystalline Lithium Niobate. *Radiat. Eff. Defects Solids* **2003**, *158*, 197–201. [CrossRef]
163. Diaz-Moreno, C.A.; Farías, R.; Galindo, J.T.E.; González, J.; Hurtado-Macias, A.; Bahena, D.; José-Yacamán, M.; Ramos, M. Structural Aspects $LiNbO_3$ Nanoparticles and Their Ferromagnetic Properties. *Materials* **2014**, *7*, 7217–7225. [CrossRef]
164. Luo, J.H. Preparation of Lithium Niobate Powders by Mechanochemical Process. *Appl. Mech. Mater.* **2011**, *121–126*, 3401–3405. [CrossRef]
165. Takacs, L. Multiple combustion induced by ball milling. *Appl. Phys. Lett.* **1996**, *69*, 436–438. [CrossRef]
166. Xue, J.; Wan, D.; Lee, S.-E.; Wang, J. Mechanochemical Synthesis of Lead Zirconate Titanate from Mixed Oxides. *J. Am. Ceram. Soc.* **1999**, *82*, 1687–1692. [CrossRef]
167. Pitcher, M.W.; He, Y.; Bianconi, P.A. Facile in situ synthesis of oriented $LiNbO_3$ single crystals in a polymer matrix. *Mater. Chem. Phys.* **2005**, *90*, 57–61. [CrossRef]
168. Zhenxiang, C.; Kiyoshi, O.; Akimitsu, M.; Hideo, K. Formation of Niobates from Aqueous Peroxide Solution. *Chem. Lett.* **2004**, *33*, 1620–1621.
169. Zeng, H.C.; Tung, S.K. Synthesis of Lithium Niobate Gels Using a Metal Alkoxide–Metal Nitrate Precursor. *Chem. Mater.* **1996**, *8*, 2667–2672. [CrossRef]
170. Camargo, E.R. Low temperature synthesis of lithium niobate powders based on water-soluble niobium malato complexes. *Solid State Ionics* **2002**, *151*, 413–418. [CrossRef]
171. Niederberger, M.; Pinna, N.; Polleux, J.; Antonietti, M. A General Soft-Chemistry Route to Perovskites and Related Materials: Synthesis of $BaTiO_3$, $BaZrO_3$, and $LiNbO_3$ Nanoparticles. *Angew. Chem. Int. Ed.* **2004**, *43*, 2270–2273. [CrossRef] [PubMed]
172. An, C.; Tang, K.; Wang, C.; Shen, G.; Jin, Y.; Qian, Y. Characterization of $LiNbO_3$ nanocrystals prepared via a convenient hydrothermal route. *Mater. Res. Bull.* **2002**, *37*, 1791–1796. [CrossRef]
173. Yoshikawa, Y.; Tsuzuki, K. Fabrication of Transparent Lead Lanthanum Zirconate Titanate Ceramics from Fine Powders by Two-Stage Sintering. *J. Am. Ceram. Soc.* **1992**, *75*, 2520–2528. [CrossRef]
174. Oren, E.E.; Taspinar, E.; Tas, A.C. Preparation of Lead Zirconate by Homogeneous Precipitation and Calcination. *J. Am. Ceram. Soc.* **1997**, *80*, 2714–2716. [CrossRef]
175. Camargo, E.R.; Frantti, J.; Kakihana, M. Low-temperature chemical synthesis of lead zirconate titanate (PZT) powders free from halides and organics. *J. Mater. Chem.* **2001**, *11*, 1875–1879. [CrossRef]
176. Blum, J.B.; Gurkovich, S.R. Sol-gel-derived $PbTiO_3$. *J. Mater. Sci.* **1985**, *20*, 4479–4483. [CrossRef]

177. Kim, S.; Jun, M.-C.; Hwang, S.-C. Preparation of Undoped Lead Titanate Ceramics via Sol-Gel Processing. *J. Am. Ceram. Soc.* **2004**, *82*, 289–296. [CrossRef]
178. Tartaj, J.; Moure, C.; Lascano, L.; Duran, P. Sintering of dense ceramics bodies of pure lead titanate obtained by seeding-assisted chemical sol-gel. *Mater. Res. Bull.* **2001**, *36*, 2301–2310. [CrossRef]
179. Santos, I.; Loureiro, L.; Silva, M.; Cavaleiro, A.M.V. Studies on the hydrothermal synthesis of niobium oxides. *Polyhedron* **2002**, *21*, 2009–2015. [CrossRef]
180. Sato, S.; Murakata, T.; Yanagi, H.; Miyasaka, F.; Iwaya, S. Hydrothermal synthesis of fine perovskite $PbTiO_3$ powders with a simple mode of size distribution. *J. Mater. Sci.* **1994**, *29*, 5657–5663. [CrossRef]
181. Peterson, C.R.; Slamovich, E.B. Effect of Processing Parameters on the Morphology of Hydrothermally Derived $PbTiO_3$Powders. *J. Am. Ceram. Soc.* **1999**, *82*, 1702–1710. [CrossRef]
182. Chen, D.-R.; Jiao, X.-L.; Xu, R.-R. Hydrothermal synthesis of $PbZr_xTi_{1-x}O_3$ (x = 0.45–0.65) powders without using alkaline mineralizer. *J. Mater. Sci. Lett.* **1998**, *17*, 53–56. [CrossRef]
183. Narendar, Y.; Messing, G.L. Kinetic Analysis of Combustion Synthesis of Lead Magnesium Niobate from Metal Carboxylate Gels. *J. Am. Ceram. Soc.* **2005**, *80*, 915–924. [CrossRef]
184. Arendt, R.; Rosolowski, J.; Szymaszek, J. Lead zirconate titanate ceramics from molten salt solvent synthesized powders. *Mater. Res. Bull.* **1979**, *14*, 703–709. [CrossRef]
185. Chiu, C.C.; Li, C.C.; Desu, S.B. Molten Salt Synthesis of a Complex Perovskite, $Pb(Fe_{0.5}Nb_{0.5})O_3$. *J. Am. Ceram. Soc.* **1991**, *74*, 38–41. [CrossRef]
186. Kuz'Minov, Y.S.; Osiko, V.V. Nonstoichiometric composition of lithium niobate crystal. *Ferroelectrics* **1993**, *142*, 105–113. [CrossRef]
187. Vyalikh, A.; Zschornak, M.; Köhler, T.; Nentwich, M.; Weigel, T.; Hanzig, J.; Zaripov, R.; Vavilova, E.; Gemming, S.; Brendler, E.; et al. Analysis of the defect clusters in congruent lithium tantalate. *Phys. Rev. Mater.* **2018**, *2*, 013804. [CrossRef]
188. Spaldin, N.A. Analogies and Differences between Ferroelectrics and Ferromagnets. In *Physics of Ferroelectrics: A Modern Perspective*, 1st ed.; Rabe, K.M., Ahn, C.H., Triscone, J.-M., Eds.; Springer: Berlin/Heidelberg, Germany, 2007; pp. 175–218.
189. Toyoura, K.; Ohta, M.; Nakamura, A.; Matsunaga, K. First-principles study on phase transition and ferroelectricity in lithium niobate and tantalate. *J. Appl. Phys.* **2015**, *118*, 064103. [CrossRef]
190. Sirenko, A.A.; Bernhard, C.; Golnik, A.; Clark, A.M.; Hao, J.; Si, W.; Xi, X.X. Soft-mode hardening in $SrTiO_3$ thin films. *Nature* **2000**, *404*, 373–376. [CrossRef] [PubMed]
191. Gehring, P.M.; Wakimoto, S.; Ye, Z.-G.; Shirane, G. Soft Mode Dynamics above and below the Burns Temperature in the Relaxor $Pb(Mg_{1/3}Nb_{2/3})O_3$. *Phys. Rev. Lett.* **2001**, *87*, 277601. [CrossRef] [PubMed]
192. Cohen, R.E. Origin of ferroelectricity in perovskite oxides. *Nature* **1992**, *358*, 136–138. [CrossRef]
193. Opik, U.; Pryce, M.H.L. Studies of the Jahn-Teller effect. I. A survey of the static problem. *Proc. R. Soc. Lond. Ser. A, Math. Phys. Sci.* **1957**, *238*, 425–447. [CrossRef]
194. Halasyamani, P.S.; Poeppelmeier, K.R. Noncentrosymmetric Oxides. *Chem. Mater.* **1998**, *10*, 2753–2769. [CrossRef]
195. Bersuker, I.B. *The Jahn–Teller Effect*; Cambridge University Press: Cambridge, UK, 2006.
196. Bersuker, I.B. Modern Aspects of the Jahn–Teller Effect Theory and Applications to Molecular Problems. *Chem. Rev.* **2001**, *101*, 1067–1114. [CrossRef]
197. Burdett, J.K. Use of the Jahn-Teller theorem in inorganic chemistry. *Inorg. Chem.* **1981**, *20*, 1959–1962. [CrossRef]
198. Bersuker, I.B. *The Jahn-Teller Effect and Vibronic Interactions in Modern Chemistry*; Springer: New York, NY, USA, 1984.
199. Atanasov, M.; Reinen, D. Density Functional Studies on the Lone Pair Effect of the Trivalent Group (V) Elements: I. Electronic Structure, Vibronic Coupling, and Chemical Criteria for the Occurrence of Lone Pair Distortions in AX_3 Molecules (A = N to Bi; X = H, and F to I). *J. Phys. Chem. A* **2001**, *105*, 5450–5467. [CrossRef]

200. Troiler-McKinstry, S. Chapter 3. Crystal Chemistry of Piezoelectric Materials. In *Piezoelectric and Acoustic Materials for Transducer Applications*, 1st ed.; Safari, A., Koray Akdogan, E., Eds.; Springer Science+Business Media, LLC: New York, NY, USA, 2008; pp. 39–56.
201. Ringwood, A.E. Mineralogical constitution of the deep mantle. *J. Geophys. Res. Space Phys.* **1962**, *67*, 4005–4010. [CrossRef]
202. Mitchell, R.H. *Perovskites Modern and Ancient*; Almaz Press: Ontario, CA, USA, 2002.
203. Extance, A. The reality behind solar power's next star material. *Nature* **2019**, *570*, 429–432. [CrossRef] [PubMed]
204. Snaith, H.J. A decade of perovskite photovoltaics. *Nat. Energy* **2019**, *4*, 1. Available online: https://ui.adsabs.harvard.edu/abs/2019NatEn...4....1./abstract (accessed on 31 July 2020). [CrossRef]

Publisher's Note: MDPI stays neutral with regard to jurisdictional claims in published maps and institutional affiliations.

crystals

MDPI

Review

Lithium Niobate Single Crystals and Powders Reviewed—Part II

Oswaldo Sánchez-Dena [1,*], Sergio David Villalobos-Mendoza [1], Rurik Farías [1] and Cesar David Fierro-Ruiz [2,*]

1 Instituto de Ingeniería y Tecnología, Universidad Autónoma de Ciudad Juárez, Av. Del Charro 450 Norte, Ciudad Juárez 32310, Chihuahua, Mexico; sergio.d.v.m@hotmail.com (S.D.V.-M.); rurik.farias@uacj.mx (R.F.)
2 Departamento de Mecatrónica y Energías Renovables, Universidad Tecnológica de Ciudad Juárez, Avenida Universidad Tecnológica 3051, Colonia Lote Bravo II, Ciudad Juárez 32695, Chihuahua, Mexico
* Correspondence: ossdena@gmail.com (O.S.-D.); cesar_fierro@utcj.edu.mx (C.D.F.-R.); Tel.: +52-656-308-5850 (C.D.F.-R.)

Received: 2 September 2020; Accepted: 27 October 2020; Published: 31 October 2020

check for updates

Abstract: A review on lithium niobate single crystals and polycrystals has been prepared. Both the classical and recent literature on this topic is revisited. It is composed of two parts with several sections. The current part discusses the available defect models (intrinsic), the trends found in ion-doped crystals and polycrystals (extrinsic defects), the fundamentals on dilute magnetic oxides, and their connection to ferromagnetic behavior in lithium niobate.

Keywords: lithium niobate; lithium tantalate; intrinsic defects; extrinsic defects; elemental doping; ferromagnetism; diluted-magnetic oxides

1. Introduction

The paramountcy of lithium niobate ($LiNbO_3$, LN) to the applied sciences has already been outlined in the first part of this review. Its versatility in what respects technology has promoted it in science for almost 60 years. Perhaps an effective way to quantitatively assess the success of a researching field is to analyze the number of peer-reviewed publications in scientific journals in a period. This approach was taken by Volk and Wöhlecke (2008): the Web Of Science database was used to track the number of publications between 1965 and 2007 devoted to LN and $BaTiO_3$ (BT) annually [1]. Their plot shows a shoulder-to-shoulder competition of our protagonist with BT, a material that has been the "heart and soul of several multibillion-dollar industries" and which importance in history can be advertised with nothing less than wartime secrecy and a lack of open publications right at its birth around 1945 [2,3]. Today, BT dominates the market of multilayered ceramic capacitors, nearly $6 billion [4]. The bibliographical survey done in 2008 has been retaken only for the case of LN. The actualized data are shown in Figure 1.

With the present contribution, we aim to complete a formal justification to the paradigm that defines the researching field of LN, explicitly stated in *Part I*. A central idea there is that of the inherence of this material in forming solid solutions (SSs) with a significant number of intrinsic defects which at the macroscale define the congruent (CG) composition. The perturbation from the stoichiometric (ST) composition under ordinary growth circumstances is substantiated by knowing the crystal structure (described in *Part I*). However, on a deeper level, it essentially relies on understanding the available defect models that provide a solution to the electrical charge neutrality in stable off-ST SSs. No consensus can be found today in this subject. So far, it appears that both experimental and simulation approaches provide conclusive evidence on the type of intrinsic defects being present within Li-deficient crystals, namely Li vacancies (V_{Li}) and Nb antisites (Nb_{Li}), in turn favoring the *lithium vacancy defect* model. However, debate still prevails on their intrinsic distribution, that is, the available models on defect clusters. Less conclusive arguments can be given in this respect for doped LN. Only site occupancy trends have been established depending on the dopant characteristics and doping extent. One can often find discrepancies or irregularities on this point. However, recently, most dopant species seem to adjust to a rule or, better stated, a generalized trend [5]. We discuss the three most widely accepted defect models through LN research (Section 2). Then, the subject of extrinsic defects naturally follows. The role of the ubiquitous presence of hydrogen in the crystal structure and the recent developments on ion-doped LN single crystals are thus revisited (Section 3). Reviews on this subject and focused on other distinct aspects have been recently published [6–8]. Lastly, a connection is made between magnetism, ferromagnetism, and ion-doped LN powders through a formal discussion on diluted magnetic oxides (Section 4); this is a relatively new topic regarding LN, not discussed in detail before.

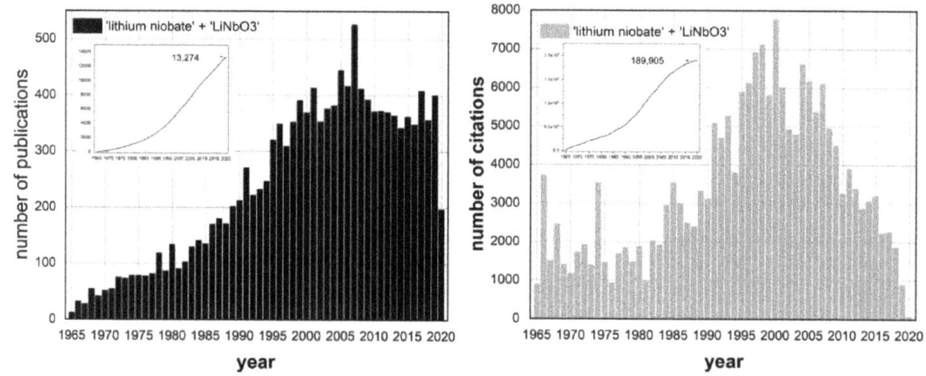

Figure 1. Statistics of peer-reviewed scientific reports on $LiNbO_3$ from 1965 to mid-2020. According to the bibliographical survey done using The Web of Science database, data correspond to the sum of the in-title results obtained under typing '$LiNbO_3$' and 'lithium niobate' in the search engine. The insets show the cumulative distributions. **Left**: publications per year. **Right**: citations per year.

2. Intrinsic Defects

A compromise between the increasing (decreasing) concentration of Nb (Li) atoms exists for stable off-stoichiometric SSs of LN. Electrical charge compensation occurs through an adequate accommodation of both cation species and their respective vacancies (V_{Nb} and V_{Li}). Since the latter are defects with a net electrical charge, other defects with countercharges are required to guarantee overall charge neutrality [9]. The occurrence of a stacking fault envisioned as the relatively excess Nb partially occupying the available V_{Li}s is proposed. It is denoted as an *Nb antisite* (Nb_{Li}) and is the carotid upon which rests the framework of defect models in LN (see the fourth figure in *Part I* for visualization aid) [10]. Several defect models have been constructed, and debate still prevails on which of these indeed represents a "solution of the off-ST LN crystal structure with a stable charge equilibrium" [7]. The discrepancies observed over time might be addressed to intrinsic factors persistent on the different crystal growth technologies or suppliers, as recently reported for isostructural lithium tantalate ($LiTaO_3$, LT) [9], a situation that also leads to the irreproducibility of results in at least two cases of doped LN crystals [5,11]. The main characteristics of the three most popular models ever devised are now described in the lines that follow.

Oxygen vacancy model: proposed by Fay et al. (1968) [12], it can be illustrated by the following constitutional formula: $[Li_{1-2x}V_{2x}]Nb[O_{3-x}V_x]$, $x = 0.028$. Supported by Prokhorov and Kuzminov (1990) [13], it assumes the formation of Li vacancies due to the lack of Li_2O (used as a raw material in the initial melt, Czochralski method) at the CG point. Then, charge compensation comes from the formation of oxygen vacancies, which is characteristic of all oxides. Thus, at the earliest stages of the CG crystal structure discussion, it was reasonable to assume such vacancies as the most probable defect. However, nowadays, it is tacitly known that LN "is to some extent a unique oxide containing no oxygen vacancies" [10]; based on theoretical calculations, Donnerberg et al. (1989) ruled out the existence of V_Os [14]. In addition, the model predicts the crystal density to decrease with greater Li_2O deficiency (that is, increasing concentration of V_{Li}), which is inconsistent with the observations obtained in the experimental studies by Lerner et al. (1968) [15]. The existence of Nb_{Li}s copes with this situation, whereas the inexistence of V_Os in LN single crystals has been demonstrated by detailed structure measurements and supported by electronic-structure calculations [16–20].

Niobium vacancy model: proposed by Peterson and Carnevale (1972) [21], it can be illustrated by the following constitutional formula: $[Li_{1-5x}Nb_{5x}][Nb_{1-4x}V_{4x}]O_3$, where $x \approx 0.012$ and the Nb in the first bracket denotes the Nb antisite defect. It was once widely accepted based on detailed structural studies by Abrahams and Marsh (1986) [16]. It successfully describes the relationship between the density and the Li content in the crystal. Oxygen vacancies are neglected, and the Li sites are filled with an excess of Nb. Likewise, some Nb vacancies are generated at the former 5+ sites for charge compensation. However, according to this model in congruent LN (cLN), there are 5.9 mol% Nb_{Li} and 4.7 mol% V_{Nb}. Such charged Nb antisite concentration is believed to be structurally unstable. In principle, the existence of Nb_{Li} is in agreement with the Hume–Rothery rules for SSs, given that the measured ionic radii for Li^+ and Nb^{5+} are 74 and 64 pm (6-fold coordination), respectively [22,23]. In the analysis of the atomic size effects and their influence within a structure, data are usually taken from the 'effective' ionic radii for this coordination number, as given by Shannon and Prewitt [23].

Lithium vacancy model: proposed by Lerner et al. (1968) [15] and extended by Kim et al. (2001) [24], it can be illustrated by the following constitutional formula: $[Li_{1-5x}V_{4x}Nb_x]NbO_3$, $x = 0.01$. It also includes Nb_{Li} and neglects V_O. The Li vacancy model differs from the Nb one in that no Nb vacancies are formed. Instead, four V_{Li} stands for compensation of a single Nb_{Li}. The existence of a high concentration of Li vacancies in cLN was confirmed by the structure analysis reported in Refs. [18,19,25,26], as well as by Nuclear Magnetic Resonance (NMR) studies [27,28]. Nowadays, the Li vacancy model is commonly accepted as valid, and the considerations of all defect reconstructions in LN are based on this framework. Reports supporting this model based on computational simulations can be consulted [14,29–31]. Nonetheless, a new model has recently been proposed based on the coexistence of all three cation stacking faults Nb_{Li}, V_{Nb},, and V_{Li} [32,33]. According to this model, the Nb and Li vacancy models are simply the two opposite ultimate states. In 1998, Wilkinson et al. had already indicated this possibility [19]. In this respect, Schirmer et al. (1991) pointed out that the existence of ilmenite-type stacking faults in the regular LN crystal structure would support these two ultimate states [17]. On the other hand, refined studies of NMR spectra have concluded that a combination of the oxygen vacancy and lithium vacancy models in a ratio 1.1:1.0, respectively, also gives a satisfactory defect framework [34–36].

The formation of oxygen vacancies is the least probable event, whereas the combined formation of four Li vacancies for each Nb antisite is the most probable one in our defect scenario, which can be observed in Figure 2, and results from a first-principles study on the formation energies of intrinsic defects in off-ST LN [31]. It is crucial to notice that such a framework can only be considered realistic at moderate temperatures. For high Ts (starting from about 300 °C above room T), it cannot apply mainly to a tendency of a small fraction of Li cations to occupy the vacant octahedral sites, as discussed by Volk and Wöhlecke and references therein cited [10,37–39]. On the other hand, quoting Kong et al. (2020): "a variety of further defect configurations and defect clusters have been investigated based on the defect models above, especially the Li vacancy model" [6]. Here, a defect cluster means a specific configuration or distribution of closely spaced defects. The formation of defect clusters subdues energetic misfits and was proposed ever since the earlier defect investigations on LN [15,17,18,40].

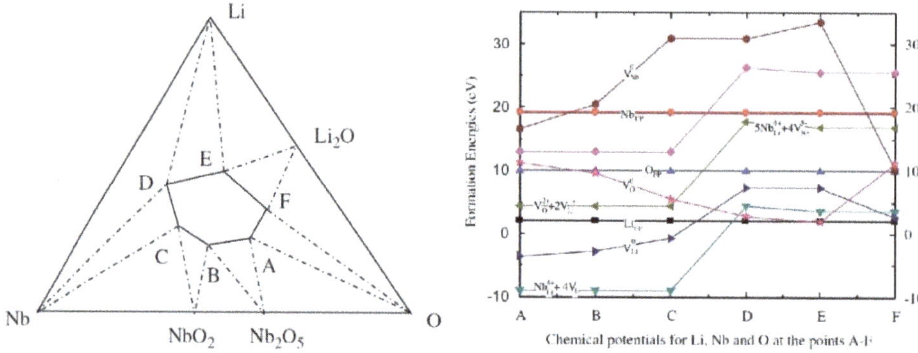

Figure 2. Simulations are showing the viability of the lithium vacancy model based on first-principles calculations. **Left**: constructed phase diagram of the ternary system Li–Nb–O showing the points of interest A–F (different electronic environments). **Right**: formation energies for intrinsic defects in LiNbO3, as a function of the atomic chemical potential)—doi.org/10.1016/j.jpcs.2007.02.035 (Reproduced from Li, Q.; Wang, B.; Woo, C.H.; Wang, H.; Wang, R. First-principles study on the formation energies of intrinsic defects in $LiNbO_3$. *J. Phys. Chem. Solids* **2007**, *68*, 1336–1340, with permission from Elsevier.)

Recently, Vyalikh et al. (2018) have performed a detailed study based on Density Functional Theory (DFT) electronic structure calculations and measurement of the spectral characteristics under three different techniques, and, among six proposed cluster models (often found in literature), they found out that two are the most probable to describe cLT [9]; these are therein labeled as models M1 and M3. The first—the least probable among the two according to this study, model M1, with calculated energy formation of 1.94 eV—was initially proposed by Kim et al. (2001), and it assumes that a Nb_{Li} antisite (or Ta_{Li}) is surrounded by three V_{Li}s in the nearest neighborhood and fourth V_{Li} located along the polar *c*-axis [24], as shown on the left side of Figure 3. According to the same study, the most probable one—model M3, with calculated energy formation of 1.61 eV—locates all the Nb excess interstitially within the structural vacancy, provoking five VLis instead of four as shown in the right side of Figure 3. Accordingly, Zotov et al. (1994) became supporters of the Nb antisite scheme. They proposed the empty site scheme as a viable alternative to describe stoichiometry disturbance in LN, presenting similar goodness-of-fit factors [18].

Debate still prevails on Nb ions occupying interstitial structural sites [8,41,42]. Other alternatives of intrinsic defect clusters have been proposed, perhaps more intricated [43–46]; most have been proposed to describe the effect of the localized polarization caused by the defect clusters on the overall crystal polarization of the observed features of domain stabilization and reversal [6]. Given the bilateral acceptance in what respects the defect scenario of undoped or pure LN, it is reasonable to adopt the Li vacancy model and its associated defect cluster (left side of Figure 3) as a feasible physical instance describing both the type of intrinsic defects and their ordering in off-stoichiometric LN, respectively. As will be discussed in the next section, this framework is usually adopted to describe the substitution mechanisms in doped LN—concerning at least the type of intrinsic defects. However, the main result presented by Vyalikh et al. must not be overlooked [9], mostly because of the overall confidence recently put on the results obtained with ever more advanced simulations [6,8,47,48].

Indeed, computational power and numerical simulation techniques have made significant progress in the last two decades. Even to the extent that the analysis of extrinsic defects—so far limited to establish substitution mechanisms of dopants entering either a Li site or Nb one, by assumption a preconceived model of intrinsic defects—could soon enable the construction of more sophisticated models. For example, the specific configuration of *extrinsic defect clusters*. Apropos Kong et al. have made a relevant statement regarding the evolution of simulation methods over time [6]: compared to earlier works, today results can be obtained by the direct insertion of the stacking fault defects into a designed supercell standing for the actual crystal structure. Usually, the extended supercell size is 2 × 2 × 2 hexagonal unit cells comprising 240 atoms. This has been enough, for example, to obtain satisfactory results (2011) on the features shown by a more realistic LN single crystal in which hydrogen has inevitably entered the structure in a low ratio of 10–100 ppm [49,50]. More recently, the above-discussed results reported by Vyalikh et al. (2018) were obtained using 2 × 2 × 1 (120 atoms) and 2 × 2 × 2 supercells. In the same year, Li et al. simulated doping effects with various elements using a basal intrinsic defect supercell of 120 and 540 atoms [51]. Earlier works used to emulate the existence of intrinsic defects by changing the chemical potential reference state respect to that corresponding to the non-defective system (under Li-rich conditions). The enterprise of combined calculation and theoretical analysis is undoubtedly changing.

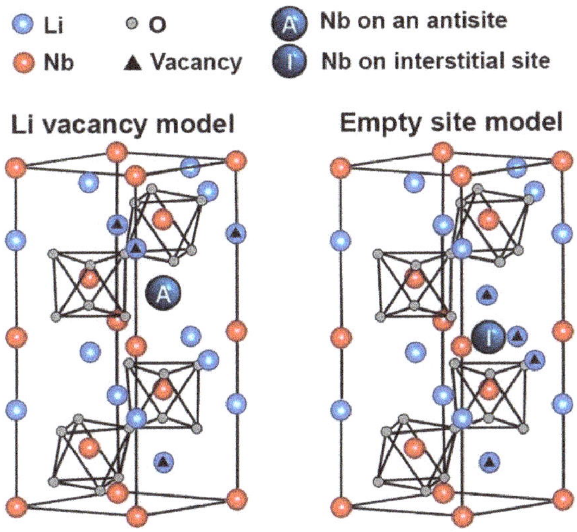

Figure 3. Scheme of the $LiNbO_3$ crystal structures according to the defect cluster models discussed above **Left**: Nb antisite surrounded by three first-neighbor Li vacancies and fourth one located at the *c*-polar axis, supporting the Li vacancy defect model. **Right**: interstitial Nb in the structural vacancies, compensated by five Li vacancies, supporting the empty site model—redrawn from the publication by Vyalikh et al. [9].

Lastly, regarding LT, it is also essential to notice that its properties are often discussed in terms of an identical defect scenario, that is, the existence of Ta antisites and the inexistence of oxygen vacancies. No evidence of the presence of such stacking faults in LT is available yet; thus, in principle, the "considerations based on this defect model are speculative in contrast to LN" [10]. However, Li vacancies are undoubtedly present in LT and based on its structural isomorphism. It is permissible to describe one in terms of the defect models proposed for the other. That a Nb_{Li} (Ta_{Li}) forms instead of a Li_{Nb} (Li_{Ta}) is related to the larger dimensions of the octahedrons occupied by Li cations compared to those occupied by Nb (Ta) cations, as discussed in *Part I*. In general, the incorporation of impurity atoms is predominant in Li sites.

3. Extrinsic Defects

Nowadays, it is common to regard this subject as a synonym of elemental doping. Historically, the doping of the (Li-deficient) LN matrix with several ions belonging to distinct *families* or classes has been highly motivated by either the suppression or enhancement of photorefraction (PR). Also termed 'optical damage' or 'optical breakdown', PR stands for the process whereby an inhomogeneity in the refractive index is optically induced. It is characteristic of most insulators and, regularly, the irreversible change is generated by applying intense laser-light [52]. The case of LN, however, is peculiar since low irradiance of the order of 20 Wcm^{-2} (λ = 532 nm) is enough for its manifestation in a congruent crystal [6]; even the use of incoherent visible/ultraviolet light suffices, as reported by Ashkin et al. (1966, considered the discovery of the effect) [53]. Wavefront distortion often follows PR, which is detrimental for applications involving high-intensity laser light and undisturbed beam propagation in media such as nonlinear-optical frequency conversion, Q-switching, optical waveguides, and electro-optical modulation [54–59]. Single-crystal LN outstands in such and, if not for its rather ubiquitous PR, it would satisfactorily serve for such technological approaches. Hence, a demand exists for suppressing optical breakdown in LN, or at least to lessen it.

The supply is nowadays found by elemental doping (above certain threshold concentrations) of off-ST single crystals with a particular group of impurities termed 'optical-damage-resistant ions' (ODRI): divalent (Mg, Zn), trivalent (In, Sc), and tetravalent metals (Hf, Zr, Sn) [5,6,8,10,60]. Increment of the crystal density and phase transition temperature, together with a blue shift of the fundamental absorption edge, are the main characteristic effects that all these dopants share when they occupy a given site in the defect structure [10]. The term family has recently been coined regarding ODRI, given the similarity of their properties when doped into LN, which have their origin from a shared structural factor despite their state of valence differences. These impurities are optically inactive; "doping with any member of this family induces no changes within the whole transparency range 0.35–4 μm except for a slight shift" of the UV fundamental absorption edge [10]. It is the main reason behind the outstanding interest for optical applications.

Photorefraction (PR) is enhanced by doping with another type of ions—mainly transition metals. In this case, the applications favored by charge transport and trapping phenomena are exploited, such as holography storage, beam coupling, information processing, and computation [61–65]. The bulk photovoltaic effect has been acknowledged as the primary source of optical damage. Ion Fe has been considered for a long time the strongest photovoltaic-active impurity due to the tendency of Li-deficient LN to host the two charged states Fe^{2+} and Fe^{3+} [66,67]; ions Cu, Mn, Ni, Co also belong to this group of multi-charge transition metals of valence below +5 used to enhance PR in LN. Thus, despite specific difficulties (low response speed and volatility), Fe-doped LN (LN:Fe) has been "the mainstay of holographic data storage" [61]. Double and triple co-doping (such as LN:Hf,Fe, and LN:Zr,Ru,Fe) and single-doping with ions of valence 5+, 6+, namely LN:V and LN:Mo, have been done to cope with the shortcomings present in LN:Fe and without compromising significantly one or the other, or either the strength of the PR effect [61,68–72].

The use of dopants not only entails a modification of the intrinsic capabilities of the crystal, but also new ones that could arise, as stated by Kovács et al. [73]. Laser emission is the best example of such, accomplished when lanthanide trivalent rare-earth ions are the dopants: Pr, Tb, Er, Nd, Yb, Eu, Dy, Ho, Tm [73,74]. These may also be photorefractive ions; for example, the development of two-color holographic recording by doping with Pr^{3+} and Tb^{3+} has been highlighted [10,75,76]. This limits to some extent their usage in designing new sources of laser light, for which LN:Nd^{3+} is remarkable; even detection of the coupled self-doubled frequency in LN:Nd^{3+} based materials have been reported [77–79]. In this respect, transition metal Cr^{3+} has been extensively studied since, apart from not presenting PR properties, LN:Cr^{3+} also shows a "strong broad-band luminescence spanning the 770–1170 nm range" [80–82]: both features make LN:Cr^{3+} a suitable material for room temperature tunable solid-state lasing. The versatility of LN in connection to elemental doping and the correspondent derived capabilities is synthesized in Figure 4, according to the consulted literature; because research on elemental doping of LN is merely abundant, it may be incomplete. Still, the subject of extrinsic defects is not entirely covered by discussing elemental doping only.

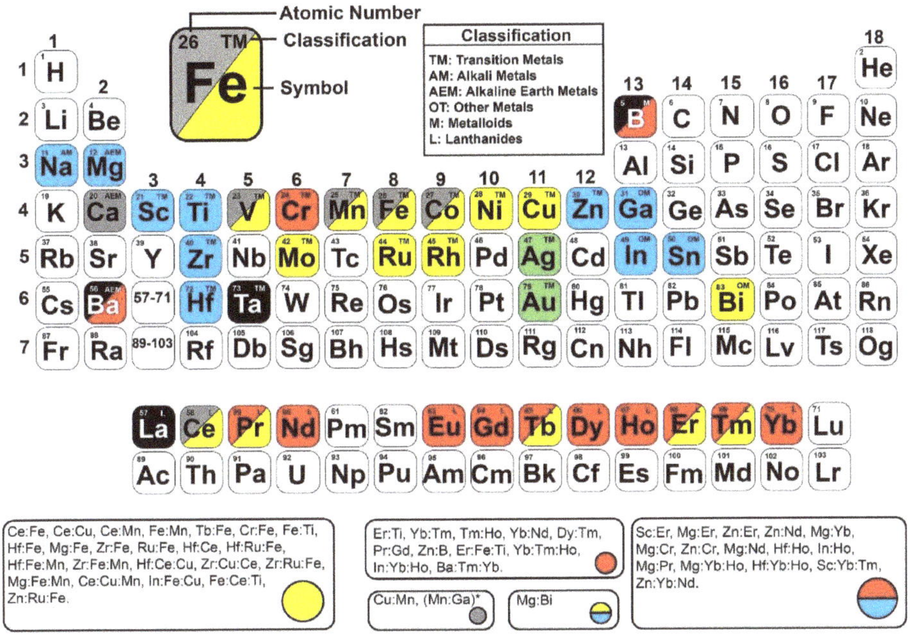

Figure 4. The Periodic Table of Elements showing single elemental doping in $LiNbO_3$, according to the consulted literature and classified in terms of the sought functionality: optical damage resistance (blue), photorefraction enhancement (yellow), laser activation (red), and ferromagnetism (gray). In black color Ta, La (new functionality), and the elements that have not been used alone but do have been used in double or triple co-doping (combined with the color of its corresponding functionality). Green color for interstitial impurities, usually created by ion-implantation techniques. Cases of double and triple co-doping are also shown below the table, using the same color code (the symbol * stands for simulation). By convention, from higher to lowest concentration within the crystal, the elements are written from left to right; some might be inaccurate.

Thermal treatment or annealing in air and at moderate temperatures (600–700 °C), combined with applying a small electric field on the surface of the crystal, precedes ODRI-doping in what respects the efforts to reduce the degree of index inhomogeneity. Such a strategy was implemented first by Levinstein et al. [83], as early as 1967 (just one year after the discovery of PR), and soon afterward by Smith et al. (1968) [84]. In contrast, the first report of a significant reduction of the optical damage in Mg-doped LN was published until 1980 [85]. In the annealing approach of 1968, Smith et al. addressed the partial PR suppression to the diffusion hydrogen in the crystal, a result that has been confirmed by more recent and detailed investigations [86–88].

Thus, hydrogen, or better said the hydrogen-based defect hydroxyl ion (OH^-), represents the first-ever appointed extrinsic defect in LN. Today, it is almost 70 years ago since OH^- molecular ion has been routinely found in a large variety of oxide compounds [89,90]. It occupies a regular oxygen site, and the promoted "positive charge with respect to the lattice compensates some other kinds of intrinsic or extrinsic defects, allowing one to probe the defect structure by studying the spectroscopic properties of the hydroxyl ions affected by their surroundings" [10]. The easiest way to trace it is by measuring the infrared (IR) absorption band associated with the stretching vibration, the frequency of which is much higher respect to the vibrational frequencies of the host, regularly falling in the 3200–3700 cm^{-1} range; after Smith et al. (1968), whose work is considered the first report of hydrogen incorporating in LN [84]. Hydroxyl ion concentrations in cLN have been determined by this approach (10^{18} to 10^{19}

cm^{-3}) [87]. The measurement of the diffraction efficiency of a photorefractive grating has also been used for this purpose [91].

The characteristic IR absorption spectra of a regular cLN crystal consist of a strong peak at 3482 cm^{-1} (2.87 μm) and a weaker one at 3467 cm^{-1} (2.88 μm), which together account for a full-width at half maximum of 32 cm^{-1} [8,10]; the IR band dramatically shifts and broadens to 3300 cm^{-1} (3.03 μm) and (about) 400 cm^{-1}, respectively, for a congruent crystal saturated with hydrogen [92]. Nowadays, OH^- Vibrational Spectroscopy is a popular term used in the context of LN to denote the use of IR absorption measurements to probe its defect structure, either it is undoped (intrinsic defects) or ion-doped [5,8,73,93,94]. The sensitivity of this positively charged extrinsic defect (the hydroxyl ion) to its surroundings may be easily understood by observation of Figure 5: for a Li-enriched crystal, the local environment tends to homogenize due to a proper cationic interchange with Nb antisites and Li vacancies, manifested in OH^- vibrational spectra as a doubled-tendency to band narrowing and blue shifting of the peak position (also observed for isostructural LT [95]).

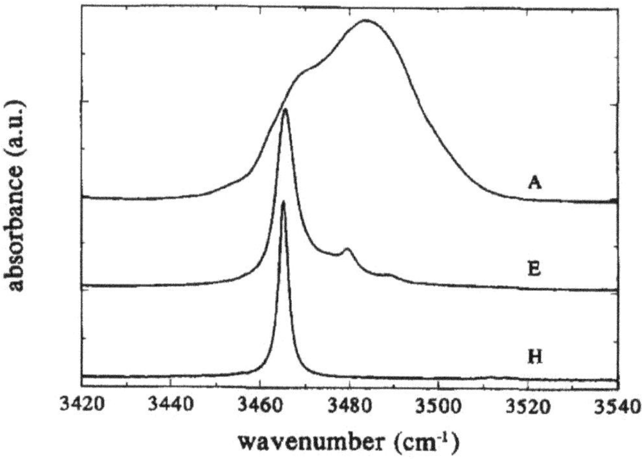

Figure 5. Infrared absorption bands from OH^- ions in LiNbO3 with different crystal compositions measured at 300 K (room temperature): [Li]/[Nb] in (A, E, H) = (0.950, 0.988, 1.000)—doi.org/10.1016/S0022-0248(96)01098-6. (Reproduced from Polgár, K. Péter, Á.; Kovács, L.; Corradi, G.; Szaller, Zs. Growth of stoichiometric $LiNbO_3$ single crystals by top seeded solution growth method. *J. Cryst. Growth* **1997**, *177*, 211–216, with permission from Elsevier.)

Hydrogen enters the crystal structure from the ambient atmosphere either during the growth process or after it. The standard Czochralski method of crystal growth is effectuated in an air atmosphere. The OH^- centers form within the crystal since a certain degree of humidity is always present in the air [87,96]. The diffusion rate can be optimized by subsequent annealing treatment in a water-vapor-rich or hydrogen atmosphere at temperatures between 400 and 700 °C and pressures within the range 1–30 bar [87]. A reduction treatment is subsequent annealing under a hydrogen-enriched atmosphere: reduced LN crystals exhibit a loss of transparency by the formation of color centers that translate into absorption coefficients in the visible region of the order of 50–200 cm^{-1} (use of water vapor prevents this, which can also be reverted by subsequent annealing in a dry oxidizing atmosphere) [10,87]. The use of temperatures above 700 °C in the reduction treatment is contraindicated because it could distort the cation sublattice, and the formation of oxygen vacancies would also occur during this process [37,38,97]. Embedding the crystal in a Li salt reduces Li loss probability during the reduction treatment [10].

From the early 1970s to the late 1980s, reduction treatment played a decisive role in the final acceptance of the polaron model to describe charge transport and trapping phenomena in (undoped)

LN, in terms of photoexcitation and diffusion of small polarons and bipolarons. There is a consensus nowadays about this model satisfactorily describing the origin of optical absorption bands observed in reduced and non-reduced LN crystals, and which together cover an energy range of 0.2–3.5 eV, in terms of the light-induced transfer of various polaron carriers (Nb^{4+}_{Nb}, Nb^{4+}_{Li} and $(Nb_{Nb}–Nb_{Li})^{2-}$) out of their potential well to one of the neighboring Nb ions. It also allows one to predict the existence/absence of Nb antisites based on the spectral characteristics of chemically reduced crystals: after reduction, the observation of an absorption band around 0.95 eV implies a lack, whereas a redshift in the absorption to 1.6 eV combined with a photoinduced Electron Paramagnetic Resonance (EPR) structure (observation of a characteristic 10-line EPR spectrum) stands for the evidence of their existence [10].

Rigorously, the type of point defects induced by chemical reduction, as well as the small polarons and bi-polarons themselves, are not extrinsic defects. They instead convey a redistribution (and formation of new complexes) of (with) the already existent intrinsic defects. Such concepts have been barely mentioned in this section due to their high relevance to LN subject in general, doped and undoped. Schirmer et al. (2009) and Imlau et al. (2015) have written comprehensive reviews on this topic [98,99]. A good introduction can be found in the comprehensive monograph written by Volk and Wöhlecke (2008), and it is also recommended to read the section devoted to it in work by Schirmer et al. (1991) [17]. On the other hand, we have noticed (according to the consulted bibliography) that, to follow a train of the earlier concepts, ideas, and facts that led to the development and acceptance of the polaron model concerning LN is not an easy task. While not being synthesized in a sole publication, some space and ink are next dedicated to it.

To the best of our knowledge, the first-ever mention of a small polaron related to reduced LN:Fe was made in 1973 by Clark et al. [100]. The report was written after a decade of theoretical developments to describe the electrical conductivity and to establish an absorption mechanism based on small polaron interactions in ionic crystals [101–105], extended for the case of small bipolarons in 1983 [106], and revisited in the early 1990s [107,108]. The polaron concept in connection with LN:Fe was refined in 1976 to describe thermally activated electron capture, where, for the first time, the localization of the small polaron in undoped LN was assigned to the Nb site [109]. In 1978, Schirmer and von der Linde observed a broad and complex absorption band extending from the near-infrared up to the absorption edge after irradiation of undoped and *non-reduced* crystals with X-rays or two-photon excitation at low temperature [110,111]. This work confirmed previous results obtained at room temperature by Karaseva et al. [112], and in conjunction with the report by Bernhardt (1976) [113], it also showed that the effect is more significant at lower temperatures. According to Schirmer et al. (1991) and references therein cited, by then, such broad absorptions in the given range of 1.5–3 eV were already correlated to O^- trapped holes having the character of a bound small polaron [17,114]. Schirmer and von der Linde confirmed it, based on two different EPR spectra and assigned the localization of the trapped holes at the Li vacancies, whereas that of the trapped electrons to the Nb sites [110,111].

In 1984, Arizmendi et al. refined the study described above and added elemental doping and reduction treatment to it: from the comparison of the effects of low-T X-ray irradiation and reduction treatment on the absorption spectra of undoped LN crystals, they found that the broadband resolves into three prominent bands with peaks at 1.6, 2.5, and 3.2 eV [115]. They concluded, in terms of the previous assignment of polaron states, that a self-trapped electron by ion Nb^{5+}_{Nb} (the small polaron Nb^{4+}_{Nb}) is responsible for the band at 1.6 eV, whereas holes trapped by Li vacancies or *other lattice defects* are responsible for the other two bands. According to the detailed analysis they made, by 'other lattice defects,' the formation of oxygen vacancies under the reduction treatment is implied, which acts as electron traps, in agreement with the conclusions obtained by similar studies a year earlier [116,117]. Under such an assumption, the band at 2.5 (3.2) eV is assigned to F (F^+) centers. It also copes with the fact reduced LN crystals are essentially diamagnetic, that is, no EPR signal can be correlated to the absorption band at 2.5 eV, which is intensified after reduction: the existence of oxygen vacancies being able to capture two electrons would fulfill these criteria, noticing that the absorption band itself cannot be assigned to electron capture by the Nb_{Nb} sites only, which would form a paramagnetic Nb^{4+}_{Nb}. Back then, there was no consensus regarding a model for intrinsic defects. The oxygen vacancy model was not overruled yet; recall that the Li vacancy model was proposed as early as 1968 by Lerner et al. [15], but it was not until the early 1990s that was widely accepted, mainly due to the works by Zotov [18], Iyi [25], and Blumel et al. [27].

Mainly based on the earlier report by Smyth (1983) [118], Jhans et al. (1986) gave solid arguments to disregard the neutral oxygen vacancy model and concluded that no single small polaron could be assigned to the 2.5 eV band [119]. In 1987, Schirmer and Koppitz (among others) proposed in two separate works that this band is ascribed to the dissociation of a bipolaron localized at a newly formed neighboring complexes Nb_{Nb}–Nb_{Li}, whereas the band at 1.6 eV was ascribed to a small polaron localized at Nb_{Li} antisite instead of the regular Nb_{Nb} site [120,121]. For such assignations, the existence of the Nb_{Li} intrinsic defect was taken into consideration, as well as the report by Chakraverty et al. (1978), in which it was demonstrated that in transition metal oxides, bipolaron formation is more energetically favorable if the electrons to be paired are localized on neighboring sites rather than on the same cation [122]. The previous success of the bipolaron scheme explaining the diamagnetism and optical properties of reduced WO_3 also played a role in constructing this model [123]. It was soon supported by computer simulations and model calculations [20,29,124]. The dissociation of the bipolaron $(Nb_{Nb}–Nb_{Li})^{2-}$ can be induced by thermal or optical means (about 0.3 eV), and physically it stands for "the transfer of one of the paired electrons from the bipolaron and further capturing by an isolated Nb_{Li}" [10].

The other important topics mentioned in this section (hydroxyl ions, ODRI, elemental doping for PR-enhancement, and laser-emission) have also been discussed in detail by Volk and Wöhlecke (2008) [10]. These can be further consulted in References [86,87,94], [6,17,60], [17,61,125], and [73,74,126], respectively. In the lines that follow, we express our belief about the top 10 facts (or sets of facts) regarding extrinsic defects in LN. Excluding induced ferromagnetism by elemental doping (treated in the next section) and neither chronologically nor hierarchically:

(1) Since 1973, a well-known spectral characteristic of the IR absorption of undoped LN is dichroism. In other words, optical absorption is prominently measured for a given linear polarization state of the incident light, which results in being perpendicular to the polar *c*-axis of the crystal [127]. It was concluded that the O–H dipole vibration essentially lies in the oxygen plane perpendicular to that axis. On a recent theoretical approach, it has been shown that, in ST crystals, the stretch mode is "tilted weakly out of the oxygen plane by about 4.3°" [50].

- Polarized Raman Spectroscopy of CG LN shows a very low local symmetry of OH^- related defect, assigned to point group C_1 [128].
- There is a tendency of increasing tilting angle (between the O–H stretch vibration and the oxygen plane) as the valence of the ODRI in doped crystals (above their respective threshold concentration) decreases [94]. As shown in Figure 6 (left side), for Mg^{2+}, the effect is pronounced with an angle of roughly 15°.

(2) For ODRI doping, the *threshold concentration* is the critical value in mol% describing the amount of incorporated ion into the crystal and "above which the optical damage drastically falls off by more than two orders of magnitude" [10]. It fundamentally depends on the stoichiometry of the crystal and in the valence state of the dopant: as a rule of thumb, less dopant is needed for the obtention of an optical-damage-resistant (ODR) crystal for higher valences ($n+ < 5+$). Independently of the dopant, it is less than 1 mol% for a near-ST composition. On the other hand, in a congruent crystal, it is roughly (5, 4, 2) mol% for a (di-, tri-, tetra-) valent ODRI [8].

- The first ODR composition found is LN:Mg^{2+} by doping with 4.6mol% MgO [85]. Later studies settled the threshold value at 5.0 mol% MgO for the congruent melt [129–131]. The dependence of the content of Nb_2O_5, Li_2O, and MgO in CG LN:Mg^{2+} has been plotted by Liu et al. [132].

- According to proposed models on the defect structure of LN:Mg, it is usually assumed that the Mg ions compete for Li sites with antisite Nb_{Li} ions up to the threshold concentration where all Nb antisites are eliminated [29,132,133]. Hence, regarding ODRI in general, the concentration threshold is also defined as the physical instance in which all Nb_{Li} antisites have been replaced by the doping ions, which above the threshold keep replacing regular Li sites and start to replace Nb sites.
- The threshold concentration for Mg can be calculated in mol% units, by use of Equation (1) [132]:

$$c_{th,Mg} = \frac{1000x}{3(100+2x)} \tag{1}$$

x is the same parameter used to describe the chemical composition of undoped LN, the formula $(Li_2O)_{50-x}(Nb_2O_5)_{50+x}$. Such extension of the x-values to define the [Li]/[Nb] ratio in LN:Mg comes from the lack of a direct method for this purpose, and it assumes that not only this ratio is equal in the melts but also the growing conditions of both crystals, undoped and Mg-doped [8,134]. Precautions must be taken on the ambiguity regarding reported concentration values and their dependence on properties due to the existence of two ways in specifying the impurity content (distinct chemical formulas), as noted by Donnerberg [135].
- LN:Mg has been extensively studied, and similar to undoped LN, there exist various alternatives to determine the dopant concentration (c_{Mg} mol%) using linear relationships with a corresponding measured parameter related to a specific experimental technique: IR absorption (OH^- vibrational spectroscopy), UV absorption edge, or polarized Raman spectroscopy. The results obtained using these three techniques are consistent for a given crystal; the equations can be consulted in Refs. [8,134,136]. Qualitative results on the effect of c_{Mg} on the PR effect presented by the crystals also have been studied through the z-scan method, a standard method often used to assess (quantitatively) third-order optical nonlinearities [137,138].

(3) Regarding other spectral characteristics of the IR absorption, the band components tend to narrow, and the amplitudes of those of higher frequency approach zero as the Li content increases in the crystal (see Figure 5). This tendency culminates at the stoichiometric (ST) point, characterized by a single-mode located at 3465 cm^{-1} with a Full Width at Half Maximum (FWHM) about equal to 3 cm^{-1} at room temperature (RT) [93]:

- In doped crystals with ODRI and below the threshold concentration, the IR absorption spectra are essentially the same as undoped crystals: the peak position appears near the 3465 cm^{-1}, and no new OH^- bands are detected. In the case of Mg, only the linewidth of the band is affected, decreasing while the doping concentration increases [139].
- Above the threshold, the dopant ions start to occupy regular Nb sites, "opening the way for the formation of M_{Nb}^{n+}–OH^- type complexes resulting in new absorption bands in the infrared spectral range" [94]. These bands are well separated from that of the undoped ST crystal (3465 cm^{-1}) and always are of higher frequency, as shown in the center and the right side of Figure 6. Their linewidths are narrower than the OH^- of undoped and congruent LN (cLN), for all ODRI doping cases. These characteristics resemble undoped and stoichiometric LN (sLN) so that *the existence of different classes of sLN crystals might be postulated.* In this regime of concentrations, the intensity ratio between the new and the 3465 cm^{-1} band increases with the dopant amount [94,140].
- According to a recent investigation (2017), the IR absorption bands of the lanthanide trivalent rare earth Er^{3+}, Nd^{3+}, Yd^{3+} feature similar characteristics to those of the trivalent ODRI (Sc and In), including the tilting angle out of the oxygen plane, which is around 8° [73]. More recently (2019), similar conclusions have been obtained for transition metal ions Fe^{3+}, Cr^{3+}, and Ti^{4+} [5].

- Thus, based on these substantial and systematic studies by Kovács et al. [5,73,94], it appears to be that all type of ions follow a generalized trend regarding the substitution mechanisms in doped LN, where the dominant factor is the valence state of the dopant. Even a partial solution to the configuration or distribution of the extrinsic defects (a model of extrinsic defect clusters) is acknowledged locally in the surroundings of a hydroxyl ion. As stated by Kovács et al., "only a small proportion of the incorporated dopants are involved in the $M_{Nb}^{n+} - OH^-$ type defect complexes while the others are assumed to occupy either Li or Nb sites," due to significant less concentration of OH^- ions compared to that of the dopant [5].

(4) Any given chemical composition, "the bandwidths, peak positions and intensities of different components are almost temperature-independent within the range of liquid-helium temperature to RT. Above RT, the maximum of some bands increase, and those of other bands decrease; the integrated intensity (i.e., the total amount of hydrogen) remains constant up to 200–250 °C" [87]. Conversely, they change slowly with time on a scale of 10 h; an equilibrium state of the OH^- vibrational spectra is reached only after some years [141,142]:

- The time-varying characteristic of the spectra is accounted for by a non-equilibrium distribution of the ocuppancy of the hydroxyl ions over four nonequivalent positions [10,88,143]. Accordingly, the broad OH band should be resolved into at least four overlapping components corresponding to different O–H bonds. In near-ST crystals, up to five distinct Lorentzian components have been resolved (Figure 8): **3466**, *3470*, **3481**, *3490*, and 3499 cm^{-1} [143]. Significantly, with time, the intensities of those in italics increase at the expense of those in bold (slight differences in the numerical assignments scattered throughout the literature), the integrated intensity being essentially unchanged [10].
- The intensity ratio (R) of the two components at 3480 and 3465 cm^{-1} (above in bold), I_{3480}/I_{3465}, slowly grows after annealing near-ST crystals, reaching a saturation point in between 16 and 18 months (Figure 9) [142].
- The chemical composition of an undoped crystal can be assessed with an absolute accuracy of 0.01 mol%, using the estimation of the Li content c_{Li} in terms of this ratio measured at RT and at a time t after its growth (in units of months) [8]:

$$R(t) = \left(2.01 - 1.27e^{-t/5.9}\right)(50 - c_{Li}) \quad (2)$$

(5) Recent advances in OH^- Vibrational Spectroscopy have led to a detailed RT study on the dynamics of strongly localized and ultrashort-living lattice distortions in the neighborhood of an O–H dipole. A novel achievement is the reconstruction of the electrostatic potential change in the oxygen plane of LN, based on the local polaronic distortion correlated to a pump-induced (scale of femtoseconds) –3 cm^{-1} shift of the frequency of the OH^- stretch vibration [144]. • This result stands for the first visualization of the important observation made about 15 years earlier (2000–2006, mainly by two German groups) on the Nb_{Li}^{5+} antisite being a shallow electron trap, based on transient absorption of LN:Fe crystals induced by ultrashort-pulsed green illumination [145–148]. By use of λ = 388 nm light with 240 fs pulse duration in undoped CG crystal, Beyer et al. estimated the formation of the small polarons or Nb_{Li}^{4+} centers in less than 400 fs [148].

(6) "The basic advantage of choosing higher valence ODR dopants comes from their lower damage threshold concentrations; lower built-in dopant content facilitates the growth of more homogeneous crystals resulting in the high-quality sample for device applications" [8]. Recently, research efforts on LN:Zr^{4+} have been intensified for various practical reasons, which make it an optimal choice for applications at RT:

- *Low doping concentration and high optical-damage resistance:* a concentration threshold of 2.0 mol% ZrO_2 holds for cLN:Zr [6,149], whereas 0.1 mol% in near- sLN:Zr [140], compared to those of 5 mol% and 0.2 mol% for MgO, respectively [46]. A crystal doped with 2 mol% ZrO_2 can withstand a light intensity (λ = 514.5 nm) about 40 times higher (around 20 MW cm^{-2}) than a 6.5 mol% MgO doped crystal [6]. Apropos, only a near- sLN:Mg crystal (1 mol% Mg), competes with the former, withstanding close to 26 MW cm^{-2} at λ = 488 nm [150]; the obtention of this crystal involves an expensive yet unpractical vapor-transport-equilibria process at 1000 °C for about 120 h. It has recently been demonstrated that near-ST LN:Zr crystals can be illuminated with 80 GW cm^{-2} of pulsed 10 ns green light without optical breakdown [151].

- *Segregation and distribution coefficient:* The segregation between solvent and solute and the distribution of dopants between Li and Nb sites influence the incorporation of ODRI into flux-grown LN crystals, the quality (i.e., homogeneity) of the latter per se. Segregation is a severe problem even for the growth of pure or undoped LN crystals since the CG and ST compositions do not coincide. It remains in melt systems for the obtention of ODRI-based compositions, causing the scattering in values of the distribution coefficient (K_{eff}) from unity, except perhaps for the case of Li_2O–Nb_2O_5–ZrO_2. Most ODRI are characterized by K_{eff} falling in the range 1.2–1.3 for the low concentration regime [10]. At higher concentrations, they decrease close to 1 for MgO [133] and ZnO [152] in the range 5–6 mol%, and to about 0.9 for In_2O_3 in the range 1.5–2 mol% [153]. In contrast, $K_{eff} \approx 1$ for LN:Zr^{4+}, independently of the concentration range of 1–5 mol% ZrO_2 [149]. In addition, "it should be noted that in crystals prepared for device applications, ODR dopant concentrations slightly above the critical threshold are preferred to suppress the formation of unwanted microdomains" [8].
- *Working in the ultraviolet range:* compared to the rest of ODRI, LN:Zr maintains its ODR feature in an extended region from the visible to the UV [6,154]. Interestingly, PR enhancement has been determined for LN:Mg, LN:Zn, and LN:In crystals excited by UV laser light [155,156].
- Similar characteristics of Hf^{4+} makes it also a promising ODRI [157–160].

(7) Recently, a Japanese group reported the novel result of overcoming segregation in the pseudo-ternary system Li2O–Nb2O5–MgO using in melt compositions 45.30:50.00:4.70, respectively [161,162]. As shown in Figure 7 (left side), the CG and ST compositions coincide with the obtained single crystal. For this reason, it has been labeled as 'cs-MgO:LN'. Because of a higher compositional homogeneity, its optical properties are expected to be superior to those featured by CG, ST, and LN:Mg. Indeed, optical second harmonic generation (SHG) measurements show comparable conversion efficiencies and a more uniform in-plane distribution of the non-critical phase-matching wavelength (right side of Figure 7) [162]. Its ODR, however, has been elsewhere stated to be inferior to that of cLN:Mg—5 mol% [6].

(8) Upon elemental doping with different ions, several defect-related structures and possible charge misfits are compensated by the formation of suitable intrinsic defect complexes. Given the similarity of the octahedral surroundings at the Li and Nb sites, one of the most challenging questions to answer since earlier studies in this field is the dopant-site occupancy, whether a given doped-ion substitutes Li or Nb or both. No consensus exists on these grounds. Depending on the classification of the dopants (ODRI, PR-active, or laser activation, i.e., lanthanides), among other factors, different trends on site occupancy and doping mechanisms (the dynamics) can be stated. As of today, and in general terms, that is in what respects for *all kinds of impurity ions* and independently of their charge state, a strong statement, say undisputable, is that dopants are "localized dominantly on the Li sites" [10]. It has been qualitatively explained according to the larger dimension of the Li octahedra respect to Nb one [163,164]. According to a recent review by Kang et al. (and references therein), the dopant concentration, ionic radii, and valence states are among the most influential factors [7]. In general:

- The valence state induces priority substitution: a +2 valence dopant substitutes Li^+ ions, whereas one with a valence state greater or equal to +5 replaces Nb^{+5} ions. No explicit priority substitution can be noticed for dopants of valence +3 and +4 [165].
- Longer (smaller or equal) dopant–O bonds than the mean length of the Nb–O bond reflects occupancy at the Li (either Li or Nb) site; a small shift from the regular Li site takes place if the bond is also longer than the mean length of the Li–O bond [163].
- Low (high) doping concentrations occupy Li or (and) Nb sites.
- The doping process has also been shown to influence the site occupancy. Recently, it was found that the OH^- vibration spectra are significantly different for two LN:Cr^{3+} crystals sharing the same dopant concentration (0.5 mol%) but obtained by two distinct growth techniques, thus implying different substitution mechanisms; the IR absorption spectral characteristics of one crystal shows the fingerprint of a heavily doped crystal while those of the other do not [5]. A similar observation has been made for Cu^{2+} doping [11], although it is not well supported by the bulk of the literature: copper is normally assumed to enter a Li site, irrespective of the charge state [8,115].

(9) The central role of ODRI is to vary the amount of Nb antisites in the crystal. Extensive studies by several experimental methods and model calculations have been motivated by explaining the microscopic origin of the threshold concentrations in ODRI. In other words, the incorporation mechanisms of the dopants and their influence on the threshold values. Among others, a striking result from such is that *Mg and Zn at relatively low concentration values may be regarded as controllers of the intrinsic defects* Nb_{Li} *and* V_{Li} *contents, so their effect is qualitatively analogous a Li enrichment* [10].

- Although still being debated, the current most accepted description on the microscopic origin threshold concentrations for ODRI is that its existence is associated with *partial* incorporation of the dopants onto *regular* Nb_{Nb} *sites*. In contrast, below these critical values, the incorporation mechanism is described by applying a *simultaneous* substitution of Nb_{Li} antisites and regular Li_{Li} sites by the dopants. This framework is constructed upon results reported for Mg^{2+} [17,132,135,166,167]. The same conclusions were obtained for later detailed structure studies with X-ray and neutron diffraction on LN:Zn [168–170]. Of course, this discussion holds for Li-deficient LN-doped crystals (congruent). As the crystal composition is closer to the ST one, less Nb_{Li} antisites will be available, so that the incorporation will be split mainly into regular Li and Nb sites. It explains that significantly less dopant content is needed to reach the threshold concentration in near-ST crystals.
- Impurity Zn possesses the largest threshold concentration among ODRI, above 7 mol% [171,172], which is particularly important regarding the reliability of the results obtained by detailed structural studies. The extension of this framework onto the rest of the ODRI seems quite natural; it is also supported by further evidence found for In^{3+} and Hf^{4+} [173–175]. Based on the study of the structural properties of doped crystals, the available information has been recently updated for In^{3+} and Zn^{2+} [60,176], and extended for Zr^{4+} [177].
- All lanthanide trivalent rare-earth ions were long thought to occupy Li sites only, independently of the [Li]/[Nb] ratio in the crystal [10]. However, based on detailed OH^- Vibrational Spectroscopy studies in near-ST LN-doped crystals with Er, Nd, and Yd ions, Kovács et al. have inferred partial occupation at the Nb sites above a critical concentration value, denoted as 'threshold' given the similarity of the trends observed in connection to crystals doped with ODRI [94]. Quoting them: "for rare-earth ions, the term 'threshold' possibly does not mean a photorefractive damage threshold, but implies the concentration of the dopant above which the rare-earth ions can occupy Nb sites in the lattice" [94]. Still, most of the dopant ions are assumed to enter Li sites; the amount of those entering an Nb site has been estimated to be very small, of the order of the OH^- content in the samples

(10–100 ppm). The same conclusions were recently obtained for transition metal ions (Fe^{3+}, Cr^{3+}, Ti^{4+}), as discussed in bulletpoint 3 within the present section.

- An intermediate stage of the substitution mechanism has also been suggested, where Li_{Li} substitution alone lies in between the simultaneous occupation of Li_{Li} sites and Nb_{Li} antisites, and the partial incorporation of the dopant into regular Nb_{Nb} sites [178–180]. This has led elsewhere to construct the *first threshold* concept, a lower concentration threshold than the beforehand discussed [10]. Thus, under such a framework, the first threshold would stand for the complete removal of Nb_{Li} antisites, whereas the second threshold for the beginning of Nb_{Nb} site occupancy.

(10) Simultaneous enhancement for both photorefraction (PR) and optical damage (OD) resistance has been reported for MgO and Bi_2O_3 co-doped LN (LN:Mg, Bi) [181]. This observation conflicts with the mainstream opinion found in literature: for a long time, these terms have been used indistinctly to address the same process: the optically induced inhomogeneities in the refractive indexes. Hence, it had been tacitly conceived that by making a crystal more resistant to OD, it could also hardly be PR enhanced. They had been thought all this time to "be just opposite sides of the same medal" [6]. Inspired from a rather old result, Zhang et al. did experiments involving the coupling of two beams of equal intensity and came to conclude that OD is not the same as PR, but only a form of expression of it [181]:

- Just a year ago, the same group proved this material to efficiently display real-time dynamic holograms with a 30 Hz refresh time when concentrations of 6 mol% Mg and 1 mol% Bi are used [182]. A video of the display of a running leopard is available in the supplementary material of that contribution. Regarding the labeling of co-doped crystals, recall that, in the present contribution, we use the convention of writing first the dopant with higher concentration, after the colon, followed by a comma, and then the dopant of less concentration, so that LN:Mg, Bi, and LN:Bi, Mg stand for two distinct crystals (see Figure 4).
- Doping with more than one ion species, i.e., co-doping, clearly provides some advantages of technological relevance, particularly in the field of photorefraction. The substitution mechanisms in co-doped LN is out of the scope of the present contribution. Nonetheless, it is expected to behave similarly to single yet heavily doping [5]. Further details on this subject can be consulted in Reference [61].

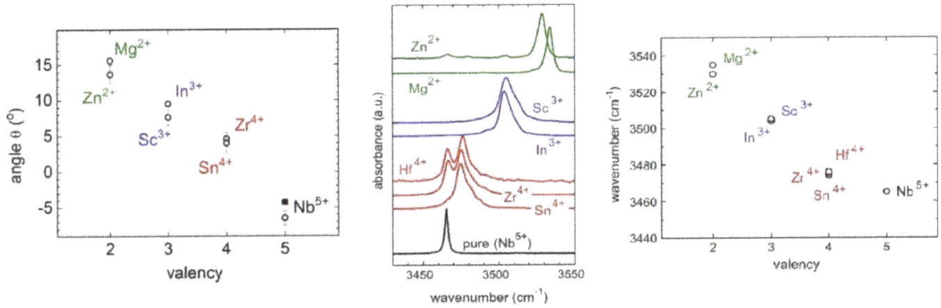

Figure 6. Trends observed in the OH^- absorption bands of ODRI doped-$LiNbO_3$ (above the threshold) in terms of the valence state of the dopant. **Left**: angle θ between the polar *c*-axis of the crystal and the plane containing the stretch vibration of the O–H dipole. **Center**: IR absorption spectra for the seven known ODRI dopants, in the cases where multiple bands are observed, presumably the doping concentration is close to the threshold. **Right**: the vibrational frequency of the stretch vibration of the O–H dipole—doi.org/10.1016/j.optmat.2014.04.043 (Reproduced from Kovács, L.; Szaller, Zs.; Lengyel, K.; Corradi, G. Hydroxyl ions in stoichiometric $LiNbO_3$ crystals doped with optical damage resistant ions. *Opt. Mater* **2014**. *37*, 55-58, with permission from Elsevier).

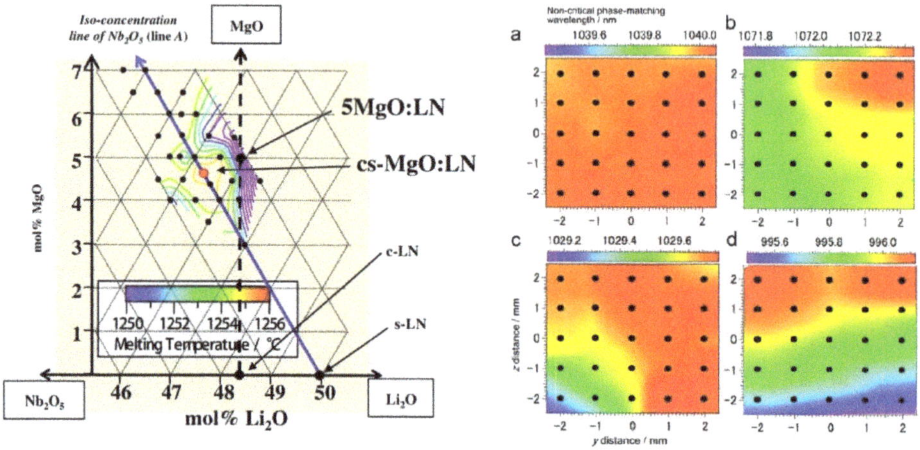

Figure 7. cs-MgO:LN crystal. **Left:** equilibria phase diagram of the pseudo-ternary system Li_2O–Nb_2O_5–MgO showing isothermal lines of the liquidus plane around the zone where the stoichiometric and congruent points coincide. **Right:** in-plane distribution of the non-critical phase-matching wavelength for cs-MgO:LN (a), cLN (b), LN:Mg—5mol% (c) and sLN (d)—doi.org/10.1016/j.jcrysgro.2010.08.056. (Reproduced from Kimura, H.; Taniuchi, T.; Iida, S.; Uda, S. Bulk crystal growth of congurent MgO-doped $LiNbO_3$ crystal with stoichiometric structure and its second-harmonic-generation properties. *J. Cryst. Growth* **2010**, *312*, 3425–3427, with permission from Elsevier.)

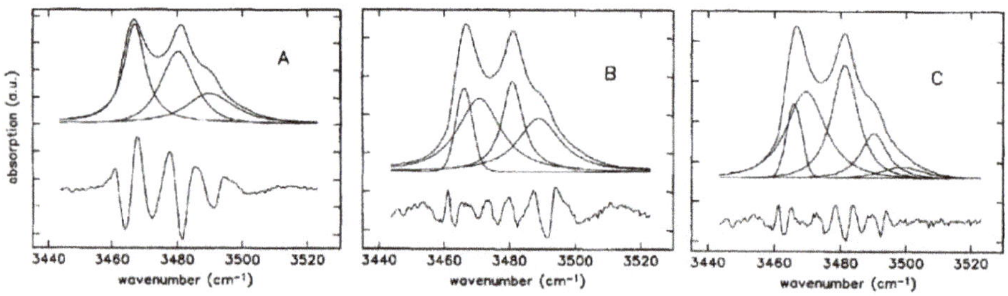

Figure 8. Lorentzian-based resolution of the IR absorption band of a LiNbO3 crystal grown from a melt with [Li]/[Nb] = 1.2, using three (**A**), four (**B**), and five (**C**) components. The residue is enlarged by a factor of 10—doi.org/10.1016/022-3697(91)90078-E. (Reproduced from Kovács, L.; Wöhlecke, M.; Jovanovic, A.; Polgár, K.; Kapphan, S. Infrared absorption study of the OH vibrational band in $LiNbO_3$ crystals. *J. Phys. Chem. Solids* **1991**, *52*, 797–803, with permission from Elsevier.)

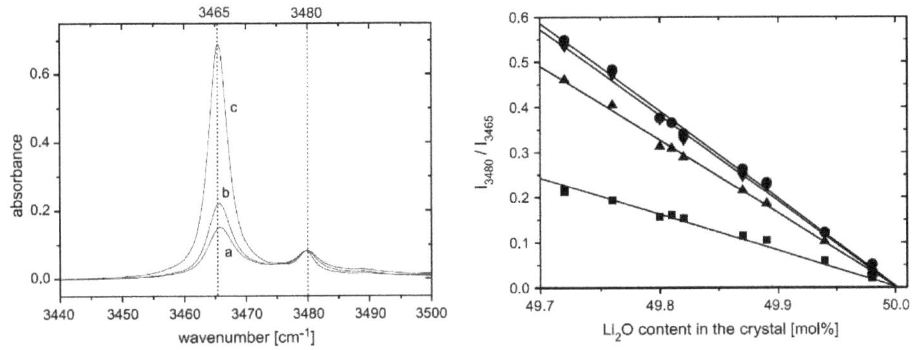

Figure 9. IR absorption spectra of near-stoichiometric $LiNbO_3$ crystals, normalized to the band located at 3480 cm^{-1}. **Left**: Li_2O content of 49.72 (a), 49.80 (b), and 49.94 mol% (c). **Right**: intensity ratio of the two components at 3480 and 3465 cm^{-1} in terms of the crystal composition and measured at 0, 7.5, 16, and 18 months after crystal growth (squares, up-triangles, down-triangles, and circles, respectively)—doi.org/10.1007/s00340-007-2704-9. (Reproduced from Dravecz, G.; Kovács, L. Determination of the crystal composition from OH^- vibrational spectrum in lithium niobate. *Appl. Phys. B* **2007**, *88*, 305–307, with permission of Springer Nature.)

4. Dilute Magnetic Oxides and Ferromagnetism in $LiNbO_3$ Doped with Magnetic *3d* Cations

Historically, the study of the magnetic properties of materials has been of great importance. Novel technological advances can be ascribed to this physical phenomenon, such as ultra-strong permanent magnets, information storage devices with high magnetic density, nano-biomagnetic sensors, and spintronics [183]. Over the years, ferromagnetism has been reported in metals such as Fe, Ni, and Co [184]; in insulators, where the EuO, $YTiO_3$, $SeCuO_3$, and $BiMnO_3$ are some examples [185]; and in diluted magnetic semiconductors (DMSs) [185]. In the classification of DMSs, there are three subgroups: the II–VI, III–V, and the rare-earth-V groups, where the (Zn, Mn)Se, (Ga, Mn)As and ErAs are excellent representatives, respectively [185]. In the same framework, the diluted magnetic oxides (DMOs) is another critical group in the classification of ferromagnetic materials, with the general formula [186]:

$$(A_{1-x}M_x)(O\square_\delta)_n \tag{3}$$

A corresponds to a nonmagnetic cation, *M* is the magnetic cation, □ representing a donor defect, $n = 1$ or 2, and δ corresponds to the donor concentration. Systems as SnO_2 (conductor), ZnO (semiconductor), and TiO_2 (insulator) doped with magnetic *3d* cations are some examples. In the search for new magnetic materials, some non-magnetic ABO_3-type oxides doped with *3d* magnetic cations have attracted much attention due to the possibility of inducing ferromagnetism in them, expanding this way the range of technological applications. Nowadays, there are reports already of systems such as LN:Mn, LN:Fe, $LiTaO_3$(LT):Fe, $BaTiO_3$:Fe, and $SrTiO_3$:Co that exhibit ferromagnetism at room temperature (RT) [187–192].

As for DMOs, Coey defines these as nonmagnetic oxides doped with small percentages of magnetic *3d* cations (generally under 10 at.%), which show signs of ferromagnetism at high temperatures (the general formula is shown in Equation (3)) [186,193,194]. These are often synthesized as thin films. Regarding the reported magnetic curves *M–H*, scattered throughout the literature, these are not temperature-dependent below RT and are practically anhysteretic [194]. The fact that the magnetic coercivities observed in these materials are not dependent on temperature implies that the magnetization process that favors the manifestation of ferromagnetism is not dominated by magneto-crystalline anisotropy, but instead by magnetic dipole interactions [194]. In DMOs, the magnetization curve is described by Equation (4) [194]:

$$M = M_s tanh\left(\frac{H}{H_0}\right) \quad (4)$$

where M_s is the saturation magnetization, H the applied magnetic field, and H_0 is an effective magnetic field constant that prevents saturation. The observed ferromagnetism in DMOs can be explained from the bound magnetic polaron (BMP) model, proposed by Coey [186]. It is based on the formation of point defects and the association of an electron to each defect. In turn, each electron is confined in a 1s hydrogenic orbital of radius $r_H = \varepsilon(m/m^*)a_0$, ε is the high-frequency dielectric constant, m the mass of the electron, m^* the effective mass of the donor electrons, and a_0 the B–r radius (~53 pm). Hence, defects—often induced by heat treatments under certain atmospheres—play an essential role in the origin of magnetism. As the donor concentration increases, so does the number of *1s* orbitals, which overlap and form the so-called impurity band. When the hydrogenic electrons interact with the magnetic cations M in the impurity band, magnetic polarons tend to form, coupling the *3d* orbitals of the ions within their orbits [186]. A ferromagnetic coupling between the spin of the donors and the spin of the *3d* orbitals exists if $\gamma > 4$, where $\gamma = \varepsilon(m/m^*)$. This ferromagnetic interaction is defined by the exchange parameter J_{sd} in Equation (5) [186]:

$$-J_{sd}S\cdot S|\Psi(r)|^2\Omega \quad (5)$$

S corresponds to the spin of the 3d cations with a volume Ω, and **s** is the spin of the donor electron. Finally, long-range ferromagnetic order is favored in the DMOs if the amount of magnetic cations is large enough, and $\delta > \delta_p$, $x < x_p$. Parameters δ_p and x_p are the percolation thresholds of the polaron and the cation, respectively; x is the doping and $\gamma^3\delta_p \simeq 4.3$ [186]. The BMP model is summarized in Figure 10. The ferromagnetism observed in DMOs samples is reported as intrinsic. In the cases of thin films or nanocrystals, the formation of secondary magnetic phases must be ruled out by a convenient characterization. Regarding the magnetic hysteresis loops (measured at RT), these present very small coercivities on the order of 5–10 mT, and a remanence ratio (M_r/M_s, the numerator being the remanence magnetization) of 5 to 10% [193]; there are instances where the magnetic moment exceeds the spin-only moment of the doping cation for small concentrations of doping. Conversely, these kinds of oxides present some disadvantages, the most important being the fact that the magnetic moment is frequently unstable in time, and the reproducibility of the samples is poor [193]. The BMP model fits well with the experimental results reported over the years. Nevertheless, other models cannot be discarded, namely those based on whether the magnetic moment in the samples is localized or delocalized. The nature of the electrons involved in the magnetic interactions also plays a role. In this context, the diluted magnetic semiconductor, the split-spin impurity band, and the charge-transfer ferromagnetism models may be consulted [195–197]. With DMOs as a framework, the magnetic properties of LN doped with magnetic *3d* cations will now be introduced. As mentioned above, the impurification of LN with transition metals (Fe, Cu, Mn, Ni) has been extensively explored in the study of the photorefractive effect; however, there is an important opportunity area in terms of magnetic properties.

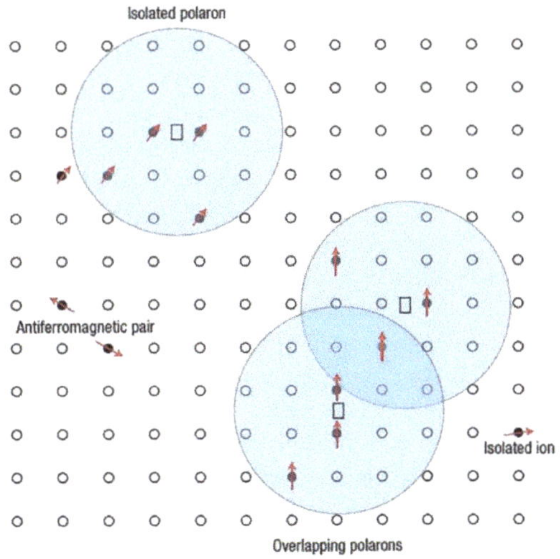

Figure 10. Schematic representation of magnetic polarons: the spin of a donor electron in its hydrogenic orbit couples to antiparallel half-full (or more than half-full) *3d* shells of the magnetic impurities. In this case, x = 0.1 and γ = 12. The small circles represent the cations and the squares the unoccupied oxygen sites (oxygen is not shown)—doi.org/10.1038/nmat1310. (Reproduced from Coey, J.M.D.; Venkatesan, M.; Fitzgerald, C.B. Donor impurity band exchange in dilute ferromagnetic oxides. *Nature Mater.* **2005**, *4*, 173–179, with permission from Springer Nature.)

In 2006, Song et al. prepared $Co_{0.05}(LiNb)_{0.95}O_{3-\delta}$ by Co implantation [198]. Their analysis performed by energy dispersive spectroscopy reported that Co was uniformly distributed to a depth of 220 nm on $LiNbO_3$ wafers. On the other hand, through ab initio calculations and X-ray absorption near-edge structure (XANES), it was demonstrated that the Co substitutes the Li in the LN crystal lattice. The $Co_{0.05}(LiNb)_{0.95}O_{3-\delta}$ sample showed ferromagnetism at RT with an atomic magnetic moment (AMM) of 1.3 μ_B/Co and a high Curie temperature (T_c) of 710 K. In terms of the BMP model, also within the context of the discussion given by Song et al. [198], the implantation of ions is a nonequilibrium process that makes possible the formation of defects located at aleatory distances respect to Co sites in the $Co_{0.05}(LiNb)_{0.95}O_{3-\delta}$ layer. In this sense, a donor spin of a particular defect interacts with Co ions within its orbit, forming a BMP. The orbits overlap and interact with adjacent BMPs, creating a magnetic order. Thus, Song et al. associated the ferromagnetism in Co-doped $LiNbO_3$ with the formation of BMPs. However, other mechanisms have been found to cause ferromagnetism in $LiNbO_3$ doped with magnetic *3d* cations. For example, Chen et al. (2011) developed an interesting work where LN wafers implanted with a dose of Mn (1, 2, 3, 4, 5 at.%) were prepared, having a thickness of 200 nm [192]. The samples exhibited ferromagnetism, a maximum AMM of 5.83 μ_B/Mn was determined for the sample at 3 at.%. XANES ruled out the formation of secondary phases and oxygen vacancies, and it was concluded that the Mn ions substitute the Li ions in the crystal lattice. In addition, they found that, above this concentration, the crystal lattice began to present damages, translating into a significant decrease in the AMM. The origin of ferromagnetism (half-metallic ferromagnetism) was attributed to a strong *d–d* interaction between the dopant and its neighboring atoms Nb, inducing a long-range interaction through the whole crystal. The experimental results were supported by electronic structure calculations, where a magnetic moment of 5.93 μ_B per cell was calculated for an Mn concentration of 3.3 at.% [192]. Finally, their theoretical analysis led them to conclude that the distances between the Mn and two neighboring Nb atoms (Nb1 and Nb2) are in consideration of the Li–Mn cationic interchange d_{Mn-Nb1} = 277 pm and d_{Mn-Nb2} = 300 pm, whereas for the undoped crystals d_{Li-Nb} = 312 pm.

In a similar study, Zeng et al. (2012) prepared LN:Fe by ion-beam implantation at different Fe concentrations (1, 2, and 3 at.%); the obtained implanted layers were about 200 nm thick and showed ferromagnetism at RT [190]. A maximum AMM of 3.33 μ_B/Fe was determined for the sample doped with 1 at.% [190]. It was concluded that Fe substitutes Li for a concentration of 1 at.%, the interaction between the Fe and its neighbor Nb enables strong spin coupling throughout the whole crystal, resulting in an effective enhancement of the ferromagnetic behavior. Here, it is important to note that the Fe cation environment into the LN structure has been extensively studied using X-ray absorption fine structure (EXAFS) and XANES, where the results indicate that Fe^{3+} ions unambiguously occupy Li sites [199–201]. Zeng et al. also found a significant decrease in the AMM measured values (0.87 and 0.30 μ_B/Fe for concentration 2 and 3 at.%, respectively), this being attributed to the possible formation of FeO_x clusters not coupled in a long-range order [190]. The theoretically calculated distances between the dopant Fe and its two neighboring Nb atoms were $d_{Fe\text{–}Nb1}$ = 274 pm and $d_{Fe\text{–}Nb2}$ = 299 pm, in agreement with the results reported by Chen et al. (2011) [192].

Figure 11 shows the plots of magnetic moment per transition metal-cation of LN doped with Fe (1-LN:Fe), Mn (2-LN:Mn), V (7-LN:V), as well as four different systems of LN doped with Co (labeled as 3-LN:Co, 4-LN:Co, 5-LN:Co, and 6-LN:Co). All samples were prepared as layers by different methods [190,192,198,202–205]. As can be seen, for 2-LN:Mn and 7-LN:V, the AMM increases as the x increases until reaching a maximum; after that, the AMM begins to decrease as x increases. On the other hand, 1-LN:Fe presents a maximum AMM for small values of x. As previously explained, the AMM decrease is mainly associated with the formation of secondary phases or lattice damage due to excess dopant, which destroys the magnetic coupling. Finally, the LN:Co samples (4-LN:Co, 5-LN:Co and 6-LN:Co) exhibit a low AMM, except for 3-LN:Co, around 5.5 μ_B/Co.

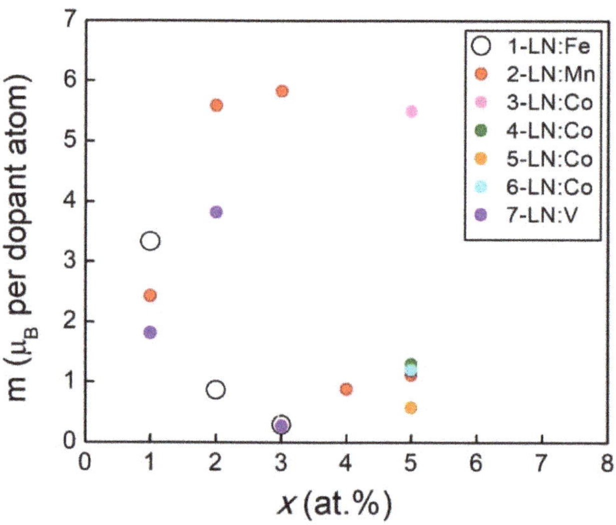

Figure 11. Plots of magnetic moment per transition metal-cation in doped LN. The transition metal-cation substitutes Li in 1-LN:Fe, 2-LN:Mn, 3-LN:Co, and 4-LN:Co when x = 1 at.%, 3 at.%, 5 at.%, and 5 at.%, respectively. On the other hand, the transition metal-cation substitutes Nb in 5-LN:Co, 6-LN:Co, and 7-LN:V when x = 5 at.%, 5 at.%, and 2 at.%, respectively.

There is a tendency in what respects doped-LN to study single-crystal and polycrystal thin films, in comparison to polycrystalline powders; further results on the Fe impurity and reports on the doping with Co, Cu, and V—based on similar studies to those above just described—can be consulted in the literature [198,202–207]. However, recently some efforts have been put on as-obtained coarse LN particles that show ferromagnetism at RT. On an investigation carried out in 2018 by our group, it was found that the generation of oxygen vacancies—by thermal treatment in reducing atmospheres: thermal treatment in a reducing atmosphere (TTRA)—at the surfaces of LN:Fe powders is a mechanism that favors RT ferromagnetism at low doping concentrations of 0.44%, 0.89%, 1.47%, and 2.20% by mass of Fe_2O_3 (labeled as nLN:Fe-0.44-R, nLN:Fe-0.89-R, nLN:Fe-1.47-R, nLN:Fe-2.20-R, respectively) [208]. A saturation magnetization of 0.96 $Am^2\ kg^{-1}$ was determined for the sample of higher concentration. Structural analysis by X-ray diffraction and Raman spectra led to the conclusion that Fe enters Li sites, in agreement with the work of Zeng et al. (2012). However, in contrast, no decrease in the magnetization values as the concentration increases gradually was found this time. It suggests that the crystal structure was not strongly distorted and also that the formation of secondary magnetic phases in the form of clusters is unlikely, in the studied concentration range. Figure 12 shows the ferromagnetic curves for the chemically reduced samples with different Fe-doped concentrations. In this case, all the systems presented a constant coercive field (H_c) of $2 \times 10^4\ Am^{-1}$. In addition, two samples, oxidized LN doped at 2.20% (labeled as nLN:Fe-2.20-O) and a chemically reduced sample of undoped LN (labeled as nLN-R), are shown. It can be seen that the saturation magnetization of the reduced samples (doped at different Fe concentrations) is about two orders of magnitude higher than the oxidized samples.

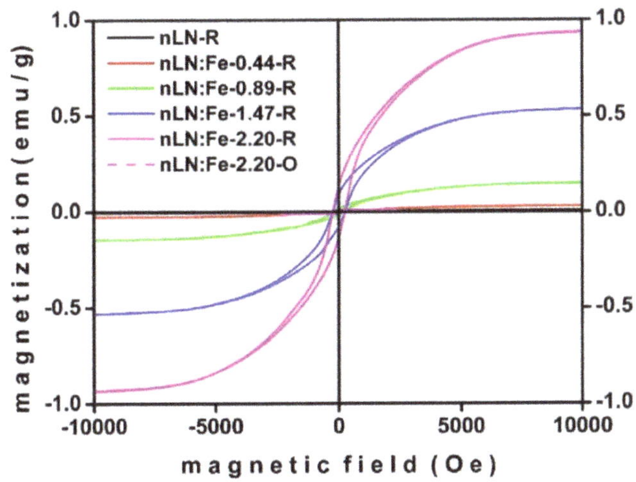

Figure 12. Comparison between the *M–H* curves of the chemically reduced samples (doped at different Fe concentrations), nLN-R, and nLN:Fe-2.20-O—doi.org/10.3390/cryst8030108. (Reproduced from Fierro-Ruíz, C.D.; Sánchez-Dena, O.; Cabral-Larquier, E.M.; Elizalde-Galindo, J.T.; Farías, R. Structural and Magnetic Behavior of Oxidized and Reduced Fe Doped $LiNbO_3$ Powders. *Crystals*, **2018**, *8*, 108, with permission from MDPI.)

The connection between ferromagnetism and the generation of oxygen vacancies through TTRA was previously reported (2014), also by our group [209]. In that instance, the generation of the oxygen vacancies at the surfaces of the nanocrystals was confirmed by the measuring of oxygen depletion across the nanoparticles using transmission electron microscopy (TEM). Experimentally, a magnetic moment of $5.24 \times 10^{-3}\ \mu_B$ was determined. Although this value is minimal and did not agree with ab initio calculations, this contribution is relevant because it is one of the first reports on the magnetic properties of undoped LN: see also the reports in 2012–2013 by Díaz et al. [210], Ishii et al. [211], and Cao et al. [212], and in 2016 by Yan et al. [213].

In the work by Cao et al. (2013), based on ab initio calculations, it is stated that the formation of neutral V_Os is favorable and that the formation of two intrinsic defects of this type (in specific coordinates) induces ferromagnetism by the spin polarization of the Nb *s* electron [212]. The work by Yan et al. (2016) is particularly important because it represents the first experimental study on undoped *single-crystal* LN (and $LiTaO_3$) in connection to induced ferromagnetism by TTRA [213]. They measured a magnetization saturation, a magnetic remanence, and a coercive field in reduced CG LN of 7.0×10^{-3} emu g^{-1}, 0.65×10^{-3} emu g^{-1}, and 0.050 kOe, respectively. The induced ferromagnetism is therein associated with the formation Nb^{4+} (Ta^{4+}) with non-zero net spin and oxygen vacancies on the surface due to the chemical reduction.Here, an important fact regarding the implications made in Reference [209] is discussed. According to the paramount observation in the peer reviewing process of this review, the depletion of oxygen does not neccesarily implies the formation of oxygen vacancies. The existence of oxygen vacancies in LN from TTRA is still on debate. The studies just discussed above attribute the manifestation of ferromagnetism to mechanisms where oxygen vacancies play an essential role, while others have entirely overruled their existence. Before entering the decade of the 1990s, Smyth (1983) and Donnerberg et al. (1989) separately proposed an acceptable model that explains the defect structure of CG crystals after chemical reduction (see discussion of the polaron model in the previous section) [14,118]:

$$3O_O + V'_{Li} + \mathrm{Nb}_{\mathrm{Nb}} \leftrightarrow \frac{3}{2}O_2 + \mathrm{Nb}^{4+}_{\mathrm{Li}} + 6e^- \quad (6)$$

where a loss of oxygen on the surface of a molecule promotes a rearrangement of the constitutional elements of LN, causing Nb^{5+} ions to move into empty Li sites, in turn leading to a rearrangement of Li vacancies and Nb antisites; the released electrons are captured by traps [10]. As already discussed, this model predicts the crystal density to decrease with greater Li_2O deficiency. Thus, the TTRA could in principle be held for long times to the extent where the secondary Li-deficient phase, $LiNb_3O_8$, starts to form at the expense of the pure LN phase. Indeed, this is the case recently reported by Kocsor et al. [214]: pure LN nanocrystals were prepared by ball-milling the crucible residues of a Czochralski grown congruent crystal, however, $LiNb_3O_8$ crystallized after reduction and oxidation thermal treatments at 800 °C for 180 min. There, the oxidizing treatments induce the formation of the secondary phase more effectively as a consequence of Li_2O evaporation. The formation of oxygen vacancies was also excluded from the mechanisms picture.

After Schirmer and Koppitz (1987), it is acknowledged that the trapped electrons create small polarons and bipolarons, localized at the Nb_{Li} antisites and newly formed neighboring complexes Nb_{Nb}–Nb_{Li}, respectively [120,121]. Thus, TTRA plays a role in the generation, or at least the redistribution, of point defects in LN, which in turn induce changes in its physical properties. Because it has been elsewhere stated that "the oxygen leaves the crystal surface, so no oxygen vacancies appear at the end" [10], the implications made by our group in Reference [209] need to be revisited. A set of studies shall soon be done in our laboratories in order to study such implications more carefully and coherently. The permanent existence of vacancies in the surfaces of nanocrystals cannot be assured (or denied) with the current available literature. Nonetheless, at least it can be argued that during the TTRA, the in situ formation of Vo's could be responsible for the construction of permanent electron trapps that eventually lead to strong (weak) magnetism in LN:Fe (undoped LN) powders.

Other significant conclusions can be obtained from the studies analyzed in this section:

- Based on XANES, EXAFS, and Raman Spectroscopy measurements and ab initio calculations, there is a trend of Fe, Mn, and Co cations to occupy Li sites at low doping concentrations.
- The substitution of Li with Fe and Mn cations modifies the interatomic distances between their closest neighbors, favoring the interaction between the *3d* cations and their neighbors.
- When LN is doped with a small amount of *3d* magnetic cations such as Fe, Mn, and Co, it manifests ferromagnetism at RT. However, in the high-doping concentration regime, the magnetization

saturation tends to decrease, mostly due to the destruction of the magnetic coupling around the magnetic *3d* cation and sometimes to the formation of secondary magnetic phases.

- The formation of point defects usually addressed as oxygen vacancies after TTRA in either doped or undoped LN serves as a mechanism for ferromagnetism enhancement.

5. Conclusions

Lithium niobate ($LiNbO_3$, LN) is a promising material in fundamental and applied sciences. A decreasing tendency to respect fundamental research in this topic for about the last 10 years can be noticed in Figure 1. This fact might be interpreted as LN becoming indeed practical on a large scale, and perhaps its normal science activities have diminished in compensation for recent efforts in the realization of tractable technologies.

Research at the fundamental level still needs to be done to tie up loose ends. Remarkably, the study of the intimate relationship between the chemistry and crystal structure. The flexibility of the latter to support several chemistries has been summarized in Figure 4; potassium (K) is not highlighted, recalling that it is not considered a dopant according to the conceptual refinement given by Polgár et al. [93] and within the context of the high-temperature top-seeded solution (HTTSSG) growth technique. On the other hand, recently, Kang et al. have explicitly stated the powerful idea of obtaining stoichiometric LN crystals by the route of elemental doping, that is, by an effective elimination of intrinsic defects by filling of the sites [7]. This simple yet appealing concept rests upon trends observed in most doped LN systems, particularly those with divalent optical damage resistant ions (ODRI) Mg and Zn. For a given doping concentration below the threshold values, these have shown effects qualitatively analogous to a Li enrichment, and overall, doped LN crystals show improved physical properties compared to pure and congruent crystals [7]. However, it is crucial to notice that, by doping with any kind of element, the structure is distorted to some extent; for ODRI in general, the unit cell volume increases with doping concentration [60,133,176], whereas, in the case of pure LN, it decreases in the elimination process of Nb antisites and Li vacancies. Moreover, in pure LN, the blue shifting of the fundamental absorption edge in going from off-stoichiometric compositions to the stoichiometric one is a well-known fact. Congruent crystals doped with Mg have shown the same trend, as reported for concentrations below the threshold value; it reverses by doping further [139]. Change in the substitution mechanism in passing from Li site occupancy to Nb site occupancy influences the tendency reversal. However, it is important to recall that, under extreme Li enrichment conditions, the formation of the parasitical phase Li_3NbO_4 occurs instead of showing the characteristics of a partial Li-Nb exchange. Still, such a dissimilarity in the structure relaxation/hardening tendencies should be enough to state with confidence that no doping can be entirely analogous to the process of Li enrichment. Thus, in the pursuit of obtaining high-quality large single LN crystals, the chemical composition is to be adjusted, and the structural distortion or deformation should be minimized.

The best example of structural distortion affecting the properties is found by comparing LN and lithium tantalate ($LiTaO_3$, LT). Recall that, in pure LN, the Curie temperature is higher for crystals of ever-increasing quality (less intrinsic defects). Although the unit cell volume of LT is smaller than that of LN by a factor of about 0.5% [215–217], a dramatic difference between their corresponding Curie points holds: in going from off-stoichiometric compositions to the stoichiometric one, these have been reported to vary linearly in the range 1037–1207 °C for LN, and in the range 507–692 °C for LT, respectively [217]. Nearly the double notwithstanding that both phases are isostructural and Ta possesses identical chemical/physical properties to Nb, such as the state of valence and ionic radii. Hence, this can be attributed to lattice deformation, shown to be pronounced in LN single crystals doped with Ta, by polarized Raman Spectroscopy [218]. In contrast, the fundamental absorption edge can also be tuned with chemical composition in LT, but these are blue-shifted ca. 40 nm with respect to those of LN [219,220]; for this reason, the extension of transparency to the UV region has been proposed by a partial Nb-Ta exchange [8].

According to a quick survey, few experimental works on LNT (LN doped with Ta) have been done [217,218,221–223]. The study of this system could clarify some aspects regarding the substitution mechanisms in doped LN, as well as to pave the way for a close and quantitative investigation of the structural distortion in equivalent stoichiometric LN materials, accounting for proper doping (choice of element and concentration) for the adjustment of the chemical composition. In this respect, the recent advances in state-of-the-art simulation approaches could be useful. However, some peculiarities are still left to be solved. For example, the electronic bandgap is reported to be significantly below or above compared to the bulk of reports based on experiments [6,224]. Another situation to be considered is that of the role of hydrogen within the structure. Recently, LN crystals with a hydroxyl ion have been simulated with DFT calculations and by the direct insertion of the stacking fault defect into a designed supercell standing for the actual crystal structure [49,50]. However, the extended supercell size is limited to computational power, where $2 \times 2 \times 2$ hexagonal unit cells (240 atoms) are usually used in such approaches. The question then arises regarding whether such simulated crystals represent real LN crystals for which the concentration of hydrogen is very low, of the order of 10–100 ppm, or if crystals saturated with hydrogen are modeled in this way. A similar argument can be stated concerning the population of intrinsic and extrinsic defects (the chemical composition).

A final comment is given regarding the concept of substitution mechanisms. In writing this contribution, it has been noticed that, particularly in those publications related to the elemental doping in connection with ferromagnetism, *Li substitution* may have been used as a synonym of *Li site occupancy*. Rigorously, Li substitution disregards the filling of Li vacancies and dopant substitution for an Nb atom located at the Li site (antisite filling). Both processes have a higher probability of occurrence than Li substitution in off-stoichiometric crystals and under low doping concentrations, as is the case being described for diluted magnetic oxides.

Author Contributions: Conceptualization, O.S.-D. and S.D.V.-M.; validation, R.F. and C.D.F.-R.; investigation, O.S.-D., S.D.V.-M. and C.D.F.-R.; writing—original draft preparation, O.S.-D. and S.D.V.-M.; writing—review and editing, O.S.-D. and R.F.; visualization, S.D.V.-M.; supervision, R.F.; project administration, R.F.; funding acquisition, C.D.F.-R. All authors have read and agreed to the published version of the manuscript.

Funding: The publishing fee was covered by Universidad Tecnológica de Ciudad Juárez.

Acknowledgments: Special thanks to Omar Alejandro Velasco Cortez for his help with the bibliographical survey in the preparation of Figure 1. O.S.-D. is thankful for the post-doctoral grant PRODEP Oficio No. 511-6/2020-2054.

Conflicts of Interest: The authors declare no conflict of interest.

References

1. Volk, T.; Wöhlecke, M. Introduction. In *Springer Series in Materials Science 115. Lithium Niobate. Defects, Photorefraction and Ferroelectric Switching*, 1st ed.; Hull, R., Osgood, R.M., Jr., Parisi, J., Warlimont, H., Eds.; Springer: Berlin/Heidelberg, Germany, 2009; Volume 115, pp. 1–8.
2. Haertling, G.H. Ferroelectric Ceramics: History and Technology. *J. Am. Ceram. Soc.* **1999**, *82*, 797–818. [CrossRef]
3. Bhalla, A.S.; Guo, R.; Roy, R. The perovskite structure—A review of its role in ceramic science and technology. *Mat. Res. Innovat.* **2000**, *4*, 3–26. [CrossRef]
4. Troiler-McKinstry, S. Impact of ferroelectricity. *Am. Cer. Soc. Bull.* **2020**, *99*, 22–23.
5. Kovács, L.; Kocsor, L.; Tichy-Rács, É.; Lengyel, K.; Béncs, L.; Corradi, G. Hydroxyl ion probing transition metal dopants occupying Nb sites in stoichiometric $LiNbO_3$. *Opt. Mater. Express* **2019**, *9*, 4506–4516. [CrossRef]
6. Kong, Y.; Bo, F.; Wang, W.; Zheng, D.; Liu, H.; Zhang, G.; Rupp, R.; Xu, J. Recent Progress in Lithium Niobate: Optical Damage, Defect Simulation, and On-Chip Devices. *Adv. Mater.* **2020**, *32*, 1806453. [CrossRef] [PubMed]
7. Kang, X.; Liang, L.; Song, W.; Wang, F.; Sang, Y.; Liu, H. Formation mechanism and elimination methods for anti-site defects in $LiNbO_3/LiTaO_3$ crystals. *CrystEngComm* **2016**, *18*, 8136–8146. [CrossRef]

8. Lengyel, K.; Péter, Á.; Kovács, L.; Corradi, G.; Pálfalvi, L.; Hebling, J.; Unferdorben, M.; Dravecz, G.; Hajdara, I.; Szaller, Z.; et al. Growth, defect structure, and THz application of stoichiometric lithium niobate. *Appl. Phys. Rev.* **2015**, *2*, 040601. [CrossRef]
9. Vyalikh, A.; Zschornak, M.; Köhler, T.; Nentwich, M.; Weigel, T.; Hanzing, J.; Zaripov, R.; Vavilova, E.; Gemming, S.; Brendler, E.; et al. Analysis of the defect clusters in congruent lithium tantalite. *Phys. Rev. Mater.* **2018**, *2*, 013804. [CrossRef]
10. Volk, T.; Wöhlecke, M. Point defects in LiNbO3. In *Springer Series in Materials Science 115. Lithium Niobate. Defects, Photorefraction and Ferroelectric Switching*, 1st ed.; Hull, R., Osgood, R.M., Jr., Parisi, J., Warlimont, H., Eds.; Springer: Berlin/Heidelberg, Germany, 2009; Volume 115, pp. 9–50.
11. Pracka, I.; Bajor, A.L.; Kaczmarek, S.M.; Swirkowickz, M.; Kaczmarek, B.; Kieselewski, J.; Lukasiewicz, T. Growth and Characterization of LiNbO3 Single Crystals Doped with Cu and Fe Ions. *Cryst. Res. Technol.* **1994**, *34*, 627–634. [CrossRef]
12. Fay, H.; Alford, W.J.; Dess, H.M. Dependence of second-harmonic phase-matching temperature in $LiNBO_3$ crystals on melt composition. *Appl. Phys. Lett.* **1968**, *12*, 89–92. [CrossRef]
13. Prokhorov, A.M.; Kuz'minov, Y.S. *Physics and Chemistry of Crystalline Lithium Niobate*, 1st ed.; Hilger: New York, NY, USA, 1990.
14. Donnerberg, H.; Tomlinson, S.M.; Catlow, C.R.A.; Schirmer, O.F. Computer-simulation studies of intrinsic defects in $LiNbO_3$ crystals. *Phys. Rev. B* **1989**, *40*, 11909–11916. [CrossRef] [PubMed]
15. Lerner, P.; Legras, C.; Dumas, J.P. Stoechiometrie des monocristaux de metaniobate de lithium. *J. Cryst. Growth* **1968**, *3*, 231–235. [CrossRef]
16. Abrahams, S.C.; Marsh, P. Defect Structure Dependence on Composition in Lithium Niobate. *Acta Cryst.* **1986**, *B42*, 61–68. [CrossRef]
17. Schirmer, O.F.; Thiemann, O.; Wöhlecke, M. Defects in $LiNbO_3$—I Experimental aspects. *J. Phys. Chem. Solids* **1991**, *52*, 185–200. [CrossRef]
18. Zotov, N.; Boysen, H.; Frey, F.; Metzger, T.; Born, E. Cation substitution models of congruent $LiNbO_3$ investigated by X-ray and neutron powder diffraction. *J. Phys. Chem. Solids* **1994**, *55*, 145–152. [CrossRef]
19. Wilkinson, A.P.; Cheetham, A.K.; Jarman, R.H. The defect structure of congruently melting lithium niobate. *J. Appl. Phys.* **1998**, *74*, 3080–3083. [CrossRef]
20. DeLeo, G.G.; Dobson, J.L.; Masters, M.F.; Bonjack, L.H. Electronic structure of an oxygen vacancy in lithium niobate. *Phys. Rev. B* **1988**, *37*, 8394–8400. [CrossRef]
21. Peterson, G.E.; Carnevale, A. ^{93}Nb NMR Linewidths in Nonstoichiometric Lithium Niobate. *J. Chem. Phys.* **1972**, *56*, 4848–4851. [CrossRef]
22. Ye, N.; Wang, J.-Y.; Boughton, R.I.; Hong, M.-C. Chapter 20. Functional Crystals. In *Modern Inorganic Synthethic Chemistry*, 2nd ed.; Xu, R., Xu, Y., Eds.; Elsevier, B. V.: Amsterdam, The Netherlands, 2017; pp. 575–611.
23. Shannon, R.D.; Prewitt, C.T. Effective Ionic Radii in Oxides and Flourides. *Acta Cryst.* **1969**, *B25*, 925–946. [CrossRef]
24. Kim, S.; Gopalan, V.; Kitamura, K.; Furukawa, Y. Domain reversal and nonstoichiometry in lithium tantalate. *J. Appl. Phys.* **2001**, *90*, 2949–2963. [CrossRef]
25. Iyi, N.; Kitamura, K.; Izumi, F.; Yamamoto, J.K.; Hayashi, T.; Asano, H.; Kimura, S. Comparative of defect structures in lithium niobate with different compositions. *J. Soild State Chem.* **1992**, *101*, 340–352. [CrossRef]
26. Zotov, N.; Frey, F.; Boysen, H.; Lehnert, H.; Hornsteiner, A.; Strauss, B.; Sonntag, R.; Mayer, H.M.; Güthoff, F.; Hohlwein, D. X-ray and neutron diffuse scattering in $LiNbO_3$ from 38 to 1200 K. *Acta Crystallogr. B* **1995**, *51*, 961–972. [CrossRef]
27. Blumel, J.; Born, E.; Metzger, T. Solid State NMR study supporting the lithium vacancy defect model in congruent lithium niobate. *J. Phys. Chem. Solids* **1994**, *55*, 589–593. [CrossRef]
28. Yatsenko, A.V.; Ivanova-Maksimova, H.M.; Sergeev, N.A. NMR study of intrinsic defects in congruent $LiNbO_3$. 2. "Overlapping" defects. *Phys. B Condens. Matter* **1998**, *254*, 256–259. [CrossRef]
29. Donnerberg, H.; Tomlinson, S.M.; Catlow, C.R.A. Defects in $LiNbO_3$—II Computer simulation. *J. Phys. Chem. Solids* **1991**, *52*, 201–210. [CrossRef]
30. Safaryan, F.P.; Feigelson, R.S.; Petrosyan, A.M. An approach to the defect structure analysis of lithium niobate single crystals. *J. Appl. Phys.* **1999**, *85*, 8079–8082. [CrossRef]

31. Li, Q.; Wang, B.; Woo, C.H.; Wang, H.; Wang, R. First-principles study on the formation energies of intrinsic defects in $LiNbO_3$. *J. Phys. Chem. Solids* **2007**, *68*, 1336–1340. [CrossRef]
32. Abdi, F.; Fontana, M.D.; Aillerie, M.; Bourson, P. Coexistence of Li and Nb vacancies in the defect structure of pure $LiNbO_3$ and its relationship to optical properties. *Appl. Phys. A* **2006**, *83*, 427–434. [CrossRef]
33. Li, Y.; Sanna, S.; Schmidt, W.G. Modeling intrinsic defects in $LiNbO_3$ within the Slater-Janak transition state model. *J. Chem. Phys.* **2014**, *140*, 234113. [CrossRef]
34. Ivanova, E.M.; Sergeev, N.A.; Yatsenko, A.V. Analysis of Intrinsic Defects in the Lithium Niobate Structure by the NMR ^7Li Method. *Krist. Crystallogr. Rep.* **1998**, *43*, 303–306.
35. Yatsenko, A.V.; Ivanova, E.N.; Sergeev, N.A. NMR study of intrinsic defects in congruent $LiNbO_3$. 1. "Unoverlapping" defects. *Phys. B Condens. Matter* **1997**, *240*, 254–262. [CrossRef]
36. Gopalan, V.; Dierolf, V.; Scrymgeour, D.A. Defect-Domain Wall Interactions in Trigonal Ferroelectrics. *Annu. Rev. Mater. Res.* **2007**, *37*, 449–489. [CrossRef]
37. Boysen, H.; Altorfer, F. A Neutron Powder Investigation of the High-Temperature Structure and Phase Transition in $LiNbO_3$. *Acta Cryst.* **1994**, *B50*, 405–414. [CrossRef]
38. Lehnert, H.; Boysen, H.; Frey, F.; Hewat, A.; Radaelli, P. A neutron powder investigation of the high-temperature structure and phase transition in stoichiometric $LiNbO_3$. *Z. Krist. Cryst. Mater.* **1997**, *212*, 712–719. [CrossRef]
39. Etschmann, B.; Ishizawa, N.; Streltsov, V.; Oishi, S. A synchrotron X-ray diffraction analysis of near-stoichiometric $LiNbO_3$. *Z. Krist. Cryst. Mater.* **2001**, *216*, 455–461. [CrossRef]
40. Räuber, A. Chemistry and physics of lithium niobate. In *Current Topics in Materials Science*, 1st ed.; Kaldis, E., Ed.; North-Holland: Amsterdam, The Netherlands, 1978; Volume 1, pp. 481–601.
41. Araujo, R.M.; Lengyel, K.; Jackson, R.A.; Valerio, M.E.G.; Kovács, L. Computer modelling of intrinsic and substitutional defects in $LiNbO_3$. *Phys. Status Solidi A* **2007**, *4*, 1201–1204. [CrossRef]
42. Araujo, R.M.; Lengyel, K.; Jackson, R.A.; Kovács, L.; Valerio, M.E.G. A computational study of intrinsic and extrinsic defects in $LiNbO_3$. *J. Phys. Condens. Matter* **2007**, *19*, 046211. [CrossRef]
43. Wang, W.; Zheng, D.; Hu, M.; Saeed, S.; Liu, H.; Kong, Y.; Zhang, L.; Xu, J. Effect of Defects on Spontaneous Polarization in Pure and Doped LiNbO3: First-Principles Calculations. *Materials* **2019**, *12*, 100. [CrossRef]
44. Köhler, T.; Mehner, E.; Hanzig, J.; Gärtner, G.; Stöcker, H.; Leisegang, T.; Meyer, D.C. Real structure influencing the hydrogen defect chemistry in congruent $LiNbO_3$ and $LiTaO_3$. *J. Solid State Chem.* **2016**, *244*, 108–115. [CrossRef]
45. Li, Y.; Schmidt, W.G.; Sanna, S. Defect complexes in congruent $LiNbO_3$ and their optical signatures. *Phys. Rev. B* **2015**, *91*, 174106. [CrossRef]
46. Xu, H.; Lee, D.; Sinnott, S.B.; Dierolf, V.; Gopalan, V.; Phillpot, S.R. Structure and diffusion of intrinsic defect complexes in $LiNbO_3$ from density functional theory calculations. *J. Phys. Condens. Matter* **2010**, *22*, 135002. [CrossRef] [PubMed]
47. Lejaeghere, K.; Bihlmayer, G.; Björkman, T.; Blaha, P.; Blügel, S.; Blum, V.; Caliste, D.; Castelli, I.E.; Clark, S.J.; Dal Corso, A.; et al. Reproducibility in density functional theory calculations. *Science* **2016**, *351*, aad3000. [CrossRef] [PubMed]
48. Jones, R.O. Density functional theory: Its origins, rise to prominence and future. *Rev. Mod. Phys.* **2015**, *87*, 897–923. [CrossRef]
49. Szalay, V.; Lengyel, K.; Kovács, L.; Timón, V.; Hernández-Laguna, A. Vibrations of H+ (D+) in stoichiometric $LiNbO_3$ single crystal. *J. Chem. Phys.* **2011**, *135*, 124501. [CrossRef] [PubMed]
50. Lengyel, K.; Timón, V.; Hernández-Laguna, A.; Szalay, V.; Kovács, L. Structure of OH^- defects in $LiNbO_3$. *IOP Conf. Ser. Mater. Sci. Eng.* **2010**, *15*, 012015. [CrossRef]
51. Li, L.; Li, Y.; Zhao, X. Doping stability of nonphotorefractive ions in stoichiometric and congruent $LiNbO_3$. *Phys. Chem. Chem. Phys.* **2018**, *20*, 17477–17486. [CrossRef]
52. Volk, T.; Wöhlecke, M. General Introduction to Photorefraction in $LiNbO_3$. In *Springer Series in Materials Science 115. Lithium Niobate. Defects, Photorefraction and Ferroelectric Switching*, 1st ed.; Hull, R., Osgood, R.M., Jr., Parisi, J., Warlimont, H., Eds.; Springer: Berlin/Heidelberg, Germany, 2009; Volume 115, pp. 51–74.
53. Ashkin, A.; Boyd, G.D.; Dziedzic, J.M.; Smith, R.G.; Ballman, A.A.; Levinstein, J.J.; Nassau, K. Optically-induced refractive index inhomogeneities in $LiNbO_3$ and $LiTaO_3$. *Appl. Phys. Lett.* **1966**, *9*, 72–74. [CrossRef]

54. Furukawa, Y.; Sato, M.; Nitanda, F.; Ito, K. Growth and characterization of MgO-Doped $LiNbO_3$ for electro-optic devices. *J. Cryst. Growth* **1990**, *99*, 832–836. [CrossRef]
55. Furukawa, Y.; Yokotani, A.; Sasaki, T.; Yoshida, H.; Yoshida, K.; Nitanda, F.; Sato, M. Investigation of bulk laser damage threshold of lithium niobate single crystals by Q-switched pulsed laser. *J. Appl. Phys.* **1991**, *69*, 3372–3374. [CrossRef]
56. Dunn, M.H.; Ebrahimzadeh, M. Parametric Generation of Tunable Light from Continuous-Wave to Femtosecond Pulses. *Science* **1999**, *286*, 1513–1517. [CrossRef]
57. Ilchenko, V.S.; Savchenkov, A.A.; Matsko, A.B.; Meleki, L. Nonlinear Optics and Crystalline Whispering Gallery Mode Cavities. *Phys. Rev. Lett.* **2004**, *92*, 043903. [CrossRef] [PubMed]
58. Guarino, A.; Poberaj, G.; Rezzonico, D.; Degl'Innocenti, R.; Günter, P. Electro-optically tunable microring resonators in lithium niobate. *Nat. Photon.* **2007**, *1*, 407–410. [CrossRef]
59. Canalias, C.; Pasiskevicius, V. Mirrorles optical parametric oscillator. *Nat. Photon.* **2007**, *1*, 459–462. [CrossRef]
60. Sulyanov, S.; Volk, T. Lattice Parameter of Optical Damage Resistant In-Doped $LiNbO_3$ Crystals. *Crystals* **2018**, *8*, 210. [CrossRef]
61. Kong, Y.; Liu, S.; Xu, J. Recent Advances in the Photorefraction of Doped Lithium Niobate Crystals. *Materials* **2012**, *5*, 1954–1971. [CrossRef]
62. Dhar, L.; Curtis, K.; Fäcke, T. Coming of age. *Nat. Poton.* **2008**, *2*, 403–405. [CrossRef]
63. Haw, M. The light fantastic. *Nature* **2003**, *422*, 556–558. [CrossRef]
64. Staebler, D.L.; Phillips, W. Fe-Doped for Read-Write Applications. *Appl. Opt.* **1974**, *13*, 788–794. [CrossRef]
65. von der Linde, D.; Glass, A.M.; Rodgers, K.F. Multiphoton photorefractive processes for optical storage in $LiNbO_3$. *Appl. Phys. Lett.* **1974**, *25*, 155–157. [CrossRef]
66. Keune, W.; Date, S.K.; Dézsi, I.; Gonser, U. Mösbauser-effect study of Co^{57} and Fe^{57} impurities in ferroelectric $LiNbO_3$. *J. Appl. Phys.* **1975**, *46*, 3914–3924. [CrossRef]
67. Falk, M.; Buse, K. Thermoelectric-method for nearly complete oxidation of highly iron-doped lithium niobate crystals. *Appl. Phys. B* **2005**, *81*, 853–855. [CrossRef]
68. Li, S.; Liu, S.; Kong, Y.; Xu, J.; Zhang, G. Enhanced photorefractive properties of $LiNbO_3$:Fe crystals by HfO_2 codoping. *Appl. Phys. Lett.* **2006**, *89*, 101126. [CrossRef]
69. Xu, C.; Yang, C.; Zhu, C.; Sun, T.; Wang, R.; Xu, Y. Improved nonvolatile holographic storage properties in Zr:Ru:Fe:$LiNbO_3$ crystal by blue light recording. *Mater. Lett.* **2012**, *67*, 320–322. [CrossRef]
70. Dong, Y.; Liu, S.; Li, W.; Kong, Y.; Chen, S.; Xu, J. Improved ultraviolet photorefractive properties of vanadium-doped lithium niobate crystals. *Opt. Lett.* **2011**, *36*, 1779–1781. [CrossRef]
71. Dong, Y.; Liu, S.; Kong, Y.; Chen, S.; Rupp, R.; Xu, J. Fast photorefractive response of vanadium-doped lithium niobate in the visible region. *Opt. Lett.* **2012**, *37*, 1841–1843. [CrossRef]
72. Tian, T.; Kong, Y.; Liu, S.; Li, W.; Wu, L.; Chen, S.; Xu, J. Photorefraction of molybdenum-doped lithium niobate crystals. *Opt. Lett.* **2012**, *37*, 2679–2681. [CrossRef] [PubMed]
73. Kovács, L.; Kocsor, L.; Szaller, Z.; Hajdara, I.; Dravecz, G.; Lengyel, K.; Corradi, G. Lattice Site of Rare-Earth Ions in Stoichiometric Lithium Niobate Probed by OH^- Vibrational Spectroscopy. *Crystals* **2017**, *7*, 230. [CrossRef]
74. García-Solé, J.; Bausá, L.E.; Jaque, D.; Montoya, E.; Murrieta, H.; Jaque, F. Rare earth and transition metal ion centers in $LiNbO_3$. *Spectrochim. Acta Part A* **1998**, *54*, 1571–1581. [CrossRef]
75. Lande, D.; Orlov, S.S.; Akella, A.; Hesselink, L.; Neurgaonkar, R.R. Digital holographic storage system incorporating optical fixing. *Opt. Lett.* **1997**, *22*, 1722–1724. [CrossRef]
76. Lee, M.; Takekawa, S.; Furukawa, Y.; Kitamura, K.; Hatano, H.; Tanaka, S. Nonvolatile two-color holographic recording in Tb-doped $LiNbO_3$. *Appl. Phys. Lett.* **2000**, *76*, 1653–1655. [CrossRef]
77. Kaminow, I.P.; Stulz, L.W. Nd:$LiNbO_3$ laser. *IEEE J. Quantum Electron.* **1975**, *11*, 306–308. [CrossRef]
78. Fan, T.Y.; Cordova-Plaza, A.; Digonnet, M.J.F.; Byer, R.L.; Shaw, H.J. Nd:MgO:$LiNbO_3$ spectroscopy and laser devices. *J. Opt. Soc. Am. B* **1986**, *3*, 140–148. [CrossRef]
79. Yamamoto, J.K.; Sugimoto, A.; Yamagishi, K. Self-frequency doubling in Nd,Sc_2O_3:$LiNbO_3$ at room temperature. *Opt. Lett.* **1994**, *19*, 1311–1313. [CrossRef] [PubMed]
80. Lhomme, F.; Bourson, P.; Fontana, M.D.; Malovichko, G.; Aillerie, M.; Kokanyan, E. Luminescence of Cr^{3+} in lithium niobate: Influence of the chromium concentration and crystal composition. *J. Phys. Condens. Matter* **1998**, *10*, 1137–1146. [CrossRef]

81. Qiu, Y. Assesment of tunable laser crystal $LiNbO_3:Cr^{3+}$. *J. Phys. Condens. Matter* **1993**, *5*, 2041–2044. [CrossRef]
82. Babadjanyan, V.G.; Kokanyan, E.P.; Kostanyan, R.B.; Nikogosyan, V.R. Spectral and kinetic properties of $LiNbO_3:Cr^{3+}$ crystals. *Opt. Commun.* **1990**, *78*, 247–249. [CrossRef]
83. Levinstein, H.J.; Ballman, A.A.; Denton, R.T.; Ashkin, A.; Dziedzic, J.M. Reduction of the Susceptibility to Optically Induced Index Inhomogeneities in $LiTaO_3$ and $LiNbO_3$. *J. Appl. Phys.* **1967**, *38*, 3101–3102. [CrossRef]
84. Smith, R.G.; Fraser, D.B.; Denton, R.T.; Rich, T.C. Correlation of Reduction in Optically Induced Refractive-Index Inhomogeneity with OH Content in $LiTaO_3$ and $LiNbO_3$. *J. Appl. Phys.* **1968**, *39*, 4600–4602. [CrossRef]
85. Zhong, G.G.; Jin, J.; Wu, Z.K. Measurements of Optically Induced Refractive-Index Damage of Lithium Niobate Doped with Different Concentrations of Mg. In Proceedings of the 11th International Quantum Electronics Conference, Boston, MA, USA, 23–26 June 1980; Volume 70, p. 631.
86. Wöhlecke, M.; Kovács, L. OH^- ions in Oxide Crystals. *Crit. Rev. Solid State Mater. Sci.* **2001**, *26*, 1–86. [CrossRef]
87. Cabrera, J.M.; Olivares, J.; Carrascosa, M.; Rams, J.; Müller, R.; Diéguez, E. Hydrogen in lithium niobate. *Adv. Phys.* **1996**, *45*, 349–392. [CrossRef]
88. Kovács, L.; Szalay, V.; Capelleti, R. Stoichiomnetry dependence of the OH^- absorption band in $LiNbO_3$ crystals. *Solid State Commun.* **1984**, *52*, 1029–1031. [CrossRef]
89. Thomas, D.G.; Lander, J.J. Hydrogen as a Donor in Zinc Oxide. *J. Chem. Phys.* **1956**, *25*, 1136–1142. [CrossRef]
90. Norby, T.; Wideroe, M.; Glöckner, R.; Larring, Y. Hydrogen in oxides. *Dalton Trans.* **2004**, 3012–3018. [CrossRef]
91. Müller, R.; Arizmendi, L.; Carrascosa, M.; Cabrera, J.M. Determination of H concentration in $LiNbO_3$ by photorefractive fixing. *Appl. Phys. Lett.* **1992**, *60*, 3212–3214. [CrossRef]
92. Rice, C.E. The Structure and Properties of $Li_{1-x}H_xNbO_3$. *J. Solid State Chem.* **1986**, *64*, 188–199. [CrossRef]
93. Polgár, K.; Péter, Á.; Kovács, L.; Corradi, G.; Szaller, Z. Growth of stoichiometric $LiNbO_3$ single crystals by top seeded solution growth method. *J. Cryst. Growth* **1997**, *177*, 211–216. [CrossRef]
94. Kovács, L.; Szaller, Z.; Lengyel, K.; Corradi, G. Hydroxyl ions in stoichiometric $LiNbO_3$ crystals doped with optical damage resistant ions. *Opt. Mater* **2014**, *37*, 55–58. [CrossRef]
95. Bäumer, C.; David, C.; Betzler, K.; Hesse, H.; Lengyel, K.; Kovács, L.; Wöhlecke, M. Composition dependence of the OH-stretch-mode spectrum in lithium tantalate. *Phys. Stat. Solidi A* **2004**, *201*, R13–R16. [CrossRef]
96. Bollmann, W.; Gernand, M. On the disorder of $LiNbO_3$ crystals. *Phys. Stat. Solidi A* **1972**, *9*, 301–308. [CrossRef]
97. Bredikhin, S.; Scharner, S.; Klingler, M.; Kveder, V.; Red'kin, B.; Weppner, W. Nonstoichiometry and electrocoloration due to injection of Li^+ and O^{2-} ions into lithium niobate crystals. *J. Appl. Phys.* **2000**, *88*, 5687–5694. [CrossRef]
98. Schirmer, O.F.; Imlau, M.; Merschjann, C.; Schoke, B. Electron small polarons and bipolarons in $LiNbO_3$. *J.Phys. Condens. Matter* **2009**, *21*, 123201. [CrossRef] [PubMed]
99. Imlau, M.; Badorreck, H.; Merschjann, C. Optical nonlinearities of small polarons in lithium niobate. *Appl. Phys. Rev.* **2015**, *2*, 040606. [CrossRef]
100. Clark, M.G.; DiSalvo, F.J.; Glass, A.M.; Peterson, G.E. Electronic structure and optical index damage of iron-doped lithium niobate. *J. Chem. Phys.* **1973**, *59*, 6209–6219. [CrossRef]
101. Eagles, D.M. Optical Absorption in Ionic Crystals Involving Small Polarons. *Phys. Rev.* **1963**, *130*, 1381–1400. [CrossRef]
102. Klinger, M. Quantum theory of non-steady-state conductivity in low mobility solids. *Phys. Lett.* **1963**, *7*, 102–104. [CrossRef]
103. Reik, H.G. Optical properties of small polarons in the infrared. *Solid State Commun.* **1963**, *1*, 67–71. [CrossRef]
104. Reik, H.G.; Heese, D. Frequency dependence of the electrical conductivity of small polarons for high and low temperatures. *J. Phys. Chem. Solids* **1967**, *28*, 581–596. [CrossRef]
105. Austin, I.G.; Mott, N.F. Polarons in crystalline and non-crystalline materials. *Adv. Phys.* **1969**, *18*, 41–102. [CrossRef]
106. Bryksin, V.V.; Voloshin, V.S.; Raitsev, A.V. *Sov. Phys. Solid State* **1983**, *25*, 820.

107. Emin, D. Optical properties of large and small polarons and bipolarons. *Phys. Rev. B* **1993**, *48*, 13691–13702. [CrossRef]
108. Alexandrov, A.S.; Mott, N.F. Bipolarons. *Rep. Prog. Phys.* **1994**, *57*, 1197–1288. [CrossRef]
109. Zylbersztejn, A. Thermally activated trapping in Fe-Doped $LiNbO_3$. *Appl. Phys. Lett.* **1976**, *29*, 778–780. [CrossRef]
110. Schirmer, O.F.; von der Linde, D. Two-photon- and x-ray-induced Nb^{4+} and O^- small polarons in $LiNbO_3$. *Appl. Phys. Lett.* **1978**, *33*, 35–38. [CrossRef]
111. von der Linde, D.; Schirmer, O.F.; Kurz, H. Intrinsic Photorefractive Effect of $LiNbO_3$. *Appl. Phys.* **1978**, *15*, 153–156. [CrossRef]
112. Karaseva, L.G.; Bondarenko, G.P.; Gromov, V.V. Investigation of optical active centres of the irradiated lithium niobate. *Radiat. Phys. Chem.* **1977**, *10*, 241–245. [CrossRef]
113. Bernhardt, H.J. Comparative studies of the LNT coloration of $Y_3Al_5O_{12}$, $YAlO_3$, and $LiNbO_3$ crystals. *Phys. Status Solidi A* **1976**, *33*, 211–216. [CrossRef]
114. Schirmer, O.F. Optical Absorption of Small Polarons Bound in Octahedral Symmetry: V^- Type Centers in Alkaline Earth Oxides. *Z. Physik B* **1976**, *24*, 235–244. [CrossRef]
115. Arizmendi, L.; Cabrera, J.M.; Agulló-López, F. Defects induced in pure and doped $LiNbO_3$ by irradiation and thermal reduction. *J. Phys. C Solid State Phys.* **1984**, *17*, 515–529. [CrossRef]
116. Ketchum, J.L.; Sweeney, K.L.; Halliburton, L.E. Vacuum Annealing Effects in Lithium Niobate. *Phys. Lett.* **1983**, *94A*, 450–453. [CrossRef]
117. Sweeney, K.L.; Halliburton, L.E. Oxygen vacancies in lithium niobate. *Appl. Phys. Lett.* **1983**, *43*, 336–338. [CrossRef]
118. Smyth, D.M. Defects and transport in $LiNbO_3$. *Ferroelectrics* **1983**, *50*, 93–102. [CrossRef]
119. Jhans, H.; Honig, J.M.; Rao, C.N.R. Optical properties of reduced $LiNbO_3$. *J. Phys. C Solid State Phys.* **1986**, *19*, 3649–3658. [CrossRef]
120. Schirmer, O.F.; Juppe, S.; Koppitz, J. Electron-spin-resonance, optical and photovoltaic studies of reduced undoped $LiNbO_3$. *Cryst. Latt. Def. Amorph. Mater.* **1987**, *16*, 353–357.
121. Koppitz, J.; Schirmer, O.F.; Kuznetsov, A.I. Thermal Dissociation of Bipolarons In Reduced Undoped $LiNbO_3$. *Europhys. Lett.* **1987**, *4*, 1055–1059. [CrossRef]
122. Chakraverty, B.K.; Sienko, M.J.; Bonnerot, J. Low-temperature specific heat and magnetic susceptibility of nonmetallic vanadium bronzes. *Phys. Rev. B* **1978**, *17*, 3781–3789. [CrossRef]
123. Schirmer, O.F.; Salje, E. Conduction bipolarons in low-temperature crystalline WO_{3-x}. *J. Phys. C Solid State Phys.* **1980**, *13*, L1067–L1072. [CrossRef]
124. Dutt, D.A.; Feigl, F.J.; DeLeo, G.G. Optical absorption and electron paramagnetic resonance studies of chemically reduced congruent lithium niobate. *J. Phys. Chem. Solids* **1990**, *51*, 407–415. [CrossRef]
125. Krätzig, E.; Schirmer, O.F. Photorefractive centers in electro-optic crystals. In *Photorefractive Materials and Their Applications, I. Topics in Applied Physics*; Günter, P., Huignard, J.P., Eds.; Springer: Berlin/Heidelberg, Germany, 1988; Volume 61, pp. 131–166.
126. Tsonev, L. Luminiscent activation of planar optical waveguides in $LiNbO_3$ with rare earth ionsLn^{3+}—A review. *Opt. Mater.* **2008**, *30*, 892–899. [CrossRef]
127. Herrington, J.R.; Dischler, B.; Räuber, A.; Schneider, J. An optical study of the stretching absorption band near 3 microns from OH^- defects in $LiNbO_3$. *Solid State Commun.* **1973**, *12*, 351–354. [CrossRef]
128. Klauer, S.; Wöhlecke, M. Incorporation of hydrogen in cubic and uniaxial oxidic crystals deduced from polarized Raman scattering. *Phys. Rev. B* **1994**, *49*, 158–181. [CrossRef]
129. Bryan, D.A.; Gerson, R.; Tomaschke, H.E. Increased optical damage resistance in lithium niobate. *Appl. Phys. Lett.* **1984**, *44*, 847–849. [CrossRef]
130. Bryan, D.A.; Rice, R.R.; Gerson, R.; Tomaschke, H.E.; Sweeney, K.L.; Halliburton, L.E. Magnesium-Doped Lithium Niobate For Higher Optical Power Applications. *Opt. Eng.* **1985**, *24*, 241138. [CrossRef]
131. Sweeney, K.L.; Halliburton, L.E.; Bryan, D.A.; Rice, R.R.; Gerson, R.; Tomaschke, H.E. Point defects in Mg-doped lithium niobate. *J. Appl. Phys.* **1985**, *57*, 1036–1044. [CrossRef]
132. Liu, J.; Zhang, W.; Zhang, G. Defect Chemistry Analysis of the Defect Structure in Mg-Doped $LiNbO_3$ Crystals. *Phys. Status Solidi A* **1996**, *156*, 285–291. [CrossRef]
133. Grabmaier, B.C.; Otto, F. Growth and investigation of MgO-doped $LiNbO_3$. *J. Cryst. Growth* **1986**, *79*, 682–688. [CrossRef]

134. Lengyel, K.; Kovács, L.; Péter, Á.; Polgár, K.; Corradi, G. The effect of stoichiometry and Mg doping on the Raman spectra of $LiNbO_3$:Mg crystals. *Appl. Phys. B* **2007**, *87*, 317–322. [CrossRef]
135. Donnerberg, H. Comments on the Defect Chemistry of Magnesium-Doped Lithium Niobate ($LiNbO_3$). *J. Solid State Chem.* **1996**, *123*, 208–214. [CrossRef]
136. Lengyel, K.; Péter, Á.; Polgár, K.; Kovács, L.; Corradi, G. UV and IR absorption studies in $LiNbO_3$:Mg crystals below and above the photorefractive threshold. *Phys. Status Solidi C* **2005**, *2*, 171–174. [CrossRef]
137. Pálfalvi, L.; Hebling, J.; Almási, G.; Péter, A.; Polgár, K.; Lengyel, K.; Szipöcs, R. Nonlinear refraction and absorption of Mg doped stoichiometric and congruent $LiNbO_3$. *J. Appl. Phys.* **2004**, *95*, 902–908. [CrossRef]
138. Pálfalvi, L.; Almási, G.; Hebling, J.; Péter, Á.; Polgár, K. Measurement of laser-induced refractive index changes of Mg-doped congruent and stoichiometric $LiNbO_3$. *Appl. Phys. Lett.* **2002**, *80*, 2245–2247. [CrossRef]
139. Polgár, K.; Kovács, L.; Földvári, I.; Cravero, I. Spectroscopic and electrical conductivity investigation of Mg doped $LiNbO_3$ single crystals. *Solid State Commun.* **1986**, *59*, 375–379. [CrossRef]
140. Kovács, L.; Szaller, Z.; Lengyel, K.; Péter, Á.; Hajdara, I.; Mandula, G.; Pálfvalvi, L.; Hebling, J. Photorefractive damage resistance threshold in stoichiometric $LiNbO_3$:Zr crystals. *Opt. Lett.* **2013**, *38*, 2861–2864. [CrossRef]
141. Kovács, L.; Polgár, K.; Capelletti, R. IR absorption study of OH^- in pure and Mg-doped $LiNbO_3$ single crystals. *Cryst. Latt. Def. Amorph. Mater.* **1987**, *15*, 115–121.
142. Dravecz, G.; Kovács, L. Determination of the crystal composition from OH^- vibrational spectrum in lithium niobate. *Appl. Phys. B* **2007**, *88*, 305–307. [CrossRef]
143. Kovács, L.; Wöhlecke, M.; Jovanovic, A.; Polgár, K.; Kapphan, S. Infrared absorption study of the OH vibrational band in $LiNbO_3$ crystals. *J. Phys. Chem. Solids* **1991**, *52*, 797–803. [CrossRef]
144. Freytag, F.; Corradi, G.; Imlau, M. Atomic insight to lattice distortions caused by carrier self-trapping in oxide materials. *Sci. Rep.* **2016**, *6*, 36929. [CrossRef] [PubMed]
145. Berben, D.; Buse, K.; Wevering, S.; Herth, P.; Imlau, M.; Woike, T. Lifetime of small polarons in iron-doped lithium-niobate crystals. *J. App. Phys.* **2000**, *87*, 1034–1041. [CrossRef]
146. Herth, P.; Schaniel, D.; Woike, T.; Graznow, T.; Imlau, M.; Krätzig, E. Polarons generated by laser pulses in doped $LiNbO_3$. *Phys. Rev. B* **2005**, *71*, 125128. [CrossRef]
147. Herth, P.; Graznow, T.; Schaniel, D.; Woike, T.; Imlau, M.; Krätzig, E. Evidence for Light-Induced Hole Polarons in $LiNbO_3$. *Phys. Rev. Lett.* **2005**, *95*, 067404. [CrossRef]
148. Beyer, O.; Maxein, D.; Woike, T.; Buse, K. Generation of small bound polarons in lithium niobate crystals on the subpicosecond time scale. *Appl. Phys. B* **2006**, *83*, 527–530. [CrossRef]
149. Kong, Y.; Liu, S.; Zhao, Y.; Liu, H.; Chen, S.; Xu, J. High optical damage resistant crystal: Zirconium-oxide-doped lithium niobate. *Appl. Phys. Lett.* **2007**, *91*, 081908.
150. Chen, S.; Liu, H.; Kong, Y.; Huang, Z.; Xu, J.; Zhang, G. The resistance against optical damage of near-stoichiometric $LiNbO_3$:Mg crystals prepared by vapor transport equilibration. *Opt. Mater.* **2007**, *29*, 885–888.
151. Liu, H.; Liang, Q.; Zhu, M.; Li, W.; Liu, S.; Zhang, L.; Chen, S.; Kong, Y.; Xu, J. An excellent crystal for high resistance against optical damage in visible-UV range: Near-stoichiometric zirconium-doped lithium niobate. *Opt. Exp.* **2011**, *19*, 1743–1748.
152. Schlarb, U.; Wöhlecke, M.; Gather, B.; Reichert, A.; Betzler, K.; Volk, T.; Rubinina, N. Refractive indices of Zn-doped lithium niobate. *Opt. Mater.* **1995**, *4*, 791–795.
153. Schlarb, U.; Matzas, B.; Reichert, A.; Betzler, K.; Wöhlecke, M.; Gather, B.; Volk, T. Refractive indices of Zn/In-co-doped lithium niobate. *Ferroelectrics* **1996**, *185*, 269–272.
154. Liu, F.; Kong, Y.; Li, W.; Liu, H.; Liu, S.; Chen, S.; Zhang, X.; Rupp, R.; Xu, J. High resistance against ultraviolet photorefraction in zirconium-doped lithium niobate crystals. *Opt. Lett.* **2010**, *35*, 10–12. [PubMed]
155. Xu, J.; Zhang, G.; Li, F.; Zhang, X.; Sun, Q.; Liu, S.; Song, F.; Kong, Y.; Chen, X.; Qiao, H.; et al. Enhancement of ultraviolet photorefraction in highly magnesium-doped lithium niobate crystals. *Opt. Lett.* **2000**, *25*, 129–131.
156. Qiao, H.; Xu, J.; Zhang, G.; Zhang, X.; Sun, Q.; Zhang, G. Ultraviolet photorefractive features in doped lithium niobate crystals. *Phys. Rev. B* **2004**, *70*, 094101.
157. Razzari, L.; Minzioni, P.; Cristiani, I.; Degiorgio, V.; Kokanyan, E.P. Photorefractivity of Hafnium-doped congruent lithium–niobate crystals. *Appl. Phys. Lett.* **2005**, *86*, 131914.
158. Li, S.; Liu, S.; Kong, Y.; Deng, D.; Gao, G.; Li, Y.; Gao, H.; Zhang, L.; Hang, Z.; Chen, S.; et al. The optical damage resistance and absorption spectra of LiNbO3:Hf crystals. *J. Phys. Condens. Matter* **2006**, *18*, 3527–3534.

159. Minzioni, P.; Cristiani, I.; Degiorgio, V.; Kokanyan, E.P. Strongly sublinear growth of the photorefractive effect for increasing pump intensities in doped lithium-niobate crystals. *J. Appl. Phys.* **2007**, *101*, 116105.
160. Minzioni, P.; Cristiani, I.; Yu, J.; Parravicini, J.; Kokanyan, E.P.; Degiorgio, V. Linear and nonlinear optical properties of Hafnium-doped lithium-niobate crystals. *Opt. Express* **2007**, *15*, 14171–14176. [CrossRef]
161. Kimura, H.; Uda, S. Conversion of non-stoichiometry of LiNbO3 to constitutional stoichiometry by impurity doping. *J. Cryst. Growth* **2009**, *311*, 4094–4101. [CrossRef]
162. Kimura, H.; Taniuchi, T.; Iida, S.; Uda, S. Bulk crystal growth of congurent MgO-Doped $LiNbO_3$ crystal with stoichiometric structure and its second-harmonic-generation properties. *J. Cryst. Growth* **2010**, *312*, 3425–3427.
163. Rebouta, L.; Smulders, P.J.M.; Boerma, D.O.; Agúllo-López, F.; da Silva, M.F.; Soares, J.C. Ion-beam channeling yields of host and impurity atoms in $LiNbO_3$: Computer simulations. *Phys. Rev. B* **1993**, *48*, 3600–3610.
164. Kling, A.; Soares, J.C.; da Silva, M.F. Channeling Investigations of Oxide Materials for Optoelectronic Applications. In *Insulating Materials for Optoelectronics: New Developments*; Agúllo-López, F., Ed.; World Scientific: Singapore, 1995; pp. 175–200.
165. Xue, D.; He, X. Dopant occupancy and structural stability of doped lithium niobate crystals. *Phys. Rev B* **2006**, *73*, 064113. [CrossRef]
166. Donnerberg, H.; Tomlinson, S.M.; Catlow, C.R.A.; Schirmer, O.F. Computer-simulation studies of extrinsic defects in $LiNbO_3$ crystals. *Phys. Rev. B* **1991**, *44*, 4877–4883.
167. Feng, X.; Wang, D.; Zhang, J. NMR Spectra of Mg Nuclei in Mg-Doped $LiNbO_3$ Crystals. *Phys. Stat. Solidi B* **1990**, *157*, K127–K130.
168. Chernaya, T.S.; Maksimov, B.A.; Volk, T.R.; Rubinina, N.M.; Simonov, V.I. Zn atoms in lithium niobate and mechanism of their insertion into crystals. *JETP Lett.* **2000**, *73*, 103–106.
169. Volk, T.; Maximov, B.; Chernaya, T.; Rubinina, N.; Wöhlecke, M.; Simonov, V. Photorefractive properties of $LiNbO_3$:Zn crystals related to the defect structure. *Appl. Phys. B* **2001**, *72*, 647–652.
170. Sulyanov, S.; Maximov, B.; Volk, T.; Boysen, H.; Schneider, J.; Rubininca, N.; Hansen, T. Neutron and X-ray study of stoichiometric and doped $LiNbO_3$:$Zn_{0.08}$. *Appl. Phys. A* **2001**, *74*, s1031–s1033.
171. Volk, T.R.; Rubinina, N.M.; Pryalkin, V.I.; Krasnikov, V.V.; Volkov, V.V. Optical and non-linear optical investigations in $LiNbO_3$:Mg and $LiNbO_3$:Zn. *Ferroelectrics* **1990**, *109*, 345–350.
172. Volk, T.R.; Pryalkin, V.I.; Rubinina, N.M. Optical-damage-resistant $LiNbO_3$:Zn crystal. *Opt. Lett.* **1990**, *15*, 996–998. [PubMed]
173. Volk, T.; Maximov, B.; Sulyanov, S.; Rubinina, N.; Wöhlecke, M. Relation of the photorefraction and optical-damage resistance to the intrinsic defect structure in $LiNbO_3$ crystals. *Opt. Mater.* **2003**, *23*, 229–233. [CrossRef]
174. Marques, J.G.; Kling, A.; Soares, J.C.; Rebouta, L.; da Silva, M.F.; Diéguez, E.; Agulló-López, F. Lattice site location of Hf in $LiNbO_3$: Influence of dopant concentration and crystal stoichiometry. *Nuclr. Instr. Meth. B* **1998**, *136–138*, 431–435. [CrossRef]
175. Marques, J.G.; Kling, A.; Rebouta, L.; da Silva, M.F.; Melo, A.A.; Soares, J.C.; Serrano, M.D.; Diéguez, E.; Agulló-López, F. Lattice Location of Hf in Near-Stoichiometric $LiNbO_3$: RBS/Channeling and PAC Studies. *Mater. Sci. Forum* **1997**, *248–249*, 395–398.
176. Chia, C.-T.; Lee, C.-C.; Chang, P.-J.; Hu, M.-L.; Hu, L.-J. Substitution mechanism of ZnO-doped lithium niobate crystal determined by powder x-ray diffraction and coercive field. *Appl. Phys. Lett.* **2005**, *86*, 182901. [CrossRef]
177. Argiolas, N.; Bazzan, M.; Ciampolillo, M.V.; Pozzobon, P.; Sada, C.; Saoner, L.; Zaltron, A.M.; Bacci, L.; Minzioni, P.; Nava, G.; et al. Structural and optical properties of zirconium doped lithium niobate crystals. *J. Appl. Phys.* **2010**, *108*, 093508. [CrossRef]
178. Iyi, N.; Kitamura, K.; Yajima, Y.; Kimura, S.; Furukawa, Y.; Sato, M. Defect Structure Modelo f MgO-Doped $LiNbO_3$. *J. Solid State Chem.* **1995**, *118*, 148–152.
179. Hammoum, R.; Fontana, M.D.; Gilliot, M.; Bourson, P.; Kokanyan, E.P. Site spectroscopy of Hf doping in Hf-doped $LiNbO_3$ crystals. *Solid State Comm.* **2009**, *149*, 1967–1970.
180. Kokanyan, N.; Chapron, D.; Kokanyan, E.; Fontana, M.D. Zr doping on lithium niobate crystals: Raman spectroscopy and chemometrics. *J. Appl. Phys.* **2017**, *121*, 095103. [CrossRef]

181. Zheng, D.; Kong, Y.; Liu, S.; Chen, M.; Chen, S.; Zhang, L.; Rupp, R.; Xu, J. The simultaneous enhancement of photorefraction and optical damage resistance in MgO and Bi_2O_3 co-doped $LiNbO_3$ crystals. *Sci. Rep.* **2016**, *6*, 20308. [CrossRef] [PubMed]
182. Zheng, D.; Wang, W.; Wang, S.; Qu, D.; Liu, H.; Kong, Y.; Liu, S.; Chen, S.; Rupp, R.; Xu, J. Real-time dynamic holographic display realized by bismuth and magnesium co-doped lithium niobate. *Appl. Phys. Lett.* **2019**, *114*, 241903. [CrossRef]
183. Bader, S.D. Colloquium: Opportunities in nanomagnetism. *Rev. Mod. Phys.* **2006**, *78*, 1–15. [CrossRef]
184. Cullity, B.D.; Graham, C.D. *Introduction to Magnetic Materials*, 2nd ed.; IEEE/Wiley: Hoboken, NJ, USA, 2009.
185. Spaldin, N.A. *Magnetic Materials*, 2nd ed.; Cambridge University Press: Cambridge, UK, 2010.
186. Coey, J.M.D.; Venkatesan, M.; Fitzgerald, C.B. Donor impurity band exchange in dilute ferromagnetic oxides. *Nat. Mater.* **2005**, *4*, 173–179. [CrossRef]
187. Villalobos-Mendoza, S.D.; Holguín-Momaca, J.T.; Elizalde-Galindo, J.T.; Carrillo-Flores, D.M.; Olive-Méndez, S.F.; Farías-Mancilla, J.R. Iron-Doped Lithium Tantalate Thin Films Deposited by Magnetron Sputtering: A Study of Iron Role in the Structure and the Derived Magnetic Properties. *Crystals* **2020**, *10*, 50. [CrossRef]
188. Song, Y.-J.; Zhang, Q.-H.; Shen, X.; Ni, X.-D.; Yao, Y.; Yu, R.-C. Room-Temperature Magnetism Realized by Doping Fe into Ferroelectric $LiTaO_3$. *Chin. Phys. Lett.* **2014**, *31*, 017501. [CrossRef]
189. Verma, K.C.; Gupta, V.; Kaur, J.; Kotnala, R.K. Raman spectra, photoluminescence, magnetism and magnetoelectric coupling in pure and Fe doped $BaTiO_3$ nanostructures. *J. Alloy. Compnd.* **2013**, *578*, 5–11. [CrossRef]
190. Zeng, F.; Sheng, P.; Tang, G.S.; Pan, F.; Yan, W.S.; Hu, F.C.; Zou, Y.; Huang, Y.Y.; Jiang, Z.; Guo, D. Electronic structure and magnetism of Fe-doped $LiNbO_3$. *Mater. Chem. Phys.* **2012**, *136*, 783–788. [CrossRef]
191. Yao, D.; Zhou, X.; Ge, S. Raman scattering and room temperature ferromagnetism in Co-doped $SrTiO_3$ particles. *Appl. Surf. Sci.* **2011**, *257*, 9233–9236. [CrossRef]
192. Chen, C.; Zeng, F.; Li, J.H.; Sheng, P.; Luo, J.T.; Yang, Y.C.; Pan, F.; Zou, Y.; Huang, Y.Y.; Jiang, Z. Strong *d–d* electron interaction inducing ferromagnetism in Mn-doped $LiNbO_3$. *Thin Film Solids* **2011**, *520*, 764–768. [CrossRef]
193. Coey, J.M.D. Dilute Magnetic Oxides. *Curr. Opin. Solid State Mater. Sci.* **2006**, *10*, 83–92. [CrossRef]
194. Coey, J.M.D.; Mlack, J.T.; Venkatesan, M.; Stamenov, P. Magnetization Process in Dilute Magnetic Oxides. *IEEE Trans. Magn.* **2010**, *46*, 2501–2503. [CrossRef]
195. Dietl, T.; Ohno, H.; Matsukura, F.; Cibert, J.; Ferrand, D. Zener model description of ferromagnetism in zinc-blende magnetic semiconductors. *Science* **2000**, *287*, 1019–1022. [CrossRef] [PubMed]
196. Edwards, D.M.; Katsnelson, M.I. High-temperature ferromagnetism of sp electrons in narrow impurity bands: Application to CaB6. *J. Phys. Condens. Matter* **2006**, *18*, 7209–7225. [CrossRef]
197. Coey, J.M.D.; Wongsaprom, K.; Alaria, J.; Venkatesan, M. Charge-transfer ferromagnetism in oxide nanoparticles. *J. Phys. D Appl. Phys.* **2008**, *41*, 134012. [CrossRef]
198. Song, C.; Zeng, F.; Shen, Y.X.; Geng, K.W.; Xie, Y.N.; Wu, Z.Y.; Pan, F. Local Co structure and ferromagnetism in ion-implanted Co-doped $LiNbO_3$. *Phys. Rev. B* **2006**, *73*, 172412. [CrossRef]
199. Vitova, T.; Hormes, J.; Falk, M.; Buse, K. Site-selective investigation of site symmetry and site occupation of iron in Fe-doped lithium niobate crystals. *J. Appl. Phys.* **2009**, *105*, 013524. [CrossRef]
200. Olimov, K.; Falk, M.; Buse, K.; Woike, T.; Hormes, J.; Modrow, H. X-ray absorption near edge spectroscopy investigations of valency and lattice occupation site of Fe in highly iron-doped lithium niobate crystals. *J. Phys. Condes. Matter* **2006**, *18*, 5135–5146. [CrossRef]
201. Bush, T.S.; Catlow, C.R.A.; Chadwick, A.V.; Cole, M.; Geatches, R.M.; Greaves, G.N.; Tomlinson, S.M. Studies of cation dopant sites in metal oxides by EXAFS and computer-simulation techniques. *J. Mater. Chem.* **1992**, *2*, 309–316. [CrossRef]
202. Ye, J.; Sun, X.; Wu, Z.; Liu, J.; An, Y. Evidence of the oxygen vacancies-induced room temperature ferromagnetism in multiferroic Co-doped $LiNbO_3$ films. *J. Alloy. Compnd.* **2018**, *768*, 750–755. [CrossRef]
203. Sheng, P.; Zeng, F.; Tang, G.S.; Pan, F.; Yan, W.S.; Hu, F.C. Structure and ferromagnetism in vanadium-doped $LiNbO_3$. *J. Appl. Phys.* **2012**, *112*, 033913. [CrossRef]
204. Song, C.; Wang, C.; Liu, X.; Zeng, F.; Pan, F. Room Temperature Ferromagnetism in Cobalt-Doped $LiNbO_3$ Single Crystalline Films. *Cryst. Growth Des.* **2009**, *9*, 1235–1239. [CrossRef]

205. Song, C.; Wang, C.Z.; Yang, Y.C.; Liu, X.J.; Zeng, F.; Pan, F. Room temperature ferromagnetism and ferroelectricity in cobalt-doped $LiNbO_3$ film. *Appl. Phys. Lett.* **2008**, *92*, 262901. [CrossRef]
206. Yang, D.-S.; Sung, N.; Yeom, T.H. Local Structure of Fe-Doped $LiNbO_3$ Crystal Measured by X-ray Absorption Fine Structure. *J. Phys. Soc. Jpn.* **2009**, *78*, 114605. [CrossRef]
207. Bu, D.; Fu, Y.; Sun, N.; Li, C.; Li, Y.; An, Y.; Liu, J. Effect of Cu doping on room temperature ferromagnetic behavior of Mn doped $LiNbO_3$ films. *J. Cryst. Growth* **2016**, *453*, 198–202. [CrossRef]
208. Fierro-Ruíz, C.D.; Sánchez-Dena, O.; Cabral-Larquier, E.M.; Elizalde-Galindo, J.T.; Farías, R. Structural and Magnetic Behavior of Oxidized and Reduced Fe Doped $LiNbO_3$ Powders. *Crystals* **2018**, *8*, 108. [CrossRef]
209. Díaz-Moreno, C.A.; Farías-Mancilla, R.; Matutes-Aquino, J.A.; Elizalde-Galindo, J.; Espinosa-Magaña, F.; Hernández-González, J.; Hurtado-Macías, A. Magnetic behavior in $LiNbO_3$ nanocrystallites caused by oxygen vacancies. *J. Magn. Magn. Mater.* **2014**, *356*, 82–86. [CrossRef]
210. Díaz-Moreno, C.; Farias, R.; Hurtado-Macias, A.; Elizalde-Galindo, J.; Hernandez-Paz, J. Multiferroic response of nanocrystalline lithium niobate. *J. Appl. Phys.* **2012**, *111*, 07D907. [CrossRef]
211. Ishii, M.; Ohta, D.; Uehara, M.; Kimishima, Y. Ferromagnetism of Nano-$LiNbO_3$ with Vacancies. *Trans. Mat. Res. Soc. Jpn.* **2012**, *37*, 443–446. [CrossRef]
212. Cao, E.; Zhang, Y.; Qin, H.; Zhang, L.; Hu, J. Vacancy-induced magnetism in ferroelectric $LiNbO_3$ and $LiTaO_3$. *Phys. B Condens. Matter* **2013**, *410*, 68–73. [CrossRef]
213. Yan, T.; Ye, N.; Xu, L.; Sang, Y.; Chen, Y.; Song, W.; Long, X.; Wang, J.; Liu, H. Ferromagnetism in chemically reduced $LiNbO_3$ and $LiTaO_3$ crystals. *J. Phys. D Appl. Phys.* **2016**, *49*, 195005. [CrossRef]
214. Kocsor, L.; Péter, L.; Corradi, G.; Kis, Z.; Gubicza, J.; Kovács, L. Mechanochemical Reactions of Lithium Niobate Induced by High-Energy Ball-Milling. *Crystals* **2019**, *9*, 334. [CrossRef]
215. Abrahams, S.C.; Reddy, J.M.; Bernstein, J.L. Ferroelectric Lithium Niobate—3. Single crystal X-ray diffraction study at 24 °C. *J. Phys. Chem. Solids* **1966**, *27*, 971–1012.
216. Abrahams, S.C.; Reddy, J.M.; Bernstein, J.L. Ferroelectric Lithium Tantalate—1. Single crystal X-ray diffraction study at 24 °C. *J. Phys. Chem. Solids* **1967**, *28*, 1685–1692. [CrossRef]
217. Huband, S.; Keeble, D.S.; Zhang, N.; Glazer, A.M.; Bartasyte, A.; Thomas, P.A. Crystallographic and optical study of $LiNb_{1-x}Ta_xO_3$. *Acta Cryst.* **2017**, *B73*, 498–506. [CrossRef] [PubMed]
218. Sidorov, N.V.; Serebryakov, Y.A. Investigation of structural peculiarities of impure lithium niobate crystals by Raman spectroscopy. *Vib. Spectrosc.* **1994**, *6*, 215–223. [CrossRef]
219. Bäumer, C.; David, C.; Tunyagi, A.; Betzler, K.; Hesse, H.; Krätzig, E.; Wöhlecke, M. Composition dependene of the ultraviolet absorption edge in lithium tantalite. *J. Appl. Phys.* **2003**, *93*, 3102–3104. [CrossRef]
220. Kovács, L.; Ruschhaupt, G.; Polgár, K.; Corradi, G.; Wöhlecke, M. Composition dependence of the ultraviolet absorption edge in lithium niobate. *Appl. Phys. Lett.* **1997**, *70*, 2801–2803. [CrossRef]
221. Young, S.-L.; Kao, M.-C.; Chen, H.-Z. The properties of tantalum modified lithium niobate thin films prepared by a diol-based sol-gel process. *Eur. Phys. J. Appl. Phys.* **2006**, *36*, 5–10. [CrossRef]
222. Kao, M.C.; Chen, H.Z.; Yang, S.L.; Chen, Y.C.; Hsieh, P.T.; Yu, C.C. Pyroelectric Ta-modified $LiNbO_3$ thin films and devices for thermal infrared detection. *Thin Film Solids* **2008**, *516*, 5518–5522. [CrossRef]
223. Irzaman; Sitompul, H.; Masitoh; Misbakhusshudur, M.; Mursyidah. Optical and structural properties of lanthanum doped lithium niobate thin films. *Ferroelectrics* **2016**, *502*, 9–18. [CrossRef]
224. Thierfelder, C.; Sanna, S.; Schindlmayr, A.; Schmidt, W.G. Do we know the band gap of lithium niobate? *Phys. Status Solidi C* **2010**, *7*, 362–365. [CrossRef]

crystals

Review

Structures of Impurity Defects in Lithium Niobate and Tantalate Derived from Electron Paramagnetic and Electron Nuclear Double Resonance Data

Valentin G. Grachev * and **Galina I. Malovichko** *

Physics Department, Montana State University, Bozeman, MT 59717, USA
* Correspondence: valentin.g.grachev@gmail.com (V.G.G.); galina.i.malovichko@gmail.com (G.I.M.)

Abstract: Point intrinsic and extrinsic defects, especially paramagnetic ions of transition metals and rare-earth elements, have essential influence on properties of lithium niobate, LN and tantalate, LT, and often determine their suitability for numerous applications. Discussions about structures of the defects in LN/LT have lasted for decades. Many experimental methods facilitate progress in determining the structures of impurity centers. This paper gives current bird's eye view on contributions of Electron Paramagnetic Resonance (EPR), and Electron Nuclear Double Resonance (ENDOR) studies to the determination of impurity defect structures in LN and LT crystals for a broad audience of researchers and students. Symmetry and charge compensation considerations restrict a number of possible structures. Comparison of measured angular dependences of ENDOR frequencies with calculated ones for Li and Nb substitution using dipole–dipole approximation allows unambiguously to determine the exact location of paramagnetic impurities. Models with two lithium vacancies explain angular dependencies of EPR spectra for Me^{3+} ions substituting for Li^+ like Cr, Er, Fe, Gd, Nd, and Yb. Self-compensation of excessive charges through equalization of concentrations of $Me^{3+}(Li^+)$ and $Me^{3+}(Nb^{5+})$ and appearance of interstitial Li^+ in the structural vacancy near $Me^{3+}(Nb^{5+})$ take place in stoichiometric LN/LT due to lack of intrinsic defects.

Keywords: impurity; intrinsic defect; paramagnetic ion; lithium niobate; lithium tantalate; electron paramagnetic resonance; electron nuclear double resonance; lithium vacancy

Citation: Grachev, V.G.; Malovichko, G.I. Structures of Impurity Defects in Lithium Niobate and Tantalate Derived from Electron Paramagnetic and Electron Nuclear Double Resonance Data. *Crystals* **2021**, *11*, 339. https://doi.org/10.3390/cryst11040339

Academic Editor: Gábor Corradi

Received: 28 February 2021
Accepted: 23 March 2021
Published: 27 March 2021

1. Introduction

Point intrinsic and extrinsic defects, especially paramagnetic ions of transition metals and rare-earth elements, belong to the most important defects in lithium niobate (LN, $LiNbO_3$) and tantalate (LT, $LiTaO_3$), because of their essential influence on properties of this material, such as domain structure, electro-optical coefficients, light absorption, refractive indices, birth and evolution of wave-front dislocations ([1,2] and references there), and their consequences for present and potential applications [3–7]. A lot of effort was spent to establish a correlation between the observable data and the crystal composition, and to develop experimentally supported models of the defects: ion charges, identities, and position of the ions in the lattice, their nearest surroundings, ways of charge compensation and recharge mechanisms. Discussions about structures of intrinsic and extrinsic defects in LN/LT have lasted for decades [1]. With time, the proposed structures were evolved and detailed. Some early models were rejected. For instance, after a supposition that effective net charges are about 2.0+ [8], or 1.59+ for Nb and 1.21+ for Ta [9], it was natural to assume that divalent and trivalent impurities should preferably substitute for Nb, but not for Li. Later, numerous investigations have shown that the real picture is more complicated and richer.

In the course of the investigation of defect structures in LN/LT the following difficulties take place:

- high quality conventional samples with crystal composition x_C are usually grown from the congruent melt with the composition $x_m \approx 48.4\%$, for LN and $x_m \approx 48.7\%$ for LT (x = [Li]/([Li] + [Nb/Ta]); this means that the congruent crystals with $x_m = x_C$ contain many intrinsic (non-stoichiometric) defects, causing a broadening of the observable spectral lines and ambiguities in their interpretation;
- the crystal composition x_C of the undoped samples depends on both melt composition and growth conditions;
- the most probable positions for impurity incorporation, the Li and Nb sites as well as the octahedral structural vacancy, have the same local symmetry $\mathbf{C}_3$; this means that they are not distinguishable by many spectroscopic techniques.

Techniques for the defect study can be conditionally divided into two groups: direct (Mössbauer spectroscopy [10], Rutherford Backscattering Spectrometry (RBS) [11], Extended X-ray Absorption Fine Structure Analysis (EXAFS), Particle Induced X-ray Emission (PIXE) combined with channeling, X-ray standing wave (XSW) [12], Electron Paramagnetic Resonance (EPR), and Electron Nuclear Double Resonance (ENDOR) [13]) and indirect (optical absorption and luminescence, measurements of electro-optical coefficients, etc.).

Attempts to determine impurity positions by indirect methods often gave contradicting information. Direct methods also have some shortcomings. For instance,

- Mössbauer spectroscopy demands the presence of special nuclei,
- channeling investigations are more successful in the case of heavy, many-electron ions,
- channeling methods are not sensitive to the charge state of the impurity and do not distinguish centers with $\mathbf{C}_3$ and $\mathbf{C}_1$ symmetry,
- due to the relatively low sensitivity the EXAFS needs high levels of crystal dopants (about 3–5 mol.%), for which clustering and occupation of both Li and Nb positions become very probable,
- EPR/ENDOR techniques are applicable to systems with non-zero spins only.

Investigations of impurity defects were carried out by all direct and indirect techniques in parallel and together, and made invaluable contributions to the problem solution. Let's mention briefly a few of them. The positions of many impurity ions were determined by the EXAFS, PIXE, XSW, and channeling methods. Lithium substitution was found for most elements including Sc, Ti, Mn, Ni, In, Pr, Gd, Ho, Er, Tm, Yb, Lu ([11,14–18], and references there), Nd [19,20], Co, Fe [21–28]. A few ions substitute for Nb in LN (Hf [29], Er [30]) and in LN heavily co-doped by Mg or Zn (Sc, In, Nd, Lu [31]). Some of the ions can occupy both positions (Cr [32], Gd, Nd). Mössbauer spectroscopy [33,34] gave indirect evidence of Fe^{2+} and Fe^{3+} substitution for Li^+. Raman spectroscopy data [35] were interpreted on the base of Fe substitution for Li. Magneto-optical and luminescence studies [36,37] identified non-paramagnetic pairs Cr_{Li}-Cr_{Nb} and established a correlation between optical bands and Cr^{3+} positions.

Convincing arguments of Fe^{3+} substitution for Li^+ were obtained by ENDOR [38–40]: it was shown that spectrum angular dependencies calculated on the base of dipole-dipole interactions of Fe^{3+} electrons with surrounding Li nuclei are qualitatively different for Li and Nb substitution; quantitative agreement calculated and experimental data was achieved for Li substitution.

The huge amount of nonstoichiometric intrinsic defects in congruent LN/LT grown by the conventional Czochralski method [41,42] (we shall call them cLN and cLT) has often undesirable effects on crystal properties. Several techniques were invented in order to reduce the concentration of these defects. Some decrease of their concentration was achieved by using melts with Li excess (with x_m up to 60%) [43,44]. Thin LN/LT samples were enriched with Li by vapor transport equilibrium treatment (VTE [45–50], we shall call them vLN and vLT). Using double crucible growth [51–55] with excess of Li allowed one to obtain nearly stoichiometric samples (nsLN). Crystals grown by the Czochralski method from a melt to which potassium oxide K_2O has been added ([56–67], and references there) have composition $x_C \approx 50\%$ (LN_K, stoichiometric LN, sLN). The method is scientifically called the High Temperature Top Seeded Solution Growth (HTTSSG).

The defect elimination in sLN crystals has led to a tremendous narrowing of the resonance lines and a corresponding increase of spectral resolution, and resulted in improved precision of obtained spectroscopic characteristics. Discovery of new centers substituting for Nb (or Ta in $LiTaO_3$) and a change of charge compensation mechanism in sLN was accompanied by a significant progress in understanding the nature and structures of intrinsic and extrinsic defects [68–76]. Several methods of determination of real crystal composition were developed [68,77–80]. It should be mentioned that even crystals with x_C = 50% are not completely free of intrinsic defects. There is a class of intrinsic defects which break the regular order of the LN lattice without a change of x_C: the permutations Nb_{Li} and Li_{Nb}, and cyclical permutation of Nb in Li site, Li in structural vacancy, and vacancy of Nb. Such defects can cause broadening of spectral lines.

EPR is resonance absorption of microwave quanta by electron spins placed in a swept external magnetic field at definite value of magnetic field B_{res}. Depending on measurement conditions it has usual sensitivity about 10^{12}–10^{15} spins/cm^3 or higher. Equipment with wave lengths 3cm (X-band, microwave frequency $\nu \approx 9.5$ GHz) and 8 mm (Q-band, $\nu \approx 34$ GHz) are usually used. EPR lines have typical widths about 1–100 MHz. This defines the resolution ability of the EPR.

Nuclear magnetic resonance (NMR) is resonance absorption of radiofrequency quanta by nuclear spins. It has worse sensitivity due to thousand time smaller magnetic moments of nuclei. However, NMR has significantly higher resolution ability about 1–100 KHz or less. Only nuclei with large natural abundance of magnetic isotopes (like ^{7}Li—92.5%, ^{93}Nb—100%, ^{181}Ta—99.99%) can be studied by NMR. Due to interactions with electron spins, nuclei in the impurity neighborhood have energy levels and resonance frequencies different from those of bulk nuclei. As their concentrations, which are about impurity concentration, are significantly smaller than concentrations of bulk nuclei, their NMR signals are too weak to be registered directly.

ENDOR is the indirect way to register NMR of nuclei in impurity surrounding via a stimulated change of intensity of the EPR signal in magnetic field B_{res}. ENDOR has sensitivity that is worse than EPR (about one-two orders of magnitude), but is better than NMR. Its resolution is comparable with NMR resolution. Characteristics of all nuclei (natural abundance of isotopes, their nuclear spins, magnetic and quadrupole moments) are well known and tabulated. This allows to relate observed ENDOR frequencies to interactions of impurity electrons with own nucleus and surrounding nuclei.

For a paramagnetic impurity these techniques are capable of determining: its element, charge, spin state, nearest surrounding (i.e., its location), and many spectroscopic characteristics. Fortunately, most impurities which are useful for optoelectronic application (including transition and rare-earth elements) enter LN and LT in a paramagnetic state. Li, Nb and Ta nuclei have significant (93–100%) natural abundance of magnetic isotopes. The main isotope ^{16}O has no magnetic moment and is invisible in EPR/ENDOR/NMR experiments.

The aim of this paper is to give current bird's eye view on contributions of EPR/ENDOR studies to the determination of impurity defect structures in LN and LT crystals for a broad audience of researchers and students. The name "center" will be used for a complex consisting of the paramagnetic impurity and its surrounding that can have zero, one or more defects besides regular ions.

2. Materials and Methods

Most of our measurements presented below were carried out on samples grown by E.P.Kokanyan from Li-rich melt or from the melt with the addition of K_2O [56,57].

The EPR/ENDOR spectra were measured at microwave frequencies $v \approx 9.4$ GHz (X-band) and $v \approx 34.4$ GHz (Q-band) on the Bruker ELEXSYS EPR/ENDOR spectrometer at temperatures between 4.2 and 300 K at Montana State University (Bozeman, MT, USA).

The treatment and simulation of all spectra and their angular dependencies, as well as the preparation of corresponding figures were carried out with the help of the "Visual EPR" and "Visual ENDOR" program packages [81].

All lattice images were generated using CrystalMaker®program [82] and crystallographic data [83–86]. Sizes of oxygen balls were artificially reduced, since nonmagnetic O^{2-} ions are invisible for EPR/ENDOR. As $LiNbO_3$ and $LiTaO_3$ are isostructural crystals, all figures with lattice and structures of impurity centers are equally applicable to both LN and LT. However, dipole-dipole interactions should be calculated taking into account a difference of their lattice constants.

3. Basics of EPR and ENDOR Spectra Interpretation

3.1. *Symmetry Considerations*

The ideal lattice of LN and LT crystals (Figure 1, see also [87]) has two molecules in its rhombohedral elementary unit cell and the space group symmetry is R3c ($C_{3v}{}^6$, 3m) at room temperature [8,88–90]. If ionic radii are taken into account, the LT/LN lattices look like consisting mainly of planar layers of closely packed oxygen ions with small voids between them, filled by Li and Nb/Ta. There are several electrically non-equivalent positions in the lattice. Some of them—the sites on the **z** (or optical **c**) axis of the crystal, including the sites of Li, Nb/Ta and the octahedral structural void (vacancy), v_{oct}—have the symmetry of the point group C_3. An isolated defect in any of these positions creates a C_3 (in the following also labeled "axial") center. The tetrahedral structural void, oxygen sites and all other off axis positions have the lowest possible symmetry, C_1.

A complex of two defects has C_3 symmetry, if both are located on the crystal **z** axis, and C_1 symmetry in all other cases. If the difference in positions of the nearest oxygen ions is ignored, then each unit cell has the following site sequence along the **z**-axis: Li, Ta, v_{oct}, Li, Ta, v_{oct}. However, oxygen octahedrons and next neighbors for two positions of one type cation are not identical, therefore, the correct assignment of the sequence should be Li_L, Ta_L, $v_{oct,L}$, Li_R, Ta_R, $v_{oct,R}$. The surrounding of the "Right" (R) position can be transformed to the "Left" (L) one by a reflection $\mathbf{x} \Leftrightarrow -\mathbf{x}$ and a shift by c/2, because **zy** is a glide mirror plane in the R3c lattice (Figure 1b). The same consideration is applicable to C_1 positions also. This means that each axial (C_3) or low-symmetry (C_1) L center has a corresponding R partner. The L and R partners are electrically identical and are not distinguishable by optical or channeling methods, but they are magnetically non-equivalent and can be resolved from each other in favorable cases (high spin value, small line width) by magnetic resonance techniques.

Each C_1 center in the R3c lattice has two additional magnetically nonequivalent partners, which can be transformed into each other by a rotation around the **z**-axis of the crystal by 120° and 240°. They can be distinguished by the EPR at arbitrary orientations of the magnetic field.

As any crystal characteristic, the EPR spectra must reflect the crystal symmetry and symmetry of all present paramagnetic centers. The best way to determine the defect symmetry is to study angular dependence of the spectra in external magnetic field **B** under rotation of the crystal sample around **z**-axis (shortly speaking, in **xy** plane). For a C_3 paramagnetic defect with electron spin $S = 1/2$ the EPR spectrum consists of one resonance at the same position B_{res} at any orientation of **B** with respect to crystallographic axes. A set of up to six lines can be observed at arbitrary orientation of **B** for C_1 centers; however, some of them coincide for directions of **B** along **x**, **y**, or **z**. If two or more centers present in the sample studied, the spectrum has two or more sets of the lines. If a center has electron spin $S > 1/2$ multiplets of $2S$ lines can be observed for every center.

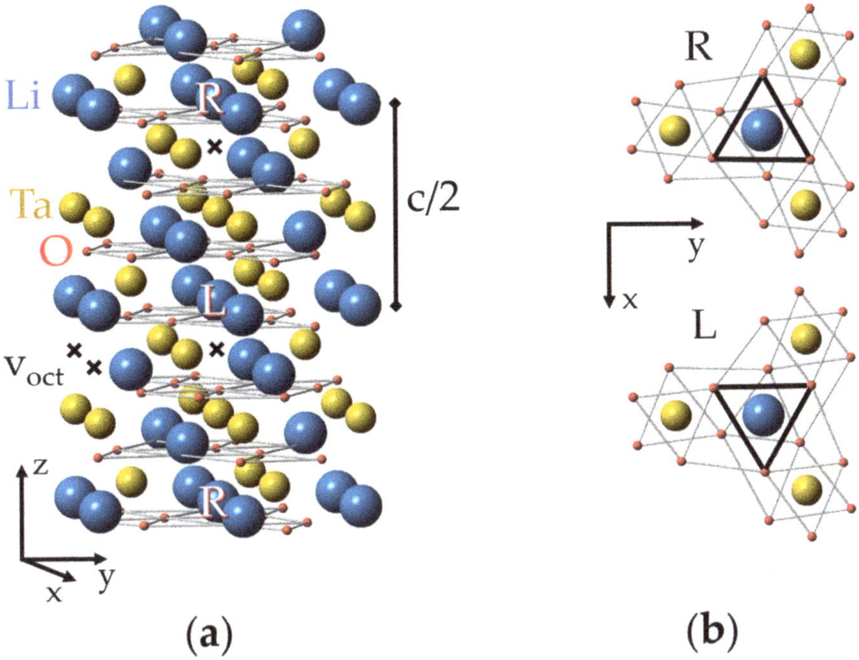

Figure 1. (**a**) Ideal lattice of lithium tantalite (LT). To simplify the representation of the lattice and defect structures in it, the sizes of "balls" imitating ions were made intentionally different. The L (Left) and R (Right) positions are distinguished by their different oxygen surroundings. (**b**) An illustration of local C_3 symmetry for the nearest surroundings of cation sites and glide mirror **zy** plane that transforms the L center to the R partner; a projection of one Li, Ta layer between two oxygen layers on the **xy** plane is shown.

For centers with $S = 1/2$ the resonance line positions can be described using spin-Hamiltonian

$$H = \mu_B \mathbf{B}\mathbf{g}\mathbf{S} \tag{1}$$

Here μ_B is the Bohr magneton, **B** is the vector of static magnetic field, **g** is the tensor of spectroscopic splitting, and **S** is the vector of electron spin.

As an example of the multi-center spectra let's consider ion Nd^{3+} in LN [91–93] (Figure 2). Every spectrum consists of a single dominant line labeled with the number 1 and many satellite lines of smaller intensities. The positions of the dominant line have the same value of the resonance field for **B**| |**x** and **B**| |**y** (φ = 90 deg), whereas positions of other lines do not coincide. We have to suppose that the single line #1 belongs to the axial C_3 symmetry center, Nd_1.

To make correct labeling of all observed lines a detailed study of angular dependence of EPR line positions is required. For instance, we can rotate magnetic field in **xy** crystallographic plane, changing azimuthal angle φ between **x**-axis and **B**. Circular diagram of measured spectra (Figure 3) show that dominant line draw a circle, i.e., has no dependence of $B_{res}(1)$ on φ. Angular patterns of other lines demonstrate the presence of all elements of the 3m group, namely, C_3 symmetry and the mirror with respect to **y**-axis at φ = 90 deg. Measurement of such a diagram is redundant, since **xy** plane angular patterns simply repeat themselves with a period 60 deg. Due to the glide mirror plane, the 30 degree dependence measured from **x** axis contains all information.

Usually, 90 deg of the measured dependence is plotted (Figure 4). If spectra are measured with small angular steps (said 1 deg), it is easy to trace every line. The line tracing allows to put labels on every resonance line on Figure 2. Without doubt, at least four groups of angular branches are clearly distinguished on Figure 4: one straight branch of axial Nd_1 and three sets of curved branches, which correspond to low-symmetry centers Nd_2, Nd_3, and Nd_4. Therefore, we can conclude that Nd^{3+} in Nd_1 center occupies one of three possible positions on the **z**-axis. If Nd^{3+} has additional defect or defects (charge compensators) in its neighborhood, the defect is located on the same axis. In the case of Nd_2, Nd_3, and Nd_4 centers with the C_1 symmetry, the additional defects are located off the **z**-axis. It is unlikely that Nd^{3+} ions occupy sites with C_1 symmetry (tetrahedral void or O^{2-}) due to large charge misfit.

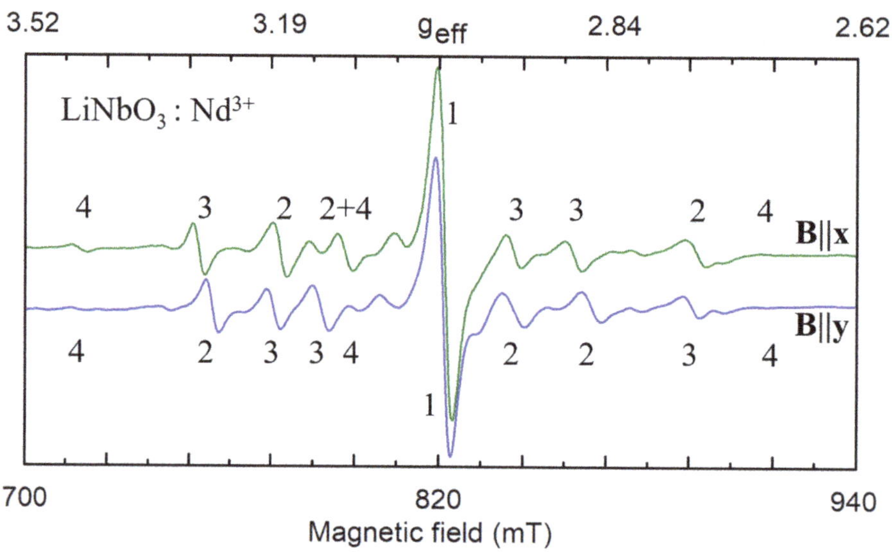

Figure 2. Fragments of electron paramagnetic resonance (EPR) spectra of Nd^{3+} impurity ions in nearly stoichiometric lithium niobate (LN) for microwave frequency ν = 34.445 GHz, T = 19 K. The numbers 1–4 correspond to four different electrically non-equivalent defects. Lines of magnetically nonequivalent partners have the same number.

Symmetry of any center reflects the symmetry of lattice sites occupied by impurity and charge compensators (if any) and their relative locations. To distinguish the C_3 and C_1 symmetries the measurement of angular dependence in **xy** plane is sufficient. However, to obtain a full set of spectroscopic characteristics of the center (like six components of **g**-tensor) by fitting measured angular dependence a study of spectra under three rotations in perpendicular planes (road map) is required.

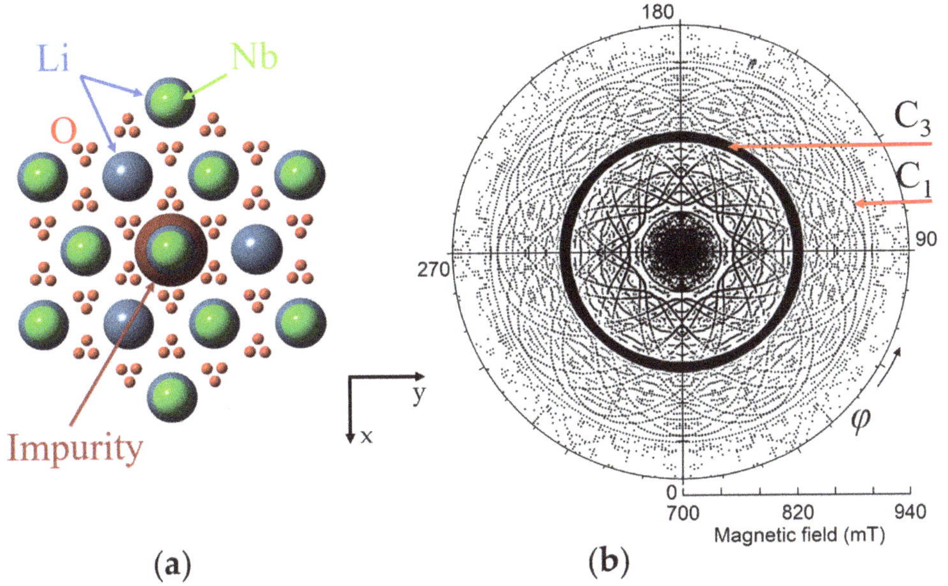

Figure 3. (**a**) Projection of ideal lithium niobate lattice with Nd^{3+} impurity ion on **xy** plane (artificial sizes of balls are used). (**b**) Circular diagram of measured angular dependence of electron paramagnetic resonance (EPR) lines of Nd^{3+} ions in Figure 3. Magnetic field was rotated in **xy** plane from 0 to 360 deg. ν = 34.445 GHz, T = 19 K. Symbol sizes reflect line intensities.

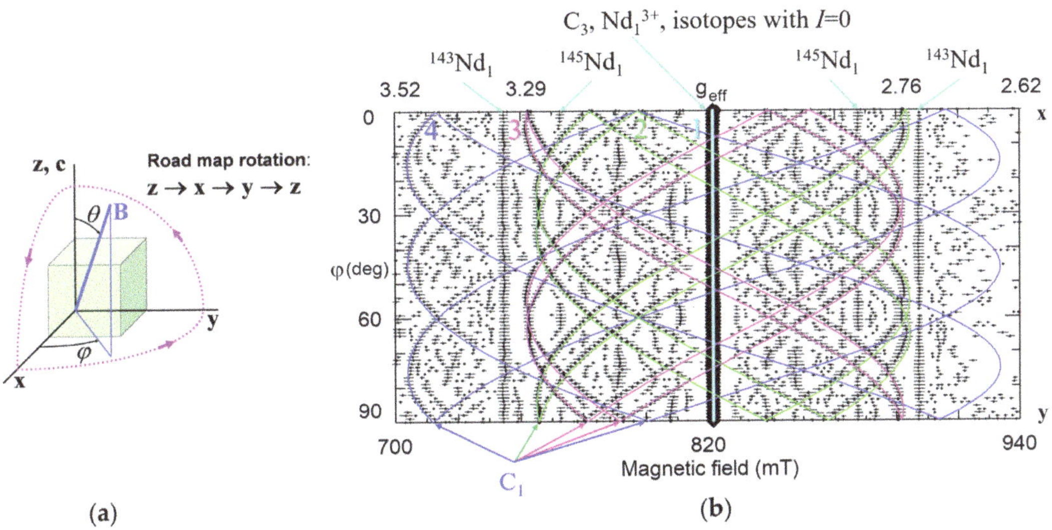

Figure 4. (**a**) The road map rotation corresponds to the following changes of the θ, ϕ angles: θ = 0–90° at ϕ = 0°; θ = 90°, ϕ = 0–90°; θ = 0–90° at ϕ = 90°. (**b**) Angular dependence of the EPR spectra in **xy** plane for nearly stoichiometric LN:Nd^{3+}. Rhombs represent experimental line positions for Q-band measurements (ν = 34.445 GHz), their sizes reflect line intensities. Horizontal whiskers near rhombs represent line widths. Cyan, lime, fuchsia, and blue curves represent simulated dependencies for axial Nd_1, and low-symmetry Nd_2, Nd_3, and Nd_4 centers, respectively. Curves for isotopes with nuclear spin $I \neq 0$ are not shown.

3.2. Isotropic and Dipole-Dipole Interactions

Calculations of the ENDOR frequencies and transition probabilities (i.e., ENDOR line positions and relative intensities) for i-th nucleus is usually based on nuclear spin-Hamiltonians H_i

$$H_i = -\mu_n g_n^{(i)} \mathbf{B}\mathbf{I}^{(i)} + \mathbf{S}\mathbf{A}^{(i)}\mathbf{I}^{(i)} + \mathbf{I}^{(i)}\mathbf{Q}^{(i)}\mathbf{I}^{(i)} \quad (2)$$

where μ_n—nuclear magneton, $g_n^{(i)}$—nuclear g-factor; $\mathbf{A}^{(i)}$, $\mathbf{Q}^{(i)}$—tensors of hyperfine and quadrupole interactions.

In many cases there are two dominant contributions to the hyperfine tensor $\mathbf{A}^{(i)}$: isotropic (contact) hyperfine interaction, $a^{(i)}\mathbf{S}\mathbf{I}^{(i)}$ and dipole-dipole interaction of vectors of electron $\mu_B g\mathbf{S}$ and nuclear $-\mu_n g_n^{(i)}\mathbf{I}^{(i)}$ magnetic moments. The dipole–dipole interaction can be described by

$$A_{jm}^{(i)dd} = \frac{\mu_B \mu_n}{\left(R^{(i)}\right)^3} g_n^{(i)} \left[\frac{3}{\left(R^{(i)}\right)^2} \left(\sum_{p\,=\,x,y,z} g_{jp} R_p^{(i)} \right) \left(\sum_{l\,=\,x,y,z} R_l^{(i)} \delta_{lm} \right) - g_{jm} \right] \quad (3)$$

where $\mathbf{R}^{(i)}$ is the vector from paramagnetic impurity to the i-th nucleus. Most characteristics in Equation (3) are known: μ_B, μ_n, $g_n^{(i)}$ are tabulated, g_{jm} components of $\mathbf{g}$-tensor are determined from EPR measurements. Therefore, comparison of measured angular dependences of ENDOR frequencies with calculated ones on the base of Equations (2) and (3) can be used for determination of $\mathbf{R}^{(i)}$, i.e., the position of the paramagnetic impurity relative to surrounding nuclei.

Nuclei at the same distance from impurity ions are usually called a shell. The first shell of an impurity in a Li site has six nearest Li nuclei off the **z**-axis at the distance $R^{(1)}$ (Figure 5a). Three of them are located above the impurity (the subshell 1a), other three—below the impurity (the sub-shell 1b); if the impurity is shifted from regular Li site then $R^{(1a)} \neq R^{(1b)}$. The second shell consist of six Li nuclei in the **xy** plane of the impurity at the distance $R^{(2)}$ (subshells 2a, 2b).The first shell of an impurity in a Nb site has one Li nucleus on **z**-axis, the second shell has three Li nuclei (the nuclei are labelled 1 and 2 on Figure 5b). As directions from the impurity to the nuclei in a shell are different, their hyperfine interactions are magnetically non-equivalent, i.e., produce different branches in angular dependences of ENDOR spectra.

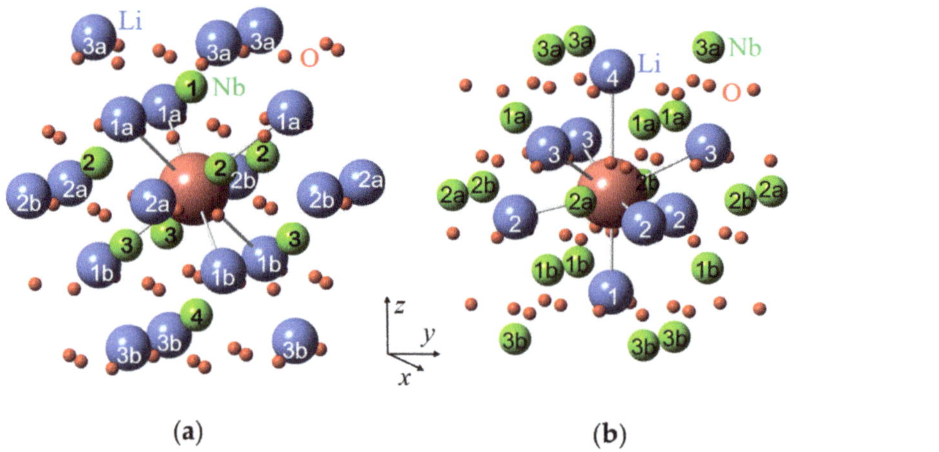

Figure 5. Surroundings of impurity ions substituted for Li (**a**) and Nb (**b**) in LN crystal lattice with corresponding shell numbers. Large blue balls represent Li^+ ions, medium green balls—Nb^{5+} ions, small red balls—O^{2-} ions, and the largest magenta ball—the impurity ion.

The simplest way to determine the impurity lattice site is:

- to calculate $A_{jm}^{(i)dd}$ for the several nearest nuclei around the impurity taking lattice distances from X-ray data and g_{jm} from EPR measurements,
- to calculate ENDOR frequencies, to plot patterns of their angular dependencies, and
- to compare the patterns with measured angular dependencies of ENDOR for a definite EPR line [13,38,39].

Due to different surroundings for impurities in the Li and Nb sites, the calculated patterns are completely different (Figure 6). ENDOR frequencies for the first shell of Li nuclei for the Li site vary with rotation of magnetic field in **xy** plane(Figure 6a), whereas for the single Li nucleus of the first shell for Nb site two straight branches should be observed in angular dependence (Figure 6c). As Li nuclei of the all corresponding shells are closer to the impurity ion in the Nb site than in the Li site, the range of angular variation for Nb site is larger than for Li site. The branches of the 2nd and 3rd shells for Li site practically coincide with measured angular dependencies for Nd_1 center (Figure 6b), whereas no branch for Nb site is close to observed one. The branches of the 1st shell for Li substitution also agree with observed ones after small correction due to isotropic hyperfine interaction (Figure 6b). Based on clear agreement of hundred measured values of ENDOR frequencies for the dominant EPR line #1 with calculated frequencies for Li site and obvious disagreement with calculated ones for Nb site, we can definitely conclude that the Nd_1 line in EPR spectra (Figure 2) belongs to Nd^{3+} ion substituted for Li.

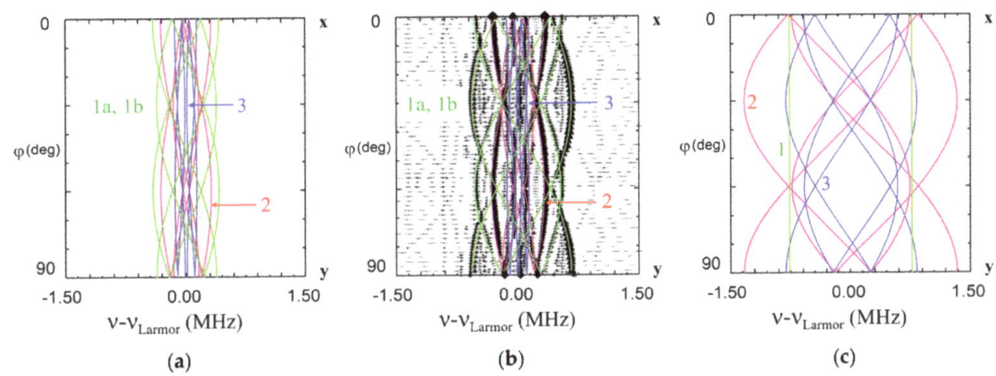

Figure 6. Angular dependence of electron nuclear double resistance (ENDOR) frequencies of Li nuclei in **xy** plane for the $Nd_1{}^{3+}$ center in $LiNbO_3$. The green, red and blue curves—calculated frequencies for Li nuclei of the 1st, 2nd and 3rd shells based on dipole-dipole interaction for $Nd^{3+}{}_{Li}$ (**a**,**b**), and $Nd^{3+}{}_{Nb}$ (**c**). ENDOR frequencies represented on the part (**b**) were measured at magnetic fields corresponding to the position of dominant EPR line #1 on Figure 2 (B = 820 mT for ν = 34.445 GHz, rhombs, and B = 228 mT for ν = 9.500 GHz, triangles). ν_{Larmor} is Larmor frequency of ^{7}Li nuclei.

The simulation of dipole-dipole interactions for the Li and Nb sites has no fitting parameter. Therefore, the described qualitative and quantitative approach for the determination of impurity positions gives reliable results. Note that hyperfine interactions for Nb/Ta nuclei have additional contributions due to a polarization of inner electron shells of oxygen and niobium ions (transferred hyperfine interaction).

The characteristic of isotropic hyperfine interaction $a^{(i)}$ is related to a density of electron cloud at the location of i-nucleus $\mathbf{R}^{(i)}$. For the isotropic g-factor:

$$a^{(i)} = \frac{8\pi}{3} g\mu_B g_n^{(i)} \mu_n \left|\psi\left(\mathbf{R}^{(i)}\right)\right|^2 \tag{4}$$

The isotropic hyperfine interaction often (but not always) exponentially decreases with the distance from impurity. The largest value $a^{(i)}/g_n^{(i)}$ should correspond to nuclei nearest to the impurity. It allows to determine nuclei of the nearest surrounding by comparison of $a^{(i)}/g_n^{(i)}$ for Li and Nb nuclei, and therefore, to find the impurity location. Impurities substituted for Li have Nb nucleus on **z**-axis in the nearest surrounding, whereas ions substituted for Nb have Li nucleus (Figure 6, shell #1).

Another way to use contact and dipole-dipole interactions is:

- to measure the road map of angular dependencies in three perpendicular planes (Figure 4a),
- to determine all components of $\mathbf{A}^{(i)}$ tensors by fitting observed angular dependencies,
- to separate isotropic and anisotropic parts,
- to find principal values of the anisotropic part, and finally,
- to compare these principal values with calculated ones on the base of Equation (3).

3.3. *Charge Compensation and Intrinsic Defects*

The structure of a center, in which a lattice site is occupied by an extrinsic impurity ion having a charge different from that of the respective lattice ion (in the following labeled "non-isocharged replacement"), depends on the charge of the impurity and the mechanism of charge compensation. Because of non-stoichiometry, the real lattice of conventional congruent LN contains many intrinsic defects, the relative concentrations of which have not yet been determined reliably. The following entities have been considered (their charges with respect to the lattice being given by Kröger-Vink notation in brackets): $Nb^{5+}{}_{Li}{}^{+}$ ($Nb_{Li}{}^{4\bullet}$) antisite defect, $v_{Li}{}^{+}$ ($v_{Li}{}^{'}$) lithium vacancy, $v_{Nb}{}^{5+}$ ($v_{Nb}{}^{5'}$) niobium vacancy, $Nb^{5+}{}_{v}$ ($Nb_{v}{}^{5\bullet}$) niobium on structural vacancy, $Li^{+}{}_{v}$ ($Li_{v}{}^{\bullet}$) lithium on structural vacancy, v_O oxygen vacancy. During recent years the existence of the following charge compensated complexes has been postulated most often: Nb_{Li} + $4v_{Li}$ [90,94,95] (including three-dimensional complexes $3v_{Li}$ + Nb_{Li} + v_{Li} [96,97]), $5Nb_{Li}$ + $4v_{Nb}$ [98,99], $2Nb_{Li}$ + $2Nb_v$ + $3v_{Li}$ + $3v_{Nb}$ [96] and some others [100]. Some features accompanying the crystal growth (Li_2O evaporation, variation of oxygen deficiency in Nb_2O_{5-x} [101]) and specific changes of some crystal properties after thermal oxidation and/or reduction definitely indicate that the oxygen sub-lattice is not always perfect and stable as well; therefore the oxygen non-stoichiometry has been also discussed for a long time [102–105]. The intrinsic defects by themselves or complexes of them, which are not charge compensated, can furthermore serve as local or distant charge compensators for non-isocharged extrinsic or radiation defects. Due to the high concentration of the intrinsic defects the congruent LN and LT crystals are very tolerant to substitutional or interstitial impurities, including non-controlled ones, because the necessary charge compensators (local or distant) can be easily found among the non-stoichiometric defects. Real crystals often contain H, Cu, Co, Mn, Fe, etc. in concentrations about 0.00X–0.0X.

Non-isovalent cation Me^{n+} ($n > 1$) substituted for Li^+ requires negative charge compensator like lithium vacancy or interstitial O^{2-}. If Me^{n+} ($n < 5$) substituted for Nb^{5+} or Ta^{5+} the $n - 5$ negative charge can be compensated by antisite ions, oxygen vacancies, v_O or interstitial H^+ and Li^+.

In some cases, a self-compensation of impurity charges takes place. For instance, no additional charge is required if two Me^{3+} ions substitute for both Li^+ and Nb^{5+} in nearest neighbor or next neighbor sites (local self-compensation) or even relatively far one from another (distant self-compensation). There are two critical parameters, which stimulate the self-compensation: total concentration of possible charge compensation defects, [D] and the total concentration of Me^{3+} ions in the sample, [Me]. The [D] includes a dominant contribution due to a deviation of the crystal composition from stoichiometry, concentrations of H^+ ions and v_O, as well as contributions due to non-controlled impurities (Mg, C, Si, Cl, …), which are always present in real crystal in concentrations about 1–500 ppm. If [D] >> [Me] the isolated centers with additional intrinsic defects are dominating. If [D] << [Me] then the self-compensation is preferable due to lack of charge compensators.

Distant charge compensators produce small distortions of the crystal field at the impurity site, what normally causes EPR line broadening, and an asymmetry of their shapes, but do not change the center symmetry revealed in the observed positions and splitting of EPR lines. Defects in the immediate neighborhood (local charge compensation) cause strong changes of the center characteristics (**g** and **A** tensors in the case of Nd^{3+}) and lowering of center symmetry.

3.4. Di-Vacancy Models for Trivalent Impurities

Many impurities in LN/LT crystals create a family of paramagnetic centers that includes dominant axial center and several satellite low-symmetry centers with EPR lines of smaller intensities (like presented on Figure 2). It would be reasonable to look for center models which are able to describe the whole family.

Low-symmetry C_1 centers appear in three cases:

1. Impurity ion like Nd^{3+} has an off-axis lattice defect (charge compensator) in the immediate neighborhood.
2. Impurity ion substitutes for O^{2-} (this is very improbable for cations).
3. Impurity ion incorporates into a small tetrahedral void (this is possible, but often unlikely due to larger charge misfit than for Li substitution).

For axial centers, the directions of principal axes of the **g**-tensor are dictated by crystal symmetry: the 3rd axis of the center coincides with the crystal axis and the directions of the 1st and 2nd axes are arbitrary. For low-symmetry centers, there are no symmetry restrictions on the orientation of principal axes. However, if the symmetry lowering is related to an off-axis lattice defect, it is reasonable to expect that directions of the center axes are related to the distortion created by the off-axis defect. The 3rd axis will have some inclination from the **z**-axis to the defect, and projections of the 1st or 2nd axis will be close to the projection of line from the impurity ion to the defect. The distortion should decrease with the distance from the impurity ion to the defect.

ENDOR data for the dominant EPR line have confirmed that Nd^{3+} substitutes for Li^+ and only Li and Nb nuclei were found in the neighborhood. Therefore, intrinsic defects without nuclear spin should be considered for the required 2-charge compensation. The size O^{2-} is comparatively large to be placed into small octahedral or tetrahedral voids. Nearly stoichiometric crystals have significantly reduced concentration of v_{Li}. Nevertheless, their concentration often exceeds the impurity concentration. Sufficient concentration of v_{Li} can be also created in the process of growth of samples with non-isocharged impurities.

A key to the identification of models which describe all varieties of Nd^{3+} centers with minimal assumptions was obtained from the analysis of angular dependencies of the Nd_4 center. The extrema in the **xy** plane and, correspondingly, projections of one principal axis of the **g**-tensor, are close to ϕ values of 15, 45, and 75 degrees (Figure 4b). In the projection of the LN lattice onto the **xy** plane (Figure 7, after [93]) there are no similar values of ϕ from Nd^{3+} to any ion in the lattice: directions from the Nd^{3+} site to all cation sites have azimuthal angles equal to $n \times 30°$. Therefore, the simplest model, that has a single lattice defect, like a Li^+ vacancy, v_{Li}, cannot explain the observed angular dependencies of Nd_4 centers. However, if two Li^+ vacancies are located in the first and second shells of the surroundings of an Nd^{3+} ion substituted for a Li^+, the three defects are organized into a triangle. The longest side of the triangle connected two v_{Li} has a perpendicular oriented in the required direction (Figure 7).

The hypothesis that Nd^{3+} substituted for Li^+ has two v_{Li} as charge compensators allows for consistent models for the whole family of Nd^{3+} centers in LN crystals with a low concentration of neodymium doping. After analysis of principal values and axes of **g**-tensors it was concluded [92,93] that Nd^{3+} centers have lithium vacancies in sites described in Table 1. Some of possible structures are presented on Figures 7 and 8.

The di-vacancy models can explain EPR spectra of many other (but not all) trivalent impurities $Me^{3+}{}_{Li}$.

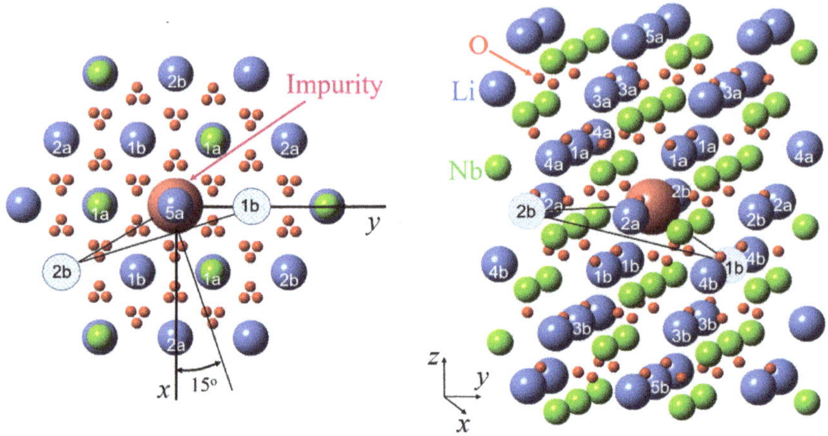

(a) (b)

Figure 7. A projection of lithium niobate crystal lattice on **xy** plane (**a**) and 3D model (**b**) with shell numbers for an impurity substituted for Li^+ ion (sizes of all ions were artificially changed in order to make clear positions of cation ions). Hatched circles represent two vacancies of Li^+ ions for Nd_4 center.

Table 1. Sites for lithium vacancies, v_{Li} in the surroundings of $Me^{3+}{}_{Li}$ centers in $LiNbO_3$.

Center v_{Li} Site	Me_1	Me_2	Me_3	Me_4	Me_5	Me_6	Me_7	Me_8
First	5a/5b	1a/1b	2a/2b	1a	1b	1a/1b	2a/2b	5a
Second	distant	5a/5b	5b/5b	2a	2b	3a/3b	3a/3b	5b

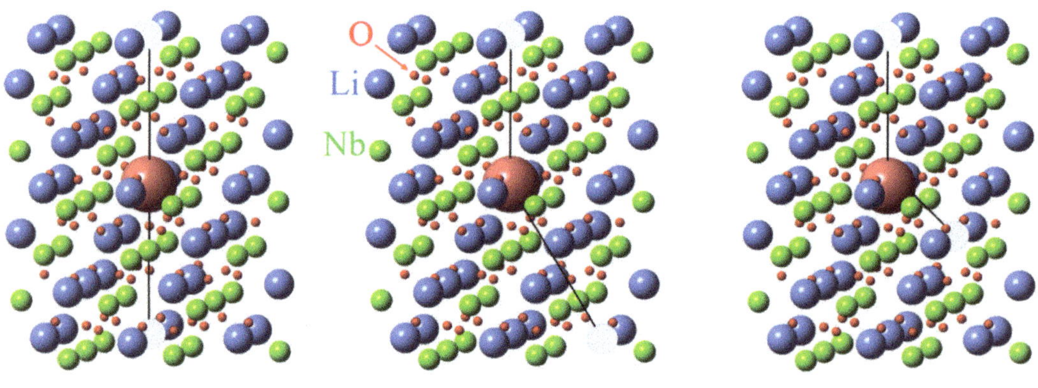

(a) (b) (c)

Figure 8. Divacancy models for an impurity substituted for Li^+ ion in LN crystal lattice. Hatched circles represent two v_{Li}. (**a**) Axial center, (**b**) nearly axial center, (**c**) low-symmetry center.

4. Structures of Impurity Centers

4.1. Monovalent Cations

The most probable incorporations of Me^+ is the substitution for Li^+. No charge compensation is required in this case. The $Me_{Li}{}^+$ center should have axial C_3 symmetry.

Proton H⁺. This is an example of off-site position: protons occupy positions between two oxygen ions in an oxygen plane. The non-paramagnetic OH^- centers were studied by infrared and NMR spectroscopies [80,106–108].

A hydrogen associated paramagnetic center (g = 2.0028, A = 3 mT at 77 K) was identified as an OH^{2-} ion, produced because of an electron capture by a diamagnetic OH^- ion, substituting the O^{2-} ion in LN [109].

Ni^+ ($3d^9$, $S = 1/2$). Observed EPR spectra of the Ni^+ have axial symmetry at room temperature [110,111]. That excludes interstitial position in tetrahedral structural vacancies. However, at low temperatures the Ni^+ center has C_1 symmetry, as up to six lines were observed at arbitrary orientation of magnetic field (Figure 9a, after [111]).

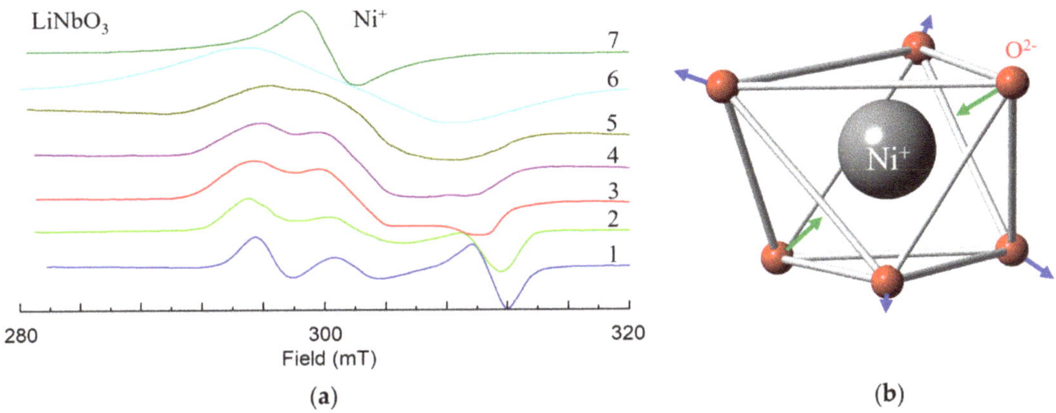

Figure 9. (**a**) X-band EPR spectra of Ni^+ in $LiNbO_3$ at θ = 25 deg in **zy** plane at various temperatures T: 1–40 K, 2–90 K, 3–151 K, 4–206 K, 5–244 K, 6–300 K. The spectrum 7 was measured at θ = 0 deg, T = 300 K. (**b**) Distortion of oxygen octahedron due to the Jahn-Teller effect.

The center was characterized by anisotropic **g**-tensor with principal values $g_1 = 2.246$, $g_2 = 2.217$ and $g_3 = 2.061$; principal axes of the **g**-tensor are rotated with respect to crystallographic axes by Euler angles $\alpha = \gamma \approx 0$, $\beta \approx 55°$. As the 3rd principal axis is directed approximately to one of the nearest oxygen ions, the reason for the low symmetry is a static Jahn-Teller effect for $3d^9$ ions in $Ni^+O_6{}^{2-}$ complexes (Figure 9b, [111,112]), and not a presence of an intrinsic defect in the neighborhood. Dynamic averaging due to center reorientation leads to the axial symmetry of observed EPR spectra at room temperatures [111].

Mg^+ ($3s^1$, $S = 1/2$). Following vacuum reduction at 1000 °C, LN crystals heavily-doped with Mg exhibit an optical absorption spectrum that can be decomposed into two bands peaking near 760 and 1200 nm, and a broad EPR spectrum with $g_c = 1.82$ [113]. The 1200-nm band and ESR signal are associated with an electron trap (identical to the one produced during the irradiations). This electron trap is suggested to be a Mg^+ complex. There is an alternative interpretation of this spectrum [114].

4.2. Divalent Cations

For divalent Me^{2+} impurities a substitution for Li^+ ions and incorporation in structural vacancies has essentially less charge misfit than a substitution for Nb^{5+}. Since the Li-Li distance is much larger than Li-Nb or v_{oct}-Li, v_{oct}-Nb, the $Me^{2+}{}_{Li} + v_{Li}$ centers should be slightly distorted by the presence of a local charge compensator. For v_{Li} located on C_3 axis and for distant cation vacancies (local and distant charge compensation) axial centers should be observed.

Co^{2+} ($3d^7$, $S = 1/2$, $I = 7/2$). Dominant axial Co^{2+} center with $g_\parallel = 2.6$ and $g_\perp = 4.96 \div 5.04$, $A_\parallel \approx 0$, $|A_\perp| = 0.0154\ cm^{-1}$, as well as low-intensity low-symmetry satellite centers were observed in cLN [115,116] and vLN [117]. Similar **g**- and **A**-values were reported for $LiTaO_3:Co^{2+}$ [116]. The picture agrees with Co^{2+} substitution for Li^+ in the dominant axial center (Co_{Li}) and excess charge compensation by v_{Li} (the $Co^{2+} \Leftrightarrow 2Li^+$ substitution mechanism). A small EPR line of axially symmetric cluster of Co^{2+} ions appeared in sLN [118] (Figure 10a). To explain it the substitution mechanism $4Co^{2+} \Leftrightarrow 3Li^+ + Nb^{5+}$ [119] was considered. The four Co^{2+} ions can occupy nearest possible cation sites by occupying one Nb site and three neighbor Li sites, creating a trigonal pyramid with C_3 symmetry (Figure 10b).

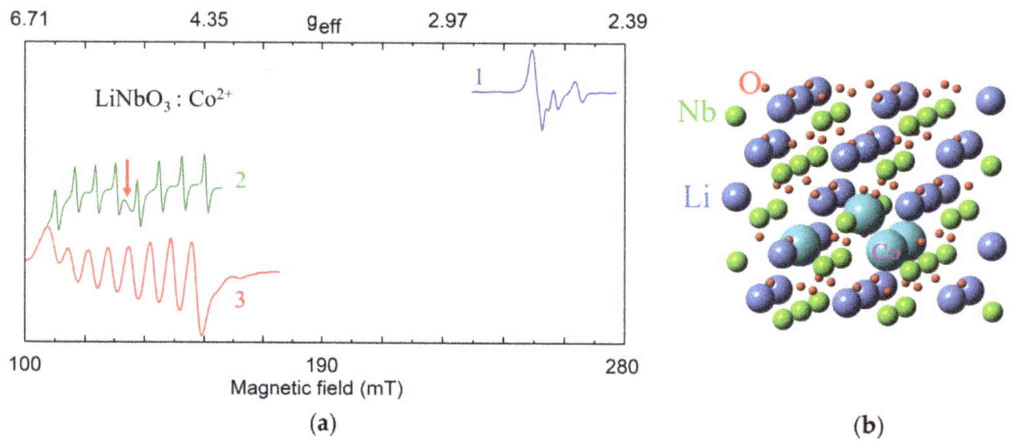

Figure 10. (**a**) Measured EPR spectra of Co^{2+}, X-band, T = 5 K. 1—sLN, **B** | | **z**; 2—sLN, **B** | | **x**; 3—cLN, **B** | | **x**. The red arrow indicates an additional line assigned to clusters of Co^{2+} ions. (**b**) Possible model for a cluster of four Co^{2+} ions substituted for one Nb^{5+} and three Li^+ ions.

From comparison of measured and calculated characteristics it was found that Co^{2+} does not occupy exactly the host Li^+ site but undergoes an off-center displacement 0.006 nm away from the oxygen octahedron center in $LiNbO_3$ (or $LiTaO_3$) [120,121].

Cu^{2+} ($3d^9$, $S = 1/2$). Copper nuclei have two isotopes ^{63}Cu ($I = 3/2$, $g_n = 1.484$, natural abundance 69.2%) and ^{65}Cu ($I = 3/2$, $g_n = 1.588$, 30.8%). Hyperfine interaction of Cu^{2+} electrons with their own nucleus leads to splitting of its single EPR line into a quartet ([110,111], and references there; [122–124]). As magnetic moments of ^{63}Cu and ^{65}Cu are very close, the quartets from the two isotopes overlap in cLN (Figure 11a). The Cu^{2+} center in LN was characterized by anisotropic **g**-tensor ($g_1 = 2.095$, $g_2 = 2.111$ and $g_3 = 2.428$; $\alpha = \gamma \approx 0$, $\beta \approx 51°$) [111]. At low temperatures a static Jahn-Teller effect for $3d^9$ ions in $Cu^{2+}O_6{}^{2-}$ complexes reduces the center symmetry to C_1(Figure 11b). At room temperature, the Cu^{2+} center has axial symmetry due to center reorientation and motional averaging. The EPR parameters of the impurity Ni^+, Cu^{2+}, and Ni^{3+} in $LiNbO_3$ were theoretically studied from the perturbation formulas for $3d^9$ ions [112,122].

Ni^{2+} ($3d^8$, $S = 1$). Several additional terms, which describe zero field splitting (ZFS) of energy levels, should be added to the spin-Hamiltonian (1) for paramagnetic centers with $S > 1/2$:

$$H_{ZFS} = \sum_{k=2,4,6} f_k \left[\sum_{q=0}^{k} b_k^q O_k^q(\mathbf{S}) + \sum_{q=1}^{k} c_k^q \Omega_k^q(\mathbf{S}) \right] \quad (5)$$

Here $f_2 = 1/3, f_4 = 1/60, f_6 = 1/1260$; O_k^q, $\Omega_k^q(\mathbf{S})$—Stevens operators, which are non-zero for $k \geq 2S$. For C_3 symmetry only q equal to 0, 3, and 6 are allowed. For $S = 1$ only terms with $k = 2$ are present in Equation (5).

The Ni^{2+} centers in LN exhibit the EPR spectra of C_3 symmetry. Therefore, the sum in (5) turns into one term $b_2^0 O_2^0/3$. It was found that $b_2^0 = -5.31$ cm^{-1} and $\Delta g = g_{\parallel} - g_{\perp} = 0.04$ [115,125]. Since EXAFS data supports Ni^{2+} substitution for Li^+ [24], the reasonable choice for the charge compensator is one v_{Li} that is located on the C_3 axis or very far of Ni^{2+}.

A cluster substitution $4Ni^{2+} \Leftrightarrow Ta^{5+} + 3Li^{+}$ was considered for LT [126]. Note that an agreement of measured and calculated spin-Hamiltonian parameters [127] was obtained for Ni^{2+} substitution for Nb^{5+} in LN without a charge compensator.

The charge excess of one interstitial Me^{2+} or two $Me^{2+}{}_{Li}$ could be exactly compensated by an additional O^{2-} ion. However, such a compensation looks unlikely, as the ionic radius of O^{2-} (about 1.4 Å) is larger than the sizes of octahedral or tetrahedral vacancies.

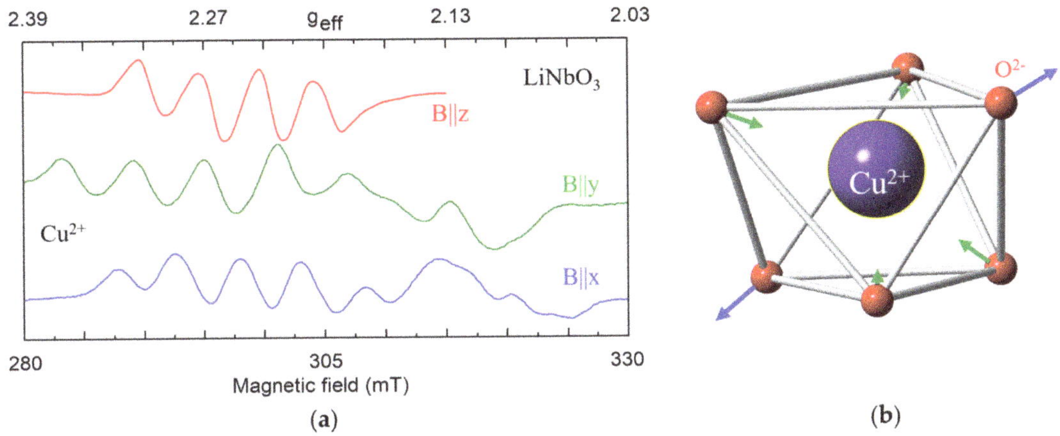

Figure 11. (**a**) Measured EPR spectrum of Cu^{2+} in LN, X-band, T = 7 K. (**b**) Distortion of oxygen octahedron due to Jahn-Teller effect.

Mn^{2+} ($3d^5$, $S = 5/2$, $I = 5/2$). For $S = 5/2$ the ZFS causes splitting into $2S = 5$ components of fine structure, each of them additionally splits into $2I + 1 = 6$ lines due to the hyperfine interaction of Mn^{2+} electrons with their own nucleus (Figure 12).

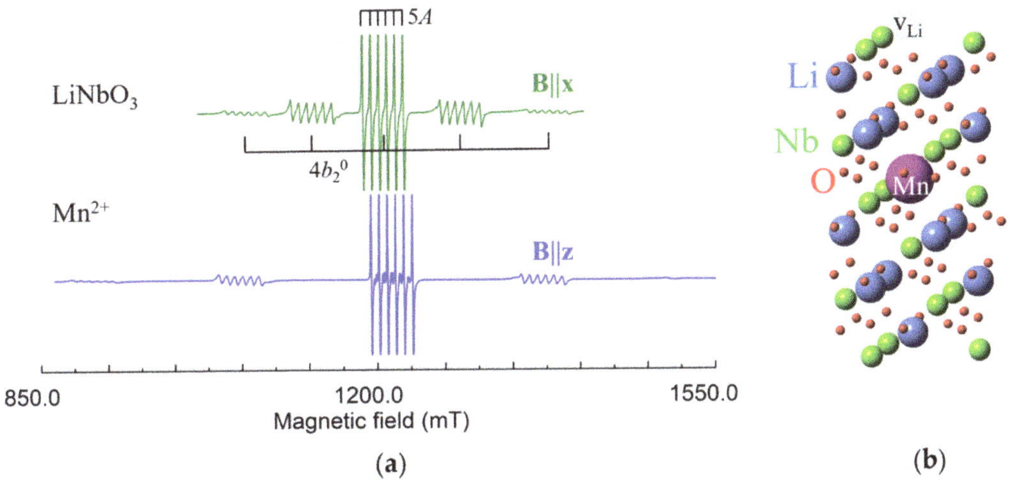

Figure 12. (**a**) Measured EPR spectrum of Mn^{2+} in LN_K, ν = 34.45 GHz. (**b**) Model of axial $Mn^{2+}{}_{Li}$ center with v_{Li}.

Angular dependences of Mn^{2+} in both LN and LT are described by spin-Hamiltonian of axial symmetry with $g \approx 2.0$, and $A \approx -0.008$, $b_2^0 = 0.0731$ (LN), $b_2^0 = 0.1694$ (LT), all in cm^{-1}, [128–138]. Our spectra (Figure 12) were fitted with $b_2^0 = 0.074$ (LN), $b_2^0 = 0.175$ (LT).

Hyperfine interactions with four Li and two Nb shells of surrounding nuclei were determined by ENDOR study in LN [139]. For isotropic **g**-tensor, the principal values of dipole-dipole interaction (3) can be described by:

$$b_{dd}^{(i)} = \frac{g\mu_B g_n^{(i)} \mu_n}{\left(R^{(i)}\right)^3} \tag{6}$$

Comparison of measured values of $b_{dd}^{(i)}$ with values calculated by Equation (6) for Li and Nb substitution has definitely shown that Mn^{2+} ions occupy Li site [139].

4.3. Trivalent Cations

Most transition metals (including iron, titanium, and chromium) and rare-earth elements enter LN in this valence. If Me^{3+} substitutes Li^+ there are three possibilities to compensate its 2+ excess charge: v_{Li}, v_{Nb} and self-compensation with ${Me^{3+}}_{Nb}{}^{5+}$ in the nearest or distant neighborhood. Every ${Me^{3+}}_{Li}$ can be compensated by two v_{Li}; every five ${Me^{3+}}_{Li}{}^{+}$ ions—by two niobium vacancies. The positive antisite defect Nb_{Li} can serve as a charge compensator for Me^{3+} replacing Nb^{5+} (but not ${Me^{3+}}_{Li}{}^{+}$ or ${Me^{3+}}_{v}$). It is remarkable that one ${Nb^{5+}}_{Li}{}^{+}$ exactly compensates the excess charge of two ${Me^{3+}}_{Nb}{}^{5+}$ ions. Since ${v_{Li}}^{+}$ has only one negative charge relative to the ideal lattice, it produces a 4–5 times weaker perturbation of the crystal field than ${Nb^{5+}}_{Li}{}^{+}$ or ${v_{Nb}}^{5+}$. Therefore the centers with ${v_{Li}}^{+}$ at distances of about 6 Å should probably are not distinguishable from axial centers with non-local charge compensation.

Interstitial Li^+ ions should be considered as charge compensators for ${Me^{3+}}_{Nb}{}^{5+}$ in Li-rich, VTE treated, and stoichiometric crystals. The association of ${Me^{3+}}_{Nb}{}^{5+}$ with one Li^+ ion in the nearest vacancy (partial local charge compensation) leads to an axial center, the second Li^+ in the next vacancies can decrease symmetry to C_1, if located off center axis and near the impurity. Mg^{2+} or Zn^{2+} ions substituted for Li^+ can be also suitable compensators for ${Me^{3+}}_{Nb}$.

Distances between the replaced ion and shells of possible location of compensators, various configurations of Me and charge compensators, as well as the symmetries of the corresponding complexes are given in References [87,140–142].

Cr^{3+} ($3d^3$, $S = 3/2$). In congruent and Li-rich LN samples, the EPR lines of dominant axial Cr^{3+} center, characterized with ZFS $b_2^0 \approx 0.39\ cm^{-1}$, are accompanied with small satellite lines (Figure 13). Initial discussion with plausible but contradictory arguments about Li^+ or Nb^{5+} substitution [134,143–148] should be ended after PIXE [32] and detailed ENDOR [140] studies have shown that Cr^{3+} substitutes for Li^+ and slightly shifted from regular Li site. ENDOR measurements confirmed that Cr^{3+} substitutes for Li^+ also in all satellite centers. Therefore, the whole family of these Cr^{3+} centers can be described as ${Cr^{3+}}_{Li}$ with location of charge compensator on C_3 axis for axial or off it for low-symmetry centers.

The ENDOR measurements [140] found that hyperfine interactions with Nb nuclei of the 2nd and 3rd shells (Figure 5a) are stronger than with Li nuclei (Figure 14, bottom). However, lines of Nb nuclei on the center axis (1st and 4th shells) were not identified. It can be caused by unfortunate conditions of their observation, petal distribution of electron density for the $3d^3$ ion or absence of Nb ion in one of these sites, i.e., v_{Nb}. Two v_{Nb} can serves as the charge compensator for five ${Cr^{3+}}_{Li}$. Although the presence of v_{Nb} in undoped LN looks unlikely, the charge compensation defects in doped crystals (especially, if dopant concentrations exceed 0.X%) can differ from dominant intrinsic defects in undoped LN or LT. During the growth process, the required compensators can organize themselves around impurities or enter from air in order to minimize the creation energy for the impurity center. This is why structures with v_{Nb} were proposed for satellite centers of ${Cr^{3+}}_{Li}$ [87].

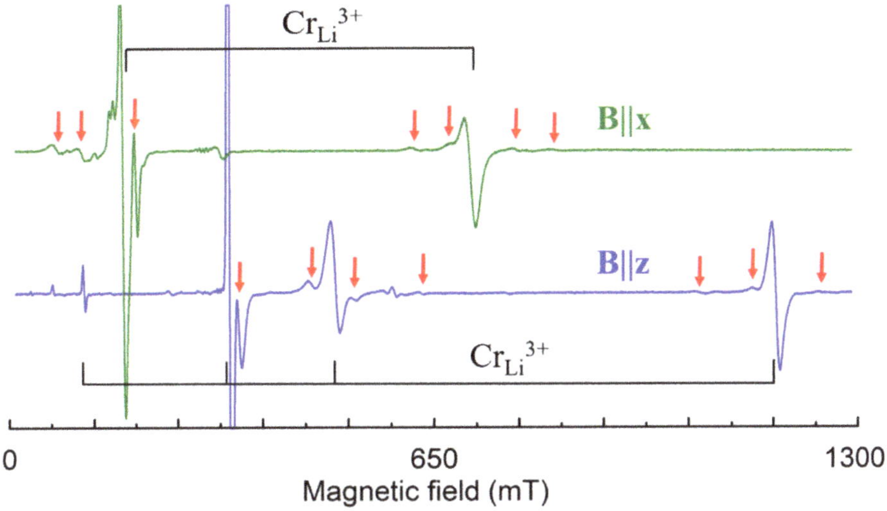

Figure 13. Measured EPR spectrum of Cr^{3+} in LN grown from the melt with $x_m \approx 60\%$ and 0.1 wt.% of Cr (room temperature, ν = 9.84 GHz). Black braces indicate lines of dominant Cr^{3+} center, red arrows—lines of satellite centers.

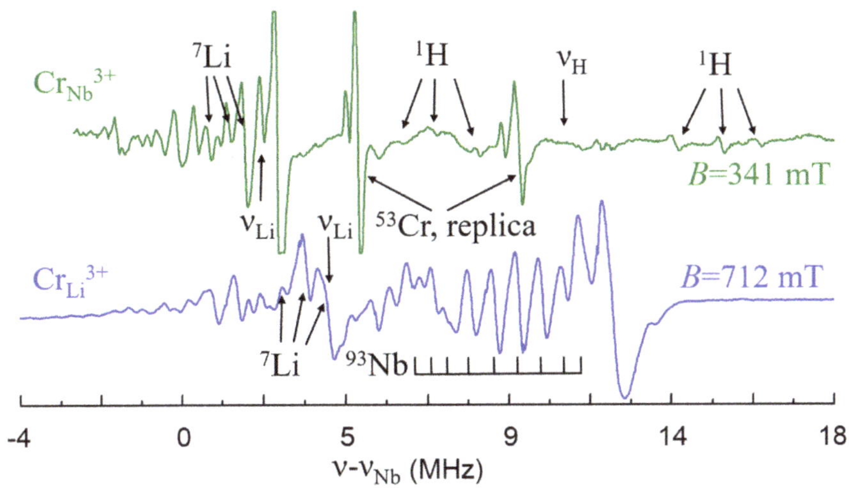

Figure 14. The ENDOR spectra of $Cr^{3+}{}_{Li}$ (high-field EPR transition) and $Cr^{3+}{}_{Nb}$ (central EPR transition) at **B** | | **x**, T = 5 K. To facilitate a comparison of the spectra, they were shifted to the ^{93}Nb Larmor frequencies.

On the other hand, v_{Li} are considered as dominant intrinsic defects in LN and LT. Angular dependencies of EPR spectra for the dominant (C_3) and satellite (C_1 symmetry) Cr^{3+} centers (Figure 15) are pretty similar to observed patterns for Nd^{3+} (Figure 4). Therefore, two vacancy models for trivalent impurities (Table 1) and structures on Figures 7 and 8 can be viable alternatives for Cr^{3+} family.

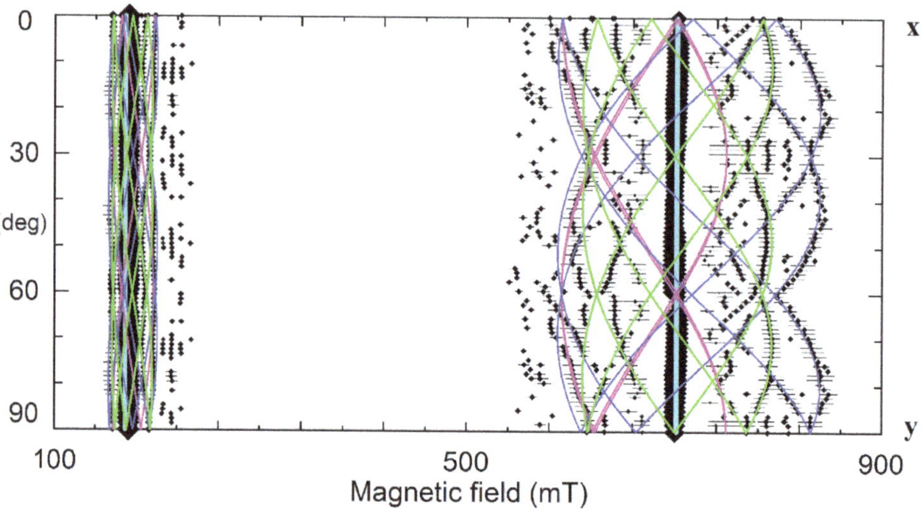

Figure 15. The angular dependencies of EPR spectra in the **xy** plane for LN:Cr^{3+} grown from the melt with $x_m \approx 60\%$, ν = 9.26 GHz. Rhombs represent line positions; their sizes are proportional to line intensities. Horizontal whiskers near rhombs represent line widths. Cyan, lime, fuchsia, and blue curves represent simulated dependencies for axial Cr_1, and low-symmetry Cr_2, Cr_3, and Cr_4 centers, respectively.

A lack of intrinsic defects in stoichiometric samples leads unavoidably to a change of charge compensation mechanism for trivalent impurities, and substitution for Nb^{5+} becomes possible. An axial Cr^{3+} center with significantly smaller ZFS b_2^0 = 0.0215 cm^{-1} (Figure 16a) was found in LN_K samples [142]. ENDOR study has shown that hyperfine interactions with Li nuclei significantly larger than with Nb nuclei for this center, i.e., the nearest surrounding consist of Li nuclei. This means that Cr^{3+} substitutes for Nb^{5+} in this center. As lines of protons, H^+, were found in the ENDOR spectra (Figure 14, top), they compensate the negative charge of ${Cr^{3+}}_{Nb}{}^{5+}$ (Figure 16b).

Exchange interaction $\mathbf{S}^A\mathbf{J}\mathbf{S}^B$ between spins of Cr^{3+} ions ($S^A = S^B = 3/2$) leads to gaps between states with values of total spin $S = S^A + S^B$ equal to 0, 1, 2, and 3. The state with $S = 0$ is non paramagnetic. For pairs at a close distance the gaps can exceed energies of microwave quantum (36 GHz ≈ 1.2 cm^{-1}). It was found by magneto-optical study that for ${Cr^{3+}}_{Li}$–${Cr^{3+}}_{Nb}$ substituted for nearest Li and Nb sites (at the distance about 0.3 nm) the exchange interaction is antiferromagnetic and $J \approx 480$ cm^{-1}. As 1 K × $k_B \approx$ 0.7 cm^{-1} the upper states of such pairs with non-zero S are not populated even at room temperatures, and the pairs are EPR silent. However, the pairs of the next orders with the isotropic exchange coupling parameter $J \approx 1.5$ cm^{-1} were observed by EPR at relatively low concentration of chromium in LN (less than 0.1 at.%) [87,134,148–150].

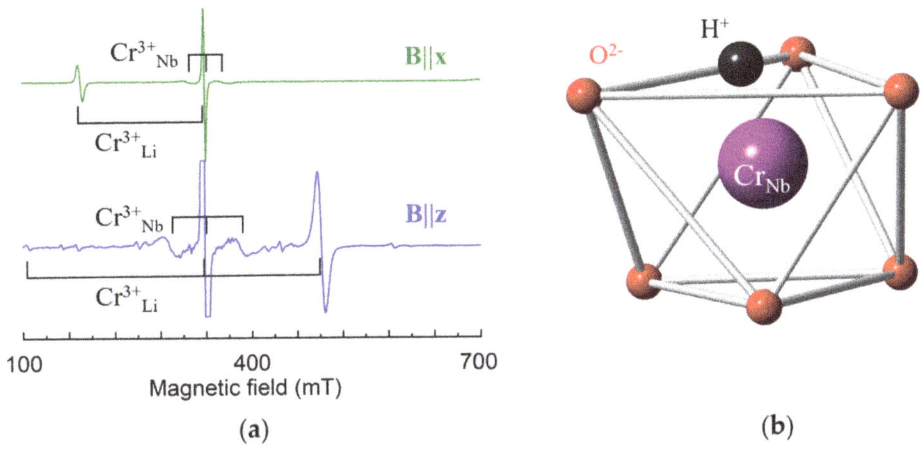

Figure 16. (**a**) Measured EPR spectra of Cr^{3+} in LN grown from the melt with addition of K_2O and 1 wt.% of Cr; room temperature, ν = 9.445 GHz. (**b**) Model of $Cr^{3+}{}_{Nb}$ center.

PIXE/channeling study [32] revealed that chromium ions occupy both regular cation sites (60% on Li sites and 40% on Nb sites) in congruent LN doped with 0.1 mol.% of Cr. This means that the majority of chromium ions enter cLN as non-paramagnetic pair centers. The EPR observes only a top of iceberg in cLN: dominant $Cr^{3+}{}_{Li}$ and small signals of non-nearest paramagnetic pairs.

ENDOR study of Cr^{3+} centers in LN heavily doped with Mg and co-doped with Cr has unambiguously shown that Cr^{3+} in the dominant center has ZFS b_2^0 close to zero (nearly isotropic case) and substitute for Nb [151–155]. Measured anisotropic hyperfine interactions of Cr^{3+} with for four Li shells were close to $b_{dd}^{(i)}$ for Nb site. From comparison of data obtained by EPR, ENDOR, optical absorption, fluorescence, fluorescence line narrowing, selective excitation and radiative lifetime measurements [156–162] it has been concluded that the addition of Mg^{2+} ions to LN does not create new Cr^{3+} complexes, but changes the relative concentrations of the $Cr^{3+}{}_{Li}$ and $Cr^{3+}{}_{Nb}$ centers.

The measured value of ZFS for dominant Cr^{3+} in cLT $b_2^0 \approx 0.444$ cm^{-1} [144] is very close to the value for $Cr^{3+}{}_{Li}$ in LN. EPR spectra of this center together with signals of weaker intensities of a second center [163,164] were explained in a supposition that they originate from Cr^{3+} ions located at Li^+ sites and that two v_{Li} play the role of a divalent charge compensator for both centers. EPR study of Cr^{3+} in nsLT and superposition model analysis [165] are in good agreement with the Cr^{3+} substitution for Li. The temperature dependence of $b_2{}^0$ term showed a non-monotonic behavior in the region of 40 K.

Finally, a lot of studies were devoted to various properties of Cr^{3+} in LN and LT crystals of different compositions in order to clarify relations of optical characteristics with structures of chromium centers [166–175] etc.

Dy^{3+} ($4f^9$). The observed Zeeman splitting [143] was described with g-tensor of axial symmetry: $g_{\|} = 8.7$ and $g_{\perp} = 1.3$. The single EPR line had width about 8–10 mT at **B** | | **z** and became broader at **B**⊥**z**. Such a behavior can be related to unresolved splitting due to satellite centers, if Dy^{3+} occupies Li position and its charge is compensated by v_{Li}. Two Dy^{3+} centers with $g_{xx}(1) = 2.56(1)$, $g_{zz}(1) = 4.43(1)$, and $g_{xx}(2) = 6.67(1)$, $g_{zz}(2) = 1.23(1)$ and linewidth about 20 mT were registered in LN after γ-irradiation [176]. Both centers were attributed to $Dy^{3+}{}_{Li}$. A broad line of the third center with $g_{zz}(3) \approx 1.2$ appeared only at **B** close to the **z**-axis. Weak hyperfine lines due to isotopes ^{161}Dy (natural abundance 19%), ^{163}Dy (2.49%) was observed in single crystal of LN [177].

Er^{3+} ($4f^{11}$). Due to fast spin-lattice relaxation, the EPR signals of Er^{3+} are observable at low-temperatures only. Earlier studies claimed that Er^{3+} ions in cLN create an axial center with $g_{\|} \approx 15.1$–15.4 and $g_{\perp} \approx 2.1$ [143,178] or $g_{zz} \approx 15.5$ and $g_{xx} \approx g_{yy} \approx 0.8$ [179]. A proposed model with lithium vacancies statistically distributed around $Er^{3+}{}_{Li}$ [179] supposed that the center with no vacancies in surrounding, i.e., the center with axial C_3 symmetry, should give a dominant (54%) line in the EPR spectra. Note that angular dependences of EPR spectra in **xy**-plane were not measured in these studies. Later measurements in all three principal planes [180–183] have shown that there is no line without angular dependence in the **xy**-plane, i.e., dominant Er^{3+} center in cLN has C_1 symmetry. This does not agree with statistically distributed v_{Li} around $Er^{3+}{}_{Li}$ [179].

Significant narrowing of EPR lines in LN_K (Figure 17a) allowed us to trace two different $Er_1{}^{3+}$ and $Er_2{}^{3+}$ centers with extrema at about 15, 45 and 75 degrees in **xy** plane (Figure 17b) [184,185]. The divacancy model (Figure 7) gives a possible explanation if a shift of $Er^{3+}{}_{Li}$ from regular Li site is taken into account: RBS, XSW and ion-beam/channeling studies have determined that Er occupies Li sites, but is shifted from the ferroelectric Li position by 0.03 [186], 0.046 [187], and 0.02 nm [188]. In this case, distances from Er_{Li} to the v_{Li} in the 1a and 1b shells (Figures 5a and 7b) are completely different, and these centers have one charge compensating v_{Li} in the nearest neighborhood (the shell 1a for Er_1, and 1b for Er_2), and the second v_{Li} in the next nearest neighborhood (shells 2a, 2b, Figure 7).

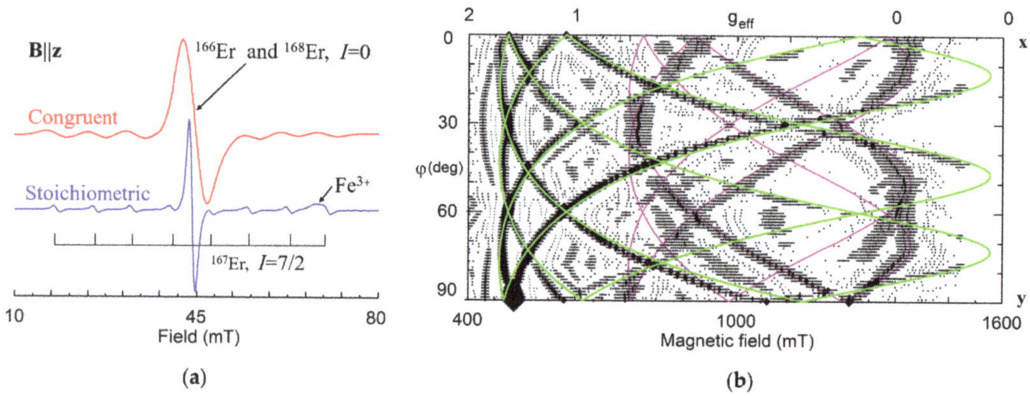

Figure 17. (**a**) EPR spectra of Er^{3+} in congruent and stoichiometric LN at **B** | | **z**, X-band. (**b**) Calculated angular dependence of EPR spectra in **xy** plane for the $Er_1{}^{3+}$ (green lines) and $Er_2{}^{3+}$ (fuchsia lines) centers in sLN. Symbols with horizontal whiskers represent line position and widths.

Magnetic moments $\mu_B g\mathbf{S}$ for both Er^{3+} centers are very large and strongly anisotropic. At low temperatures their interactions lead to magnetic ordering for Er concentration about 0.5 at.% in sLN [189].

Various models with two v_{Li} were also extensively discussed in papers devoted to site-selective spectroscopy [190–196] and references there. Note that models with one of two v_{Li} on **z**-axis (Figure 8) do not agree with the EPR spectra for dominant lines and hyperfine satellites on Figure 17a. However, such centers can probably be associated with weaker lines or may have no EPR lines at all, if they are non-paramagnetic.

EPR spectra detected in cLN heavily doped with Mg or Zn and co-doped with Er were described with $g_{\|} = 4.3$, $g_{\perp} = 7.6$ for the Mg-doped samples and $g_{\|} = 4.26$, $g_{\perp} = 7.8$ for the Zn-doped ones [197]. The spectra can be attributed to Er^{3+} located at the Nb^{5+} site of LN, as they are compared to additional centers observed for some trivalent transition metal ions (particularly Cr^{3+}) in LN: Mg or LN:Zn.

Fe^{3+} ($3d^5$, $S = 5/2$). EPR studies have shown that the dominant iron center, Fe_1 in cLN has axial symmetry with the ZFS b_2^0 about 0.1680 cm^{-1} [131,133,134,198–208]. Calculations of optical and EPR characteristics based on the superposition models or generalized crystal field theory gave a preference for Nb substitution [209], no definite conclusion for Li or Nb substitution [210–213] or some preference for Li site [214,215]. Mössbauer [33,34] and Raman [35] spectroscopy data, Stark effect [216] were interpreted on the base of Fe substitution for Li. Iron clusters were observed at high Fe concentration [33]. Comparison of calculated dipole-dipole interactions of Fe^{3+} electrons with surrounding Li nuclei for Li and Nb substitution (Equation (6)) with measured ENDOR spectra [38–40] has shown that Fe^{3+} is undoubtedly located in Li site in both cLN and cLT (Figure 5a). Independently, XSW [12,28], EXAFS [21–26], and ion beam/channeling [18,27] confirmed Fe^{3+} substitution for Li^+. It was concluded also that Fe^{3+} ions can be shifted from regular Li-site; however, the obtained values for the shift were very different: 0.006 nm [21] or 0.05 nm from octahedron center [24], less than 0.012 nm [25], about 0.01 nm [26], 0.018 $\pm$ 0.007 nm [28], less than 0.005 nm [39]. Fitting observed ENDOR data in LN crystals grown from Li-rich melt ($x_m \approx 55\%$) was obtained with the shift 0.009 nm from Li site [217].

Fe^{3+} center with small ZFS, Fe_2 observed in congruent LN doped with Mg [104,218–222], In [223], and Zn [224] was assigned to $Fe^{3+}{}_{Nb}$.

Comparison of EPR spectra of congruent (Figure 18a, 1), near-stoichiometric grown from Li enriched melts (Figure 18a, 2) and grown from congruent melt with the addition of K_2O, LN_K samples (Figure 18a, 3) shows[a]:

- Lines of allowed transitions in LN_K become symmetric—intensities of left (up) and right (down) wings become equal.
- Lines of forbidden transitions (see yellow box on Figure 18) disappear in LN_K.

In LN_K the EPR lines become narrower (up to dozen times at some magnetic field orientation). The narrowing strongly increases spectral resolution. This allows to register even trace impurities in undoped (nominally pure) LN_K samples (see lines of Fe^{3+} and Mn^{2+} on Figure 18a, 3).

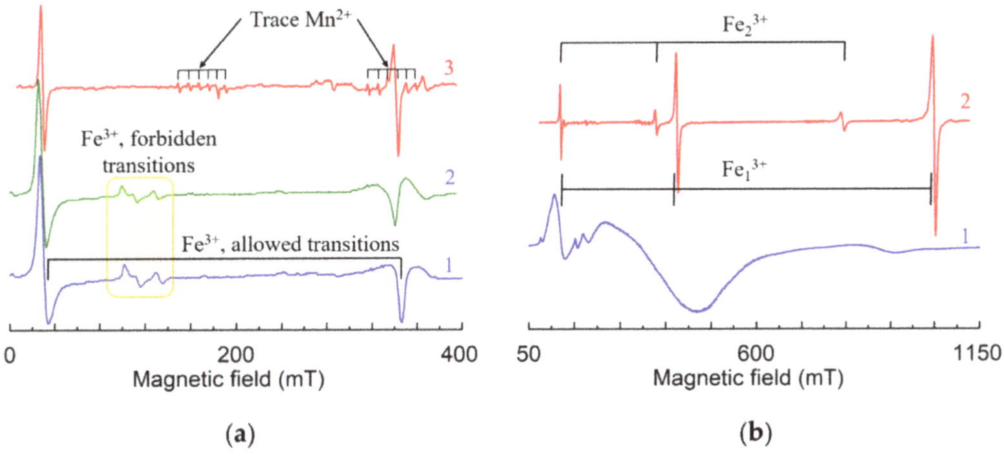

Figure 18. Normalized spectra of Fe^{3+} in LN (**a**) and LT (**b**), X-band. (**a**) Congruent (1), grown with x_m = 0.545 (2), undoped sample grown from congruent the melt with 6 wt% of K_2O (3), **B** | | **z**. (**b**) Congruent (1), vapor transport equilibrium (VTE) treated sample (2), **B** | | **x**.

[a] A brief glance on history of sLN for curios researchers. Looking for better crystals, G. I. Malovichko asked crystal growers for different samples of LN and LT. Dr. E. P. Kokanyan (at that time, one of the engineers in the group of Dr. V. T. Gabrielyan) was interested to grow crystals under various conditions (melt composition x_m, growth under applied electric field, variation of temperature, etc.). He has prepared for her a set of non-stoichiometric LN samples grown from melts with x_m from 43% to 60%. During her PhD research of LN and LT of different compositions, Malovichko found that EPR lines become narrower and more symmetric in samples grown with larger x_m. Dr. V. G. Grachev has simulated

EPR spectra with random distributions of non-axial components of ZFS, b_2^q ($q \neq 0$) for Cr^{3+} and Fe^{3+} and confirmed that line width and asymmetry, as well as intensities of forbidden transitions are related to intrinsic defects in non-stoichiometric samples [43]. In 1991, on Malovichko's request Dr. Kokanyan has grown several congruent samples with 2, 4, and 6% of K_2O in the melt (LN_K). First EPR measurement has surprisingly shown narrow symmetrical lines in LN_K with traces of Fe^{3+} (Figure 18a, 3). Based on the experience of studies of Li-rich samples, it was concluded that the concentration of intrinsic defects in LN_K is smaller than in LN grown from the melt with x_m = 60%, and that K_2O may serve as a catalyst in electrochemical reaction of crystal growth [56]. As abilities of research techniques in Ukraine were limited, Dr. Malovichko asked Prof. O. F. Schirmer (Osnabrueck University, Germany) for an international collaboration. Prof. Schirmer was very enthusiastic and has quickly managed to involve many of his colleagues in the investigation of LN_K properties. K. Betzler, B. Faust, B. Gather, F. Jermann, S. Klauer, U. Schlarb, M. Wesselmann, M. Woehlecke and others participated in the LN_K study by different techniques resulting in publications [57,69,77].

All these features are the result of significant reduction of intrinsic defects in LN_K samples. Estimations of crystal composition x_C made by different methods [59,68,77,79] have shown that x_C can exceed 49.8–49.9%, i.e., LN_K is really a stoichiometric crystal.

As for other impurities, the lack of intrinsic defects in LN_K:Fe has led to the appearance of centers where impurities substitutes for Nb. Two additional axial Fe^{3+} centers named Fe_3 and Fe_4 were observed in LN_K [69]. Their ZFS are b_2^0 = 0.0495 and b_2^0 = 0.0688 cm^{-1}. Compared with Fe_2, the Fe_3 and Fe_4 centers were also assigned to Fe^{3+} in Nb sites with different charge compensation. The Fe^{3+} center with b_2^0 = 0.0656 cm^{-1} (that is very close to b_2^0 for Fe_4) was observed in VTE treated stoichiometric LN [76]. Therefore, Fe_4, Fe_3, and Fe_2 were attributed to Fe_{Nb} with different charge compensator ions in nearest structural vacancy v_{oct}: Fe_4—non-regular, interstitial $Li^+{}_v$ (Figure 19a), Fe_3–$K^+{}_v$, and Fe_2–$Mg^{2+}{}_v$. Two $Li^+{}_v$ (Figure 19b), or interstitial $Mg^{2+}{}_v$ or two $Mg^{2+}{}_{Li}{}^+$ are required for full charge compensation of $Fe^{3+}{}_{Nb}{}^{5+}$. Our measurements of EPR in LN:Mg show angular patterns with extrema at 15, 45 and 75 degrees in **xy** plane. It is why we think that models with two differently located $Mg^{2+}{}_{Li}{}^+$ (Figure 19c,d) are more suitable for Fe_{Nb} (similar to models with two v_{Li} for Fe_{Li}). ENDOR measurements could confirm some of these reasonable assignments.

Dominant $Fe_1{}^{3+}$ center in cLT has b_2^0 = 0.33 cm^{-1} [38,225] (Figure 18b, 1), and according to ENDOR data [38] it is definitely $Fe^{3+}{}_{Li}$. Line narrowing in nsLT grown by double crucible Czochralski method from an Li rich melt composition (about 60 mol.% Li_2O) allowed to determined $b_2{}^0$ more accurately [226].

The main features of the spectra in sLT obtained by VTE treatment are strong line narrowing, higher resolution, and a clear presence of the second axial $Fe_2{}^{3+}$ center with b_2^0 = 0.205 cm^{-1} (Figure 18b, 2) [75]. Comparison of calculated angular dependences of ENDOR frequencies for Li and Nb substitution using Equation (6) with measured ones (Figure 20) obviously shows that in the case of Fe_2 center the Fe^{3+} ion substitutes for Nb. Both Fe_1 and Fe_2 centers in sLT have C_3 symmetry. No foreign nuclei in the nearest neighborhood were detected for Fe_1. As sLT still have some residual concentration of v_{Li}, the charge compensators for Fe_1 centers are one v_{Li} on center axis (shells 5a, 5b on Figure 7) and probably one other distant v_{Li} (any vacancy cannot be directly detected by ENDOR).

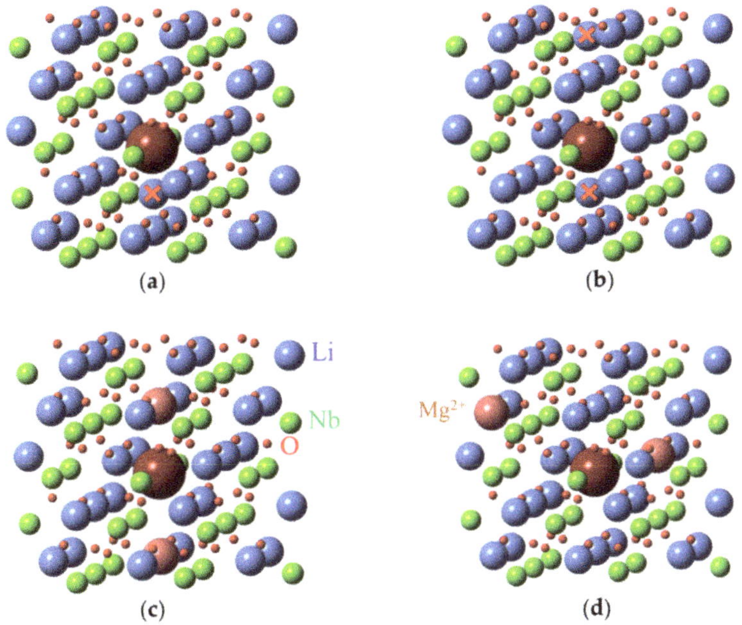

Figure 19. Possible structures of $Me^{3+}{}_{Nb}{}^{5+}$ (large magenta ball) centers in LN. Structure with one (**a**) and two (**b**) additional Li ions in v_{oct} (marked with red crosses). Axial (**c**) and low-symmetry (**d**) structures with two $Mg^{2+}{}_{Li}{}^{+}$ ions.

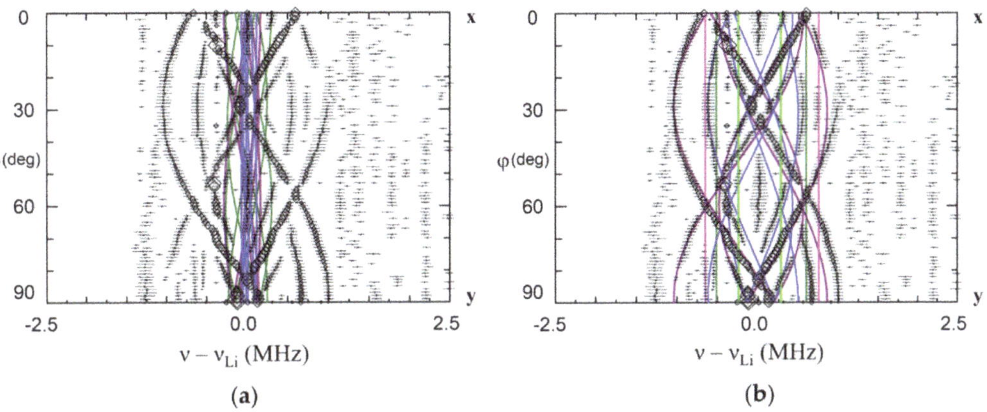

Figure 20. Angular dependence of ENDOR frequencies of Li nuclei in **xy** plane for the $Fe_2{}^{3+}$ center in $LiTaO_3$. The green, magenta, blue and lime curves—calculated frequencies for Li nuclei of the 1st, 2nd, 3rd and 4th shells based on dipole-dipole interaction for $Fe^{3+}{}_{Li}$ (**a**), and $Fe^{3+}{}_{Nb}$ (**b**). Fuchsia lines on (**b**)—calculated frequencies for the additional Li nucleus in the nearest v_{oct} on the axis of $Fe^{3+}{}_{Nb}$ center. ENDOR frequencies represented by rhombs were measured on the EPR line at B = 1162 mT ($\nu \approx 34.5$ GHz).

A pair of ENDOR lines for Fe_2 (fuchsia line on Figure 20b) were attributed to the additional $Li^{+}{}_{v}$ in the nearest v_{oct} at the distance about 0.277 nm from $Fe^{3+}{}_{Nb}$. It is difficult to make a choice between two models on Figure 19a,b as lines of the second $Li^{+}{}_{i}$ in the next or next-next v_{oct} are not identified yet. The ratio of concentrations of Fe_2 and Fe_1 centers changed from less than 0.2 for Fe concentrations 1.1×10^{19} cm^{-3} to about 1 for 6.7×10^{19} cm^{-3} in the crystals. Therefore, there are three different mechanisms for excessive charge compensation: distant v_{Li} for small concentrations of $Fe^{3+}{}_{Li}{}^{+}$, partial self-compensation of charges of $Fe^{3+}{}_{Li}{}^{+}$ by $Fe^{3+}{}_{Nb}{}^{5+}$ and partial compensation of $Fe^{3+}{}_{Nb}{}^{5+}$ by Li^+ in v_{oct} for Fe^{3+} concentrations, which exceed the concentration of v_{Li}.

Gd^{3+} ($4f^7$, $S = 7/2$). The EPR spectrum for every Gd^{3+} center consists of $2S + 1 = 8$ strong lines of fine structure and many low intensity lines of so called forbidden transitions due to nearly equal values of Zeeman splitting and ZFS. Results of the first EPR study [227], were interpreted as a presence of two axial Gd_1 and Gd_2 centers with $b_2^0 = 0.118$ and $b_2^0 = 0.126\ cm^{-1}$. From the viewpoint of charge compensation, a preference was given to Gd^{3+} substitution for Nb^{5+}. A small rhombic distortion ($b_2^2 \approx 0.004$) determined for Gd_2 [228] was attributed to an off axis charge compensator, while for Gd_1 the charge compensator may either be absent or must lie on the same threefold axis as Gd^{3+}. However, no decision on whether Gd^{3+} substitutes for Li^+ or Nb^{5+} or for both was made. Similar results with slightly smaller $b_2^2 \approx 0.002$ were also reported [229].

At least four different Gd^{3+} centers were identified in our study of Li-rich LN doped with 1 wt.% Gd_2O_3 in the melt. As patterns for every EPR transition in **xy**-plane (Figure 21) are very similar to presented on Figures 4b and 15, the divacancy model is suitable for Gd^{3+} in LN. Gd^{3+} substitutes for Li^+ in all centers. The dominant axial Gd_1 center and low-symmetry Gd_2, Gd_3, and Gd_4 centers have v_{Li} in positions described in Table 1. This assignment agrees with the Gd substitution for Li found by RBS and PIXE channeling [31].

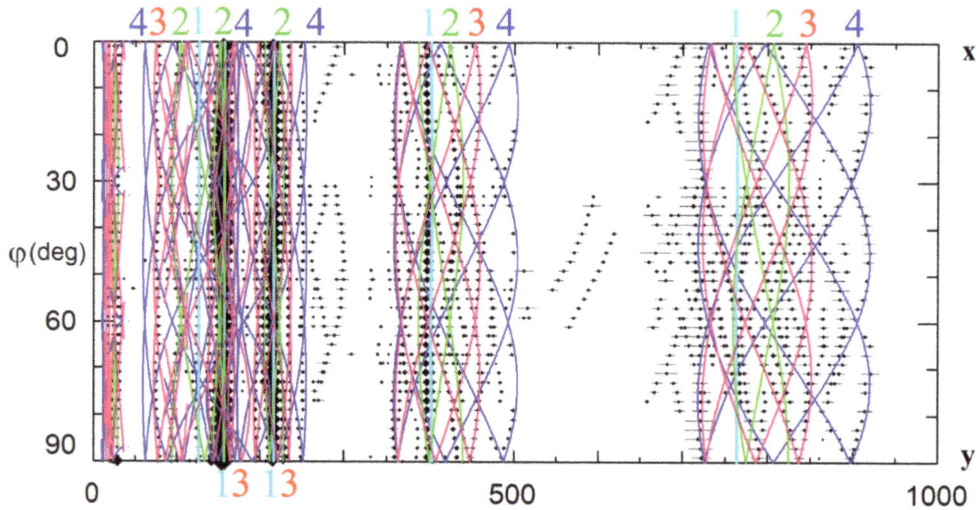

Figure 21. The angular dependencies of EPR spectra in **xy** plane for Li-rich LN:Gd^{3+}, $\nu = 9.26$ GHz. Rhombs represent line positions; horizontal whiskers—line widths. Cyan, lime, fuchsia, and blue curves are simulated dependencies for axial Gd_1, and low-symmetry Gd_2, Gd_3, and Gd_4 centers, respectively.

Nd^{3+} ($4f^3$) First EPR spectra have shown that in cLN Nd^{3+} creates an axial center with $g_{\|} \approx 1.43$ and $g_{\perp} \approx 2.95$ (Nd_1) [143,144,230] and second center with $g_{\|} \approx 1.33$ and $g_{\perp} \approx 2.95$ (Nd_2) [143]. EPR/ENDOR studies in sLN [91–93] has established that Nd^{3+} substitutes for Li^+, and resolved eight different Nd^{3+} centers. Divacancy models explain angular dependencies of EPR spectra for the whole family of Nd^{3+} centers [93], and Section 3.1 of this paper.

It is supposed that Nd^{3+} center found in LN:Mg and LN:Zn belong to Nd^{3+} substituted for Nb [231].

Ti^{3+} ($3d^1$, $S = 1/2$) EPR spectrum of Ti^{3+} consist of one line with $g_{\|} \approx 1.961$ and $g_{\perp} \approx 1.840$ in LN [232–234], and $g_{\|} = 1.948$, and $g_{\perp} = 1.827$ in reduced LT [235]. An axial

EPR signal observed in vacuum annealed $LiNbO_3$ single crystals doped with 8 mol.% Mg and 0.05 mol.% Ti has $g_{\|} = 1.760$ and $g_{\perp} = 1.786$ for $T = 5$ K and $g_{\|} = g_{\perp} = 1.893$ for $T = 74$ K. The signal has been attributed to Ti^{3+} on Nb site [236–240]. The **g** tensor components of these centers were explained by a model calculation involving a dynamic pseudo Jahn-Teller effect. Spin-orbit coupling, lattice vibration, pseudo Jahn-Teller interaction, and the Zeeman term were treated on equal footing. Electron transfer from the observed $Ti^{3+}{}_{Nb}$ center to lattice niobiums, resulting in Nb^{4+} trapped polarons, has been stimulated by illumination in the near UV region.

Yb^{3+} ($4f^{13}$) The EPR lines of Yb^{3+} in congruent LN are very broad (Figure 22a,c) [241]. The axial center Yb_1 with $g_{\|} \approx 4.7$–4.86 and $g_{\perp} \approx 2.7$ was observed in cLN with 0.5–1.2 wt.% Yb_2O_3 in the melt [144,230,241]. The second $Yb_2{}^{3+}$ center with $g_{\|} \approx 1.9$ and $g_{\perp} \approx 2.8$ was found in LN:Mg [241]. By comparison of these observations with EPR data for Cr^{3+}, Er^{3+}, Fe^{3+}, and Ti^{3+} in LN and LN:Mg the Yb_1 center was tentatively assigned to $Yb^{3+}{}_{Li}$ and Yb_2 center to $Yb^{3+}{}_{Nb}$. A variation of lattice parameters found in cLN:Yb supported Yb^{3+}substitution for Li compensated by Li^{+} vacancy [242].

Complicated EPR spectra without 60 deg repetitions in **xy** plane were registered in cLN doped with 1wt.% Yb, and cLN doubly doped with 0.8 wt.% of Yb and 0.1 wt.% of Pr [243–247]. Some of the spectra were attributed to Yb^{3+} pairs with the parameter of isotropic exchange interaction $J = -0.0283$ cm^{-1}.

Our study of LN_K crystals doped with 0.02 wt.% Yb_2O_3 has shown that the broad line observed in cLN belongs in reality to a family of at least 8 different centers (Figure 22b,d) [92,248,249]. Hyperfine structures from isotopes ^{171}Yb ($I = 1/2$, natural abundance 14.4%) and 173 ($I = 5/2$, 16.6%) that are barely distinguished in cLN, were well resolved in LN_K (see Figures 22 and 23). The dominant Yb_1 line that represents even isotopes with $I = 0$ (^{even}Yb, 69%) was described with $g_{\|} = 4.46$, $g_{\perp} = 2.706$. Intensities of the Yb^{3+} lines were proportional to natural abundances of isotopes and their spins: $\mathcal{I}$(even):$\mathcal{I}$(171):$\mathcal{I}$(173) = 0.69:0.144/2:0.166/6 = 0.69:0.072:0.028. It was found for Yb_1 that $^{171}A_{\|} = 0.119$ cm^{-1}, and $^{171}A_{\perp} = 0.0715$ cm^{-1}. Branches of angular dependencies of the Yb_5 center (fuchsia lines on Figure 22) were described with spin-Hamiltonian for $S = 1$ and anisotropic $\mathbf{S}^A\mathbf{J}\mathbf{S}^B$ interaction ($J_{ik} \approx 0.012$–0.066 cm^{-1}). They were assigned to low-symmetry Yb^{3+}–Yb^{3+} pairs. The presence of additional lines of hyperfine structure with similar angular dependence (indicated by red arrows in Figure 23) supports the assignment, as their intensities are proportional to probabilities to meet two ^{even}Yb, or one ^{even}Yb and one ^{171}Yb, or one ^{even}Yb and one ^{173}Yb in such pairs. A self-compensated pair consisted of $Me^{3+}{}_{Li}$ and $Me^{3+}{}_{Nb}$ in the nearest sites (Nb shell 1 at a distance 0.3 nm on the crystal axis, Figure 5a) creates an axial center with rather strong exchange interaction (Jiso > 300 cm^{-1}). Therefore, the observed low-symmetry pairs were attributed to the Yb^{3+}–Yb^{3+} ions in next neighbor or next-next neighbor positions (shells 2 and 3 on Figure 5a).

One of the satellite centers (Yb_6) has axial symmetry; all others are low-symmetry centers. ENDOR measurements has revealed that Yb^{3+} substitutes for Li: at the first, comparison of measured angular dependencies with calculated ones on the base of dipole-dipole interactions (Equation (3)) gave undisputable preference for Li site, and at the second, the strongest axial hyperfine interaction was found for ^{93}Nb on the **z**-axis (the first shell on Figure 5a). The most reasonable explanation for the existence of the whole family of ytterbium centers is that $Yb^{3+}{}_{Li}$ is compensated by one or two v_{Li} in different configurations.

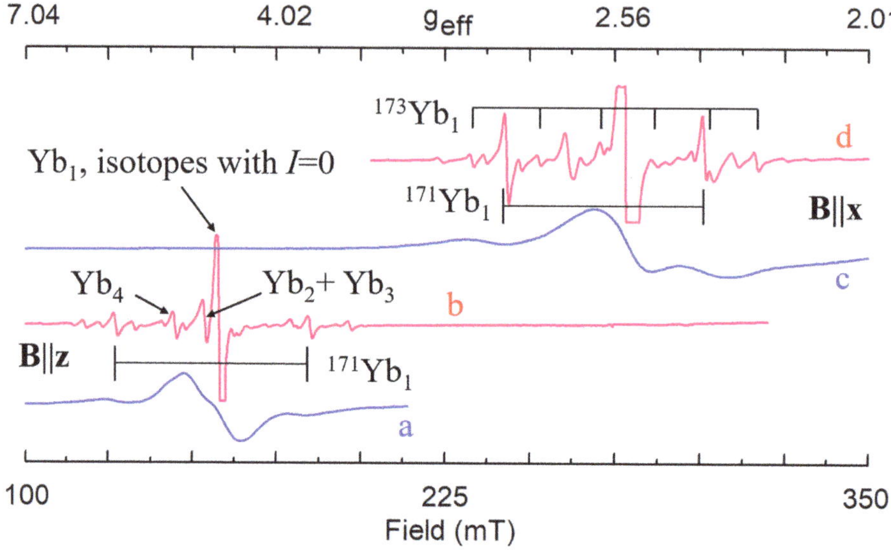

Figure 22. EPR spectra of Yb^{3+} in congruent (a, c) and stoichiometric (b, d) LN.

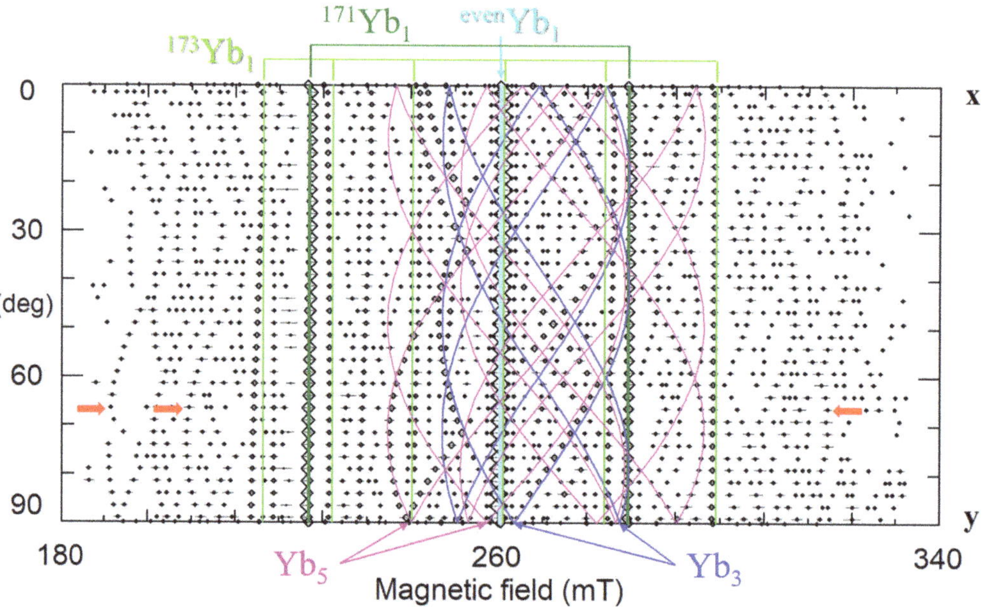

Figure 23. Angular dependence of EPR spectra of Yb^{3+} in stoichiometric LN, **xy** plane, ν = 9.86 GHz. Red arrows—lines of hyperfine structure for Yb_5 centers.

4.4. Tetra-, Penta- and Heptavalent Cations

The non-paramagnetic tetravalent impurities (C and Si) are always present in LN in rather high concentrations (about 50–500 ppm). Since determined concentrations of chlorine Cl (50–500 ppm) and manganese Mg (1–100 ppm) have the same order of magnitudes, no other charge compensators are necessary if C or Si substitutes for Nb creating $C^{4+}{}_{Nb}$ – $Cl^{-}{}_{O}{}^{2-}$ or $C^{4+}{}_{Nb}$ – $Mg^{2+}{}_{Li}$. The additional possibilities for the charge compensation supply H^{+} and $Li^{+}{}_{V}$ ions. It is supposed that Ti^{4+} substitutes for Nb^{5+}; however, at present a mechanism of its charge compensation is not well established.

Due to the lithium deficiency of congruent lithium niobate crystals, v_{Li} have been considered as possible charge compensators for $Nb^{5+}{}_{Li}{}^{+}$ or $Ta^{5+}{}_{Li}{}^{+}$ antisites for a long time. Models with plane and space configurations of v_{Li} for non-stoichiometric defects were proposed [96,97] and used for calculations of stability of intrinsic defects and defect clusters in LN [250–253] and LN:Mg [254]. The antisites become paramagnetic ions $Nb^{4+}{}_{Li}{}^{+}$ or $Ta^{4+}{}_{Li}{}^{+}$ after irradiation (see Section 4.5 below).

The Ta^{5+} substitution for Nb^{5+} in LN causes minor lattice distortions only. The heptavalent ions (Mo, W) probably substitute for Nb^{5+} having lithium vacancies as charge compensators.

U^{5+} ($5f^1$). EPR spectra of U^{5+} were studied in $LiNbO_3$ powders doped with natural U_3O_8 and $^{233}U_3O_8$. A hyperfine sextet of EPR line for ^{233}U (I = 5/2) was described with $A_{\|}$ = 0.0145 and $A_{\perp}$ = 0.0128 cm^{-1}. $g_{\|}$ = 0.71, $g_{\perp}$ = 0.724 were determined for the line of natural U^{5+} (even isotopes with I = 0 have total natural abundance about 99%). It was observed that U^{5+} takes part in photoinduced valence change which is the basic mechanism for photorefraction.

4.5. Radiation and Reduction Defects

A radiation usually recharges of regular lattice, interstitial and impurity ions or produces interstitial ions. Two kinds of recharged defects were observed in LN: electron traps like Nb^{4+} and hole traps like O^{-} ions [236,255,256].

Nb^{4+} ($4d^1$). Hyperfine interaction of the unpaired $4d^1$ electron with the ^{93}Nb nuclear spin I = 9/2 splits its EPR line into ten components. The ten-line EPR spectrum ($g_{\|}$ = 1.90 and $g_{\perp}$ = 1.72) has been described for congruent $LiNbO_3$ after ionizing radiation [257]. Later this spectrum has been reproduced in vacuum-reduced and UV bleached crystals [258–261] (Figure 24, 1) and ascribed to antisite Nb [255]). It is remarkable that at least part of the Nb^{4+} centers has C_1 symmetry [262], although the main possible positions for Nb (regular Nb and Li site, v_{oct}) have C_3 symmetry. It means that compensating defects for Nb^{4+}, most probably lithium vacancies, are located in the nearest neighborhood (Figure 25a,b).

Similar spectra were observed in $LiNbO_3$ doped with 6 mol.% Mg (Figure 24, 2) after X-irradiation or vacuum reduction treatments and were related to Nb^{4+} centers on niobium sites [104,113,114], obviously with nearby defects. The Nb^{4+} center in $LiNbO_3$ doped with 10 mol.% Mg belongs to Nb in regular position, but with $Mg^{2+}{}_{Li}$ in neighborhood (Figure 24, 3) [238,263]. The neighboring Mg^{2+} ion redistributes a cloud of Nb^{4+} electrons and changes the hyperfine splitting.

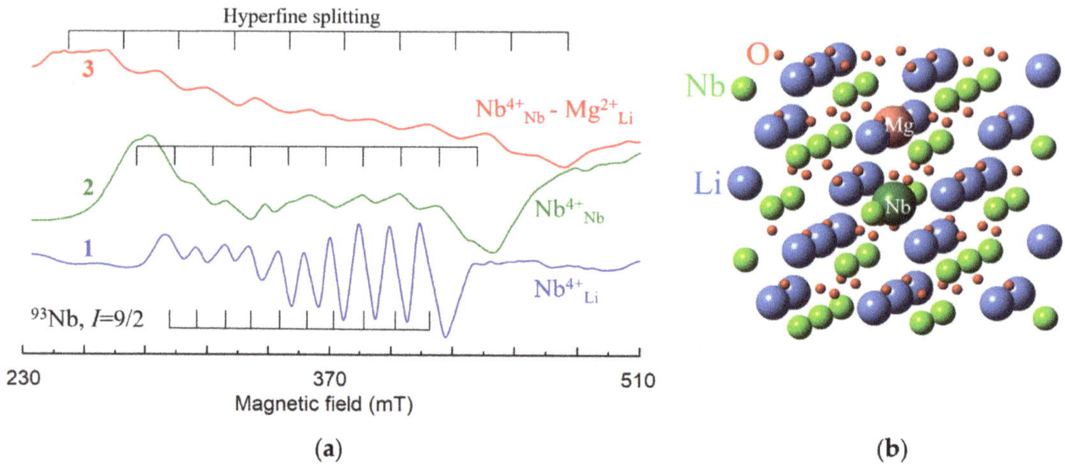

Figure 24. (**a**) EPR spectra of Nb^{4+} centers in cLN (1) and cLN doped with 6 mol.% Mg (2) and 10 mol.% Mg (3), X band. (**b**) Model for axial $Nb^{4+}{}_{Nb} - Mg^{2+}{}_{Li}$ center.

Ta^{4+} ($5d^1$). The isotope ^{181}Ta has $I = 7/2$ and 100% natural abundance. The eight-line axial EPR spectrum with $g_{\|} = 1.503$ and $g_{\perp} = 1.172$, $A_{\|} = 0.0023$, and $A_{\perp} = 0.0234$ cm^{-1} has been observed in $LiTaO_3$ after reduction in argon and attributed to axial $Ta^{4+}{}_{Li}$ [257]. Other ways to obtain Ta^{4+} are an irradiation of as-grown crystals with X-rays or optical bleaching of crystals that had been previously reduced [264]. A possible model for axial Ta^{4+} center is similar to the presented on Figure 25a.

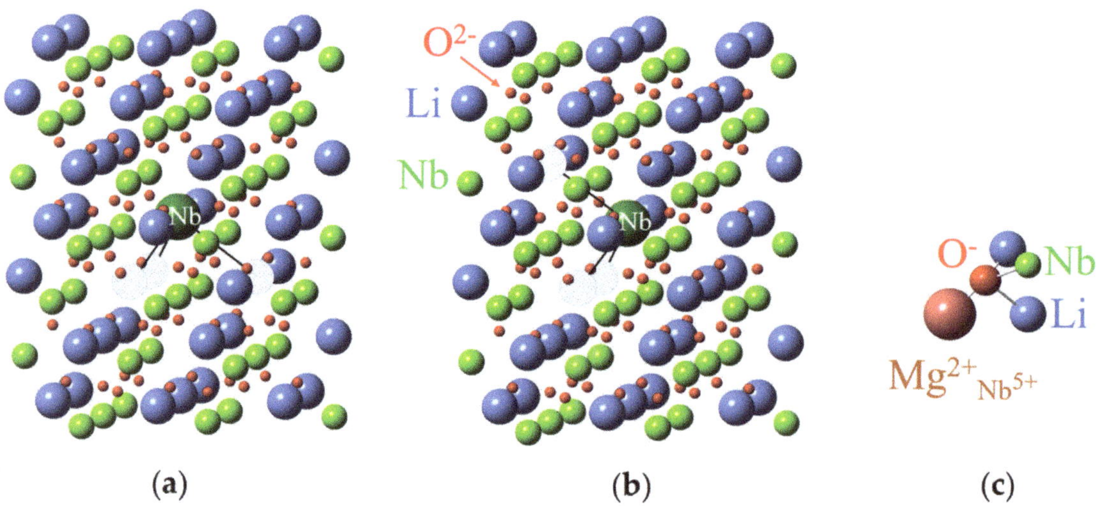

Figure 25. Models for $Nb^{4+}{}_{Li}$ (large green ball) centers with three v_{Li} in LN. Hatched circles represent lithium vacancies. (**a**) Axial center. (**b**) Low-symmetry center. (**c**) The center with O^- near Mg^{2+} substituted for Nb^{5+}.

Tb^{4+} ($4f^7$). EPR study at 15 K revealed a signal of $g \approx 2.0$ appearing after UV irradiation with a simultaneous decrease in the Fe^{3+} signal intensity in near-stoichiometric $LiNbO_3$:Tb and $LiNbO_3$:Tb:Fe [265]. This implies that the Fe^{3+} ions act as electron traps. Irradiation by UV light induced an absorption band extending from $\lambda \approx 650$ nm to the absorption edge caused by the charge transfer from UV-sensitive absorption centers to Fe^{3+} ions via the conduction band.

Other electron and hole traps. A trapped-hole center with $S = 1/2$ was produced in $LiNbO_3$ by ionizing radiation [266]. Its ESR spectrum contains at least 26 equally spaced lines with 1.54 mT separation at $\mathbf{B} \| \mathbf{z}$. This hyperfine pattern was explained as one "hole" interacting equally with three ^{93}Nb nuclei ($I = 9/2$ and 100% abundant). The hole is equally shared by three equivalent oxygen ions adjacent to a cation vacancy. A center with 25 lines of hyperfine interaction with two ^{7}Li and two ^{93}Nb nuclei was ascribed to O^- in regular O^{2-} site with unclear stabilizing factor for the hole trap [236].

An OH^{2-} ion was identified in undoped and weakly doped with Mg LN samples after γ-irradiation and subsequent partial UV bleaching [109]. Specific hyperfine structure with ^{93}Nb was observed for a hole trap in LN doped with 6–8 mol.% Mg. The center was attributed to O^- near Mg^{2+} ion substituted for Nb^{5+} (Figure 25c) [263].

5. Impurity Identification

At an investigation of a sample grown from a melt with an addition of a paramagnetic spice in concentration 0.00X–0.X at.%, which warrants that EPR signal significantly exceeds noise, it would be reasonable to expect EPR signal from this spice. Often, studied samples (especially, commercial ones) contain so-called non-controlled or trace impurities like Cr, Mn, Fe, Cu etc. in a slightly smaller or comparable concentration. Fortunately, most paramagnetic impurities are already investigated by EPR/ENDOR techniques and their characteristics—electron and nuclear spins, zero-field splitting, hyperfine and quadrupole interactions, kinds of charge compensators and their locations, i.e., "passports" of the impurities—are known. Comparison of published and observed spectra (see, for instance, figures with EPR spectra and their angular dependencies above) allows to identify the impurities in films, epitaxial layers, and fibers, or to evaluate profile of impurity distribution in bulk samples. Spectra of LN/LT powders and ceramics can be simulated using determined spin-Hamiltonian parameters.

6. Conclusions

Very detailed information about structures of impurity defects (charge state, point symmetry, hyperfine interactions with neighbor nuclei, charge compensation mechanisms etc.) was obtained with the help of EPR and ENDOR.

The necessity of a charge compensation for non-isovalent substitution usually leads to the creation of families of electrically and magnetically non-equivalent impurity centers. The families of satellite centers exist due to the different relative locations of the impurity ion and its charge compensator. Two or more different centers were observed for Co^{2+}, Cr^{3+}, Cu^{2+}, Er^{3+}, Fe^{3+}, Gd^{3+}, Mn^{2+}, Nd^{3+}, Yb^{3+}, and other ions. Since the relative concentrations of satellite centers are comparable with the concentration of the main center, both kinds of centers generally are equally responsible for many of the properties of LN/LT crystals, and they should both be taken into consideration, especially in non-stoichiometric crystals.

Several different mechanisms for a compensation of excessive charge of Me impurity were found:

- the compensation by nonstoichiometric defects (v_{Li}, v_{Nb}) for small concentrations of Me_{Li} in crystals,
- the partial self-compensation of charges of Me_{Li} by Me_{Nb} mainly for the Me_{Li} concentration that exceeds the concentration of v_{Li},
- the partial compensation of Me_{Nb} by interstitial Li^+ in v_{oct},
- the compensation by other impurities (H, co-dopants like Mg, Zn, etc.).

The presence of non-stoichiometric defects is one of the reasons why LN tolerates a strong incorporation of dopants non-isovalent to Li^+ and Nb^{5+}. As long as the impurity concentration [Me] is smaller than $\delta x_C = |50\% - x_C|$, the number of intrinsic defects is large enough to compensate the corresponding charge excess. However, for stoichiometric or nearly stoichiometric samples with high impurity concentrations (when [Me] > δx_C) and with the lack of charge compensators a decrease of the distribution coefficient of impurities is observed in comparison with congruent material. A further increase of the [Me]/δx_C

ratio up to [Me] >> δx_C can result in a change of the charge compensation mechanism. This can reveal itself in the appearance of new impurity centers.

The charge compensation by v_{Li} works well for most of the non-isovalent ions. Vacancies in the nearest neighborhood of Me_{Li} decrease symmetry of centers from C_3 to C_1, whereas distant vacancies cause spectral line broadening. For $Me^{3+}{}_{Nb/Ta}$ in stoichiometric LN/LT the charge compensation by interstitial H^+ and $Li^+{}_v$ ions was found. The interstitial $Li^+{}_v$ should be considered as a concurrent charge compensation mechanism in VTE treated samples. The compensation of Me_{Nb} by Mg_{Li}, Zn_{Li} and other impurities occurs when the co-dopant concentration exceeds some threshold that depends on δx_C. Typically, the threshold is about 6–8% for congruent samples, but it is lower than 1% for nearly stoichiometric samples.

The use of stoichiometric or nearly stoichiometric crystals with $\delta x_C \approx 0$ presents many advantages for the investigation of impurity centers by spectroscopic techniques. The decreased concentration of intrinsic defects causes a tremendous narrowing of the spectral lines. This is accompanied with the increase of spectral resolution and sensitivity, facilitates the analysis of the spectra, and simplifies the interpretation of the data. However, together with the disappearance of intrinsic defects the satellite centers disappear also. For a detailed investigation of such additional centers, the crystals with high $x_C \approx 49.5$–49.85% are more suitable than others: they have more narrow spectral lines than congruent samples, but the satellite centers are still present. Further study of nearly stoichiometric and stoichiometric LN and LT samples should help to eliminate some disagreements in published data and to clarify intimate details of structures of impurity defects in materials, which are important for both physics and applications.

Derived structures and obtained by EPR/ENDOR characteristics of Zeeman and zero-field splitting, quadrupole and hyperfine interactions of impurity electrons with own and surrounding nuclei can be reliable corner stones for modelling of the structures.

Author Contributions: V.G.G.—theorist (yin), and G.I.M.—experimentalist (yang) made inseparable complementary contributions. All authors have read and agreed to the published version of the manuscript.

Funding: The publishing fee is covered by V.G.Grachev, SimuMag, Bozeman, MT, USA.

Institutional Review Board Statement: Not applicable.

Informed Consent Statement: Not applicable.

Data Availability Statement: The data presented in this study are available on request from the corresponding author.

Conflicts of Interest: The authors declare no conflict of interest.

References

1. Räuber, A. Chemistry and Physics of Lithium Niobate. In *Current Topics in Materials Science*, 1st ed.; Kaldis, E., Ed.; North-Holland: Amsterdam, The Netherlands, 1978; Volume 1, pp. 481–601.
2. Volk, T.; Wöhlecke, M. Point defects in $LiNbO_3$. In *Lithium Niobate. Defects, Photorefraction and Ferroelectric Switching*, 1st ed.; Hull, R., Osgood, R.M., Jr., Parisi, J., Warlimont, H., Eds.; Springer Series in Materials Science; Springer: Berlin/Heidelberg, Germany, 2009; Volume 115, pp. 1–50.
3. Ilyenkov, A.V.; Khiznyak, A.I.; Kreminskaya, L.V.; Soskin, M.S.; Vasnetsov, M.V. Birth and evolution of wave-front dislocations in a laser beam passed through a photorefractive $LiNbO_3$: Fe crystal. *Appl. Phys. B* **1996**, *62*, 465–471. [CrossRef]
4. Prokhorov, A.M.; Kuz'minov, Y.S. *Physics and Chemistry of Crystalline Lithium Niobate*; CRC Press: Boca Raton, FL, USA, 1990.
5. Wong, K.K. (Ed.) *Properties of Lithium Niobate*; EMIS Data Review Series No. 5; The Institution of Electrical Engineers, INSPEC: London, UK, 2002.
6. Toney, J.E. *Lithium Niobate Photonics*; Artech House Publishers: London, UK, 2015; pp. 1–265.
7. Streque, J.; Aubert, T.; Kokanyan, N.; Bartoli, F.; Taguett, A.; Polewczyk, V.; Kokanyan, E.; Hage-Ali, S.; Boulet, P.; Elmazria, O. Stoichiometric Lithium Niobate Crystals: Towards Identifiable Wireless Surface Acoustic Wave Sensors Operable up to 600 °C. *IEEE Sens. Lett.* **2019**, *3*, 2501204. [CrossRef]

8. Abrahams, S.C.; Reddy, J.M.; Bernstein, J.L. Ferroelectric lithium niobate. 3. Single crystal X-ray diffraction study at 24 °C. *J. Phys. Chem. Solids* **1966**, *27*, 997–1012. [CrossRef]
9. Peterson, G.E.; Bridenbaugh, P.M.; Green, P. NMR Study of Ferroelectric $LiNbO_3$ and $LiTaO_3$. *J. Chem. Phys.* **1967**, *46*, 4009–4014. [CrossRef]
10. Vianden, R. Hyperfine techniques for defects studies. In *Insulating Materials for Optoelectronics. New Developments*; Agulló-López, F., Ed.; World Scientific Publishing: Singapore; Hackensack, NJ, USA; London, UK; Hong Kong, China, 1995; pp. 125–146.
11. Kling, A.; Soares, J.C.; da Silva, M.F. Channeling Investigation of Oxide materials for Optoelectronic Applications. In *Insulating Materials for Optoelectronics. New Development*; Agulló-López, F., Ed.; World Scientific Publishing: Singapore; Hackensack, NJ, USA; London, UK; Hong Kong, China, 1995; pp. 175–200.
12. Gog, T.; Marerlik, G. X-ray Standing Wave Determination of Lattice Positions of Impurities in Lithium Niobate. In *Insulating Materials for Optoelectronics. New Developments*; Agulló-López, F., Ed.; World Scientific Publishing: Singapore; Hackensack, NJ, USA; London, UK; Hong Kong, China, 1995; pp. 201–220.
13. Spaeth, J.-M.; Niklas, J.R.; Bartram, R.H. *Structural Analysis of Point Defects in Solids: An Introduction to Multiple Magnetic Resonance Spectroscopy*; Springer: Singapore, 1992; pp. 1–367.
14. Prieto, C. Influence of lithium niobate stoichiometry on the lattice position of Nd^{3+} in Nd:$LiNbO_3$. *Opt. Mater.* **1999**, *12*, 135–142. [CrossRef]
15. Lorenzo, A.; Jaffrezic, H.; Roux, B.; Boulon, G.; Garcia-Sole, J. Lattice location of rare-earth ions in $LiNbO_3$. *Appl. Phys. Lett.* **1995**, *67*, 3735–3737. [CrossRef]
16. Lorenzo, A.; Loro, H.; Muñoz Santiuste, J.E.; Terrile, M.C.; Boulon, G.; Bausa, L.E.; Garcia Sole, J. RBS/channeling to locate active ions in laser materials: Application to rare earth activated $LiNbO_3$. *Opt. Mater.* **1997**, *8*, 55–63. [CrossRef]
17. Rebouta, L.; Smulders, P.J.M.; Boerma, D.O.; Agullo-Lopez, F.; da Silva, M.F.; Soares, J.C. Ion-beam channeling yields of host and impurity atoms in $LiNbO_3$: Computer simulations. *Phys. Rev. B* **1993**, *48*, 3600–3610. [CrossRef]
18. Rebouta, L.; da Silva, M.F.; Soares, J.C.; Santos, M.T.; Dieguez, E.; Agulló-López, F. Ion-beam/channeling characterization of $LiNbO_3$: Interaction between impurity sites. *Opt. Mater.* **1995**, *4*, 174–178. [CrossRef]
19. Garcia Sole, J.; Petith, T.; Jaffrez, H.; Boulon, G. Lattice Location of the Non-Equivalent Nd^{3+} Ions in $LiNbO_3$: Nd and $LiNbO_3$: MgO:Nd. *Europhys. Lett.* **1993**, *24*, 719–724. [CrossRef]
20. Vila, M.; de Bernabe, A.; Prieto, C. EXAFS determination of the Nd lattice position in Nd:$LiNbO_3$: Influence of lithium niobate stoichiometry and Mg and Zn co-doping. *J. Alloys Compd.* **2001**, *323–324*, 331–335. [CrossRef]
21. Catlow, C.R.A.; Chadwick, A.V.; Cole, M.; Tomlinson, S.M. An EXAFS study of dopant substitution in $LiNbO_3$ and $LiTaO_3$. *Radiat. Eff. Defects Solids* **1991**, *119–121*, 565–570. [CrossRef]
22. Bush, T.S.; Catlow, C.R.A.; Chadwick, A.V.; Cole, M.; Geatches, R.M.; Greaves, G.N.; Tomlinson, S.M. Studies of cation dopant sites in metal oxides by EXAFS and computer-simulation techniques. *J. Mater. Chem.* **1992**, *2*, 309–316. [CrossRef]
23. Prieto, C.; Zaldo, C. Determination of the lattice site of Fe in photorefractive $LiNbO_3$. *Solid State Commun.* **1992**, *83*, 819–821. [CrossRef]
24. Zaldo, C.; Prieto, C. Lattice site of transition metal and rare-earth impurities in $LiNbO_3$ single crystals: An exafs study. *Ferroelectrics* **1992**, *134*, 47–51. [CrossRef]
25. Corradi, G.; Chadwick, A.V.; West, A.R.; Cruickshankand, K.; Paul, M. On the substitution site of Cr and Fe in $LiNbO_3$: An exafs study. *Radiat. Eff. Defects Solids* **1995**, *134*, 219–222. [CrossRef]
26. Sanson, A.; Zaltron, A.; Argiolas, N.; Sada, C.; Bazzan, M.; Schmidt, W.G.; Sanna, S. Polaronic deformation at the $Fe^{2+/3+}$ impurity site in Fe:$LiNbO_3$ crystals. *Phys. Rev. B* **2015**, *91*, 094109. [CrossRef]
27. Rebouta, L.; da Silva, M.F.; Soares, J.C.; Hage-Ali, M.; Stoquert, J.P.; Siffert, P.; Sanz-García, J.A.; Diéguez, E.; Agulló-López, F. Lattice Site of Iron in $LiNbO_3(Fe^{3+})$ by the PIXE/Channelling Technique. *Europhys. Lett.* **1991**, *14*, 557–561. [CrossRef]
28. Gog, T.; Schotters, P.; Falta, J.; Materlik, G.; Grodzicki, M. The lattice position of Fe in Fe-doped $LiNbO_3$. *J. Phys. Condens. Matter* **1995**, *7*, 6971–6980. [CrossRef]
29. Marques, J.G.; Kling, A.; Soares, J.C.; Rebouta, L.; da Silva, M.F.; Diéguez, E.; Agulló-López, F. Lattice site location of Hf in $LiNbO_3$: Influence of dopant concentration and crystal stoichiometry. *Nucl. Instrum. Methods Phys. Res. B* **1998**, *136–138*, 431–435. [CrossRef]
30. Mignotte, C. Determination of lattice site locations of erbium ions implanted into $LiNbO_3$ single crystals after annealing at moderate and high temperature. *Appl. Surf. Sci.* **2001**, *185*, 11–26. [CrossRef]
31. Kovacs, L.; Rebouta, L.; Soaresh, J.C.; da Silva, M.F.; Hage-Ali, M.; Stoquert, J.P.; Siffert, P.; Sanz-Garcia, J.A.; Corradi, G.; Szaller, Z. On the lattice site of trivalent dopants and the structure of Mg^{2+}-OH^--M^{3+} defects in $LiNbO_3$:Mg crystals. *J. Phys. Condens. Matter* **1993**, *5*, 781–794. [CrossRef]
32. Kling, A.; Soares, J.C.; da Silva, M.F.; Sanz-Garcia, J.A.; Dieguez, E.; Agullo-Lopez, F. Lattice site determination of Cr in low doped lithium niobate single crystals using PIXE/channeling. *Nucl. Instrum. Methods Phys. Res. B* **1998**, *136–138*, 426–430. [CrossRef]
33. Engelmann, H.; Mouahid, F.; Dezsi, I.; Molnar, B.; Gonser, U.; Siebert, D.; Dahlem, J.; Tuczek, F. Mössbauer and ESR study of $LiNbO_3$-Fe_2O_3 for low Fe_2O_3 concentrations. *Appl. Phys. A* **1989**, *48*, 211–217. [CrossRef]

34. Tomov, T.; Engelmann, H.; Dezci, I.; Gonser, U. Investigation of the ferroelectric phase transition in $LiNbO_3$: Fe by Mössbauer spectroscopy. *Solid State Commun.* **1989**, *69*, 41–44. [CrossRef]
35. Mignoni, S.; Fontana, M.D.; Bazzan, M.; Ciampolillo, M.V.; Zaltron, A.M.; Argiolas, N.; Sada, C. Micro-Raman analysis of Fe-diffused lithium niobate waveguides. *Appl. Phys. B* **2010**, *101*, 541. [CrossRef]
36. Kamińska, A.; Suchocki, A.; Kobyakov, S.; Arizmendi, L.; Potemski, M.; Teran, F.J. High-pressure and magneto-optical studies of Cr-related defects in the lithium-rich $LiNbO_3$:Cr, Mg crystal. *Phys. Rev. B* **2007**, *76*, 144117. [CrossRef]
37. Basun, S.; Kaplyanskii, A.; Kutsenko, A.; Dierolf, V.; Troester, T.; Kapphan, S.E.; Polgar, K. Optical characterization of Cr^{3+} centers in $LiNbO_3$. *Appl. Phys. B* **2001**, *73*, 453–461. [CrossRef]
38. Sothe, H.; Rowan, L.G.; Spaeth, J.-M. Sites of Fe^{3+} impurities in a congruent $LiTaO_3$ crystal. *J. Phys. Condens. Matter* **1989**, *1*, 3591–3600. [CrossRef]
39. Sothe, H.; Spaeth, J.-M. The site and local environment of Fe^{3+} in $LiNbO_3$ investigated with ENDOR. *J. Phys. Condens. Matter* **1992**, *4*, 9901–9906. [CrossRef]
40. Spaeth, J.-M. Determination of atomic positions of point defects in solids by ENDOR. *Appl. Magn. Reson.* **1992**, *3*, 257–281. [CrossRef]
41. Lerner, P.; Legras, C.; Dumas, J.P. Stoechiometrie des monocristaux de metaniobate de lithium. *J. Cryst. Growth* **1968**, *3*, 231–235. [CrossRef]
42. Byer, R.L.; Young, J.F.; Feigelson, R.S. Growth of High-Quality $LiNbO_3$ Crystals from the Congruent Melt. *J. Appl. Phys.* **1970**, *41*, 2320–2325. [CrossRef]
43. Malovichko, G.I.; Grachev, V.G.; Gabrielyan, V.T.; Kokanyan, E.P. Widths and intensities of ESR lines of iron-group impurities in nonstoichiometric lithium niobate crystals. *Sov. Phys. Solid State* **1986**, *28*, 1453–1458.
44. Iyi, N.; Kitamura, K.; Izumi, F.; Yamamoto, J.; Hayashi, T.; Asano, H.; Kimura, S. Comparative of defect structures in lithium niobate with different compositions. *J. Solid State Chem.* **1992**, *101*, 340–352. [CrossRef]
45. Holman, R.L. *Processing of Crystalline Ceramics*; Palmour, H., Davis, R.F., Eds.; Plenum: New York, NY, USA, 1978; p. 343.
46. Jundt, D.H.; Fejer, M.M.; Byer, R.L. Optical properties of lithium-rich lithium niobate fabricated by vapor transport equilibration. *IEEE J. Quantum Electron.* **1990**, *26*, 135–138. [CrossRef]
47. Bordui, P.F.; Norwood, R.G.; Bird, C.D.; Calvert, G.D. Compositional uniformity in growth and poling of large-diameter lithium niobate crystals. *J. Cryst. Growth* **1991**, *113*, 61–68. [CrossRef]
48. Bordui, P.F.; Norwood, R.G.; Jundt, D.H.; Fejer, M.M. Preparation and characterization of off-congruent lithium niobate crystals. *J. Appl. Phys.* **1992**, *71*, 875–879. [CrossRef]
49. Jundt, D.H.; Fejer, M.M.; Norwood, R.G.; Bordui, P.F. Composition dependence of lithium diffusivity in lithium niobate at high temperature. *J. Appl. Phys.* **1992**, *72*, 3468–3473. [CrossRef]
50. Fischer, C.; Kapphan, S.; Feng, X.-Q.; Cheng, N. Sharp R-lines in absorption and emission of Cr^{3+} in stoichiometric (VTE) $LiNbO_3$. *Radiat. Eff. Defects Solids* **1995**, *135*, 199–202. [CrossRef]
51. Furukawa, Y.; Sato, M.; Kitamura, K.; Yajima, Y.; Minakata, M. Optical damage resistance and crystal quality of $LiNbO_3$ single crystals with various [Li]/[Nb] ratios. *J. Appl. Phys.* **1992**, *72*, 3250–3254. [CrossRef]
52. Kitamura, K.; Yamamoto, J.; Iyi, N.; Kirnura, S.; Hayashi, T. Stoichiometric $LiNbO_3$ single crystal growth by double crucible Czochralski method using automatic powder supply system. *J. Cryst. Growth* **1992**, *116*, 327–332. [CrossRef]
53. Kitamura, K.; Furukawa, Y.; Iyi, N. Progress in single crystal growth of $LiNbO_3$ using double crucible Czochralski method. *Ferroelectrics* **1997**, *202*, 21–28. [CrossRef]
54. Kitamura, K.; Furukawa, Y.; Ji, Y.; Zgonik, M.; Medrano, C.; Montemezzani, G.; Guenter, P. Photorefractive effect in $LiNbO_3$ crystals enhanced by stoichiometry control. *J. Appl. Phys.* **1997**, *82*, 1006–1009. [CrossRef]
55. Furukawa, Y.; Kitamura, K.; Suzuki, E.; Niwa, K. Stoichiometric $LiTaO_3$ single crystal growth by double crucible Czochralski method using automatic powder supply system. *J. Cryst. Growth* **1999**, *197*, 889–895. [CrossRef]
56. Malovichko, G.I.; Grachev, V.G.; Yurchenko, L.P.; Proshko, V.Y.; Kokanyan, E.P.; Gabrielyan, V.T. Improvement of $LiNbO_3$ microstructure by crystal growth with potassium. *Phys. Status Solidi (a)* **1992**, *133*, K29–K33. [CrossRef]
57. Malovichko, G.I.; Grachev, V.G.; Kokanyan, E.; Schirmer, O.F.; Betzler, K.; Gather, B.; Jermann, F.; Klauer, S.; Schlarb, U. Characterization of stoichiometric $LiNbO_3$ grown from melts containing K_2O. *Appl. Phys. A* **1993**, *56*, 103–108. [CrossRef]
58. Bermudez, V.; Dutta, P.S.; Serrano, M.D.; Dieguez, E. The effect of native defects on the domain structures of Fe—A case study by means of the addition of MgO and K_2O to the congruent melt. *J. Phys. Condens. Matter* **1997**, *9*, 6097–6101. [CrossRef]
59. Polgar, K.; Peter, A.; Kovacs, L.; Corradi, G.; Szaller, Z. Growth of stoichiometric $LiNbO_3$ single crystals by top seeded solution growth method. *J. Cryst. Growth* **1997**, *177*, 211–216. [CrossRef]
60. Balasanyan, R.N.; Vartanyan, E.S.; Gabrielyan, V.T.; Kazaryan, L.M. Method for Growing Lithium Niobate Single Crystals. Patent SU 845506 A1, 29 March 1979. Available online: https://www.elibrary.ru/item.asp?id=41087794 (accessed on 27 February 2000).
61. Serrano, M.D.; Bermudez, V.; Arizmendi, L.; Dieguez, E. Determination of the Li/Nb ratio in $LiNbO_3$ crystals grown by Czochralski method with K_2O added to the melt. *J. Cryst. Growth* **2000**, *210*, 670–675. [CrossRef]
62. Kokanyan, E.; Dieguez, E. New perspectives of lithium niobate crystals. *J. Optoelectron. Adv. Mater.* **2000**, *2*, 205–214.

63. Polgar, K.; Peter, A.; Foldvari, I.; Szaller, Z. Structural defects in flux-grown stoichiometric $LiNbO_3$ single crystals. *J. Cryst. Growth* **2000**, *218*, 327–333. [CrossRef]
64. Polgár, K.; Péter, Á.; Földvári, I. Crystal growth and stoichiometry of $LiNbO_3$ prepared by the flux method. *Opt. Mater.* **2002**, *19*, 7–11. [CrossRef]
65. Polgár, K.; Péter, Á.; Pöppl, L.; Ferriol, M.; Földvári, I. Chemical and thermal conditions for the formation of stoichiometric $LiNbO_3$. *J. Crystal Growth* **2002**, *237–239*, 682–686. [CrossRef]
66. Grunskij, O.S.; Denisov, A.V.; Badmaev, T.V. A Method for Growing Doped Crystals of Lithium Niobate with a Composition close to Stoichiometric, and a Device for Its Implementation. Patent RU 2367730 C2, 29 November 2007. Available online: https://www.elibrary.ru/item.asp?id=37690005 (accessed on 20 September 2009).
67. Lengyel, K.; Péter, Á.; Kovács, L.; Corradi, G.; Pálfalvi, L.; Hebling, J.; Unferdorben, M.; Dravecz, G.; Hajdara, I.; Szaller, Z.; et al. Growth, defect structure, and THz application of stoichiometric lithium niobate. *Appl. Phys. Rev.* **2015**, *2*, 040601. [CrossRef]
68. Malovichko, G.; Grachev, V.; Schirmer, O. Interrelation of intrinsic and extrinsic defects—Congruent, stoichiometric, and regularly ordered lithium niobate. *Appl. Phys. B* **1999**, *68*, 785–793. [CrossRef]
69. Malovichko, G.I.; Grachev, V.G.; Schirmer, O.F.; Faust, B. New axial Fe^{3+} centers in stoichiometric lithium niobate crystals. *J. Phys. Condens. Matter* **1993**, *5*, 3971–3976. [CrossRef]
70. de Bernabe, A.; Prieto, C.; de Andres, A. Effect of stoichiometry on the dynamic mechanical properties of $LiNbO_3$. *J. Appl. Phys.* **1995**, *79*, 143–148. [CrossRef]
71. Garret, M.H.; Mnushkina, I.; Furukawa, Y.; Kitamura, K.; Halliburton, L.E.; Giles, N.C.; Setzler, S.D. Photorefractive Properties of Iron-Doped Congruent Lithium Niobate, and Iron-Doped K_2O Flux Grown Stoichiometric Lithium Niobate. Proceedings of 1997 Topical Meeting on Photorefractive Materials, Effects and Devices, Chiba, Japan, 11–13 June 1997.
72. Abdi, F.; Aillerie, M.; Bourson, P.; Fontana, M.D.; Polgar, K. Electro-optic properties in pure $LiNbO_3$ crystals from the congruent to the stoichiometric composition. *J. Appl. Phys.* **1998**, *84*, 2251–2254. [CrossRef]
73. Choi, Y.N.; Choh, S.H.; Park, I.W.; Koh, E.K.; Kim, S.S. Characterization of Stoichiometry of $LiNbO_3$ Crystals by EPR and Raman Scattering Measurements. *J. Korean Phys. Soc.* **1998**, *32*, S643.
74. Malovichko, G. Nonstoichiometry as a powerful tool for photorefractive material optimization. Lithium Niobate Crystals. In *Advances in Photorefractive Materials, Effects and Devices*; Andersen, P.E., Johansen, P.M., Pedersen, H.C., Petersen, P.M., Saffman, M., Eds.; Trends in Optics and Photonics; Optical Society of America: Washington, DC, USA, 1999; Volume 27. [CrossRef]
75. Malovichko, G.; Petersen, R.; Bäuman, C.; Grachev, V. Second axial Fe^{3+} center in stoichiometric lithium tantalate. *J. Appl. Phys.* **2006**, *100*, 023911. [CrossRef]
76. Razdobarin, A.G.; Basun, S.A.; Bursian, V.É.; Sochava, L.S.; Evans, D.R. A Fe[Nb]-Li center in stoichiometric $LiNbO_3$ crystals: Mechanism of formation. *Phys. Solid State* **2010**, *52*, 706–711. [CrossRef]
77. Schlarb, U.; Klauer, S.; Wesselmann, M.; Betzler, K. Determination of the Li/Nb ratio in lithium niobate by means of birefringence and Raman measurements. *Appl. Phys. A* **1993**, *56*, 311–315. [CrossRef]
78. Kling, A.; Marques, J.G.; Correia, J.G.; da Silva, M.F.; Dieguez, E.; Agullo-Lopez, F.; Soares, J.C. Study of structural differences between stoichiometric and congruent lithium niobate. *Nucl. Instrum. Methods Phys. Res. B* **1996**, *113*, 293–295. [CrossRef]
79. Kovacs, L.; Ruschhaupt, G.; Polgar, K.; Corradi, G.; Woehlecke, M. Composition dependence of the ultraviolet absorption edge in lithium niobate. *Appl. Phys. Lett.* **1997**, *70*, 2801–2803. [CrossRef]
80. Dravecz, G.; Kovács, L. Determination of the crystal composition from the OH^- vibrational spectrum in lithium niobate. *Appl. Phys. B* **2007**, *88*, 305–307. [CrossRef]
81. Grachev, V.G. Visual EPR, Visual ENDOR Packages, Version 20.11. Available online: www.visual-epr.com (accessed on 1 June 2013).
82. *CrystalMaker Program, Version 2.7.7*; CrystalMaker Software Ltd.: Oxford, UK; Available online: www.crystalmaker.com (accessed on 31 December 2011).
83. Boysen, H.; Altorfer, F. A neutron powder investigation of the high-temperature structure and phase transition in $LiNbO_3$. *Acta Crystallogr. B* **1994**, *50*, 405–414. [CrossRef]
84. Hsu, R.; Maslen, E.N.; du Boulay, D.; Ishizawa, N. Synchrotron X-ray Studies of $LiNbO_3$ and $LiTaO_3$. *Acta Crystallogr. B* **1997**, *53*, 420–428. [CrossRef]
85. Malovichko, G.; Cerclier, O.; Estienne, J.; Grachev, V.; Kokanyan, E.; Boulesteix, C. Lattice constants of K- and Mg-doped $LiNbO_3$. Comparison with nonstoichiometric lithium niobate. *J. Phys. Chem. Sol.* **1995**, *56*, 1285–1289. [CrossRef]
86. Lehnen, H.; Boysen, H.; Frey, F.; Hewat, A.; Radaelli, P. A neutron powder investigation of the high-temperature structure and phase transition in stoichiometric $LiNbO_3$. *Zeitschrift für Kristallographie* **1997**, *212*, 712–719. [CrossRef]
87. Malovichko, G.; Grachev, V.; Kokanyan, E.; Schirmer, O. Axial and low-symmetry centers of trivalent impurities in lithium niobate. Chromium in congruent and stoichiometric crystals. *Phys. Rev. B* **1999**, *59*, 9113–9125. [CrossRef]
88. Abrahams, S.C.; Hamilton, W.C.; Reddy, J.M. Ferroelectric lithium niobate. 4. Single crystal neutron diffraction study at 24 °C. *J. Phys. Chem. Solids* **1966**, *27*, 1013–1018. [CrossRef]
89. Megaw, H.D. A note on the structure of lithium niobate, $LiNbO_3$. *Acta Crystallogr. A* **1968**, *24*, 583–588. [CrossRef]

90. Abrahams, S.C.; Levinstein, H.J.; Reddy, J.M. Ferroelectric lithium niobate. 5. Polycrystal X-ray diffraction study between 24° and 1200 °C. *J. Phys. Chem. Solids* **1966**, *27*, 1019–1026. [CrossRef]
91. Malovichko, G.; Grachev, V.; Okulov, S.; Kokanyan, E.; Henecker, F.; Hofstaetter, A.; Schirmer, O. EPR of Nd^{3+} in congruent and nearly stoichiometric lithium niobate. *Phys. Status Solidi (b)* **2005**, *243*, 1–7. [CrossRef]
92. Malovichko, G.; Bratus, V.; Munro, M.; Kokanyan, E.; Okulov, S.; Grachev, V. Multifrequency spectroscopy of laser active centers Nd^{3+} and Yb^{3+} in nearly stoichiometric $LiNbO_3$. *Phys. Status Solidi (c)* **2007**, *4*, 1346–1351. [CrossRef]
93. Grachev, V.; Munro, M.; Kokanyan, E.; Malovichko, G. Determination of g-tensors of low-symmetry Nd^{3+} centers in $LiNbO_3$ by rectification of angular dependence of electron paramagnetic resonance spectra. *J. Appl. Phys.* **2015**, *118*, 044103. [CrossRef]
94. Blümel, J.; Born, E.; Metzger, T. Solid state NMR study supporting the lithium vacancy defect model in congruent lithium niobate. *J. Phys. Chem. Solids* **1994**, *55*, 589–593. [CrossRef]
95. Zotov, N.; Boysen, H.; Frey, F.; Metzger, T.; Born, E. Cation substitution models of congruent $LiNbO_3$ investigated by X-ray and neutron powder diffraction. *J. Phys. Chem. Solids* **1994**, *55*, 145–152. [CrossRef]
96. Leroux, C.; Nihoul, G.; Malovichko, G.; Grachev, V.; Boulesteix, C. High resolution electron microscopy investigation of correlated defects in non-stoichiometric lithium niobate. *J. Phys. Chem. Solids* **1998**, *59*, 311–319. [CrossRef]
97. Kim, S.; Gopalan, V.; Kitamura, K.; Furukawa, Y. Domain reversal and nonstoichiometry in lithium tantalate. *J. Appl. Phys.* **2001**, *90*, 2949–2963. [CrossRef]
98. Abrahams, S.C.; Marsh, P. Defect structure dependence on composition in lithium niobate. *Acta Crystallogr. B* **1986**, *42*, 61–68. [CrossRef]
99. Peterson, G.E.; Carnevale, A. ^{93}Nb NMR linewidths in nonstoichiometric lithium niobate. *J. Chem. Phys.* **1972**, *56*, 4648–4851. [CrossRef]
100. Araujo, R.M.; dos Santos Mattos, E.F.; Valerio, M.E.G.; Jackson, R.A. Computer Simulation of the Incorporation of V^{2+}, V^{3+}, V^{4+}, V^{5+} and Mo^{3+}, Mo^{4+}, Mo^{5+}, Mo^{6+} Dopants in $LiNbO_3$. *Crystals* **2020**, *10*, 457. [CrossRef]
101. Kuzminov, Y.; Osiko, V.V. Disorder of stoichiometry in lithium niobate crystals. *Cryst. Rep.* **1994**, *39*, 471–476.
102. Fay, H.; Alford, W.J.; Dess, H.M. Temperature in $LiNbO_3$ Crystals on Melt Composition. *Appl. Phys. Lett.* **1968**, *12*, 89. [CrossRef]
103. Suzuki, T. Oxygen partial pressure dependence of optical absorption in lithium niobate. *J. Cryst. Growth* **1996**, *163*, 403–410. [CrossRef]
104. Sweeney, K.L.; Halliburton, L.E.; Bryan, D.A.; Rice, R.R.; Gerson, R.; Tomaschke, H.E. Point defects in Mg-doped lithium niobate. *J. Appl. Phys.* **1985**, *57*, 1036–1044. [CrossRef]
105. Erdei, S.; Ainger, F.W. Trends in the growth of stoichiometric single crystals. *J. Cryst. Growth* **1997**, *174*, 293–300. [CrossRef]
106. Kovács, L.; Wohlecke, M.; Jovanović, A.; Polgár, K.; Kapphan, S. Infrared absorption study of the OH vibrational band in $LiNbO_3$ crystals. *J. Phys. Chem. Solids* **1991**, *52*, 797–803. [CrossRef]
107. Engelsberg, M.; de Souza, R.E.; Pacobahyba, L.H.; do Nascimento, G.C. Structural determination of hydrogen site occupation in proton-exchanged $LiNbO_3$ by nuclear magnetic resonance. *Appl. Phys. Lett.* **1995**, *67*, 359–361. [CrossRef]
108. Yevdokimov, S.V.; Yatsenko, A.V. Investigation of the Localization of H^+ Ions in Stoichiometric $LiNbO_3$. *Crystallogr. Rep.* **2003**, *48*, 542–546. [CrossRef]
109. Rakitina, L.G.; Zaritskii, I.M.; Polgar, K. New hydrogen associated paramagnetic centre in $LiNbO_3$. *Appl. Magn. Reson.* **1990**, *1*, 149–154. [CrossRef]
110. Rosa, J.; Polak, K.; Kubatova, J. ESR and optical studies of impurity centres in γ- and X-irradiated $LiNbO_3$. *Phys. Status Solidi (b)* **1982**, *111*, K85–K87. [CrossRef]
111. Corradi, G.; Polgar, K.; Bugai, A.A.; Grachev, V.G.; Deryugina, N.I.; Rakitina, L.M.; Zaritskii, I.M. Models of Ni^+ and Cu^{2+} impurity centers in $LiNbO_3$. *Sov. Phys. Solid State* **1986**, *28*, 739–748.
112. Zhang, H.-M.; Xiao, W.-B. Investigations on the EPR parameters and defect structures due to Jahn-Teller effect for the Cu^{2+} and Ni^+ centers in $LiNbO_3$. *J. Alloys Compd.* **2018**, *745*, 586–591. [CrossRef]
113. Zaritskii, I.M.; Rakitina, L.G.; Polgar, K. Annealing and gamma irradiation effects on the ESR in $LiNbO_3$:Mg crystals. *Sov. Phys. State Solid* **1995**, *37*, 1073–1978.
114. Faust, B.; Müller, H.; Schirmer, O.F. Free small polarons in $LiNbO_3$. *Ferroelectrics* **1994**, *153*, 297–302. [CrossRef]
115. Mirzakhanyan, A.A.; Petrosyan, A.K. EPR and optical absorption of Co^{2+} impurity ions in single crystal α-$LiIO_3$ and $LiNbO_3$. *Sov. Phys. Solid State* **1986**, *904–906*, 1593–1595.
116. Donnerberg, H.J.; Shirmer, O.F. The paramagnetic properties of Co^{2+} in $LiNbO_3$ and $LiTaO_3$. *Solid State Commun.* **1987**, *63*, 29–32. [CrossRef]
117. Choi, Y.N.; Park, I.-W.; Kim, S.S.; Park, S.S.; Choh, S.H. Electron paramagnetic resonance studies of Co^{2+} ions in congruent and nearly stoichiometric $LiNbO_3$ single crystals. *J. Phys. Condens. Matter* **1999**, *11*, 4723–4730. [CrossRef]
118. Grachev, V.G.; Hansen, K.; Meyer, M.; Kokanyan, E.P.; Malovichko, G.I. Substitution mechanisms and location of Co^{2+} ions in congruent and stoichiometric lithium niobate crystals derived from electron paramagnetic resonance data. *Mater. Res. Express* **2017**, *4*, 036204. [CrossRef]
119. Paul, M.; Tabuchi, M.; Weast, A.R. Defect Structure of Ni, Co-doped $LiNbO_3$ and $LiTaO_3$. *Chem. Mater.* **1997**, *9*, 3206–3214. [CrossRef]

120. Mei, Y.; Wu, X.-X.; Zheng, W.-C. EPR parameters and defect structures for the Co^{2+} ions in $LiNbO_3$ and $LiTaO_3$ crystals. *Spectrochim. Acta Part A Mol. Biomol. Spectrosc.* **2008**, *71*, 714–719. [CrossRef]
121. Wu, S.-Y.; Lu, G.D.; Zhang, Z.H.; Wei, L.H.; Hu, Y.-X. Theoretical investigations of the local structures and spin Hamiltonian parameters for the trigonal Co^{2+} centers in $LiNbO_3$ and $LiTaO_3$. *J. Alloys Compd.* **2009**, *1–2*, 1–5. [CrossRef]
122. Setser, G.G. An Investigation by Electron Paramagnetic Resonance of the Jahn-Teller Effect in $LiNbO_3$:Ni^{3+} and $LiNbO_3$:Cu^{2+}. Ph.D. Thesis, Rice University, Houston, TX, USA, 1974; pp. 1–237.
123. Palatnikov, M.N.; Sidorov, N.V.; Skiba, V.I.; Biryukova, I.V.; Serebryakov, Y.A.; Kravchenko, O.E.; Balabanov, Y.I.; Kalinnikov, V.T. Effects of nonstoichiometry and doping on the curie temperature and defect structure of lithium niobate. *Inorg. Mater.* **2000**, *36*, 489–493. [CrossRef]
124. Pape, M.; Reyher, H.-J.; Schirmer, O.F. Optically detected magnetic resonance of Cu, Fe and Mn defects in $LiNbO_3$. *J. Phys. Condens. Matter* **2005**, *17*, 6835–6847. [CrossRef]
125. Mirzakhanyan, A.A. Zero-field splitting of energy levels of Ni^{2+} ground state in α-$LiIO_3$ and $LiNbO_3$. *Sov. Phys. State Solid* **1981**, *23*, 2452.
126. Khalil, A.; Masaif, N.; Jennane, A.; Maaider, K. Substitution Mechanism of Ni:$LiTaO_3$. *J. Mater. Environ. Sci.* **2011**, *2*, 196–200.
127. Yang, Z.Y.; Rudowicz, C.; Yeung, Y.Y. Microscopic spin-Hamiltonian parameters and crystal field energy levels for the low C_3 symmetry Ni^{2+} centre in $LiNbO_3$ crystals. *Physica B* **2004**, *348*, 151–159. [CrossRef]
128. Petrov, M.P. Spectrum of Electron Paramagnetic Resonance of Mn^{2+} in ferroelectric $LiNbO_3$. *Sov. Phys. Solid State* **1968**, *10*, 2574.
129. Takeda, T.; Watanabe, A.; Sugihara, K. Spacing of the hyperfine sextet in Mn^{2+} ESR in $LiNbO_3$. *Phys. Lett. A* **1968**, *27*, 114–115. [CrossRef]
130. Danner, J.C.; Ranon, U.; Stamires, D.N. ESR of Mn^{2+} in ferroelectric $LiTaO_3$ and $LiNbO_3$ single crystals. *Chem. Phys. Lett.* **1968**, *2*, 605–606. [CrossRef]
131. Herrington, J.B.; Dischler, B.; Schneider, J. An EPR investigation of Fe^{3+} and Mn^{2+} in $LiNbO_3$. *Solid State Commun.* **1972**, *10*, 509–511. [CrossRef]
132. Rexford, D.J.; Kim, Y.M. Electron-Spin Resonance Studies of Crystal Field Parameters in Mn^{2+}:$LiNbO_3$. *J. Chem. Phys.* **1972**, *57*, 3094–3098. [CrossRef]
133. Malovichko, G.I.; Grachev, V.G. Use of nuclear quadrupole splitting in determination of positions of iron group impurities in ferroelectric $LiNbO_3$. *Sov. Phys. Solid. State* **1985**, *27*, 1678–1679.
134. Malovichko, G.I.; Grachev, V.G.; Lukin, S.N. EPR of axial and low symmetry Cr^{3+}, Fe^{3+}, and Mn^{2+} centers in $LiNbO_3$. *Sov. Phys. Solid State* **1986**, *28*, 553–557.
135. Malovichko, G.I.; Karmazin, A.A.; Bykov, I.P.; Laguta, V.V.; Yarunichev, V.P. Electron spin resonance investigation of the temperature dependence of the axial crystal field parameters in Mn^{2+} and Cr^{3+} doped $LiNbO_3$. *Sov. Phys. Solid State* **1983**, *25*, 2038–2041.
136. Park, I.W.; Choh, S.H.; Song, K.J. The hyperfine structure of Mn^{2+} in $LiNbO_3$ and the sign of the second-order axial zero-field splitting. *J. Korean Phys. Soc.* **1993**, *26*, 77–80.
137. Park, I.-W. Three-fold Symmetry Effect on Mn^{2+} Centers in a $LiNbO_3$ Crystal. *J. Korean Magn. Res. Soc.* **2008**, *12*, 103–110. [CrossRef]
138. Yeom, T.H.; Lee, S.H. Temperature Dependence of Mn^{2+} Paramagnetic Ion in a Stoichiometric $LiNbO_3$ Single Crystal. *J. Magn.* **2013**, *18*, 221–224. [CrossRef]
139. Corradi, G.; Söthe, H.; Spaeth, J.-M.; Polgar, K. Mn^{2+} defects in $LiNbO_3$: An electron nuclear double resonance (ENDOR) investigation of the Mn^{2+} site and the local disorder. *J. Phys. Condens. Matter.* **1990**, *2*, 6603–6618. [CrossRef]
140. Malovichko, G.; Grachev, V.; Hofstaetter, A.; Kokanyan, E.; Scharmann, A.; Schirmer, O. ENDOR study of Cr^{3+} centers substituting for lithium in lithium niobate. *Phys. Rev. B* **2002**, *65*, 224116. [CrossRef]
141. Grachev, V.; Malovichko, G. EPR, ENDOR, and optical-absorption study of Cr^{3+} centers substituting for niobium in Li-rich lithium niobate crystals. *Phys. Rev. B* **2000**, *62*, 7779–7790. [CrossRef]
142. Grachev, V.; Malovichko, G.; Schirmer, O. Structures of point defects in lithium niobate. *Ukr. J. Phys.* **2004**, *49*, 438–448.
143. Evlanova, N.F.; Kornienko, L.S.; Rashkovich, L.N.; Rybaltovskii, A.O. EPR of some rare-earth ions and Cr^{3+} in $LiNbO_3$. *Sov. JETP* **1967**, *53*, 1920–1926.
144. Burns, G.; O'Kane, D.F.; Title, R.S. Optical and Electron-Spin-Resonance Spectra of $\mathbf{Yb^{3+}}$, $\mathbf{Nd^{3+}}$, and $\mathbf{Cr^{3+}}$ in $LiNbO_3$ and $LiTaO_3$. *Phys. Rev.* **1968**, *167*, 314–318. [CrossRef]
145. Rexford, D.J.; Kim, Y.M.; Story, H.S. Electron-Spin resonance studies of Cr^{3+} in $LiNbO_3$. *J. Chem. Phys.* **1970**, *52*, 860–863. [CrossRef]
146. Choh, S.H.; Kim, H.T.; Choh, H.K.; Han, C.S.; Choi, D.; Kim, J.N. Characterization of Stoichiometry of $LiNbO_3$ Crystals by EPR and Raman Scattering Measurements. *Bull. Magn. Res.* **1989**, *11*, 371.
147. Salley, G.M.; Basun, S.A.; Imbusch, G.F.; Kaplyanskii, A.A.; Kapphan, S.; Meltzer, R.S.; Happek, U. Chromium centers in $LiNbO_3$ revisited. *J. Luminescence* **1999**, *83–84*, 423–427. [CrossRef]
148. Grachev, V.; Malovichko, G.; Schirmer, O. Single, dimer and trimer chromium centers in lithium niobate. *Ferroelectrics* **1996**, *185*, 5–8. [CrossRef]

149. Grachev, V.G.; Malovichko, G.I.; Troitskii, V.V. Investigation of the charge compensation mechanisms of Cr^{3+} ions in $LiNbO_3$ by EPR and ENDOR methods. *Sov. Phys. Solid State* **1987**, *29*, 349–350.
150. Siu, G.G.; Zhao, M.-G. Electron-spin-resonance studies of Cr-ion pairs in $LiNbO_3$: Cr^{3+}. *Phys. Rev. B* **1991**, *43*, 13575–13578. [CrossRef] [PubMed]
151. Corradi, G.; Sothe, H.; Spaeth, J.-M.; Polgar, K. Electron spin resonance and electron-nuclear double-resonance investigation of a new Cr^{3+} defect on an Nb site in $LiNbO_3$:Mg:Cr. *J. Phys. Condens. Matter* **1991**, *3*, 1901–1908. [CrossRef]
152. Corradi, G.; Sothe, H.; Spaeth, J.-M.; Polgar, K. ENDOR for characterizing transition metal centres in $LiNbO_3$. *Radiat. Eff. Defects Solids* **1991**, *119–121*, 583–587. [CrossRef]
153. Corradi, G.; Söthe, H.; Spaeth, J.-M.; Polgar, K. Local environment of Mn^{2+} and Cr^{3+} centres in $LiNbO_3$: Discussion of recent ENDOR results. *Ferroelectrics* **1992**, *125*, 295–299. [CrossRef]
154. Diazcaro, J.; Garcia-Sole, J.; Bravo, D.; Han, T.P.J.; Jaque, F.; Henderson, B. EPR and infrared absorption study in MgO Co-Doped $LiNbO_3$ crystals. *Ferroelectr. Lett. Sect.* **1997**, *23*, 27–35. [CrossRef]
155. Thiemann, O.; Corradi, G.; Reyher, H.-J. ODMR-investigations of transition metals in $LiNbO_3$. *Ferroelectrics* **1992**, *125*, 283–288. [CrossRef]
156. Camarillo, E.; Garcia-Sole, J.; Cusso, F.; Agullo-Lopez, F.; Sanz-Garcia, J.A.; Han, T.P.J.; Jaque, F.H.; Henderson, B. Polarization spectroscopy of Cr^{3+} ions in $LiNbO_3$ single crystals: Effect of Mg^{2+} ions. *Chem. Phys. Lett.* **1991**, *185*, 505–510. [CrossRef]
157. Camarillo, E.; Tocho, J.; Vergara, I.; Dieguez, E.; Garcia-Sole, J.; Jaque, F. Optical bands of Cr^{3+} induced by Mg^{2+} ions in $LiNbO_3$:Cr,Mg. *Phys. Rev. B* **1992**, *45*, 4600–4604. [CrossRef]
158. Martin, A.; Lopez, F.J.; Agullo-Lopez, F. Cr^{3+} in pure and Mg-doped $LiNbO_3$: Analysis of the EPR and optical spectra. *J. Phys. Condens. Matter* **1992**, *4*, 847–853. [CrossRef]
159. Macfarlane, P.I.; Holliday, K.; Nicholls, J.F.H.; Henderson, B. Characterization of Cr^{3+} centres in $LiNbO_3$ using fluorescence line narrowing. *J. Phys. Condens. Matter.* **1995**, *7*, 9643–9656. [CrossRef]
160. Diaz-Caro, J.; Garcia-Sole, J.; Bravo, D.; Sanz-Garcia, J.A.; Lopez, F.J.; Jaque, F. MgO codoping-induced change in the site distribution of Cr^{3+} ions in $LiNbO_3$. *Phys. Rev. B* **1996**, *54*, 13042–13046. [CrossRef] [PubMed]
161. Diaz-Caro, J.; Garcia-Sole, J.; Martinez, J.L.; Henderson, B.; Jaque, F.; Han, T.P.J. Redistribution of Cr^{3+} defect centres in $LiNbO_3$ crystals: The MgO effect. *Opt. Mater.* **1998**, *10*, 69–77. [CrossRef]
162. Torchia, G.A.; Martinez Matos, O.; Vaveliuk, P.; Tocho, J.O. Influence of the electron-lattice coupling for Cr^{3+} ions in Nb^{5+} site into congruent co-doped $LiNbO_3$: Cr^{3+}: ZnO crystal. *Solid State Commun.* **2003**, *127*, 535–539. [CrossRef]
163. Ahn, S.W.; Rudowicz, C.; Choh, S.H.; Han, S.Y. EPR study of two Cr^{3+}defect centers in $LiTaO_3$ single crystals. *J. Korean Phys. Soc.* **1997**, *30*, 99–102.
164. Ahn, S.W.; Kim, J.S.; Choh, S.H.; Yeom, T.H. An Induced Cr^{3+} Center in gamma-Irradiated $LiTaO_3$. *J. Korean Phys. Soc.* **1994**, *27*, 535–538.
165. Loyo-Menoyo, M.; Keeble, D.J.; Furukawa, Y.; Kitamura, K. Electron paramagnetic resonance of Cr^{3+} in near-stoichiometric $LiTaO_3$. *J. Appl. Phys.* **2005**, *97*, 123905. [CrossRef]
166. Bourson, P.; Malovichko, G.; Ridah, A.; Kokanyan, E. Effect of chromium concentration on site selective luminescence in nearly stoichiometric lithium niobate crystals. *Ferroelectrics* **1996**, *155*, 273–276. [CrossRef]
167. Lhomme, F.; Bourson, P.; Fontana, M.D.; Malovichko, G.; Aillerie, M.; Kokanyan, E. Luminescence of Cr^{3+} in lithium niobate: Influence of the chromium concentration and crystal composition. *J. Phys. Condens. Matter* **1998**, *10*, 1137–1146. [CrossRef]
168. Lhomme, F.; Bourson, P.; Moncorge, R.; Burlot, R.; Fontana, M.D.; Aillerie, M.; Boulon, G.; Malovichko, G. New study of the 720–750 nm range in the emission spectra of $LiNbO_3$:Cr^{3+} crystals with different compositions. *Radiat. Eff. Defects Solids* **1999**, *150*, 259–263. [CrossRef]
169. Kaminska, A.; Suchocki, A.; Grinberg, M.; Garcia-Sole, J.; Jaque, F.; Arizmendi, L. High-pressure spectroscopy of $LiNbO_3$:MgO, Cr^{3+} crystals. *J. Lumin.* **2000**, *87–89*, 571–573. [CrossRef]
170. Salley, G.M.; Basun, S.A.; Kaplyanskii, A.A.; Meltzer, R.S.; Polgar, K.; Happek, U. Chromium centers in stoichiometric $LiNbO_3$. *J. Lumin* **2000**, *87–89*, 1133–1135. [CrossRef]
171. Dierolf, V.; Kaplyanskii, A.A.; Kapphan, S.; Kutsenko, A.B. The fine structure of spectral response of Cr^{3+} sites in $LiNbO_3$. In *Radiat. Eff. Defects Solids*; 2001; Volume 155, pp. 241–246. [CrossRef]
172. Basun, S.A.; Salley, G.M.; Kaplyanskii, A.A.; Gallagher, H.G.; Polgar, K.; Lu, L.; Happek, U. A novel luminescent center in $LiNbO_3$:Cr:Mg crystals. *J. Lumin.* **1999**, *83–84*, 435–439. [CrossRef]
173. Lhomme, F.; Bourson, P.; Boulon, G.; Guyot, Y.; Fontana, M.D. Comparative analysis of the Cr^{3+} centre spectroscopic properties in $LiNbO_3$ crystals from congruent to nearly stoichiometric compositions. *Eur. Phys. J. AP* **2002**, *20*, 29–40. [CrossRef]
174. Han, T.P.J.; Jaque, F.; Bermudez, V.; Dieguez, E. Luminescence of the Cr^{3+} R-lines in pure and MgO co-doped near stoichiometric $LiNbO_3$:Cr crystals. *Chem. Phys. Lett.* **2003**, *369*, 519–524. [CrossRef]
175. Han, T.P.J.; Jaque, F. Optical stability of the Cr^{3+} centres in codoped stoichiometric and congruent $LiNbO_3$:Cr:Mg. *Opt. Mater.* **2007**, *29*, 1041–1043. [CrossRef]
176. Kaczmarek, S.M.; Lukasiewicz, T.; Pracka, I.; Jablonski, R.; Boulon, G.; Kaczmarek, B.; Warchol, S. Radiation defects in Dy^{3+} doped $LiNbO_3$ single crystals. *J. Alloys Compd.* **1998**, *275–277*, 105–108. [CrossRef]

177. Jablonski, R.; Pracka, I.; Swirkowicz, M. Electron spin resonance spectra of Nd^{3+}, Dy^{3+}, Er^{3+} and Yb^{3+} in $LiNbO_3$. In *Solid State Crystals: Growth and Characterization, Proceedings of the XII Conference on Solid State Crystals: Materials Science and Applications, Zakopane, Poland, 7–11 October 1996*; SPIE: Bellingham, WA, USA, 1997. [CrossRef]
178. Milori, D.M.B.P.; Moraes, I.J.; Hernandes, A.C.; de Souza, R.R.; Li, M.S.; Terrile, M.C.; Barberis, G.E. Optical and ESR study of Er^{3+} in $LiNbO_3$. *Phys. Rev. B* **1995**, *51*, 3206—3209. [CrossRef] [PubMed]
179. Nolte, T.; Pawlik, T.; Spaeth, J.-M. EPR study of Er^{3+} in congruent $LiNbO_3$. *Solid State Commun.* **1997**, *104*, 535–539. [CrossRef]
180. Bodziony, T.; Kaczmarek, S.M. EPR study of low symmetry Er centers in congruent lithium niobate. *Phys. Status Solidi (b)* **2008**, *245*, 998–1002. [CrossRef]
181. Kaczmarek, S.M.; Bodziony, T. Low symmetry centers in $LiNbO_3$ doped with Yb and Er. *J. Non-Cryst. Solids* **2008**, *354*, 4202–4210. [CrossRef]
182. Bodziony, T.; Kaczmarek, S.M. Temperature dependence of the EPR spectra and optical measurements of $LiNbO_3$: Er, Tm single crystal. *J. Alloys Compd.* **2009**, *468*, 581–585. [CrossRef]
183. Bodziony, T.; Kaczmarek, S.M.; Kruk, R. Low temperature magnetic measurements of $LiNbO_3$ single crystal weakly doped with Er and codoped with Tm ions. *J. Spectrosc. Dyn.* **2011**, *1*, 1–8.
184. Vrable, I.; Malovichko, G.; Grachev, V.; Meyer, M. EPR/ENDOR Studies of Erbium Centers in Stoichiometric Lithium Niobate Crystals. In Proceedings of the 49th Rocky Mountain Conference on Analytical Chemistry, Breckenridge, CO, USA, 22–26 July 2007; Abstract C 143. p. 71. Available online: http://rockychem.com/links/past-conferences.html (accessed on 1 March 2007).
185. Malovichko, G.; Grachev, V.; Jorgensen, J.; Meyer, M.; Munro, M.; Todt, B.; Vrable, I.; Kokanyan, E.; Bratus, V.; Okulov, S. Magnetic resonance study of non-equivalent centers created by 4f-ions in congruent and stoichiometric lithium niobate. *MRS Online Proc. Libr.* **2008**, *1111*, 103. [CrossRef]
186. Mackova, A.; Groetzschel, R.; Eichhorn, F.; Nekvindova, P.; Spirkova, J. Characterization of Er:$LiNbO_3$ and APE:Er:$LiNbO_3$ by RBS–channeling and XRD techniques. *Surf. Interface Anal.* **2004**, *36*, 949–951. [CrossRef]
187. Gog, T.; Griebenow, M.; Materlik, G. X-ray standing wave determination of the lattice location of Er diffused into $LiNbO_3$. *Phys. Lett. A* **1993**, *181*, 417–420. [CrossRef]
188. Rebouta, L.; da Silva, M.F.; Soares, J.C.; Serrano, D.; Diéguez, E.; Agulló-López, F.; Tornero, J. Nonaxial sites for Er in $LiNbO_3$. *J. Appl. Phys. Lett.* **1997**, *70*, 1070–1072. [CrossRef]
189. Malovichko, G.; Vrable, I.; Meyer, M.; Kokanyan, E.; Grachev, V. Ordering magnetic moments of Er^{3+} in ferroelectric lithium niobate. In Proceedings of the International Conference "Oxide Materials for Electronic Engineering—Fabrication, Properties and applica$tion" (OMEE-2014), Lviv, Ukraine, 26–30 May 2014; pp. 35–36.
190. Gill, D.M.; Wright, J.C.; McCaughan, L. Site characterization of rare-earth-doped $LiNbO_3$ using total site selective spectroscopy. *J. Appl. Phys. Lett.* **1994**, *64*, 2483–2485. [CrossRef]
191. Gill, D.M.; McCaughan, L.; Wright, J.C. Spectroscopic site determinations in erbium-doped lithium niobate. *Phys. Rev. B* **1996**, *53*, 2334–2344. [CrossRef] [PubMed]
192. Muntoz, J.A.; Cantelar, E.; Sanz-garcia, J.A.; Duchowicz, R.; Serrano, D.; Dieguez, E.; Lifante, G.; Cusso, F. Site-selective spectroscopy of Er^{3+} and Er^{3+} Yb^{3+} doped stoichiometric and congruent lithium niobate. *Radiat. Eff. Defects Solids* **1999**, *150*, 221–225. [CrossRef]
193. Dierolf, V.; Koerdt, M. Combined excitation-emission spectroscopy of Er^{3+} ions in stoichiometric $LiNbO_3$: The site selectivity of direct and up conversion excitation processes. *Phys. Rev. B* **2000**, *61*, 8043–8048. [CrossRef]
194. Dierolf, V.; Kutsenko, A.B.; Ostendorf, A.; von der Osten, W.; Sohler, W.; Suche, H. Site-selective spectroscopy of Er^{3+}:Ti:$LiNbO_3$ waveguides. *Appl. Phys. B* **2001**, *72*, 803–810. [CrossRef]
195. Dierolf, V.; Kutsenko, A.B.; Sandmann, C.; Troester, T.; Corradi, G. High-resolution site selective optical spectroscopy of rare earth and transition metal defects in insulators. *J. Lumin.* **2000**, *87–89*, 989–991. [CrossRef]
196. Dierolf, V.; Sandmann, C. Combined excitation emission spectroscopy of defects for site-selective probing of ferroelectric domain inversion in lithium niobate. *J. Lumin.* **2007**, *125*, 67–79. [CrossRef]
197. Bravo, D.; Martin, A.; Lopez, F.J. A new centre of Er^{3+} in MgO or ZnO co-doped $LiNbO_3$ single crystals. *Solid State Commun.* **1999**, *112*, 541–544. [CrossRef]
198. Mehran, F.; Scott, B.A. Electron paramagnetic resonance of $LiNbO_3$: Fe^{3+}. *Solid State Commun.* **1972**, *11*, 15–19. [CrossRef]
199. Towner, H.H.; Kim, Y.M.; Story, H.S. EPR studies of crystal field parameters in Fe^{3+}: $LiNbO_3$. *J. Chem. Phys.* **1972**, *56*, 3676–3679. [CrossRef]
200. Dikantova, Z. EPR study of impurities centers in $LiNbO_3$. *Acta Phys. Slov.* **1973**, *23*, 245–250.
201. Keune, W.; Date, S.K.; Gonser, U.; Bunzel, H. Mossbauer effect study of Fe^{57} doped $LiNbO_3$ and $LiTaO_3$. *Ferroelectrics* **1976**, *13*, 443–445. [CrossRef]
202. Kurz, H.; Krätzig, E.; Keune, W.; Engelmann, H.; Gonser, U.; Dischler, B.; Räuber, A. Photorefractive centers in $LiNbO_3$, studied by optical-, Mössbauer- and EPR-methods. *Appl. Phys.* **1977**, *12*, 355–368. [CrossRef]
203. Grachev, V.G.; Malovichko, G.I. Determination of the point symmetry of defects in crystals exhibiting structural phase transitions, using the temperature dependencies of the EPR spectra. Impurities in $LiNbO_3$. *Sov. Phys. Solid State* **1985**, *27*, 686–689.
204. Pfannes, H.D.; Putzka, A.; Sampaio, J.F. Electronic structure and spin relaxation of Fe(III) in $LiNbO_3$. *Hyperfine Interact.* **1986**, *28*, 785–788. [CrossRef]

205. Jablonski, R.; Kaczmarek, S.M.; Praska, I.; Surma, B.; Swirkowicz, M.; Lukasiewicz, T. ESR and optical measurements of $LiNbO_3$ and $LiTaO_3$ single crystals doped with ions of the first transition series. *Spectrochim. Acta A* **1998**, *54*, 1701–1709. [CrossRef]
206. Yeom, T.H.; Choh, S.H. Magnetic Resonance Investigations of $LiNbO_3$ and $LiTaO_3$ Single Crystals. *J. Korea Phys. Soc.* **1998**, *32*, S672–S674.
207. Santana, R.C.; Terrile, M.C.; Hernandes, A.C.; Andreeta, M.R.B.; Barberis, G.E. Electron spin resonance study of Fe^{3+} in $LiNbO_3$ single crystals: Bulk and fibres. *Solid State Commun.* **1997**, *103*, 61–64. [CrossRef]
208. Keeble, D.J.; Loyo-Menoyo, M.; Furukawa, Y.; Kitamura, K. Electron paramagnetic resonance of $\mathbf{Fe^{3+}}$ in $\mathbf{LiNbO_3}$. *Phys. Rev. B* **2005**, *71*, 224111. [CrossRef]
209. Agullò-López, F.; Müller, K.A. Superposition Model Calculations for Fe^{3+} in $LiNbO_3$. *Cryst. Latt. Def. Amorph. Mater.* **1987**, *15*, 89.
210. Zhao, M.G.; Chiu, M. Substitution site of the Fe^{3+} impurity in crystalline $LiNbO_3$. *Phys. Rev. B* **1994**, *49*, 12556–12558. [CrossRef]
211. Wang, H.; Kuang, X.Y.; Die, D.; Tan, X.-M.; Yang, X. EPR spectra and local lattice structure of Fe^{3+} impurity ions in ferroelectric $LiNbO_3$. *Chem. Phys.* **2006**, *330*, 212–215. [CrossRef]
212. Wang, H.; Kuang, X.Y.; Die, D.; Yang, X.; Zhou, K.-W. EPR investigation of substitution position for Fe^{3+} in $LiNbO_3$:Fe^{3+} system. *Physica B* **2005**, *367*, 53–60. [CrossRef]
213. Yeom, T.H.; Chang, Y.M.; Choh, S.H.; Rudowicz, C. Experimental and Theoretical Investigation of Spin-Hamiltonian Parameters for the Low Symmetry Fe^{3+} Centre in $LiNbO_3$. *Phys. Status Solidi (b)* **1994**, *185*, 409–415. [CrossRef]
214. Xue, D.; He, X. Dopant occupancy and structural stability of doped lithium niobate crystals. *Phys. Rev. B* **2006**, *73*, 064113. [CrossRef]
215. Zhang, Z.; Xue, D. Local Lattice Structure and Dopant Occupancy Of Doped Lithium Niobate Crystals. *Mod. Phys. Lett. B* **2009**, *23*, 3687–3694. [CrossRef]
216. Basun, S.A.; Bursian, V.E.; Evans, D.R.; Kaplyanskii, A.A.; Razdobarin, A.G.; Sochava, L.S. Ferroelectric-Specific Stark Effect in Stoichiometric $LiNbO_3$:Fe at Room Temperature. *Phys. Rev. Lett.* **2008**, *100*, 057602. [CrossRef]
217. Grachev, V.G.; Petersen, R.T.; Kokanyan, E.P.; Schirmer, O.F.; Malovichko, G.I. Structural analysis of the dominant axial Fe^{3+} center in $LiNbO_3$ crystal by electron nuclear double resonance. *J. Appl. Phys.* **2016**, *120*, 193901. [CrossRef]
218. Boker, A.; Donnerberg, H.; Schirmer, O.F.; Feng, X. Two sites of Fe^{3+} in highly Mg-doped $LiNBO_3$. *J. Phys. Condens. Matter* **1990**, *2*, 6865–6868. [CrossRef]
219. Volk, T.R.; Rubinina, N.M.; Pryalkin, V.I.; Krasnikov, V.V.; Volkov, V.V. Optical and non-linear optical investigations in $LiNbO_3$:Mg and $LiNbO_3$:Zn. *Ferroelectrics* **1990**, *109*, 345–350. [CrossRef]
220. Yeom, T.H.; Lee, S.H.; Choh, S.H.; Choi, D. Electron paramagnetic resonance study of Fe^{3+} in $LiNbO_3$:Mg:Fe crystal. *J. Korean Phys. Soc.* **1998**, *32*, S647.
221. Malovichko, G.I.; Grachov, V.G.; Kokanyan, E.P. Low-symmetry Fe-Mg complexes in $LiNbO_3$:Mg. *Ferroelectrics* **1992**, *125*, 289–294. [CrossRef]
222. Grachev, V.; Malovichko, G.; Kokanyan, E. Optimization of lithium niobate for advanced applications by variation of extrinsic and intrinsic defect subsystems. *Ferroelectrics* **2001**, *258*, 131–140. [CrossRef]
223. Volk, T.; Wöhlecke, M.; Rubinina, N.; Razumovskii, N.V.; Jermann, F.; Fischer, C.; Böwer, R. $LiNbO_3$ with the damage-resistant impurity indium. *Appl. Phys. A* **1995**, *60*, 217–225. [CrossRef]
224. Volk, T.R.; Rubinina, N.M. Non photorefractive impurities in lithium niobate: Magnesium and zinc. *Sov. Phys. Solid State* **1991**, *33*, 674–680.
225. Vazhenin, V.A.; Guseva, V.B.; Artyomov, M.Y.; Route, R.K.; Fejer, M.M.; Byer, R.L. Fine structure of the electron paramagnetic resonance spectrum of Fe^{3+} centres in $LiTaO_3$. *J. Phys. Condens. Matter* **2003**, *15*, 275–280. [CrossRef]
226. Loyo-Menoyo, M.; Keeble, D.J.; Furukawa, Y.; Kitamura, K. Electron paramagnetic resonance of Fe^{3+} in near-stoichiometric $LiTaO_3$. *J. Phys. Condens. Matter* **2004**, *16*, 9047–9057. [CrossRef]
227. McDonald, P.F.; Tam, C.P.; Mok, Y.W. EPR Gd^{3+} in $LiNbO_3$. *J. Chem. Phys.* **1972**, *56*, 1007–1008. [CrossRef]
228. Dischler, B.; Herrington, J.R.; Räuber, A.; Schneider, J.; Urban, W. An EPR study of different Gd^{3+} centers in $LiNbO_3$. *Solid State Communs.* **1973**, *12*, 737–740. [CrossRef]
229. Park, I.-W.; Kim, M.; Choh, S.H.; Kim, J.N. EPR study of Gd^{3+} Paramagnetic Centers in Congruent $LiNbO_3$ Single Crystal. *New Phys. (Korean Phys. Soc.)* **1994**, *34*, 464–469.
230. Bonardi, C.; Carvalho, R.A.; Basso, H.C.; Terrile, M.C.; Cruz, G.K.; Bausa, L.E.; Sole, J.G. Magnetic circular dichroism of Nd^{3+} and Yb^{3+} ions in $LiNbO_3$ crystals. *J. Chem. Phys.* **1999**, *111*, 6042–6046. [CrossRef]
231. Camarillo, E.; Hernandez, J.; Garcia Sole, J.; Caldino, U.; Munoz Santiuste, J.E.; Loro, H.; Voda, M.; Jaque, F.; Murrieta, H. Nd^{3+} centres induced by ZnO or MgO codoping $LiNbO_3$. *J. Phys. Condens. Matter* **1995**, *7*, 96359641. [CrossRef]
232. Ziling, K.K.; Nadolinnii, V.A.; Shashkin, V.V. *Sov.:* Reports of AS of USSR. *Inorg. Mater.* **1980**, *16*, 701.
233. Juppe, S.; Schirmer, O.F. EPR Ti^{3+} in $LiNbO_3$. *Phys. Lett. A* **1986**, *117*, 150–155. [CrossRef]
234. Rakitina, L.G.; Zaritskii, I.M.; Corradi, G.; Polgar, K. Mechanisms of broadening of ESR lines of polaron centers in $LiNbO_3$: Ti subjected to irradiation and heat treatments. *Sov. Phys. Solid State* **1990**, *32*, 654–660.
235. Thiemann, O.; Donnerberg, G.H.; Wöhlecke, M.; Schirmer, O. Vibronic structure, energy level, and incorporation mechanism of Ti^{3+} in $LiNbO_3$ and $LiTaO_3$. *Phys. Rev. B* **1994**, *49*, 5845–5851. [CrossRef]

236. Corradi, G.; Polgar, K.; Zaritskii, I.M.; Rakitina, L.G.; Deryugina, N.I. Defect formation under γ-irradiation and heat treatment in pure and doped $LiNbO_3$ monocrystals. *Sov. Phys. State Solid* **1989**, *31*, 115–122.
237. Corradi, G.; Zaritskii, I.M.; Hofstaetter, A.; Polgar, K.; Rakitina, L.G. Ti^{3+} on Nb site: A paramagnetic Jahn-Teller center in vacuum-reduced $LiNbO_3$:Mg:Ti single crystals. *Phys. Rev. B* **1998**, *58*, 8329–8337. [CrossRef]
238. Corradi, G. Nb^{4+} Polaron and Ti^{3+} Shallow Donor Jahn-Teller Centers in $LiNbO_3$ Systems. In *Defects and Surface-Induced Effects in Advanced Perovskites*; Borstel, G., Krumins, A., Millers, D., Eds.; NATO Science Series (Series 3. High Technology); Springer: Dordrecht, The Netherlands; Boston, MA, USA; London, UK, 2000; Volume 77, pp. 89–100. [CrossRef]
239. Corradi, G.; Meyer, M.; Kovács, L.; Polgar, K. Gap levels of Ti^{3+} on Nb or Li sites in $LiNbO_3$:(Mg):Ti crystals and their effect on charge transfer processes. *Appl. Phys. B* **2004**, *78*, 607–614. [CrossRef]
240. Corradi, G.; Meyer, M.; Polgár, K. Bipolarons localised by Ti dopants in reduced $LiNbO_3$ crystals double-doped by Ti and Mg. *Phys. Status Solidi (c)* **2005**, *2*, 132–135. [CrossRef]
241. Bonardi, C.; Magon, C.J.; Vidoto, E.A.; Terrile, M.C.; Bausa, L.E.; Montoya, E.; Bravo, D.; Martin, A.; Lopez, F.J. EPR spectroscopy of Yb^{3+} in $LiNbO_3$ and Mg:$LiNbO_3$. *J. Alloys Compd.* **2001**, *323*, 340–343. [CrossRef]
242. Dohnke, I.; Trusch, B.; Klimm, D.; Hulliger, J. A study of influence of ytterbium and impurities on lattice parameters and phase transition temperature of Czochralski grown $LiNbO_3$. *J. Phys. Chem. Solids* **2004**, *65*, 1297–1305. [CrossRef]
243. Bodziony, T.; Kaczmarek, S.M. New low symmetry centres of Yb^{3+} impurities in lithium niobate single crystal. *Opt. Mater.* **2007**, *29*, 1440–1446. [CrossRef]
244. Bodziony, T.; Kaczmarek, S.M. EPR and optical study of coupled Yb^{3+} ion pairs in weakly doped $LiNbO_3$:Yb single crystal. *Res. Chem. Intermed.* **2007**, *33*, 885–899. [CrossRef]
245. Bodziony, T.; Kaczmarek, S.M.; Rudowicz, C. Temperature dependence of the EPR lines in weakly doped $LiNbO_3$:Yb—Possible evidence of Yb3+ ion pairs formation. *Physica B* **2008**, *403*, 207–218. [CrossRef]
246. Bodziony, T.; Kaczmarek, S.M.; Hanuza, J. EPR and optical studies of $LiNbO_3$:Yb and $LiNbO_3$:Yb, Pr single crystals. *J. Alloys Compd.* **2008**, *451*, 240–247. [CrossRef]
247. Bodziony, T.; Kaczmarek, S.M.; Kruk, R. Magnetic properties of $LiNbO_3$ single crystals weakly doped by Yb and/or codoped by Pr. *Rev. Adv. Mater. Sci.* **2010**, *23*, 1–7.
248. Malovichko, G.; Grachev, V.; Kokanyan, E.; Schirmer, O. EPR, NMR and ENDOR study of intrinsic and extrinsic defects in disordered and regularly ordered lithium niobate crystals. *Ferroelectrics* **2000**, *239*, 357–366. [CrossRef]
249. Malovichko, G.; Bratus, V.; Grachev, V.; Kokanyan, E. Electron paramagnetic resonance and electron-nuclear double resonance study of nonequivalent Yb^{3+} centers in congruent and nearly stoichiometric lithium niobate. *Phys. Status Solidi (b)* **2009**, *246*, 215–225. [CrossRef]
250. Xu, H.; Lee, D.; He, J.; Sinnott, S.B.; Gopalan, V.; Dierolf, V.; Phillpot, S.R. Stability of intrinsic defects and defect clusters in $LiNbO_3$ from density functional theory calculations. *Phys. Rev. B* **2008**, *78*, 174103. [CrossRef]
251. Xu, H.; Chernatynskiy, A.; Lee, D.; Sinnott, S.B.; Gopalan, V.; Dierolf, V.; Phillpot, S.R. Stability and charge transfer levels of extrinsic defects in $LiNbO_3$. *Phys. Rev. B* **2010**, *82*, 184109. [CrossRef]
252. Xu, H.; Lee, D.; Sinnott, S.B.; Dierolf, V.; Gopalan, V.; Phillpot, S.R. Structure and diffusion of intrinsic defect complexes in $LiNbO_3$ from density functional theory calculations. *J. Phys. Condens. Matter* **2010**, *22*, 135002. [CrossRef]
253. Vyalikh, A.; Zschornak, M.; Köhler, T.; Nentwich, M.; Weigel, T.; Hanzing, J.; Zaripov, R.; Vavilova, E.; Gemming, S.; Brendler, E.; et al. Analysis of the defect clusters in congruent lithium tantalate. *Phys. Rev. Mater.* **2018**, *2*, 013804. [CrossRef]
254. Zhang, Q.-R.; Feng, X.-Q. Defect structures and the MgO-doping-level-threshold effect on the optical absorption of reduced MgO-doped lithium niobate. *Phys. Rev. B* **1991**, *43*, 12019–12024. [CrossRef] [PubMed]
255. Schirmer, O.F.; Thiemann, O.; Wohlecke, M. Defects in $LiNbO_3$—I. experimental aspects. *J. Phys. Chem. Solids* **1991**, *52*, 185–200. [CrossRef]
256. Schirmer, O.F.; Imlau, M.; Merschjann, C.; Schoke, B. Electron small polarons and bipolarons in $LiNbO_3$. *J. Phys. Condens. Matter* **2009**, *21*, 123201. [CrossRef] [PubMed]
257. Schirmer, O.F.; von der Linde, D. Two photon and x-ray induced Nb^{4+} and O^- small polarons in $LiNbO_3$. *Appl. Phys. Lett.* **1978**, *33*, 35–38. [CrossRef]
258. Ketchum, J.L.; Sweeney, K.L.; Halliburton, L.E.; Armington, A.F. Vacuum annealing effects in lithium niobate. *Phys. Lett. A* **1983**, *94*, 450–453. [CrossRef]
259. Sweeney, K.L.; Halliburton, L.E. Oxygen vacancies in lithium niobate. *Appl. Phys. Lett.* **1983**, *43*, 336–338. [CrossRef]
260. Halliburton, L.E.; Sweeney, K.L.; Chen, C.Y. Electron spin resonance and optical studies of point defects in lithium niobate. *Nucl. Instrum. Methods Phys. Res. B* **1984**, *1*, 344–347. [CrossRef]
261. Dutt, D.A.; Feigl, F.J.; DeLeo, G.G. Optical absorption and electron paramagnetic resonance studies of chemically reduced congruent lithium niobate. *J. Phys. Chem. Solids* **1990**, *51*, 407–415. [CrossRef]
262. Müller, H.; Schirmer, O.F. Microscopic structure of Nb_{Li} related defects in reduced undoped $LiNbO_3$. *Ferroelectrics* **1992**, *125*, 319–324. [CrossRef]
263. Zaritskii, I.M.; Rakitina, L.G.; Corradi, G.; Polgar, K.; Bugai, A.A. A new trapped-hole radiation defect in heavily Mg-doped $LiNbO_3$. *J. Phys. Condens. Matter* **1991**, *3*, 8457–8465. [CrossRef]
264. Halliburton, L.E.; Sweeney, K.L.; Kappers, L.A. Self-trapped electrons in lithium tantalate. *Phys. Lett. A* **1986**, *116*, 81–84. [CrossRef]

265. Lee, M.; Gyoo Kim, I.; Takekawa, S.; Furukawa, Y.; Uchida, Y.; Kitamura, K.; Hatano, H. Electron paramagnetic resonance investigation of the photochromic effect in near-stoichiometric **$LiNbO_3$** with applications to holographic storage. *J. Appl. Phys.* **2001**, *89*, 5311. [CrossRef]
266. Miki, T.; Hantehzadeh, M.R.; Halliburton, L.E. A new trapped-hole center in irradiated $LiNbO_3$. *J. Phys. Chem. Solids* **1989**, *50*, 1003–1007. [CrossRef]

crystals

MDPI

Review

Unveiling the Defect Structure of Lithium Niobate with Nuclear Methods

Andreas Kling * and José G. Marques

Centro de Ciências e Tecnologias Nucleares, Instituto Superior Técnico, Universidade de Lisboa, Estrada Nacional 10, km 139.7, P-2695-066 Bobadela, Portugal; jmarques@ctn.tecnico.ulisboa.pt
* Correspondence: akling@ctn.tecnico.ulisboa.pt; Tel.: +351-219-946-154

Abstract: X-ray and neutron diffraction studies succeeded in the 1960s to determine the principal structural properties of congruent lithium niobate. However, the nature of the intrinsic defects related to the non-stoichiometry of this material remained an object of controversial discussion. In addition, the incorporation mechanism for dopants in the crystal lattice, showing a solubility range from about 0.1 mol% for rare earths to 9 mol% for some elements (e.g., Ti and Mg), stayed unresolved. Various different models for the formation of these defect structures were developed and required experimental verification. In this paper, we review the outstanding role of nuclear physics based methods in the process of unveiling the kind of intrinsic defects formed in congruent lithium niobate and the rules governing the incorporation of dopants. Complementary results in the isostructural compound lithium tantalate are reviewed for the case of the ferroelectric-paraelectric phase transition. We focus especially on the use of ion beam analysis under channeling conditions for the direct determination of dopant lattice sites and intrinsic defects and on Perturbed Angular Correlation measurements probing the local environment of dopants in the host lattice yielding independent and complementary information.

Keywords: lithium niobate; intrinsic defects; extrinsic defects; lattice location; radiation damage; ion beam analysis; hyperfine interactions

Citation: Kling, A.; Marques, J.G. Unveiling the Defect Structure of Lithium Niobate with Nuclear Methods. *Crystals* **2021**, *11*, 501. https://doi.org/10.3390/cryst11050501

Academic Editors: Gábor Corradi and László Kovács

Received: 31 March 2021
Accepted: 21 April 2021
Published: 2 May 2021

1. Introduction

Present opto-electronic and future quantum-process based communication technologies rely strongly on optical materials with their inherent or artificially modified properties. Lithium niobate ($LiNbO_3$, henceforth abbreviated LN) is one of the materials that has drawn most attention in this process. Its excellent linear and non-linear optical properties, facility to form waveguide using a variety of methods, ability to incorporate optically active dopants and the applications based on periodically poled crystals make it a favorite target for basic and applied research [1–4].

In order to achieve the modification of its properties with a certain goal in mind, the basic properties of LN have to be understood. Among them, the intrinsic defect structure of the congruent lithium-deficient crystals in comparison to stoichiometric ones, phenomena related with the phase ferroelectric-paraelectric phase transition, incorporation of dopants and their interaction with the lattice and lattice damage production and removal are topics for fundamental research fostering future application directed research.

Methods based on nuclear techniques played and will play an important role in materials research. Their application can clarify or—in combination with other methods—assist to clarify the topics mentioned in the preceding paragraph. While ion beam analysis methods are ideal for research on composition as well as lattice defect concentration and distribution, hyperfine interaction methods can yield complementary information on local environments in the lattice and dopant charge states. As will be demonstrated in the course of this review the combination of several of these techniques can unveil the nature of defects in LN. For illustration, Figure 1 shows a "periodic table of elements" for which nuclear techniques have been applied in the case of LN.

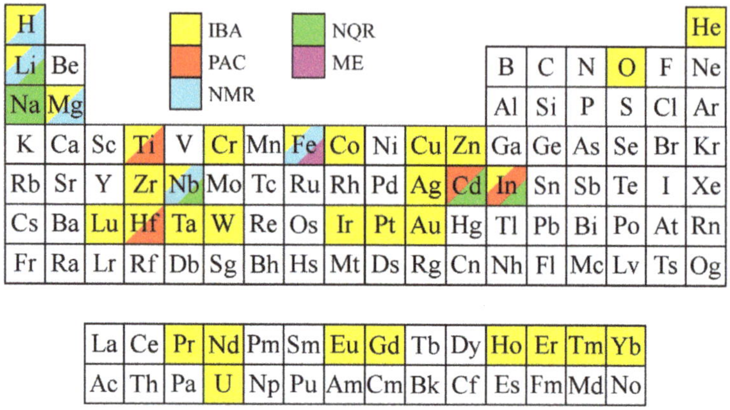

Figure 1. "Periodic table of elements" in LN for which one or several nuclear techniques have been applied in the investigation of their properties. IBA—Ion Beam Analysis, PAC—Perturbed Angular Correlation, NMR—Nuclear Magnetic Resonance, NQR—Nuclear Quadrupole Resonance, ME—Mössbauer Effect.

2. Short Overview on Nuclear Methods for Materials Research

The nuclear techniques addressed in this review are characterized shortly indicating some peculiarities with regard to their application in research on LN. References to general introductory books and reviews for further in depth study are supplied for each of the methods.

2.1. Ion Beam Analysis

Ion beam analysis is a group of nuclear methods that are based on the interaction of energetic ion beams (typically few keV to tens of MeV) with materials [5–8]. Its main applications are:

- Composition determination in bulk materials, surface layers and multilayer systems. Some of the methods attain a depth resolution down to sub-monolayer range;
- Detection and quantitation of lattice damage in monocrystalline materials with high depth resolution up to a few µm from the surface;
- Characterization of intrinsic and extrinsic defects in terms of lattice site location and defect type.

The basic ion beam analysis methods applied in LN are Rutherford Backscattering Spectrometry (RBS) and Resonant Elastic Scattering (RES) [9], Nuclear Reaction Analysis (NRA) [10], Particle Induced X-ray Emission (PIXE) [11], Channeling (C) [12,13], Secondary Ion Mass Spectroscopy (SIMS) [14], Elastic Recoil Detection (ERD) and Heavy Ion Elastic Recoil Detection (HIERD) [15,16].

RBS relies on elastic backscattering of the incoming ions by the target atoms. Its origins can be traced back to the 1909 experiments performed by Geiger and Marsden [17] which led to the first modern model of the atom. The energy spectra of backscattered light ions (protons, He^{+} and He^{++}) in the MeV range provide information on the composition of the material and its depth dependence up to several µm into the bulk. Since the backscattering yield is proportional to the square of the target atomic number, this method works best for targets composed of heavy elements. Therefore, lighter elements in heavy matrices are often difficult or only indirectly determinable while the method excels in the detection of small quantities of heavy elements in light targets. Further, the difference in the maximum backscattering energy decreases rapidly with increasing target atom mass making it difficult to discern elements that are neighbors in the periodic table. In order to overcome the restrictions for light elements, resonant elastic scattering, i.e., deviations from the Rutherford cross section, are employed to improve the sensitivity for their detection. This technique is known as Resonant Elastics Scattering or Elastic Backscattering Spectrometry (EBS). Analysis of RBS and RES spectra is done with dedicated software, e.g., RUMP [18] and NDF [19]. As an example, Figure 2 shows a backscattering spectrum for congruent LN irradiated with 3.1 MeV He^+ demonstrating Nb-RBS and O-RES while Li remains invisible.

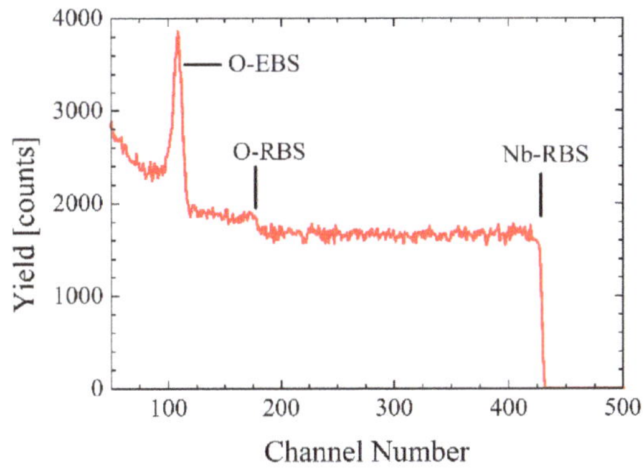

Figure 2. Backscattering spectrum (3.25 MeV He^+) for congruent LN with Rutherford Backscattering for niobium and Resonant Elastic Backscattering for oxygen using the $^{16}O(\alpha,\alpha)^{16}O$ resonance (about 30 times higher than Rutherford scattering) at 3.045 MeV.

Another possibility for the detection and quantification of light elements in heavy matrices is to look for the light charged nuclei produced by ion beam induced reactions in the target material, within the so called NRA technique. Long measuring times are frequently necessary, since the reaction cross sections are typically much smaller than for backscattering. The analysis of the spectra is analogous to that of RBS and EBS. As an example, Figure 3 depicts a spectrum for LN irradiated with 1.75 MeV protons. Contributions of backscattering by the target and by the alpha particles produced in the $^7Li(p,\alpha)^4He$ reaction are clearly distinguishable due to the high Q-value (17.35 MeV) of the reaction employed.

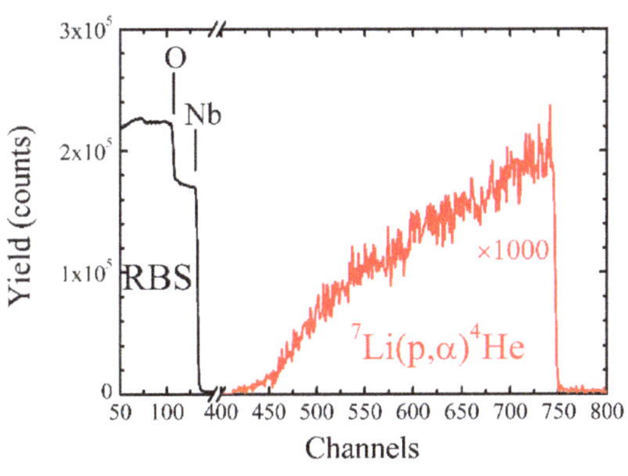

Figure 3. Backscattering spectrum for 1.75 MeV protons on LN and the spectrum of the alpha particles produced in the $^7Li(p,\alpha)^4He$ reaction.

The PIXE technique relies on the production of characteristic X-rays (mainly K- and L-lines) by interaction of the incoming ion (in most cases protons, for special purposes He^+)

with inner shell electrons of the target atoms. It is most useful for the detection of medium and heavy elements in compound materials. The quantitative analysis of the elemental concentrations is significantly more complicated than in the previous cases since the technique integrates over the whole ion path in the material. This lack of depth resolution can be partially overcome by measuring the sample at different angles of incidence. Figure 4 shows an example of an X-ray energy spectrum for LN irradiated with 3.1 MeV He^+ ions.

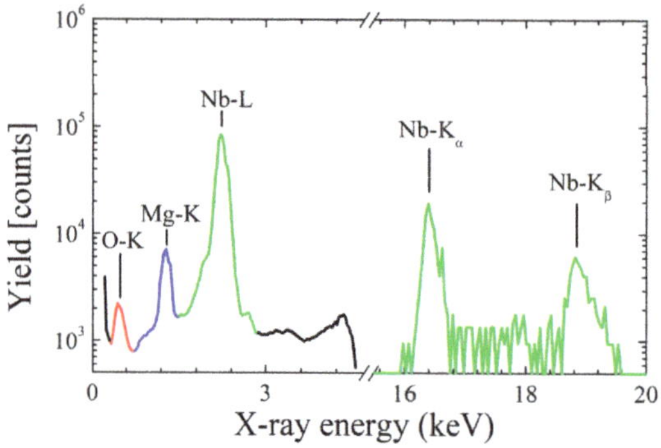

Figure 4. Characteristic X-ray spectrum induced by irradiation of LN by 3.1 MeV He^+ ions.

Ions incident along major crystallographic directions (axes or planes) of monocrystalline materials show the channeling effect. Along a crystallographic axis or plane, the atoms of the crystal atoms appear perfectly aligned in strings or plates, respectively. These form, due to their repellent Coulomb potential, a "channel" in which the ions are guided and the ion fluence is concentrated in the center of the channel. Due to this guiding effect the probability for close encounters with lattice atoms are reduced diminishing the yield for ion scattering, nuclear reactions and X-ray emission.

In a typical channeling experiment the yield of backscattered ions, nuclear reactions products or characteristic X-rays is measured tilting the crystal stepwise from a major crystallographic direction to one of random ion beam incidence (in which the atomic arrangement appears disordered). These "scans" are essentially characterized by two parameters, depicted in Figure 5:

- The minimum yield, χ_{min}, i.e., the ratio of the reaction yield at perfect alignment with a crystallographic direction to that at random incidence. The lower χ_{min} the better is the crystalline quality of the sample.
- The half-width of the scan, $\psi_{1/2}$, that is larger the less defective the material is.

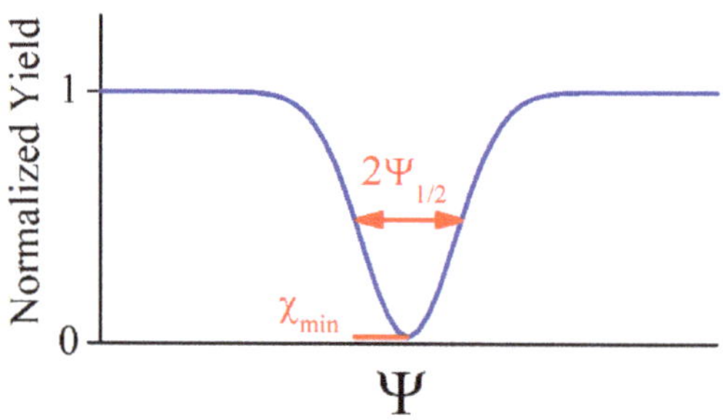

Figure 5. Typical shape and characteristic parameters of a channeling scan.

Possible applications of the channeling effect are [20]:

- Determination of intrinsic lattice defect concentration (typically, genuine defects of the material or radiation induced defects) with high depth resolution. For the analysis, spectra are taken for the undamaged and the damaged sample aligned with a major crystallographic axis and for a random ion incidence angle. From their comparison a depth-resolved defect profile can be derived using appropriate calculation methods or programs, e.g., DICADA [21];
- Determination of intrinsic lattice defect type. Spectra are recorded in the same manner as in the item above but for different ion energies yielding a dependence of the minimum yield difference on E^{κ}, where E is the ion energy. Determining the constant κ allows to distinguish e.g., between point defects, dislocations, inclusions, etc.
- Lattice location of dopants in the crystal. While an angular scan for a dopant located on a regular lattice site of the crystal will follow that one of the host atoms, foreign atoms on interstitial sites will show scans with distinctly different shapes, depending on their relative position in the channel (e.g., a peak instead of a dip if located in the center). Measurements for different crystallographic directions allow to determine the lattice site by triangulation.

The channeling method was initially developed for the study of simple mono- or bi-elemental cubic crystals. Therefore, an extension to materials as complex as LN was not straightforward and it was only attempted in the 1980s. In contrast to simple materials, in most axial directions of LN the channels are highly asymmetric or split in several subchannels. In all cases they are dominated by the heaviest constituent (Nb). Figure 6 shows the projections of five important axial and a planar directions in LN indicating the regular and some interstitial lattice sites together with the respective calculated continuum potentials governing the channels.

The evaluation of the effect of a dopant on a non-Nb site—regular or interstitial—by simple reasoning can lead to serious errors. Therefore, the use of simulation software is mandatory for the lattice site determination of intrinsic and extrinsic defects in LN. Results presented and discussed in this review are based on two channeling simulation programs. FLUX [22,23] is a long-known code for RBS/C and NRA/C in cubic crystals that has been adapted to the trigonal LN lattice. CASSIS [24–27] is a multi-purpose code, i.e., it is applicable for all Bravais lattices and can also deal realistically with PIXE/C and high dopant concentrations. Figure 7 shows results of CASSIS for two-dimensional RBS/C calculations for the *y*-axis.

The SIMS method uses low energy heavy ions (mostly Ar^+) to remove and ionize surface atoms by sputtering and analyzes these secondary ions using a mass spectrometer. This allows determining the composition of materials with very good depth resolution. The sensitivity of the method—ranging from ppm to ppb depending on the material—is its main advantage. However, this method is destructive, in contrast to those mentioned above.

ERD uses typically He-ions with MeV energies to remove hydrogen from a sample surface. An absorber foil in front of the detection stops the backscattered He-ions and allows pass only the protons for analysis. In contrast, HIERD uses heavy ions with specific energy of several MeV/amu in order to sputter even heavy atoms from the surface. Various methods exist for the detection and analysis of the secondary ions emitted.

For better readability, technical details on the ion beam analysis such as ion species, charge state, current, detector geometry, installation used and provenience of samples are mostly omitted in this review. However, the information can always be recovered from the cited works.

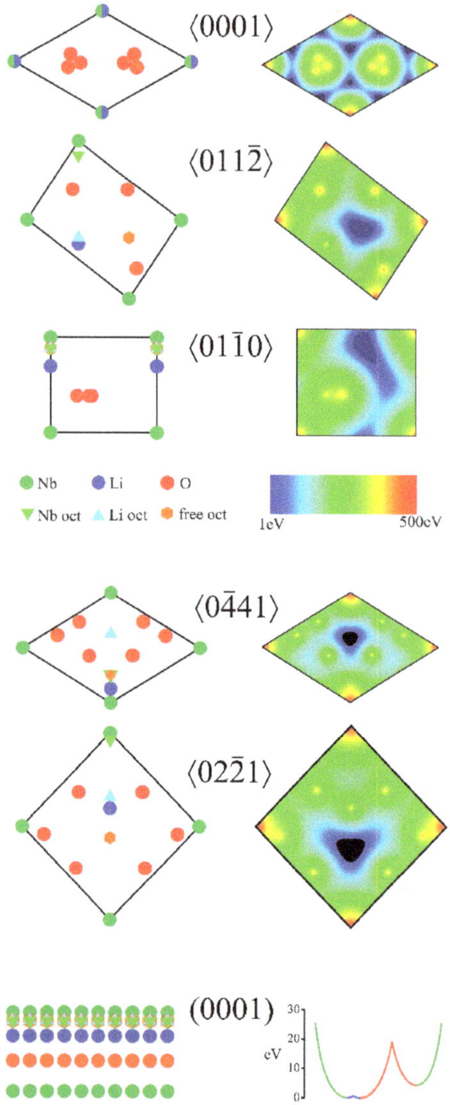

Figure 6. Channel projections with regular and some interstitial sites for important crystallographic directions and the corresponding calculated continuum potentials. The two-dimensional potentials for axial directions are plotted in logarithmic scale; the channeling occurs in the blue regions. The projection of the (0001)-plane and the corresponding one-dimensional continuum potential can be seen on the right side down.

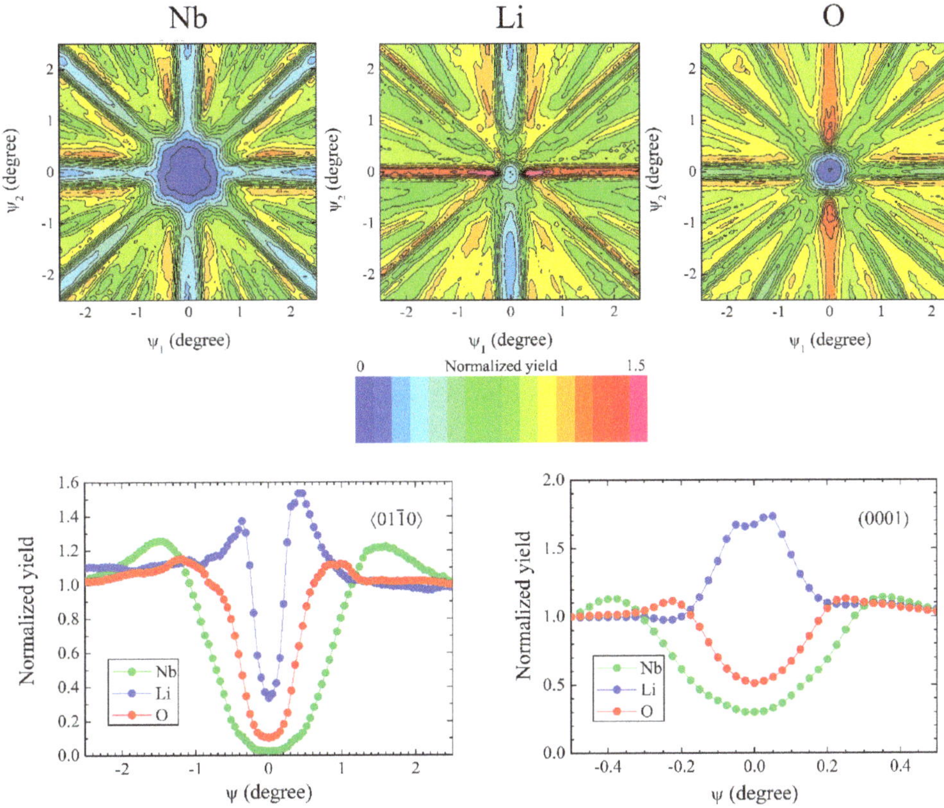

Figure 7. RBS/C calculations with CASSIS. Upper row: Two-dimensional scans for the $\langle 01\bar{1}0\rangle$-axis, ψ_1 corresponds to a rotation around the c-axis and ψ_2 around the a-axis of LN. The calculated yields for Li, Nb and O are shown. Li and O exhibit a yield enhancement for the (0001)- or the $(11\bar{2}0)$-plane respectively. Lower row: One-dimensional scans through $\langle 01\bar{1}0\rangle$ close to $(11\bar{2}0)$, in which Li shows the typical asymmetric yield curve, and through the (0001)-plane, in which Li acts similar to a host atom located in the center of the planar channel.

2.2. Hyperfine Interaction Methods

Hyperfine interaction methods such as Perturbed Angular Correlations (PAC), Mössbauer Effect (ME), Nuclear Magnetic Resonance (NMR) and Nuclear Quadrupole Resonance (NQR) are based on the interaction of a nucleus with an electric (or magnetic) field produced in a solid by the electrons and other nuclei near the nucleus. In several cases, techniques from the ion beam analysis and hyperfine interaction groups are used to study the same samples, obtaining complementary information and enlarging significantly the range of sensitivity for characterization of defects [28].

The PAC probe is an unstable nucleus decaying through a γ-γ cascade that has an intermediate state with suitable half-life and nuclear quadrupole moment (or dipole moment, in case one is interested in magnetic interactions). Conversion electrons from one of the transitions of the γ-γ cascade can also be used [29]. In the case of Hf-doped LN, the PAC probe is usually produced by the neutron capture reaction $^{180}Hf(n,\gamma)^{181}Hf$ on natural Hf introduced in the samples during growth as HfO_2. A typical activity of $\approx$15 µCi is obtained after reactor irradiation. Although thermal neutrons are the most important in the reactor irradiation, the LN crystal is also irradiated with fast neutrons, at a fluence 1–2 orders of magnitude smaller than the thermal flux [30] and these higher energy neutrons produce defects distinct from the ones produced by the thermal neutrons.

Annealing in air at 700 °C is enough to restore the lattice completely [31]. The ^{111}In probe was implanted to a typical fluence of $\approx 10^{13}$ cm^{-2} from which more than 90% were due to an unavoidable stable Cd contamination, resulting in a typical activity of 50 μCi. The crystals were annealed at temperatures in the 700–900 °C range under flowing wet oxygen to remove the defects introduced by ion implantation. PAC measurements were made using standard spectrometers with four BaF_2 detectors in a plane at 90° intervals for γ-γ PAC or two BaF_2 detectors with two magnetic lenses in a plane at 90° intervals for e$^-$-γ PAC [29].

Equation (1) is used to calculate the time differential anisotropy $R(t)$ from the coincidence spectra $N(\theta,t)$, where θ is the angle between detectors and t is the time delay between events.

$$R(t) = 2\frac{N(180^o,t) - N(90^o,t)}{N(180^o,t) + 2N(90^o,t)} \approx A_{22}G_{22}(t). \quad (1)$$

$G_{22}(t)$ is the perturbation function which describes the modulation of the angular correlation and A_{22} is a parameter which depends on the nature of the radiations of the cascade [32]. For a cascade having an intermediate state with spin I = 5/2, such as the ones from the decay of ^{111}In and ^{181}Hf, three frequencies $\omega_n = C_n(\eta)\nu_Q$ are observed, corresponding to:

$$G_{22}(t) = \sum_{n=0}^{3} S_{2n}\cos\left(C_n(\eta)\nu_Q t\right)e^{-\delta C_n(\eta)\nu_Q t}, \quad (2)$$

The quadrupole interaction frequency (QIF), $\nu_Q = eQV_{zz}/h$, and the asymmetry parameter $\eta = (V_{xx} - V_{yy})/V_{zz}$, which can be deduced from the frequency factors $C_n(\eta)$, contain information about the magnitude of the principal component V_{zz} and the asymmetry of the EFG, respectively. The exponential factor allows for a Lorentzian distribution of EFGs around a mean value. Such a distribution can be caused, e.g., by different lattice defects in the vicinity of the probe atoms. The S_{2n} coefficients, which give the amplitudes for each frequency, can be calculated for polycrystalline as well as single crystalline samples. If fractions f_i of the total number of radioactive probes are in different but unique sites experiencing different EFGs the observed spectra is described by a sum over several perturbation functions, $f_i G^i_{22}(t)$, with the sum of the f_i normalized to unity.

Figure 8 shows the PAC coincidence count rates $N(\theta,t)$ obtained with detectors at angles of θ = 90° and θ = 180° and the corresponding spectrum obtained in near-stoichiometric LN (doped with 4.6 wt % K_2O) implanted with ^{111}In. The coincidence count rates $N(\theta,t)$ show the half-life of the intermediate state of the γ-γ cascade from the decay of ^{111}In ($T_{1/2}$ = 80 ns) with the perturbation resulting from the interaction of the quadrupole moment of the intermediate state of the cascade and the EFG produced by the surrounding electric charges. The anisotropy ratio $R(t)$ (usually designated "PAC spectrum") calculated according to eq. 1 extracts the perturbation, exploiting the difference between minima and maxima in the experimental count rates at 90° and 180°.

Details on the other hyperfine interaction techniques can be found, e.g., in references [8,32]. In general one can extract information on the principal component V_{zz} and on the asymmetry of the EFG, although the observable frequencies are dependent on the particular nuclear method and on the nuclear properties of the probe nucleus. One must emphasize that the knowledge of the EFG is not yet enough for a direct assignment of the probe to a given lattice site. Although much progress has been made in the calculation of EFG [33] this is still time-consuming and far from generalized. Nevertheless, one can compare EFG values obtained with different probes in LN [34].

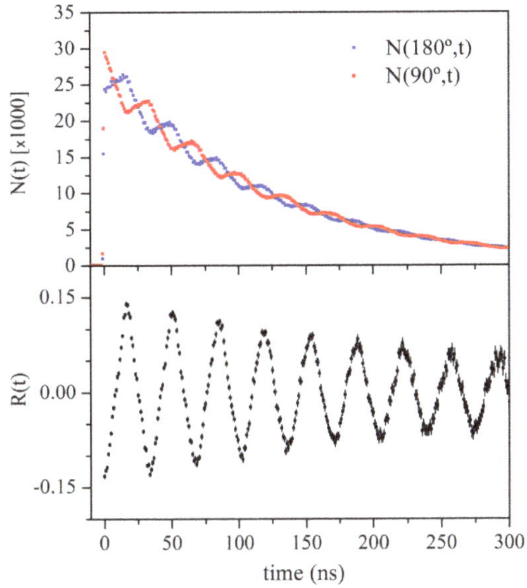

Figure 8. PAC coincidence count rates $N(\theta,t)$ obtained with detectors at angles of $\theta = 90°$ and $\theta = 180°$ (**top**) and corresponding spectrum obtained in near-stoichiometric LN (4.6 wt% K_2O) implanted with ^{111}In (**bottom**).

3. Investigation of Intrinsic Defects

3.1. Congruent Versus Stoichiometric Lithium Niobate

Already in the first growth studies it was noted that LN does not melt congruently with the first phase diagram published by Reisman and Holtzberg [35] in 1958. This non-stoichiometry implies and intrinsic defect structure and influences strongly many material properties which triggered intense research activities on the nature of the defects and the production of stoichiometric material.

Initial NMR measurements with ^{7}Li and ^{93}Nb in LN and lithium tantalate ($LiTaO_3$, henceforth abbreviated LT) were reported in the late 1960s [36–39]. Measurements were performed with the then available magnetic fields below 1 T [40]. These measurements were characterized by wide resonance lines, whose width could be empirically related to the material's departure from stoichiometry. Measurements with NQR [41,42] and ME [43] have also shown broader than expected interaction lines in LN. Later, Peterson et al. [44] found a second ^{93}Nb NMR signal in congruent LN crystals, with a near zero QIF and an intensity of approximately 6% of the total Nb signal that was assigned to Nb in Li sites (5.9%) thus providing a strong support to the Nb vacancy model, even if the authors assumed later that the interpretation of their NMR spectra was not so straightforward [45]. More recent measurements by Blümel et al. [40] using a higher magnetic field of ≈7 T did not confirm a second ^{93}Nb signal with such high intensity, thus favoring instead the Li vacancy model. A series of experimental studies complemented with simulations assuming O-, Li- or Nb-vacancies (but unfortunately omitting models based on ilmenite stacking faults) was published by Yatsenko et al. [46–48], favoring the formation of Li-vacancies in the vicinity of Nb located on regular Li sites.

Rebouta et al. [49] reported RBS/C and ^{181}Hf PAC measurements in LN doped with 1 mol% HfO_2 in the melt. The RBS/C measurements unambiguously determined that Hf replaced Li, but the obtained PAC spectra could only be fitted considering two QIF around 1200 MHz. Two frequencies of the same magnitude were also observed in ^{181}Hf PAC measurements done in polycrystalline LN samples [50]. These QIF were later detailed as ν_{Q1} = 1154(12) MHz and ν_{Q2} = 1213(12) MHz, with asymmetry parameters in the

0.2–0.3 range, from new measurements in single-crystals [34]. NMR studies [51,52] confirm an asymmetry parameter of 0.2 for congruent material. The observed frequency doublet was unexpected, as Hf was found only replacing Li and a single QIF should suffice. Significant progress was made when higher precision PAC measurements were performed using the ^{111}In probe, also replacing Li in congruent LN crystals [53]. Again, the new spectra could not be described by a single frequency even if considering an asymmetric EFG or a broad frequency distribution, both already indicative of crystal imperfections [54]. The frequency doublet now found was ν_{Q1} = 192(2) MHz and ν_{Q2} = 205(2) MHz, with asymmetry parameters in the 0.1–0.2 range. Since the frequencies were again very close, the simplest explanation is that the probes were in slightly different Li sites, one of them possibly linked to a defect located in the Li–Octahedron. Since the fraction of PAC probe atoms located at this additional site is significant (about one third) it should be energetically favorable for the inclusion of atoms causing a charge excess relative to Li. It is therefore unlikely that this defect is identical to a Li-position distorted by a nearby excess Nb atom, since the occupation of such a site by a dopant with a higher charge state than Li would increase further the local charge imbalance. A far more reasonable assumption is to identify this defect with a site, shifted form the regular lithium site that actually incorporates the excess Nb (or dopants) in congruent material. The broadening of quadrupole interaction lines seen by other techniques is also compatible with the existence of two nearby frequencies that cannot be resolved.

Growth procedures developed in the 1990s [55–58] made available high-quality crystals with stoichiometric composition, allowing a direct comparison of the two compositions.

The possible impacts of the Li- and the Nb-vacancy models on axial and planar channeling can be easily estimated qualitatively from the channel projections presented above. Eventually existing vacancies on regular Nb-sites should have only small effects on the Nb-channeling since even in the case of 5.9 mol% vacancies the steering effect is hardly diminished. On the other hand, anti-site Nb-atoms located on regular Li-sites should lower the minimum yields for Li-NRA measurements for congruent material in axial directions clearly since the channeling effect on the Li-rows is enhanced by the Nb. For a 1.2% fraction the average atomic number of the row increases from 3 (stoichiometric) to 3.5 (Nb-vacancy model). For planar channeling in the (0001)-direction Li-NRA-scans should show clear peaks in the aligned direction; but for the congruent case, it should be lower and narrower than for the stoichiometric.

Measurements with congruent and near-stoichiometric LN, grown with 4.6 wt % K_2O in the melt ([Li]/[Nb] = 0.979 [59]), reported by Kling et al. [60,61] showed contrary effects. NRA/C using the ^{7}Li(p,α)^{4}He nuclear reaction and RBS/C scans were performed simultaneously in order to study differences in Li- and Nb-sublattice between the two crystals. Several crystallographic axes and planes were investigated, as shown in Figure 9. The Li-NRA data corresponding to the (0001) plane, perpendicular to the c-axis, were particularly interesting. They showed a prominent flux peak that is considerably broadened in the congruent crystal in comparison to the near-stoichiometric one. The NRA-Li broadening suggests either a disordered Li sublattice or several well-defined defect sites for the Li ions. For the $\langle 02\bar{2}1\rangle$-axis, in which Li is located in the center of the channel, the minimum yield for Li-NRA/C drops drastically ruling out the possibility of simple Nb_{Li} antisites, i.e., the Li- and the Nb-vacancy model [60] Besides the stronger reduction of the ^{7}Li(p,α)^{4}He reaction yield at the surface the spectra for the stoichiometric extend also to higher depth indicating a reduced stopping power due to a diminished number of lattice defects in the channel compared to congruent material [61].

Further, a strong broadening of Nb-RBS/C in the $\langle 01\bar{1}0\rangle$-axis of the near-stoichiometric sample has been observed indicating an improvement of the Nb-sublattice structural ordering with regard to the congruent case [60]. These observations indicate that the excess Nb cannot be located in regular Li lattice sites, favoring the ilmenite stacking fault model but not ruling out the formation of small non-crystalline or Li-deficient (e.g., ilmenite-type

LN or monoclinic lithium $LiNb_3O_8$) regions in the material and on the other extreme, point defects with inversed cation occupation.

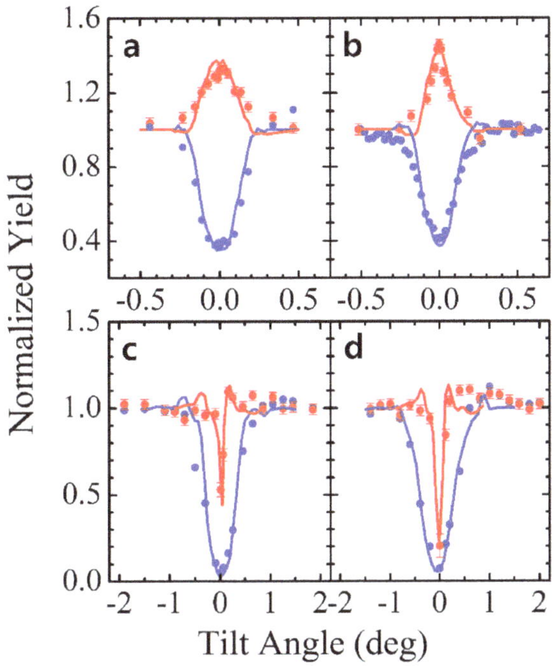

Figure 9. RBS/C dips obtained for the (0001) plane in (**a**) congruent and (**b**) near-stoichiometric $LiNbO_3$, as well as for the $\langle 02\bar{2}1\rangle$ axis in (**c**) congruent and (**d**) near-stoichiometric $LiNbO_3$. The Li data are plotted as red circles and were obtained using NRA via the $^7Li(p,\alpha)^4He$ reaction and the Nb data are plotted as blue circles and were obtained using Rutherford Backscattering. The uncertainty bars for Nb-RBS are smaller than symbols. Adapted from Kling et al. [59,61].

This point can be clarified by studying the Nb sublattice measuring the energy dependence of axial dechanneling induced by defects along several crystallographic directions. The difference in the minimum yields between near-stoichiometric and congruent, ΔX_{min}, at a fixed depth interval yields information about the type and size of defects in a non-perfect crystal [62]. The observed values for κ range between 0.69 to 0.78 falls between the values for true point defects (κ = 1) and extended defect clusters (κ = 0.5) which indicates defects should have a size in the order of nanometers (as expected for local stacking fault sequences). However, a small admixture of true point defects cannot be excluded. In addition, the possibility that microcrystals of ilmenite-type or $LiNb_3O_8$ (κ = 0) or dislocations (κ= −0.5) exist can be rejected.

A comparison was also made with PAC. Figure 10 shows PAC spectra obtained with the ^{111}In probe in the same near-stoichiometric crystal grown with 4.6 wt % K_2O after implantation and annealing [63], compared with a spectrum previously obtained in a congruent crystal [53]. It is clear that the spectrum with the highest attenuation is the one obtained in the congruent crystal. The spectrum obtained in the near-stoichiometric crystal is described by a single QIF, ν_Q = 191(2) MHz, i.e., only one of the two QIF previously found, and with a zero asymmetry parameter. It can also be seen that the pattern from the congruent crystal is not in phase with the one from the near-stoichiometric crystal, with the maxima in the congruent crystal occurring progressively a few nanoseconds after the corresponding ones for the stoichiometric crystal, due to the modulation imposed by the frequency doublet, which is also responsible for a higher attenuation. As the QIF have almost identical values, the PAC measurements suggest the existence of two non-equivalent Li sites in congruent crystals and only one Li site in stoichiometric material. No conclusions can be drawn on the Nb sublattice from these measurements, as ^{111}In replaces Li only [53].

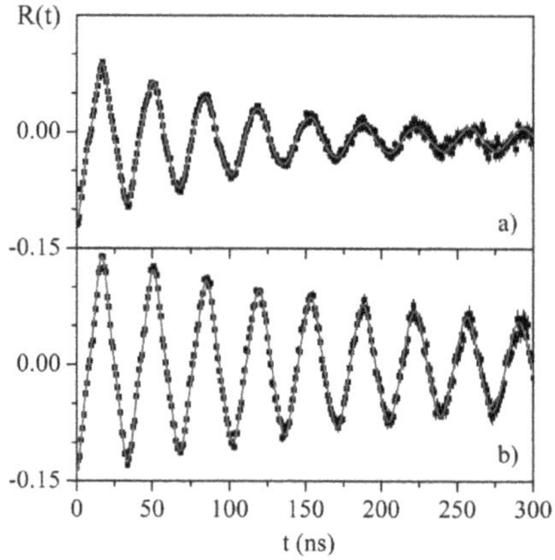

Figure 10. Perturbed Angular Correlation spectra obtained (**a**) in a congruent LN crystal, and (**b**) a near-stoichiometric LN crystal grown with 4.6 wt % K_2O in the melt. Drawn using data from references [53,63].

The joint analysis of the NRA and PAC data thus suggest the existence of two distinct Li sites in a congruent crystal, while only one of them remains in a stoichiometric crystal. Natural candidates for such defects are stacking faults. Several models have been proposed for LN in the literature. Both the Li and the Nb vacancy models include Nb on Li sites and thus a sequence of three Nb consecutive atoms along the c-axis would be created. Nassau et al. [64] proposed different stacking sequences in order to avoid such improbable arrangements and Smyth [65,66] proposed short stacking sequences with the ilmenite structure, which were later found energetically plausible [67]. The stacking sequence along the c-axis of ilmenite-like LN would be Nb, Li, V, Li, Nb, . . . , where V denotes the intrinsic vacancy. Since these inverse stacking sequences accommodate no Nb excess themselves they have to accompanied by Li vacancies. The "inverted" Li site in the ilmenite sequence is generally assumed to be one of this vacancies leading to the "complex defect" model. This defect structure can be regarded as a Li-vacancy model in which the Nb is shifted from the regular Li-site and the inconvenient stacking of three consecutive Nb atoms along the c-axis is avoided. Further, it furnishes a straightforward explanation for the occurrence of the second electric field gradient in the Li–Octahedron observed in hyperfine interaction measurements. Figure 11 compares the regular LN stacking with other sequences, including short ilmenite stacking sequences and the "complex defect".

The channel projections of the lattice sites corresponding to the "inverted" Li and Nb-sites in ilmenite stacking faults are plotted in Figure 12. The Nb on inverted locations shows up close but shifted from the regular Li sites and therefore should induce dechanneling near the Li-rows instead of an enhancement of the channeling effect for regular Li-site occupation. This effect is expected to be most visible in the case of the $\langle 02\bar{2}1\rangle$-axis and the (0001)-plane where these sites are located close to center of the channel.

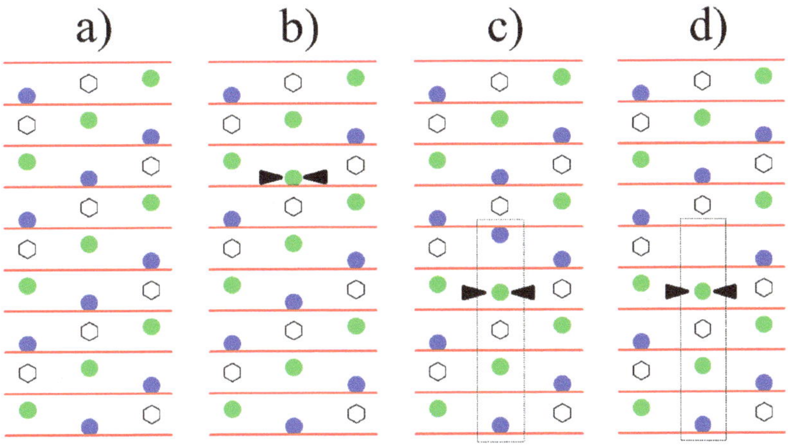

Figure 11. Stacking models for $LiNbO_3$: (**a**) regular stacking, (**b**) Nb-antisites (extra Nb indicated by arrows), (**c**) ilmenite stacking faults (boxed area) and (**d**) like (**c**) but with Li vacancy ("complex defect"). Nb is represented by green, Li by blue and free octahedral positions by white symbols and the oxygen layers are depicted as red lines.

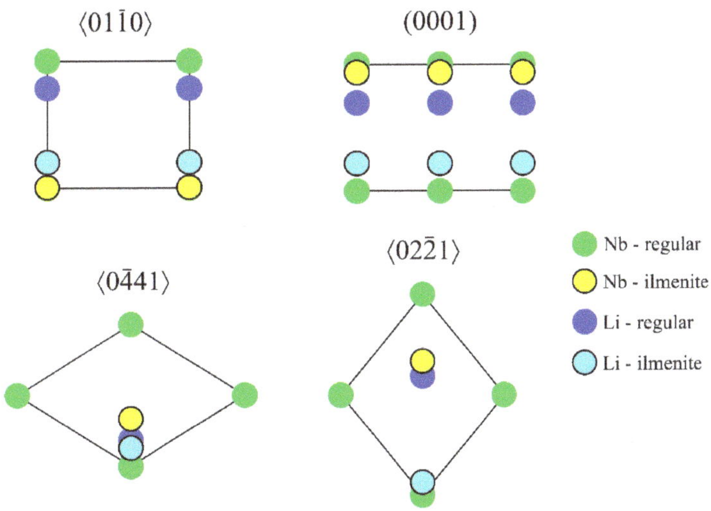

Figure 12. Channel projections with regular and ilmenite type lattice site for Li and Nb for important crystallographic direction.

These results can be related to the lattice structure looking at the interaction potential in the (0001) plane seen by the incoming ions, shown in Figure 13, which was calculated using Molière's approximation to the Thomas–Fermi potential under the assumption of a static lattice [68]. Two pronounced minima are visible, showing the existence of two channels for the incoming ions, separated by a potential barrier associated to O ions. One of those minima, designated as Li-I, corresponds to the regular Li position, whereas the other one, named as Li-II, lies close to a regular Nb location. It could be considered as a Li that substitutes for an Nb and then shifts towards the nearest oxygen plane up to a position symmetrical to that of Li. The existence of those defective Li sites is a key result that provides a confirmation of theoretical analyses [67] proposing the occurrence of inverted Li-Nb sequences as they appear in the related ilmenite structure.

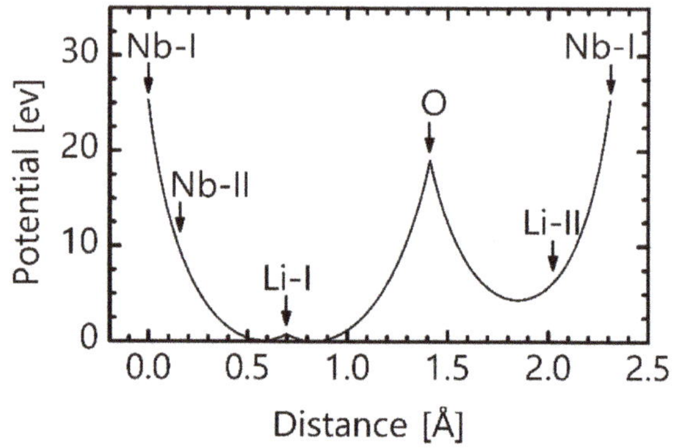

Figure 13. Interaction potential in the (0001) plane of $LiNbO_3$ seen by the incoming protons, calculated using a Molière's approximation to the Thomas–Fermi potential. The positions of Li and Nb in the regular lattice as Li-I and Nb-I as well as in the ilmenite stacking fault as Li-II and Nb-II are marked.

Taking as a basis this local ilmenite structure, the broadened flux peak in the NRA/ channeling spectra has been simulated using the CASSIS code. The peak shape and particularly the broadening of the Li-NRA signal is reasonably well explained. The best fit is obtained assuming 2% of dispersed ilmenite stacking faults, Li—Nb—Nb—Li. On the contrary, other models exclusively involving Nb antisites and cation vacancies cannot account for this key experimental feature, as shown in Figure 14.

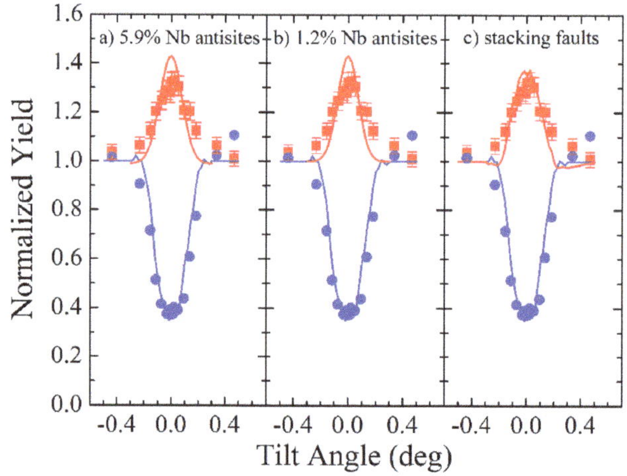

Figure 14. Comparison of experimental results and Monte Carlo simulations for the (0001) Li-NRA angular scan assuming different defect models for congruent lithium niobate: (**a**) 5.9% niobium antisites, (**b**) 1.2% niobium antisites, and (**c**) 2% ilmenite type stacking faults. The Li data are plotted as red circles and were obtained using NRA via the $^{7}Li(p,\alpha)^{4}He$ reaction and the Nb data are plotted as blue circles and were obtained using Rutherford Backscattering. The uncertainty bars for Nb-RBS are smaller than symbols. Redrawn from Kling et al. [61].

The energy dependence of the axial dechanneling reported in [60] has also been simulated using CASSIS. Figure 15 shows a comparison of the experimentally observed energy dependence of the minimum yield X_{min} in several axes for protons together with CASSIS simulations for (a) a defect free stoichiometric crystal; (b) congruent crystal containing 5.9 and 1.2% Nb antisites and (c) congruent cristal containing 2.0% ilmenite stacking inversions with and without 1.2% Nb antisites, from Kling at al. [26]. The best fit to the Nb data is obtained when the ilmenite stacking inversions with 1.2% Nb antisites, i.e., within the Li vacancy model.

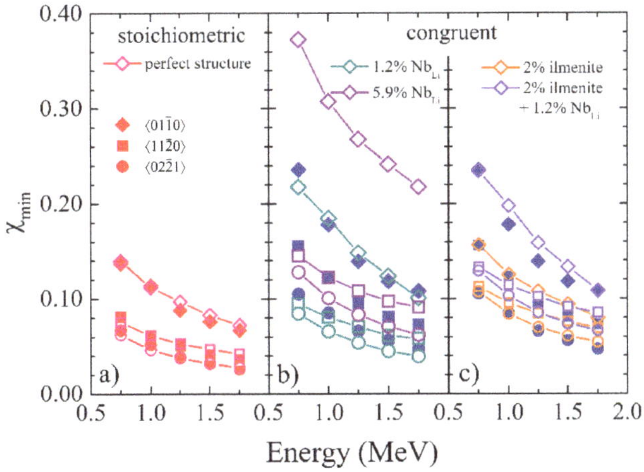

Figure 15. Comparison of the experimentally observed energy dependence of χ_{min} for protons (full symbols) with CASSIS simulations (open symbols) for (**a**) a defect free stoichiometric crystal; (**b**) congruent crystal containing 5.9 and 1.2% Nb antisites and (**c**) congruent cristal containing 2.0% ilmenite stacking inversions with and without 1.2% Nb antisites. Redrawn from Kling et al. [26].

The crystal structures of $LiNbO_3$, perovskite and ilmenite are interrelated and some transitions between them are possible [69,70]. The ilmenite form of LN has been synthetized by ion-exchange from the ilmenite form of sodium niobate, $NaNbO_3$ [71–73]. This LN form transforms back to the regular structure above 900 K. The estimated enthalpy difference between the two LN forms [73] is in good agreement with the calculated difference of 0.1 eV in their lattice energies [67].

The channeling studies on near-stoichiometric LN clearly indicate the absence of any intrinsic disorder—including the occurrence of ilmenite stacking faults—suggesting that in this case the small energy difference is sufficient to suppress their existence. However, LN has a strong tendency to grow non-stoichiometric with the congruent melt showing an excess of Nb. The resulting need for an intrinsic defect structure may locally outbalance the extra energy needed for the formation of ilmenite-type stacking to accommodate the excess Nb and the Li vacancies leaving the "stoichiometric" remainder of the crystal virtually defect-free. This arrangement may result in a global energy minimum for the congruent material.

The current knowledge on the properties of ilmenite-like LN is still limited. Wiegel et al. [74] have reported the UV optical absorption edge in ilmenite-like LN moved towards longer wavelengths when compared to a congruent crystal. For comparison purposes, the UV absorption edge of LN moves towards shorter wavelengths when approaching stoichiometric composition [75]. Baran et al. [76] have reported a very strong Raman scattering peak at about 735 cm^{-1} for ilmenite-like LN. Kong et al. [77] have reported a weak peak at 738 cm^{-1} in congruent LN that is absent in near-stoichiometric crystals, in congruent crystals doped with MgO above 2 mol%, and in congruent crystals doped with 2.5 mol% TiO_2 and suggested that it comes from an ilmenite-like stacking defect in nonstoichiometric LN. Lengyel et al. [78] confirmed the initial decrease of the intensity of a band at around 740 cm^{-1} with MgO doping, but have seen this band reappear for heavily doped crystals.

3.2. Ferroelectric to Paraelectric Phase Transition

The high-temperature phase of LN, above $T_C \approx 1470$ K for material with the congruent composition [79], is paraelectric and belongs to space group $R\bar{3}c$. For LT, the transition temperature is significantly lower, $T_C \approx 878$ K [80]. In this phase, Nb (or Ta) a moves along the c-axis to the center of the octahedral oxygen cages, while for Li two models were proposed in the literature [81]. In the displacive model, Li moves into a triangularly coordinated site within the oxygen layer below the former Li site [82], while in the order-disorder model, the Li atoms are randomly distributed on both sides of the oxygen plane [83,84], as depicted in Figure 16.

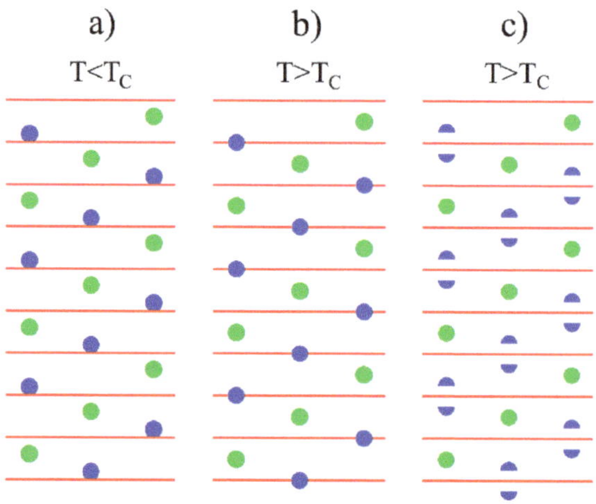

Figure 16. Projection of the $LiNbO_3$ structure, showing two models for the phase transition: (**a**) Li and Nb locations in the ferroelectric phase below T_C; (**b**) Li and Nb locations in the paraelectric phase above T_C for a displacive model; and (**c**) Li and Nb locations in the paraelectric phase for a disordering model, where half-circles indicate a split occupancy with a probability of 50%. Nb is represented in green, Li in blue and the oxygen layers by red lines.

There is no agreement among the theoretical calculations on the phase transition. The calculations of Inbar et al. [85] suggest order-disorder features for the oxygen atoms, while the Li atoms are passive players. More recent molecular dynamics calculations of Phillpot at al. [86] and Sanna et al. [87] suggest that the phase transition presents displacive features for Nb/Ta and order-disorder features for Li. In contrast, the theoretical calculations of Toyoura et al. [88] suggest order-disorder features for both Li and Nb/Ta.

There are few measurements of the temperature dependence of the nuclear quadrupole interaction at the Nb/Ta site in LN and LT. Schempp et al. [41] reported ^{181}Ta NQR measurements in polycrystalline near-stoichiometric LN in a reduced temperature range from 21 to 515 K, where a continuous decrease of the EFG was observed. Moreover, the linewidths of the signal from near-stoichiometric samples were markedly narrower than the ones from Li-deficient material and became narrower with the increase in temperature. Figure 17 plots ^{181}Ta NQR and ^{181}Ta ME results in polycrystalline LT reported by Zhukov et al. [89] and by Löhnert et al. [90], respectively. The variation of the EFG derived from the NQR and ME measurements shows a continuous decrease with temperature in an extended temperature range which includes T_C. This variation can be essentially reproduced using the point-charge model of De Wette et al. [91], taking into account thermal lattice expansion and atom displacement [90], using data from Abrahams et al. [83]. As thermal lattice expansion alone would make the EFG increase moderately with temperature, the main contribution to the EFG comes from the displacement of the atoms. More advanced ab initio cluster calculations of Shelyapina et al. [92] reproduced the essential features of the EFG temperature dependence up to T_C.

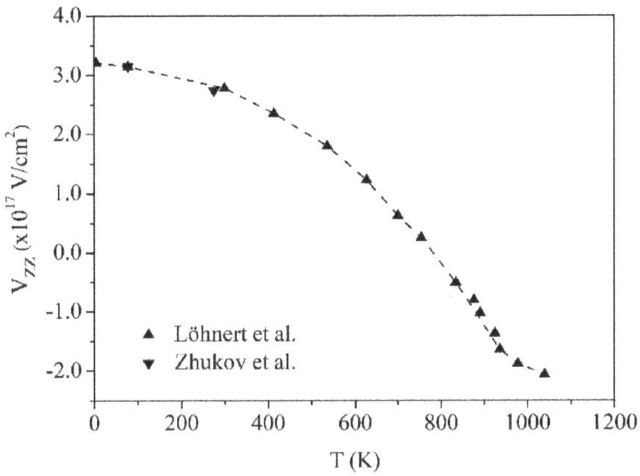

Figure 17. Temperature dependence of the nuclear quadrupole interaction at the Ta site in lithium tantalate measured with ^{181}Ta NQR and ME. Drawn using data from references [89,90]. The dashed line was drawn to guide the eye.

The temperature dependence of the nuclear quadrupole interaction at the Li site was measured using NMR, MS and PAC methods in LN and LT. ^{7}Li NMR measurements in single-crystalline LN were reported by Halstead [39] below T_C, up to ≈950 K. A linear increase of the QIF was observed in this temperature range. The NMR signal had broad lines at low temperature, which became narrower with the increase in temperature, especially after ≈800 K. ^{7}Li NMR measurements in polycrystalline LT were reported by Slotfeldt-Ellingsen et al. [93,94], below and above T_C. A linear increase of the QIF was observed until close to T_C and then a decrease. PAC measurements in polycrystalline LN and LT doped with ^{181}Hf were reported by Catchen et al. [50,95]. They observed extensive line broadening at temperatures well below T_C and significantly less line broadening above T_C. The anisotropy coefficient was non-zero below T_C and dropped to zero above T_C. This result was not expected, because the axial symmetry at the Li site implies that η should vanish at temperatures both below and above T_C. A second QIF was reported in the measurements above T_C, with a value about an order of magnitude lower than the one of Hf probes in Li sites, which was tentatively attributed to Hf probes in Nb/Ta sites [95]. Moreover, in the case of a previous MS study in Fe doped LN, a second QIF was found at high temperature, but no more details were given [96]. Marques et al. [97] reported high precision PAC measurements in single-crystalline LT implanted with ^{111}In. While at RT two nearby QIF were found, ν_{Q1} = 230(2) MHz and ν_{Q2} = 242(2) MHz, with nonzero asymmetry parameters, above T_C only one QIF and $\eta = 0$ was found. The observation of two nearby QIF is consistent with the previous measurements of Hauer et al. [53] in congruent LN and with the extensive line broadening reported in other studies well below T_C. Ohkubo et al. [98] reported PAC measurements in polycrystalline LT with the ^{111m}Cd and ^{117}Cd probes. Figure 18 shows the temperature dependence of the QIF for the measurements in LT.

The temperature dependence of the QIF doublet reported by Marques et al. [97] is shown in Figure 19. The temperature dependence of ν_{Q1} is close to the observed with ^{7}Li in LT [93], as expected since the ^{111}In probe atoms replace Li. The same is applicable for the ^{181}Hf, ^{117}Cd and ^{111m}Cd probes which also replace Li. In contrast, ν_{Q2} that is associated to probes in Li-II sites, shows only a very weak temperature dependence. Close to and above T_C, the derived ν_{Q1} and ν_{Q2} values are no longer distinguishable. This suggests that the differences between the Li-I and Li-II sites vanish in the high-temperature paraelectric phase.

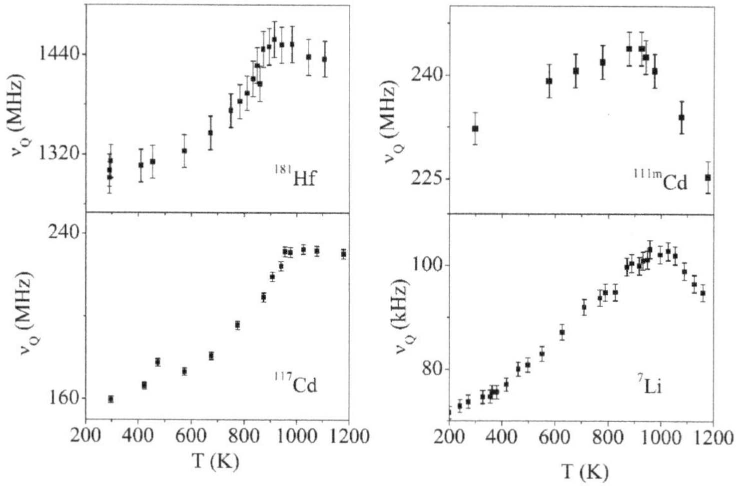

Figure 18. Temperature dependence of the nuclear quadrupole interaction at the Li site measured with the ^{181}Hf, ^{117}Cd, ^{111m}Cd and ^{7}Li probes in lithium tantalate. Drawn using data from references [50,93,95,98].

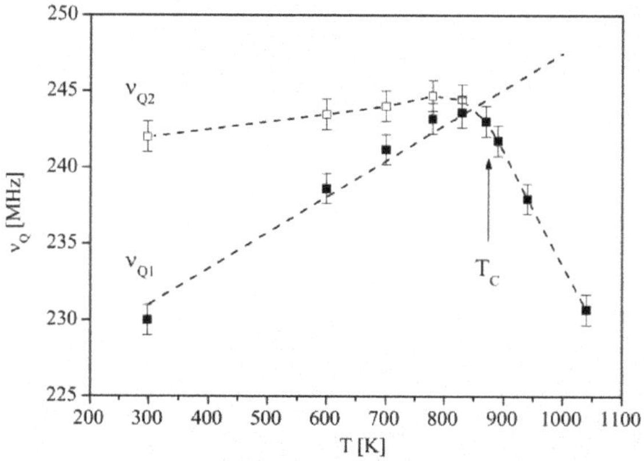

Figure 19. Temperature dependence of the quadrupole interaction frequencies measured with the ^{111}In probe at the Li site in lithium tantalate. Redrawn from ref. [97].

Point-charge model EFG calculations reproduce the observed linear increase of ν_{Q1} up to T_C taking into account thermal lattice expansion only [97]. If the corresponding Li/O displacements of Abrahams et al. [83] are taken into account, then the EFG would be too large by a factor of two close to T_C. The main contribution to the EFG at a Li site comes from the six surrounding O atoms, distributed in two planes, with Li–O distances of 2.08 Å and 2.29 Å at room temperature. At 940 K these distances are 2.05 Å and 2.51 Å [83], which results in a significantly increased EFG. To explain the observed decrease above T_C it is necessary to decrease the difference between the two Li–O distances. Inbar and Cohen [85] proposed that the driving mechanism for the phase transition in LT is a displacement of the O atoms towards the Ta atoms. Since a displacement of the O only along the c-axis would result in too short Ta–O distances, they further proposed that the O atoms move also in the plane perpendicular to the c-axis, toward Li and lead to a coupled Li–O motion. EFG calculations using the above-mentioned point charge model show that such a Li–O coupled movement explains qualitatively the decrease of the EFG above T_C [97].

3.3. *Ion Implantation Induced Defects*

Ion implantation matured to a standard technique for defect engineering and doping of all kinds of materials. In the case of LN the main applications that will be discussed in their respective section are the formation of optical waveguides (Section 5.3) as well as the introduction of optical active (Section 4.3) and insoluble dopants (Section 4.5). This section is dedicated solely to investigation of basic effects of ion implantation leading to the production of intrinsic defects and their removal by annealing.

The production of defects occurs in two ways: (i) by collisions of the incoming ion with a lattice atom that cause its displacement from the lattice site, eventually initiating a collision cascade (nuclear damage); (ii) by excitation of electrons in the material which, in the case of swift heavy ions, can cause the formation of amorphization tracks on the nanometer scale (electronic damage). Since the nuclear damage is mainly occurring at the end of the ion track, it is the dominating damage producer at low energy irradiation. On the other hand, the electronic damage dominates in swift ion irradiations on the major part of the trajectory. In both cases, increasing fluence causes the defects to accumulate until full amorphization of the material is achieved. Defects cause dechanneling similar to interstitials allowing to determine their concentration and depth distribution by ion channeling.

3.3.1. Nuclear Damage Dominated Cases

For irradiation, already the first systematic studies on ion implantation in LN by Jetschke et al. [99] and Götz and Karge [100,101] showed a clear anisotropy in the generation of lattice defects. For irradiation with 150 keV N^+ ions at room temperature amorphization at the surface was reached with fluences of 5×10^{15} cm^{-2} and 2×10^{16} cm^{-2} for x- and y-cut material. In contrast, z-cut material sustained a heavily damaged but not amorphized surface layer even at a fluence of 4×10^{16} cm^{-2}. Three different stages in the damage evolution were distinguished:

(i) In the pre-damage stage (fluence $< 1 \times 10^{15}$ cm^{-2}) mainly point defects were produced of which a significant fraction was attributed to Nb displaced from its regular lattice site to the free octahedron. This explains the lower dechanneling along the $\langle 0001 \rangle$-axis in which these displaced Nb is still aligned with the guiding atomic rows, while for the $\langle 11\bar{2}0 \rangle$-axis this lattice site appears in the center of the channel enhancing the dechanneling.

(ii) The heavy damage stage (1×10^{15} cm^{-2} < fluence < 5×10^{15} cm^{-2}) defect clusters form that cause $\langle 11\bar{2}0 \rangle$- and $\langle 01\bar{1}0 \rangle$-directions to be amorphous but still maintain some alignment with the $\langle 0001 \rangle$-axis.

(iii) In the final stage (fluence $> 5 \times 10^{15}$ cm^{-2}) saturation of defect formation is observed along the $\langle 11\bar{2}0 \rangle$- and $\langle 01\bar{1}0 \rangle$-axis. A recent study [102] revisited the issue employing 350 keV Ar^+ implantation into samples with x- and z-cut as well as a sample cut in a direction equally distanced from $\langle 11\bar{2}0 \rangle$- and $\langle 0001 \rangle$-axis by 45° which allowed to study both axial directions in a single sample. Plotting the mean maximum damage concentration versus the ion fluence yielded perfect overlaps for $\langle 11\bar{2}0 \rangle$- and $\langle 0001 \rangle$ channeling measurements for the x-z cut sample with those of the respective standard cuts. The same work presents an additional confirmation of displaced Nb occupying the intrinsic vacant by angular-resolved RBS/C for a 1 MeV I^+ implanted x-cut LN.

A study on ion irradiation at low temperature (15 K) was performed by Gischkat et al. [103] for various ions (50 keV H^+, 160 keV Li^+, 160 keV O^+, 350 keV Ar^+) at fluences ranging from 5×10^{11} cm^{-2} and 2×10^{17} cm^{-2}. The analysis by RBS/C was done in situ at the same low temperature. Nevertheless, the analyzing He^+-ion beam induced an annealing effect attributed to energy deposition by electronic processes. Taking into account this effect, the undisturbed damage concentrations were calculated indicating an enhanced defect formation with regard to room temperature.

Bianconi et al. [104] deduced, based on own experimental (C^+, 0.5–5.0 MeV, O^+, 0.44–8.0 MeV) and literature data, a universal curve for the Nb-defect fraction in the end-of-track region as function of the energy deposited by nuclear stopping. The relation holds for all ions with $Z \leq 14$ independent of the fluence.

While the Nb-related damage is easily detected by RBS/C the other two constituents of LN are more difficult to access. In a work by Schmidt et al. [105] the $^7Li(p,\alpha)^4He$ reaction was employed to study the Li-related damage in x- and z-cut material irradiated with 1 MeV I^+ (fluences between 2×10^{13} cm^{-2} and 1×10^{14} cm^{-2}). The evolution of the Li-damage concentration with the fluence rate was found to accompany closely that for Nb in the z-cut material. Due to the low contribution of Li to the channeling effect in LN other crystallographic directions than along the ⟨0001⟩-axis are difficult to investigate in terms of defect formation.

Götz and Karge [100] also found an enhancement of the etching rate in diluted hydrofluoric acid for room temperature implantation fluences above 5×10^{14} N^+ cm^{-2} that increased with increasing fluence. Schrempel et al. [106] investigated the etching rates for z-cut LN implanted at 15 K and room temperature with 40 keV He^+ and 350 keV Ar^+ with fluences ranging from 5×10^{12} to 5×10^{16} cm^{-2}. The larger amount of damage created by low temperature irradiation as well as the higher nuclear damage produced by the irradiation by heavier ions increase the etching rate strongly.

Jentschke and Hehl [107] studied the influence of annealing procedures in air on the implantation induced by in P^+ (250 keV) and N^+ (150 keV) implanted LN. For the case of a partial damaged (5×10^{14} N^+ cm^{-2}) x-cut sample, a virtually complete recovery of the lattice was achieved after annealing at 300 °C for 30 min. On the other hand, x- and z-cut samples with RBS channeling spectra indicating full amorphization show only a moderate reduction of the dechanneling even for annealing temperatures up to 1100 °C. This is related to the loss of Li by Li_2O outdiffusion from the amorphous layer during the annealing process. Random RBS spectra of these samples show a surface region with a higher yield of Nb than for the virgin crystal due to the formation of the Li-deficient $LiNb_3O_8$ compound. The loss of Li during annealing at high temperature can be avoided by substituting air by flowing wet oxygen resulting in high-quality recrystallization of the material. This is demonstrated in Figure 20 for Cr-implanted LN after annealing in the two different atmospheres.

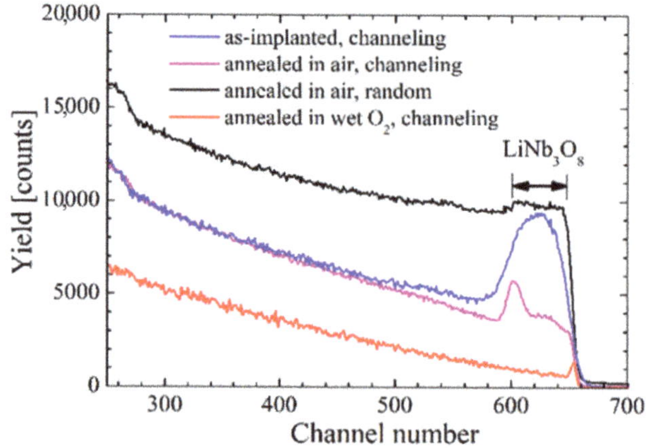

Figure 20. RBS spectra in random and channeling conditions for Cr-implanted LN in the as-implanted state and after annealing in air and wet oxygen. For the sample annealed in air the recrystallization is incomplete and shows the formation of $LiNb_3O_8$ at the surface.

3.3.2. Electronic Damage Dominated Cases

For ions with specific energies above 0.1 MeV/amu, the electronic stopping power outweighs the nuclear along most of the path. For the formation of damage consisting of amorphous tracks the electronic stopping power of the ions has to exceed a material dependent threshold value.

The effect of electronic damage is seen most perfectly separated from nuclear damage for irradiation of materials with GeV ions. Canut et al. [108–110] performed a series of irradiation of LN using Sn-, Gd- and U-ions with energies in the GeV range and fluences ranging between 1×10^{11} to 1×10^{12} cm^{-2}. RBS/C investigations reveal even for the lowest fluences a significant damage in the observable depth range. The lattice disorder was found to increase with ion energy and fluence showing some saturation tendency for the latter at the maximum fluence. Plotting the damage fraction against the electronic stopping power for the different ions reveals a linear dependence.

Irradiations with lower ion energies, a few to tens of MeV, are of interest for opto-electronic applications and have therefore experienced an immense attention in the last two decades. In the following only a selective overview on the observed features can be given. Bentini et al. [111] investigated the implantation of 5 MeV O^{3+} ions with fluences of 1×10^{14} cm^{-2} and 6×10^{14} cm^{-2} into x-cut LN. RBS/C spectra of the as-implanted samples show for all fluences a damage peak at the end-of-range (2.7 μm) of the implantation, while a second peak appears at the surface region for fluences above 4×10^{14} cm^{-2} that has to be attributed to electronic damage. Samples annealed for 2 h at 235 °C in dry oxygen show a significant for the low fluence irradiation but a poor effect for higher fluences. The literature reports further similar RBS/C investigations on electronic damage for C [112], N [113], O [112–115], F [113,115,116], Si [117] and Mg [115].

Determination of the threshold electronic stopping power induced damage was the aim of several works. Rivera et al. [118] reported a value of about 6 keV/nm based on measurements for very low fluences (1×10^{11} cm^{-2} to 1×10^{12} cm^{-2}) investigating a set of six different ion species. Ramos et al. [119] found a smaller value, 3.5 keV/nm, for irradiation with Cl- and Br-ions in the adjacent fluence range (1×10^{12} cm^{-2} to 1×10^{13} cm^{-2}). An extensive study for the higher fluence range (1×10^{13} cm^{-2} to 1×10^{15} cm^{-2}) using O- and Si-ions was recently presented by Wesch et al. [120]. A threshold value of 2.5 keV was extracted using the RBS/C data, in good agreement with the 2.2 keV reported in ref. [113] for 1×10^{14} cm^{-2} O implantation. These data evidence the reduction of the threshold value with increasing fluence. A plot of the threshold electronic stopping power versus the logarithm of the ion fluence can be fitted with a straight line as demonstrated by Agulló-Lopez et al. [121].

The theoretical modelling of the processes leading to amorphization experimentally observed in the electronic stopping power regime is a very active field, steadily generating new approaches. A detailed discussion is therefore out of the scope of this paper. In brief, the most used description is based on thermal spike model that assumes that local melting of the crystal structure occurs above the threshold value [122]. However, this cannot explain the damage production observed even below the threshold. Other effects as the exciton based mechanisms have been invoked as an explanation for this effect [123,124]. Finally, the cumulative character of defect generation has been invoked to explain the damage enhancement and amorphization in sub-threshold irradiation [120,125].

The defect formation by Si^+ ion irradiation of LN along crystallographic directions has been investigated by Schrempel et al. [126,127]. For the case of low ion energies (<1 MeV for Si^+) it was observed that the nuclear damage maxima were shifted to higher depth for aligned irradiation with respect to those of random incidence. This effect is related to the reduced stopping power in axial direction due to the channeling effect that keeps the ion in regions with lower electron density and consequently less energy loss. Additionally, the aligned irradiation creates significantly less defects for the same fluence. For higher energies the electronic stopping passes the threshold. Surface region defect formation is observed and the shift of the damage maxima for channeling irradiation decreases again.

3.4. Neutron Irradiation Induced Defects

The interest in the behavior of LN-based acoustical devices close to the core of a nuclear fission reactor started in the 1970s [128–131]. Materials close to a reactor core are exposed to neutrons from the fission of ^{235}U and to gamma radiation [132]. While the neutrons directly released from fission have an average energy of ≈2 MeV (thus being considered fast neutrons), neutron moderation in most reactors extends the spectrum down to the meV energy range (thermal neutrons). Thus, most irradiations are done with neutrons with a wide energy range, unless specific filters are interposed [133]. Besides structural characterization performed with neutron scattering, LN crystals mostly with congruent composition were irradiated with fast and/or thermal neutrons with fluences up to 10^{20} cm^{-2} [31,49,134–141], although with scarce details on the actual neutron spectrum in many cases. An expansion along the c-axis and a shift towards longer wavelengths of the UV absorption edge were reported by Dowell et al. [128] for neutron fluences up to 4×10^{17} cm^{-2}. Primak et al. [130] reported that irradiated LN becomes highly disordered and shows loss of piezoelectric response and optical birefringence at a fluence of 8×10^{19} cm^{-2}. This process of destruction of LN was found to be significantly reduced when the LN sensors were kept at ≈800 K during the irradiation [131].

The basic physics of neutron damage production is well understood [142,143]. Fast neutrons with energy up to a few MeV produce displacement damage mostly through elastic collisions. In contrast, thermal neutrons produce displacement damage mostly through (n,γ) reactions, due to the recoil induced by the emission of the γ photons. In the case of LN one has to consider also the $^{6}Li(n,\alpha)^{3}H$ nuclear reaction, with a high cross-section for thermal neutrons [144] and whose products, α and ^{3}H, also produce displacement damage. Although in general fast neutrons are expected to create more damage than thermal neutrons, this is not always the case [145].

Figure 21 shows the number of collisions per neutron incident on a thin slab of LN (100 μm thick) as function of energy, covering a range from 0.1 meV to 20 MeV in 69 groups, calculated with the MCNPX Monte Carlo code [146]. Besides the total number of collisions, the contributions coming from collisions with Li, O and Nb are also shown. In the thermal neutron region, until about 1 eV, the number of collisions is essentially driven by reactions with Li (and within Li, with its less abundant isotope, ^{6}Li). After this energy, the number of collisions per incident neutron is essentially constant until about 1 MeV, when it starts to decrease.

In order to evaluate the damage caused by the above collisions, it is necessary to know the threshold energy for displacement of the atoms in LN. Seitz estimated the threshold energy for displacement of an ion or atom in a tightly bound solid, E_d, to be 25 eV [147]. Calculations of E_d values have been done using molecular dynamics for several technologically-relevant materials (see, e.g., references [148–151]), but are scarce for LN. The displacement energy values calculated by Petersen [152] are $E_d(Li) = 30$ eV, $E_d(Nb) = 34$ eV and $E_d(O) = 22$ eV. The value of E_d for oxygen has been estimated from irradiations with electrons as 53 eV [153,154], which is well within the normal range for oxides [155], but significantly higher than obtained by calculation. An estimate for the value of E_d for Hf in Nb and Li sites can be obtained from the PAC measurements of Marques et al. [31]. The typical neutron fluence (all energies) for the preparation of the PAC measurements was less than 5×10^{16} cm^{-2}, well below the fluence for which amorphization was observed [130]. Annealing in air at 573 K was enough to remove the coloration induced by the irradiation and annealing at 973 K was enough to remove all defects, as seen by RBS [31].

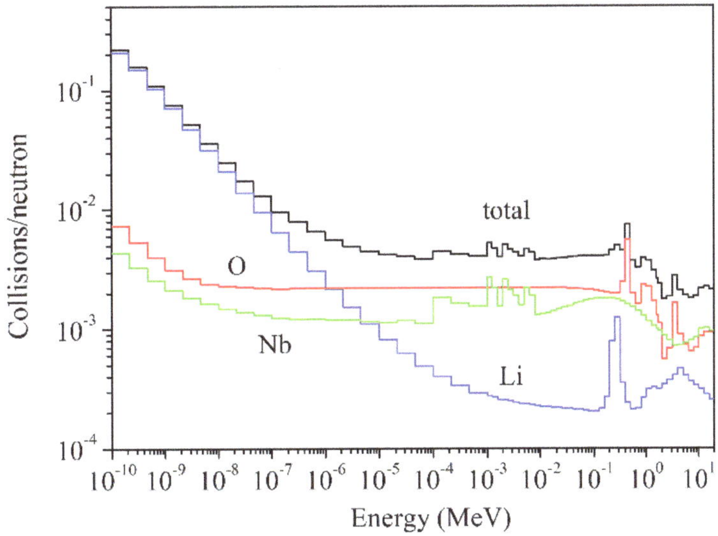

Figure 21. Number of collisions per neutron incident on a thin slab of LN (100 μm thick) as function of energy from 0.1 meV to 20 MeV calculated with the MCNPX Monte Carlo code.

In a congruent LN crystal doped with 1 mol% HfO_2 the spectra before and after thermal annealing show the same characteristic QIF around 1200 MHz, associated with Hf probes in Li sites. In contrast, in a near-stoichiometric crystal doped with 1 mol% HfO_2 where Hf is distributed between Li (45%) and Nb (55%) sites, the spectrum after irradiation shows that the fraction of Hf probes in Nb sites is strongly reduced to 18%, with a full recovery only after annealing at 973 K. However, RBS data shows no change on the lattice site of Hf after irradiation [31]. Thus the change of lattice location after irradiation can only affect the radioactive ^{181}Hf atoms produced via the (n,γ) reaction. Since these atoms are much less than 1 ppm of the total amount of Hf in the crystal, no effect can be detected by RBS. It thus seems reasonable to assume that the displacement is due to the recoil induced by emission of the γ photons from the ^{180}Hf(n,γ)^{181}Hf reaction, characterized by Alenius et al. [156]. As the maximum recoil that can be induced is 96 eV, corresponding to the emission of a 5.6 MeV γ photon, only one displacement per recoil is expected.

The transfer of Hf form Nb to Li sites upon irradiation can be qualitatively explained by the existence of Li vacancies in the congruent and near-stoichiometric crystals. In this way, a recoiled ^{181}Hf from a Nb position finds easily a vacant Li site, leading to the observed change. In the case of a recoiled ^{181}Hf from a Li site, which finds another Li site, no change is seen. If the energy necessary to displace Hf from Li and Nb lattice sites is approximately the same and x represents the fraction of ^{181}Hf recoiled from Li and Nb sites, we can write:

$$x = 1 - \frac{[Hf_{Nb}]'}{[Hf_{Nb}]}, \tag{3}$$

where $[Hf_{Nb}]$ and $[Hf_{Nb}]'$ are, respectively, the fraction of Hf probes in Nb sites before and after irradiation. Using the $[Hf_{Nb}]$ and $[Hf_{Nb}]'$ values and respective uncertainties reported by Marques et al. [31], we obtain $x = 0.37 \pm 0.07$. From the intensity of the discrete values of the γ photons emitted by the ^{180}Hf(n,γ)^{181}Hf reaction, a recoiled fraction of 37% would correspond to the emission of photons with energy of 3979 keV and above, imposing a minimum recoil energy of 47 eV to Hf. Thus, our estimate for the required energy to displace Hf in Nb and Li sites is 47 ± 8 eV, in reasonable agreement with the E_d value for Li and Nb sites calculated by Petersen [152].

Displacement of other dopants as a result of (n,γ) reactions is possible. Table 1 compiles data on the neutron capture cross section at 0.025 eV (1 b = 100 fm^2) [157], main activation reactions(s), respective cross section [158], maximum γ energy released [159] and maximum recoil energy.

Table 1. Recoil resulting from neutron activation of several dopants in LN.

Dopant	$\overline{\sigma_C}$ [b]	Main Activation Reaction(s)	σ_C[b]	Relative Weight [%]	Max γ Energy [MeV]	Max Induced Recoil [eV]	Importance
Er	159.4	^{167}Er(n,γ)^{168}Er	642.3	92	7.691	188	High
Fe	2.56	^{56}Fe(n,γ)^{57}Fe	2.59	93	7.646	551	Moderate
Hf	104.1	^{177}Hf(n,γ)^{178}Hf	375.5	67	6.451	126	High
In	193.8	^{115}In(n,γ)^{116}In	201.0	99	6.561	199	High
Mg	0.063	^{24}Mg(n,γ)^{25}Mg	0.05	63	7.329	1153	Low
Nd	50.5	^{143}Nd(n,γ)^{144}Nd	325.0	78	6.502	158	High
Sc	27.14	^{45}Sc(n,γ)^{46}Sc	27.14	100	8.760	896	High
Ti	6.09	^{48}Ti(n,γ)^{49}Ti	7.84	95	6.760	501	Moderate
Yb	34.8	^{174}Yb(n,γ)^{175}Yb	69.4	63	5.308	86	Moderate
Zn	1.11	^{64}Zn(n,γ)^{65}Zn	0.93	41	7.864	511	Moderate
		^{67}Zn(n,γ)^{68}Zn	6.8	28	9.120	657	Moderate

For the estimation of displacement damage, Dale et al. [160] have shown that is convenient to use a quantity named Non-Ionizing Energy-Loss (NIEL) to quantify the portion of energy lost by a particle that is used for displacement damage, with the remainder of the energy being dissipated as heat. It was shown that NIEL can be used to correlate displacement damage effects in materials induced by different types of radiation [160], even if some exceptions are known [161]. For a particle with energy E_0 incident on a material with a single type of atom with atomic weight A, it is defined as:

$$NIEL(E_0) = \frac{N_A}{A} \int_{T_{min}}^{T_{max}} \left(\frac{d\sigma}{dT}\right)_{E_0} Q(T) T dT, \tag{4}$$

where is N_A is Avogadro's number, $d\sigma/dT$ is the total differential cross section (elastic and inelastic) for energy transfer to an atom of the material, T_{min} and T_{max} are the minimal and maximal energies that can be transferred to a lattice atom, and Q(T) is a partition function which gives the fraction of T that is lost to NIEL. T_{min} is usually taken as $T_{min} = 2E_d$, while $T_{max} = 4E_0A/(A + A')^2$, where A′ is the mass of the incident particle. The partition function Q(T) has a classical formulation given by Lindhard et al. [162] and fitted by Robinson et al. [143], as well as a more recent formulation by Akkerman et al. [163] with improved handling of low energy ions. Total NIEL for LN results from the application of Equation (4) to Li, Nb and O, taking stoichiometry in account, as:

$$NIEL(LiNbO_3, E_0) = \frac{A(Li)NIEL(Li, E_0) + A(Nb)NIEL(Nb, E_0) + 3A(O)NIEL(O, E_0)}{A(Li) + A(Nb) + 3A(O)}, \tag{5}$$

The product of the NIEL and the particle fluence gives the displacement damage energy deposition per unit mass of material. NIEL plays the same role to the displacement damage energy deposition as the stopping power to the total ionizing dose. Figure 22 shows the values of NIEL for electrons, neutrons, protons and alphas incident in LN, with energy up to 100 MeV, calculated with the tools available at the SR-NIEL platform [164], using the Akkerman partition function. Displacement energy values of 30 eV, 35 eV and 25 eV were used for Li, Nb and O, as the nearest to the values calculated by Petersen [152], due to the constraints of the calculations for neutrons. NIEL for incident neutrons has a weak dependence of the value of E_d: a change from 20 to 50 eV in the E_d of the 3 species leads to a neutron NIEL average decrease of 3% in the range 10^{-4}–10^{-2} MeV and less than 1% outside this range. Table 2 shows discrete values, for reference. The tabled values for neutrons were obtained by a simple linear interpolation between the two nearest values. It is worth noting that at 1 MeV protons are about 20 times more effective than neutrons in producing displacement damage in LN.

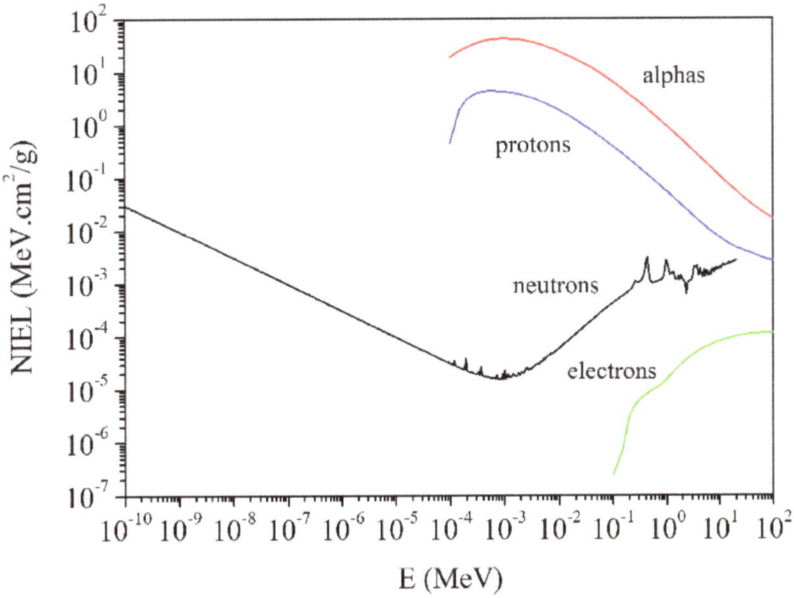

Figure 22. Non-Ionizing Energy-Loss (NIEL) for electrons, neutrons, protons and alphas incident in lithium niobate, with energy up to 100 MeV, calculated with the tools available at the SR-NIEL website (www.sr-niel.org; acessed on 31 March 2021).

Table 2. Non-Ionizing Energy-Loss (NIEL) values for electrons, neutrons, protons and alphas incident in lithium niobate, calculated with the tools available at the SR-NIEL website (www.sr-niel.org; acessed on 31 March 2021).

Energy [MeV]	NIEL e^- [MeV·cm^2/g]	NIEL n [MeV·cm^2/g]	NIEL p [MeV·cm^2/g]	NIEL α [MeV·cm^2/g]
0.001		1.886×10^{-5}	4.255	$4.285 \times 10^{+1}$
0.01		5.984×10^{-5}	1.914	$2.450 \times 10^{+1}$
0.1	2.346×10^{-7}	4.229×10^{-4}	3.932×10^{-1}	6.528
1	1.506×10^{-5}	2.568×10^{-3}	5.537×10^{-2}	9.452×10^{-1}
10	8.037×10^{-5}	1.962×10^{-3}	7.368×10^{-3}	1.090×10^{-1}

4. Extrinsic Point Defects

This section is dedicated to the determination of lattice sites occupied by foreign atoms in LN. The influence of dopant concentration, co-doping and stoichiometry of the material are addressed. The subsections are mainly organized according to the functionality of the dopant.

4.1. Dopants to Increase Photorefractive Damage Resistance

Although congruent LN exhibits excellent optical, electro-optical and non-linear optical properties, the relatively low optical damage resistance hampers its application in optical devices [165]. Besides the attempt to improve the optical damage resistance by growing stoichiometric LN, the introduction of divalent (Mg^{2+} [166,167], Zn^{2+} [168]), trivalent (Sc^{3+} [169], In^{3+} [170]) and tetravalent (Hf^{4+} [171], Zr^{4+} [172]) dopants has been proven to be successful. The observed minimum concentration for suppression of the optical damage (4.5 mol% for doping with MgO, 6 mol% for ZnO, 1.5 mol% for Sc_2O_3 and

In_2O_3 and 4 mol% for HfO_2 and 6 mol% for ZrO_2) indicate that it is important to clarify the lattice location and interaction of these dopants with intrinsic defects of LN.

As the doping with Mg is by far the most technically most important way innumerous studies about the characteristics and application extensively studied and attempts were made. From the more fundamental viewpoint, the main interest is to understand the incorporation of this dopant in dependence on its concentration and to elucidate the underlying mechanisms.

Most of the theoretical approaches published so far depend largely on the model adopted to explain the non-stoichiometry of the congruent material. Zhang and Feng [173] proposed a series of overlapping concentration dependent mechanisms, based on the Nb-vacancy model. At low concentration (<2.7 mol%) Mg substitutes Nb on antisites until they are consumed; for higher concentration up to the threshold Mg is incorporated at the Li site and accompanied by a trapped electron for charge compensation; above the threshold the incorporation occurs on both Li and Mg sites. These ideas were readopted by Liu et al. [174] and more recently by Abdi et al. [175]. Based on the Li-vacancy model, Iyi et al. [176] propose that Mg at first replaces Nb at Li sites up to 3 mol% MgO at which the removal of Nb_{Li} is completed. In the range between 3 and 8 mol% MgO magnesium will only occupy regular Li sites while at higher concentration both cation sites are incorporating Mg. The studies by Donnerberg et al. [177,178] (further refined in [179,180]) are based on the occurrence of ilmenite-type stacking faults with Li vacancies but are formally treated as Nb antisites and Nb vacancies. In the following, we also adopt this procedure for simplicity. According to this model, Mg is replacing in the two concentration ranges < 1.5 mol% and 1.5 to 5.0 mol% simultaneously Nb on antisites and Li on regular sites employing different mechanisms. The preferential substitution of Nb_{Li} by Mg in the second range results in their complete removal at the threshold level. For even higher Mg concentrations both Li and Nb sites should be occupied.

Since a direct determination of the lattice site of magnesium in dependence of its concentration has to be regarded as essential to decide which of the proposed models is the most adequate, an extended channeling study on a set of Mg- doped crystals was performed employing He^+ in order to improve the attainable minimum yield and minimize bremsstrahlung background [181]. As no adequate nuclear reaction for the study of magnesium is available, PIXE/C has been applied. Due to very low energy of the characteristic Mg-K X-rays a Si(Li)-detector with an extremely thin Be window was used to limit absorption. This set of crystals, containing one undoped and six doped (0.5, 1.0, 2.0, 4.0, 6.0 and 9.0 mol% MgO in melt), was grown at the Siemens Research Laboratory in Munich and thoroughly characterized regarding crystal composition, lattice constants, density, temperature, piezoelectricity and various optical properties [182–187].

Angular scans crossing through the $\langle 0001 \rangle$-axis showed for all concentration dips with a width and minimum yield for the Mg-PIXE between those of Nb-RBS and Li-NRA. Computer simulations with CASSIS confirmed that this is consistent with the absence of non-axial lattice sites for Mg.

For MgO concentrations of 0.5 and 1.0 mol% the most striking observation is the strong peak in the Mg-PIXE yield for alignment with the $\langle 11\bar{2}0 \rangle$-axis. This indicates that Mg is not occupying a regular cation lattice site but some interstitial site. The fact that the Mg- PIXE scans for the other axial directions show clear dips and a similar behavior to the Li-scans one can further exclude that Mg is located in the intrinsically vacant or in the Nb-octahedron. The computer simulation with the CASSIS code allows to identify the occupied position as the center of the Li–Octahedron which yields for Mg an oxygen environment similar to that of magnesium oxide. The epitaxial growth of LN on MgO substrates it is known that the (111)-plane of MgO and (001)-plane of LN have the same oxygen framework [188] which might explain this behavior.

It is tempting to identify the observed behavior with a substitution of Nb in the stacking but this would require a substitution of the following type:

$$2\,MgO + 3\,Li_2O + 3\,V_{Li} + 2\,Nb_{Li} \rightarrow 2\,Mg_{Li} + 6\,Li_{Li} + Nb_2O_5 \quad (6)$$

incorporating so much lithium that a near-stoichiometric crystal would be obtained. However, the measurements on the lithium fraction and Curie temperature for these samples exclude this scenario [182]. It is more likely that the Mg occupied sites associated to the ilmenite stacking faults (Li vacancies or Li sites in the vicinity) and the movement to the center of the Li–Octahedron can be interpreted as a relaxation mechanism.

The Mg-scans along the $\langle 11\bar{2}0\rangle$-axis for concentrations above 2.0 mol% MgO in melt differ clearly from those at low concentration. For all axial directions clear minima are observed with scan curve that resemble those of Li. However, computer simulations demonstrate that Mg solely occupying regular Li sites would result in minimum yields that are significantly lower than the observed ones. The best fit to the experimental data was obtained assuming 2/3 of the Mg to occupy regular Li sites and 1/3 sites in the center of the Nb oxygen cage (occupation of regular Nb sites yielded even lower minima). Interestingly this observation holds for the whole concentration range from 2.0 to 9.0 mol% MgO, without any change at the threshold concentration for photorefractive damage suppression. Indeed, there exists a substitution reaction for divalent dopants that would be compatible with these findings:

$$3MgO + V_{Li} + Li_{Li+}\,Nb_{Nb} \rightarrow 2\,Mg_{Li} + Mg_{Nb} + LiNbO_3, \tag{7}$$

which on the other hand does not remove any antisites. A possible solution is that the reactions described by Equations (6) and (7) occur simultaneously until the threshold of 6 mol% Mg in the crystal is reached, causing a deviation in the Mg_{Li}/Mg_{Nb} ratio that is too small to be detectable by the PIXE/C measurements.

A further study on this set of crystals focused on the influence of the Mg incorporation on the Li-NRA channeling minimum yield [189] providing further insight on the lattice modifications. The minimum yield for an atomic string in channeling directions depends strongly on repelling force of its continuum potential. The replacement of Li (Z = 3) by atoms with higher atomic number Nb (Z = 41) or Mg (Z = 12) increases the continuum potential of the Li-string and therefore reduces the minimum yield. Figure 23 shows the dependence of the Li-NRA minimum yield in various axial directions in dependence of the MgO-concentration. The main observation is that the minimum yield increases steeply for all but the $\langle 0001\rangle$-axis (in which all cations are aligned with each other) strongly for MgO melt concentrations up to 1 mol%. This indicates an increase in the disturbance of the lattice due to the interstitial location of magnesium rather than changes of the continuum potential due to removal of excess Nb. In the investigated axial directions ilmenite-like and regular Li site occupying Nb are located very close and do not produce distinguishable effects. For further increasing Mg concentrations the $\langle 11\bar{2}0\rangle$- and $\langle 0\bar{4}41\rangle$-axis no systematic alteration of the minimum yield is observed since the Li atomic strings are in close neighborhood to the—in terms of channeling—dominating Nb-rows. In addition, for both axial directions χ_{min} does not decrease anymore to the value of the congruent case. On the other hand, Li is isolated in the $\langle 11\bar{2}0\rangle$-axis within the channel and the incorporation of the Mg reinforces the repelling force of the Li strings. This leads to a steady decrease of the Li-NRA minimum yields for this axial direction reaching at 9 mol% MgO a value below that of the congruent material.

In addition, one NMR study on ^{25}Mg (nuclear spin 5/2+) has been reported by Feng et al. [190] for LN doped with MgO with concentration ranging from 1.0 to 6.0 mol%. The NMR spectra for ^{25}Mg show clear differences for the various concentration but surprisingly no new line, that would indicate change in the lattice site, appears for concentrations above the threshold. Yatsenko et al. [191] used an indirect approach to obtain information on the incorporation of Mg using ^{7}Li- and ^{93}Nb-NMR in samples doped with 2 to 7 mol%. They conclude that the amount of vacancies is steadily increasing up to the threshold at which on the other hand no more Nb-antisites are present. Mg is supposed to occupy Li sites forming next-neighbor Mg_{Li}-V_{Li} complexes.

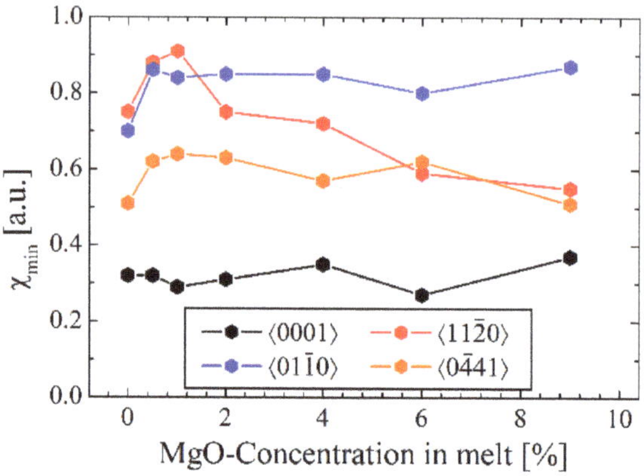

Figure 23. The dependence of the Li-NRA minimum yield on the MgO doping concentration for various axial directions. Redrawn from Kling et al. [189].

Although less popular as dopant for photorefractive suppression in LN, Hf is highly interesting for studies on the basic incorporation mechanism of dopants since it can be accessed simultaneously by RBS/C and PAC. The advantage use of the combination of these two nuclear techniques in investigating has already been demonstrated in the case of intrinsic defects.

Hf-doped crystals LN were grown with 1.0 mol% [49,192] and 6.0 mol% [138] of HfO_2 in congruent melt. For comparison, samples of near-stoichiometric material, to which 4.6 wt % K_2O was added, were doped with 0.2 and 1.0 mol% [138,193] HfO_2.

RBS enabled to determine the actual concentrations of the melt grown samples and derive the corresponding distribution coefficients (Table 3). Channeling measurements for various axial direction proved that in the 1.0 mol% congruent crystal Hf occupies exclusively the Li site [49,192]. For the samples doped with 6.0 mol% RBS measurements revealed that the Hf distribution in the samples is inhomogeneous (3.5–5.1 mol%) and the distribution coefficient lower than one [138]. Further, the higher minimum yield (10%) at ⟨02$\bar{2}$1⟩-axis for Nb and Hf in comparison to the 1.0 mol% sample (2.5%) indicates an inferior crystal quality. Both findings indicate that the maximum solubility of Hf in LN has been exceeded in the sample. However, the quality of angular scans was still sufficient to obtain meaningful results for the incorporated Hf-fraction: the best fit was obtained assuming 55% of the Hf to occupy regular or slightly shifted (0.1 Å towards the octahedron center) Nb lattice sites and 45% to occupy regular Li sites. This ratio is not compatible with a charge compensation mechanism suggested in theoretical studies for which 25% of the Hf has to reside on Li sites and 75% on Nb sites.

Table 3. Incorporation of Hf into crystals grown from congruent and stoichiometric melts.

Melt Type	Concentration in Melt (mol%)	Concentration in Crystal (mol%)	Distribution Coefficient
congruent	1.0	1.3	1.3
congruent	6.0	3.5–5.1	0.58–0.85
stoichiometric	0.2	0.6	3.0
stoichiometric	1.0	1.7	1.7

Besides the already known frequencies for Hf in Li sites, the Fourier analyses of the PAC spectra obtained in the crystal doped with 6 mol% HfO_2 and in the near-stoichiometric crystals has shown another peak at a frequency about 3.5 times smaller. The relative intensity agreed with the fraction for Hf in Nb sites determined by RBS and the EFG scaled with the ^{93}Nb NMR measurements previously reported by Peterson et al. [44]. It can thus be safely assigned to Hf probes occupying Nb sites at room temperature. A frequency doublet for probes in Nb sites, ν_{Q1} = 353 MHz and ν_{Q2} = 494 MHz, was necessary for a good fit of the PAC spectra [193]. However, the best fits for the 6 mol% congruent and 0.2 mol% near stoichiometric crystals, for which the fraction of probes in Nb sites was close to 50%, were obtained without a fraction of probes in Li II sites. This suggests that the defect associated Li II site has disappeared or is of too small importance to make an effect.

The influence of stoichiometry on the incorporation of Hf into LN was studied by RBS/C on samples grown from stoichiometric melt [138,193]. The sample with higher Hf concentration (1.0 mol%) shows clearly a mixture of Li- and Nb-lattice sites occupied by Hf. Computer simulations with FLUX yield a best fit for 66% Nb sites and 34% Li sites, i.e., roughly a 2:1 ratio. This deviates from the 3:1 ratio expected in the case of charge compensation during the incorporation of Hf into LN. From angular scans of 0.2 mol% HfO_2 sample, which can be regarded closer to a stoichiometric case, a fraction of more than 85% of Hf is located on the Li site. It can be concluded that Hf, despite being a tetravalent ion, shows a strong preference to incorporation on Li sites.

In order to demonstrate the influence of co-doping in LN on the incorporation mechanisms in LN, the case of Hf-Mg will be discussed in detail. A set of single crystals of congruent LN doped with 1 mol% of HfO_2 in melt and MgO concentrations ranging between 2.0 and 4.0 mol% in steps of 0.5 mol% as well as one with 6.0 mol% have been studied by RBS/C and PAC [194–197]. The evaluation of the first set of co-doped samples showed that the fraction of Hf located on Li sites decreases constantly with increasing MgO concentration from about 90% at 2.0 mol% MgO to about 20% at 4.0 mol% and changes to full occupation of Nb sites at 6.0 mol%. This reveals that the tendency of Mg to occupy Li sites is stronger than that of Hf but falls short on being a charge compensation effect. Unfortunately, only for the case of 6 mol% MgO data on the lattice site occupation by Mg are available from an additional study [198]. In this case the mixed occupation of regular Li and Nb-octahedron sites as in the single Mg-doped case has been confirmed indicating that Mg may keep its incorporation mechanisms despite co-doping.

PAC measurements were performed in the same crystals after irradiation with neutrons and annealing. The spectrum obtained in the 6 mol% co-doped crystal could only be fitted considering a doublet ν_{Q1} = 327(10) MHz and ν_{Q2} = 398(10) MHz for probes in Nb sites. This frequency doublet is slightly different from the one observed for Hf probes in Nb sites of near-stoichiometric crystals, although no shift was observed for probes in Li sites, ν_{Q1} = 1154(12) MHz and ν_{Q2} = 1213(12) MHz. Figure 24 shows the dependence of the fractions of Hf on the different Li and Nb sites on the MgO concentration. The fraction of Hf occupying the Nb antisite (Li-II site) is constant up to 2.5 mol% MgO in the crystal. In this concentration range, Mg is exclusively located at the center of the Li octahedron. It therefore appears that Mg substitutes in Li_{Li} or vacant Li sites, decreasing the number of Li sites available for Hf. Further, full suppression of Hf on the Li-II occurs between 4 and 6 mol%, the region in which the suppression of the photorefractive damage is completed.

For indium only one PIXE/C study for co-doping with 6 mol% MgO in melt is available [199]. The results indicate that In mainly substitutes Li in its regular lattice site but a fraction occupying regular Nb sites cannot be ruled completely.

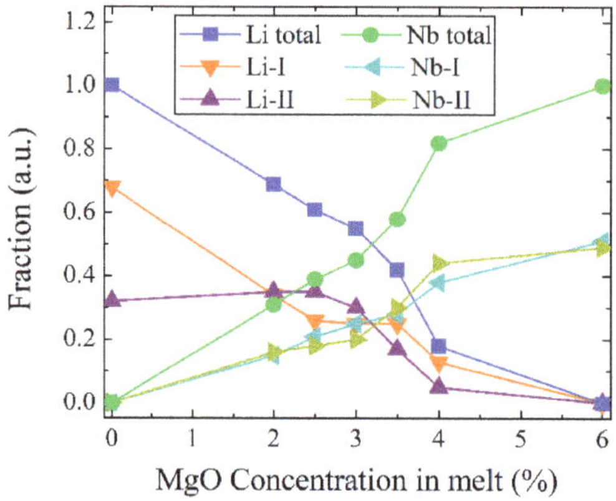

Figure 24. Dependence of the fractions of Hf on the different Li and Nb sites on the MgO concentration as determined by PAC.

4.2. Dopants to Enhance Photorefractive Damage

Despite the photorefractive effect being undesired for most optical applications it has an interesting application in the field of holographic recording of data [200]. Therefore also dopants that increase this effect have been the target of intense research. So far Fe has been identified as the single dopant with the most promising properties but other single dopants as Mn, Co, Cu, Ni, Cr, Mo, Ru, Rh, Ce and Tb [201–203] were investigated. Moreover, combinations of dopants, e.g., $LiNbO_3$:Fe,Mn [204] or $LiNbO_3$:Zr,Cu,Ce [205] were investigated intensively.

In contrast to the suppressing dopants the incorporation of photorefractive elements has gotten rather little attention by nuclear methods. Studies exist on the technically most important dopant, iron, on cobalt and on titanium. The latter enhances the photorefractive effect only very slightly but is of importance for optical waveguide applications.

It is generally presumed that the charge state of iron (Fe^{2+} or Fe^{3+}) is of importance for the photorefractive efficiency. Various teams studied this topic applying Mössbauer Spectroscopy with ^{57}Co as radiation source. The first study by Vayashko et al. [206] used lithium deficient LN heavily Fe-doped (2.4 to 5.1 mol%) crystals and detected both charge states in the material. They also reported that the fraction of Fe^{2+} increases with increasing Li-deficiency. Further studies with diffusion-doped [43] (0.5 to 1.0 mol% of Fe enriched to 86% in ^{57}Fe) and melt-doped [207,208] (0.22 mol% Fe_2O_3 enriched to 90.7% in ^{57}Fe) by and confirmed that a major fraction (90%) of the iron is present as Fe^{3+} in the virgin material. Annealing under reducing atmosphere (in Ar at 1000 °C for 140 h or under vacuum at 635 °C for 24 h), changes the charge state of the majority of the dopant to Fe^{2+}. The measurements also indicate that the Fe dopant is located on an axial site [209].

Additional Mössbauer studies addressed the electron trapping in congruent, near-stoichiometric, Mg-doped and Fe-doped LN [210,211]. It was found that the electron trapping effect by Nb-antisite outweighs that attributed to Fe^{3+} despite the larger cross section of the latter. For crystals doped with Mg up to threshold concentration the trapping is strongly reduced. Due to Li-loss during ^{57}Co diffusion at high temperature no quantitative results could be obtained for the near-stoichiometric material.

A first study on the lattice location of Fe was performed by Rebouta et al. [212] using samples of congruent LN doped in melt with 0.5 mol% Fe_2O_3. Due to the growth method, Fe was mainly incorporated as Fe^{2+} (as was confirmed by optical absorption spectrometry) rather than in the form of Fe^{3+} preferred in applications using the photorefractive effect However, previous studies suggest that the lattice location should not be affected by this [213]. PIXE/C with protons was used to determine the lattice location. Although the use of protons produces less pronounced dips and peaks in the angular scans the results definitely show that Fe is located in the Li site.

Zaltron et al. [214] presented a PIXE/C study on Fe-diffused samples. The material was produced by coating the LN substrate with a metallic Fe film and performing a diffusion treatment for 5 h at 1000 °C in oxygen atmosphere. Since the resulting samples contained Fe^{3+}, an additional treatment under reducing atmosphere (96% Ar + 4% H_2) for different times at 500 °C was performed in order to obtain up to 100% Fe^{2+}. The depth profiles of Fe before and after reduction were measured by SIMS revealing a distribution extending up μm into the bulk with a maximum concentration of about 0.9 mol% Fe at the surface. The distribution remained unaltered after reduction. Angular scans measured for the as-diffused (1% Fe^{2+}) and two reduced (20% and 100% Fe^{2+}) samples show fully overlapping Fe-PIXE curves. Despite the poor crystalline quality of the samples (Nb-RBS minimum yield of ≈45% in the $\langle 02\bar{2}1 \rangle$-axis) the data indicate that Fe is located on the Li site. This confirms also the assumption that the lattice site does not depend iron being incorporated in the divalent or trivalent state.

The lattice site of Co in LN was studied by Szilagy et al. [215] in melt-doped (0.2 mol% CoO) congruent material. In this case, a He^+ ion beam was used to obtain PIXE/C scans for Co- and Nb-PIXE with pronounced dips for five major axial directions. The simultaneous recording of Nb-RBS axial scans allowed to demonstrate the high quality of the crystals grown ($\chi_{min} \approx 2\%$). Comparison of the experimental results with simulations using CASSIS demonstrated that Co is located exclusively on the regular Li site.

In contrast to its homologues Zr and Hf, Ti does not suppress the photorefractive effect. It is therefore of interest if its behavior is linked to some different incorporation mechanism. A first study was made by Buchal et al. [216] with a sample implanted with Ti at a fluence of 2.5×10^{17} cm^{-2} and 360 keV energy and subjected to a heat treatment at 1000 °C. This caused an almost complete recovery of the crystal structure and a diffusion of Ti in the bulk material. The maximum concentration of Ti in the samples was reported to 12 mol%. The PIXE/C measurements for Ti-K characteristic X-ray showed an almost perfect overlap with that of Nb leading to the conclusion that Ti occupies the Nb lattice site. Kollewe et al. [217] studied two different cases: melt-doped material with 3 mol% Ti and Ti-indiffused waveguides with a peak concentration of about 9 mol%. PIXE/C scans were recorded for various axial directions. For measurements along the $\langle 0001 \rangle$-axis both samples evidence the absence of off-axis lattice sites while for the other axial directions clear differences are visible. While for the melt-doped crystal an almost complete overlap of the Ti-PIXE scan with those of the Li-NRA is found, the indiffused sample shows a behavior that resembles mainly the Nb-PIXE signal. Finally, Hauer et al. [218] reported a combined PIXE/C and PAC study for LN doped in melt with 1 mol% of Ti. The angular PIXE/C scans for various axial directions confirm clearly that Ti occupies only Li sites at this concentration level and that no off-axis sites are occupied. The PAC measurements show a single QIF, ν_Q = 15.4(2) MHz, which scales up very well with the results obtained with other probes occupying Li sites [34]. These findings led to the conclusion that Ti occupies solely Li-sites for lower concentrations while for higher concentrations a mixture of Li- and Nb-sites has to be assumed. Ti and Hf have obviously similar tendencies in the occupation of lattice sites although with opposite influence on the photorefractive index.

4.3. *Optically Active Dopants*

The excellent linear and non-linear properties turn LN into a highly desirable host material for optically active ions. Coherent stimulated emission from the rare earths Nd, Tm and Ho was observed in this material already in 1969 [219]. The interest on the laser applications of rare earths and Cr has led to innumerous studies on their optical properties in the last decades. Since these are strongly dependent on the dopant's crystallographic environment the determination of its lattice site is of outmost interest. Although some pertinent information can be obtained by spectroscopic methods the direct determination of the lattice location by ion beam methods is indispensable. The study of rare earths in LN by RBS/C is fostered by their larger mass in comparison to Nb as the heaviest component of the host material while chromium has to be investigated using PIXE/C.

For optical communications applications Er^{3+} has attracted interest in a wide range of laser host materials due to its IR emission at about 1.5 μm coinciding with a minimum in the absorption in optical fibers. Erbium has been incorporated into LN by melt-doping, indiffusion and ion-implantation. RBS/C studies have been reported only for melt-doped LN. The first investigation used two crystals co-doped with 0.5 mol% Er_2O_3 and 5.8 and 8.0 mol% MgO in melt (0.79 mol% Er, 4.8 and 7.5 mol% Mg in crystal), respectively [220,221]. Angular scans for axial directions show an exclusive occupation of the regular Li lattice site. For a second study, a series of singly-doped crystals with concentrations of 0.5, 1, 2, 3, 4 and 5 mol% in a congruent melt was grown and investigated [222]. The comparison of the experimental angular scans with simulations using the modified FLUX program confirmed that the main fraction Er is located in the Li–Octahedron but appears to be shifted form the regular site by 0.2 Å towards the closest oxygen layer. The Er concentration of these sites increases at first with increasing total Er concentration but reaches a saturation value at about 2.5 mol%. The high minimum yields for Er in axial directions point out that an additional fraction of Er occupies non-axial sites but cannot be identified with the known tetrahedral site in LN. The authors claim that these sites off the c-axis cannot be seen as a new crystalline structure but are rather amorphous precipitates or defect clusters.

Another important rare earth in optical communication application is praseodymium. Crystals doped with 0.2 and 0.5 mol% Pr^{3+} in melt were studied by RBS/C [223–225]. The comparison of experimental scans and simulations with the FLUX code shows that as in the case of erbium Pr is located in the Li oxygen-cage but with a larger shift (0.45 Å) towards the closest oxygen layer. On the other hand, site selective spectroscopy performed on the same samples suggests that more than one site for Pr^{3+} exists. This discrepancy can be easily understood taking into account the defect structure of the congruent material that causes (due to vacancies and stacking faults) different local surroundings for the Pr^{3+} ion.

Also the lattice site of the probably most popular laser active ion, Nd, has been investigated using RBS/C. The history of the investigation of this impurity highlights the importance of computer simulations for the interpretation of lattice location measurements in complex non-cubic crystal structures like LN. The first studies of the Nd^{3+} lattice sites for a 1 mol% Nd^{3+} in melt (0.33 mol% in the crystal) where first interpreted based on assumptions adopted from the study of simpler crystal lattice and pointed out a mixed occupation of Li- and Nb-sites [192]. Another study even claimed the occupation of three different axial lattice sites within the Li–Octahedron [226,227]. A new interpretation of the experimental results using the FLUX code revealed that Nd^{3+} is actually located in the Li–Octahedron displaced by 0.4 Å towards the closest oxygen layer [23].

In order to complete the overview on the lanthanides it has to be mentioned that three other rare earths were also found to occupy positions shifted from the Li site towards the closest oxygen plane: Eu (0.4 Å) [23,192,228], Ho (0.38 Å) [223,224] and Yb (0.3 Å) [223,224]. Plotting the five known displacement values against the ionic radius as suggested by Lorenzo et al. [223,224], shows an apparent linear dependence, depicted in Figure 25. The measured position of the main fraction of Er [222] (0.2 Å) has been added to the Lorenzo's data.

For Nd, Gd, Er, Tm, and Lu (1 mol% in melt) RBS/C studies on samples co-doped with Mg (6 mol%) have been performed [199,221]. For Nd, Gd and Tm the angular scans show only a partial occupation of the Li octahedron. In contrast to the singly-doped crystals, remarkably large fractions of randomly incorporated ions (almost complete for Nd, 70% for Gd, 20% for Tm) are observed. Rare earths with low or no random fractions are Er and Lu. Interestingly, also the distribution coefficient of Nd in the MgO co-doped crystal, k = 0.03, is an order of magnitude lower than in the singly-doped material [192]. This indicates that the incorporation of lanthanides with large ionic radius into the LN lattice is strongly hampered by Mg. The lattice location study for Mg on the $LiNbO_3$:Lu,Mg sample by Kling et al. [198] demonstrates that, in contrast to Hf-co-doping, the distribution of Mg in the lattice is identical to the singly 6 mol% MgO-doped crystal. In the competition for the occupation of sites in the Li–Octahedron the divalent Mg may be favored with regard to the trivalent lanthanides since it requires less charge compensation by formation of additional Li vacancies.

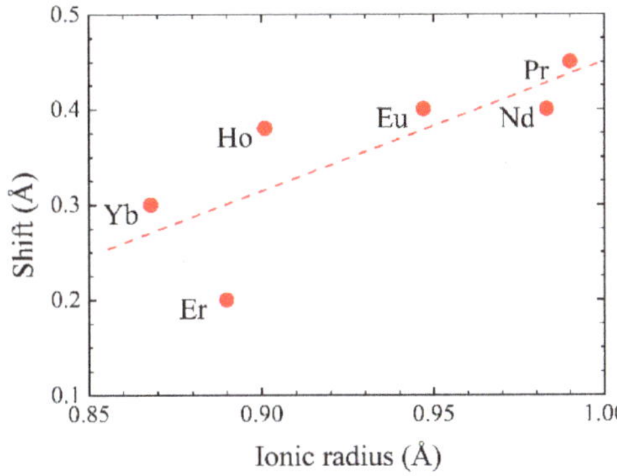

Figure 25. Dependence of the shift of the rare earth lattice sites with respect to the regular Li-site on the ionic radius.

Trivalent chromium has been already used in the very first laser as optically active ion. The potential application as a compact tunable laser initiated an intense research on the properties of Cr-doped LN. The introduction of chromium in LN with chromium can be achieved by melt doping [229,230], ion implantation [231] and diffusion [232]. Since chromium as a transition metal shows the tendency to increase the photorefractive effect the co-doping with Mg at high concentration is desired. However, the co-doping alters the absorption (Cr-doped LN is green, Cr-Mg-co-doped pink) and luminescence properties. Optical [233], ESR [234] and EXAFS [235] studies suggest that these effects are related to changes in the lattice site occupied by the Cr—only Li-sites for singly-doped material, Li- and Nb-sites for Mg-co-doped.

The lattice location of Cr was studied by PIXE/C in a congruent sample doped with 0.1 mol% Cr and one with 6 mol% Mg and 0.1 mol% Cr in melt [236]. The very low Cr dopant concentration turned to out to be extremely challenging for the experiment and a copper monolayer had to be deposited on the samples to suppress low energy X-ray emission due to charging effects by the ion beam. Nevertheless, it was possible to determine the lattice location by comparing the angular scans obtained with computer simulations using CASSIS. Surprisingly, it was found that already the singly doped crystal contains about 40% of the Cr on niobium sites and 60% on Li sites. This implies that chromium, in contrast to all other trivalent dopants studied so far, shows a tendency for charge compensation Cr^{3+}_{Li}–Cr^{3+}_{Nb} replacing Li^{1+}–Nb^{5+}). The results align well with those of the ESR study by Jaque et al. [234] suggesting that in singly-doped material mainly Cr^{3+}_{Li}–Cr^{3+}_{Nb} occur with some Cr^{3+} on Li-sites. For the case of the co-doped material the fraction of Cr located on Li sites is reduced to about 20% while 80% occupy the Nb octahedron although shifted about 0.1 Å towards the center of the octahedron.

Waveguide lasers require only a thin near-surface with the optically active ion, instead of bulk-doping. Ion implantation is an advantageous tool to dope a material with a controllable concentration and depth distribution. In order to verify the applicability of this method y-cut LN samples were implanted with 100 keV Cr to fluences of 1×10^{16} and 5×10^{16} cm^{-2}, respectively [115]. This resulted in a peak concentration of 2.2 mol% Cr for the latter fluence with a projected range of about 55 nm. In order to remove the implantation induced lattice damage annealing was performed under wet oxygen atmosphere for various temperatures and times. The success of the annealing process was verified by RBS/C showing that for 1000 °C and 4 h the almost completely amorphized surface layer was reconstructed and the RBS/C spectra hardly distinguishable from those for virgin material. In order to verify the incorporation of chromium in the crystal lattice angular scans were taken with PIXE/C. An almost perfect overlap with the results for the melt-doped case demonstrate the successful incorporation of Cr into the LN crystal.

4.4. Pentavalent and Hexavalent Dopants

Dopants in the pentavalent state are of interest since they can replace Nb without need of charge compensation. However, none of the elements in the pnictogen group incorporates into the LN lattice. An alternative are the homologues of Nb, V and Ta. The latter is known to be able to form mixed crystals covering the full stoichiometry range [237].

Since Ta has about double the atomic mass of Nb it was easy to study this impurity by RBS/C. The results of congruent crystal doped with 1 mol% of Ta showed that the angular scans for Nb and Ta fully overlap for all investigated axial and planar direction evidencing that Ta fully replaces Nb on its lattice site [228].

Similar to the case of pnictogens, also no doping of LN with chalcogenides could be achieved. Nevertheless, three elements that readily form hexavalent oxides—Mo [238], W [239–241] and U [242] have been successfully incorporated into LN. The lattice location of latter two was subject to studies by RBS/C.

Tungsten doping was performed in three different ways: ion implantation, doping in melt and diffusion. In the first case a fluence of 1×10^{14} cm^{-2} of ^{186}W was implanted into x-cut congruent LN resulting in maximum W concentration of 1.25 mol% peaking at about 50 nm depth [239]. In order to remove the implantation induced damage the sample was subjected to a two-step annealing process (600 °C for 1 h, 800 °C for 4 h) in a wet flowing oxygen atmosphere. The angular scans of the recrystallized material showed a complete overlap of the Nb- and W signals for all crystallographic directions proving the complete incorporation of tungsten on the niobium site. In a second study, LN crystals were grown from melt doped with 1 and 2.5 mol% WO_3 [240]. The concentration in the grown samples were determined by RBS to be 0.2 and 0.5 mol%, respectively, fixing the distribution coefficient of tungsten to 0.2. The RBS/C measurements confirmed the results of the ion implantation work on W replacing solely Nb. In the third approach a y-cut LN crystal was placed on a bed of metallic tungsten and kept at 1000 °C for 2.5 h under vacuum which yielded an extremely W-rich (about 35 mol%) layer at the surface [241]. A subsequent annealing treatment at 1000 °C for 2 h reduced the W concentration to 0.75 mol% and improved the crystal quality sufficiently to perform RBS/C scans. As in the previous two cases, the exclusive occupation of Nb sites by W was observed.

For the investigation of the lattice site of uranium a crystal grown from a congruent melt doped with uranylacetate-dihydrate (0.5 at.% U equivalent to roughly 2.5 mol%) was used, in which uranium should be incorporated in the hexavalent state. The samples obtained contained about 0.18 mol% U (distribution coefficient ≈ 0.07) as determined by RBS. Optical absorption spectroscopy confirmed that uranium had been incorporated as U^{6+} [242]. As in the case of W the angular scans for U-RBS and W-RBS overlapped fully confirming also for uranium the occupation of the Nb lattice site [240]. Calvo et al. [242] observed that a heat treatment under vacuum, i.e., reducing conditions for LN, turns the U-doped crystals from orange to red. Optical absorption studies revealed the existence of trivalent (and eventually tetravalent) uranium after this treatment. In order to check if this alteration of the valence state has any influence on the lattice site of uranium RBS/C with a sample reduced for 16 h at 1000 °C under high vacuum were performed. The angular scans, shown in Figure 26 for various crystallographic directions, show that part of the U has been pushed out of the Nb site but is still on an axial position [243]. Employing simulations using CASSIS suggest that about 20% of the U is now on Li sites while the remainder still occupies the Nb site. These findings corroborate the assumption that the cation lattice site occupied by ad dopant is mainly governed by its valence state.

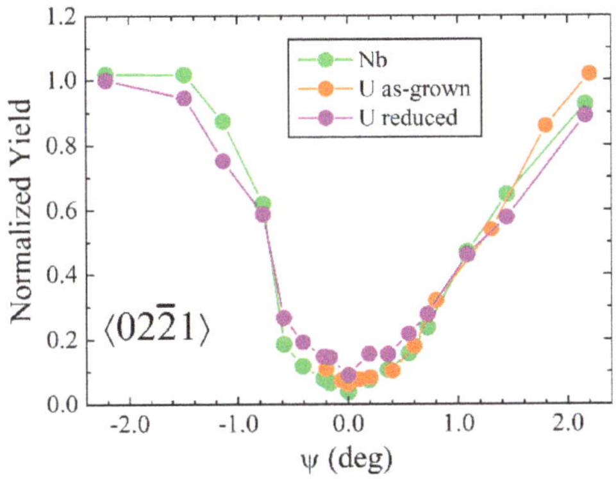

Figure 26. Angular RBS/C scans for the ⟨02$\bar{2}$1⟩-axis for Nb and U in the as-grown and reduced sample. In the reduced case the narrowing of the uranium-dip near the axial direction is due to the partial occupancy of Li-sites by U.

4.5. Noble Metals

The introduction of noble metals into glass in order to change its color is a method applied since Roman times [244]. The coloration is due the formation of noble metal nanoparticles for which the existence of surface plasmon resonances causes strong optical absorption peaks. The associated non-linear optical properties raised the interest on formation of such nanoparticles in modern optical materials. LN can incorporate noble metals as Cu and Au only at very low concentrations in melt-doped crystals causing the absence of surface plasmon related effects. Since the first successful formation of Cu-nanoparticles and the observation of their typical optical absorption behavior by Saito et al. [245] in 1987, this method has been applied for the production of Cu [246], Ag [247], Au [248] and mixed [249] nanoparticles. The influence of ion energy (20 keV–3 MeV), fluence (4×10^{13}–2×10^{17} cm^{-2}), implantation temperature (room temperature—700 K), post-implantation treatment (annealing at various temperatures and atmospheres) on the optical properties has been studied in detail [178–181]. RBS and RBS/C was employed in several studies to investigate the distribution of the implanted species, the implantation-induced damage and its removal by annealing as well as some structural characteristics of the implanted elements.

The best investigated case is that of Au. Due its high mass it can be easily detected by RBS and the implantation-related peak in the spectrum is well separated from niobium. Shi et al. [250] demonstrated that an implantation at 1 MeV with a fluence of 4×10^{13} Au^{+} cm^{-2} only partially damages the implanted layer while for 5×10^{14} Au^{+} cm^{-2} already full amorphization is achieved. Taking RBS/C spectra they were able to calculate damage depth profiles with nanometric resolution. A comparison with the popular TRIM code [251] for damage profile computation revealed that the experimentally determined damage reaches far deeper into the crystal than computed slowing down recovery of the crystal lattice during annealing. The study by Takahira et al. [252] was the first one to investigate the recrystallization of an Au-implanted (550 keV, 1×10^{16} to 1×10^{17} cm^{-2}) LN crystals in detail. They showed that annealing at 600 °C for 1 h in air efficiently removes a large part of the damage for the sample implanted with 3.7×10^{16} Au^{+} cm^{-2}—corresponding to a peak concentration of 7 at.%. It is noteworthy that the Au implantation profile is not affected by the heat treatment.

Dev et al. [253] performed RBS/C measurements for the ⟨0001⟩-axis of LN implanted with 1×10^{15} cm^{-2} Au (1.43 MeV) and annealed at 450 °C in flowing oxygen atmosphere for 40 min. The curves for the Nb- and Au-RBS overlap mostly but show a distinct kink at about 0.5° from the axial direction, ruling out a formation of nanoclusters due to a low Au concentration (0.5 mol%). This indicates that a major part of the Au is in axial lattice site but in contrast to the claims of the authors, not necessarily the Nb-site. Due to its high atomic mass the Au-RBS scan would also overlap for Li-, free octahedron or other aligned sites. The kink-related fraction is definitely related to a fraction on a non-axial site, possibly the tetrahedral site (although this would have to be verified by additional studies).

For a higher fluence (1×10^{16} Au^+ cm^{-2}, 800 keV, implanted at room temperature) Kling et al. [254] investigated the recrystallization of the sample after subsequent annealing at 600 °C (2 h) and 800 °C (1 h) in flowing wet oxygen. RBS/C measurements showed that even at 800 °C only a partial recovery of the crystal lattice was achieved. Further, it was found that the yield for Au-RBS is reduced for the $\langle 01\bar{1}0 \rangle$-axial direction. Since optical absorption demonstrated the formation of Au-nanoparticles in the sample an incorporation into the LN lattice was ruled out leaving only the hypothesis that the Au-cluster are aligned the with the LN lattice. Complementary studies by TEM of this sample revealed that gold did not form spherical nanoparticles as deduced from various optical investigations but thin flakes parallel to the surface with lateral dimension of up to 200 nm. An electron diffraction measurement confirmed that these flakes are highly coherent with the LN lattice.

In another study by the same authors, the interaction between Au- and He-implants was in the focus of the interest [255]. All samples were implanted with 20 keV He^+ to a fluence of 1×10^{16} cm^{-2}. Au^+-implantations with 50 keV (5×10^{15} cm^{-2}) were used to obtain two well-separated profiles while others using 800 keV Au^+ (1×10^{16} cm^{-2}) achieved an overlap with the profile of He. In RBS/C spectra for the samples with separate profiles the two damage-related peaks for He and Au appear separately even after annealing at 270 °C. Annealing at 600 °C causes the Au to migrate towards the bulk and broadens the profile while at 800 °C a movement of the gold towards the surface is observed. This mobility of Au contrast clearly with the static Au profiles observed for single implantation at the same temperature discussed above. Most likely the diffusion of Au at 600 °C is facilitated by the presence of He in the material but not showing a real gettering effect as in Si [256]. As in the previous case, diffusion towards the surface and an alignment of the Au-nanoparticles was found for the samples with overlapping profiles. TEM observations revealed again the existence of Au-flakes but with drastically reduced lateral sizes (<100 nm) compared to singly-implanted LN opening a path for the control of nanoparticle properties by co-implantation of noble gases.

For the case of Ag-implantation, Milz et al. [257] studied the influence of implantation and annealing temperatures on the damage production and removal. X-cut LN was implanted with 380 keV Ag^+ to a high fluence (1×10^{17} cm^{-2}) at liquid nitrogen (−196 °C) temperature, room temperature and 700 K (423 °C). RBS/C spectra show that in the first two cases the implanted surface layer was completely amorphized while in the high-temperature case the layer was heavily damaged and thinner retaining some crystallinity. Rapid thermal annealing at temperatures ranging from 300–900 °C was employed for recrystallization. In the sample implanted at room temperature an increasing reduction of the amorphized layer with increasing temperature was observed but only at 900 °C a partial recovery of the crystal lattice at the surface was observed.

The interaction of Ag and O implants was the topic of a study by Williams et al. [258]. Z-cut samples were implanted with Ag and O at energies of 160 and 35 keV and fluences of 4×10^{16} cm^{-2} and 8×10^{16} cm^{-2}, respectively at RT and 500 °C. Both implantation sequences were used and additionally a singly Ag^+ implanted samples (1.5 MeV, 2×10^{16} cm^{-2} and 1.7×10^{17} cm^{-2}, same temperatures) were produced. The implanted material was annealed for 1 h at 500 °C in air. From RBS/C spectra recorded for the Ag-implanted sample it was concluded that in case of high-temperature implantation is located closer to the surface than in that for room temperature. On the other hand, during the annealing the Ag moves towards the bulk in the high-temperature sample while it approaches the surface in the RT-sample. Co-implanted material shows some difference between the two implantation sequences when the high-temperature procedure is applied. For Ag- then O-implantation the Ag-profile stays practically unaltered with most of the Ag remaining at the surface.

For the opposite sequence, a significant Ag-diffusion towards the bulk was observed that seems to stabilize the Ag-nanoparticles formed.

The incorporation of Pt into LN was achieved by 50 keV Pt^+ ion implantation at room temperature at a fluence of 5×10^{15} cm^{-2} [259]. A maximum concentration of about 1.5 mol% of Pt was attained. The samples were annealed under flowing wet oxygen atmosphere at 600 °C and 800 °C for 1–6 h in steps of one hour or at 1000 °C for 60, 90 and 120 min. Further, an incremental annealing was done at 300 °C, 400 °C, 450 °C, 500 °C, 600 °C and 700 °C for 30 min followed by one at 1000 °C for up to 5 h. Studying the Pt-RBS yield in the surface region (0–10 nm) showed no effect for 600 °C, a slight diffusion towards the bulk after 6 h at 800 °C and a very rapid one at 1000 °C leaving the surface region virtually without Pt. RBS/C spectra showed no alignment effect for Pt except for the sample that underwent incremental annealing. A minimum yield of 53% is observed for alignment with $\langle 01\bar{1}0 \rangle$-axis while for all other crystallographic directions random behavior is found. It can be concluded that Pt forms precipitates that are preferentially aligned with the $\langle 01\bar{1}0 \rangle$-axis of LN as in the case of Au.

Iridium implanted into LN (130 keV, 5×10^{16} cm^{-2}) with a high peak concentration has been studied by Kling et al. [260] The samples were annealed under flowing wet oxygen at 500 °C, 800 °C and 1000 °C for 30 min. In contrast to Pt, significant diffusion towards the bulk was found by RBS already at 500 °C. After heat treatment at 1000 °C a practically uniform concentration of 0.015 mol%, i.e., about two orders of magnitude lower than in the as-implanted material, is reached. RBS/C scans show an alignment effect with the LN for all investigated axial directions with a width comparable to that of Nb-RBS. However, the high minimum yields (40–80%) do not allow attributing positively an occupation of the Nb-site by Ir.

5. Optical Waveguides

Photonic applications using optical materials like LN receive a constantly increasing interest. A key part in any photonic device is an optical waveguide. It consists of a thin surface layer (in the order of the wavelength of the light guided) with a refractive index larger than the underlying substrate. Total reflection at the interfaces with the bulk and the air confines the light propagation to the waveguide. Many different approaches for the fabrication of waveguides in LN and other optical materials have been developed in last decades. For a comprehensive review see Bazzan and Sada [261].

5.1. Indiffusion

The formation of LN waveguides by in-diffusion of Ti from a thin metallic layer deposited on the surface was reported for the first time by Kaminow and Carruthers [262] in 1973 and became with time the most common method. It combines a good guiding behavior with the preservation of the electro-optic properties. The downside is that Ti, in contrast to its homologues Zr and Hf, it does not suppress the photo-refractive effect.

For applications, it is of important to determine the depth profile of the in-diffused Ti that controls the refractive index profile of the optical waveguide. The method of choice is in this case the SIMS. The first Ti-profile measurement was performed in 1979 by Burns et al. [263]. The samples were prepared from x-cut LN coated with 17.5 nm of Ti and diffusing at 900–1100 °C for 6h in Ar atmosphere. SIMS concentration profiles for Ti and Li ions for a sample treated at 1000 °C show pronounced peaks with half-width of 200–300 nm near the sample surface. A clear correlation between decreasing Ti- and increasing Li-concentration with depth is observed. These results suggest that the diffusion treatment time was too short for a sufficient redistribution of the Ti. Arizmendi et al. [264,265] investigated in detail this first phase of the diffusion process combining ion beam analysis, X-ray diffraction and electron microscopy techniques for samples in-diffused in Ar and O_2 atmospheres between 500 and 950 °C. RBS/C measurements for samples annealed at 500 °C showed the formation of an amorphous TiO_2-layer on the substrate and a lack of oxygen in the LN surface. For the technically more important case of 950 °C (30 min, dry oxygen) a $Ti_xNb_{1-x}O_2$ compound with x = 0.65 in the surface region is inferred from the RBS results in accordance with x-ray diffraction. Angular scans for several axial directions revealed that this new compound is aligned along the c-axis of the LN substrate but not with other axial directions. The authors argue that the compound formed in this step constitutes the actual source for the further in-diffusion of Ti.

Based on these observations for technical applications the diffusion treatment at high temperature and for extended time periods (up to 16 h) has been adopted. Several groups reported SIMS studies on the Ti-diffusion process. Bremer et al. [266] reported a Gaussian profile peaking at the surface for Ti after indiffusion for 16 h under wet Ar atmosphere. A vast investigation of the main determining factors that govern Ti-diffusion (Ti-film thickness, diffusion temperature) has been presented by Caccavale et al. [267,268]. With the help of SIMS measurements on samples annealed at 900–1000 °C they could confirm that only for the highest temperature a complete indiffusion of Ti takes place while for lower temperatures an enhanced Ti-concentration near the surface is found. Further, they were able to detect a significant anisotropy in the diffusion process leading to a diffusion coefficient twice as high for z-cut material than for x-cut. Caccavale et al. [269] have given an impressive example of the lateral resolution of the SIMS technique by studying the lateral diffusion of Ti in waveguides. It was possible to determine the lateral Ti-profiles of two waveguides separated only by a few µm. For near-stoichiometric LN produced by the vapor transport equilibration Zhang et al. [270] observed a diffusion profile that can rather be described by an error-function than by a Gaussian as in the congruent material.

To overcome the photorefractive effect in Ti-diffused waveguides diffusion into Mg-doped material or Mg co-diffusion can be applied. The first SIMS study by Bremer et al. [271] used already Ti-diffused waveguides as described above [266]. MgO was deposited as Mg-source on the waveguide and indiffused at 950 °C in order to avoid further Ti-diffusion and taking into account the more rapid diffusion of Mg. A single Mg-diffusion treatment leads to a Gaussian distribution of the Mg peaking at the surface with some reduction of the Li content at the waveguide surface. Repetition of the Mg-indiffusion causes a drastic (20%) loss of Li in the surface region accompanied by high Mg-peak in this region. The authors interpreted this as hint to the substitution of Li by Mg during doping. Caccavale et al. [272] concluded that the Ti-profile suffers only minor additional diffusion by Mg co-diffusion but that on the other hand Mg-diffusion is clearly enhanced in comparison to an undoped crystal [273].

A comparative study on the Ti diffusion in undoped and Mg-melt-doped congruent and near-stoichiometric LN was presented by Kumar et al. [274]. Samples were indiffused at 1080 °C for 23 h in dry air atmosphere. The values of the Ti-diffusion constant determined using SIMS profiles show that Ti-diffusion is slowed down in MgO-doped material for near-stoichiometric (1 mol% MgO) and congruent material (5 mol% MgO) with respect to their undoped counterparts.

An important alternative is the indiffusion of Zn from vapor phase into LN, reported for the first time by Herreros and Lifante [275]. Nevado et al. [276,277] produced samples indiffused at 550 °C and 700 °C with a diffusion times of 2 h and 8 h, respectively. The samples were subsequently annealed for 4 h at the 700 °C or 800 °C, respectively, to form the actual waveguide. As-diffused and annealed samples were investigated by SIMS and RBS. In both cases the as-diffused samples showed Zn-rich surface layers followed by a zone with decreasing Zn concentration. After the annealing Zn diffused from the Zn-rich superficial layer several µm deep into the bulk forming the actual waveguide. A further quantitative analysis was performed using HIERD [278]. This high depth-resolution quantitative technique allowed to confirm that in the as-diffused case the top layer consists of LN and $ZnNb_2O_6$ phases, i.e., a partial substitution of Li by Zn occurs. On the other hand, in the annealed sample a homogenous mixture of ZnO and LN was found on top and a Zn-diffusion profile for the remainder of the waveguide.

For completeness, it should be mentioned, that also other elements, Sc [279] and Zr [280], which are candidates for waveguide formation were studied with regard to their diffusion properties using SIMS. Further, several studies focused on the co-diffusion of two elements: Zn/Ni [281], Zr/Ti [282,283], Sc/Ti [283] as well as for waveguide laser applications, the important Er/Ti [283,284].

5.2. Proton-Exchange

A simple alternative to metal-diffusion based waveguide production is the proton-exchange (PE) technique reported in 1981 by Jackel and Rice [285], trailed by the production of the first optical waveguide in the following year [286]. Proton exchange is achieved by placing the LN crystal in a liquid source (most commonly melted benzoic acid) that acts as hydrogen source at elevated temperature. For a review see Cabrera et al. [287]. Unfortunately, the as-exchanged layer suffers from lattice stress and damage, as found in early RBS/C studies [288,289], which degrades the optical properties. Therefore, a subsequent annealing procedure at about 300 °C in air is employed in order to obtain a high-quality waveguide. Figure 27 shows the effect for LN proton-exchanged using benzoic acid.

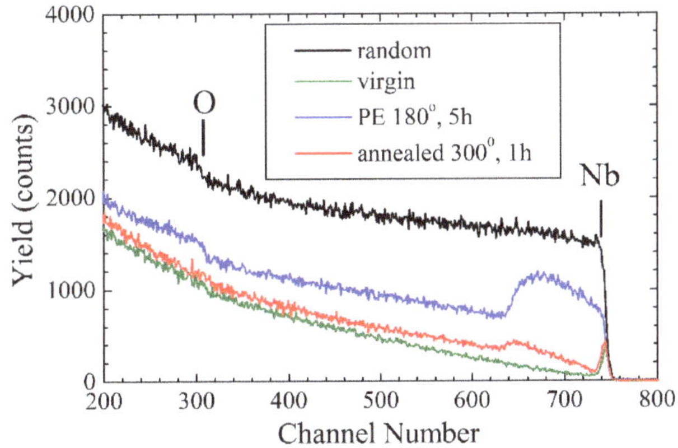

Figure 27. RBS spectra for the virgin LN (random and $\langle 01\bar{1}0\rangle$-axis) and after proton-exchange in benzoic acid ($\langle 01\bar{1}0\rangle$ -axis, as-exchanged and annealed). The clearly visible surface damage layer has a thickness of 270 nm.

Studies of the as-exchanged layer thickness are performed by analyzing the damage profile in axially aligned directions by RBS. Changing process temperatures and times as well as the cut of the substrate allowed to determine their influence on the exchange-depth and the hydrogen diffusion coefficient. Results were reported for benzoic acid in the low (110–220 °C) [290] and high temperature range (200–300 °C) [291,292]. In accordance with the expectations, the exchange depth was found to be proportional to the square root of the exchange time while the diffusion constant follows an Arrhenius law [290] Analogue findings were reported for samples treated with sulfuric [293] and pyrophosphoric acid [294]. The comparison of channeling spectra taken before and after annealing shows a strong but incomplete recovery of the LN crystal structure [291,292,294,295].

Nb-disorder profiles extracted from the RBS/C spectra show differences between x- and z-cut LN [292]. The as-exchanged z-cut material shows a low damage (up to 25% of the Nb-atoms displaced) increasing from surface to bulk with an abrupt end. On the other hand, for x-cut material the damage was found to be constant at a high level (80%). After annealing the damage profiles extended for both cases deeper into the bulk and interestingly, showed a smaller degree of recovery for the z-cut substrate. Olivares et al. [291] measured RBS/C spectra and angular scans for Nb-RBS along several axes in as-exchanged waveguides demonstrating high-temperature treatment leads to a superior crystal quality. Moreover, virtually no differences between samples prepared under slightly different conditions (addition of lithium benzoate, annealing atmosphere) were observed. The authors also performed ERD angular scans for the lattice location of hydrogen at several planar directions [291,296]—due to the need of large ion beam incidence angles ($\theta > 70°$) for this technique no appropriate axial directions were available. A comparison of the experimental results with computer simulations based on a theoretical model proposed by Kóvacs et al. [297] (shift of H in c-direction by 0.1 Å, in y-direction $\approx$ 1Å) revealed a good agreement. Although somewhat ambiguous, these results support clearly support the formation of O-H suggested by theory.

As in the case of Ti-diffusion the actual elemental profiles within the waveguide are of outmost interest for establishing a correspondence with refractive index profile. Suitable nuclear depth profiling techniques employed in proton-exchanged LN for hydrogen are ERD, NRA and SIMS, for lithium NRA and SIMS and for oxygen SIMS.

Canali et al. [289] measured the H and Li depth profiles of as-exchanged layers using nuclear reactions. In the case of hydrogen profiling was done with the $^{15}N(p,\alpha\gamma)^{12}C$ reaction that allows due to its resonant character a very high depth resolution; Li-profiles were taking using the non-resonant $^{7}Li(p,\alpha)^{4}He$ reaction that allows only moderate depth resolution. The hydrogen profiles overlap with the damage depth observed in the RBS/C. The Li-NRA spectra reveal a 70% reduction of lithium in the treated layer corroborating that indeed an exchange reaction between Li and H takes place. Additional SIMS studies by the authors confirmed these results. Hsu et al. [295] verified the depletion in as-exchanged layers and demonstrated that after annealing a partial recovery of the Li concentrations occurs although a thin surface layer stays poor in lithium. Resonant NRA depth profiling by Ito and Kawamoto [298] revealed a reduction of the H-concentration in the surface after annealing and further diffusion into the bulk out of the depth region accessible by this method.

Rottschalk et al. [299] employed ERD for the H depth profiling of PE waveguides (benzoic acid, 180 °C, 5 h or 250 °C, 1 min) in as exchanged state and after annealing at various temperatures. Despite the extremely short treatment time, it was clearly visible that higher temperature yielded a higher hydrogen surface concentration. As expected, the hydrogen diffuses more rapidly with increasing annealing temperature reducing its concentration at the surface. The work of Hagner and Bachmann [300] investigated the H-concentration in the as-exchanged waveguide in dependence of the amount of lithium benzoate admixture to the benzoic acid used in the process. The ERD spectra showed a reduction of the hydrogen incorporation into the layer with otherwise identical conditions.

Hydrogen profiles in PE-LN layers (benzoic acid, 185 °C, 70 min) annealed at 400 °C for 6–180 min were obtained by Casey et al. [301] and Zavada et al. [302] using SIMS. The measured H concentration profiles had to be fit using two Gaussian diffusion profiles which suggest the existence of two different types of hydrogen diffusion with distinct diffusion coefficients. The authors suggest that the fast component is related to interstitial diffusion while the slower represents substitution of lithium by hydrogen. De Souza et al. [303,304] confirmed and quantified the fractions of these two different species of hydrogen with NMR measurements for several exchange conditions. In a second study [305] the authors succeeded to separate the signal origination from the "static" hydrogen and compared the results with calculations for four defect models. The best agreement was found for H occupying a position within an oxygen plane along a side of an oxygen triangle equidistant from two oxygen atoms.

Yamamoto and Taniuchi [306] as well as Cheng et al. [307] reported SIMS-based hydrogen profiles for the alternative pyrophosphoric acid exchange method and show that the profiles obtained are very similar to those of samples using the common benzoic acid.

5.3. Ion Implantation

Among the waveguide fabrication techniques ion implantation is the one that can be applied to largest variety of materials. For comprehensive reviews see Chen et al. [308] and Peña-Rodriguez et al. [309]. As discussed in Section 3.3 defects are produced by electronic and nuclear interactions. In many cases, these defects cause an alteration of the refractive index forming an optical waveguide. The first successful waveguide formation using ion implantation was reported by Destefanis et al. [310]. For waveguides based on nuclear damage induced defects the quality of the waveguides is improved by an annealing procedure to remove electronic damage in the light guiding layer but leaving the buried damage layer intact. In the meantime, waveguide formation based on electronic damage production gained importance with many ions reported to be useful: B, C, N, O, F, Si, P, Ti, Cu, Ni, and Ag.

Barfoot et al. [311] studied the damage built-up during implantation with 1 MeV He^+ (2.0×10^{15} and 2.4×10^{16} cm^{-2}) and annealing induced recovery by in-situ RBS/C. Varying the implantation temperature between 40 and 140 °C resulted in a decrease of the Nb-lattice disorder with increasing temperature. In all cases a saturation of the damage was observed at about 1×10^{16} He^+ cm^{-2} coinciding with the value found for optimal waveguide performance. Further, the best annealing temperature was determined to be about 180 °C. Williams et al. [312] investigated similar waveguides (implanted with 1.7 MeV He^+, 1.5×10^{15} and 2.0×10^{16} cm^{-2}, RT; annealed at 175 °C and 400 °C) by NRA using the $^7Li(p,\alpha)^4He$ reaction. A comparison indicated among the NRA/C spectra indicated a lithium movement towards the nuclear damage region during annealing. The authors suggest that this effect is responsible for the increase in the extraordinary index in the surface region.

Low-energy He^+-implantation (50 keV, 2.0×10^{15} and 2.4×10^{16} cm^{-2}) performed at temperatures of −80 °C, −30 °C, 25 °C and 80 °C has been studied by Al-Chalabi et al. [313] employing RBS/C and HIERD. The damage in the sample implanted at −80 °C appeared strongly enhanced while no difference between those implanted at −30 °C and 80 °C was detected. Li- and H-profiles obtained from the HIERD data indicate a loss of lithium in the implanted layer that is compensated by the incorporation of hydrogen. The lithium concentration reduction was found to increase with increasing fluence and implantation temperature.

Kling et al. [314] studied the annealing temperature dependence of the lattice damage in y-cut LN using RBS/C. Samples were implanted at room temperature with 20 keV He^+ to a fluence of 1×10^{16} cm^{-2} or with 25 keV Li^+ (1×10^{16} cm^{-2})—that yield the same damage profile—in order to check for the chemical influence of the implant. Annealing was performed under wet oxygen atmosphere for 10 min at five different temperatures (150–350 °C). At first, an increase of the damage during the heat treatment was observed with increasing temperature reaching a maximum at 270 °C with an apparent activation energy of 0.14 eV, as depicted in Figure 28. For Li-implanted samples an almost complete recovery of the lattice has been observed at this temperature indicating that He-implantation induces fundamentally different damage structures. A further increase of temperature induced a decrease in the number of defects with a much larger activation energy of 1.3 eV. This two-phase behavior resembles closely that observed for He-implantation in silicon [315]. An additional TEM study on the same samples annealed at 270 °C revealed the existence of platelet voids in the depth region at which the He ions were stopped. This region forms naturally a layer of low refractive index.

The annealing behavior of LN implanted with 800 keV He^+ ions at very low temperature (−173 °C) to fluences of 5×10^{15} and 1×10^{16} cm^{-2} was the topic of a study of Gischkat et al. [316]. RBS/C measurements were performed at implantation temperature, at room temperature and after heat treatments at 250 °C and 300 °C in air. An apparently higher defect concentration observed in x-cut samples with respect to z-cut material was observed. As discussed in the Section 3.3 this anisotropy is due to the dislocation of Nb to the free octahedron site during implantation leading to different defect geometries in terms of channeling for the two directions. The dechanneling effect due to defects was observed to decrease with heat treatment for the low-fluence case and to increase for the high-fluence case. In the latter case, the enhancement was attributed to the formation of dislocation loops.

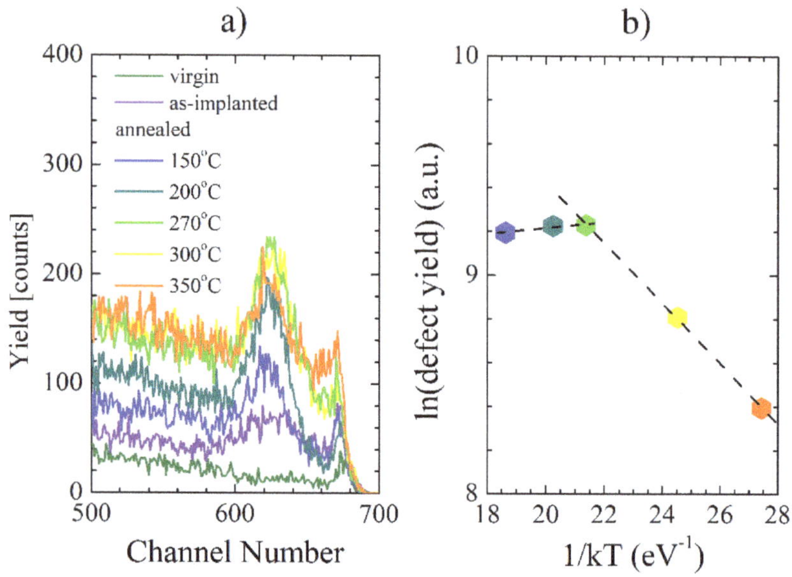

Figure 28. (**a**) Channeling spectra of virgin, He-implanted and annealed LN in the defect area. (**b**) Arrhenius plot for the determination of the apparent activation energies.

Besides helium, the implantation of nitrogen has already been explored in 1980s with view to its applications in optical waveguides [99,100]. Damage profiles for samples implanted with 200 keV N^+ with fluences between 5×10^{14} and 2×10^{16} cm^{-2} were determined by RBS/C and correlated with the refractive index of the material. The formation of optically isotropic waveguide by amorphization due to electronic damage formation of a surface layer by implantation of high-energy (5 MeV and 7.5 MeV) Si^+ ions has been studied by Olivares et al. [117,317]. From RBS/C spectra for samples implanted with fluences ranging from 5×10^{13} to 1×10^{15} cm^{-2} the authors determined threshold values for the amorphization and the associated electronic stopping power. Similar studies have also been performed for H [318], O [111,319–321] and F [116].

Optical waveguides can also be formed by high-fluence implantation of Ti with subsequent annealing/diffusion treatment. Bremer et al. [322] implanted y-cut LN with 200 keV Ti to fluences of 2.5×10^{17} cm^{-2} and 4×10^{17} cm^{-2} at liquid nitrogen temperature, room temperature and 350 °C. The samples underwent subsequent heat treatment at 1000 °C under wet oxygen atmosphere for 2 to 8 h. RBS/C studies demonstrated a good crystal lattice recovery was achieved in all cases for long annealing times but that a higher implantation temperature favors the recrystallization process. Optical waveguide properties were observed for all samples.

A technique related to implantation is ion beam mixing. A LN sample plated with a 40 nm metallic Ti layer has been irradiated with 3 MeV Ti^+ ions at a fluence of 1×10^{16} cm^{-2} by Bremer et al. [266]. After removal of excess Ti from the surface the sample was annealed at 1000 °C under wet oxygen atmosphere to achieve recrystallization. The SIMS profile measurement showed a Ti profile with a steady slope reaching deep into the bulk material which differs strongly from the Gaussian profiles found in indiffused waveguides.

5.4. Combination of Techniques

The combination of the proton-exchange and the implantation technique produces interesting results. For samples subjected first to a He^+ implantation (200 keV, 3×10^{16} cm^{-2}) and then undergoing a proton exchange treatment the hydrogen indiffusion is strongly enhanced [323]. It appears that the damage introduced by ion implantation process facilitates the incorporation of the hydrogen. For the inverse treatment sequence, it was observed that a proton-exchanged sample subsequently implanted with 6 MeV O^+ at a fluence of 6×10^{14} cm^{-2} is fully amorphized while PE-only and implantation-only samples show only minor damage [324]. Obviously, the proton-exchange process leads to a more fragile lattice increasing the susceptibility to lattice disorder.

5.5. Interaction with Optical Active Dopants

For optical waveguide lasers the absence of interferences between the waveguide formation process and the optically active dopants is mandatory. Therefore lattice site location and crystal structure studies for several rare earths in bulk and processed waveguide materials have been performed. Herreros et al. [325,326] compared the lattice sites for $LiNbO_3$:Tm,Mg and $LiNbO_3$:Er,Mg in as-grown and ion-implanted (2.6 and 2.0 MeV He^+, respectively, 2×10^{16} cm^{-2}; annealing: 250 °C, 30 min, air) samples. Angular scans for an axial direction demonstrated that the lattice sites of Tm and Er stay unaltered after waveguide fabrication. The same authors also studied the structural differences in waveguides fabricated by proton-exchange (benzoic acid, 300 °C, 20 h) and by He^+ implantation (2 MeV, 2×10^{16} cm^{-2}) in $LiNbO_3$:Nd,Mg using RBS/C. Again, the results show that ion-implanted waveguides essentially maintain their properties while proton-exchanged present a larger deterioration [327]. Cajzl et al. [328] reported on the influence of the three waveguide fabrication methods (Ti-diffusion, proton-exchange and ion implantation) on the lattice site of erbium. Angular RBS scans confirm that none of the methods has any measurable influence on the position of the Er. These results indicate that rare-earth ions are successfully incorporated of into LN waveguides.

6. Conclusions

The relevance of nuclear techniques in the investigation of defect structures in modern materials has been demonstrated for the case of LN, presently one of the most important optoelectronic materials. It has been shown that the techniques originally developed for a simple crystal system can be successfully transferred to complex cases.

In the case of the intrinsic defect structure of congruent LN the combination of Ion Beam Analysis and Perturbed Angular Correlation studies delivered evidence for an additional lattice site in the Li–Octahedron. The features of this site observed by nuclear techniques support the interpretation that the excess Nb is incorporated through ilmenite-type stacking faults dispersed in the crystal lattice.

Channeling measurements are found to be the method of choice for the characterization of ion implantation induced damage and its recovery laying the basis for the theoretical interpretation of damage formation in oxides.

The determination of lattice sites for a large number of elements, including the role of co-doping furnishes pertinent information for future engineering of LN by introduction of foreign atoms. It was demonstrated that the main factor determining the occupation of a Li- or a Nb-octahedron by a dopant is the valence state. Divalent, trivalent and tetravalent ions occupy—with the single exception of Cr, the Li–Octahedron, pentavalent and hexavalent the Nb.

Depth profiling and defect characterization in optical waveguides by ion beam methods is essential for the development of opto-electronic devices.

Author Contributions: Conceptualization, writing, review and editing: A.K., J.G.M. All authors have read and agreed to the published version of the manuscript.

Funding: This research was partially funded by Fundação para a Ciência e Tecnologia, Portugal, through project UIDB/04349/2020 of Centro de Ciências e Tecnologias Nucleares.

Institutional Review Board Statement: Not applicable.

Informed Consent Statement: Not applicable.

Data Availability Statement: Not applicable.

Acknowledgments: The authors gratefully acknowledge José Carvalho Soares (University of Lisbon, Portugal) and Fernando Agulló-López (Universidad Autónoma de Madrid, Spain) for their guidance and support. Juan Antonio Sanz-García, Maria Dolores Serrano and Ernesto Diéguez (Universidad Autónoma de Madrid, Spain) and Barbara Christa Grabmaier (Siemens AG, Germany), Ágnes Péter and Katalin Polgár (Hungarian Academy of Sciences, Hungary) are gratefully acknowledged for the many high-quality lithium niobate crystals supplied over the years. Reiner Vianden (Helmholtz-Institut für Strahlen- und Kernphysik, University of Bonn, Germany) is gratefully acknowledged for the local support to perform PAC measurements. Pier Giorgio Rancoita (Istituto Nazionale di Fisica Nucleare, Milano-Bicocca, Italy) is gratefully acknowledged for granting access to the SR-NIEL web platform.

Conflicts of Interest: The authors declare no conflict of interest.

References

1. Kong, Y.; Bo, F.; Wang, W.; Zheng, D.; Liu, H.; Zhang, G.; Rupp, R.; Xu, J. Recent progress in lithium niobate: Optical damage, defect simulation, and on-chip devices. *Adv. Mater.* **2020**, *32*, 1806452. [CrossRef]
2. Volk, T.; Wöhlecke, M. *Lithium Niobate: Defects, Photorefraction and Ferroelectric Switching*; Springer: Berlin, Germany, 2008.
3. Shur, V.Y.; Akhmatkhanov, A.R.; Baturin, I.S. Micro- and nano-domain engineering in lithium niobate. *Appl. Phys. Rev.* **2015**, *2*, 040604. [CrossRef]
4. Arizmendi, L. Photonic applications of lithium niobate crystals. *Phys. Stat. Sol.* **2004**, *201*, 253–283. [CrossRef]
5. Alford, T.L.; Feldman, L.C.; Mayer, J.W. *Fundamentals of Nanoscale Film Analysis*; Springer Science and Business Media Inc.: Berlin/Heidelberg, Germany, 2007; ISBN 978-0-387-29260-1.
6. Nastasi, M.; Mayer, J.W.; Wang, Y. (Eds.) *Ion Beam Analysis—Fundamentals and Applications*; CRC Press: Boca Raton, FL, USA; Taylor & Francis Group: Boca Raton, FL, USA, 2014; ISBN 978-1-4398-4639-1.
7. Nastasi, M.; Wang, Y. (Eds.) *Handbook of Modern Ion Beam Materials Analysis*, 2nd ed.; Cambridge University Press: Cambridge, UK, 2010; Volume 2, ISBN 9781605112176.
8. Verma, H.R. *Atomic and Nuclear Analytical Methods*; Springer: Berlin, Germany, 2007; ISBN 3-540-30277-8.
9. Chu, W.-K.; Mayer, J.W.; Nicolet, M.A. *Backscattering Spectrometry*; Elsevier Inc.: Amsterdam, The Netherlands, 1978; ISBN 978-0-12-173850-1.
10. Amsel, G.; Lanford, W.A. Nuclear Reaction Techniques in Materials Analysis. *Ann. Rev. Nucl. Part. Sci.* **1984**, *34*, 435–460. [CrossRef]
11. Johansson, S.A.E.; Campbell, J.L. (Eds.) *Particle-Induced X-Ray Emission Spectrometry (PIXE)*; Wiley & Sons: Hoboken, NJ, USA, 1995; ISBN 978-0-471-58944-0.
12. Gemmell, D.S. Channeling and Related Effects in the Motion of Charged Particles through Crystals. *Rev. Mod. Phys.* **1974**, *46*, 129–235. [CrossRef]
13. Morgan, D.V. *Channeling: Theory, Observation and Applications*; Wiley & Sons: Hoboken, NJ, USA, 1973; ISBN 978-0471615101.
14. Van der Heide, P. *Secondary Ion Mass Spectrometry: An Introduction to Principles and Practices*; Wiley & Sons: Hoboken, NJ, USA, 2014; ISBN 978-1-118-48048-9.
15. Lanford, W.A. Analysis for Hydrogen by Nuclear Reaction and Energy Recoil Detection. *Nucl. Instr. Meth. B* **1992**, *66*, 65–82. [CrossRef]
16. Bik, W.M.A.; Habraken, F.H.P.M. Elastic Recoil Detection. *Rep. Prog. Phys.* **1983**, *56*, 859–902. [CrossRef]
17. Geiger, H.; Marsden, E. On a diffuse reflection of the α-particles. *Proc. R. Soc. Lond. A* **1909**, *82*, 495–500. [CrossRef]
18. Doolittle, L. Algorithms for the rapid simulation of Rutherford backscattering spectra. *Nucl. Instr. Meth. B* **1985**, *9*, 344–351. [CrossRef]
19. Barradas, N.P.; Jeynes, C.; Webb, R.P. Simulated annealing analysis of Rutherford backscattering data. *Appl. Phys. Lett.* **1997**, *71*, 291–293. [CrossRef]
20. Swanson, M. The Study of Lattice Defects by Channeling. *Rep. Prog. Phys.* **1982**, *45*, 47–93. [CrossRef]
21. Gärtner, K. Modified master equation approach of axial dechanneling in perfect compound crystals. *Nucl. Instr. Meth. B* **2005**, *227*, 522–530. [CrossRef]
22. Smulders, P.J.M.; Boerma, D.O. Computer Simulation of Channeling in Single Crystals. *Nucl. Instr. Meth. B* **1987**, *29*, 471–489. [CrossRef]
23. Rebouta, L.; Smulders, P.J.M.; Boerma, D.O.; Agulló-López, F.; da Silva, M.F.; Soares, J.C. Ion-Beam Channeling Yields of Host and Impurity Atoms in $LiNbO_3$: Computer Simulation. *Phys. Rev. B* **1993**, *48*, 3600–3610. [CrossRef]
24. Kling, A. CASSIS—A New Monte-Carlo Computer Program for Channeling Simulation of RBS, NRA and PIXE. *Nucl. Instr. Meth. B* **1995**, *102*, 141–144. [CrossRef]
25. Kling, A.; Soares, J.C.; da Silva, M.F. Computer Simulation of Channeling Spectra with the CASSIS Program. *Rad. Eff. Def. Solids* **1997**, *141*, 53–61. [CrossRef]
26. Kling, A.; Soares, J.C.; da Silva, M.F. Simulation of Channeling in Crystals with Defects Using the CASSIS Code. *Nucl. Instr. Meth. B* **1999**, *153*, 457–460. [CrossRef]

27. Kling, A. The refurbished CASSIS code for channeling simulations. *Nucl. Instr. Meth. B* **2012**, *273*, 88–90. [CrossRef]
28. Soares, J.C. Microscopic characterization of materials by ion beam and hyperfine interaction analysis. *Nucl. Instr. Meth. B* **1992**, *64*, 215–220. [CrossRef]
29. Marques, J.G.; Correia, J.G.; Melo, A.A.; da Silva, M.F.; Soares, J.C. A four-detector spectrometer for e^--γ PAC on-line with the ISOLDE-CERN isotope separator. *Nucl. Instr. Meth. B* **1995**, *99*, 645–648. [CrossRef]
30. Fernandes, A.C.; Santos, J.P.; Marques, J.G.; Kling, A.; Ramos, A.R.; Barradas, N.P. Validation of the Monte Carlo model supporting core conversion of the Portuguese Research Reactor (RPI) for neutron fluence rate determinations. *Ann. Nucl. Ener.* **2010**, *37*, 1139–1145. [CrossRef]
31. Marques, J.G.; Kling, A.; de Jesus, C.M.; Soares, J.C.; da Silva, M.F.; Diéguez, E.; Agulló-López, F. Annealing recovery of neutron irradiated $LiNbO_3$:Hf single crystals. *Nucl. Instr. Meth. B* **1998**, *141*, 326–331. [CrossRef]
32. Schatz, G.; Weidinger, A. *Nuclear Condensed Matter Physics*; John Wiley and Sons: Chichester, UK, 1996; ISBN 0471954799.
33. Schwarz, K.; Blaha, P. Electronic structure of solids and surfaces with WIEN2k. In *Practical Aspects of Computational Chemistry I*; Leszczynski, J., Shukla, M., Eds.; Springer: Berlin, Germany, 2012; pp. 191–207.
34. Marques, J.G.; Lorenz, K. Lattice location of Hf and its interaction with other impurities in $LiNbO_3$: A review. *Opt. Eng.* **2014**, *53*, 060901. [CrossRef]
35. Reisman, A.; Holtzberg, F. Heterogeneous Equilibria in the Systems Li_2O-, Ag_2O-Nb_2O_5 and Oxide-Models. *J. Am. Chem. Soc.* **1958**, *80*, 6503–6507. [CrossRef]
36. Peterson, G.E.; Bridenbaugh, P.M.; Green, P. NMR study of ferroelectric $LiNbO_3$ and $LiTaO_3$ I. *J. Chem. Phys.* **1967**, *46*, 4009–4014. [CrossRef]
37. Peterson, G.E.; Bridenbaugh, P.M. NMR study of ferroelectric $LiNbO_3$ and $LiTaO_3$ II. *J. Chem. Phys.* **1968**, *48*, 3402–3406. [CrossRef]
38. Peterson, G.E.; Carruthers, J.R. ^{93}Nb NMR as a sensitive and accurate probe of stoichiometry in $LiNbO_3$ crystals. *J. Sol. State Chem.* **1969**, *1*, 98–99. [CrossRef]
39. Halstead, T.K. Temperature dependence of the Li NMR spectrum and atomic motion in $LiNbO_3$. *J. Chem. Phys.* **1970**, *53*, 3427–3435. [CrossRef]
40. Blümel, J.; Born, E.; Metzger, T. Solid state NMR study supporting the lithium vacancy defect model in congruent lithium niobate. *J. Phys. Chem. Sol.* **1994**, *55*, 589–593. [CrossRef]
41. Burns, G. Nuclear quadrupole resonance of Li in ferroelectric compounds. *Phys. Rev.* **1962**, *127*, 1193–1197. [CrossRef]
42. Schempp, E.; Peterson, G.E.; Carruthers, J.R. ^{93}Nb nuclear quadrupole resonance investigation of $LiNbO_3$. *J. Chem. Phys.* **1970**, *53*, 306–311. [CrossRef]
43. Keune, W.; Date, S.K.; Dézsi, I.; Gonser, U. Mössbauer-effect study of Co^{57} and Fe^{57} impurities in ferroelectric $LiNbO_3$. *J. Appl. Phys.* **1975**, *46*, 3914–3924. [CrossRef]
44. Peterson, G.E.; Carnevale, A. ^{93}Nb NMR linewidths in nonstoichiometric lithium niobate. *J. Chem. Phys.* **1972**, *56*, 4848–4851. [CrossRef]
45. Douglass, D.C.; Peterson, G.E.; McBrierty, V.J. Reexamination of the local electric field gradients in $LiNbO_3$. *Phys. Rev. B* **1989**, *40*, 10694–10703. [CrossRef] [PubMed]
46. Yatsenko, A.Y.; Ivanova, E.N.; Sergeev, N.A. NMR Study of Defects in Congruent $LiNbO_3$. 1. "Unoverlapping" Defects. *Phys. B* **1997**, *240*, 254–262. [CrossRef]
47. Yatsenko, A.Y.; Ivanova-Maksimova, H.M.; Sergeev, N.A. NMR Study of Intrinsic Defects in Congruent $LiNbO_3$. 2. "Overlapping" Defects. *Physica B* **1998**, *254*, 256–259. [CrossRef]
48. Yatsenko, A.; Maksimova, H.; Sergeev, N.A. NMR Study of Intrinsic Defects in Congruent Lithium Niobate. *Cryst. Res. Technol.* **1999**, *34*, 709–713. [CrossRef]
49. Rebouta, L.; Da Silva, M.F.; Soares, J.C.; Sanz-Garcia, J.A.; Diéguez, E.; Agulló-López, F. Combined RBS/Channeling and PAC studies of hafnium doped $LiNbO_3$. *Nucl. Instr. Meth. B* **1990**, *45*, 495–498. [CrossRef]
50. Catchen, G.L.; Spaar, D.M. Order-disorder effects in the phase transitions of $LiNbO_3$ and $LiTaO_3$ measured by perturbed-angular-correlation spectroscopy. *Phys. Rev. B* **1991**, *44*, 12137–12145. [CrossRef] [PubMed]
51. Davis, J.; Tinet, D.; Fripiat, J.J.; Amarilla, J.M.; Casal, B.; Ruis-Hitzky, E. ^{51}V and ^{93}Nb high resolution NMR study of $NbVO_5$. *J. Mat. Res.* **1991**, *6*, 393–400. [CrossRef]
52. Lapina, O.B.; Khabibulin, D.F.; Romanenko, K.V.; Gan, Z.; Zue, M.G.; Krasil'nikov, V.N.; Federov, V.E. ^{93}Nb NMR chemical shift scale for niobia systems. *Sol. State Nucl. Magn. Resonance* **2005**, *28*, 204–224. [CrossRef]
53. Hauer, B.; Vianden, R.; Marques, J.G.; Barradas, N.P.; Correia, J.G.; Melo, A.A.; Soares, J.C.; Agulló-López, F.; Diéguez, E. Electric-field gradients at the ^{111}In and ^{111m}Cd in undoped and Mg-doped $LiNbO_3$. *Phys. Rev. B* **1995**, *51*, 6208–6214. [CrossRef]
54. Forker, M. The problematic of the derivation of the electric field gradient asymmetry parameter from TDPAC measurements or Mössbauer spectroscopy in imperfect crystal lattices. *Nucl. Instr. Meth.* **1973**, *106*, 121–126. [CrossRef]
55. Kitamura, K.; Yamamoto, J.K.; Iyi, N.; Kimura, S.; Hayashi, T. Stoichiometric $LiNbO_3$ single crystal growth by double crucible Czochralski method using automatic powder supply system. *J. Cryst. Growth* **1992**, *116*, 327–332. [CrossRef]
56. Bordui, P.F.; Norwood, R.G.; Jundt, D.H.; Fejer, M.M. Preparation and characterization of off-congruent lithium niobate crystals. *J. Appl. Phys.* **1992**, *71*, 875–879. [CrossRef]
57. Polgár, K.; Péter, Á.; Kovács, L.; Corradi, G.; Szaller, Z. Growth of stoichiometric $LiNbO_3$ single crystals by top seeded solution growth method. *J. Cryst. Growth* **1997**, *177*, 211–216. [CrossRef]

58. Serrano, M.D.; Bermúdez, V.; Arizmendi, L.; Diéguez, E. Determination of the Li/Nb ratio in $LiNbO_3$ crystals grown by Czochralski method with K_2O added to the melt. *J. Cryst. Growth* **2000**, *210*, 670–676. [CrossRef]
59. Földvári, I.; Polgár, K.; Voszka, R.; Balasanyan, R.N. A simple method to determine the real composition of $LiNbO_3$ crystals. *Cryst. Res. Tech.* **1984**, *19*, 1659–1661. [CrossRef]
60. Kling, A.; Marques, J.G.; Correia, J.G.; Da Silva, M.F.; Diéguez, E.; Agulló-López, F.; Soares, J.C. Study of structural differences between stoichiometric and congruent lithium niobate. *Nucl. Instr. Meth. B* **1996**, *113*, 293–295. [CrossRef]
61. Kling, A.; Rebouta, L.; Marques, J.G.; Correia, J.G.; Da Silva, M.F.; Diéguez, E.; Agulló-López, F.; Soares, J.C. Ion beam channeling and hyperfine interaction analysis for the characterization of stoichiometry and anti-site population in $LiNbO_3$. *Nucl. Instr. Meth. B* **1996**, *118*, 622–625. [CrossRef]
62. Gärtner, K.; Hehl, K.; Schlotzhauer, G. Axial dechanneling: II. Point defects. *Nucl. Instr. Meth. B* **1984**, *4*, 55–62. [CrossRef]
63. Marques, J.G.; Kling, A.; Soares, J.C.; Da Silva, M.F.; Vianden, R.; Polgár, K.; Diéguez, E.; Agulló-López, F. Structural defects in congruent and near-stoichiometric $LiNbO_3$. *Rad. Eff. Def. Solids* **1999**, *150*, 233–236. [CrossRef]
64. Nassau, K.; Lines, M.E. Stacking-fault model for stoichiometry deviations in $LiNbO_3$ and $LiTaO_3$ and the effect on the Curie temperature. *J. Appl. Phys.* **1970**, *41*, 533–537. [CrossRef]
65. Smyth, D.M. Defects and transport in $LiNbO_3$. *Ferroelectrics* **1983**, *50*, 93–102. [CrossRef]
66. Smyth, D.M. Defect Chemistry of $LiNbO_3$. In Proceedings of the Sixth IEEE International Symposium on Applications of Ferroelectrics, Bethlehem, PA, USA, 8–11 June 1986; Wood, V.E., Ed.; IEEE: Piscataway, NJ, USA, 1986; pp. 115–117. [CrossRef]
67. Donnerberg, H.; Tomlinson, S.M.; Catlow, C.R.A.; Schirmer, O.F. Computer simulation studies of intrinsic defects in $LiNbO_3$ crystals. *Phys. Rev. B* **1989**, *40*, 11909–11916. [CrossRef]
68. Molière, G. Theorie der Streuung schneller geladener Teilchen I. Einzelstreuung am abgeschirmten Coulomb-Feld. *Z. Natur. A* **1947**, *2*, 133–145. [CrossRef]
69. Navrotsky, A. Energetics and crystal chemical systematics among ilmenite, lithium niobate, and perovskite structures. *Chem. Mat.* **1998**, *10*, 2787–2793. [CrossRef]
70. Yusa, H.; Akaogi, M.; Sata, N.; Kojitani, H.; Yamamoto, R.; Ohishi, Y. High-pressure transformations of ilmenite to perovskite, and lithium niobate to perovskite in zinc germanate. *Phys. Chem. Min.* **2006**, *33*, 217–226. [CrossRef]
71. Kumada, N.; Ozawa, N.; Muto, F. $LiNbO_3$ with ilmenite-type structure prepared via ion-exchange reaction. *J. Sol. St. Chem.* **1985**, *57*, 267–268. [CrossRef]
72. Kumada, N.; Kinomura, N.; Muto, F. Crystal-structures of ilmenite type $LiNbO_3$ and $NaNbO_3$. *J. Ceram. Soc. Japan* **1990**, *98*, 384–388. [CrossRef]
73. Mehta, A.; Navrotsky, A.; Kumada, N.; Kinomura, N. Structural transitions in $LiNbO_3$ and $NaNbO_3$. *J. Solid State Chem.* **1993**, *102*, 213–225. [CrossRef]
74. Wiegel, M.; Blasse, G.; Navrotsky, A.; Mehta, A.; Kumada, N.; Kinomura, N. Luminescence of the ilmenite phase of $LiNbO_3$. *J. Sol. St. Chem.* **1994**, *109*, 413–415. [CrossRef]
75. Corradi, G.; Kovács, L.; Zaritskii, I.M. Optical absorption edge and some shallow donor levels in $LiNbO_3$ systems. *Rad. Eff. Def. Sol.* **1999**, *150*, 211–219. [CrossRef]
76. Baran, E.J.; Botto, I.L.; Muto, F.; Kumada, N.; Kinomura, N. Vibrational spectra of the ilmenite modifications of $LiNbO_3$ and $NaNbO_3$. *J. Mat. Sci. Lett.* **1986**, *5*, 671–672. [CrossRef]
77. Kong, Y.; Xu, J.; Chen, X.; Zhang, C.; Zhang, W.; Zhang, G. Ilmenite-like stacking defect in nonstoichiometric lithium niobate crystals investigated by Raman scattering spectra. *J. Appl. Phys.* **2000**, *87*, 4410–4414. [CrossRef]
78. Lengyel, K.; Kovács, L.; Péter, Á.; Polgár, K.; Corradi, G. The effect of stoichiometry and Mg doping on the Raman spectra of $LiNbO_3$:Mg crystals. *Appl. Phys. B* **2007**, *87*, 317–322. [CrossRef]
79. Abrahams, S.C.; Marsh, P. Defect structure dependence on composition in lithium niobate. *Acta Crystal.* **1986**, *42*, 61–68. [CrossRef]
80. Glass, A.M. Dielectric, thermal, and pyroelectric properties of ferroelectric $LiTaO_3$. *Phys. Rev.* **1968**, *172*, 564–571. [CrossRef]
81. Birnie, D.P. Model for the ferroelectric transition in nonstoichiometric lithium niobate and lithium tantalate. *J. Amer. Ceram. Soc.* **1991**, *74*, 988–993. [CrossRef]
82. Johnston, W.D.; Kaminow, I.P. Temperature dependence of Raman and Rayleigh scattering in $LiNbO_3$ and $LiTaO_3$. *Phys. Rev.* **1968**, *168*, 1045–1054. [CrossRef]
83. Abrahams, S.C.; Buehler, E.; Hamilton, W.C.; Laplaca, S.J. Ferroelectric lithium tantalate-III. Temperature dependence of the structure in the ferroelectric phase and the paraelectric phase at 940 K. *J. Phys. Chem. Sol.* **1973**, *34*, 521–532. [CrossRef]
84. Boysen, H.; Altdorfer, F. A neutron powder investigation of the high-temperature structure and phase transition in $LiNbO_3$. *Acta Crystal. B* **1994**, *50*, 405–414. [CrossRef]
85. Inbar, I.; Cohen, R.E. Comparison of the electronic structures and energetics of ferroelectric $LiNbO_3$ and $LiTaO_3$. *Phys. Rev. B* **1996**, *53*, 1193–1204. [CrossRef]
86. Phillpot, S.R.; Gopalan, V. Coupled displacive and order-disorder dynamics in $LiNbO_3$ by molecular-dynamics simulation. *Appl. Phys. Lett.* **2004**, *84*, 1916–1918. [CrossRef]
87. Sanna, S.; Schmidt, W.G. Ferroelectric phase transition in $LiNbO_3$: Insights from molecular dynamics. *IEEE Trans. Ultrason. Ferroel. Freq. Control* **2012**, *59*, 1925–1928. [CrossRef]
88. Toyoura, K.; Ohta, M.; Nakamura, A.; Matsunaga, K. First-principles study on phase transition and ferroelectricity in lithium niobate and tantalate. *J. Appl. Phys.* **2015**, *118*, 064103. [CrossRef]

89. Zhukov, A.P.; Soboleva, L.V.; Belyaev, L.M.; Volkov, A.F. Niobium-93, Lanthanum-139, and Tantalum-181 pure NQR in rare earth orthoniobates and lithium tantalate. *Ferroelectrics* **1978**, *21*, 601–604. [CrossRef]
90. Löhnert, M.; Kaindl, G.; Wortmann, G.; Salomon, D. Temperature-induced changes of the electronic structure of ferroelectric $LiTaO_3$. *Phys. Rev. Lett.* **1981**, *47*, 194–197. [CrossRef]
91. De Wette, F.W.; Schacher, G.E. Electric field gradients in point-ion and uniform-background lattices II. *Physical Review* **1965**, *137*, A92–A94. [CrossRef]
92. Shelyapina, M.G.; Kasperovich, V.S.; Shchegolev, B.F.; Charnaya, E.V. Cluster calculations of electric-field-gradients at the Ta Site for the ferroelectric $LiTaO_3$ crystal. *Ferroelectrics* **2003**, *282*, 1–7. [CrossRef]
93. Slotfeldt-Ellingsen, D. NMR study of the temperature dependence of the 7Li quadrupole coupling constant above and below the Curie temperature in ferroelectric $LiTaO_3$. In *Proceedings of the 17th Congress Ampere*; Hovi, V., Ed.; North Holland Publishing Company: Amsterdam, The Netherlands, 1972; pp. 350–352.
94. Slotfeldt-Ellingsen, D.; Pedersen, B. The structural phase transition in $LiTaO_3$ studied by pulsed ^{7}Li-NMR. *Phys. Stat. Sol.* **1974**, *24*, 191–195. [CrossRef]
95. Catchen, G.L.; Adams, J.M.; Rearick, T.M. High-temperature partitioning of ^{181}Hf-probe impurities between Li and group-V sites in $LiNbO_3$ and $LiTaO_3$. *Phys. Rev. B* **1992**, *46*, 2743–2749. [CrossRef]
96. Tomov, T.; Engelmann, H.; Dézsi, I.; Gonser, U. Investigation of the ferroelectric phase transition in $LiNbO_3$:Fe by Mössbauer spectroscopy. *Solid State Commun.* **1989**, *69*, 41–44. [CrossRef]
97. Marques, J.G.; Kling, A.; De Jesus, C.M.; Soares, J.C.; Friedsam, P.; Freitag, K.; Vianden, R. Electric field gradients in the ferroelectric and paraelectric phases of $LiTaO_3$. *Hyperfine Interact.* **1999**, *120–121*, 485–489. [CrossRef]
98. Ohkubo, Y.; Murakami, Y.; Saito, T.; Yokohama, A.; Uerhara, S.; Kawase, Y. Mechanism of the ferroelectric phase transitions in $LiNbO_3$ and $LiTaO_3$. *Phys. Rev. B* **2002**, *65*, 052107. [CrossRef]
99. Jetschke, S.; Karge, H.; Hehl, K. Anisotropic Effects in N+-Implanted $LiNbO_3$. *Phys. Stat. Sol.* **1983**, *77*, 207–214. [CrossRef]
100. Götz, G.; Karge, H. Ion Implantation into $LiNbO_3$. *Nucl. Instr. Meth.* **1983**, *209/210*, 1079–1088. [CrossRef]
101. Götz, G. Radiation Effects in Optoelectronic Materials. *Rad. Eff.* **1986**, *98*, 189–209. [CrossRef]
102. Wendler, E.; Becker, G.; Rensberg, J.; Schmidt, E.; Wolf, S.; Wesch, W. Direction-Dependent RBS Channelling Studies in Ion Implanted $LiNbO_3$. *Nucl. Instr. Meth. B* **2016**, *379*, 195–199. [CrossRef]
103. Gischkat, T.; Schrempel, F.; Wesch, W. Ion-beam induced effects at 15 K in $LiNbO_3$. *Nucl. Instr. Meth. B* **2008**, *266*, 2906–2909. [CrossRef]
104. Bianconi, M.; Argiolas, N.; Bazzan, M.; Bentini, G.G.; Cerutti, A. Quantification of Nuclear Damage in High Energy Ion Implanted Lithium Niobate. *Nucl. Instr. Meth. B* **2007**, *257*, 597–600. [CrossRef]
105. Schmidt, E.; Ritter, K.; Gärtner, K.; Wendler, E. Investigation of Li/Nb-Sublattices in Ion Implanted $LiNbO_3$ by RBS and NRA in Channelling Configuration. *Nucl. Instr. Meth. B* **2017**, *409*, 126–132. [CrossRef]
106. Schrempel, F.; Gischkat, T.; Hartung, H.; Kley, E.-B.; Wesch, W. Ion Beam Enhanced Etching of $LiNbO_3$. *Nucl. Instr. Meth. B* **2006**, *250*, 164–168. [CrossRef]
107. Jetschke, S.; Hehl, K. Diffusion and Recrystallization Processes during Annealing of N^+- and P^+-Implanted $LiNbO_3$. *Phys. Stat. Sol.* **1985**, *88*, 193–205. [CrossRef]
108. Canut, B.; Brenier, R.; Meftah, A.; Moretti, P.; Ould Salem, S.; Ramos, S.M.M.; Thevenard, P.; Toulemonde, M. Damage Induced in $LiNbO_3$ Single Crystals by GeV Gadolinium Ions. *Nucl. Instr. Meth. B* **1994**, *91*, 312–316. [CrossRef]
109. Canut, B.; Brenier, R.; Meftah, A.; Moretti, P.; Salem, S.O.; Pitaval, M.; Ramos, S.M.M.; Thevenard, P.; Toulemonde, M. Latent track formation in $LiNbO_3$ single crystals irradiated by GeV uranium ions. *Rad. Eff. Def. Sol.* **1995**, *136*, 307–310. [CrossRef]
110. Canut, B.; Ramos, S.M.M.; Brenier, R.; Thevenard, P.; Loubet, J.L.; Toulemonde, M. Surface Modifications of $LiNbO_3$ Single Crystals Induced by Swift Heavy Ions. *Nucl. Instr. Meth. B* **1996**, *107*, 194–198. [CrossRef]
111. Bentini, G.G.; Bianconi, M.; Chiarini, M.; Correra, L.; Sada, C.; Mazzoldi, P.; Argiolas, N.; Bazzan, M.; Guzzi, R. Effect of Low Dose High Energy O^{3+} Implantation on Refractive Index and Linear Electro-Optic Properties in X-Cut $LiNbO_3$: Planar Optical Waveguide Formation and Characterization. *J. Appl. Phys.* **2002**, *92*, 6477–6483. [CrossRef]
112. Bentini, G.G.; Bianconi, M.; Cerutti, A.; Chiarini, M.; Pennestrì, G.; Sada, C.; Argiolas, N.; Bazzan, M.; Mazzoldi, P.; Guzzi, R. Structural and Compositional Characterization of X-Cut $LiNbO_3$ Crystals Implanted with High Energy Oxygen and Carbon Ions. *Nucl. Instr. Meth. B* **2005**, *240*, 174–177. [CrossRef]
113. Bentini, G.G.; Bianconi, M.; Correra, L.; Chiarini, M.; Mazzoldi, P.; Sada, C.; Argiolas, N.; Bazzan, M.; Guzzi, R. Damage Effects Produced in the Nearsurface Region of X-cut $LiNbO_3$ by Low Dose, High Energy Implantation of Nitrogen, Oxygen, and Fluorine Ions. *J. Appl. Phys.* **2004**, *96*, 242–247. [CrossRef]
114. Bianconi, M.; Argiolas, N.; Bazzan, M.; Bentini, G.G.; Chiarini, M.; Cerutti, A.; Mazzoldi, P.; Pennestrì, G.; Sada, C. On the Dynamics of the Damage Growth in 5 MeV Oxygen-Implanted Lithium Niobate. *Appl. Phys. Lett.* **2005**, *87*, 072901. [CrossRef]
115. Olivares, J.; García-Navarro, A.; García, G.; Agulló-López, F.; Agulló-Rueda, F.; García-Cabañes, A.; Carrascosa, M. Buried Amorphous Layers by Electronic Excitation in Ion-Beam Irradiated Lithium Niobate: Structure and Kinetics. *J. Appl. Phys.* **2007**, *101*, 033512. [CrossRef]
116. Olivares, J.; García, G.; García-Navarro, A.; Agulló-López, F.; Caballero, O.; García-Cabanes, A. Generation of High-Confinement Step-Like Optical Waveguides in $LiNbO_3$ by Swift Heavy Ion-Beam Irradiation. *Appl. Phys. Lett.* **2005**, *86*, 183501. [CrossRef]

117. Olivares, J.; García, G.; Agulló-López, F.; Agulló-Rueda, F.; Kling, A.; Soares, J.C. Generation of Amorphous Surface Layers in $LiNbO_3$ by Ion Beam-Irradiation: Thresholding and Boundary Propagation. *Appl. Phys. A* **2005**, *81*, 1465–1469. [CrossRef]
118. Rivera, A.; Olivares, J.; Crespillo, M.L.; García, G.; Bianconi, M.; Agulló-López, F. Assessment of Swift-Ion Damage by RBS/C: Determination of the Amorphization Threshold. *Nucl. Instr. Meth. B* **2009**, *267*, 1460–1463. [CrossRef]
119. Ramos, S.M.M.; Canut, B.; Ambri, M.; Bonardi, N.; Pitaval, M.; Bernas, H.; Chaumont, J. Defect creation in $LiNbO_3$ irradiated by medium masses ions in the electronic stopping power regime. *Rad. Eff. Def. Sol.* **1998**, *143*, 299–309. [CrossRef]
120. Wesch, W.; Rensberg, J.; Schmidt, M.; Wendler, E. Damage evolution in $LiNbO_3$ due to Electronic Energy Deposition below the Threshold for Direct Amorphous Track Formation. *J. Appl. Phys.* **2019**, *126*, 125105. [CrossRef]
121. Agulló-López, F.; García, G.; Olivares, J. Lattice Preamorphization by Ion Irradiation: Fluence Dependence of the Electronic Stopping Power Threshold for Amorphization. *J. Appl. Phys.* **2005**, *97*, 093514. [CrossRef]
122. Toulemonde, M.; Dufour, C.; Meftah, A.; Paumier, E. Transient thermal processes in heavy ion irradiation of crystalline inorganic insulators. *Nucl. Instr. Meth. B* **2000**, *166–167*, 903–912. [CrossRef]
123. Agulló-López, F.; Climent-Font, A.; Muñoz-Martín, A.; Olivares, J.; Zucchiatti, A. Ion beam modification of dielectric materials in the electronic excitation regime: Cumulative and exciton models. *Progr. Mat. Sci.* **2016**, *76*, 1–58. [CrossRef]
124. Agulló-López, F.; Mendez, A.; García, G.; Olivares, J.; Cabrera, J.M. Synergy between thermal spike and exciton decay mechanisms for ion damage and amorphization by electronic excitation. *Phys. Rev. B* **2006**, *74*, 174109. [CrossRef]
125. García, G.; Rivera, A.; Crespillo, M.L.; Gordillo, N.; Olivares, J.; Agulló-López, F. Amorphization Kinetics under Swift Heavy Ion Irradiation: A Cumulative Overlapping-Track Approach. *Nucl. Instr. Meth. B* **2011**, *269*, 492–497. [CrossRef]
126. Schrempel, F.; Steinbach, T.; Gischkat, T.; Wesch, W. Channeling irradiation of $LiNbO_3$. *Nucl. Instr. Meth. B* **2008**, *266*, 2958–2961. [CrossRef]
127. Steinbach, T.; Schrempel, F.; Gischkat, T.; Wesch, W. Influence of ion energy and ion species on ion channeling in $LiNbO_3$. *Phys. Rev. B* **2008**, *78*, 184106. [CrossRef]
128. Dowell, M.B.; Lefkowitz, I.; Taylor, G.W. Radiation damage in lithium niobate. Proc. 2nd Int. Meeting on Ferroelectricity, Kyoto, 1969. *J. Phys. Soc. Japan* **1970**, *28*, 442–444.
129. Berg, N.; Speulstra, J. The operation of acoustic surface wave delay lines in a nuclear environment. *IEEE Trans. Nucl. Sci.* **1973**, *20*, 137–143. [CrossRef]
130. Primak, W.; Anderson, T.T. Metamictization of lithium niobate by thermal neutrons. *Nucl. Technol.* **1976**, *28*, 235–248. [CrossRef]
131. Primak, W.; Gavin, A.P.; Anderson, T.T.; Monahan, E. Stability of lithium niobate on irradiation at elevated reactors temperature. *Nucl. Technol.* **1977**, *36*, 79–84. [CrossRef]
132. Marques, J.G. Evolution of nuclear fission reactors: Third generation and beyond. *Energy Convers. Manag.* **2010**, *51*, 1774–1780. [CrossRef]
133. Giuliani, F.; Oliveira, C.; Collar, J.I.; Girard, T.A.; Morlat, T.; Limagne, D.; Marques, J.G.; Ramos, A.R.; Waysand, G. Response of SIMPLE SDDs to monochromatic neutron irradiations. *Nucl. Instr. Meth. A* **2004**, *526*, 348–358. [CrossRef]
134. González, R.; Chen, Y.; Abraham, M.M. Transmutation-induced tritium in $LiNbO_3$ single crystals. *Phys. Rev. B* **1988**, *37*, 6433–6435. [CrossRef]
135. González, R.; Ballesteros, C.; Chen, Y.; Abraham, M.M. Diffusion of tritons, deuterons, and protons in $LiNbO_3$ crystals. *Phys. Rev. B* **1989**, *39*, 11085–11092. [CrossRef] [PubMed]
136. Verma, R.; Arunachalam, J.; Gangadharan, S.; Vyas, J.C.; Kothiyal, G.P.; Gupta, M.K. Trace characterisation of lithium niobate by neutron activation analysis. *Fresenius J. Anal. Chem.* **1992**, *344*, 261–264. [CrossRef]
137. Brannon, P.J. Transient radiation-induced absorption at 1061 nm in $LiNbO_3$ and MgO:$LiNbO_3$. *IEEE Trans. Nucl. Sci.* **1994**, *41*, 642–647. [CrossRef]
138. Marques, J.G.; Kling, A.; Soares, J.C.; Rebouta, L.; Da Silva, M.F.; Diéguez, E.; Agulló-López, F. Lattice site location of Hf in $LiNbO_3$: Influence of dopant concentration and crystal stoichiometry. *Nucl. Instr. Meth. B* **1998**, *136–138*, 431–435. [CrossRef]
139. Orlova, A.N.; Ped'ko, B.B.; Filinova, A.V.; Franko, N.Y.; Prokhorova, A.Y. Influence of gamma and gamma–neutron irradiation on the optical properties of $LiNbO_3$ single crystals. *Phys. Sol. State* **2006**, *48*, 544–547. [CrossRef]
140. Mandula, G.; Rupp, R.A.; Balaskó, M. Erasure of elementary holograms in $LiNbO_3$:Fe by neutron irradiation. In *Proceedings of the Holography 2005: International Conference on Holography, Optical Recording and Processing of Information*; SPIE-International Society for Optics and Photonics: Bellingham, WA, USA, 2006; Volume 6252, p. 62520N. [CrossRef]
141. Sha, G.; Harlow, C.; Chernatynskiy, A.; Daw, J.; Khafizov, M. In-situ measurement of irradiation behavior in $LiNbO_3$. *Nucl. Instr. Meth. B* **2020**, *472*, 46–52. [CrossRef]
142. Coltman, R.R.; Klabunde, C.E.; McDonald, D.L.; Redman, J.K. Reactor damage in pure metals. *J. Appl. Phys.* **1962**, *33*, 3509–3522. [CrossRef]
143. Robinson, M.T. Basic physics of radiation damage production. *J. Nucl. Mat.* **1994**, *216*, 1–28. [CrossRef]
144. Pritychenko, B.; Mughabghab, S.F. Neutron thermal cross sections, Westcott factors, resonance integrals, Maxwellian averaged cross sections and astrophysical reaction rates calculated from the ENDF/B-VII.1, JEFF-3.1.2, JENDL-4.0, ROSFOND-2010, CENDL-3.1 and EAF-2010 evaluated data libraries. *Nucl. Data Sheets* **2012**, *113*, 3120–3144. [CrossRef]
145. Lorenz, K.; Marques, J.G.; Franco, N.; Alves, E.; Peres, M.; Correia, M.R.; Monteiro, T. Defect studies on fast and thermal neutron irradiated GaN. *Nucl. Instr. Meth. B* **2008**, *266*, 2780–2783. [CrossRef]
146. Pelowitz, D.B. *MCNPX User's Manual Version 2.7, LA-CP-11-00438*; Los Alamos National Laboratory: Los Alamos, NM, USA, 2011.

147. Seitz, F. On the disordering of solids by action of fast massive particles. *Discuss. Faraday Soc.* **1949**, *5*, 271–282. [CrossRef]
148. Nord, J.; Nordlund, K.; Keinonen, J. Molecular dynamics study of damage accumulation in GaN during ion beam irradiation. *Phys. Rev. B* **2003**, *68*, 184104. [CrossRef]
149. Tsuchihira, H.; Oda, T.; Tanaka, S. Molecular-dynamics simulation of threshold displacement energies in lithium aluminate. *Nucl. Instr. Meth. B* **2011**, *269*, 1707–1711. [CrossRef]
150. Lucas, G.; Pizzagalli, L. Ab initio molecular dynamics calculations of threshold displacement energies in silicon carbide. *Phys. Rev. B* **2005**, *72*, 161202. [CrossRef]
151. Liu, B.; Xiao, H.Y.; Zhang, Y.; Aidhy, D.S.; Weber, W.J. Ab initio molecular dynamics simulations of threshold displacement energies in $SrTiO_3$. *J. Phys. Cond. Matt.* **2013**, *25*, 485003. [CrossRef]
152. Petersen, B.A. Low Energy Recoil Simulations in MgO, $LiNbO_3$, and $LiTaO_3$ Using Ab Initio Molecular Dynamics. PhD Thesis, University of Tennessee, Knoxville, TN, USA, 2017.
153. Hogdson, E.R.; Agulló-López, F. Oxygen vacancy centres induced by electron irradiation in $LiNbO_3$. *Solid State Commun.* **1987**, *64*, 965–968. [CrossRef]
154. Hogdson, E.R.; Agulló-López, F. Displacement damage in $LiNbO_3$. *Nucl. Instr. Meth. B* **1988**, *32*, 42–44. [CrossRef]
155. Popov, A.I.; Kotomin, E.A.; Maier, J. Basic properties of the F-type centers in halides, oxides and perovskites. *Nucl. Instr. Meth. B* **2010**, *268*, 3084–3089. [CrossRef]
156. Alenius, G.; Arnell, S.E.; Schale, C.; Wallander, E. Low spin states in ^{181}Hf from the thermal neutron capture reaction. *Phys. Scrip.* **1971**, *3*, 105–110. [CrossRef]
157. Sears, V.F. Neutron scattering lengths and cross sections. *Neutron News* **1992**, *3*, 26–37. [CrossRef]
158. Nakagawa, T.; Shibata, K.; Kawasaki, H. *Curves and Tables of Neutron Cross Sections in JENDL-3.3. JAERI-Data/Code 2002-020*; Japan Atomic Energy Research Institute: Tokaimura, Japan, 2002.
159. Molnar, G. (Ed.) *Handbook of Prompt Gamma Activation Analysis*; Springer: Berlin, Germany, 2003; ISBN 978-1402013041.
160. Dale, C.J.; Marshall, P.W.; Summers, G.P.; Wolicki, E.A.; Burke, E.A. Displacement damage equivalent to dose in silicon devices. *Appl. Phys. Lett.* **1989**, *54*, 451–453. [CrossRef]
161. Angelescu, T.; Vasilescu, A. Comparative radiation hardness results obtained from various neutron sources and the NIEL problem. *Nucl. Instr. Meth. A* **1996**, *374*, 85–90. [CrossRef]
162. Lindhard, J.; Nielsen, V.; Scharff, M.; Thonsen, P.V. Integral equations governing radiation effects (Notes on atomic collisions III). *Matematisk-Fysiske Meddelelser* **1963**, *33*, 1–42.
163. Akkerman, A.; Barak, J. New partition factor calculations for evaluating the damage of low energy ions in silicon. *IEEE Trans. Nucl. Sci.* **2006**, *53*, 3667–3674. [CrossRef]
164. Boschini, M.J.; Rancoita, P.G.; Tacconi, M. *SR-NIEL Calculator: Screened Relativistic (SR) Treatment for Calculating the Displacement Damage and Nuclear Stopping Powers for Electrons, Protons, Light- and Heavy- Ions in Materials (version 6.9.0)*; INFN sez Bicocca: Milan, Italy, 2020; Volume 10. Available online: http://www.sr-niel.org/ (accessed on 31 March 2021).
165. Ashkin, A.; Boyd, G.D.; Dziedzic, J.M.; Smith, R.G.; Ballman, A.A.; Levinstein, J.J.; Nassau, K. Optically-Induced Refractive Index Inhomogenities in $LiNbO_3$ and $LiTaO_3$. *Appl. Phys. Lett.* **1966**, *9*, 72–74. [CrossRef]
166. Zhong, G.G.; Jian, J.; Wu, Z.-K. *Proceedings of the 11th International Quantum Electronics Conference*; Cat. No. 80 CH 1561-0; IEEE: New York, NY, USA, 1980; p. 631.
167. Bryan, D.A.; Gerson, R.; Tomaschke, H.E. Increased Optical Damage Resistance in Lithium Niobate. *Appl. Phys. Lett.* **1984**, *44*, 847–849. [CrossRef]
168. Volk, T.R.; Pryalkin, V.I.; Rubinina, N.M. Optical-Damage-Resistant $LiNbO_3$:Zn Crystal. *Opt. Lett.* **1990**, *15*, 996–998. [CrossRef]
169. Yamamoto, J.K.; Kitamura, K.; Iyi, N.; Kimura, S.; Furukawa, Y.; Sato, M. Increased Optical Damage Resistance in Sc_2O_3-Doped $LiNbO_3$. *Appl. Phys. Lett.* **1992**, *61*, 2156–2158. [CrossRef]
170. Volk, T.R.; Rubinina, N.M. A New Optical Damage Resistant Impurity in Lithium Niobate Crystals: Indium. *Ferroelectrics Lett.* **1992**, *14*, 37–43. [CrossRef]
171. Kokanyan, E.P.; Razzari, L.; Cristiani, I.; Degiorgio, V.; Gruber, J.B. Reduced Photorefraction in Hafnium-Doped Single-Domain and Periodically Poled Lithium Niobate Crystals. *Appl. Phys. Lett.* **2004**, *84*, 1880–1882. [CrossRef]
172. Sun, L.; Guo, F.; Lv, Q.; Yu, H.; Li, H.; Cai, W.; Xu, Y.; Zhao, L. Increased Optical Damage Resistance of Zr:$LiNbO_3$ Crystals. *Cryst. Res. Technol.* **2007**, *42*, 1117–1122. [CrossRef]
173. Zhang, Q.; Feng, X. Defect Structures and Densities of Mg-Doped Lithium Niobate. *Phys. Stat. Sol.* **1990**, *121*, 429–435. [CrossRef]
174. Liu, J.; Zhang, W.; Zhang, G. Defect Chemistry Analysis of the Defect Structure in Mg-Doped $LiNbO_3$ Crystals. *Phys. Stat. Sol.* **1996**, *156*, 285–291. [CrossRef]
175. Abdi, F.; Aillerie, M.; Bourson, P.; Fontana, M.D. Defect structure in Mg-doped $LiNbO_3$: Revisited study. *J. Appl. Phys.* **2009**, *106*, 033519. [CrossRef]
176. Iyi, N.; Kitamura, K.; Yajima, Y.; Kimura, S.; Furukawa, Y.; Sato, M. Defect Structure Model of MgO-Doped $LiNbO_3$. *J. Solid State Chem.* **1995**, *118*, 148–152. [CrossRef]
177. Donnerberg, H.J.; Tomlinson, S.M.; Catlow, C.R.A. Defects in $LiNbO_3$, II. Computer Simulation. *J. Phys. Chem. Solids* **1991**, *52*, 201–210. [CrossRef]
178. Donnerberg, H.J.; Tomlinson, S.M.; Catlow, C.R.A.; Schirmer, O.F. Computer-Simulation Studies of Extrinsic Defects in $LiNbO_3$ Crystals. *Phys. Rev. B* **1991**, *44*, 4877–4883. [CrossRef]

179. Donnerberg, H. Comments on the Defect Chemistry of Magnesium-Doped Lithium Niobate ($LiNbO_3$). *J. Solid State Chem.* **1996**, *123*, 208–214. [CrossRef]
180. Wöhlecke, M.; Volk, T.; Donnerberg, H. On the Role of Intrinsic Clusters in Damage-Resistant $LiNbO_3$. *Ferroelectr. Lett. Sect.* **1997**, *22*, 53–58. [CrossRef]
181. Kling, A.; Kollewe, D.; Grabmaier, B.C. Lattice site Investigations for Mg in $LiNbO_3$ by Combined RBS-PIXE-NRA-Channeling Experiments. *Nucl. Instr. Meth. B* **1992**, *64*, 232–236. [CrossRef]
182. Grabmaier, B.C.; Otto, F. Growth and Investigation of MgO-doped $LiNbO_3$. *J. Cryst. Growth* **1986**, *79*, 682–688. [CrossRef]
183. Grabmaier, B.C.; Otto, F. Growth of $LiNbO_3$ Single Crystals for Optical Applications. *Proc. SPIE* **1986**, *651*, 2–6. [CrossRef]
184. Sommerfeldt, R.; Holtmann, L.; Krätzig, E.; Grabmaier, B.C. Influence of Mg Doping and Composition on the Light-Induced Charge Transport in $LiNbO_3$. *Phys. Stat. Sol.* **1988**, *106*, 89–98. [CrossRef]
185. Koppitz, J.; Schirmer, O.F.; Wöhlecke, M.; Kuznetsov, A.I.; Grabmaier, B.C. Threshold Effects in $LiNbO_3$:Mg Caused by Change of Electron-Lattice Coupling. *Ferroelectrics* **1989**, *92*, 233–241. [CrossRef]
186. Grabmaier, B.C.; Wersing, W.; Koestler, W. Properties of Undoped and MgO-Doped $LiNbO_3$; Correlation to the Defect Structure. *J. Cryst. Growth* **1991**, *110*, 339–347. [CrossRef]
187. Schmidt, N.; Betzler, K.; Grabmaier, B.C. Composition dependence of the second harmonic phase-matching temperature in $LiNbO_3$. *Appl. Phys. Lett.* **1991**, *58*, 34–35. [CrossRef]
188. Nunomura, K.; Ishitani, A.; Matsubara, T.; Hayashi, I. Second Harmonic Generation in a Sputtered $LiNbO_3$ Film on MgO. *J. Crystal Growth* **1978**, *45*, 355–360. [CrossRef]
189. Kling, A.; Kollewe, D.; Grabmaier, B.C. Dependence of the $^7Li(p,\alpha)^4He$ Minimum Yield on the Mg Concentration for Proton Channeling. *Nucl. Instr. Meth. B* **1994**, *85*, 490–493. [CrossRef]
190. Feng, X.; Wang, D.; Zhang, J. NMR Spectra of Mg Nuclei in Mg-Doped $LiNbO_3$ Crystals. *Phys. Stat. Sol.* **1990**, *157*, K127–K130. [CrossRef]
191. Yatsenko, A.V.; Yevdokimov, S.V.; Sugak, D.Y.; Solskii, I.M. NMR Analysis of Mg Ion Localization in $LiNbO_3$ Crystal. *Acta Phys. Pol. A* **2010**, *117*, 166–169. [CrossRef]
192. Rebouta, L.; Soares, J.C.; da Silva, M.F.; Sanz-Garcia, J.A.; Diéguez, E.; Agulló-López, F. Lattice Sites for Eu, Hf and Nd in $LiNbO_3$ by RBS/Channeling Experiments. *Nucl. Instr. Meth. B* **1990**, *50*, 428–430. [CrossRef]
193. Marques, J.G.; Kling, A.; Rebouta, L.; da Silva, M.F.; Melo, A.A.; Soares, J.C.; Serrano, M.D.; Diéguez, E.; Agulló-López, F. Lattice Location of Hf in Near-stoichiometric $LiNbO_3$: RBS/Channeling and PAC Studies. *Mat. Sci. Forum* **1997**, *248/249*, 395–398. [CrossRef]
194. Rebouta, L.; da Silva, M.F.; Soares, J.C.; Santos, M.T.; Diéguez, E.; Agulló-López, F. Ion-Beam/Channeling Characterization of $LiNbO_3$: Interaction between Impurity Sites. *Rad. Eff. Def. Sol.* **1995**, *136*, 137–139. [CrossRef]
195. Rebouta, L.; da Silva, M.F.; Soares, J.C.; Santos, M.T.; Diéguez, E.; Agulló-López, F. Ion-Beam/Channeling Characterization of $LiNbO_3$: Interaction between Impurity Sites. *Opt. Mat.* **1995**, *4*, 174–178. [CrossRef]
196. Marques, J.G.; de Jesus, C.M.; Melo, A.A.; Soares, J.C.; Diéguez, E.; Agulló-López, F. PAC Study of the Interaction between Impurity Sites in $LiNbO_3$ Doped with Hf and Mg. *Hyperfine Interact.* **1996**, *1*, 348–351.
197. Dias, S.A.; Marques, J.G.; Correia, J.G.; Sanz, J.A.; Soares, J.C. The $^{181}Hf/^{181}Ta$ Probe in the Li and Nb Sites of Congruent $LiNbO_3$ Co-Doped with Mg and Cr Ions Studied by PAC. *Hyperfine Interact.* **2004**, *158*, 323–328. [CrossRef]
198. Kling, A.; Soares, J.C.; da Silva, M.F.; Rebouta, L.; Kollewe, D.; Krause, H.; Flagmeyer, R.-H.; Vogt, J. Determination of Impurity Lattice Sites in Single Crystals Using PIXE/channeling. *X-Ray Spectrom.* **1999**, *28*, 105–109. [CrossRef]
199. Kovács, L.; Rebouta, L.; Soares, J.C.; da Silva, M.F.; Hage-Ali, M.; Stoquert, J.P.; Siffert, P.; Zaldo, C.; Szaller, Z.; Polgár, K. Lattice Site of Trivalent Impurities in Mg-Doped Lithium Niobate Crystals. *Mat. Sci. Eng. B* **1991**, *9*, 505–508. [CrossRef]
200. Hesselink, L.; Orlov, S.S.; Bashaw, M.C. Holographic Data Storage Systems. *Proc. IEEE* **2004**, *92*, 1231–1280. [CrossRef]
201. McMillen, D.K.; Hudson, T.D.; Wagner, J.; Singleton, J. Holographic Recording in Specially Doped Lithium Niobate Crystals. *Opt. Express* **1998**, *2*, 491–502. [CrossRef]
202. Yue, X.; Adibi, A.; Hudson, T.; Buse, K.; Psaltis, D. Role of Cerium in Lithium Niobate for Holographic Recording. *J. Appl. Phys.* **2000**, *87*, 4051–4055. [CrossRef]
203. Kong, Y.; Liu, S.; Xu, J. Recent Advances in the Photorefraction of Doped Lithium Niobate Crystals. *Materials* **2012**, *5*, 1954–1971. [CrossRef]
204. Buse, K.; Adibi, A.; Psaltis, D. Non-Volatile Holographic Storage in Doubly Doped Lithium Niobate Crystals. *Nature* **1998**, *393*, 665–668. [CrossRef]
205. Liu, F.; Kong, Y.; Ge, X.; Liu, H.; Liu, S.; Chen, S.; Rupp, R.; Xu, J. Improved Sensitivity of Nonvolatile Holographic Storage in Triply Doped $LiNbO_3$:Zr,Cu,Ce. *Opt. Express* **2010**, *18*, 6333–6339. [CrossRef] [PubMed]
206. Valyashko, E.G.; Varina, T.M.; Kuz'min, R.N.; Rubinina, N.M.; Smirnov, V.A.; Shagdarov, V.M. Spectroscopic properties of single crystals of lithium metaniobate containing iron and neodymium ions as impurity. *J. Appl. Spectrosc.* **1974**, *21*, 877–880. [CrossRef]
207. Keune, W.; Date, S.K.; Gonser, U.; Bunzel, H. Mössbauer Effect Study of Fe^{57} Doped $LiNbO_3$ and $LiTaO_3$. *Ferroelectrics* **1976**, *13*, 443–445. [CrossRef]
208. Kurz, H.; Krätzig, E.; Keune, W.; Engelmann, H.; Gonser, U.; Dischler, B.; Räuber, A. Photorefractive Centres in $LiNbO_3$ Studied by Optical-, Mössbauer- and EPR-Methods. *Appl. Phys.* **1977**, *12*, 355–368. [CrossRef]

209. Gonser, U.; Sakai, H.; Keune, W. Mössbauer Spectroscopy of $^{57}Fe^{2+}$ Impurities in $LiNbO_3$. *J. Electrochem. Soc.* **1976**, *123*, 1915–1917. [CrossRef]
210. Becze-Deák, T.; Bottyán, L.; Corradi, G.; Korecz, L.; Nagy, D.L.; Polgár, K.; Sayed, S.; Spiering, H. Electron Trapping Centres and Cross Sections in $LiNbO_3$ Studied by ^{57}Co Mössbauer Emission Spectroscopy. *J. Phys. Condens. Matt.* **1999**, *11*, 6239–6250. [CrossRef]
211. Becze-Deák, T.; Bottyán, L.; Corradi, G.; Korecz, L.; Nagy, D.L.; Polgár, K.; Sayed, S.; Spiering, H. ^{57}Co Mössbauer Emission Study of $LiNbO_3$, Fe:$LiNbO_3$ and Mg:$LiNbO_3$ in Various Thermochemical Reduction States. *J. Radioanalyt. Nucl. Chem.* **2000**, *246*, 33–37. [CrossRef]
212. Rebouta, L.; da Silva, M.F.; Soares, J.C.; Hage-Ali, M.; Stoquert, J.P.; Siffert, P.; Sanz-García, J.A.; Diéguez, E.; Agulló-López, F. Lattice Site of Iron in $LiNbO_3(Fe^{3+})$ by the PIXE/Channeling Technique. *Europhys. Lett.* **1991**, *14*, 557–561. [CrossRef]
213. Malovichko, G.I.; Grachev, V.G. Determination of Iron Group Impurity Positions in Ferroelectric $LiNbO_3$ using nuclear-quadrupole splitting. *Sov. Phys. Sol. State* **1985**, *27*, 1678–1679.
214. Zaltron, A.; Argiolas, N.; De Salvador, D.; Bazzan, M.; Ciampolillo, M.V.; Bacci, L.; Sada, C. Iron Site Location in Fe-Diffused Lithium Niobate Crystals by Combined RBS-PIXE-NRA Analysis. *Nucl. Instr. Meth. B* **2012**, *275*, 11–15. [CrossRef]
215. Szilágy, E.; Becze-Deák, T.; Bottyán, L.; Kocsonya, A.; Kótai, E.; Nagy, D.L.; Kling, A.; Battistig, G.; Khan, N.Q.; Polgár, K. Lattice Site Determination of Co in Low Doped Congruent $LiNbO_3$ Single Crystal Using PIXE/Channeling. *Sol. State Commun.* **2000**, *115*, 535–538. [CrossRef]
216. Buchal, C.; Mantl, S.; Thomas, D.K. Channeling Investigation of the Lattice Location of Ti in Ti-Implanted Optical Waveguides in $LiNbO_3$. *Mat. Res. Soc. Symp. Proc.* **1987**, *100*, 317–324. [CrossRef]
217. Kollewe, D.; Kling, A.; Grabmaier, B.C.; Bremer, T.; Heiland, W.; Zimmermann, W. Lattice Site Location of Ti Diffused and Doped in $LiNbO_3$. *Phys. Lett. A* **1992**, *169*, 177–180. [CrossRef]
218. Hauer, B.; Vianden, R.; da Silva, M.F.; Rebouta, L.; Soares, J.C.; Diéguez, E.; Agulló-López, F. The Lattice Site of Ti in $LiNbO_3$. *J. Phys. Condens. Matter* **1994**, *6*, 267–274. [CrossRef]
219. Johnson, L.F.; Ballman, A.A. Coherent Emission from Rare Earth Ions in Electro-Optic Crystals. *J. Appl. Phys.* **1969**, *40*, 297–302. [CrossRef]
220. Kovács, L.; Rebouta, L.; Soares, J.C.; da Silva, M.F. Lattice Site of Er in $LiNbO_3$:Mg,Er Crystals. *Rad. Eff. Def. Sol.* **1991**, *119–121*, 445–450. [CrossRef]
221. Kovács, L.; Rebouta, L.; Soares, J.C.; da Silva, M.F.; Hage-Ali, M.; Stoquert, J.P.; Siffert, P.; Sanz-García, J.A.; Corradi, G.; Szaller, Z.; et al. On the Lattice Site of Trivalent Dopants and the Structure of Mg^{2+}–OH–M^{3+} Defects in $LiNbO_3$:Mg Crystals. *J. Phys.: Condens. Matter* **1993**, *5*, 781–794. [CrossRef]
222. Rebouta, L.; Soares, J.C.; da Silva, M.F.; Serrano, D.; Díeguez, E.; Agulló-López, F.; Tornero, J. Nonaxial Sites for Er in $LiNbO_3$. *Appl. Phys. Lett.* **1997**, *70*, 1070–1072. [CrossRef]
223. Lorenzo, A.; Jaffrezic, H.; Roux, B.; Boulon, G.; Bausá, L.E.; García Solé, J. Lattice Location of Pr^{3+} Ions in $LiNbO_3$. *Phys. Rev. B* **1995**, *52*, 6278–6284. [CrossRef] [PubMed]
224. Lorenzo, A.; Jaffrezic, H.; Roux, B.; Boulon, G.; García Solé, J. Lattice Location of Rare-Earth Ions in $LiNbO_3$. *Appl. Phys. Lett.* **1995**, *67*, 3735–3737. [CrossRef]
225. Lorenzo, A.; Loro, H.; Muñoz Santiuste, J.E.; Terrile, M.C.; Boulon, G.; Bausá, L.E.; García Solé, J. RBS/Channeling to Locate Active Ions in Laser Materials: Application to Rare Earth Activated $LiNbO_3$. *Opt. Mat.* **1997**, *8*, 55–63. [CrossRef]
226. García Solé, J.; Petit, T.; Jaffrezic, H.; Boulon, G. Lattice Location of the Non-Equivalent Nd^{3+} Ions in $LiNbO_3$:Nd and $LiNbO_3$:MgO:Nd. *Europhys. Lett.* **1993**, *24*, 719–724. [CrossRef]
227. García-Solé, J.; Lorenzo, A.; Petit, T.; Boulon, G.; Roux, B.; Jaffrezic, H. Site Selective Spectroscopy and RBS/Channeling in Optically Active Ion Doped $LiNbO_3$ crystals. *J. Phys. IV* **1994**, *4*, C4-293–C4-296.
228. Rebouta, L.; Soares, J.C.; da Silva, M.F.; Sanz-García, J.A.; Diéguez, E.; Agulló-López, F. Lattice Site for Transition and Rare-Earth Impurities in $LiNbO_3$ by Ion-Beam Methods. *J. Mater. Res.* **1992**, *7*, 130–135. [CrossRef]
229. Camarillo, E.; Tocho, J.; Vergara, I.; Díeguez, E.; García-Solé, J.; Jaque, F. Optical Bands of Cr^{3+} induced by Mg^{2+} Ions in $LiNbO_3$:Cr,Mg. *Phys. Rev. B* **1992**, *45*, 4600–4604. [CrossRef] [PubMed]
230. Díaz-Caro, J.; García-Solé, J.; Bravo, D.; Sanz-García, A.; Lopez, F.J.; Jaque, F. MgO Codoping-Induced Change in the Site Distribution of Cr^{3+} Ions in $LiNbO_3$. *Phys. Rev. B* **1996**, *54*, 13042–13046. [CrossRef] [PubMed]
231. Kling, A.; Soares, J.C.; da Silva, M.F. Optical and Structural Properties of Chromium Implanted Lithium Niobate—Cluster Formation and Substitutional Incorporation. *Nucl. Instr. Meth. B* **1999**, *148*, 1044–1048. [CrossRef]
232. Almeida, J.M.; Boyle, G.; Leite, A.P.; De La Rue, R.M.; Ironside, C.N.; Caccavale, F.; Chakraborty, P.; Mansour, I. Chromium Diffusion in Lithium Niobate for Active Optical Waveguides. *J. Appl. Phys.* **1995**, *78*, 2193–2197. [CrossRef]
233. Macfarlane, P.I.; Holliday, K.; Nicholls, J.F.H.; Henderson, B. Characterization of Cr^{3+} Centres in $LiNbO_3$ Using Fluorescence Line Narrowing. *J. Phys. Condens. Matter* **1995**, *7*, 9643–9656. [CrossRef]
234. Jaque, F.; García-Solé, J.; Camarillo, E.; López, F.J.; Murrieta, S.H.; Hernández, A.J. Detection of Cr^{3+} sites in $LiNbO_3$:MgO,Cr^{3+} and $LiNbO_3$:Cr^{3+}. *Phys. Rev. B* **1993**, *47*, 5432–5434. [CrossRef]
235. Corradi, G.; Chadwick, A.V.; West, A.R.; Cruickshank, K.; Paul, M. On the substitution site of Cr and Fe in $LiNbO_3$: An EXAFS study. *Rad. Eff. Def. Solids* **1995**, *134*, 219–222. [CrossRef]

236. Kling, A.; Soares, J.C.; da Silva, M.F.; Sanz-García, J.A.; Diéguez, E.; Agulló-López, F. Lattice Site Determination of Cr in Low Doped Lithium Niobate Single Crystals Using PIXE/channeling. *Nucl. Instr. Meth. B* **1998**, *136-138*, 426–430. [CrossRef]
237. Fernández-Ruiz, R.; Bermúdez, V. Determination of the Ta and Nb ratio in $LiNb_{1-x}Ta_xO_3$ by total reflection X-ray fluorescence spectrometry. *Spectrochim. Acta Part B At. Spectrosc.* **2005**, *60*, 231–235. [CrossRef]
238. Tian, T.; Kong, Y.F.; Liu, S.G.; Li, W.; Wu, L.; Chen, S.L.; Xu, J.J. Photorefraction of molybdenum-doped lithium niobate crystals. *Opt. Lett.* **2012**, *37*, 2679–2681. [CrossRef] [PubMed]
239. Kling, A.; Marques, J.G.; Soares, J.C.; da Silva, M.F.; Diéguez, E.; Agulló-López, F. Valence Effect on the Incorporation of Tungsten Implanted into Lithium Niobate. *Nucl. Instr. Meth. B* **1997**, *127/128*, 520–523. [CrossRef]
240. Kling, A.; Marques, J.G.; da Silva, M.F.; Serrano, M.D.; Diéguez, E.; Sanz-García, J.A.; García-Solé, J.; Agulló-López, F.; Soares, J.C. Incorporation of Hexavalent Impurities into $LiNbO_3$. *Rad. Eff. Def. Solids* **1999**, *150*, 249–253. [CrossRef]
241. Kling, A.; Valdrez, C.; Marques, J.G.; da Silva, M.F.; Soares, J.C. Incorporation of Tungsten in Lithium Niobate by Diffusion. *Nucl. Instr. Meth. B* **2002**, *190*, 524–527. [CrossRef]
242. Calvo, J.A.; Bausá, L.E.; Sanz, J.A.; Voda, M.; García-Solé, J. Spectroscopy of Uranium Ions in $LiNbO_3$. *Ferroelectrics* **1996**, *185*, 41–44. [CrossRef]
243. Kling, A.; da Silva, M.F.; Soares, J.C.; Sanz-García, J.A.; García-Solé, J. Change of Uranium Lattice Site in $LiNbO_3$ Induced by Thermal Reduction Treatment. *Rad. Eff. Def. Solids* **2001**, *155*, 229–233. [CrossRef]
244. Freestone, I.; Meeks, N.; Sax, M.; Higgitt, C. The Lycurgus Cup—A Roman Nanotechnology. *Gold Bull.* **2007**, *40*, 270–277. [CrossRef]
245. Saito, Y.; Suganomata, S. Coloration of $LiNbO_3$ by Metal Ion Implantation. *Jpn. J. Appl. Phys.* **1987**, *26*, 1941–1942. [CrossRef]
246. Saito, Y.; Shang, D.Y.; Kitsutaka, R.; Kitahara, A. Optical Properties of Cu-Implanted $LiNbO_3$. *J. Appl. Phys.* **1997**, *81*, 3621–3626. [CrossRef]
247. Shang, D.Y.; Saito, Y.; Kittaka, R.; Taniguchi, S.; Kitahara, A. Optical Properties of $LiNbO_3$ Implanted with Ag Ions. *J. Appl. Phys.* **1996**, *80*, 6651–6654. [CrossRef]
248. Williams, E.K.; Ila, D.; Sarkisov, S.; Curley, M.; Cochrane, J.C.; Poker, D.B.; Hensley, D.K.; Borel, C. Study of the Effects of MeV Ag and Au Implantation on the Optical Properties of $LiNbO_3$. *Nucl. Instr. Meth. B* **1998**, *141*, 268–273. [CrossRef]
249. Saito, Y.; Kitahara, A. Absorption in the Visible Region of $LiNbO_3$ Sequentially Implanted with Ag and Cu Ions. *J. Appl. Phys.* **2000**, *87*, 1276–1279. [CrossRef]
250. Shi, B.-R.; Lu, F.; Meng, M.-Q.; Wang, F.-X.; Li, W.; Liu, X.-D.; Wang, K.-M.; Wang, Z.-L. Damage Behaviours in $LiNbO_3$ Crystals Induced by MeV Au^+ Irradiation. *Nucl. Instr. Meth. B* **1998**, *135*, 275–279. [CrossRef]
251. Ziegler, J.F.; Ziegler, M.D.; Biersack, J.P. SRIM—The stopping and range of ions in matter. *Nucl. Instr. Meth. B* **2010**, *268*, 1818–1823. [CrossRef]
252. Takahiro, K.; Kunimatsu, A.; Nagata, S.; Yamaguchi, S.; Yamamoto, S.; Aoki, Y.; Naramoto, H. Crystal Structure and Optical Absorption of Au Implanted MgO, $SrTiO_3$ and $LiNbO_3$. *Nucl. Instr. Meth. B* **1999**, *152*, 314–318. [CrossRef]
253. Dev, B.N.; Kuri, G.; Satyam, P.V.; Sundaravel, B.; Gog, T.; Materlik, G. X-ray Standing Wave and Ion Scattering Studies on Au-Implanted $LiNbO_3$ (001) Single Crystals. *Appl. Surf. Sci.* **1998**, *125*, 163–172. [CrossRef]
254. Kling, A.; da Silva, M.F.; Soares, J.C.; Fichtner, P.F.P.; Amaral, L.; Zawislak, F.C. Formation of Coherent Gold Nanoclusters in Lithium Niobate. *Nucl. Instr. Meth. B* **2002**, *191*, 478–481. [CrossRef]
255. Kling, A.; da Silva, M.F.; Soares, J.C.; Fichtner, P.F.P.; Amaral, L.; Zawislak, F.C. Influence of Helium Co-Implantation on the Formation of Gold Nanoclusters in Lithium Niobate. *Mod. Phys. Lett. B* **2001**, *15*, 1348–1354. [CrossRef]
256. Wong-Leung, J.; Williams, J.S.; Eliman, R.G.; Nygren, E.; Eaglesham, D.J.; Jacobson, D.C.; Poate, J.M. Proximity Gettering of Au to Ion Beam Induced Defects in Silicon. *Nucl. Instr. Meth. B* **1995**, *96*, 253–256. [CrossRef]
257. Milz, S.; Rensberg, J.; Ronning, C.; Wesch, W. Correlation between damage evolution, cluster formation and optical properties of silver implanted lithium niobate. *Nucl. Instr. Meth. B* **2012**, *286*, 67–71. [CrossRef]
258. Williams, E.K.; Ila, D.; Darwish, A.; Poker, D.B.; Sarkisov, S.S.; Curley, M.J.; Wang, J.-C.; Svetchnikov, V.L.; Zandbergen, H.W. Characterization of Silver Colloids Formed in $LiNbO_3$ by Ag and O Implantation at Room and Elevated Temperatures. *Nucl. Instr. Meth. B* **1999**, *148*, 1074–1078. [CrossRef]
259. Kling, A.; Soares, J.C.; da Silva, M.F. Platinum Implanted Lithium Niobate—Annealing Behavior and Dopant Redistribution. *Nucl. Instr. Meth. B* **1998**, *141*, 436–440. [CrossRef]
260. Kling, A.; Soares, J.C.; da Silva, M.F. Redistribution and Incorporation of Iridium Implanted into Lithium Niobate. *Nucl. Instr. Meth. B* **2000**, *166–167*, 280–283. [CrossRef]
261. Bazzan, M.; Sada, C. Optical Waveguides in Lithium Niobate: Recent Developments and Applications. *Appl. Phys. Rev.* **2015**, *2*, 040603. [CrossRef]
262. Kaminow, I.P.; Carruthers, J.R. Optical waveguiding layers in $LiNbO_3$ and $LiTaO_3$. *Appl. Phys. Lett.* **1973**, *22*, 326–328. [CrossRef]
263. Burns, W.K.; Klein, P.H.; West, E.J.; Plew, L.E. Ti diffusion in Ti:$LiNbO_3$ planar and channel optical waveguides. *J. Appl. Phys.* **1979**, *50*, 6175–6182. [CrossRef]
264. Armenise, M.N.; Canali, C.; De Sario, M.; Carnera, A.; Mazzoldi, P.; Celotti, G. Evaluation of the Ti diffusion process during fabrication of Ti:$LiNbO_3$ optical waveguides. *J. Non-Cryst. Sol.* **1982**, *47*, 255–257. [CrossRef]
265. Armenise, M.N.; Canali, C.; De Sario, M.; Carnera, A.; Mazzoldi, P.; Celotti, G. Characterization of $(Ti_{0.65}Nb_{0.35})O_2$ Compound as a Source for Ti Diffusion During Ti:$LiNbO_3$ Optical Waveguides Fabrication. *J. Appl. Phys.* **1983**, *54*, 62–70. [CrossRef]

266. Bremer, T.; Kollewe, D.; Koschmieder, H.; Heiland, W. SIMS Investigations of Titanium Profiles in $LiNbO_3$ Produced by Ion Beam Mixing and Diffusion. *Fresenius Z. Anal. Chem.* **1989**, *333*, 485–487. [CrossRef]
267. Caccavale, F.; Chakraborty, P.; Quaranta, A.; Mansour, I.; Gianello, G.; Bosso, S.; Corsini, R.; Mussi, G. Secondary-Ion-Mass Spectrometry and Near-Field Studies of Ti:$LiNbO_3$ Optical Waveguides. *J. Appl. Phys.* **1995**, *78*, 5345–5350. [CrossRef]
268. Caccavale, F.; Gonella, F.; Quaranta, A.; Mansour, I. Analysis of Ti:$LiNbO_3$ Waveguides Using Secondary Ion Mass Spectrometry and Near Field Method. *Elect. Lett.* **1995**, *31*, 1054–1056. [CrossRef]
269. Caccavale, F.; Sada, C.; Segato, F.; Cavuoti, F. Secondary Ion Mass Spectrometry and Optical Characterization of Ti:$LiNbO_3$ Optical Waveguides. *Appl. Surf. Sci.* **1999**, *150*, 195–201. [CrossRef]
270. Zhang, D.-L.; Yang, Z.; Wong, W.H.; Pun, E.Y.B. Secondary-ion-mass spectrometry study on nearstoichiometric $LiNbO_3$ strip waveguide fabricated by vapour transport equilibration and Ti codiffusion. *Philos. Mag.* **2007**, *87*, 63–75. [CrossRef]
271. Bremer, T.; Hertel, P.; Oelschig, S.; Sommerfeldt, R.; Heiland, W. Depth Profiling of Magnesium and Titanium-Doped $LiNbO_3$ Waveguides. *Thin Sol. Films* **1989**, *175*, 235–239. [CrossRef]
272. Caccavale, F.; Chakraborty, P.; Capobianco, A.; Gianello, G.; Mansour, I. Characterization and Optimization of Ti-Diffused $LiNbO_3$ Optical Waveguides by Second Diffusion of Magnesium. *J. Appl. Phys.* **1995**, *78*, 187–193. [CrossRef]
273. Caccavale, F.; Chakraborty, P.; Mansour, I.; Gianello, G.; Mazzoleni, M.; Elena, M. A Secondary-Ion-Mass Spectrometry Study of Magnesium Diffusion in Lithium Niobate. *J. Appl. Phys.* **1994**, *76*, 7552–7558. [CrossRef]
274. Mohan Kumar, R.; Yamamoto, F.; Ichikawa, J.; Ryoken, H.; Sakaguchi, I.; Liu, X.; Nakamura, M.; Terabe, K.; Takekawa, S.; Haneda, H.; et al. SIMS-depth profile and microstructure studies of Ti-diffused Mg-doped near-stoichiometric lithium niobate waveguide. *J. Cryst. Growth* **2006**, *287*, 472–477. [CrossRef]
275. Herreros, B.; Lifante, G. $LiNbO_3$ optical waveguides by Zn diffusion from vapor phase. *Appl. Phys. Lett.* **1995**, *66*, 1449–1451. [CrossRef]
276. Nevado, R.; Cussó, F.; Lifante, G.; Caccavale, F.; Sada, C.; Segato, F. Correlation between compositional and refractive index profiles in diffused optical waveguides. *J. Appl. Phys.* **2000**, *88*, 6183–6186. [CrossRef]
277. Nevado, R.; Cantelar, E.; Segato, F.; Caccavale, F.; Kling, A.; Soares, J.C. Compositional Characterization of Zn-Diffused Lithium Niobate Waveguides. *Appl. Phys. B* **2001**, *73*, 555–558. [CrossRef]
278. Espeso-Gil, O.; García, G.; Agulló-López, F.; Climent-Font, A.; Sajavaara, T.; Domenech, M.; Cantelar, E.; Lifante, G. Characterization of surface layers in Zn-diffused $LiNbO_3$ waveguides by heavy ion elastic recoil detection. *Appl. Phys. Lett.* **2002**, *81*, 1981–1983. [CrossRef]
279. Zhang, D.-L.; Qiu, C.-X.; Wong, W.-H.; Pun, E.Y.-B. Diffusion Properties of Scandium in Lithium Niobate Crystal. *J. Am. Ceram. Soc.* **2014**, *97*, 2903–2908. [CrossRef]
280. Zhang, D.-L.; Qiu, C.-X.; Zhang, W.-Z.; Hua, P.-R.; Yu, D.-Y.; Pun, E.Y.-B. Diffusion Characteristics of Zr^{4+} in $LiNbO_3$ Single-Crystal. *J. Am. Ceram. Soc.* **2013**, *96*, 2722–2724. [CrossRef]
281. Tsai, W.-S.; Chiang, T.-Y.; Liu, L.-Y.; Chang, P.-C.; Su, Y.-K. Time and Temperature Dependent Study of Zn and Ni Codiffused $LiNbO_3$ Waveguides. *J. Light. Technol.* **2015**, *33*, 4850–4859. [CrossRef]
282. Zhang, D.-L.; Qiu, C.-X.; Wong, W.-H.; Du, W.-J.; Pun, E.Y.-B. Zr^{4+}/Ti^{4+} Codiffusion Characteristics in Lithium Niobate. *J. Am. Ceram. Soc.* **2015**, *98*, 567–573. [CrossRef]
283. Zhang, D.-L.; Zhang, Q.; Qiu, C.-X.; Wong, W.-H.; Dao, Y.-Y.; Pun, E.Y.-B. Diffusion control of an ion by another in $LiNbO_3$ and $LiTaO_3$ crystals. *Sci. Rep.* **2015**, *5*, 10018. [CrossRef] [PubMed]
284. De Almeida, J.M.M.M.; Marinho, F.; Alexandre, D.; Sada, C. Secondary Ion Mass Spectrometry Study of Erbium Titanium Codiffusion in Lithium Niobate. *IEEE Phot. Lett. Tech.* **2014**, *26*, 1307–1309. [CrossRef]
285. Jackel, J.J.; Rice, C.E. Topotactic $LiNbO_3$ to Cubic Perovskite Structural Transformation in $LiNbO_3$ and $LiTaO_3$. *Ferroelectrics* **1982**, *38*, 801–8040. [CrossRef]
286. Jackel, J.J.; Rice, C.E.; Veselka, J.J. Proton exchange for high-index waveguides in $LiNbO_3$. *Appl. Phys. Lett.* **1982**, *41*, 607–608. [CrossRef]
287. Cabrera, J.M.; Olivares, J.; Carrascosa, M.; Rams, J.; Müller, R.; Diéguez, E. Hydrogen in Lithium Niobate. *Adv. Phys.* **1996**, *45*, 349–392. [CrossRef]
288. Campari, A.; Ferrari, C.; Mazzi, G.; Summonte, C.; Al-Shukri, S.M.; Dawar, A.; De La Rue, R.M.; Nutt, A.C.G. Strain and Surface Damage Induced by Proton Exchange in Y-cut $LiNbO_3$. *J. Appl. Phys.* **1985**, *58*, 4521–4524. [CrossRef]
289. De Micheli, M.; Ostrowsky, D.B.; Barety, J.P.; Mazzi, G.; Papuchon, M. Crystalline and Optical Quality of Proton Exchanged Waveguides. *J. Light. Tech.* **1986**, *LT-4*, 743–745. [CrossRef]
290. Canali, C.; Carnera, A.; Della Mea, G.; Mazzoldi, P.; Al Shukri, S.M.; Nutt, A.C.G.; De La Rue, R.M. Structural Characterization of Proton Exchanged $LiNbO_3$ Optical Waveguides. *J. Appl. Phys.* **1986**, *59*, 2643–2649. [CrossRef]
291. Olivares, J.; Cabrera, J.M.; Agulló-López, F.; Rebouta, L.; da Silva, M.F.; Soares, J.C. Structural Characterization of $LiNbO_3$ Proton-Exchanged Waveguides by Ion-Beam Methods. *Ferroelectrics* **1995**, *174*, 93–99. [CrossRef]
292. Hu, H.; Lu, F.; Chen, F.; Wang, F.-X.; Zhang, J.-H.; Liu, X.-D.; Wang, K.-M. Channeling Study of Proton-Exchanged $LiNbO_3$ Optical Waveguides. *Phys. Stat. Sol.* **2000**, *181*, 509–513. [CrossRef]
293. Cargo, J.T.; Filo, A.J.; Hughes, M.C.; Kannan, V.C.; Stevie, F.A.; Taylor, J.A.; Holmes, R.J. Characterization of Sulfuric Acid Proton-Exchanged Lithium Niobate. *J. Appl. Phys.* **1990**, *67*, 627–633. [CrossRef]

294. Paz-Pujalt, G.R.; Tuschel, D.D.; Braunstein, G.; Blanton, T.; Tong Lee, S.; Salter, L.M. Characterization of Proton Exchange Lithium Niobate Waveguides. *J. Appl. Phys.* **1994**, *76*, 3981–3987. [CrossRef]
295. Hsu, W.-Y.; Braunstein, G.; Gopalan, V.; Willand, C.S.; Gupta, M.C. Correlation between Structural and Optical Properties in Proton-Exchanged $LiNbO_3$. *Appl. Phys. Lett.* **1992**, *61*, 3083–3085. [CrossRef]
296. Olivares, J.; Cabrera, J.M.; Agulló-López, F.; Rebouta, L.; da Silva, M.F.; Soares, J.C. Ion-Beam Channeling Characterization of Proton-Exchanged $LiNbO_3$. In Proceedings of the 6th European Conference on Integrated Optics and Technical Exhibit: ECIO 93, Neuchâtel, Switzerland, 18–22 April 1993; Roth, P., Parriaux, O.M., Eds.; ECIO: Neuchâtel, Switzerland, 1993; pp. 9-14–9-15.
297. Kovács, L.; Wöhlecke, M.; Jovanović, A.; Polgár, K.; Kapphan, S. Infrared absorption study of the OH vibrational band in $LiNbO_3$ crystals. *J. Phys. Chem. Solids* **1991**, *52*, 707–803. [CrossRef]
298. Ito, K.; Kawamoto, K. Dependence of Lattice Constant Deviation and Refractive Index on Proton Concentration in Proton-Exchanged Optical Waveguides on a Single Crystal of $LiNbO_3$. *Jpn. J. Appl. Phys.* **1992**, *31*, 3882–3887. [CrossRef]
299. Rottschalk, M.; Bachmann, T.; Witzmann, A. Investigations of Proton Exchanged Optical Waveguides in $LiNbO_3$ Using Elastic Recoil Detection. *Nucl. Instr. Meth. B* **1991**, *61*, 91–93. [CrossRef]
300. Hagner, G.; Bachmann, T. Refractive Index Profiles and Exchange Ratios of Proton-Exchanged Waveguides in Congruent and MgO-Doped $LiNbO_3$. *Phys. Stat. Sol.* **1998**, *165*, 205–212. [CrossRef]
301. Casey, H.C., Jr.; Chen, C.-H.; Zavada, J.M.; Novak, S.W. Analysis of Hydrogen Diffusion from Proton-Exchanged Layers in $LiNbO_3$. *Appl. Phys. Lett.* **1993**, *63*, 718–720. [CrossRef]
302. Zavada, J.M.; Casey, H.C., Jr.; States, R.J.; Novak, S.W.; Loni, A. Correlation of Substitutional Hydrogen to Refractive Index Profiles in Annealed Proton-Exchanged Z- and X-Cut $LiNbO_3$. *J. Appl. Phys.* **1995**, *77*, 2697–2708. [CrossRef]
303. Engelsberg, M.; do Nascimento, G.C.; Pacobahyba, L.H. ^{1}H Atomic Motion in Proton-Exchanged $LiNbO_3$. *J. Appl. Phys.* **1993**, *74*, 6427–6429. [CrossRef]
304. De Souza, R.E.; Engelsberg, M.; Pacobahyba, L.H.; do Nascimento, G.C. Nuclear Magnetic Resonance Study of Proton Exchanged $LiNbO_3$. *J. Appl. Phys.* **1995**, *77*, 408–410. [CrossRef]
305. Engelsberg, M.; de Souza, R.E.; Pacobahyba, L.H.; do Nascimento, G.C. Structural Determination of Hydrogen Site Occupation in Proton-Exchanged $LiNbO_3$ by Nuclear Magnetic Resonance. *Appl. Phys. Lett.* **1995**, *67*, 359–361. [CrossRef]
306. Yamamoto, K.; Taniuchi, T. Characteristics of Pyrophosphoric Acid Proton-Exchanged Waveguides in $LiNbO_3$. *J. Appl. Phys.* **1991**, *70*, 6663–6668. [CrossRef]
307. Cheng, C.-C.; Horng, R.-C.; Chen, Y.-C. Formation and Properties of Proton-Exchanged and Annealed $LiNbO_3$ Waveguides for Surface Acoustic Wave. *IEEE Trans. Ultras. Ferroel. Freq. Control.* **2001**, *48*, 387–391. [CrossRef]
308. Chen, F.; Wang, X.-L.; Wang, K.-M. Development of Ion-Implanted Optical Waveguides in Optical Materials: A Review. *Opt. Mat.* **2007**, *29*, 1523–1542. [CrossRef]
309. Peña-Rodríguez, O.; Olivares, J.; Carrascosa, M.; García-Cabañes, A.; Rivera, A.; Agulló-López, F. Optical Waveguides Fabricated by Ion Implantation/Irradiation: A Review. In *Ion Implantation*; Goorsky, M., Ed.; InTech: Rijeka, Croatia, 2012; pp. 267–314. ISBN 978-953-51-0634-0.
310. Destefanis, G.L.; Townsend, P.D.; Gailliard, J.P. Optical Waveguides in $LiNbO_3$ Formed by Ion Implantation of Helium. *Appl. Phys. Lett.* **1978**, *32*, 293–294. [CrossRef]
311. Barfoot, K.M.; Laursen, T.; Whitton, J.L.; Weiss, B.L.; Webb, R.P. In Situ Channeling Analysis During Thermal Annealing of $^4He^+$-Implanted $LiNbO_3$. *Nucl. Instr. Meth. B* **1989**, *44*, 141–145. [CrossRef]
312. Williams, E.K.; Ila, D.; Sarkisov, S.; Venkateswarlu, P.; Poker, D.B. Application of NRA/Channeling to Study He^+ Implanted Waveguides. *Nucl. Instr. Meth. B* **1986**, *118*, 57–61. [CrossRef]
313. Al-Chalabi, S.A.M.; Weiss, B.L.; Barfoot, K.M.; Arnold, G.W. Analysis of He-implanted $LiNbO_3$ by Elastic Recoil Detection. *J. Appl. Phys.* **1988**, *63*, 1032–1036. [CrossRef]
314. Kling, A.; da Silva, M.F.; Soares, J.C.; Fichtner, P.F.P.; Amaral, L.; Zawislak, F. Defect Evolution and Characterization in He-Implanted $LiNbO_3$. *Nucl. Instr. Meth. B* **2001**, *175–177*, 394–397. [CrossRef]
315. Fichtner, P.F.P.; Kaschny, J.R.; Kling, A.; Trinkaus, H.; Yankov, R.A.; Mücklich, A.; Skorupa, W.; Zawislak, F.C.; Amaral, L.; da Silva, M.F.; et al. Nucleation and Growth of Platelet Bubble Structures in He^+ Implanted Silicon. *Nucl. Instr. Meth. B* **1998**, *136–138*, 460–464. [CrossRef]
316. Gischkat, T.; Schrempel, F.; Höche, T.; Wesch, W. Annealing Behavior of Lithium Niobate Irradiated with He-Ions at 100 K. *Nucl. Instr. Meth. B* **2009**, *267*, 1492–1495. [CrossRef]
317. Olivares, J.; García, G.; Agulló-López, F.; Agulló-Rueda, F.; Soares, J.C.; Kling, A. Optical Investigations of the Propagation of the Amorphous-Crystalline Boundary in Ion-beam Irradiated $LiNbO_3$. *Nucl. Instr. Meth. B* **2006**, *242*, 534–537. [CrossRef]
318. Hu, H.; Lu, F.; Chen, F.; Wang, F.-X.; Zhang, J.-H.; Liu, X.-D.; Wang, K.-M.; Shi, B.-R. Optical waveguide formation by MeV H^+ implanted into $LiNbO_3$ crystal. *Opt. Comm.* **2000**, *177*, 189–193. [CrossRef]
319. Zhao, J.-H.; Huang, Q.; Wang, L.; Fu, G.; Qin, X.F.; Liu, P.; Guo, S.-S.; Liu, T.; Wang, X.-L. The Properties of Ion-Implanted $LiNbO_3$ Waveguides Measured by the RBS and Ion Beam Etching Stripping Methods. *Opt. Mat.* **2011**, *33*, 1357–1361. [CrossRef]
320. Wang, L.; Xiang, B.-X. Planar waveguides in magnesium doped stoichiometric $LiNbO_3$ crystals formed by MeV oxygen ion implantations. *Nucl. Instr. Meth. B* **2012**, *272*, 121–124. [CrossRef]
321. Song, Q.; Lu, F.; Ma, X.; Liu, H.; Liu, X.; Zhang, R.; Wang, X. MgO:$LiNbO_3$ Planar Waveguide Formed by MeV O^{2+} Implantation and Its Annealing Characteristics. *Laser Phys.* **2008**, *18*, 815–818. [CrossRef]

322. Bremer, T.; Heiland, W.; Buchal, C.; Irmscher, R.; Stritzker, B. Recrystallization and Refractive Index Profiles of Titanium-Implanted Optical Waveguides in $LiNbO_3$. *J. Appl. Phys.* **1990**, *67*, 1183–1187. [CrossRef]
323. Zhang, S.-M.; Peng, Y.-D.; Wang, P.; Liu, Q.-X. Influences on Proton Exchange by He Ion Implantation in $LiNbO_3$. *Opt. Mat. Expr.* **2015**, *5*, 1526–1531. [CrossRef]
324. Huang, Q.; Liu, P.; Liu, T.; Guo, S.-S.; Zhang, L.; Wang, X.-L. Lattice damage and waveguide properties of a proton-exchanged $LiNbO_3$ crystal after oxygen-ion implantation. *Nucl. Instr. Meth. B* **2012**, *286*, 318–321. [CrossRef]
325. Herreros, B.; Lifante, G.; Cussó, F.; Kling, A.; Soares, J.C.; da Silva, M.F.; Townsend, P.D.; Chandler, P.J. Structural and Optical Properties of Rare-earth Doped Lithium Niobate Waveguides Formed by MeV Helium Implantation. *Mat. Res. Soc. Symp. Proc.* **1996**, *396*, 355–360. [CrossRef]
326. Herreros, B.; Lifante, G.; Cussó, F.; Sanz, J.A.; Kling, A.; Soares, J.C.; da Silva, M.F.; Townsend, P.D.; Chandler, P.J. Photoluminescence and Rutherford Backscattering Spectrometry Study of Ion-implanted Er^{3+}-doped $LiNbO_3$ Planar Waveguides. *J. Phys. Condens. Matter* **1998**, *10*, 3275–3283. [CrossRef]
327. Herreros, B.; Lifante, G.; Kling, A.; Soares, J.C.; da Silva, M.F.; Townsend, P.D.; Chandler, P.J.; Olivares, J.; Cabrera, J.M. RBS/channeling Study of Ion-implanted and Proton-exchanged $LiNbO_3$:Nd^{3+}:MgO Planar Waveguides. *Opt. Mat.* **1996**, *6*, 281–286. [CrossRef]
328. Cajzl, J.; Nekvindová, P.; Macková, A.; Malinský, P.; Oswald, J.; Staněk, S.; Vytykáčová, S.; Špirková, J. Optical waveguides in Er:$LiNbO_3$ fabricated by different techniques—A Comparison. *Opt. Mat.* **2016**, *53*, 160–168. [CrossRef]

crystals

MDPI

Review

Real-time Identification of Oxygen Vacancy Centers in $LiNbO_3$ and $SrTiO_3$ during Irradiation with High Energy Particles

Miguel L. Crespillo [1,2,*], Joseph T. Graham [3], Fernando Agulló-López [2], Yanwen Zhang [4] and William J. Weber [1,4,*]

1 Department of Materials Science and Engineering, University of Tennessee, Knoxville, TN 37996, USA
2 Centro de Microanálisis de Materiales, CMAM-UAM, Cantoblanco, 28049 Madrid, Spain; fal@uam.es
3 Department of Nuclear Engineering and Radiation Science, Missouri University of Science and Technology, Rolla, MO 65409, USA; grahamjose@mst.edu
4 Oak Ridge National Laboratory, Materials Science and Technology Division, Oak Ridge, TN 37831, USA; zhangy@ornl.gov
* Correspondence: mcrespil@utk.edu (M.L.C.); wjweber@utk.edu (W.J.W.); Tel.: +1-865-360-2287(M.L.C.); +1-865-974-0415 (W.J.W.)

Abstract: Oxygen vacancies are known to play a central role in the optoelectronic properties of oxide perovskites. A detailed description of the exact mechanisms by which oxygen vacancies govern such properties, however, is still quite incomplete. The unambiguous identification of oxygen vacancies has been a subject of intense discussion. Interest in oxygen vacancies is not purely academic. Precise control of oxygen vacancies has potential technological benefits in optoelectronic devices. In this review paper, we focus our attention on the generation of oxygen vacancies by irradiation with high energy particles. Irradiation constitutes an efficient and reliable strategy to introduce, monitor, and characterize oxygen vacancies. Unfortunately, this technique has been underexploited despite its demonstrated advantages. This review revisits the main experimental results that have been obtained for oxygen vacancy centers (a) under high energy electron irradiation (100 keV–1 MeV) in $LiNbO_3$, and (b) during irradiation with high-energy heavy (1–20 MeV) ions in $SrTiO_3$. In both cases, the experiments have used real-time and in situ optical detection. Moreover, the present paper discusses the obtained results in relation to present knowledge from both the experimental and theoretical perspectives. Our view is that a consistent picture is now emerging on the structure and relevant optical features (absorption and emission spectra) of these centers. One key aspect of the topic pertains to the generation of self-trapped electrons as small polarons by irradiation of the crystal lattice and their stabilization by oxygen vacancies. What has been learned by observing the interplay between polarons and vacancies has inspired new models for color centers in dielectric crystals, models which represent an advancement from the early models of color centers in alkali halides and simple oxides. The topic discussed in this review is particularly useful to better understand the complex effects of different types of radiation on the defect structure of those materials, therefore providing relevant clues for nuclear engineering applications.

Keywords: lithium niobate; strontium titanate; self-trapped electrons; polarons; oxygen vacancies; defects; luminescence

Citation: Crespillo, M.L.; Graham, J.T.; Agulló-López, F.; Zhang, Y.; Weber, W.J. Real-time Identification of Oxygen Vacancy Centers in $LiNbO_3$ and $SrTiO_3$ during Irradiation with High Energy Particles. *Crystals* **2021**, *11*, 315. https://doi.org/10.3390/cryst11030315

Academic Editor: Robert A. Jackson

Received: 8 February 2021
Accepted: 19 March 2021
Published: 22 March 2021

1. Introduction

Oxides constitute a large family of dielectric compounds that appear in many areas of science and technology from nanoscience to geophysics and from CMOS (Complementary Metal-Oxide-Semiconductor) transistors to astronautics. An important class of oxides are the perovskites, such as cubic strontium titanate ($SrTiO_3$) and lithium niobate ($LiNbO_3$), having a distorted perovskite, with a trigonal (ilmenite-like) structure. This later material is a popular example of a photonic material, mainly due to its combination of ferroelectric, photovoltaic, and nonlinear optical properties. In particular, it is the reference material for second-harmonic generation and more specifically for charge transport (photorefractive) nonlinearities [1]. This latter property has received renewed attention in view of its function in novel applications. For example, $LiNbO_3$-functionalized surfaces have been used for trapping and manipulating nanoparticles (photovoltaic tweezers) [2]. $SrTiO_3$ and other cubic perovskites offer outstanding potential for electronic, optoelectronic, and photocatalytic devices and constitute the basis for the new growing field known as complex oxide-based microelectronics [3–7].

This review is mostly concerned with oxygen vacancies in $LiNbO_3$ and $SrTiO_3$ as representative examples of distorted and cubic perovskites, although some brief excursions into related materials are occasionally made. The oxygen vacancy centers denoted by F^+ and F represent an oxygen monovacancy with one trapped electron and an oxygen monovacancy with two trapped electrons, respectively [8]. These two centers are considered to be important point defects for many optical and transport properties and, therefore, for optoelectronic applications of $LiNbO_3$ and $SrTiO_3$. We believe that a comparative analysis of these two materials is fruitful and offers a good opportunity to examine, in depth, the role of structure on the properties and behavior of those vacancy defects.

In the basic perovskite ABO_3 structure, A is a monovalent or divalent ion, and B a tetravalent or pentavalent ion. The structure can be described as an arrangement of oxygen BO_6 octahedra, enclosing a transition metal ion B. It is useful to regard those octahedra as the basic structural units governing the electronic and optical behavior. Typical structures for $SrTiO_3$ and $LiNbO_3$ crystals are shown in Figure 1, illustrating the different octahedral arrangements. These basic structures allow us to understand many of the basic properties of perovskites. For perovskites, such as $SrTiO_3$, the oxygen octahedra are linked together at the oxygen vertices. For $LiNbO_3$, adjacent octahedra share a common face. The occurrence of one or other structure is related to the radii of ions A and B. Indeed, the Goldschmidt tolerance factor, which incorporates the ionic radii, is commonly used to predict the most stable crystal class for a particular perovskite. Note also that several phase transitions have been observed for some of these materials. A key difference between $LiNbO_3$ and the perovskites is that the ABO_3 phase of $LiNbO_3$ is stable over a wide range of Li/Nb ratios or stoichiometries. The usual or congruent composition corresponds to [Li]/[Nb] = 0.945, but other stable stoichiometries [9] can be also prepared. A large number of physical properties have been found to depend on stoichiometry, such as the optical absorption edge [10,11] and radiation-excited luminescent emission [12]. A possible reason for the wide range of stable stoichiometries in $LiNbO_3$ has to do with the close similarities of ionic radii of the host ions (0.68 $\dot{A}$ for Li^+ and 0.69 $\dot{A}$ for Nb^{5+}). In line with the central importance of the structural oxygen octahedra, the electronic band structure of $LiNbO_3$ [13], $SrTiO_3$ [14], and other oxide perovskites has some general features. The valence band (VB) is essentially associated with *p*-oxygen orbitals, whereas the conduction band (CB) is constituted by *s*-oxygen orbitals and *d*-orbitals of the B site transition metals. We will see, however, that the *d*-orbitals in the conduction band (CB) also play a key role in our understanding of color centers in oxide perovskites. The band gap, separating the oxygen *p* and *s* orbitals, lies in the range of 3 eV.

The detailed structure of defective crystals in perovskite crystals (particularly $LiNbO_3$) is a key piece of information to understand most fundamental and functional properties of those materials. However, the generation, identification, and characterization of the defect centers is a difficult subject and involves a complex set of challenges that has been a matter of controversy. A number of works have been dealing with oxygen vacancy models [15–28]. A large variety of experimental techniques have been used to generate and investigate oxygen vacancy centers. Most of these techniques use thermochemical methods (e.g., oxidation reduction treatments) or irradiation with different energetic particles. Thermochemical methods have been employed in $LiNbO_3$ [15,16], but they rely on subtle thermodynamic models whose parameters are not easy to analyze. Different models have been proposed and discussed to understand the defective structure of reduced $LiNbO_3$ samples. A very authoritative and elaborated study of those models is given in [29–31].

The defective structure strongly depends on the details of the thermal treatments and the surrounding (reducing) atmosphere and is not the objective of the present work. The major goal of the present work is to focus on the experiments that have been carried out for the production and characterization of oxygen vacancies by high energy particles, either electrons [32,33] or ions [25,26,34].

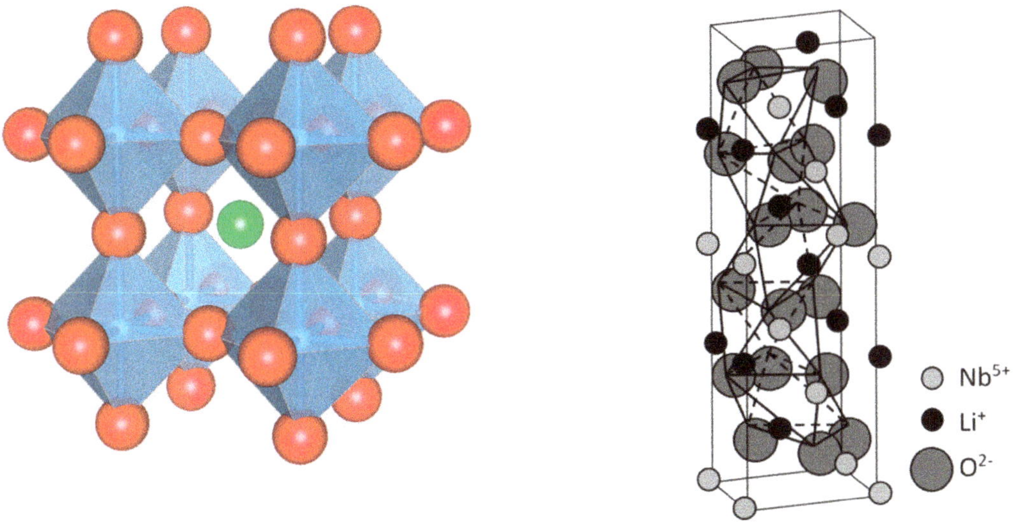

(a) (b)

Figure 1. The trigonal ilmenite-like structure of $LiNbO_3$ (**a**), and the cubic perovskite structure of $SrTiO_3$ (**b**). Schematics show the stacking of oxygen octahedra. In (**a**), note the tilting (and rotation) of the octahedral units [NbO_6] along the c-axis. In (**b**), the [TiO_6] octahedral units are clearly observed. (**a**) has been adapted from Xue et al. [35], Copyright (2003), with permission from Elsevier; (**b**) was generated using Vesta software [36].

These methods rely on the production of atomic displacements induced by momentum transfer during elastic collisions between the fast projectile particles and the atoms of the host crystal lattice. The analysis of these experiments is performed by using well-established models and available software packages [37]. A major experimental advantage of this approach is that the irradiation parameters (energy and range of the particles) can be easily modified and, therefore, the deposited energy and the effects of the damage can be tailored. Although the use of light ions, such as H and He, has been a common tool in the past [38], a growing number of studies has employed swift-heavy ions. This parallels the increased availability of laboratories with swift-heavy ion irradiation capabilities [25,26,34]. Energetic ions have the advantage of producing a broad electronic excitation spectrum that permits simultaneous access to the excited levels of all defect centers. In fact, a relevant outstanding feature of such irradiation experiments is that the production and detection of defects can be performed in situ and in real time. Unfortunately, despite these advantages, the scientific community is often not aware of the potential of the method.

Our review focuses on two representative oxide materials: (a) $LiNbO_3$, which is an example of ilmenite-like structure (distorted perovskite), a subject featured in the present issue of *Crystals*, and (b) $SrTiO_3$, a technologically important perovskite that is cubic at room temperature (RT), which may serve as an important structure for comparison. In both cases, significant experimental and theoretical progress has been achieved over the past few years.

The focus of the present review is on the production, optical identification, and electronic structure of oxygen vacancy centers produced by high-energy particles in $LiNbO_3$ and $SrTiO_3$. One important theme is the role of oxygen vacancies assumed as the main traps for electronic carriers through self-trapped electron (polaron) states and the formation of F-type centers. Our understanding of the self-trapping of electronic charges (both electrons and holes) and the further creation of polaron states, predicted by Landau [39], goes back to pioneering works by Schirmer and coworkers on $LiNbO_3$ [30,40], as well as Stoneham and coworkers in oxides [41,42]. In the case of insulators with high dielectric constants, it is often observed that free electronic carriers self-trap in the perfect lattice due to the induced local polarization, and the subsequent lowering of the energy. Oxygen vacancies and polarons turn out to be closely related and lie at the heart of this discussion. This review examines classic works in the literature, as well as very recent reports, to provide an up-to-date understanding. As a consequence of the extensive work performed on a large variety of oxide perovskites, and the new high-energy irradiation experiments, a satisfactory picture for the structure and optical behavior of oxygen vacancies in these materials is possibly emerging. It is adequate to point out that, at variance with previous investigations, the high-energy irradiation experiments are performed in situ (real-time), avoiding the processes that take place after every irradiation. Therefore, we consider it an opportune moment to present this review centered on the possible generation of oxygen vacancies in $LiNbO_3$ and $SrTiO_3$. It is expected that the comparative discussion between these two representative materials may provide a further impetus for progress in the field.

2. Oxygen Vacancy Centers in $LiNbO_3$ Created by High-Energy Electron Irradiation: Real Time in situ Detection of the Induced Optical Absorption

$LiNbO_3$ is a reference inorganic material for nonlinear optical applications, particularly in relation to charge transport (photorefractive) nonlinearities. The experiments to be discussed here involve the measurement of the optical absorption spectra induced by high-energy electron irradiation up to 2 MeV on congruent samples [32]. In these pioneer experiments, a sample is sandwiched inside a small double oven situated in a vacuum irradiation chamber, which allows for enabling in situ measurement of the optical absorption spectrum in real time. For low-electron energies, a low intensity, broad, and structureless absorption spectrum is produced. However, for higher energies (>0.3 MeV), a more well-defined absorption spectrum develops with a broad peak at about 2.6 eV (480 nm). The spectrum has been now decomposed into three main Gaussian components centered at 1.7, 2.6, and 3.2 eV, as illustrated in Figure 2. The component at 1.7 eV (730 nm) was previously associated with mobile small electron polarons such as Nb^{4+} [40] through electron spin resonance (ESR) measurements after X-ray irradiation at 20 K, and has been also detected in thermally reduced samples [15,16]. These self-trapped carriers such as Nb^{4+} constitute a localized electronic defect that can migrate through the lattice by hopping between nearby lattice sites [43]. In order to investigate the structural origin of the absorption band centered at 2.6 eV, researchers plotted its height as a function of the electron beam energy (see reference for further details). The optical density as a function of electron irradiation showed a threshold at an energy of 1.1 MeV (Figure 3); thus, in addition to the energy loss in 0.9 mm, a threshold value of 0.3 MeV can be deduced (more details can be found elsewhere [32]). Considering the energy transfer between the high-energy electrons and the $LiNbO_3$ lattice, the sharp threshold was associated with displacement of oxygen atoms by elastic electron–atom collisions. The oxygen displacement energy was determined to be 53 eV, which is similar to the values found for other oxides such as MgO (60 eV), $MgAl_2O_4$ (59 eV), and Al_2O_3 (70 eV). It is worth noting that similar absorption spectra have also been obtained after thermal reduction [15,16] where the generation of oxygen vacancies has not been demonstrated and alternative models have been invoked [29,30,44]. Pending of ESR experiments and refined theoretical calculations, a main tentative conclusion is that electron irradiation experiments can provide clear and unequivocal evidence for the generation and optical behavior of oxygen vacancy centers, i.e., F^+ and F centers. Similar irradiation experiments to those reported on congruent samples were also carried out on stoichiometric samples (Li/Nb = 1) [33]. The generated absorption spectrum is similar in both cases, indicating that it is essentially independent of the structure associated with the non-stoichiometry.

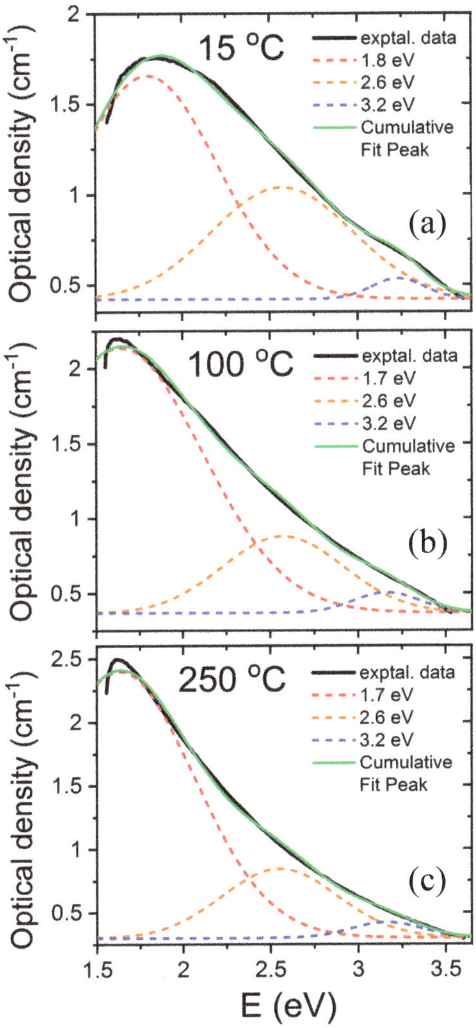

Figure 2. Absorption spectra and Gaussian decomposition obtained for congruent $LiNbO_3$ irradiated with high-energy electrons at various temperatures (**a**) 15 °C, (**b**) 100 °C, (**c**) 250 °C. This figure has been adapted from Hodgson et al. [32], Copyright (1987), with permission from Elsevier.

In the light of the available information presented in the section on $SrTiO_3$ as well as on other perovskites, one could associate the measured absorption bands to one or two Nb^{4+} small polarons trapped at the vacancy site, respectively. This model is rather different from the earlier one used to describe color centers in alkali halides, where the electrons are deeply trapped in vacancies. This is the model described in the next section on high-energy, ion-irradiated $SrTiO_3$. In order to explore the validity of such interpretation for F-type centers in $LiNbO_3$, researchers performed complementary experiments to observe the effect of heating on the absorption spectra (Figure 4). It was shown that heating in the range

RT-250 °C caused a clear conversion of the 2.6 eV band into the 1.7 eV band. This behavior was reversible on cooling the samples. It was, then, concluded that F^+-centers (one electron trapped at the vacancy) and F-centers (two electrons trapped at the vacancy) might be reasonable candidates for the 3.2 eV and 2.6 eV bands, respectively. The process can be summarized as follows. With the addition of thermal activation energy, the Nb^{4+} polarons became untrapped from the vacancy sites, giving rise to its 1.7 eV band and causing the F to F^+ conversion and the generation of free oxygen vacancies. The behavior was essentially reversible on cooling, clearly suggesting that the polarons trapped again at the vacancies. In other words, vacancies and polarons should stay close together so that the balance between the centers at 1.7 eV and 2.6 eV and the generated concentration of vacancies appear to be essentially governed by thermodynamic laws. The optical absorption bands of the F-type color centers would involve the transition between an in-gap ground state to excited Nb^{4+} *d*-levels residing in the conduction band (CB) [17]. Although, ESR experiments and theoretical calculations have yet to corroborate this model, the available data suggest that these color centers involve small Nb^{4+} polarons that are trapped in the vicinity of oxygen vacancies and form part of their electronic structure. Anyhow, one should point out that alternative models for the optical absorption spectra of the irradiated (as well as reduced) crystals have been discussed in terms of polaron and bipolaron states [30].

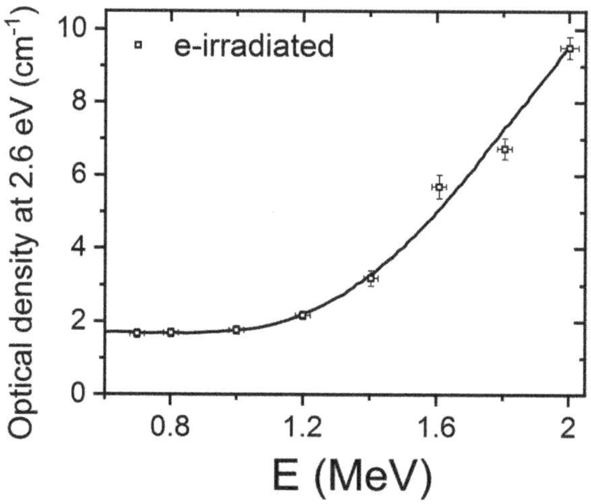

Figure 3. Normalized height of the optical absorption band at 2.6 eV as a function of the electron beam energy for congruent $LiNbO_3$. This figure has been adapted from Hodgson et al. [32], Copyright (1987), with permission from Elsevier.

It is encouraging to note that similar high-energy electron irradiation experiments have been also carried out on $KNbO_3$ (a cubic/orthorhombic perovskite) [45]. The absorption spectra are qualitatively and quantitatively similar to those obtained for $LiNbO_3$, confirming the key role of the NbO_6 octahedron in the optical response. Moreover, the experimental data also show a sharp threshold in optical density as a function of the electron energy in accordance with the impact character of the process. In our view, all these results demonstrate the potential of high-energy electron irradiation experiments to introduce oxygen vacancies in a reliable way and allow for their real-time characterization. These results could not be, in principle, achieved by any alternative methods.

At this stage, one may ask about the possible luminescence emissions associated with those color centers in $LiNbO_3$. They might provide additional insights on the responsible optical transitions, as well as on the electronic energy levels. Early luminescence experiments under X-ray irradiation showed a main emission band at around 2.9 eV, but it was found to be independent of irradiation dose [46]. Therefore, this emission was not ascribed to oxygen vacancies or any extrinsic defects, but to intrinsic electron–hole (*e-h*) recombination. The reason may lie in the fact that, after generation of free electrons (*e*) and holes (*h*) by irradiation, those pairs very rapidly couple together and form relaxed self-trapped excitons (STEs) on the sub-nanosecond time scale. Moreover, the fraction of these STEs appears to be very high (close to 1), and thus this recombination channel appears dominant.

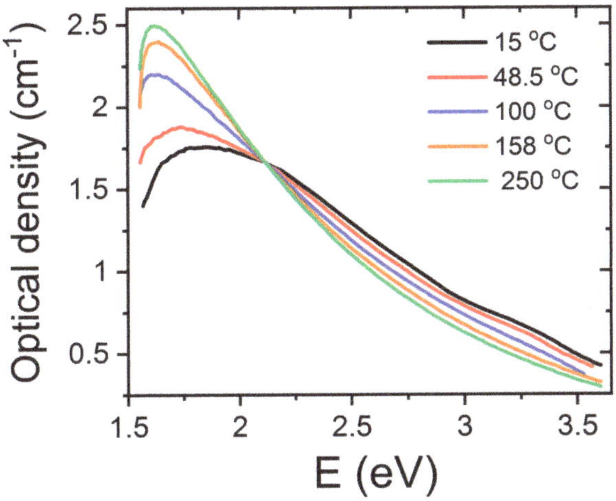

Figure 4. Effect of heating/cooling on the absorption spectra of congruent $LiNbO_3$ irradiated with high-energy electrons. This figure has been adapted from Hodgson et al. [32], Copyright (1987), with permission from Elsevier.

3. Oxygen Vacancy Centers in Cubic $SrTiO_3$ under High-Energy Ion Beam Irradiation: Real-Time Luminescence Emissions

A large fraction of the recent reports on oxide perovskites is related to $SrTiO_3$ [28], where the recent experimental and theoretical information has improved our understanding of the structure and behavior of oxygen vacancies. They are considered to be responsible for some outstanding properties of $SrTiO_3$ such as an insulating metal transition, superconductive behavior, and photocatalytic properties [3,47]. Therefore, in line with the objectives of this review, we now discuss recent experimental data obtained from irradiation experiments on $SrTiO_3$ using high-energy heavy-ion beams [25,26,48]. The use of such ions has dramatically increased in recent years. For high-energy, high-mass ions, the electronic stopping power is large and dominates the total stopping power. This large electronic stopping power gives rise to specific types of lattice damage and amorphization [34,49]. In contrast with the experiments in $LiNbO_3$ discussed in a previous section, which relied on absorbance spectra, the main detection technique with the heavy ion experiments has been luminescence. One should note that ion beam induced luminescence or ionoluminescence experiments are unique in their ability to combine high sensitivity with high spectral resolution. In such a way, the experiments offer a complementary picture to those ones described previously in Section 2 for $LiNbO_3$ and offer an experimental approach for the future.

The light emission spectra of $SrTiO_3$ under ion beam irradiation is complex and includes several overlapping bands, as shown in Figure 5, with peaks located at around 2.0, 2.5, and 2.8 eV [25–27,48]. Their relative importance strongly depends on the mass and energy of the projectile ions, which determines the electronic excitation density deposited into the material, as well as on the irradiation temperature. Moreover, heavy ions, may also introduce structural defects within the material. While these additional variables can complicate the interpretation, they also act as useful knobs to turn that allow the experimenter to gain insight into the nature of the luminescence centers. Recently, a detailed study of the luminescence emission under irradiation with a variety of high-energy heavy ions has provided unequivocal evidence supporting the assignment of the 2.0 eV emission band to Ti^{3+} polarons trapped at isolated oxygen vacancies [25,26]. In fact, the data in Figure 6 show a strong correlation between the initial growth rate of the 2.0 eV band and the production rate of isolated vacancies with irradiation fluence, as calculated by SRIM (The Stopping and Range of Ions in Matter) simulation code [37]. Density functional theory (DFT) calculations [50] indicate that the corresponding luminescence transition connects an in-gap level, having a *d*-orbital character, to Ti^{3+} *d*-levels located in the conduction band (CB). Note the similarity of this transition to the in-gap *d*-orbital to CB *d*-orbital Nb^{4+} transition associated with F-centers in electron irradiated $LiNbO_3$. One should note that X-ray irradiation of $SrTiO_3$, which does not cause lattice damage by impact collisions, does not induce the red emission associated to the proposed oxygen vacancies [51]. A schematic of the electronic structure of the oxygen vacancy center is depicted in Figure 7. The 2.0 eV band has not been often observed with pulsed laser excitation partly due to the difficulty of reaching the upper *d*-levels in the conduction band (CB) with a narrow spectral width excitation source. Other theoretical calculations have dealt with this complex problem [23,52,53], and aside from the technical details, they seem to lend support to the argument that F-type centers are small Ti^{3+} electron polarons trapped at oxygen vacancies.

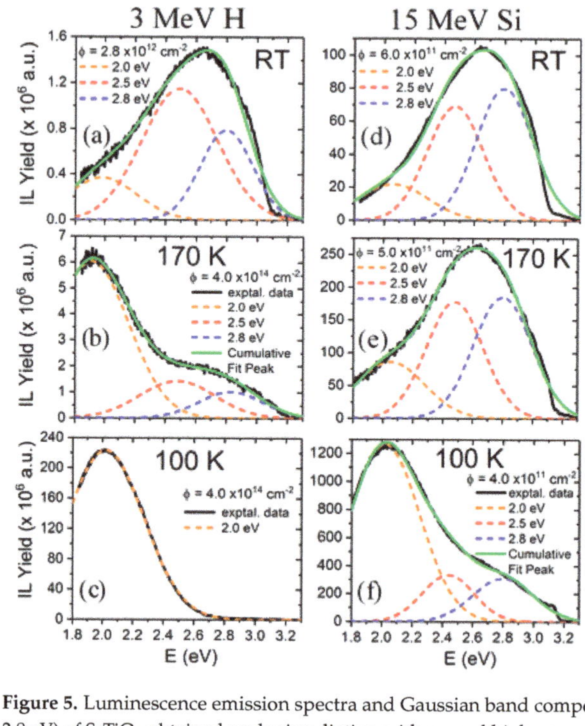

Figure 5. Luminescence emission spectra and Gaussian band components (centered at 2.0, 2.5, and 2.8 eV) of $SrTiO_3$ obtained under irradiation with several high-energy projectile ions and energies at several temperatures (100 K, 170 K and RT, see labels). Irradiation with 3 MeV H (**a–c**) and 15 MeV Si (**d–f**) ions have been selected to show the complexity of overlapping emission bands. This figure has been adapted from Crespillo et al. [27].

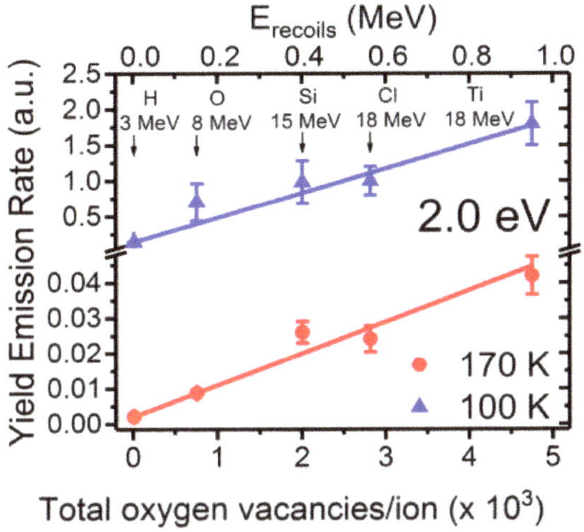

Figure 6. Normalized initial growth rate of the 2.0 eV emission band in $SrTiO_3$, associated with isolated oxygen vacancies, as a function of the vacancy production rate calculated by the SRIM code [37] under irradiation with several high-energy ions and energies (see labels). A clear linear correlation is observed for all the ions at both temperatures. This figure has been adapted from Crespillo et al. [25], Copyright (2018), with permission from Elsevier.

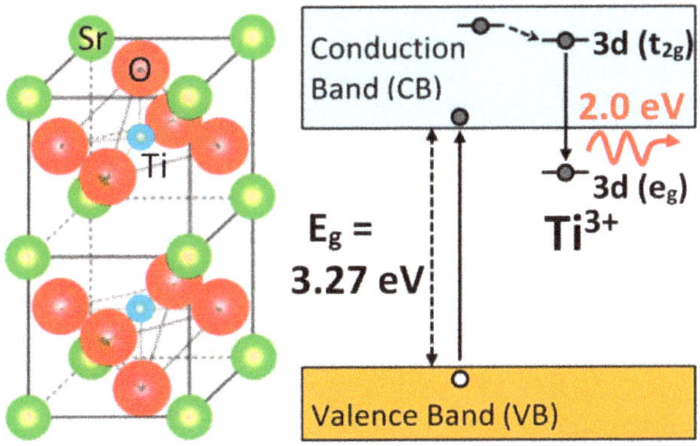

Figure 7. Schematics of the structure and electronic levels for the oxygen vacancy centers in $SrTiO_3$. Possible mechanisms of the electronic processes involving the electron polaron (Ti^{3+}) trapped adjacent to oxygen vacancies. The transition inside the TiO_6 octahedron involving the excited (t_{2g}) and ground (e_g) levels gives rise to the 2.0 eV red luminescence band. This figure has been adapted from Crespillo et al. [25], Copyright (2018), with permission of Elsevier.

Unfortunately, ESR measurements have not yet been able to provide a detailed map for the electronic structure of the oxygen vacancy centers. Nevertheless, it may be useful to comment on the detection by ESR of Ta^{4+}-polarons in $KTaO_3$ as a consequence of light irradiation with energies close to, but above, the band gap [19]. Various Ta centers have been identified and associated with oxygen vacancies. These centers break the cubic symmetry and behave as symmetry-breaking defects that are able to induce a local polar cluster at low enough temperatures (<50 K). They are responsible for second-harmonic generation (SHG) and first-order Raman scattering, among other effects. Consequently, the model derived from the ESR data indicates that the Ta^{4+} centers are located on ions nearest to the oxygen vacancy. We expect a similar defect structure in $SrTiO_3$ and $LiNbO_3$.

The centers behave as shallow electron donors with ground state energy levels close to the bottom of the conduction band (CB), around 26 meV.

4. Summary and Conclusions

From the results discussed in this review, detailed optical information has been obtained, which may contribute to understanding of the occurrence and behavior of oxygen vacancy centers in distorted ($LiNbO_3$) and cubic ($SrTiO_3$) perovskites. Irradiations with high-energy particles have provided real-time in situ information about their optical features. In particular, absorption bands in $LiNbO_3$ (1.7, 2.6, and 3.2 eV) and luminescence emissions (2.0 eV) in $SrTiO_3$ have been obtained under electron and ion beam irradiation, respectively. These spectral features can be used for the identification of the responsible centers and for monitoring their evolution and kinetics. It is important to mention that using these in situ techniques, it is possible to increase the concentration of structural defects and simultaneously monitor the optical response, thereby allowing the experimenter to associate optical spectra with lattice defect centers.

These techniques lend evidence that electrons are trapped as small polarons at host cation sites close to oxygen vacancies, although additional work is still needed. The absorption and emission bands have been ascribed to optical transitions between in-gap levels and levels within the conduction band (CB) associated with the polaron (transition metal) states. The joint analysis of these experiments together with available theoretical calculations has helped form a reasonable (although incomplete) picture of the electronic structure for oxygen vacancy centers in oxide perovskites. For $LiNbO_3$, the experimental data on high-energy electron irradiations have shown a link between the displacement of oxygen atoms and the occurrence of these optical transitions, confirming that oxygen vacancies act as efficient traps for small polarons. There are still a number of pending problems to be solved, such as determining the precise location of the corresponding electronic energy levels within the bandgap. This is not an easy task due to the many body nature of the phenomena and the need to simultaneously characterize and understand structural disorder and carrier dynamics. A specific feature to $LiNbO_3$ is the role that stoichiometry plays. The wide range of Li/Nb ratios would add an additional variable to the problem. Nevertheless, the relative insensitivity of the optical behavior to that ratio seems to reinforce the conclusion that the main electronic and optical behavior of ABO_3 perovskites is dominated by the basic octahedral BO_6 units.

As to future trends in the field, one should remark that the surfaces of perovskites have a great and largely unexplored potential, particularly $LiNbO_3$. In the area of functional materials, the surface of $LiNbO_3$ is a fertile area. Better knowledge of the surface structure and its properties (including oxygen vacancies) [22,52] would offer the possibility of nanoscale devices, molecular detection, and efficient catalysts [54], among other applications. This remark is of special relevance in relation to real-time operation of optoelectronic (photovoltaic) tweezers in $LiNbO_3$ and will advance our understanding of the trapping and untrapping cycles during their operation [55]. Furthermore, recent publications can be found in the literature showing potential applications in neuromorphic computing through the production of oxygen vacancies in $LiNbO_3$ thin films by Ar^+ irradiation [56].

Funding: This work has been supported by the U.S. Department of Energy, Office of Science, Basic Energy Sciences, Materials Sciences and Engineering Division under Contract DE-AC05-00OR22725. M.L.C. acknowledges support from the University of Tennessee Governor's Chair program.

Acknowledgments: The authors would like to pay special tribute to the pioneering role played by our colleague Ortwin Schirmer, who is responsible for many of the key achievements in the field. In particular, we deeply acknowledge and appreciate the illuminating comments and discussions we had with him in relation to some of the works cited in this review.

Conflicts of Interest: The authors declare no conflict of interest.

References

1. Günter, P.; Huignard, J.P. *Photorefractive Materials and Their Applications 1: Basic Effects*; Springer: New York, NY, USA, 2006.
2. García-Cabañes, A.; Blázquez-Castro, A.; Arizmendi, L.; Agulló-López, F.; Carrascosa, M. Recent Achievements on Photovoltaic Optoelectronic Tweezers Based on Lithium Niobate. *Crystals* **2018**, *8*, 65. [CrossRef]
3. Ogale, S.B. *Thin Films and Heterostructures for Oxide Electronics*; Springer: New York, NY, USA, 2005.
4. Reyren, N.; Thiel, S.; Caviglia, A.; Kourkoutis, L.F.; Hammerl, G.; Richter, C.; Schneider, C.W.; Kopp, T.; Ruetschi, A.-S.; Jaccard, D.; et al. Superconducting Interfaces Between Insulating Oxides. *Science* **2007**, *317*, 1196–1199. [CrossRef] [PubMed]
5. Brinkman, A.; Huijben, M.; Van Zalk, M.; Zeitler, U.; Maan, J.C.; Van Der Wiel, W.G.; Rijnders, G.; Blank, D.H.A.; Hilgenkamp, H. Magnetic effects at the interface between non-magnetic oxides. *Nat. Mater.* **2007**, *6*, 493–496. [CrossRef] [PubMed]
6. Wang, D.; Ye, J.; Kako, T.; Kimura, T. Photophysical and Photocatalytic Properties of $SrTiO_3$Doped with Cr Cations on Different Sites. *J. Phys. Chem. B* **2006**, *110*, 15824–15830. [CrossRef]
7. Crespillo, M.L.; Graham, J.T.; Agulló-López, F.; Zhang, Y.; Weber, W.J. Correlation between Cr^{3+} Luminescence and Oxygen Vacancy Disorder in Strontium Titanate under MeV Ion Irradiation. *J. Phys. Chem. C* **2017**, *121*, 19758–19766. [CrossRef]
8. Agulló-López, F.; Catlow, C.R.; Townsend, P.D. *Point Defects in Materials*; Academic Press: London, UK, 1984.
9. Polgár, K.; Péter, Á.; Kovacs, L.; Corradi, G.; Szaller, Z. Growth of stoichiometric LiNbO3 single crystals by top seeded solution growth method. *J. Cryst. Growth* **1997**, *177*, 211–216. [CrossRef]
10. Kovacs, L.; Ruschhaupt, G.; Polgár, K.; Corradi, G.; Wohlecke, M. Composition dependence of the ultraviolet absorption edge in lithium niobate. *Appl. Phys. Lett.* **1997**, *70*, 2801–2803. [CrossRef]
11. Krol, D.M.; Blasse, G.; Powell, R.C. The influence of the Li/Nb ratio on the luminescence properties of $LiNbO_3$. *J. Chem. Phys.* **1980**, *73*, 163. [CrossRef]
12. García-Cabañes, A.; Sanz-García, J.A.; Cabrera, J.M.; Agulló-López, F.; Zaldo, C.; Pareja, R.; Polgár, K.; Raksányi, K.; Fölvàri, I. Influence of stoichiometry on defect-related phenomena in$LiNbO_3$. *Phys. Rev. B* **1988**, *37*, 6085–6091. [CrossRef]
13. Javid, M.A.; Khan, Z.U.; Mehmood, Z.; Nabi, A.; Hussain, F.; Imran, M.; Nadeem, M.; Anjum, N. Structural, electronic and optical properties of $LiNbO_3$ using GGA-PBE and TB-mBJ functionals: A DFT study. *Int. J. Mod. Phys. B* **2018**, *32*, 1850168. [CrossRef]
14. Van Benthem, K.; Elsässer, C.; French, R.H. Bulk electronic structure of $SrTiO_3$: Experiment and theory. *J. Appl. Phys.* **2001**, *90*, 6156–6164. [CrossRef]
15. Sweeney, K.L.; Halliburton, L.E. Oxygen vacancies in lithium niobate. *Appl. Phys. Lett.* **1983**, *43*, 336–338. [CrossRef]
16. Garcia-Cabañes, A.; Dieguez, E.; Cabrera, J.M.; Agullo-Lopez, F. Contributing bands to the optical absorption of reduced $LiNbO_3$: Thermal and optical excitation. *J. Phys. Condens. Mat.* **1989**, *1*, 6453–6462. [CrossRef]
17. DeLeo, G.G.; Dobson, J.L.; Masters, M.F.; Bonjack, L.H. Electronic structure of an oxygen vacancy in lithium niobate. *Phys. Rev. B* **1988**, *37*, 8394–8400. [CrossRef] [PubMed]
18. Kotomin, E.A.; Eglitis, R.I.; Popov, A.I. Charge distribution and optical properties of and F centres in crystals. *J. Phys. Condens. Mat.* **1997**, *9*, L315–L321. [CrossRef]
19. Laguta, V.V.; Zaritskii, M.I.; Glinchuk, M.D.; Bykov, I.P.; Rosa, J.; Jastrabík, L. Symmetry-breakingTa4+centers in$KTaO_3$. *Phys. Rev. B* **1998**, *58*, 156–163. [CrossRef]
20. Meldrum, A.; Boatner, L.; Ewing, R. Effects of ionizing and displacive irradiation on several perovskite-structure oxides. *Nucl. Instrum. Methods Phys. Res. Sect. B* **1998**, *141*, 347–352. [CrossRef]
21. Eglitis, R.; Kotomin, E.; Borstel, G. Quantum chemical modelling of "green" luminescence in ABO_3 perovskites. *Eur. Phys. J. B* **2002**, *27*, 483–486. [CrossRef]
22. Zhukovskii, Y.F.; Kotomin, E.A.; Evarestov, R.A.; Ellis, D.E. Periodic models in quantum chemical simulations of F centers in crystalline metal oxides. *Int. J. Quantum Chem.* **2007**, *107*, 2956–2985. [CrossRef]
23. Janotti, A.; Varley, J.B.; Choi, M.; Van De Walle, C.G. Vacancies and small polarons in$SrTiO_3$. *Phys. Rev. B* **2014**, *90*, 085202. [CrossRef]
24. Wang, C.; Sun, J.; Ni, W.; Yue, B.; Hong, F.; Liu, H.; Cheng, Z. Tuning oxygen vacancy in $LiNbO_3$ single crystals for prominent memristive and dielectric behaviors. *J. Am. Ceram. Soc.* **2019**, *102*, 6705–6712. [CrossRef]
25. Crespillo, M.; Graham, J.; Agulló-López, F.; Zhang, Y.; Weber, W. Isolated oxygen vacancies in strontium titanate shine red: Optical identification of Ti^{3+} polarons. *Appl. Mater. Today* **2018**, *12*, 131–137. [CrossRef]
26. Crespillo, M.L.; Graham, J.; Agulló-López, F.; Zhang, Y.; Weber, W. Role of oxygen vacancies on light emission mechanisms in $SrTiO_3$ induced by high-energy particles. *J. Phys. D Appl. Phys.* **2017**, *50*, 155303. [CrossRef]
27. Crespillo, M.L.; Graham, J.T.; Agulló-López, F.; Zhang, Y.; Weber, W.J. The blue emission at 2.8 eV in strontium titanate: Evidence for a radiative transition of self-trapped excitons from unbound states. *Mater. Res. Lett.* **2019**, *7*, 298–303. [CrossRef]
28. Crespillo, M.L.; Graham, J.T.; Agulló-López, F.; Zhang, Y.; Weber, W.J. Recent Advances on Carrier and Exciton Self-Trapping in Strontium Titanate: Understanding the Luminescence Emissions. *Crystals* **2019**, *9*, 95. [CrossRef]
29. Schirmer, O.F.; Reyher, H.-J.; Wöhlecke, M. Characterization of point defects in photorefractive oxide crystals by para-magnetic resonance methods. In *Insulating Materials for Optoelectronics: New Developments*; Agulló-López, F., Ed.; World Scientific: London, UK, 1995; pp. 93–124.
30. Schirmer, O.F.; Imlau, M.; Merschjann, C.; Schoke, B. Electron small polarons and bipolarons in $LiNbO_3$. *J. Phys. Condens. Matter* **2009**, *21*, 123201. [CrossRef]

31. Halliburton, L.; Sweeney, K.; Chen, C. Electron spin resonance and optical studies of point defects in lithium niobate. *Nucl. Instrum. Methods Phys. Res. Sect. B* **1984**, *1*, 344–347. [CrossRef]
32. Hodgson, E.; Aguló-López, F. Oxygen vacancy centres induced by electron irradiation in $LiNbO_3$. *Solid State Commun.* **1987**, *64*, 965–968. [CrossRef]
33. Hodgson, E.R.; Agullo-Lopez, F. High-energy electron irradiation of stoichiometric $LiNbO_3$. *J. Phys. Condens. Matter* **1989**, *1*, 10015–10020. [CrossRef]
34. Agulló-López, F.; Climent-Font, A.; Muñoz-Martín, Á.; Olivares, J.; Zucchiatti, A. Ion beam modification of dielectric materials in the electronic excitation regime: Cumulative and exciton models. *Prog. Mater. Sci.* **2016**, *76*, 1–58. [CrossRef]
35. Xue, D.; Kitamura, K.; Wang, J. Atomic packing and octahedral linking model of lithium niobate single crystals. *Opt. Mater.* **2003**, *23*, 399–402. [CrossRef]
36. Momma, K.; Izumi, F. VESTA 3for three-dimensional visualization of crystal, volumetric and morphology data. *J. Appl. Crystallogr.* **2011**, *44*, 1272–1276. [CrossRef]
37. Ziegler, J.F.; Ziegler, M.; Biersack, J. SRIM—The stopping and range of ions in matter (2010). *Nucl. Instrum. Methods Phys. Res. Sect. B* **2010**, *268*, 1818–1823. [CrossRef]
38. Townsend, P.D.; Chandler, P.J.; Zhang, L. *Optical Effects of Ion Implantation (Cambridge Studies in Modern Optics)*; Cambridge University Press: Cambridge, UK, 1996.
39. Landau, L.D. Uber die Bewegung der Elektronen im Kristallgitter. *Phys. Zeit. Sowjetunion* **1933**, *3*, 664–665.
40. Schirmer, O.F.; Von Der Linde, D. Two-photon- and x-ray-induced Nb4+and O−small polarons in $LiNbO_3$. *Appl. Phys. Lett.* **1978**, *33*, 35–38. [CrossRef]
41. Stoneham, A.; Itoh, N. Materials modification by electronic excitation. *Appl. Surf. Sci.* **2000**, *168*, 186–193. [CrossRef]
42. Stoneham, A.M.; Gavartin, J.; Shluger, A.L.; Kimmel, A.V.; Ramo, D.M.; Rønnow, H.M.; Aeppli, G.; Renner, C. Trapping, self-trapping and the polaron family. *J. Phys. Condens. Matter* **2007**, *19*, 255208. [CrossRef]
43. Sturman, B.; Carrascosa, M.; Agullo-Lopez, F. Light-induced charge transport inLiNbO3crystals. *Phys. Rev. B* **2008**, *78*, 245114. [CrossRef]
44. Lengyel, K.; Péter, Á.; Kovacs, L.; Corradi, G.; Palfalvi, L.; Hebling, J.; Unferdorben, M.; Dravecz, G.; Hajdara, I.; Szaller, Z.; et al. Growth, defect structure, and THz application of stoichiometric lithium niobate. *Appl. Phys. Rev.* **2015**, *2*, 040601. [CrossRef]
45. Hodgson, E.; Zaldo, C.; Agulló-López, F. Atomic displacement damage in electron irradiated $KNbO_3$. *Solid State Commun.* **1990**, *75*, 351–353. [CrossRef]
46. Arizmendi, L.; Cabrera, J.; Agulló-López, F. X-ray induced luminescence of LiNbO3. *Solid State Commun.* **1981**, *40*, 583–585. [CrossRef]
47. Harrigan, W.L.; Michaud, S.E.; Lehuta, K.A.; Kittilstved, K.R. Tunable Electronic Structure and Surface Defects in Chromium-Doped Colloidal $SrTiO_{3-\delta}$ Nanocrystals. *Chem. Mater.* **2016**, *28*, 430–433. [CrossRef]
48. Kan, D.; Terashima, T.; Kanda, R.; Masuno, A.; Tanaka, K.; Chu, S.; Kan, H.; Ishizumi, A.; Kanemitsu, Y.; Shimakawa, Y.; et al. Blue-light emission at room temperature from Ar+-irradiated $SrTiO_3$. *Nat. Mater.* **2005**, *4*, 816–819. [CrossRef]
49. Zhang, Y.; Weber, W.J. Ion irradiation and modification: The role of coupled electronic and nuclear energy dissipation and subsequent nonequilibrium processes in materials. *Appl. Phys. Rev.* **2020**, *7*, 041307. [CrossRef]
50. Ricci, D.; Bano, G.; Pacchioni, G.; Illas, F. Electronic structure of a neutral oxygen vacancy in$SrTiO_3$. *Phys. Rev. B* **2003**, *68*, 224105. [CrossRef]
51. Aguilar, M. X-ray induced processes in $SrTiO_3$. *J. Appl. Phys.* **1982**, *53*, 9009. [CrossRef]
52. Alexandrov, V.E.; Kotomin, E.A.; Maier, J.; Evarestov, R.A. First-principles study of bulk and surface oxygen vacancies in $SrTiO_3$ crystal. *Eur. Phys. J. B* **2009**, *72*, 53–57. [CrossRef]
53. Mitra, C.; Lin, C.; Robertson, J.; Demkov, A.A. Electronic structure of oxygen vacancies in$SrTiO_3$ and $LaAlO_3$. *Phys. Rev. B* **2012**, *86*, 155105. [CrossRef]
54. Sanna, S.; Schmidt, W.G. LiNbO3surfaces from a microscopic perspective. *J. Phys. Condens. Matter* **2017**, *29*, 413001. [CrossRef]
55. Sebastián-Vicente, C.; Muñoz-Cortés, E.; García-Cabañes, A.; Agulló-López, F.; Carrascosa, M. Real-Time Operation of Photovoltaic Optoelectronic Tweezers: New Strategies for Massive Nano-object Manipulation and Reconfigurable Patterning. *Part. Part. Syst. Charact.* **2019**, *36*, 1900233. [CrossRef]
56. Pan, X.; Shuai, Y.; Wu, C.; Zhang, L.; Guo, H.; Cheng, H.; Peng, Y.; Qiao, S.; Luo, W.; Wang, T.; et al. Ar+ ions irradiation induced memristive behavior and neuromorphic computing in monolithic LiNbO3 thin films. *Appl. Surf. Sci.* **2019**, *484*, 751–758. [CrossRef]

crystals

Review

'Horror Vacui' in the Oxygen Sublattice of Lithium Niobate Made Affordable by Cationic Flexibility

Gábor Corradi * and László Kovács *

Wigner Research Centre for Physics, Institute for Solid State Physics and Optics, Konkoly-Thege M. út 29-33, H-1121 Budapest, Hungary
* Correspondence: corradi.gabor@wigner.hu (G.C.); kovacs.laszlo@wigner.hu (L.K.)

Abstract: The present review is intended to interest a broader audience interested in the resolution of the several decades-long controversy on the possible role of oxygen-vacancy defects in $LiNbO_3$. Confronting ideas of a selected series of papers from classical experiments to brand new large-scale calculations, a unified interpretation of the defect generation and annealing mechanisms governing processes during thermo- and mechanochemical treatments and irradiations of various types is presented. The dominant role of as-grown and freshly generated Nb antisite defects as traps for small polarons and bipolarons is demonstrated, while mobile lithium vacancies, also acting as hole traps, are shown to provide flexible charge compensation needed for stability. The close relationship between $LiNbO_3$ and the Li battery materials $LiNb_3O_8$ and Li_3NbO_4 is pointed out. The oxygen sublattice of the bulk plays a much more passive role, whereas oxygen loss and Li_2O segregation take place in external or internal surface layers of a few nanometers.

Keywords: $LiNbO_3$; polarons; bipolarons; defect structure and generation; Li diffusion

Citation: Corradi, G.; Kovács, L. 'Horror Vacui' in the Oxygen Sublattice of Lithium Niobate Made Affordable by Cationic Flexibility. *Crystals* **2021**, *11*, 764. https://doi.org/10.3390/cryst11070764

Academic Editor: Benoit Heinrich

Received: 4 June 2021
Accepted: 27 June 2021
Published: 29 June 2021

1. Introduction

In most simple oxides, defect generation processes by thermal reduction or irradiation are dominated by defects of the oxygen sublattice. Originally, the same was also assumed for $LiNbO_3$ (LN), and most authors did not (and many still do not) have any doubts about simply postulating vacant oxygen sites (V_O) capable of trapping one or two electrons (F^+ and F-centers, respectively) in discussions of their experiments and proposed applications (see, e.g., [1,2]). The controversy about the availability of oxygen vacancies manifests in the reviews of Sánchez-Dena et al. [3,4], is touched upon by some topical reviews [5,6], and is also closely related to the topics of other papers [7–11] of the present Special Issue, and deserves a clarifying discussion.

In this short review, aiming to integrate views on this paradigmatic material, we start by summarizing the charge compensation mechanism for Li deficiency followed by a discussion of the various defect generation procedures in congruent $LiNbO_3$ (cLN), such as thermal reduction, irradiation by ionizing or ion-displacing radiation or mechanochemical treatment. The effect of crystal stoichiometry on the defects produced is revisited before presenting a comparison with recent calculations of basic defect configurations.

2. Structure and Charge Compensation Mechanism

In contrast to ordinary perovskites with less compact anion sublattices, $LiNbO_3$ has a structure corresponding to a strongly distorted perovskite where the oxygens essentially form a hexagonally close-packed lattice offering a priori identical octahedral sites to the cations, one-third of which remain empty. In domain walls and in some defect clusters, the stacking order of the cations along the C_3-axis (Li, Nb, empty site) changes due to these liberties. The cations Li^+ and Nb^{5+} have very similar ionic radii but otherwise diametrically differing properties. Large polarizability results in ferroelectricity nearly up to the melting point and transforms all surplus charges in the lattice to small polarons by confining them via local lattice relaxation, as a rule, to a single lattice site. Small polarons still may propagate by thermally activated hopping unless sufficiently stabilized by some defect. Li readily strips an electron, whereas Nb has an intricately covalent character, as shown by the many poliforms of its oxides ranging from pentoxide Nb_2O_5 to monoxide NbO [12,13]. This explains the Li-deficient character of $LiNbO_3$ accommodated by lithium vacancies (V_{Li}) and Nb_{Li} antisites (Nb on Li site) requiring small formation energies. Compared to compensation involving anion vacancies at larger costs [13,14], compensation only by cationic disorder requires no additional volume and leads to the observed larger density of cLN compared to sLN (stoichiometric LN) [15–19]. Another characteristic property of LN is the high mobility of Li^+ ions observable already at temperatures near 350 °C (see [9,20–22] and references therein). It should be noted that LN is just an intermediary phase between Li_3NbO_4 and $LiNb_3O_8$, considered as cathode and anode materials for lithium-ion batteries, respectively [23,24]. Nb has an octahedral coordination in all these systems, the only difference being the varying Li_2O content; however, LN has a closer relationship with $LiNb_3O_8$ due to its full epitaxial compatibility with LN [25].

On the basis of extensive experimental and theoretical work, this compensation mechanism is by now fully accepted. The Nb_{Li} antisite was proven to be an efficient electron trap, forming the Nb_{Li}^{4+} small polaron state characterized by an asymmetric absorption band at ~1.6 eV (~760 nm, with a protracted high energy side) attributed to charge transfer to the conduction band, and a broad ten-line EPR spectrum reflecting hyperfine splitting due to the Nb_{Li} nucleus. The EPR spectrum has a non-axial angular dependence due to a quasi-Jahn–Teller effect or a perturbation by charge-compensating defects situated in glide-mirror (zy-type) planes containing the cation neighbors [26–28]. However, no EPR spectrum corresponding to an F^+-center or its hypothetical relaxed variant [29], a non-axial Nb_{Nb}^{4+} center next to an anion vacancy situated in zx-type planes, could be identified [27,30].

3. Defect Generation in $LiNbO_3$

3.1. Defect Generation in cLN by Thermal Reduction

Attempts to explain thermal annealing effects in LN by anionic defects [29–32] were never conclusive. Thermal reduction at high temperatures in vacuum or other reducing atmospheres in cLN results in a broad absorption band with a poorly defined maximum at ~2.5 eV (~500 nm) but having no counterpart in EPR [29–33] (see Figure 1). This band is also highly asymmetric, with its protracted high energy side merging with the UV edge and was attributed, similarly to the 1.6 eV band, to charge transfer to the conduction band [34]: asymmetry seems to be a general property of small-polaronic charge-transfer bands in LN mainly due to the structure of target states in the conduction band [35]. By illumination with photon energies above 2 eV, the ~2.5 eV band can be transferred into the ~1.6 eV band of Nb_{Li}^{4+} small polarons [29–32,34]. The same transformation can also be realized by heating the crystal; in fact, the defects responsible for these bands are in thermal equilibrium [36]. The transformation is fully reversible at least in the 122–575 K temperature range, whereby an isosbestic point near 600 nm is conserved, indicating the participation of only two kinds of centers in the transformation [36]. Some authors have also reported the participation of an additional reduction-induced absorption band near 3.2 eV (~390 nm) in these transformations [32,33]; however, due to their assumption about symmetric Gaussian bandshapes, the amplitude of this band was grossly overestimated. The interpretation of the 2.5 and 3.2 eV bands in terms of F and F^+ centers, respectively, is in immediate conflict with the fact that the 2.5 eV band is the stable in the reduced crystal corresponding to its ground state [35]; moreover, no EPR spectrum corresponding to the F^+ center could be identified, as pointed out in the previous section.

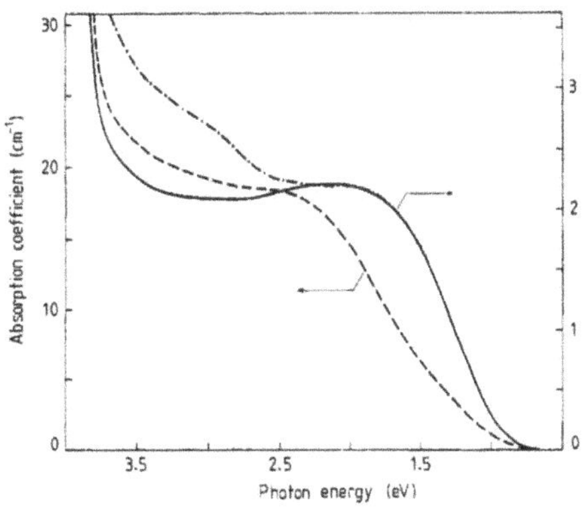

Figure 1. Reduction-induced absorption in a near-sLN sample covered by Pt foil (full line), a similar uncovered sample (dash-dotted), and a cLN sample (dashed line), measured at RT. The vacuum-reduction temperature was 880 °C (Reprinted with permission from [33]. Copyright 2021 IOP Publishing).

Possible oxygen diffusion was also checked by studies of redox kinetics and defect chemistry, including oxygen-18 tracer diffusion with the conclusion that the oxygen transport is nearly negligible compared to lithium transport and may occur via oxygen interstitials rather than via oxygen vacancies in the whole stoichiometric range [20,21]. It should be noted that the only 'free space' in LN, the octahedral structural vacancy, is an unfavorable site for anions. These results showed that the role of oxygen vacancies in the redox processes of LN is negligible, while oxygen loss/recovery is possible by the decomposition/recovery of the surface layer. For strong reduction treatments above ~650 °C, segregation and evaporation of the volatile compound Li_2O becomes important and leads to a gradual stoichiometry change in the crystal proceeding from the surface towards the bulk. This can be reversed and even overcompensated by embedding the crystal in powder mixtures serving as buffers for Li_2O. The vapor transport equilibration (VTE) method, the first technique used for the production of stoichiometric LN, is based on this procedure [37,38] and does not require the diffusion of anion vacancies either.

It took some time until the cationic interpretation of thermal reduction could be clarified. The starting point was again increased density, demonstrated for reduced crystals by Smyth [16] and his remark about the "inconsistency to propose one set of defects for Li_2O deficiency, and a different set for oxygen loss". The Nb_{Li} antisite was demonstrated to be a 'negative-U' defect capable of lowering its energy by binding a second electron, whereby a Nb_{Li}^{4+}–Nb_{Nb}^{4+} molecular ion involving a regular niobium neighbor is formed along the trigonal axis of the crystal; covalent bonding turns the interaction energy U between the two small polarons negative, despite Coulomb repulsion [35,36]. The oxygen loss was attributed to the depletion of elementary cells on the surface, while the cations left over were assumed to diffuse into the bulk, filling existing Li vacancies, according to the reaction

$$LiNbO_3 + 2V_{Li} \rightarrow Li_{Li}^{+} + Nb_{Li}^{4+} + \frac{3}{2}O_2\uparrow + 5e^- \quad (1)$$

In this scenario, the remaining electrons, transformed into small polarons, follow the cations and are trapped on the antisites. Above ~650 °C, a similar reaction with Li_2O loss (compare Equation (3)) becomes effective. The presence of Nb_{Li} point defects on the surface of reduced LN was supported by RHEED investigations [25]. Accordingly, starting from the surface, polarons and bipolarons gradually turn the crystal non-transparent. The absorption band near 2.5 eV indeed corresponds to that calculated for bipolarons [39–42], whereby further absorption of the same defect or its similar variant in the region near 3 eV is also indicated by recent calculations [10]. The transformations upon photoexcitation in the 2.5 eV band or upon heating the crystal can be explained by the reversible dissociation of the bipolaron leading to the appearance of only one defect type, the Nb_{Li}^{4+} center. Due to the absence of trapped-hole centers in reduced crystals, a thermodynamical equilibrium is established between antisite-trapped polarons and bipolarons, as observed in the range 122–575 K [36]. For higher temperatures, both kinds are liberated step by step from their bondage, leading to the appearance of free Nb_{Nb}^{4+} polarons absorbing near 1 eV [20,43,44]. The underlying kinetics have been observed and modeled in great detail in the whole temperature range, yielding activation energies in full agreement with photoexcitation, DC conductivity, and thermopower experiments [35]. Attempts to interpret the same transformation processes in the oxygen-vacancy scenario encountered difficulties due to the need to accommodate a further unfitting partner in the model: the F^+ center.

An independent, decisive argument in favor of the cationic interpretation of thermal reduction processes in LN is the lack of ~2.5 and ~3.2 eV bands in crystals where antisites are absent. Such a case, a 5% Mg-doped LN crystal grown from a melt with [Li]/[Nb] = 1:1 molar ratio and reduced at 1000 °C, was investigated by Sweeney et al. [45]. In this study, only an asymmetric band near 1 eV due to a Mg_{Li}-assisted Nb_{Nb}^{4+} polaron was found as a result of reduction (See spectrum (b) of Figure 9 in [45]; it should be noted that the decomposition of this band given in the bottom part of the same figure is incorrect, because trace (d) is already an unphysical difference-band obtained earlier by the comparison of Nb_{Nb}^{4+} and Nb_{Li}^{4+} bands in crystals above and below the photorefractive threshold, respectively. Apart from this controversy and the unconfirmed assignations to vacancy centers, the results in [45] proved to be highly reliable (see also [46])). The crystal used was beyond the photorefractive threshold where no Nb_{Li} antisites are present; part of the Mg^{2+} ions already form self-compensating complexes on both Li and Nb sites, whereas other Mg_{Li}^{2+} ions may still be charge-compensated by Li vacancies [47]. There is no reason why oxygen loss should have a different mechanism in this case, although the bands tentatively attributed by Ref. [45] to oxygen-vacancy centers in undoped LN are missing. However, a cationic mechanism, also offering a straightforward explanation for this crystal, can be readily suggested: reduction involving O_2 loss may occur by a reaction consuming elementary cells on the surface together with a transfer of nearby Mg ions from Nb to Li sites

$$LiNbO_3 + Mg_{Nb}^{2+} + 2V_{Li} \rightarrow Li_{Li}^{+} + Nb_{Nb}^{4+} + Mg_{Li}^{2+} + \frac{3}{2}O_2\uparrow + 5e^- \quad (2)$$

resulting only in Nb_{Nb}^{4+} polarons. In fact, all electrons on the right-hand side also form such small polarons overtaking the role of lithium vacancies as charge-compensators of further Mg_{Li}^{2+} defects not shown in Equation (2). Based on the above results and model calculations discussed later, the model of thermal reduction based on the formation of oxygen vacancies in cLN could be discarded.

3.2. *Defect Generation in cLN by Ionizing Irradiation at Low Temperatures ($T \leq 77$ K)*

Irradiation at low temperatures ($T \leq 77$ K) results in polaron formation in the bulk. Various kinds of radiation, including two-photon absorption [26], X-rays [26,32], Co^{60} γ-rays [48], and higher energy electrons up to 1.7 MeV [30] have been used; along with Nb_{Li}^{4+} polarons having an EPR fingerprint and an absorption band at ~1.7 eV, O^- hole polarons (also small) preferably trapped at lithium vacancies were also formed, having another EPR fingerprint and an absorption band near 2.5 eV (see Figure 2), fortuitously coinciding with that of bipolarons. In undoped $LiNbO_3$, Nb_{Li}^{4+} polarons become mobile at temperatures between 100 and 150 K and recombine with the trapped holes [26,32] in a non-radiative process [46]; therefore, the original state of the crystal is restored. Arizmendi et al. [32] also reported on the presence of a further smaller band at ~3.2 eV, ascribed to the same trapped-hole polarons in the X-irradiated crystals (again coinciding with similar structures in trapped-electron spectra). Upon heating, the three bands showed a marked parallel decrease in the 100–150 K region up to their complete disappearance below 240 K. An important achievement of this paper was an early derivation of the temperature ~312 K where $O^-(V_{Li})$ trapped holes became mobile. At the temperature ~312 K, measured in Cu-doped LN X-irradiated at low temperature, small polarons trapped as stable Cu^+ centers became spontaneously converted to Cu^{2+} centers by mobile holes.

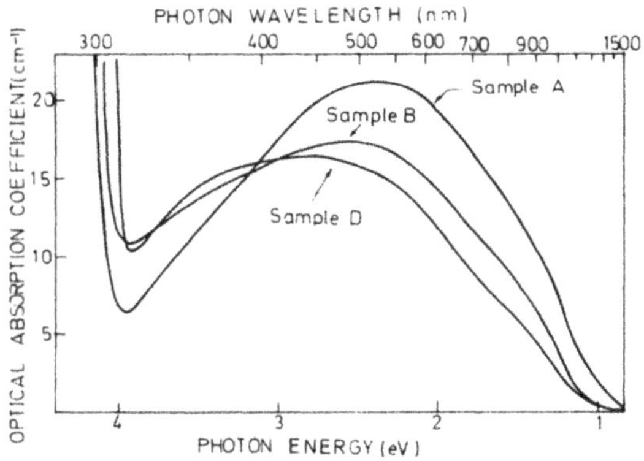

Figure 2. Optical absorption induced by 2 h 40 keV X-irradiation in LN samples with different bulk [Li]/[Nb] ratios derived a posteriori from measurements of the UV absorption edge (see Section 3.3). Sample A: 0.98 nearly stoichiometric, sample B: 0.96 nearly congruent, sample D: 0.92 sub-congruent, all irradiated and measured at 77 K. (Reprinted with the permission from Ref [49]. Copyright 2021 American Physical Society).

A peculiarity of high-energy irradiation by electrons was the small intensity of the Nb_{Li}^{4+} EPR signal compared to that of the O^- EPR signal [30], in contrast to the X-ray results where they were comparable [26]. This can be understood by assuming that the relative number of paired polaron trappings increased compared to single trapping due to the large polaron density created in the electron tracks. Therefore, the 2.5 eV band, in addition to the absorption of O^- centers, may have also contained an indistinguishable (and larger than usual) contribution of paired electron centers, i.e., bipolarons. However, the stability of the defects was found to be practically unchanged compared to experiments using lower irradiation energies, with recovery occurring for all radiation defects near 200 K. The question about missing fingerprints of F^+ centers providing deep polaron trapping sites expected by some authors [30,32] remained unanswered. Before proceeding to defect generation by high-energy irradiation at higher temperatures, we discuss similar processes caused by strong mechanic impact.

3.3. Mechanochemical Defect Generation in cLN Nanocrystals

As discussed in the review of Sánchez-Dena and coworkers [4], investigating nanocrystals is another possibility for the study of the defect structure of $LiNbO_3$. Similar defect generation phenomena as for thermal reduction have been observed upon high-energy ball-milling of polycrystalline cLN [50]. Milling resulted in sample darkening due to mechanochemical reduction via polaron and bipolaron formation detected by optical reflection spectroscopy. The obtained particles of the size of a few hundred nanometers contained monocrystalline grains sized in the range of 30–80 nm. The process was accompanied by oxygen release and Li_2O segregation:

$$2LiNbO_3 + 2V_{Li} \rightarrow 2Nb_{Li}^{4+} + 8e^- + \frac{5}{2} O_2\uparrow + (Li_2O)_{segr} \quad (3)$$

while subsequent oxidizing heat-treatments recovered the white color with the evaporation of Li_2O and crystallization of a $LiNb_3O_8$ phase on the grain surface requiring minimal adjustment, given its epitaxial compatibility with LN [25]:

$$LiNbO_3 + 2Nb_{Li}^{4+} + 8e^- + \frac{5}{2}O_2 \rightarrow 2V_{Li} + LiNb_3O_8 \quad (4)$$

The phase transformations occurring during both the grinding and the post-grinding heat treatments were also studied by Raman spectroscopy and X-ray diffraction measurements, whereas the Li_2O content of the as-ground samples was quantitatively measured by coulometric titration. The thickness of the surface layers with modified composition could be estimated to be in the orders of 2 nm and 10 nm for the as-milled and oxidized samples, respectively. The latter value is in good agreement with the results of TEM–EDS sampling measurements on similarly prepared nanoparticles of LN:Fe reduced at 650 °C [51], where the Nb and O concentrations in the grains remained unchanged during grinding with the exception of a surface layer of ~14 nm thickness: in this layer, a gradual increase in the Nb and a decrease in the O content towards the surface could be detected, in agreement with Equations (1) or (3), taking into account the evaporation of any Li_2O segregate upon thermal reduction.

These findings show the basic similarity of defect generation processes in the discussed cases: while thermal reduction promotes O_2 and Li_2O loss by lowering the respective chemical potentials, high-energy ball milling squeezes these components out of the crystal, facilitated also by the somewhat increased temperature during grinding. This indicates that the colored cores of the obtained grains have an increased density, as expected for the bipolaron model of reduction.

3.4. Defect Generation in cLN by High-Energy Irradiation at Higher Temperatures (220 °C)

Using high-energy irradiation by electrons or heavier particles at 220 °C in vacuum, absorption bands persisting well above room temperature may be detected. This was demonstrated by Hodgson et al. [52] and revisited in the present Special Issue [1]. Apart from a superficial comparison between LN and $SrTiO_3$ disregarding the deep constitutional differences between these matrices, Ref. [1] only repeats the tentative interpretation of Hodgson et al. [52] based on anion vacancies, referring to some of their figures but partly transforming them in an unexplained manner. Therefore, we confine the discussion to the original paper [52]. For gamma or electron irradiations below an energy threshold, only less structured absorption with very limited amplitude could be achieved. This changed above the threshold where the intensity of the observed spectra showed a nearly linear increase with radiation dose. Their structure and behavior upon heating/cooling strongly resembled those obtained by thermal reduction, showing broad features at ~1.6 eV, ~2.6 eV and ~3.2 eV; however, the main band showed trapped-electron character in contrast to the case of low-temperature irradiation where it was rather of trapped-hole type; upon heating there was again a reversible transformation of the main band at ~2.6 eV to the band at ~1.6 eV, whereby the same isosbestic point near 600 nm was observed as for reduced crystals [36,52]. The particle energy threshold found for increased defect formation was 0.3 MeV, corresponding to an oxygen displacement energy of 53 eV, in reasonable agreement with oxygen displacement thresholds in other oxides (see references in [6,52]). No saturation after five over-threshold doses of 2×10^{18} e/cm^2 could be observed. The authors proposed that the defects generated were oxygen vacancies trapping one or two electrons (F^+ and F-centers, respectively), and thermal excitation resulted in the reversible transformation of F-centers into Nb_{Li}^{4+} trapped polarons. Accordingly, they attributed the observed dominating absorption band near 2.6 eV to F-centers, despite its coincidence with the bipolaron band in reduced crystals, while a weak band at ~3.2 eV was again tentatively attributed to F^+-centers.

However, due to the perfect similarity with the case of thermal reduction, even in this case an interpretation in terms of bipolaron formation seems to be more appropriate.

The most important difference between the high-energy experiments in Refs. [30,51] is the irradiation temperature. Irradiation at 77 K formed both trapped-electron and trapped-hole centers, all disappearing near 200 K, whereas similar irradiation at 220 °C resulted exclusively in trapped-electron defects, this time highly stable at least up to 800 °C. As discussed earlier, above room temperature, $O^-(V_{Li})$ trapped-holes become mobile. Stable hole-trapping on oxygen interstitials may also be discarded (see Section 4 and [13]); therefore, coloration for T > 220 °C can only be stabilized by oxygen loss similarly to the case of thermal reduction and ball milling. This means that oxygen must have also left the crystal lattice during high-temperature irradiations. Note that the irradiation in [52] was carried out under vacuum. Apparently, high-energy irradiation locally stimulates processes similar to those active during thermal treatments; therefore, they can set at 220 °C, whereas normal 'unassisted' thermal reduction only starts at higher temperatures. Oxygen interstitials proved to be mobile at high temperatures [21]; the same may be assumed in local thermal spikes. For this reason, recombination of the oxygen sublattice during and immediately after irradiation is expected to be important. On the other hand, together with oxygen atoms, a large number of lithium atoms must also have been mobilized and partly left the crystal lattice. Thus, with increasing bombardment energy, local phase separation with Li_2O loss and the formation of $LiNb_3O_8$ inclusions may become possible. It can also be assumed that instead of stable oxygen vacancy generation, the formation of neutral Li_2O voids takes place, especially near the surface, and for highest doses their accumulation may turn the samples brittle. Accordingly, these internal segregation processes are expected to be similar to those observed on the surfaces of reduced crystals and nanocrystals. In this way, above a particle energy threshold, the number of Nb_{Li}^{4+} antisites will also increase, similarly to their almost unsaturable multiplication during thermal reduction. Even if present, oxygen vacancies can only offer higher-energy trapping states to polarons than antisites (see the comparison with model calculations in Section 4). Taking into account the sample thickness of ~1 mm roughly corresponding to the penetration depth of the radiation used [52], and a single defect generation event per particle, but neglecting anionic recombination, the 2×10^{18} e/cm^2 dose corresponds to an average center concentration of 2×10^{19}/cm^3 equal to that of Nb_{Li} antisites in cLN. Therefore, even without newly formed antisites, their numbers should be sufficient to trap a great part if not all of the surviving polarons. In summary, again there is no reason to deny the dominating role of cationic defects in defect production in cLN.

3.5. Defect Generation in Nearly Stoichiometric $LiNbO_3$

After the advent and spreading of techniques capable of producing stoichiometric crystals [38] (for a review see [47]), to the best of our knowledge, no studies on the coloration of sLN crystals have been published. Crystals grown with the Czochralski method from melts with Li surplus but still containing Nb_{Li} antisites (although in substantially reduced numbers) were erroneously called stoichiometric up to 1992. For such crystals, we use the term 'nearly stoichiometric LN' (near-sLN). Compared to cLN, the experimental situation in these crystals is different and can be characterized by low saturation levels, slightly shifted absorption bands (see Figures 1 and 2, [33,49]), and the absence of interpretable EPR spectra looked for in γ-irradiated crystals [48] (note that the sensitivity of EPR is reduced due to hyperfine-broadened lines). High-energy electron irradiation of near-sLN at temperatures up to 20 °C was reported by [53]. The crystals used in these papers were grown from a melt with a [Li]/[Nb] ratio of 1.2:1 by colleagues in Budapest, with an absorption edge near 311.8 nm in their records (corresponding to an absorption coefficient $\alpha = 20$ cm^{-1}) a posteriori used to derive a [Li]/[Nb] = 0.98 ratio in the bulk [54]. In [53], the same particle energy threshold (0.30 MeV) was found for increased defect production as for cLN. For a one-hour irradiation with 1.6 MeV electrons at −50 °C, a structureless absorption with a very broad maximum at ~540 nm (~2.3 eV) could be achieved. Apart from the slight redshift compared to the case of cLN (~500 nm), the spectrum showed a radically reduced stability, with two major, slightly overlapping annealing stages near −25 °C and 40 °C, and a third one near to 350 °C, where the small residual coloration consisting of a broad spectral feature near 760 nm and an even weaker one near 400 nm also disappeared (Figure 3). If, after a given annealing step, the annealing was stopped, the crystal re-cooled below the step temperature and then re-irradiated, fast (minute-timescale) reconstruction of the absorption monitored at 540 nm could be achieved. Such a fast reconstruction was not possible after annealing above 350 °C, but the whole process could be repeated. During some stages of irradiation or annealing, unspecified in [53], the spectrum showed a redistribution of bands from the short ($\lambda < 580$ nm) to the long wavelength ($\lambda > 580$ nm) part, very similar to that observed for congruent samples during heating or selective bleaching.

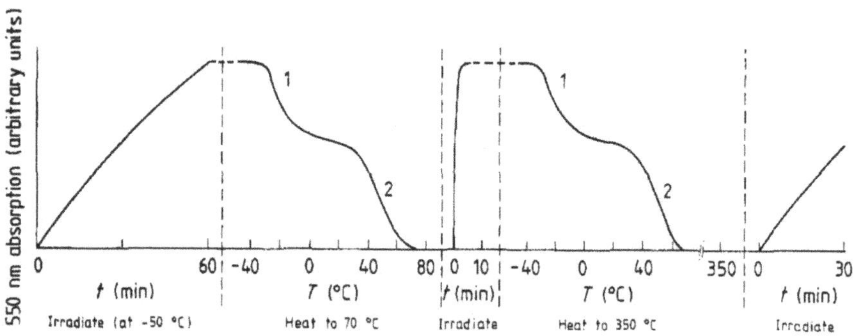

Figure 3. Irradiation and temperature annealing behavior of the 540 nm band for high-energy electron irradiation at −50 °C (Reprinted with permission from [53]. Copyright 2021 IOP Publishing).

Similarly to the case of cLN, bands at 500–540 nm and 400 nm were attributed to F- and F^+-centers generated during knock-out processes, respectively [53], disregarding the less defined position of the F band compared to cLN and the parallel behavior of both bands in the reported transformations. Although the important contribution of O^- centers to the main absorption band near 540 nm was not mentioned, the two-step annealing of this band was explained by the liberation of holes from two kinds of hole centers. Fast recovery upon re-irradiation was attributed to simple electron re-trapping at the unchanged trap, while full recovery of the lattice was assumed to occur only near 350 °C. The assumption about two kinds of hole centers is indeed supported by the reported existence of two kinds of O^- trapped holes having either unresolved or resolved hyperfine structure in their EPR spectra [48,55]. These referenced studies used cLN and photons of similarly high energy above 1 MeV. Trapped holes without distinction were shown to be stable up to ~312 K by [32] in Cu-doped cLN irradiated by 40 keV X-rays, where the lower photon energy or some other difference in preparation might have been the reason for only one annealing stage found. The decay temperature of 312 K perfectly fits the second annealing stage observed by [53].

The authors attributed the final recombination step near 350 °C in near-sLN to the release of oxygen interstitials annihilating the oxygen vacancies. However, this temperature corresponds to the well-known starting temperature of first-stage thermal reduction governed by Li mobility, also identified in ionic conductivity studies (for references see [9]), while the onset of oxygen mobility of any kind is expected at substantially higher temperatures [20–22]. In fact, in the framework of the bipolaron scenario, there is also a straightforward explanation for the observed difference between slow and fast coloration processes near room-temperature: for the stability of the doubly recharged defect, a change in its charge-compensation is required, which can be provided by the time- and energy-consuming diffusion of cation vacancies away and back to the antisite. The observed gradual blueshift of the main absorption band from 540 to 500 nm and further (see Figure 2) upon decreasing the Li/Nb ratio can also be explained by the increasing flexibility of the cation sublattices. Better charge-compensation by further neighboring defects also lowers the formation energy of the antisite complexes and possibly also that of trapped-hole defects, meaning larger distances of the trapping level from the conduction and valence bands, respectively. This fine-adjustment energy is comparable to the energy of the order of 0.5 eV calculated by [40] for the association of defects. Similar possible stoichiometry-dependent minor shifts of the ~1.6 eV small-polaron band are of the order of the experimental error.

It may be assumed that diffusion is hampered in near-sLN due to the smaller concentration of both Li vacancies and Nb_{Li} antisites. Slower diffusion in near-sLN upon thermal reduction may be the cause of the increased presence of surface coloration which could be eliminated either by polishing or by covering the samples by Pt foil [33] (see Figure 1). Absorption near to the UV edge is also caused by charge-transfer transitions involving the generation of localized excitons; these can be described as Nb^{4+}–O^- excitons having dipolar character, pinned on intrinsic and extrinsic defects of similarly strong dipolar character mainly due to their charge compensators [46].

Accordingly, the discussed experimental results in near-sLN do not support the oxygen vacancy model, whereas those involving high-energy irradiation near room temperature even provide further ingredients for the development of the bipolaron scenario. Moreover, the absence of increased defect formation at higher temperatures in near-sLN, suggested by implication by Hodgson et al. [53], might be a general argument against the oxygen vacancy scenario.

4. Comparison with Model Calculations

Earlier atomistic or cluster models did not consider the role of the ambient; however, more recent calculations also include the effect of the environment via temperature- and pressure-dependent chemical potentials, and also account for the Fermi level changing upon reduction. The formation energies of basic defect types, also including isolated oxygen vacancies and Nb_{Li} antisites, have been calculated and compared by Xu et al. [13] in the framework of density-functional theory as a function of the Fermi energy ε_F, using the perfect state of the chosen 240-atom supercell as a common reference. For values of ε_F between the valence band maximum (VBM) and the conduction band minimum (CBM), the data are shown in Figure 4 for the defect's optimal net charge state in the Li-deficient limit. In the as-grown state at low temperature, $\varepsilon_F \approx E_{gap}/2$, where E_{gap} = CBM − VBM is the gap energy and the energy zero corresponds to the VBM. A direct comparison between the formation energies of uncompensated but relaxed Nb_{Li} and V_O defects shows that the antisite underscores the oxygen vacancy for all relevant values of ε_F. In the as-grown state, the margin is large for a number of reasons:

i. The local density approximation used by [13] calculates an unphysical self-interaction for large electron densities prevailing inside the bipolaron's covalent bond between its Nb^{4+}_{Li} and its Nb^{4+}_{Nb} constituents, unduly raising its energy. Corrections within the same framework but including the use of hybrid exchange-correlation functionals [40,41] or GIPAW pseudopotentials, self-consistently calculated U values, and the Bethe–Salpeter equation [42] lower the bipolaron level (+2 net-charge state) by more than 0.5 eV, partly underscoring even the single-polaron state (+3 net charge) and thereby verifying the negative-U property of the antisite not displayed for the +3/+2 transition in the results of [13]. It should be noted that the +1 and 0 net-charge transfer levels of the antisite have been shown to correspond to a bipolaron and one or two additional free Nb^{4+}_{Nb} polarons fully separated from each other inside the supercell [42]. This means that the 'states' with more than two electrons trapped on the antisite are spurious levels inside the conduction band or closely below the CBM, but their corrected positions can be used for an independent estimate of the position of the CBM. The corrections and estimates lead to agreement with independently calculated values of E_{gap} considered realistic between 3.8 eV and 5.4 eV [56,57], while the values derived from the position of the UV absorption edge are close to 4.0 eV [54]. The charge densities of bipolaronic states have recently been calculated to extend ~1.5 eV below the CBM [10], also supporting the above estimates;

ii. Assumption (i) does not concern the position of calculated oxygen-vacancy levels. In the as-grown crystal, the preferred empty state of the oxygen vacancy is expected to have a formation energy corresponding to the CBM with an error margin of ± 0.5 eV;
iii. The Nb_{Li} antisite requires more charge compensation than the oxygen vacancy (two more lithium vacancies, if comparing defect complexes with the same net charge), which further increases the energy difference between their realized levels by several tenths of an eV. Associations with charge compensators have been shown to amount to an energy gain of up to 0.52 eV for the $Nb_{Li}^{5+}-4V_{Li}$ defect [40].

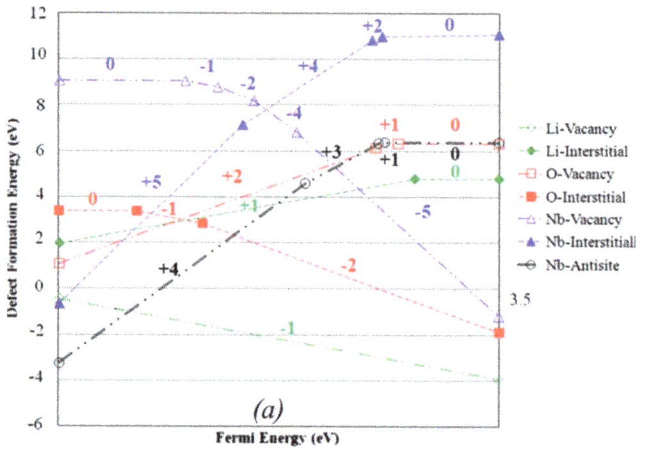

Figure 4. Defect formation energies of various point defects as a function of Fermi energy; the lowest-energy net charge state is given in each case. The Fermi energy ranges from $\varepsilon_F = 0$ (left) at the VBM to $\varepsilon_F = 3.5$ eV (right) at the CBM (without self-interaction correction and charge compensation, Reprinted with the permission from Ref [13]. Copyright 2021 American Physical Society).

Therefore, in as-grown cLN, the formation energy of the F-centers exceeds that of bipolarons by 1.4 ± 0.8 eV, making its population highly improbable. However, upon thermal reduction, the outside oxygen chemical potential μ_O may be substantially decreased: raising the temperature by 1000 °C and decreasing the oxygen partial pressure from 0.2 to 10^{-8} atm imposes changes of -2 eV and -0.8 eV on μ_O, respectively [13], adding up to $\Delta\mu_O = -2.8$ eV. This lowers the formation energy of oxygen vacancies inside the crystal which promotes their appearance, but only on the timescale of diffusion processes. Bipolarons do not involve oxygen loss; therefore, their levels remain unaffected. Therefore, a great part of the $\Delta\mu_O = 2.8$ eV difference has to be realized before the oxygen vacancy level inside the crystal may approximate that of the bipolaron. Equality between the levels may be assumed to correspond qualitatively to the end of the first annealing stage: for $T_{red} < 650$ °C, only oxygen loss from the surface together with in-diffusion of the cations and electrons occurs, raising the Fermi energy as well. For $T_{red} > 650$ °C, a second annealing stage can also be reached, where Li_2O starts to leave the crystal, meaning that the bulk chemical potential of Li_2O begins to be underscored by its outside value. This triggers surface decomposition processes described by Equation (3) and may promote also the formation of Li_2O Schottky defects. In this annealing stage, the crystal is driven towards the Li_2O–Nb_2O_5 decomposition limit corresponding to minimal possible Li_2O content in the stability quadrangle within the triangular chemical potential chart of LN (point C in [13] or point E in [40]). In fact, the Li_2O Schottky defect, the empty oxygen vacancy compensated by two empty lithium vacancies, has a near-zero formation energy already in the as-grown crystal, making it the defect of second choice after the antisite complex $Nb_{Li}-4V_{Li}$ [13]. Accordingly, for second-stage reduction followed by cool-down, instead of high-energy F-type centers, neutral Li_2O Schottky defects having much lower energy may form, while antisite-trapped bipolarons will remain the dominating trapped-electron defect. Due to their high mobilities, the movements of small polarons and charge-compensating Li vacancies are always correlated, and upon cool-down defects with zero net charge will dominate. This means that while the formation of oxygen vacancies and new lithium vacancies cannot be excluded for strong reduction treatments, this is not expected to lead to the formation of stable F^- or F^+-centers.

On the other hand, the cited calculations on electron polarons and bipolarons trapped at antisite Nb_{Li} defects reproduced the observed spectroscopic features and the values of various other properties in Li-deficient LN, including the spin-Hamiltonian parameters of Nb_{Li}^{4+} [10,42,58]. All these results give strong additional support to the established small-polaronic models and the cationic mechanism of thermal reduction.

For a refinement of the presented interpretations but also for finding minority effects related to oxygen defects in LN, further experimental and theoretical studies are required, explicitly taking into account the presence of charge-compensating defects together with diffusion and clustering as well as surface effects. In particular, modeling the structure of the modified surface layer on the nanometer scale for various treatments remains a major challenge. The rich potential of the active surfaces of LN has been presented in [59]. Oxygen vacancies as possible charge compensators of aliovalent extrinsic cation dopants, e.g., in Mo^{4+}–V_O defects found stable by cluster calculations of [60] are a separate problem requiring confirmation by experimental and comparative calculations.

5. Conclusions

After more than half a century after its first Czochralski preparation, we still have to emphasize that LN is not a 'normal' oxide, such as Al_2O_3 or $SrTiO_3$, but a compound with great freedom of redistribution, although only for its cations, which can also be considered an intermediary system between the anode and cathode materials Li_3NbO_4 and $LiNb_3O_8$ considered for lithium-ion batteries. Defect generation in non-stoichiometric LN by thermochemical, mechanochemical or irradiation methods was shown to be subject to a single principle, cationic rearrangement, in fulfillment of the remarks of Smyth [16]. We demonstrated that the repeatedly proposed oxygen-vacancy-type models cannot be sustained in Li-deficient LN. Similarly to the futile attempt to use the reversible transformation between small-polaronic bands to prove these models, the argument about non-saturating defect generation by high-energy irradiation cannot be used as a decisive proof for the generation of F-centers either. On the other hand, there are four basic arguments in favor of cationic motion and small-polaron/bipolaron formation as the main cause of coloration in cLN:

i. The practically unchanged positions of the absorption bands at ~1.6 and ~2.5 eV were obtained in cLN (attributed to small polarons and bipolarons/hole polarons, respectively) under all employed reduction, irradiation and grinding procedures up to 1050 °C; there were only small shifts of the ~2.5 eV band for changing stoichiometry which could be attributed to defect association effects both for bipolarons and hole polarons;
ii. The lack of the bands was attributed to small polarons and bipolarons in the absence of antisites in over-threshold LN:Mg [45] and the limited growth and instability of coloration observed in irradiated or reduced near-sLN;
iii. Missing of verifiable fingerprints of F^+ centers and the lack of a consequent interpretation of their role;
iv. Calculations: bipolarons at Nb_{Li} antisites are predicted to provide deeper trapping sites for small polarons than oxygen vacancies; the latter have larger calculated association energies with lithium vacancies than with electrons.

Coloration in LN doped by an over-threshold amount of Mg was also suggested to occur by cationic rearrangement, i.e., a partial change of Mg incorporation from Nb to the Li site. A key element in stable defect formation is apparently the mobility of Li^+ ions complementing small-polaronic migration and trapping processes thereby providing charge compensation for changing defects. Characteristic temperature regions below 200 K, near −25 °C/40 °C and near 350 °C, derived earlier from defect annealing experiments, have been interpreted as ranges where electron-polarons, two types of trapped hole polarons, and lithium vacancies become mobile, respectively. Another major aspect are surfaces playing a prominent role as areas for evaporation and phase segregation in thermal reduction and mechanochemical processes, and also during high-energy irradiation at elevated temperatures. Further measurements on variously treated LN crystals of bulk, thin layer, or nanocrystal forms, using optical, (para)magnetic and dielectric spectroscopies and their combinations as well as atomic resolution analytical and imaging methods (see [61] in the present Special Issue), complemented by calculations, may provide additional detail to this picture.

Author Contributions: Conceptualization, G.C.; writing—original draft preparation, G.C.; writing—review and editing, L.K. Both authors have read and agreed to the published version of the manuscript.

Funding: This research was funded by the Quantum Information National Laboratory of Hungary, supported by the Ministry of Innovation and Technology and the National Research, Development and Innovation Office (No. of Grant Agreement: NKFIH-873-3/2020).

Institutional Review Board Statement: Not applicable.

Informed Consent Statement: Not applicable.

Data Availability Statement: No new data were created or analyzed in this study.

Acknowledgments: This review and the Special Issue 'New Trends in Lithium Niobate: From Bulk to Nanocrystals' as a whole is dedicated to the memory of Ortwin F. Schirmer, who was one of the initiators of the bipolaron concept in oxides, and whose kind personality and dedicated guidance played a stimulating role in the author's careers.

Conflicts of Interest: The authors declare no conflict of interest.

References

1. Crespillo, M.L.; Graham, J.T.; Agulló-López, F.; Zhang, Y.; Weber, W.J. Real-time Identification of Oxygen Vacancy Centers in $LiNbO_3$ and $SrTiO_3$ during irradiation with High Energy Particles. *Crystals* **2021**, *11*, 315. [CrossRef]
2. Wang, C.; Sun, J.; Ni, W.; Yue, B.; Hong, F.; Liu, H.; Cheng, Z. Tuning oxygen vacancy in $LiNbO_3$ single crystals for prominent memristive and dielectric behaviors. *J. Am. Ceram. Soc.* **2019**, *102*, 6705–6712. [CrossRef]
3. Sánchez-Dena, O.; Fierro-Ruiz, C.D.; Villalobos-Mendoza, S.D.; Carillo Flores, D.M.; Elizalde-Galindo, J.T.; Farias, R. Lithium Niobate Single Crystals and Powders Reviewed—Part I. *Crystals* **2020**, *10*, 973. [CrossRef]
4. Sánchez-Dena, O.; Villalobos-Mendoza, S.D.; Farias, R.; Fierro-Ruiz, C.D. Lithium Niobate Single Crystals and Powders Reviewed—Part II. *Crystals* **2020**, *10*, 990. [CrossRef]
5. Grachev, V.G.; Malovichko, G.I. Structures of Impurity Defects in Lithium Niobate and Tantalate Derived from Electron Paramagnetic and Electron Nuclear Double Resonance Data. *Crystals* **2021**, *11*, 339. [CrossRef]
6. Kling, A.; Marques, J.G. Unveiling the Defect Structure of Lithium Niobate with Nuclear Methods. *Crystals* **2021**, *11*, 501. [CrossRef]
7. Messerschmidt, S.; Krampf, A.; Vittadello, L.; Imlau, M.; Nörenberg, T.; Eng, L.M.; Emin, D. Small-Polaron Hopping and Low-Temperature (45–225 K) Photo-Induced Transient Absorption in Magnesium-Doped Lithium Niobate. *Crystals* **2020**, *10*, 809. [CrossRef]
8. Vittadello, L.; Guilbert, L.; Fedorenko, S.; Bazzan, M. Polaron Trapping and Migration in Iron-Doped Lithium Niobate. *Crystals* **2021**, *11*, 302. [CrossRef]
9. Suhak, Y.; Roshchupkin, D.; Redkin, B.; Kabir, A.; Jerliu, B.; Ganschow, S.; Fritze, H. Correlation of Electrical Properties and Acoustic Loss in Single Crystalline Lithium Niobate-Tantalate Solid Solutions at Elevated Temperatures. *Crystals* **2021**, *11*, 398. [CrossRef]
10. Schmidt, F.; Kozub, A.L.; Gerstmann, U.; Schmidt, W.G.; Schindlmayr, A. Electron Polarons in Lithium Niobate: Charge Localization, Lattice Deformation, and Optical Response. *Crystals* **2021**, *11*, 542. [CrossRef]
11. Vasylechko, L.; Sydorchuk, V.; Lakhnik, A.; Suhak, Y.; Wlodarczyk, D.; Hurskyy, S.; Yakhnevych, U.; Zhydachevskyy, Y.; Sugak, D.; Syvorotka, I.; et al. Investigation of $LiNb_{1-x}Ta_xO_3$ Nanopowders Obtained with Mechanochemical Method. *Crystals* **2021**, *11*, 755. [CrossRef]
12. Nico, C.; Monteiro, T.; Graça, M.P.F. Niobium oxides and niobates physical properties: Review and prospects. *Progr. Mater. Sci.* **2016**, *8*, 1–37. [CrossRef]

13. Xu, H.; Lee, D.; He, J.; Sinnott, S.B.; Gopalan, V.; Dierolf, V.; Phillpot, S.R. Stability of intrinsic defects and defect clusters in $LiNbO_3$ from density functional theory calculation. *Phys. Rev. B* **2008**, *78*, 174103. [CrossRef]
14. Donnerberg, H.; Tomlinson, S.M.; Catlow, C.R.A.; Schirmer, O.F. Computer-simulation studies of intrinsic defects in $LiNbO_3$ crystals. *Phys. Rev. B* **1989**, *40*, 11909–11916. [CrossRef]
15. Lerner, P.; Legras, C.; Dumas, J.P. Stoechiométrie des monocristaux de métaniobate de lithium. *J. Cryst. Growth* **1968**, *3–4*, 231–235. [CrossRef]
16. Smyth, D.M. Defects and Transport in $LiNbO_3$. *Ferroelectrics* **1983**, *50*, 93–102. [CrossRef]
17. Kovács, L.; Polgár, K. Density measurements on $LiNbO_3$ crystals confirming Nb substitution for Li. *Cryst. Res. Technol.* **1986**, *21*, K101–K104. [CrossRef]
18. Holmes, R.J.; Minford, W.J. The effects of boule to boule compositional variations on the properties of $LiNbO_3$ electro-optic devices–An interpretation from defect chemistry studies. *Ferroelectrics* **1987**, *75*, 63–70. [CrossRef]
19. Iyi, N.; Kitamura, K.; Izumi, F.; Yamamoto, J.K.; Hayashi, T.; Asano, H.; Kimura, S. Comparative study of defect structures in lithium niobate with different compositions. *J. Sol. State Chem.* **1992**, *101*, 340–352. [CrossRef]
20. Shi, J.; Fritze, H.; Borchardt, G.; Becker, K.D. Defect chemistry, redox kinetics, and chemical diffusion of lithium deficient lithium niobate. *Phys. Chem. Chem. Phys.* **2011**, *13*, 6925–6930. [CrossRef]
21. Fielitz, P.; Schneider, O.; Borchardt, G.; Weidenfelder, A.; Fritze, H.; Shi, J.; Becker, K.D.; Ganschow, S.; Bertram, R. Oxygen-18 tracer diffusion in nearly stoichiometric single crystalline lithium niobate. *Solid State Ionics* **2011**, *189*, 1–6. [CrossRef]
22. Weidenfelder, A.; Shi, J.; Fielitz, P.; Borchardt, G.; Becker, K.D.; Fritze, H. Electrical and electromechanical properties of stoichiometric lithium niobate at high-temperatures. *Solid State Ionics* **2012**, *225*, 26–29. [CrossRef]
23. Jian, Z.; Lu, X.; Fang, Z.; Hu, Y.S.; Zhou, J.; Chen, W.; Chen, L. $LiNb_3O_8$ as a Novel Anode Material for Lithium-Ion Batteries. *Electrochem. Commun.* **2011**, *13*, 1127–1130. [CrossRef]
24. Kuganathan, N.; Kordatos, A.; Kelaidis, N.; Chroneos, A. Defects, Lithium Mobility and Tetravalent Dopants in the Li_3NbO_4 Cathode Material. *Sci. Rep.* **2019**, *9*, 2192. [CrossRef]
25. Semiletov, S.A.; Bocharova, N.G.; Rakova, E.V. Decomposition of a Solid Solution on the Surface of Lithium Niobate Crystals: Structure, Morphology, and Mutual Orientation of Phases. In *Growth of Crystals*; Givargizov, E.I., Grinberg, S.A., Eds.; Springer: Berlin/Heidelberg, Germany, 2012; Volume 17, pp. 95–103.
26. Schirmer, O.F.; Von der Linde, D. Two-photon- and x-ray-induced Nb^{4+} and O^- small polarons in $LiNbO_3$. *Appl. Phys. Lett.* **1978**, *33*, 35–38. [CrossRef]
27. Müller, H.; Schirmer, O.F. Microscopic structure of Nb_{Li} related defects in reduced undoped $LiNbO_3$. *Ferroelectrics* **1992**, *125*, 319–324. [CrossRef]
28. Reyher, H.-J.; Schulz, R.; Thiemann, O. Investigation of the optical-absorption bands of Nb^{4+} and Ti^{3+} in lithium niobate using magnetic circular dichroism and optically detected magnetic-resonance techniques. *Phys. Rev. B* **1994**, *50*, 3609–3619. [CrossRef]
29. Sweeney, K.L.; Halliburton, L.E. Oxygen vacancies in lithium niobate. *Appl. Phys. Lett.* **1983**, *43*, 336–338. [CrossRef]
30. Halliburton, L.E.; Sweeney, K.L.; Chen, C.Y. Electron spin resonance and optical studies of point defects in lithium niobate. *Nucl. Instr. Meth. Phys. Res. B* **1984**, *1*, 344–347. [CrossRef]
31. Ketchum, J.L.; Sweeney, K.L.; Halliburton, L.E.; Armington, A.F. Vacuum annealing effects in lithium niobate. *Phys. Lett. A.* **1983**, *94*, 450–453. [CrossRef]
32. Arizmendi, L.; Cabrera, J.M.; Agulló-López, F. Defects induced in pure and doped $LiNbO_3$ by irradiation and thermal reduction. *J. Phys. C Solid State Phys.* **1984**, *17*, 515–529. [CrossRef]
33. García-Cabañes, A.; Diéguez, E.; Cabrera, J.M.; Agulló-López, F. Contributing bands to the optical absorption of reduced $LiNbO_3$: Thermal and optical excitation. *J. Phys. Condens. Matter* **1989**, *1*, 6453–6462. [CrossRef]
34. Dutt, D.A.; Feigl, F.J.; DeLeo, G.G. Optical absorption and electron paramagnetic resonance studies of chemically reduced congruent lithium niobate. *J. Phys. Chem. Solids* **1990**, *51*, 407–415. [CrossRef]
35. Schirmer, O.F.; Imlau, M.; Merschjann, C.; Schoke, B. Electron small polarons and bipolarons in $LiNbO_3$. *J. Phys. Condens. Matter* **2009**, *21*, 123201. [CrossRef]
36. Koppitz, J.; Schirmer, O.F.; Kuznetsov, A.I. Thermal Dissociation of Bipolarons in Reduced Undoped $LiNbO_3$. *Europhys. Lett.* **1987**, *4*, 1055–1059. [CrossRef]
37. Holman, R.L. Novel uses of gravimetry in the processing of crystalline ceramics. In *Processing of Crystalline Ceramics*; Palmour, H., Davis, R.F., Hare, T.M., Eds.; Plenum: New York, NY, USA, 1978; pp. 343–358.
38. Bordui, P.F.; Norwood, R.G.; Jundt, D.H.; Fejer, M.M. Preparation and characterization of off-congruent lithium niobate crystals. *J. Appl. Phys.* **1992**, *71*, 875–879. [CrossRef]
39. Nahm, H.; Park, C.H. First-principles study of microscopic properties of the Nb antisite in $LiNbO_3$: Comparison to phenomenological polaron theory. *Phys. Rev. B* **2008**, *78*, 184108. [CrossRef]
40. Li, Y.; Schmidt, W.G.; Sanna, S. Intrinsic $LiNbO_3$ point defects from hybrid density functional calculations. *Phys. Rev. B* **2014**, *89*, 09411. [CrossRef]
41. Li, Y.; Sanna, S.; Schmidt, W.G. Modeling intrinsic defects in $LiNbO_3$ within the Slater-Janak transition state model. *J. Chem. Phys.* **2014**, *140*, 234113. [CrossRef]

42. Schmidt, F.; Kozub, A.L.; Biktagirov, T.; Eigner, C.; Silberhorn, C.; Schindlmayr, A.; Schmidt, W.G.; Gerstmann, U. Free and defect-bound (bi)polarons in $LiNbO_3$: Atomic structure and spectroscopic signatures from ab initio calculations. *Phys. Rev. Res.* **2020**, *2*, 043002. [CrossRef]
43. Faust, B.; Müller, H.; Schirmer, O.F. Free small polarons in $LiNbO_3$. *Ferroelectrics* **1994**, *153*, 297–302. [CrossRef]
44. Sugak, D.Y.; Syvorotka, I.I.; Buryy, O.A.; Yakhnevych, U.V.; Solskii, I.M.; Martynyuk, N.V.; Suhak, Y.; Suchocki, A.; Zhydachevskii, Y.; Jakiela, R. Spatial Distribution of Optical Coloration in Single Crystalline $LiNbO_3$ after High-Temperature H_2/Air Treatments. *Opt. Mater.* **2017**, *70*, 106–115. [CrossRef]
45. Sweeney, K.L.; Halliburton, L.A.; Bryan, D.A.; Rice, R.R.; Gerson, R.; Tomaschke, H.E. Point defects in Mg-doped lithium niobate. *J. Appl. Phys.* **1985**, *57*, 1036–1044. [CrossRef]
46. Corradi, G.; Krampf, A.; Messerschmidt, S.; Vittadello, L.; Imlau, M. Excitonic hopping-pinning scenarios in lithium niobate based on atomistic models: Different kinds of stretched exponential kinetics in the same system. *J. Phys. Condens. Matter* **2020**, *32*, 413005. [CrossRef]
47. Lengyel, K.; Péter, Á.; Kovács, L.; Corradi, G.; Pálfalvi, L.; Hebling, J.; Unferdorben, M.; Dravecz, G.; Hajdara, I.; Szaller, Z.; et al. Growth, defect structure, and THz application of stoichiometric lithium niobate. *Appl. Phys. Rev.* **2015**, *2*, 040601. [CrossRef]
48. Corradi, G.; Polgár, K.; Zaritskii, I.M.; Rakitina, L.G.; Derjugina, N.I. Characteristics of defect formation under γ-irradiation and heat treatment of pure and doped lithium niobate single crystals. *Fiz. Tverd. Tela* **1989**, *31*, 115–122. (in Russian). English translation: *Sov. Phys. Solid State* **1989**, *31*, 1540–1544.
49. García-Cabañes, A.; Sanz-García, J.A.; Cabrera, J.M.; Agulló-López, F.; Zaldo, C.; Pareja, R.; Polgár, K.; Raksányi, K.; Földvári, I. Influence of stoichiometry on defect-related phenomena in $LiNbO_3$. *Phys. Rev. B* **1988**, *37*, 6085–6091. [CrossRef]
50. Kocsor, L.; Péter, L.; Corradi, G.; Kis, Z.; Gubicza, J.; Kovács, L. Mechanochemical Reactions of Lithium Niobate Induced by High-Energy Ball-Milling. *Crystals* **2019**, *9*, 334. [CrossRef]
51. Díaz-Moreno, C.A.; Farías-Mancilla, R.; Matutes-Aquino, J.A.; Elizalde-Galindo, J.; Espinosa-Magaña, F.; Hernández-González, J.; Hurtado-Macías, A. Magnetic behavior in $LiNbO_3$ nanocrystallites caused by oxygen vacancies. *J. Magn. Magn. Mater.* **2014**, *356*, 82–86. [CrossRef]
52. Hodgson, E.R.; Agulló-López, F. Oxygen vacancy centres induced by electron irradiation in $LiNbO_3$. *Solid State Commun.* **1987**, *64*, 965–968. [CrossRef]
53. Hodgson, E.R.; Agulló-López, F. High-energy electron irradiation of stoichiometric $LiNbO_3$. *J. Phys. Condens. Matter* **1989**, *1*, 10015–10020. [CrossRef]
54. Kovács, L.; Ruschhaupt, G.; Polgár, K.; Corradi, G.; Wöhlecke, M. Composition dependence of the UV absorption edge in lithium niobate. *Appl. Phys. Lett.* **1997**, *70*, 2801–2803. [CrossRef]
55. Miki, T.; Hantehzadeh, M.R.; Halliburton, L.E. A new trapped-hole center in irradiated $LiNbO_3$. *J. Phys. Chem. Solids* **1989**, *50*, 1003–1007. [CrossRef]
56. Friedrich, M.; Riefer, A.; Sanna, S.; Schmidt, W.G.; Schindlmayr, A. Phonon dispersion and zero-point renormalization of $LiNbO_3$ from density-functional perturbation theory. *J. Phys. Condens. Matter* **2015**, *27*, 385402. [CrossRef]
57. Riefer, A.; Friedrich, M.; Sanna, S.; Gerstmann, U.; Schindlmayr, A.; Schmidt, W.G. $LiNbO_3$ electronic structure: Many-body interactions, spin-orbit coupling, and thermal effects. *Phys. Rev. B* **2016**, *93*, 075205. [CrossRef]
58. Li, Y.; Schmidt, W.G.; Sanna, S. Defect complexes in congruent $LiNbO_3$ and their optical signatures. *Phys. Rev. B* **2015**, *91*, 174106. [CrossRef]
59. Sanna, S.; Schmidt, W.G. $LiNbO_3$ surfaces from a microscopic perspective. *J. Phys. Condens. Matter* **2017**, *29*, 413001. [CrossRef]
60. Araujo, R.M.; dos Santos Mattos, E.F.; Valerio, M.E.G.; Jackson, R.A. Computer Simulation of the Incorporation of V^{2+}, V^{3+}, V^{4+}, V^{5+} and Mo^{3+}, Mo^{4+}, Mo^{5+}, Mo^{6+} Dopants in $LiNbO_3$. *Crystals* **2020**, *10*, 457. [CrossRef]
61. Reitzig, S.; Rüsing, M.; Zhao, J.; Kirbus, B.; Mookherjea, S.; Eng, L.M. "Seeing Is Believing"—In-Depth Analysis by Co-Imaging of Periodically-Poled X-Cut Lithium Niobate Thin Films. *Crystals* **2021**, *11*, 288. [CrossRef]

crystals

MDPI

Review

Epitaxy of $LiNbO_3$: Historical Challenges and Recent Success

Bill Zivasatienraj [1], M. Brooks Tellekamp [2] and W. Alan Doolittle [1,*]

[1] Georgia Institute of Technology, School of Electrical and Computer Engineering, Atlanta, GA 30332, USA; jobitron@gatech.edu

[2] National Renewable Energy Laboratory, Materials, Chemical, and Computational Science Directorate, Golden, CO 80401, USA; brooks.tellekamp@nrel.gov

* Correspondence: alan.doolittle@ece.gatech.edu

Abstract: High-quality epitaxial growth of thin film lithium niobate ($LiNbO_3$) is highly desirable for optical and acoustic device applications. Despite decades of research, current state-of-the-art epitaxial techniques are limited by either the material quality or growth rates needed for practical devices. In this paper, we provide a short summary of the primary challenges of lithium niobate epitaxy followed by a brief historical review of lithium niobate epitaxy for prevalent epitaxial techniques. Available figures of merit for crystalline quality and optical transmission losses are given for each growth method. The highest crystalline quality lithium niobate thin film was recently grown by halide-based molecular beam epitaxy and is comparable to bulk lithium niobate crystals. However, these high-quality crystals are grown at slow rates that limit many practical applications. Given the many challenges that lithium niobate epitaxy imposes and the wide variety of methods that have unsuccessfully attempted to surmount these barriers, new approaches to lithium niobate epitaxy are required to meet the need for simultaneously high crystalline quality and sufficient thickness for devices not currently practical by existing techniques.

Keywords: lithium; niobate; epitaxy; thin film; liquid phase epitaxy; molecular beam epitaxy; sputtering; pulsed laser deposition; chemical vapor deposition

check for updates

Citation: Zivasatienraj, B.; Tellekamp, M.B.; Doolittle, W.A. Epitaxy of $LiNbO_3$: Historical Challenges and Recent Success. *Crystals* **2021**, *11*, 397. https://doi.org/10.3390/cryst11040397

Academic Editors: Gábor Corradi and László Kovács

Received: 16 March 2021
Accepted: 6 April 2021
Published: 9 April 2021

1. Introduction

Lithium niobate ($LiNbO_3$: LN) is a mature material that has been pivotal in the advancement of optical and acoustic technology. With excellent ferro-electric, electro-optic, and piezoelectric properties, LN is incorporated into devices such as waveguides, modulators, frequency-doubled lasers, surface acoustic wave (SAW) devices, optical switches, and acoustic resonators [1]. These applications typically require high-quality, single-crystal LN to avoid optical and acoustic propagation losses. Presently, bulk LN material is being grown via the Czochralski pulling method to ensure the highest quality material. However, realization of LN in the miniaturization of devices, integration into hybrid systems, and the lowering of operational voltages for various electro-optical or electro-acoustic devices require thin layers of high-quality material. Thus, the deposition of high-quality, single-crystal thin films of LN has been a topic of research for at least three decades.

Traditionally, optical waveguides are formed in bulk substrates by techniques such as titanium (Ti) in-diffusion [2–4]. Compared to theoretical epitaxial counterparts, Ti-in-diffused LN has increased optical losses and the inability to form sharp index profiles as well as a top only contact structure, which leads to large voltage requirements for index modulation [5,6]. In the absence of acceptable epitaxial deposition methods, the use of ion implantation to uniformly damage and slice thin layers of LN, known as smart-cut [7] or ion-sliced LN, is presently favored. Smart-cut LN can be bonded to metal electrodes or various dielectrics, enabling lower voltage and lower loss operation, but requires a significant minimum thickness (>300 nm) [8] for wafer bonding and the transfer of the thin film, is limited to less than a few microns due to the limitations in implant energies [9], and has associated straggle-related damage, which then requires treatment for the resulting rough surface and high-temperature thermal annealing to restore optoelectrical properties [7,9]. Smart-cut LN provides a final LN wafer quality similar to bulk LN, but the thickness range available is not ideally suitable for the coupling of light on and off the chip, for example, to/from large core fibers or nanoscale integrated optics. Promising work on direct etching of LN to fabricate compact optical devices with optical propagation losses as low as 0.027 dB/cm has been shown in a 250 nm thin LN slab, achieved by a dry etch process to pattern a 600 nm smart-cut film [10,11]. Thus, high-quality epitaxial thin film LN is desirable for smaller devices with lower power consumption; lower optical losses; tighter wave confinement; integrated devices; and engineered material advances, such as doping [12,13] or strain manipulation [14], analogous to compound semiconductor electronics technology.

In this work, we will discuss the epitaxy of LN, the challenges that researchers have faced, and review the major accomplishments in LN epitaxy to date. Since some researchers target optical usage, while others merely report structural figures of merit, film quality will be detailed via metrics of optical coupling loss; deviation from single crystallinity, such as texturing of near-crystalline material; and even polycrystalline or amorphous material and the existence of rotational domains, whenever these are reported. These factors prevent or limit the practical application of LN thin epitaxial films and are the consequence of imperfections during epitaxy. Of particular interest are the elimination of rotational domains, as this has been shown to strongly effect optical quality [15,16]. There are various epitaxial methods that have successfully grown LN, including liquid phase epitaxy (LPE) [17–26], sputtering [12,14,27–40], chemical vapor deposition (CVD) [41–50], pulsed laser deposition (PLD) [13,51–65], and molecular beam epitaxy (MBE) [66–77]. Herein, we will view each of the epitaxial methods mentioned above from a historical perspective with a brief projection of prospects. When growing crystals with hexagonal symmetry, it is commonly found that the basal plane—the highest atomic density plane—grows the slowest, and thus is the most stable for epitaxy. Therefore, the epitaxy of $LiNbO_3$ is generally preferred along either the (0001) or the $(000\bar{1})$ orientation to avoid rapidly growing facets that lead to rough surfaces and domain formation. We note that MBE has achieved the highest crystalline perfection with respect to crystalline tilt, with X-ray diffraction rocking curve figures of merit as low as 8.6 arcseconds [77] for a basal plane, (0006) reflection, comparable to congruently grown bulk LN, as shown in Figure 1. However, to date, no method has simultaneously achieved the required quality needed for optical or acoustic devices along with practical deposition rates necessary for thick devices on the scale of optical (hundreds of nm) or acoustic wavelengths (microns).

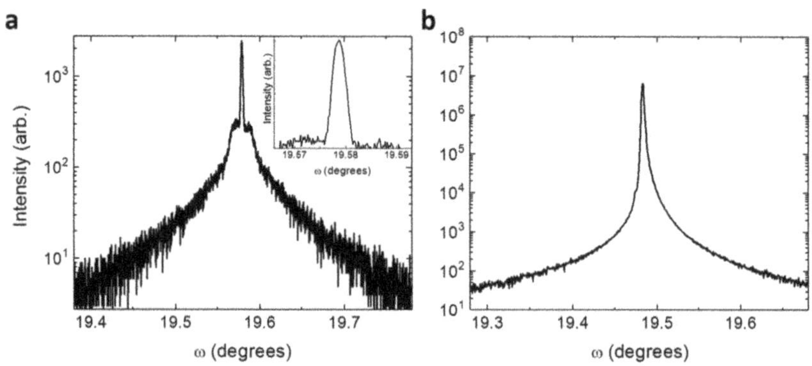

Figure 1. (**a**) X-ray diffraction rocking curve of epitaxial LN grown on sapphire via molecular beam epitaxy [77], with an 8.6 arcsec FWHM indicating a high-quality crystal strained to the sapphire substrate. The inset shows a well-resolved, narrow peak. (**b**) X-ray diffraction rocking curve of a bulk-grown LN wafer, with an FWHM of 10.4 arcsec.

2. The Technical Challenge

Lithium niobate is a remarkable crystal and, in many ways, is anomalous in the chemical world. Lithium (Li) is among the most reactive elements known, has an enormous vapor pressure compared with most elements, and readily forms stable oxides. On the other hand, niobium (Nb) is very slow to form oxides [76] and has a negligible vapor pressure except at extreme temperatures. Furthermore, Nb forms compounds of various oxidation states leading to a strong tendency to form undesirable phases, often with overlapping phase space windows resulting in undesirable multi-phase films [78]. This complexity is further exemplified by the comparison of the phase diagrams of bulk LN [78] and gallium arsenide [79] (GaAs), as shown in Figure 2. $LiNbO_3$ is not like most other crystals, as its congruent melting point favors a Li_2O-deficient stoichiometry that is difficult to reproduce outside of bulk crystal growth. Creating particular difficulty is the fact that unlike all commercially successful materials that use epitaxy, exemplified by GaAs in Figure 2b, $LiNbO_3$ does not have a line composition. Line compositions assure that regardless of the anion to cation ratio over a wide applicability range, nature takes care of forming the desired crystalline phase—GaAs in this example. In MBE, for instance, GaAs is formed whether the As to Ga flux ratio is 3:1 or 40:1. The difficulty of this lack of line composition for LN epitaxy cannot be understated. Thus, any change in the flux ratio of the cations to each other or cations to anions can result in a different stoichiometry or even multiple phases. Different ratios of Li to Nb form slightly differing material and this stoichiometry effects optical and acoustic properties as well as phase and crystal structure ($LiNbO_3$, $LiNbO_2$ [80,81], $LiNb_3O_8$ [39,49,80], Li_3NbO_4 [39,49,80]) by single digit percentage changes in fluxes. The lack of a line composition affects every aspect of LN epitaxial research, results in extreme tolerance limits on required flux ratios, and if commercial epitaxy is to ever be achieved, will dominate the challenges of scalability, uniformity, and yield.

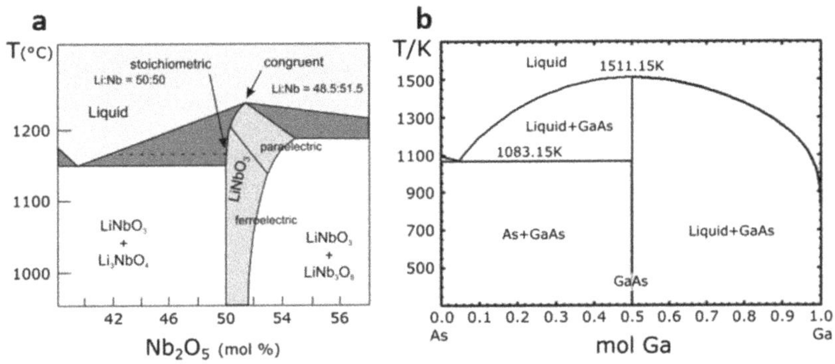

Figure 2. (**a**) The phase diagram of bulk LN adapted from Volk and Wohlecke [78] shows complex phase spaces and a narrow growth window for monophasic LN. The distinct lack of a convenient line composition creates many challenges, culminating in the need for finer control over the many growth parameters during LN epitaxy. (**b**) By comparison to LN, the phase diagram of bulk GaAs adapted from Liu et al. [79] shows that monophasic GaAs is grown, regardless of the growth ratio of As to Ga, ultimately making the epitaxial process less challenging.

In physical vapor deposition (PVD) methods, the disparity in vapor pressures between Li and Nb has not only resulted in difficult flux delivery challenges and the need for novel approaches, but the inability to keep Li on the surface without desorption typically results in the need for huge Li overpressures and surface coverages that are exponentially related to temperature, and thus, hard to control. Much of the success in recent years of the PVD methods (MBE, sputtering, and PLD) revolves around hybrid methods of generating Nb fluxes by, for example, using higher vapor pressure forms of Nb and/or using non-stoichiometric target chemistries. Contrarily, the primary challenge for CVD epitaxial methods is not generating high-vapor pressure Nb fluxes but controlling the diverse reactivity of the Nb and Li precursors that often react detrimentally with other byproducts, such as water or hydrocarbons. Each method of epitaxy deserves a full review to illuminate the subtleties in the techniques. Instead, this paper summarizes the key ideas and advances in each approach, allows the reader to compare figures of merit, and draws a conclusion on the present status of this now 48-year-old quest for high-quality $LiNbO_3$ epitaxy, a quest that unfortunately remains unfulfilled. The order of discussion of historical results follows from the earliest work to the most successful recent work.

3. Liquid Phase Epitaxy (LPE)

In 1973, Miyazawa [82] tried to melt bulk LN crystals onto a $LiTaO_3$ substrate to obtain a thin film of LN using a method he called epitaxial growth by melting (EGM). Although Miyazawa [82] was able to show some light propagation within the film, the crystal quality was poor, and optical losses were >5 dB/cm [19]. Miyazawa then decided to use LPE instead in Kondo et al. [20], looking at Li_2O-V_2O_5 as a flux material, and was able to grow high-crystallinity single-crystal LN on $LiTaO_3$ substrates, although the surface was rough and with many microcracks. Despite the film imperfections, the LPE growth in Kondo et al. [20] performed much better than EGM, with optical losses of 5 dB/cm, but at visible wavelengths, a 20–30 dB/cm optical loss was measured [19]. Ballman et al. [18] then looked at four additional flux systems, two with Li^+ rich environments and two with a Nb^+ rich environment. Ballman et al. [18] found that by using a Nb^+ rich flux system and growing at a higher temperature, they could adjust the Li/Nb ratio in the melt to obtain lattice parameters that better matched the substrate. They also looked at different crystallographic orientations of the substrate and combined those results with an annealing process in order to grow smoother films. Ballman et al. [18] were largely successful at demonstrating optical waveguiding in LPE LN films, reporting optical losses of 1–5 dB/cm at 632.8 nm. The optical losses observed decreased with a higher crystallinity film due to scattering reduction [18,19]. Baudrant et al. [21] then used the flux system in Kondo et al. [19,82], Li_2O-V_2O_5, and looked at different crystallographic orientations of the substrate to improve growth quality, thereby obtaining results that suggest a layer-by-layer growth mechanism of LN [21]. In 1991, Tamada et al. [22] used the same Li_2O-V_2O_5 flux system to grow on lattice-matched (0.08% mismatch) MgO-doped LN substrates, achieving high-quality thin films with an XRD rocking curve of 11 arcseconds, but still saw large optical losses of 25 dB/cm at 514.5 nm. However, Tamada et al. [22] were able to characterize and reduce the optical losses caused by vanadium incorporation from the flux using a 600 °C ozone anneal. Later, Tamada et al. [22] produced a high-quality LN film with a low optical loss of 1.6 dB/cm, even though they were not able to remove all of the vanadium impurities. Consequently, Yamada et al. [23] later utilized a Li_2O-B_2O_3 flux system to grow comparatively high-quality films with XRD rocking curves of 11.4 arcseconds, as shown in Figure 3, but this time without any detectable impurity inclusions and, therefore, lower optical losses of <1 dB/cm at 458 nm. Efforts toward the LPE epitaxy of LN continue and have looked at alternative growth parameters, such as using K_2O fluxes [24], doping [25], and even using LPE to grow LN waveguides in laser cut channels [26] and continue as one of the lowest optical loss sources for single crystal epitaxial LN [23].

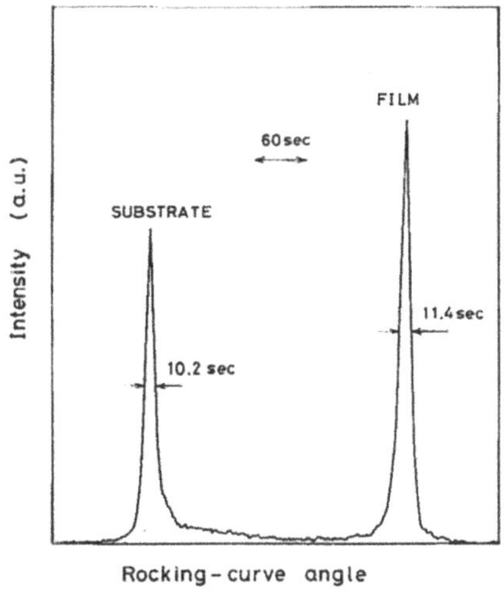

Figure 3. X-ray diffraction rocking curve with an FWHM of 11.4 arcseconds from a LN thin film grown in 1992 via liquid phase epitaxy (LPE) by Yamada et al. [23], juxtaposed with the MgO-doped LN substrate.

4. Sputtering

Following an unpublished work by Fukunishi et al. [27], Takada et al. [28] crushed LN single crystals into a powder to sinter into a sputtering target. Using an argon and oxygen gas mixture, Takada et al. [28] were able to sputter LN onto a heated sapphire substrate, and then demonstrate light propagation in the epitaxially grown c-axis-oriented film with optical losses of around 9 dB/cm. Jain and Hewig [29] later repeated the sputtering process by Fukunishi et al. [27], but onto a $Gd_3Ga_5O_{12}$ substrate to phase-match with the LN for second-harmonic generation [29]. However, the films were still rough and polycrystalline, thereby contributing to higher (unreported) propagation losses. Hewig et al. [30] studied the sputtering of LN onto both amorphous and single-crystal substrates while using different deposition parameters to investigate how the various conditions affect the LN film. Hewig et al. [30] concluded that using higher temperatures and single-crystal substrates helped avoid amorphous LN films, but the resulting film was polycrystalline of $[01\bar{1}2]$-preferred orientation. The amorphous LN films showed lower optical losses of 3 dB/cm, while the polycrystalline films measured very high losses of about 10 dB/cm due to scattering at the grain boundaries. To obtain low-loss, monocrystalline material for waveguides, Hewig et al. [30] states that the use of lattice-matched single-crystal substrates is mandatory. In the quest to obtain monocrystalline LN, Meek et al. [31] were also unsuccessful. However, during their attempts, Meek et al. [31] made two important discoveries: all films sputtered from a stoichiometric target were lithium deficient, and a Li-rich or secondary lithium source can be used to make up the lithium deficiency. Consequently, Kanata et al. [32] used a Li_2O target along with the LN target in order to compensate for the Li deficiency and succeeded in growing c-axis single-crystal LN thin films, although they do not report metrics for light propagation nor crystallinity. Similarly, Fujimura et al. [34] used a custom powder-pressed sputter target with a Li:Nb ratio of up to 3:1 instead of a sintered, stoichiometric LN target to compensate for lithium deficiency. Fujimura et al. [34] showed high-quality c-axis epitaxial LN films on sapphire without measurable rotational domains, although with twinned domains. Shimizu et al. [36] discovered that there were issues with using a powder-pressed target, and thus were able to improve on the technique used in Fujimura et al. [34] by creating a sputter target sintered from a mixture of Li_2CO_3 and Nb_2O_5 powders with a Li:Nb ratio ranging from 1 to 2. With the sintered target, Shimizu et al. [36] demonstrated monocrystalline LN films on various crystallographic orientations of Al_2O_3 and $LiTaO_3$ substrates, as well as c-axis oriented LN films on amorphous substrates using ZnO (001) and Pt (111) buffer layers, with a minimum optical loss of 6.47 dB/cm of single-crystal LN on $LiTaO_3$. In 2001, Lansiaux et al. [39] were able to grow thick c-axis monocrystalline LN films on Al_2O_3 substrates using a multi-step sputtering process from a LN target with added excess lithium of 10–20%, showing optical losses as low as 1.2 dB/cm [38]. Today, the sputtering of LN continues in research regarding doping [12], integrating and forming heterojunctions with silicon [40], tuning band gap [14], etc.

5. Chemical Vapor Deposition (CVD)

Before Curtis et al. [42], the lack of an appropriate vapor phase source of lithium prevented the epitaxial growth of LN via CVD. Following the use of 2,2,6,6-tetramethylheptane-3,5-dione (THD) for epitaxial garnet films [41], Curtis et al. [42] used the Li(THD) as a Li source and Niobium-pentamethylate [$Nb(OMe)_5$] as a Nb source to attempt the growth of LN films on various substrates. However, epitaxial growth was only seen on LN and $LiTaO_3$ substrates, and the films were a black color and amorphous. Although improvements were seen after an anneal procedure, an optical loss of about 40 dB/cm rendered this CVD technique impractical [42]. Wernberg et al. [43] used an equimolar mixture off niobium ethoxide [$Nb(OEt)_5$] and lithium dipivaloylmethanate (Li-DPM) to create a single-source organometallic liquid precursor for the growth of LN on sapphire (Al_2O_3) substrates. They were able to grow c-axis oriented films that were visually transparent and specular, with a rocking curve of 1728 arcseconds with twinned planes present. Wernberg et al. [43] report that using their chemistry, higher growth temperatures result in higher quality epitaxial LN, but this is limited by the onset of a secondary, unknown phase discovered by X-ray diffraction analysis. Wernberg et al. [43] were not able to attain smooth, featureless films. Later, Wernberg et al. [44] decided to switch to $LiNb(OEt)_6$ as a single-source precursor, thereby exchanging Li-DPM for Li(OEt) during the creation of the liquid precursor. Using the $LiNb(OEt)_6$ precursor, Wernberg et al. [44] were able to minimize ligand–exchange reactions and thus lower the FWHM of the XRD rocking curve on their single-crystal films down to 1368 arcseconds with roughly 1% twinning. In a similar fashion, Feigelson et al. [45–47] mixed unequal parts of Li(THD) and $Nb(THD)_4$ for their LN growths. A Li:Nb ratio of 65:35 was found to yield the highest quality single-phase, c-axis LN films, with FWHM values of 156 arcseconds on Al_2O_3 and 36 arcseconds on $LiTaO_3$. Optical losses of 2 dB/cm were recorded at 632.8 nm, which were later optimized to 1.8 dB/cm following the addition of an initial growth stage to control the nucleation conditions separately from the bulk growth condition [47]. Feigelson et al. [45–47] also found no twinning in the LN films grown on $LiTaO_3$ substrates, but epitaxial LN on sapphire substrates showed twinning that could be remedied by high-temperature annealing. Saulys et al. [48] performed a detailed study on the surface decomposition of many precursors used in LN MOCVD, providing enlightening results on the decomposition and dissociation of precursors due to their instability in vacuum or reaction with contaminants, such as water, which forms volatile byproducts that hinder growth. They found that water is both a reactant and a byproduct of the surface decomposition chemistry of both the Li and Nb precursors, which resulted in overall limited growth rates. Akiyama et al. [49] managed the Li and Nb feed composition by carefully controlling the flow of the precursors and varied the Li to Nb ratio with substrate temperature to find a growth window for single phase LN, as shown in Figure 4. However, Akiyama et al. [49] found polycrystals of LN within the epitaxial film and concluded that narrower operating conditions were required for high-quality single-crystal LN growth. Margueron et al. [50,83] studied the effect of deposition pressure on film quality, including a correlation with twinning. Twinning was reduced from 42% to 3% by an increase in reactor pressure from 5 to 760 Torr. A further reduction in twins was obtained by postgrowth anneals. Research on the epitaxial growth of high-quality LN via MOCVD is ongoing, including recently developed approaches, such as direct liquid injection MOCVD [84].

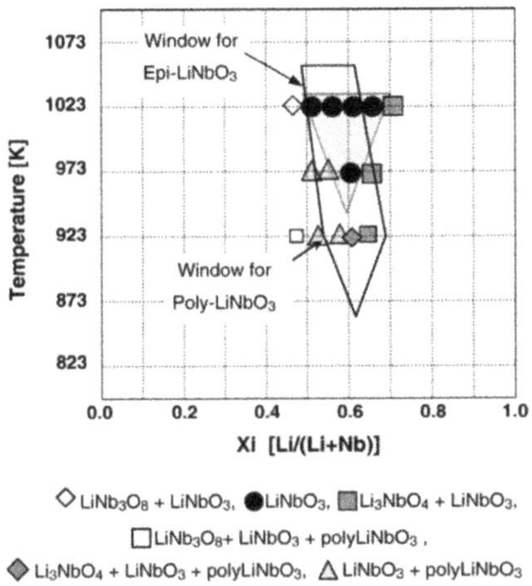

Figure 4. A pseudo-phase diagram varying the Li and Nb feed composition with growth temperature that shows growth windows for single and multiple phase LN epitaxy via MOCVD by Akiyama et al. [49].

6. Pulsed Laser Deposition (PLD)

Ogale et al. [51] were the first to use an excimer laser to ablate a stoichiometric LN sintered target for the deposition of LN onto Si. Following the sputtering work of Hewig et al. [30], Ogale et al. [51] conducted PLD in a mixed environment of argon and oxygen, making Ogale et al. [51] among the first to attempt PLD in a mixed gas ambient. By optimizing the partial pressure ratios, Ogale et al. [51] were able to grow single-phase LN, although the films were polycrystalline. Shortly after, Shibata et al. [52,56] utilized the findings from the sputtering work of Meek et al. [31] and sintered a ceramic target from Li_2CO_3 and Nb_2O_5 powders, allowing them to control the Li:Nb ratio. After tuning the Li:Nb ratio, substrate temperature, and growth pressure, Shibata et al. [52,56] were able to grow single-phase [110] LN films aligned on [110] sapphire substrates with a rocking curve FWHM of 612 arcseconds, whereas the poor-quality sapphire substrate might have affected the results, as it had an FWHM of 540 arcseconds. Without the use of a Li enriched target, many subsequent attempts resulted in polycrystalline material, twinned films, or having rotational domains [54,55,58–60]. Although Aubert et al. [55,59] were able to grow single-phase, c-axis-oriented LN with XRD rocking curve FWHM values of 360 arcseconds, they still found grain misorientations that contributed heavily to optical losses. However, it was apparent from these works [54–60] that optimal growth conditions of LN via PLD existed within a very small window amongst deposition parameters. Kakehi et al. [61] focused on the effects of laser fluence on Li incorporation into the LN films, and Son et al. [62] performed a parametric study on the effects of deposition geometries, ambient pressures, target to substrate distance, and excess Li in the target. Accounting for all of these parameters, Kilburger et al. [63,64] were able to epitaxially grow single-phase, c-axis LN on Al_2O_3 with XRD rocking curve FWHM of 165.6 arcseconds from a stoichiometric LN target, with an optical loss of about 1 ± 0.5 dB/cm, as shown in Figure 5. Recently, Paldi et al. [65] mixed LN and Au powder to create a nanocomposite seed layer before growth using both stoichiometric and Li-enriched LN targets to epitaxially grow single-domain films.

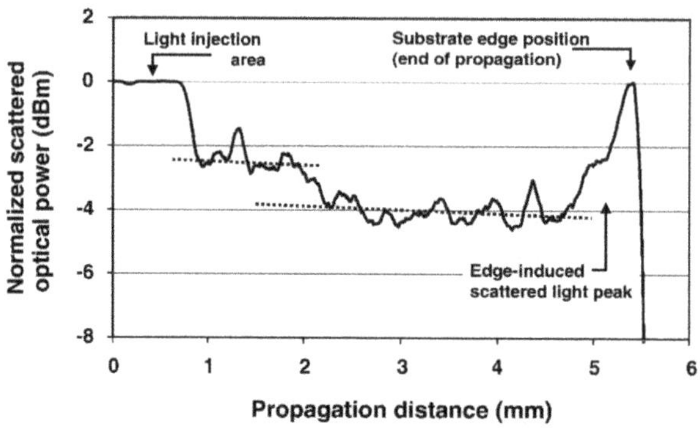

Figure 5. Light injected into a LN thin film grown on sapphire via pulsed laser deposition (PLD) by Kilburger et al. [63,64] shows low propagation losses that indicate a high-quality crystalline film.

7. Molecular Beam Epitaxy (MBE)

The first reported growth of LN via MBE was in 1985 by Betts and Pitt [66], 15 years after the first demonstration of GaAs by MBE [85,86], signifying the importance of LN as an opto-electronic material. Since Nb has a low vapor pressure and a high melting point of ~2400 °C, electron-beam evaporators were employed to produce elemental Nb and Li flux using heated sapphire or LN substrates. Upon optimization of flux values and substrate temperature, Betts and Pitt [66] were able to grow single crystal, c-axis LN films on LN substrates, but only the Li-rich phase (Li_3NbO_4) of LN grew epitaxially on sapphire substrates. The films also had a high optical propagation loss of ~15.7 dB/cm. As per the benefits of MBE, Betts and Pitt [66] characterized their growths in situ using a modified RHEED system [67]. While Sitar et al. [69] also employed RHEED, they recognized that finer control over the growth flux was necessary to grow high-quality LN films. Thus, Sitar et al. [69] interfaced a quadrupole mass spectrometer and the e-beam evaporator's emission current with a PID controller, granting finer control of the Nb flux to >1% accuracy averaged over 10 s. In addition, an elemental Li flux was instead provided by a low temperature effusion cell. Sitar et al. [69] also incorporated an oxygen plasma source. With finer control over the metal fluxes and a more reactive oxygen species, Sitar et al. [69] were able to achieve the homo- and heteroepitaxial growth of c-axis-oriented LN and $LiTaO_3$, including bilayers to form their superlattices. However, Sitar et al. [69] reported in-plane rotational domains. Matsubara et al. [70] combined the benefits of MBE with PLD to grow c-axis LN films via laser MBE, enjoying the high-vacuum growth seen in MBE, which enables fine control over the growth process and in situ characterization tools, such as RHEED, and the convenient transfer of target composition seen in PLD, avoiding the less controllable process of e-beam evaporation. Laser MBE enabled Matsubara et al. [70] to grow smooth LN thin films with an XRD rocking curve FWHM of only ~13 arcseconds. Unfortunately, Matsubara et al. [70] also reported rotation twins in their films. In 2011, Dabirian et al. [73] used a high vacuum process with chemical precursors to grow single-phase, c-axis LN thin films on sapphire. Dabirian et al. [73] evaporated $Li(OBu^t)$ and $Nb(OEt)_4$ precursors into prechambers and used Knudsen effusion sources to deliver carrier-free, collision-free, and Li-rich fluxes to the heated Al_2O_3 substrate. Dabirian et al. [73] were able to grow high-quality LN films with rocking curve FWHM values as low as 104 arcseconds. Recognizing the need for a more stable source of Nb, Doolittle et al. [71] investigated the use of $NbCl_5$ in near room temperature effusion cells, effectively replacing e-beam evaporation. By 2012, the metal-chloride growth chemistry was established as an effective means of supplying low vapor pressure metals for growth fluxes [72,74–76]. Finally, in 2016, Tellekamp et al. [77] were able to grow high-quality, c-axis LN films on Al_2O_3 and $LiTaO_3$ substrates using Li and $NbCl_5$ as solid sources. As shown in Figure 1, the heteroepitaxial growth of LN on Al_2O_3 shows XRD rocking curve FWHM values of 8.6 arcseconds, approaching the

5.6 arcsecond peaks measured from the substrate. On the lattice matched substrate of $LiTaO_3$, Tellekamp et al. [77] reported an FWHM of 193 arcseconds, closely matching the FWHM value of the substrate at 194 arcseconds. In all the films, rotational domains were not detectable [77]. This represents the best epitaxial LN grown to-date, with film quality that is comparable to bulk crystals, but the deposition rates achieved, ~75 nm/h, were limited by the ~1000 °C substrate temperature, making the thin film applications for this process limited by the practically obtainable thicknesses.

8. Conclusions

The inception of thin film epitaxial LN, the challenges faced along the way, and the results achieved by researchers were discussed, and are summarized in Table 1. Thin film LN allows for tighter wave confinement, more efficient and lower power devices, the integration of optical and acoustic systems of different scales, and even the potential for novel devices. As epitaxial techniques developed and more experiments were conducted, the phase purity, crystal quality, and optical propagation metrics improved and will likely continue to do so. Although the commercial market is currently reliant on smart-cut LN from a LN bulk crystal, the stringent thickness range of smart-cut LN obstructs some applications, and researchers have shown that epitaxial thin films of LN remain desirable. The crystal quality of state-of-the-art epitaxially grown LN has been able to match Czochralski pulled crystals using chlorinated Nb precursors and MBE. However, no present epitaxial technique has yet achieved both the film quality needed and the deposition rates (i.e., sufficient thickness in reasonable times) that make optical and acoustic devices via epitaxy practical. Although more work needs to be done on growing high-quality LN films with suitable thicknesses for system applications, the recent MBE results suggest that if specialized tooling is developed that enables higher deposition rates, the future of LN epitaxy could be promising.

Table 1. Comparison of optical loss and X-ray diffraction rocking curve full width at half maximum (FWHM) metrics for different epitaxial techniques as reported. Unless specified otherwise, optical losses are measured in the TE_0 mode at wavelength λ = 632.8 nm.

Epitaxial Technique	Optical Loss (dB/cm)	XRD FWHM (Arcseconds)	Notes	Reference
LPE	5	-	20–30 dB/cm losses at visible wavelengths.	Kondo et al. [20]
	<5	-	-	Ballman et al. [18]
	1.6	11	Significant vanadium impurities. λ = 514.5 nm.	Tamada et al. [22]
	<1	11.4	λ = 458 nm.	Yamada et al. [23]
Sputter	9	-	-	Takada et al. [28]
	3	-	Amorphous film.	Hewig et al. [30]
	6.47	-	-	Shimizu et al. [36]
	1.2	648	Multi-step process.	Lansiaux et al. [39]
CVD	-	1368	Twinning detected.	Wernberg et al. [44]
	1.8	36	No twinning detected.	Feigelson et al. [46]
PLD	-	612	-	Shibata et al. [56]
	-	360	-	Aubert et al. [59]
	<1.5	165.6	-	Kilburger et al. [64]
MBE	15.7	-	Electron-beam evaporated sources.	Betts and Pitt [66]
	-	13	Twinning detected. Laser assisted.	Matsubara et al. [70]
	-	104	Metal–organic precursors.	Dabirian et al. [73]
	-	8.6	No twinning detected. Metal–halide sources.	Tellekamp et al. [77]

Author Contributions: Data curation, B.Z. and M.B.T.; writing—review and editing, B.Z., M.B.T. and W.A.D.; supervision, W.A.D.; project administration, W.A.D.; funding acquisition, W.A.D. All authors have read and agreed to the published version of the manuscript.

Funding: This research was funded by the Air Force Office of Scientific Research (AFOSR) under award no. FA9550-18-1-0024.

Institutional Review Board Statement: Not applicable.

Informed Consent Statement: Not applicable.

Acknowledgments: Ali Sayir is the manager of the AFOSR MURI program.

Conflicts of Interest: The authors declare no conflict of interest.

References

1. Weis, R.S.; Gaylord, T.K. Lithium niobate: Summary of physical properties and crystal structure. *Appl. Phys. A* **1985**, *37*, 191–203. [CrossRef]
2. Sugii, K.; Fukuma, M.; Iwasaki, H. A study on titanium diffusion into $LiNbO_3$ waveguides by electron probe analysis and X-ray diffraction methods. *J. Mater. Sci.* **1978**, *13*, 523–533. [CrossRef]
3. Armenise, M.; Canali, C.; De Sario, M.; Carnera, A.; Mazzoldi, P.; Celotti, G. Ti Compound Formation During Ti Diffusion in $LiNbO_3$. *IEEE Trans. Compon. Hybrids Manuf. Technol.* **1982**, *5*, 212–216. [CrossRef]
4. Griffiths, G.; Esdaile, R. Analysis of titanium diffused planar optical waveguides in lithium niobate. *IEEE J. Quantum Electron.* **1984**, *20*, 149–159. [CrossRef]
5. Hall, D.G.; Hutcheson, L.D. Loss Mechanisms in Planar and Channel Waveguides. In *Integrated Optical Circuits and Components: Design and Applications*; Marcel Dekker: New York, NY, USA, 1987.
6. White, I.A.; Hutcheson, L.D.; Burke, J.J. Modal Fields and Curvature Losses in Ti-diffused $LiNbO_3$ Waveguides. *Guided Wave Opt. Surf. Acoust. Wave Devices Syst. Appl.* **1981**, *239*, 74–80. [CrossRef]
7. Bruel, M. Silicon on insulator material technology. *Electron. Lett.* **1995**, *31*, 1201. [CrossRef]
8. Thin Films—Partow Technologies. Available online: http://www.partow-tech.com/thinfilms/ (accessed on 5 January 2021).
9. Poberaj, G.; Hu, H.; Sohler, W.; Gunter, P. Lithium niobate on insulator (LNOI) for micro-photonic devices. *Laser Photon Rev.* **2012**, *6*, 488–503. [CrossRef]
10. Honardoost, A.; Abdelsalam, K.; Fathpour, S. Rejuvenating a Versatile Photonic Material: Thin-Film Lithium Niobate. *Laser Photon Rev.* **2020**, *14*, 2000088. [CrossRef]
11. Zhang, M.; Wang, C.; Cheng, R.; Shams-Ansari, A.; Lončar, M. Monolithic ultra-high-Q lithium niobate microring resonator. *Optica* **2017**, *4*, 1536–1537. [CrossRef]
12. Zhao, L.; Shi, L.; Wang, J.; Yan, J.; Chen, Y.; Zheng, Y. Effect of doping Mg on the structure and optical properties of $LiNbO_3$ films prepared by radio-frequency magnetron sputtering. *Mater. Sci. Semicond. Process.* **2020**, *108*, 104901. [CrossRef]
13. Li, W.; Cui, J.; Wang, W.; Zheng, D.; Jia, L.; Saeed, S.; Liu, H.; Rupp, R.; Kong, Y.; Xu, J. P-Type Lithium Niobate Thin Films Fabricated by Nitrogen-Doping. *Materials* **2019**, *12*, 819. [CrossRef]
14. Sumets, M.; Ovchinnikov, O.; Ievlev, V.; Kostyuchenko, A. Optical band gap shift in thin $LiNbO_3$ films grown by radio-frequency magnetron sputtering. *Ceram. Int.* **2017**, *43*, 13565–13568. [CrossRef]
15. Ballato, J.; Hawkins, T.; Foy, P.; Yazgan-Kokuoz, B.; McMillen, C.; Burka, L.; Morris, S.; Stolen, R.; Rice, R. Advancements in semiconductor core optical fiber. *Opt. Fiber Technol.* **2010**, *16*, 399–408. [CrossRef]
16. Dauter, Z.; Jaskólski, M. Crystal pathologies in macromolecular crystallography. *Postępy Biochem.* **2017**, *62*, 401–407.
17. Ballman, A.; Brown, H.; Tien, P.; Riva-Sanseverino, S. The growth of solid solution $LiNbO_3$—$LiTaO_3$ thin films for optical waveguides. *J. Cryst. Growth* **1975**, *30*, 37–41. [CrossRef]
18. Ballman, A.; Brown, H.; Tien, P.; Riva-Sanseverino, S. The growth of $LiNbO_3$ thin films by liquid phase epitaxial techniques. *J. Cryst. Growth* **1975**, *29*, 289–295. [CrossRef]
19. Miyazawa, S.; Fushimi, S.; Kondo, S. Optical waveguide of $LiNbO_3$ thin film grown by liquid phase epitaxy. *Appl. Phys. Lett.* **1975**, *26*, 8–10. [CrossRef]
20. Kondo, S.; Miyazawa, S.; Fushimi, S.; Sugii, K. Liquid-phase-epitaxial growth of single-crystal $LiNbO_3$ thin film. *Appl. Phys. Lett.* **1975**, *26*, 489–491. [CrossRef]
21. Baudrant, A.; Vial, H.; Daval, J. Liquid phase epitaxial growth of $LiNbO_3$ thin films. *J. Cryst. Growth* **1978**, *43*, 197–203. [CrossRef]
22. Tamada, H.; Yamada, A.; Saitoh, M. $LiNbO_3$ thin-film optical waveguide grown by liquid phase epitaxy and its application to second-harmonic generation. *J. Appl. Phys.* **1991**, *70*, 2536–2541. [CrossRef]
23. Yamada, A.; Tamada, H.; Saitoh, M. $LiNbO_3$ thin-film optical waveguide grown by liquid phase epitaxy using Li_2O-B_2O_3 flux. *Appl. Phys. Lett.* **1992**, *61*, 2848–2850. [CrossRef]
24. Lu, Y.; Johnston, B.; Dekker, P.; Dawes, J.M. Second Harmonic Generation in Lithium Niobate Planar Waveguides Grown by Liquid Phase Epitaxy. In Proceedings of the Conference on Lasers and Electro-Optics 2009, Baltimore, MD, USA, 31 May–5 June 2009; The Optical Society: Washington, DC, USA, 2009; pp. 1–2.

25. Lu, Y.; Dekker, P.; Dawes, J.M. Growth and characterization of lithium niobate planar waveguides by liquid phase epitaxy. *J. Cryst. Growth* **2009**, *311*, 1441–1445. [CrossRef]
26. Lu, Y.; Johnston, B.; Dekker, P.; Withford, M.J.; Dawes, J.M. Channel Waveguides in Lithium Niobate and Lithium Tantalate. *Molecules* **2020**, *25*, 3925. [CrossRef]
27. Fukunishi, S.; Kawana, S.; Uchida, N. No Title. In Proceedings of the 9th International Congress on Crystallography, Kyoto, Japan, 26 August–7 September 1972.
28. Takada, S.; Ohnishi, M.; Hayakawa, H.; Mikoshiba, N. Optical waveguides of single-crystal $LiNbO_3$ film deposited by rf sputtering. *Appl. Phys. Lett.* **1974**, *24*, 490–492. [CrossRef]
29. Jain, K.; Hewig, G. Phase-matched second-harmonic generation at 440 nm in a $LiNbO_3$ -$Gd_3Ga_5O_{12}$ waveguide with a tunable stimulated Raman source. *Opt. Commun.* **1981**, *36*, 483–486. [CrossRef]
30. Hewig, G.; Jain, K.; Sequeda, F.; Tom, R.; Wang, P.-W. R.F. Sputtering of $LiNbO_3$ thin films. *Thin Solid Film.* **1982**, *88*, 67–74. [CrossRef]
31. Meek, P.; Holland, L.; Townsend, P. Sputter deposition of $LiNbO_3$ films. *Thin Solid Film.* **1986**, *141*, 251–259. [CrossRef]
32. Kanata, T.; Kobayashi, Y.; Kubota, K. Epitaxial growth of $LiNbO_3$-$LiTaO_3$ thin films on Al_2O_3. *J. Appl. Phys.* **1987**, *62*, 2989–2993. [CrossRef]
33. Rost, T.A.; Baumann, R.C.; Stone, B.A.; Rabson, T.A. Physical characterization of RF sputtered lithium niobate films. In Proceedings of the 1990 IEEE 7th International Symposium on Applications of Ferroelectrics, Urbana-Champaign, IL, USA, 6–8 June 1990; IEEE: Piscataway, NJ, USA, 2002; pp. 125–128.
34. Fujimura, N.; Ito, T.; Kakinoki, M. Heteroepitaxy of $LiNbO_3$ and $LiNb_3O_8$ thin films on C-cut sapphire. *J. Cryst. Growth* **1991**, *115*, 821–825. [CrossRef]
35. Rost, T.A.; Lin, H.; Rabson, T.A.; Baumann, R.C.; Callahan, D.L. Deposition and analysis of lithium niobate and other lithium niobium oxides by rf magnetron sputtering. *J. Appl. Phys.* **1992**, *72*, 4336–4343. [CrossRef]
36. Shimizu, M.; Furushima, Y.; Nishida, T.; Shiosaki, T. Preparation and Optical Waveguide Properties of $LiNbO_3$ Thin Films by RF Magnetron Sputtering. *Jpn. J. Appl. Phys.* **1993**, *32*, 4111–4114. [CrossRef]
37. Kingston, J.J.; Fork, D.K.; Leplingard, F.; Ponce, F.A. C^- Outgrowths in C^+ Thin Films of $LiNbO_3$ on Al_2O_3-c. *MRS Online Proc. Libr. Arch.* **1994**, *341*, 289–294. [CrossRef]
38. Dogheche, E.-H.; Lansiaux, X.; Rémiens, D. Growth and optical waveguiding properties of rf sputtered lithium niobate thin films on sapphire substrates. *Integr. Ferroelectr.* **1999**, *25*, 47–59. [CrossRef]
39. Lansiaux, X.; Dogheche, E.; Remiens, D.; Guilloux-Viry, M.; Perrin, A.; Ruterana, P. $LiNbO_3$ thick films grown on sapphire by using a multistep sputtering process. *J. Appl. Phys.* **2001**, *90*, 5274–5277. [CrossRef]
40. Sumets, M.; Dybov, V.; Serikov, D.; Belonogov, E.; Seredin, P.; Goloshchapov, D.; Grebennikov, A.; Ievlev, V. Effect of reactive gas composition on properties of Si/$LiNbO_3$ heterojunctions grown by radio-frequency magnetron sputtering. *J. Sci. Adv. Mater. Devices* **2020**, *5*, 512–519. [CrossRef]
41. Cowher, M.E.; Sedgwick, T.O.; Landermann, J. Epitaxial garnet films by organometallic chemical vapor deposition. *J. Electron. Mater.* **1974**, *3*, 621–633. [CrossRef]
42. Curtis, B.; Brunner, H. The growth of thin films of lithium niobate by chemical vapour de position. *Mater. Res. Bull.* **1975**, *10*, 515–520. [CrossRef]
43. Wernberg, A.A.; Gysling, H.J.; Filo, A.J.; Blanton, T.N. Epitaxial growth of lithium niobate thin films from a single-source organometallic precursor using metalorganic chemical vapor deposition. *Appl. Phys. Lett.* **1993**, *62*, 946–948. [CrossRef]
44. Wernberg, A.A.; Gysling, H.J. MOCVD deposition of epitaxial lithium niobate, $LiNbO_3$, thin films using the single source precursor lithium niobium ethoxide, $LiNb(OEt)_6$. *Chem. Mater.* **1993**, *5*, 1056–1058. [CrossRef]
45. Lu, Z.; Hiskes, R.; DiCarolis, S.; Route, R.; Feigelson, R.; Leplingard, F.; Fouquet, J. Epitaxial $LiNbO_3$ thin films on sapphire substrates grown by solid source MOCVD. *J. Mater. Res.* **1994**, *9*, 2258–2263. [CrossRef]
46. Feigelson, R. Epitaxial growth of lithium niobate thin films by the solid source MOCVD method. *J. Cryst. Growth* **1996**, *166*, 1–16. [CrossRef]
47. Lee, S.; Feigelson, R. Reduced optical losses in MOCVD grown lithium niobate thin films on sapphire by controlling nucleation density. *J. Cryst. Growth* **1998**, *186*, 594–606. [CrossRef]
48. Saulys, D.; Joshkin, V.; Khoudiakov, M.; Kuech, T.; Ellis, A.; Oktyabrsky, S.; McCaughan, L. An examination of the surface decomposition chemistry of lithium niobate precursors under high vacuum conditions. *J. Cryst. Growth* **2000**, *217*, 287–301. [CrossRef]
49. Akiyama, Y.; Shitanaka, K.; Murakami, H.; Shin, Y.-S.; Yoshida, M.; Imaishi, N. Epitaxial growth of lithium niobate film using metalorganic chemical vapor deposition. *Thin Solid Film.* **2007**, *515*, 4975–4979. [CrossRef]
50. Margueron, S.; Bartasyte, A.; Plausinaitiene, V.; Abrutis, A.; Boulet, P.; Kubilius, V.; Saltyte, Z. Effect of deposition conditions on the stoichiometry and structural properties of $LiNbO_3$ thin films deposited by MOCVD. In *Oxide-based Materials and Devices IV*; International Society for Optics and Photonics: Bellingham, WA, USA, 2013; Volume 8626, p. 862612.
51. Ogale, S.B.; Nawathey-Dikshit, R.; Dikshit, S.J.; Kanetkar, S.M. Pulsed laser deposition of stoichiometric $LiNbO_3$ thin films by using O2 and Ar gas mixtures as ambients. *J. Appl. Phys.* **1992**, *71*, 5718–5720. [CrossRef]
52. Shibata, Y.; Kaya, K.; Akashi, K.; Kanai, M.; Kawai, T.; Kawai, S. Epitaxial growth of $LiNbO_3$ thin films by excimer laser ablation method and their surface acoustic wave properties. *Appl. Phys. Lett.* **1992**, *61*, 1000–1002. [CrossRef]

53. Fork, D.K.; Anderson, G.B. Epitaxial MgO on GaAs(111) as a buffer layer for z-cut epitaxial lithium niobate. *Appl. Phys. Lett.* **1993**, *63*, 1029–1031. [CrossRef]
54. Marsh, A.M.; Harkness, S.D.; Qian, F.; Singh, R.K. Pulsed laser deposition of high quality $LiNbO_3$ films on sapphire substrates. *Appl. Phys. Lett.* **1993**, *62*, 952–954. [CrossRef]
55. Aubert, P.; Garry, G.; Bisaro, R.; López, J.G. Structural properties of $LiNbO_3$ thin films grown by the pulsed laser deposition technique. *Appl. Surf. Sci.* **1995**, *86*, 144–148. [CrossRef]
56. Shibata, Y.; Kaya, K.; Akashi, K.; Kanai, M.; Kawai, T.; Kawai, S. Epitaxial growth and surface acoustic wave properties of lithium niobate films grown by pulsed laser deposition. *J. Appl. Phys.* **1995**, *77*, 1498–1503. [CrossRef]
57. Afonso, C.N.; Gonzalo, J.; Vega, F.; Dieguez, E.; Cheang Wong, J.C.; Ortega, C.; Siejka, J.; Amsel, G. Correlation between optical properties, composition, and deposition parameters in pulsed laser deposited $LiNbO_3$ films. *Appl. Phys. Lett.* **1995**, *66*, 1452–1454. [CrossRef]
58. Lee, S.-H.; Song, T.K.; Noh, T.W.; Lee, J.-H. Low-temperature growth of epitaxial $LiNbO_3$ films on sapphire (0001) substrates using pulsed laser deposition. *Appl. Phys. Lett.* **1995**, *67*, 43–45. [CrossRef]
59. Aubert, P.; Garry, G.; Bisaro, R.; Olivier, J.; López, J.G.; Urlacher, C. Epitaxial growth of $LiNbO_3$ thin films on (001) sapphire by Pulsed Laser Deposition. *Microelectron. Eng.* **1995**, *29*, 107–110. [CrossRef]
60. Veignant, F.; Gandais, M.; Aubert, P.; Garry, G. Structural evolution of lithium niobate deposited on sapphire (0001): From early islands to continuous films. *J. Cryst. Growth* **1999**, *196*, 141–150. [CrossRef]
61. Kakehi, Y.; Okamoto, A.; Sakurai, Y.; Nishikawa, Y.; Yotsuya, T.; Ogawa, S. Epitaxial growth of $LiNbO_3$ thin films using pulsed laser deposition. *Appl. Surf. Sci.* **2001**, *169–170*, 560–563. [CrossRef]
62. Son, J.-W.; Orlov, S.S.; Phillips, B.; Hesselink, L. Pulsed laser deposition of single phase $LiNbO_3$ thin film waveguides. *J. Electroceram.* **2006**, *17*, 591–595. [CrossRef]
63. Kilburger, S.; Chety, R.; Millon, E.; Di Bin, P.; Di Bin, C.; Boulle, A.; Guinebretière, R. Growth of $LiNbO_3$ thin films on sapphire by pulsed-laser deposition for electro-optic modulators. *Appl. Surf. Sci.* **2007**, *253*, 8263–8267. [CrossRef]
64. Kilburger, S.; Millon, E.; Di Bin, P.; Boulle, A.; Guinebretière, R.; Di Bin, C. Properties of $LiNbO_3$ based heterostructures grown by pulsed-laser deposition for optical waveguiding application. *Thin Solid Film.* **2010**, *518*, 4654–4657. [CrossRef]
65. Paldi, R.L.; Qi, Z.; Misra, S.; Lu, J.; Sun, X.; Phuah, X.L.; Kalaswad, M.; Bischoff, J.; Branch, D.W.; Siddiqui, A.; et al. Nanocomposite-Seeded Epitaxial Growth of Single-Domain Lithium Niobate Thin Films for Surface Acoustic Wave Devices. *Adv. Photon Res.* **2021**, 2000149. [CrossRef]
66. Betts, R.; Pitt, C. Growth of thin-film lithium niobate by molecular beam epitaxy. *Electron. Lett.* **1985**, *21*, 960–962. [CrossRef]
67. Petrucci, M.; Pitt, C. Reflection high-energy electron diffraction from lithium niobate single-crystal surfaces for application in molecular beam epitaxy. *Electron. Lett.* **1986**, *22*, 637. [CrossRef]
68. Petrucci, M.; Pitt, C.W.; Reynolds, S.R.; Milledge, H.J.; Mendelssohn, M.J.; Dineen, C.; Freeman, W.G. Growth of thin-film niobium and niobium oxide layers by molecular-beam epitaxy. *J. Appl. Phys.* **1988**, *63*, 900–909. [CrossRef]
69. Sitar, Z.; Gitmans, F.; Liu, W.; Günter, P. Homo and Heteroepitaxial Growth of $LiTaO_3$ and $LiNbO_3$ by Mbe. *MRS Online Proc. Libr. Arch.* **1995**, *401*, 255–260. [CrossRef]
70. Matsubara, K.; Niki, S.; Watanabe, M.; Fons, P.; Iwata, K.; Yamada, A. Growth of $LiNbO_3$ epitaxial films by oxygen radical-assisted laser molecular beam epitaxy. *Appl. Phys. A* **1999**, *69*, S679–S681. [CrossRef]
71. Doolittle, W.A.; Carver, A.G.; Henderson, W. Molecular beam epitaxy of complex metal-oxides: Where have we come, where are we going, and how are we going to get there? *J. Vac. Sci. Technol. B Microelectron. Nanometer Struct. Process. Meas. Phenom.* **2005**, *23*, 1272. [CrossRef]
72. Doolittle, W.; Carver, A.; Henderson, W.; Calley, W. Molecular Beam Epitaxy of Lithium Niobate Multifunctional Materials Using a Chloride Refractory Metal Chemistry. *ECS Trans.* **2006**, *2*, 103–114. [CrossRef]
73. Dabirian, A.; Harada, S.; Kuzminykh, Y.; Sandu, S.C.; Wagner, E.; Benvenuti, G.; Brodard, P.; Rushworth, S.; Muralt, P.; Hoffmann, P. Combinatorial Chemical Beam Epitaxy of Lithium Niobate Thin Films on Sapphire. *J. Electrochem. Soc.* **2011**, *158*, D72–D76. [CrossRef]
74. Henderson, W.E.; Calley, W.L.; Carver, A.G.; Chen, H.; Doolittle, W.A. A versatile metal-halide vapor chemistry for the epitaxial growth of metallic, insulating and semiconducting films. *J. Cryst. Growth* **2011**, *324*, 134–141. [CrossRef]
75. Greenlee, J.D.; Calley, W.L.; Henderson, W.; Doolittle, W.A. Halide based MBE of crystalline metals and oxides. *Phys. Status Solidi C* **2011**, *9*, 155–160. [CrossRef]
76. Tellekamp, M.B.; Greenlee, J.D.; Shank, J.C.; Doolittle, W.A. Molecular beam epitaxy growth of niobium oxides by solid/liquid state oxygen source and lithium assisted metal-halide chemistry. *J. Cryst. Growth* **2015**, *425*, 225–229. [CrossRef]
77. Tellekamp, M.B.; Shank, J.C.; Goorsky, M.S.; Doolittle, W.A. Molecular Beam Epitaxy Growth of High Crystalline Quality $LiNbO_3$. *J. Electron. Mater.* **2016**, *45*, 6292–6299. [CrossRef]
78. Volk, T.; Wöhlecke, M. *Lithium Niobate: Defects, Photorefraction and Ferroelectric Switching*; Springer: Berlin/Heidelberg, Germany, 2008; Volume 115.
79. Liu, D.; Zha, G.; Hu, L.; Jiang, W. Recovery of Gallium and Arsenic from Gallium Arsenide Semiconductor Scraps. In *TMS Annual Meeting & Exhibition*; Springer: Berlin/Heidelberg, Germany, 2018; pp. 319–330.
80. Tellekamp, M.B.; Shank, J.C.; Doolittle, W.A. Molecular Beam Epitaxy of lithium niobium oxide multifunctional materials. *J. Cryst. Growth* **2017**, *463*, 156–161. [CrossRef]

81. Zivasatienraj, B.; Tellekamp, M.B.; Weidenbach, A.S.; Ghosh, A.; McCrone, T.M.; Doolittle, W.A. Temporal versatility from intercalation-based neuromorphic devices exhibiting 150 mV non-volatile operation. *J. Appl. Phys.* **2020**, *127*, 084501. [CrossRef]
82. Miyazawa, S. Growth of $LiNbO_3$single-crystal film for optical waveguides. *Appl. Phys. Lett.* **1973**, *23*, 198–200. [CrossRef]
83. Bartasyte, A.; Plausinaitiene, V.; Abrutis, A.; Stanionyte, S.; Margueron, S.; Boulet, P.; Kobata, T.; Uesu, Y.; Gleize, J. Identification of $LiNbO_3$, $LiNb_3O_8$ and Li_3NbO_4 phases in thin films synthesized with different deposition techniques by means of XRD and Raman spectroscopy. *J. Phys. Condens. Matter* **2013**, *25*, 205901. [CrossRef]
84. Astié, V.; Oliveri, S.; Millon, C.; Margueron, S.; Rachetti, M.; Decams, J.; Boulet, P.; Bartasyte, A. Towards stoichiometric $LiNbO_3$ epitaxial thin films grown by DLI-MOCVD. In Proceedings of the GDR Oxydes Fonctionnels—des Matériaux Aux Dispositifs, Piriac sur Mer, France, 20–23 March 2018.
85. Cho, A.Y. Morphology of Epitaxial Growth of GaAs by a Molecular Beam Method: The Observation of Surface Structures. *J. Appl. Phys.* **1970**, *41*, 2780–2786. [CrossRef]
86. Cho, A.Y.; Panish, M.B.; Hayashi, I. Molecular beam epitaxy of GaAs, $Al_xGa_{1-x}As$ and GaP. *Proc. Symp. GaAs Relat. Compd.* **1970**, *2*, 18–29.

crystals

MDPI

Article

Small-Polaron Hopping and Low-Temperature (45–225 K) Photo-Induced Transient Absorption in Magnesium-Doped Lithium Niobate †

Simon Messerschmidt [1,‡], Andreas Krampf [1,‡], Laura Vittadello [1], Mirco Imlau [1,*], Tobias Nörenberg [2], Lukas M. Eng [2,3] and David Emin [4]

1 School of Physics, Osnabrueck University, Barbarastraße 7, 49076 Osnabrueck, Germany; smesserschmi@uni-osnabrueck.de (S.M.); andreas.krampf@uni-osnabrueck.de (A.K.); laura.vittadello@uni-osnabrueck.de (L.V.)

2 Institut für Angewandte Physik, TU Dresden, Nöthnitzerstr. 61, 01187 Dresden, Germany; tobias.noerenberg@tu-dresden.de (T.N.); lukas.eng@tu-dresden.de (L.M.E.)

3 ct.qmat: Dresden-Würzburg Cluster of Excellence—EXC 2147, Technische Universität Dresden, 01069 Dresden, Germany

4 Department of Physics and Astronomy, University of New Mexico, Albuquerque, NM 87131, USA; emin@unm.edu

* Correspondence: mirco.imlau@uni-osnabrueck.de

† Dedicated to the memory of Ortwin F. Schirmer (1937–2020).

‡ These authors contributed equally to this work.

Received: 20 August 2020; Accepted: 11 September 2020; Published: 14 September 2020

check for updates

Abstract: A strongly temperature-dependent photo-induced transient absorption is measured in 6.5 mol% magnesium-doped lithium niobate at temperatures ranging from 45 K to 225 K. This phenomenon is interpreted as resulting from the generation and subsequent recombination of oppositely charged small polarons. Initial two-photon absorptions generate separated oppositely charged small polarons. The existence of these small polarons is monitored by the presence of their characteristic absorption. The strongly temperature-dependent decay of this absorption occurs as series of thermally assisted hops of small polarons that facilitate their merger and ultimate recombination. Our measurements span the high-temperature regime, where small-polaron jump rates are Arrhenius and strongly dependent on temperature, and the intermediate-temperature regime, where small-polaron jump rates are non-Arrhenius and weakly dependent on temperature. Distinctively, this model provides a good representation of our data with reasonable values of its two parameters: Arrhenius small-polaron hopping's activation energy and the material's characteristic phonon frequency.

Keywords: lithium niobate; small polaron hopping; transient absorption

1. Introduction

This paper is in memoriam of Ortwin Schirmer who throughout his long career pioneered the study of lithium niobate. Lithium niobate ($LiNbO_3$, LN) is widely employed in electro-, acousto- and nonlinear optical applications. In addition, since free and trapped charge carriers of both positive and negative signs are known to form small polarons in LN, this material serves as an elegant and robust model system to study fundamental physical properties [1].

An electronic charge carrier becomes self-trapped when it is bound within the potential well produced by its displacements of the equilibrium positions of the atoms that surround it. The composite quasiparticle comprising a self-trapped electronic charge carrier taken together with the displaced atomic equilibrium positions is termed polaron [1]. This name polaron was adopted in recognition of their prevalence in polar (i.e., ionic) materials.

Lithium niobate is a perovskite-like oxide ferroelectric with especially displaceable ions [2,3]. The dependence of the energy of an electronic carrier on displacements of surrounding atoms produces both long-range and short-range electron–phonon interactions. The long-range electron–phonon interaction results from the Coulomb energy of an electronic carrier depending on displacements of distant ions. It is proportional to $(1/\epsilon_\infty - 1/\epsilon_0)$, where ϵ_0 and ϵ_∞ respectively designate a material's static and high-frequency constant. Values of this parameter in LN ($\epsilon_0 = 30 - 80$, $\epsilon_\infty = 5.5$ [4]) are comparable to those of other ionic materials (e.g., alkali halides) whose charge carriers form polarons. Short-range electron–phonon interactions result from the energies and covalency of bonds between neighboring atoms changing with their separations. Short-range electron–lattice interactions as well as disorder foster the collapse of self-trapped electronic carriers to single sites thereby forming small polarons [1,5].

In addition to free small-polarons in LN [6], defects are sites for defect-related electron and hole polarons as well as bipolarons [7,8]. Small-polarons govern many of the linear and non-linear optical properties of LN [9–11] that include the bulk photovoltaic effect [12], green-induced infrared absorption (GRIIRA), optical damage and photoconductivity (PC) [8,11,13,14]. The transient broad-band small-polaron absorptions and index changes in LN [9,10] provide a clear link between small-polaron absorption and their hopping motion. For instance, the stretched-exponential relaxation of the transient absorption is attributed to small-polarons' hopping [15–17].

Small polarons exhibit distinctive optical and electronic transport properties [1]. In particular, phonon-broadened absorption bands result from photon-assisted inter-site transfers of small-polarons' self-trapped electronic carriers. In addition, small polarons generally move with extremely low mobilities via phonon-assisted hopping. A small-polaron absorption band disappears as its small polarons are eliminated by their recombination with oppositely charged carriers. Recombination is facilitated by both photon-assisted and phonon-assisted hopping. In the former case, increasing the intensity of absorption within a small-polaron band fosters its elimination. In this manner, small-polaron absorptions are bleached. In the latter case, increasing temperature fosters disappearance of small-polaron absorptions.

Essential features of a small-polaron's phonon-assisted hopping are well established [1]. Self-trapped electronic carriers usually move fast enough to adiabatically adjust to the relatively slow vibrations of atoms. Adiabatic small-polaron hopping has been addressed at high enough temperatures for atoms' vibrations to be classical. The small-polaron jump rate is then Arrhenius both for uncorrelated hops [18,19] and when the slow transfer of vibrational energy between atoms causes hops to occur in flurries with enhanced pre-exponential factors [1,20,21]. Nonetheless, for computational simplicity, almost all calculations of phonon-assisted hopping are performed in the non-adiabatic limit in which an electronic carrier's inter-site motion is assumed to be arbitrarily slow. In particular, the elemental non-adiabatic small-polaron jump rate has been calculated for its self-trapped carrier interacting with acoustic [22], optical [23] and both acoustic and optical phonons [24].

All studies of small-polaron hopping report three distinct temperature regimes [1]. At the highest temperatures, atoms' vibrations are classical with a self-trapped carrier moving between initial and final sites when its electronic energy at these sites become coincident with each other. The jump rate's primary temperature dependence is then Arrhenius and independent of phonon energies: $\propto \exp[-(4E_a + \Delta)^2/16E_a k_B T]$, where Δ denotes the energy difference between final and initial sites with E_a being the activation energy for equivalent sites, $\Delta = 0$. At the lowest temperatures, a self-trapped carrier hops when atoms quantum-mechanically tunnel between the configurations they assume when the self-trapped carrier occupies initial and final sites. The jump rate's primary temperature dependence is then also Arrhenius and independent of phonon energies: $\propto \exp[-(\Delta + |\Delta|)/2k_B T]$. Then a hop upward in energy, $\Delta > 0$, is Arrhenius, $\propto \exp(-\Delta/k_B T)$, while a hop downward in energy, $\Delta < 0$, is temperature independent. Distinctively, the small-polaron jump rate between these high-temperature and low-temperature limits is non-Arrhenius and dependent on phonon energies. The primary temperature dependence of the small polaron jump rate is $\propto \exp[-(4E_a/\hbar\omega)\tanh(\hbar\omega/4k_BT)]\exp(-\Delta/2k_BT)$ for $|\Delta| \ll 4E_a$, where $\hbar\omega$ represents the characteristic phonon energy. Thus, measuring the temperature dependence in the high-temperature and intermediate regimes provides an estimate of this characteristic phonon energy. The reasonableness of this estimate tests the small-polaron approach.

Sufficiently separated oppositely charged small polarons will experience a mutual Coulomb attraction [1]. Recombination between oppositely charged small polarons can occur once a series of small-polaron hops enables them to move close enough to one another. These hops will tend to be downward in energy, $\Delta < 0$, since small-polaron jump rates are proportional to $\exp(-\Delta/2k_BT)$. The rate-limiting hop, the slowest of this sequence, occurs at its beginning where $|\Delta|$ tends to be smallest. Since the oppositely charged small polarons generated by our experiment's pump will have a distribution of initial separations, their recombination will be characterized by a distribution of recombination times. Thus, as observed [15], the time-dependent relaxation of the pump-induced transient absorption will be a stretched exponential.

We utilize a two-photon pump to produce electron and hole small-polarons whose recombination is then monitored by their absorption's temporal decay. To generate free small-polarons, we study an LN crystal that was congruently grown from a melt with 6.5 % Mg since it contains a negligible concentration of antisite defects. Our optical excitation then primarily produces free Nb^{4+} n-type small polarons and O^{-}–V_{Li} sites, p-type polarons bound to lithium vacancies [25,26]. We thereby avoid the profusion of other types of polaron states that occur in materials such as Fe:LN [17,27]. The transient near infra-red absorption in LN is primarily due to small polarons [6,28]. Other optical phenomena are also associated with self-trapped charge carriers and excitons [29–33]

We explore the transient absorption from 10^{-7} to 10^{5} s for temperatures between 45 K and 225 K. A simple model enables us to relate the observed decay rate of the absorption to the temperature-dependent small-polaron jump rate. Analysis of the small-polaron jump rate through the observed temperature range yields plausible estimates of the activation energy for small-polaron hopping and distinctively, the characteristic atomic-vibration frequency.

2. Methods

The lithium niobate sample (thickness d = 2 mm, Mg-doping of 6.5 mol% in the melt, z-cut) was grown, cut and polished by Kovács et al. at the WIGNER Research Centre for Physics, Budapest. The crystal was mounted in a closed-cycle cryostat which operates between 40 K and room temperature.

Optical windows on the cryostat enable nanosecond-pump and continuous-wave-probe measurements for probing small-polarons' transient absorption (TA). Small polarons were generated in Mg:LN via two-photon absorption from an ordinarily polarized, frequency-doubled Nd:YAG pulse laser ($\lambda_p = 532\,\mathrm{nm}$, τ_{FWHM} = 8 ns, $I_p \approx 170\,\mathrm{MW/cm^2}$). Transmitted polarized light of a continuous wave probe laser ($\lambda = 785$ nm) was split and detected (1) by a Si-PIN photo diode connected to a fast digital storage oscilloscope and (2) by a photometer. Thus, the TA signal was recorded (1) in a time range from nanoseconds up to five seconds and (2) in a time range from one second up to several thousands of seconds, respectively. The power of the probe laser was kept below 1 mW to minimize its effect on the transient absorption. For this purpose, a fast electronic shutter was also introduced into the probe beam's path so that the transient absorption at long times was only affected intermittently. The transient absorption $\alpha_{li}(\lambda, t)$ is determined from the ratio of the intensity of the transmitted probe light after the pulse was applied $I(\lambda, t)$ to that prior to the pulse event $I(\lambda, t \leq 0)$, giving: $\alpha_{li}(\lambda, t) = -(1/d)\ln[I(\lambda, t)/I(\lambda, t \leq 0)]$.

3. Decay of Photo-Induced Absorption

Figure 1 shows the normalized temperature-dependent results of transient absorption measurements at the probe wavelength $\lambda = 785$ nm for Mg:LN. Obviously, these data are not well described as simple temporal decays, $\exp[-(t/\tau)]$. However, our data are well fitted with the two temperature-dependent parameters [$\tau(T)$ and $\beta(T)$] of Kohlrausch–Williams–Watts (KWW) stretched-exponentials, $\exp[-(t/\tau)^{\beta}]$. The red curve accompanying the lowest-temperature data shows such a fit. The red curve accompanying the highest-temperature data illustrates that the fitting result is only slightly improved upon employing six adjustable parameters of a sum of two KWW stretched exponentials, $\alpha_{\text{li},0}(1-f)\exp[-(t/\tau_1)^{\beta_1}] + \alpha_{\text{li},0} f \exp[-(t/\tau_2)^{\beta_2}]$. We note that we used the fit parameter for the starting amplitude $\alpha_{\text{li},0}$ to account for the elevation of the measured data above the value 1, that was caused by the noise in the experimental data. We therefore illustrate our results in terms of a single KWW stretched exponential with its two temperature dependent parameters, $\tau(T)$ and $\beta(T)$.

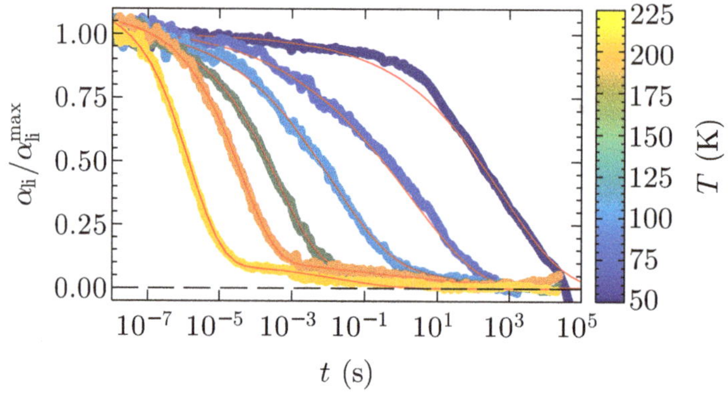

Figure 1. Normalized photo-absorptions at a wavelength of 785 nm of 6.5 mol% Mg:LN are plotted versus decay time for representative temperatures between 50 K and 225 K. The solid red curves show fits using a single Kohlrausch–Williams–Watts (KWW) stretched-exponential for $T < 100$ K and the sum of two KWW functions for $T > 100$ K, respectively.

KWW stretched-exponentials are phenomenological representations of processes with distributions of relaxation times. The mean relaxation time for a KWW stretched exponential is $\langle\tau(T)\rangle \equiv \tau(T)\Gamma\{[1/\beta(T)]+1\}$, where Γ denotes the mathematical gamma function. $\langle\tau(T)\rangle$ is found from the values of $\tau(T)$ and $\beta(T)$ obtained from stretched-exponential fits to our transient-absorption data.

The resulting values of $1/\langle\tau(T)\rangle$ plotted against reciprocal temperature in units of 1000/T are the experimental points shown in Figure 2. Distinctively, the photo-induced decay rate $1/\langle\tau(T)\rangle$ appears Arrhenius above 100 K and becomes non-Arrhenius below 100 K.

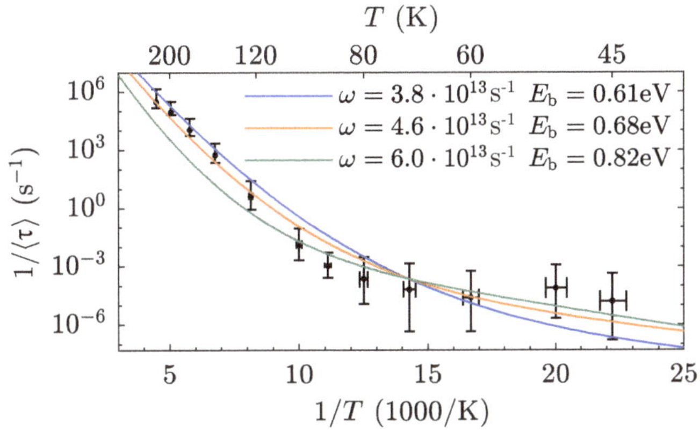

Figure 2. The temperature dependence of the inverse mean decay time of photo-induced absorption in Mg:LN is compared to the small-polaron hopping and recombination rate.

4. Small-Polaron Recombination and Phonon-Assisted Hopping

Well-separated oppositely charged small polarons attract each other through their mutual Coulomb attraction, $e^2/\epsilon_0 s$ when separated by distance s. Oppositely charged small polarons only recombine once they are close enough to one another to overlap significantly. Small polarons move toward recombination via a series of phonon-assisted hops. The rate for each phonon-assisted jump increases as the energy of its final state falls increasingly below that of its initial state except when this energy disparity is exceptionally large [1,22–24]. Thus, the recombination of well-separated oppositely charged small polarons tends to occur via a series of jumps in which each hop is progressively faster. The rate-limiting jump is the initial hop of this series, a jump with minimal energy disparity. The recombination rate of oppositely charged small polarons therefore is determined by the hop between sites of nearly equal energy.

A phonon-assisted hop occurs when an electronic carrier transfers between sites in response to atoms assuming favorable configurations [1]. A hop is termed non-adiabatic when the inter-site transfer energy of the electronic charge carrier is so small that it rarely avails itself of the opportunity to move [18,19]. Alternatively, adiabatic hopping occurs when the electronic carriers are able to adjust to changing atomic configurations. Small-polaron hopping is generally adiabatic [1]. Nonetheless, phonon-assisted jump rates are usually computed for the non-adiabatic limit, where they are perturbative to lowest-order in the hop's electronic transfer-energy.

The non-adiabatic small-polaron jump rate has been calculated exactly for the Holstein Molecular-Crystal-Model (MCM), where the electronic carrier is assumed to interact with a single vibrational mode of frequency ω, only [23]. To enable small-polaron formation in this idealized model, its small-polaron binding energy E_b is taken to exceed both the phonon energy $\hbar\omega$ and the inter-site electronic transfer energy J. An approximate formula for the adiabatic small-polaron jump rate is obtained at all but very low temperatures when dividing this non-adiabatic rate by the high-temperature probability of the electronic carrier transferring in response to an appropriate atomic configuration [1,19]. In particular, the approximate formula for the adiabatic rate for a hop between equivalent sites is found by dividing the right-hand-side of Equation (59) of Ref. [23] by $J^2/(\hbar\omega/2\pi)(2E_b k_B T/\pi)^{1/2}$:

$$R_{ad} = \sqrt{\frac{E_b k_B T}{\pi \hbar^2}} \exp\left[-\left(\frac{2E_b}{\hbar\omega}\right)\coth\left(\frac{\hbar\omega}{2k_B T}\right)\right]\left\{I_0\left[\left(\frac{2E_b}{\hbar\omega}\right)\operatorname{csch}\left(\frac{\hbar\omega}{2k_B T}\right)\right]-1\right\}. \quad (1)$$

where $I_0(x)$ denotes the zeroth-order modified Bessel function. At high and intermediate temperatures relative to $\hbar\omega$, this expression becomes

$$R_{ad} \cong \left(\frac{\omega}{2\pi}\right) \exp\left[\left(\frac{2E_b}{\hbar\omega}\right) \tanh\left(\frac{\hbar\omega}{4k_B T}\right)\right]. \tag{2}$$

Distinctively, the Arrhenius behavior with activation energy $E_b/2$ progressively disappears as $k_B T$ is lowered relative to $\hbar\omega/4$. In other words, as shown in Figure 3, the activation energy progressively shrinks from $E_b/2$ as the temperature is reduced.

Beyond the simple MCM, a small-polaron's phonon-assisted hop is primarily associated with movements of the atoms in the immediate vicinity of its self-trapped electronic carrier [1]. As such, a small-polaron hop generally involves interactions with short-wavelength phonons from both acoustic and optic vibration modes. The activation energy for an Arrhenius small-polaron jump is the sum of contributions from these modes. The pre-exponential factor of the adiabatic jump rate of Equation (2) is then the square-root of the sum of the squares of these zone edge phonons weighted by their relative contributions to the jump's activation energy [19].

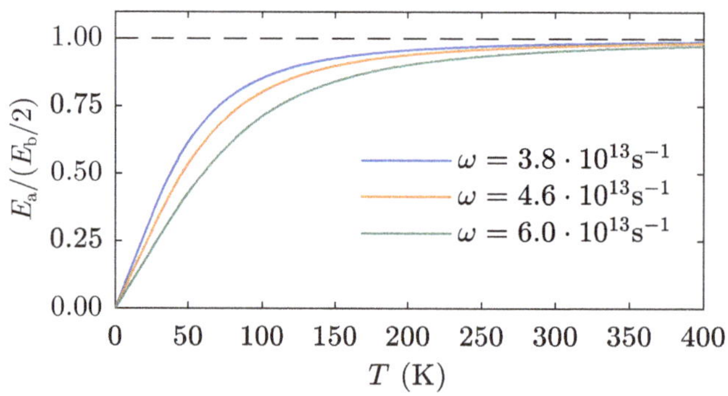

Figure 3. Normalized small-polaron hopping activation energy E_a over $E_b/2$ as a function of temperature. Note the different slopes and absolute values at different phonon frequencies, ω, when lowering the temperature.

5. Combining the Decay of Lithium Niobate's Induced Absorption with Small-Polaron Recombination

We ascribe the observed photo-induced absorption mainly to photo-generated Nb^{4+} n-type free small polarons. The decay of this photo-induced absorption is attributed to their recombination with p-type small polarons that are bound to lithium vacancies, $O^- - V_{Li}$. This attribution is supported by evidence of only negligible Fe and Nb_{Li} contamination in these crystals [17]. Indeed, room-temperature transient absorption measurements on a series of Mg-doped and nominally pure LN samples by Conradi et al. [26] also concluded that the primary signal at 785 nm stems from Nb^{4+} free small polarons. Furthermore, subsequent work only found a secondary relatively weak and slow decay that most likely results from the recombination of self-trapped excitons pinned at various defects [29,31]. As shown in Figure 1, a single KWW function provides a satisfactory fit to our low-temperature data.

Distinctively, we measure the decay of the photo-induced absorption in Mg:LN over a wide temperature range. As a result, we observe the shift from Arrhenius behavior at highest temperatures to a relatively weak temperature-dependence at lowest temperatures. This fundamental feature of small-polarons' hopping is sensitive to the phonon energies with which their self-trapped electrons interact [1,22–24]. As shown in Figure 2, the mean decay time of the photo-induced absorption is well fitted by our model for small-polaron recombination. Best fit's are obtained for $E_b = 0.68$ eV and $\omega = 4.6 \times 10^{13}\ s^{-1}$. Both of these values are typical of small-polaron hopping.

6. Discussion

Polarons are generally characterized by broad asymmetric absorption bands. These absorptions arise from exciting polarons' self-trapped electronic carriers from the potential wells that bind them [1,34]. As shown in Figures 5 and 6 of Ref. [34], small-polaron absorption bands tend to be higher on the low-frequency side of their peaks. By contrast, large-polaron absorptions tend to be higher on the high-frequency side of their peaks [34]. This distinction can be used to differentiate between small-polaron and large-polaron absorption bands. As illustrated in Figure 7 of Ref. [23], the peak of the small-polaron absorption in the MCM occurs at $2E_b$, twice the small-polaron binding energy. This figure shows that the absorption becomes broader and less symmetric with increasing temperature. An early calculation of a small-polaron absorption band [35] did not adequately describe its width and inherent asymmetry.

In the non-adiabatic limit of the MCM, the activation energy for Arrhenius small-polaron hopping E_a is simply half the small-polaron binding energy E_b. However, the MCM only envisions a short-range component of the electron–phonon interaction. As noted in this paper's introduction, LN is an ionic material with a significant long-range component of its electron–phonon interaction. As the long-range component of the electron–phonon interaction increases, E_a decreases as a fraction of the small-polaron binding energy [1]. Furthermore, the small-polaron hopping activation energy decreases upon progressing from the non-adiabatic limit into the adiabatic regime [1,18]. Thus, a small-polaron's binding energy E_b will tend to be somewhat larger than twice the activation energy of its high-temperature Arrhenius hopping E_a.

A polaron becomes small when its self-trapped electronic carrier collapses to a single site, e.g., a single Nb cation in LN. Its self-trapped electronic carrier then hops between adjacent sites by primarily interacting with short-wavelength phonons. The characteristic phonon frequency for small-polaron hopping will then be comparable to those of these short-wavelength modes. In LN these modes have wavenumbers of about 250 cm^{-1} [36–39]. The inverse of the frequencies of these modes also corresponds to the time determined for atoms to shift their equilibrium positions in response to the addition of a severely localized electronic charge carrier [40,41]. Thus, the frequencies of these short-wavelength phonons are comparable to what we measure to be the characteristic phonon frequency associated with n-type small-polaron hopping in LN, $\omega = 4.6 \times 10^{13}$ s^{-1}.

7. Summary

In summary, we measured the temporal decay of the photo-induced absorption of Mg:LN from 45 K to 225 K. The time dependencies of decays of the induced absorption at 785 nm fit well with the phenomenological Kohlrausch–Williams–Watts (KWW) stretched-exponentials, $\exp[-(t/\tau)^{\beta}]$. The effective decay rate, the inverse of the mean decay time of each KWW curve, a function of τ and β, is a function of temperature. At the highest temperatures, the decay rate for the photo-induced absorption manifests an Arrhenius temperature dependence. This temperature dependence weakens and becomes non-Arrhenius as the temperature is lowered.

These decays are ascribed to the hopping of Nb^{4+} n-type small polarons leading to their eventual recombination with O^- p-type small polarons bound to Li vacancies. The n-type and p-type small polarons are attracted to one another by their mutual Coulomb attraction reduced by the static dielectric constant of LN. We model this hopping sequence as a series of jumps that are increasingly downward in energy. The rate-limiting step in this sequence of small-polaron hops, its slowest jump, is then its first step since its energy disparity is smallest. We take the energy disparity of well-separated oppositely charged small polarons to be minimal. As such, we compare the adiabatic jump rate of small polarons between sites of equal energy with the inverse of the mean decay time for KWW decays.

The rate for adiabatic small-polaron hops between equivalent sites is Arrhenius at high temperatures. As the temperature is reduced the temperature dependence of this adiabatic small-polaron jump rate weakens as it becomes non-Arrhenius [1,22–24]. The transition between these two behaviors occurs at the temperature corresponding to a significant fraction of that characterizing the principal phonons with which a small-polaron's self-trapped electronic carrier interacts. This feature yields a more precise value for the characteristic phonon frequency than can be obtained from only fitting data from the high-temperature Arrhenius region. Fitting the decay rate of photo-induced decay in Mg:LN to the adiabatic small-polaron jump rate between equivalent states yields reasonable values of the small-polaron binding energy and the characteristic frequency of the phonons with which its self-trapped electron interacts: $E_b = 0.68\,\text{eV}$ and $\omega = 4.6 \times 10^{13}\,\text{s}^{-1}$.

Author Contributions: Conceptualization, S.M. and M.I.; Methodology, S.M. and T.N.; Software, S.M. and A.K.; Validation, S.M., A.K. and L.V.; Formal Analysis, S.M., A.K. and T.N.; Investigation, S.M., A.K. and L.V.; Writing—Original Draft Preparation, S.M., A.K., M.I. and D.E.; Writing—Review & Editing, S.M., A.K., L.V., M.I., L.M.E. and D.E.; Visualization, S.M. and A.K.; Project Administration, M.I.; Funding Acquisition, M.I. and L.M.E. All authors have read and agreed to the published version of the manuscript.

Funding: This research was funded by the Deutsche Forschungsgemeinschaft, DFG, grant numbers: IM 37/11-1, INST 190/137-1 FUGG, and INST 190/165-1 (S.M., A.K., L.V., M.I.) and the Bundesministerium für Bildung und Forschung, BMBF, grant number 05K16ODA (T.N.) and through the Center of Excellence—Complexity and Topology in Quantum Matter (ct.qmat) (L.M.E.). Support by Deutsche Forschungsgemeinschaft (DFG) and Open Access Publishing Fund of Osnabrück University is gratefully acknowledged.

Acknowledgments: The authors thank L. Kovács and coworkers at the Wigner Research Centre for Physics, Budapest, for crystal preparation.

Conflicts of Interest: The authors declare no conflict of interest.

References

1. Emin, D. *Polarons*; Cambridge University Press: Cambridge, UK, 2012.
2. Freytag, F.; Corradi, G.; Imlau, M. Atomic insight to lattice distortions caused by carrier self-trapping in oxide materials. *Sci. Rep.* **2016**, *6*, 36929, doi:10.1038/srep36929.
3. Sanson, A.; Zaltron, A.; Argiolas, N.; Sada, C.; Bazzan, M.; Schmidt, W.G.; Sanna, S. Polaronic deformation at the $Fe^{2+/3+}$ impurity site in Fe:$LiNbO_3$ crystals. *Phys. Rev. B* **2015**, *91*, 094109, doi:10.1103/PhysRevB.91.094109.
4. Weis, R.S.; Gaylord, T.K. Lithium niobate: Summary of physical properties and crystal structure. *Appl. Phys. A* **1985**, *37*, 191–203, doi:10.1007/BF00614817.
5. Emin, D.; Holstein, T. Adiabatic Theory of an Electron in a Deformable Continuum. *Phys. Rev. Lett.* **1976**, *36*, 323–326, doi:10.1103/PhysRevLett.36.323.
6. Faust, B.; Müller, H.; Schirmer, O.F. Free small polarons in $LiNbO_3$. *Ferroelectrics* **1994**, *153*, 297–302, doi:10.1080/00150199408016583.
7. Schirmer, O.F. O^- bound small polarons in oxide materials. *J. Phys. Condens. Matter* **2006**, *18*, R667–R704, doi:10.1088/0953-8984/18/43/r01.
8. Schirmer, O.F.; Imlau, M.; Merschjann, C.; Schoke, B. Electron small polarons and bipolarons in $LiNbO_3$. *J. Phys. Condens. Matter* **2009**, *21*, 123201, doi:10.1088/0953-8984/21/12/123201.
9. Imlau, M.; Brüning, H.; Schoke, B.; Hardt, R.-S.; Conradi, D.; Merschjann, C. Hologram recording via spatial density modulation of $Nb_{Li}^{4+/5+}$ antisites in lithium niobate. *Opt. Express* **2011**, *19*, 15322–15338, doi:10.1364/OE.19.015322.
10. Brüning, H.; Dieckmann, V.; Schoke, B.; Voit, K.-M.; Imlau, M.; Corradi, G.; Merschjann, C. Small-polaron based holograms in $LiNbO_3$ in the visible spectrum. *Opt. Express* **2012**, *20*, 13326–13336, doi:10.1364/OE.20.013326.
11. Imlau, M.; Badorreck, H.; Merschjann, C. Optical nonlinearities of small polarons in lithium niobate. *Appl. Phys. Rev.* **2015**, 2, 040606, doi:10.1063/1.4931396.
12. Schirmer, O.F.; Imlau, M.; Merschjann, C. Bulk photovoltaic effect of $LiNbO_3$:Fe and its small-polaron-based microscopic interpretation. *Phys. Rev. B* **2011**, *83*, 165106, doi:10.1103/PhysRevB.83.165106.
13. Furukawa, Y.; Kitamura, K.; Alexandrovski, A.; Route, R.K.; Fejer, M.M.; Foulon, G. Green-induced infrared absorption in MgO doped $LiNbO_3$. *Appl. Phys. Lett.* **2001**, *78*, 1970–1972, doi:10.1063/1.1359137.
14. Hirohashi, J.; Pasiskevicius, V.; Wang, S.; Laurell, F. Picosecond blue-light-induced infrared absorption in single-domain and periodically poled ferroelectrics. *J. Appl. Phys.* **2007**, *101*, 033105, doi:10.1063/1.2434007.

15. Berben, D.; Buse, K.; Wevering, S.; Herth, P.; Imlau, M.; Woike, T. Lifetime of small polarons in iron-doped lithium niobate crystals. *J. Appl. Phys.* **2000**, *87*, 1034–1041, doi:10.1063/1.371976.
16. Mhaouech, I.; Guilbert, L. Temperature dependence of small polaron population decays in iron-doped lithium niobate by Monte Carlo simulations. *Solid State Sci.* **2016**, *60*, 28–36, doi:10.1016/j.solidstatesciences.2016.08.002.
17. Vittadello, L.; Bazzan, M.; Messerschmidt, S.; Imlau, M. Small Polaron Hopping in Fe:$LiNbO_3$ as a Function of Temperature and Composition. *Crystals* **2018**, *8*, 294, doi:10.3390/cryst8070294.
18. Emin, D.; Holstein, T. Studies of small polaron motion IV: Adiabatic theory of the Hall Effect. *Ann. Phys.* **1969**, *53*, 439–520, doi:10.1016/0003-4916(69)90034-7.
19. Emin, D. Semiclassical small-polaron hopping in a generalized molecular-crystal model. *Phys. Rev. B* **1991**, *43*, 11720, doi:10.1103/PhysRevB.43.11720.
20. Emin, D. Generalized adiabatic polaron hopping: Meyer-Neldel compensation and Poole-Frenkel behavior. *Phys. Rev. Lett.* **2008**, *100*, 166602, doi:10.1103/PhysRevLett.100.166602.
21. Emin, D. Theory of Meyer–Neldel compensation for adiabatic charge transfer. *Monatshefte Für Chem.- Mon.* **2012**, *144*, 3–10, doi:10.1007/s00706-012-0836-z.
22. Emin, D. Phonon-assisted jump rate in noncrystalline solids. *Phys. Rev. Lett.* **1974**, *32*, 303–307, doi:10.1103/PhysRevLett.32.303.
23. Emin, D. Phonon-assisted transition rates I. Optical-phonon-assisted hopping in solids. *Adv. Phys.* **1975**, *24*, 305–348, doi:10.1080/00018737500101411.
24. Gorham-Bergeron, E.; Emin, D. Phonon-assisted hopping due to interaction with both acoustical and optical phonons. *Phys. Rev. B* **1977**, *15*, 3667–3680, doi:10.1103/PhysRevB.15.3667.
25. Volk, T.; Wöhlecke, M. *Lithium Niobate*; Springer GmbH: Berlin/Heidelberg, Germany, 2008.
26. Conradi, D.; Merschjann, C.; Schoke, B.; Imlau, M.; Corradi, G.; Polgár, K. Influence of Mg doping on the behaviour of polaronic light-induced absorption in $LiNbO_3$. *Phys. Stat. Sol. (RRL)* **2008**, *2*, 284–286, doi:10.1002/pssr.200802176.
27. Guilbert, L.; Vittadello, L.; Bazzan, M.; Mhaouech, I.; Messerschmidt, S.; Imlau, M. The elusive role of Nb_{Li} bound polaron energy in hopping charge transport in Fe:$LiNbO_3$. *J. Phys. Condens. Matter* **2018**, *30*, 125701, doi:10.1088/1361-648x/aaad34.
28. Merschjann, C.; Schoke, B.; Conradi, D.; Imlau, M.; Corradi, G.; Polgár, K. Absorption cross sections and number densities of electron and hole polarons in congruently melting $LiNbO_3$. *J. Phys. Condens. Matter* **2008**, *21*, 015906, doi:10.1088/0953-8984/21/1/015906.
29. Messerschmidt, S.; Krampf, A.; Freytag, F.; Imlau, M.; Vittadello, L.; Bazzan, M.; Corradi, G. The role of self-trapped excitons in polaronic recombination processes in lithium niobate. *J. Phys. Condens. Matter* **2019**, *31*, 065701, doi:10.1088/1361-648x/aaf4df.
30. Freytag, F.; Booker, P.; Corradi, G.; Messerschmidt, S.; Krampf, A.; Imlau, M. Picosecond near-to-mid-infrared absorption of pulse-injected small polarons in magnesium doped lithium niobate. *Opt. Mater. Express* **2018**, *8*, 1505, doi:10.1364/ome.8.001505.
31. Messerschmidt, S.; Bourdon, B.; Brinkmann, D.; Krampf, A.; Vittadello, L.; Imlau, M. Pulse-induced transient blue absorption related with long-lived excitonic states in iron-doped lithium niobate. *Opt. Mater. Express* **2019**, *9*, 2748, doi:10.1364/ome.9.002748.
32. Corradi, G.; Krampf, A.; Messerschmidt, S.; Vittadello, L.; Imlau, M. Excitonic hopping-pinning scenarios in lithium niobate based on atomistic models: Different kinds of stretched exponential kinetics in the same system. *J. Phys. Condens. Matter* **2020**, *32*, 413005, doi:10.1088/1361-648X/ab9c5b.
33. Krampf, A.; Messerschmidt, S.; Imlau, M. Superposed picosecond luminescence kinetics in lithium niobate revealed by means of broadband fs-fluorescence upconversion spectroscopy. *Sci. Rep.* **2020**, *10*, 11397, doi:10.1038/s41598-020-68376-6.
34. Emin, D. Optical properties of large and small polarons and bipolarons. *Phys. Rev. B* **1993**, *48*, 13691–13702, doi:10.1103/PhysRevB.48.13691.
35. Reik, H.G.; Heese, D. Frequency dependence of the electrical conductivity of small polarons for high and low temperatures. *J. Phys. Chem. Solids* **1967**, *28*, 581–596, doi:10.1016/0022-3697(67)90089-3.
36. Caciuc, V.; Postnikov, A.V.; Borstel, G. Ab initiostructure and zone-center phonons in $LiNbO_3$. *Phys. Rev. B* **2000**, *61*, 8806–8813, doi:10.1103/physrevb.61.8806.

37. Fontana, M.D.; Bourson, P. Microstructure and defects probed by Raman spectroscopy in lithium niobate crystals and devices. *Appl. Phys. Rev.* **2015**, *2*, 040602, doi:10.1063/1.4934203.
38. Sanna, S.; Neufeld, S.; Rüsing, M.; Berth, G.; Zrenner, A.; Schmidt, W.G. Raman scattering efficiency in $LiTaO_3$ and $LiNbO_3$ crystals. *Phys. Rev. B* **2015**, *91*, 224302, doi:10.1103/physrevb.91.224302.
39. Xin, F.; Zhai, Z.; Wang, X.; Kong, Y.; Xu, J.; Zhang, G. Threshold behavior of the Einstein oscillator, electron-phonon interaction, band-edge absorption, and small hole polarons in $LiNbO_3$:Mg crystals. *Phys. Rev. B* **2012**, *86*, 165132, doi:10.1103/physrevb.86.165132.
40. Badorreck, H.; Nolte, S.; Freytag, F.; Bäune, P.; Dieckmann, V.; Imlau, M. Scanning nonlinear absorption in lithium niobate over the time regime of small polaron formation. *Opt. Mater. Express* **2015**, *5*, 2729, doi:10.1364/ome.5.002729.
41. Sasamoto, S.; Hirohashi, J.; Ashihara, S. Polaron dynamics in lithium niobate upon femtosecond pulse irradiation: Influence of magnesium doping and stoichiometry control. *J. Appl. Phys.* **2009**, *105*, 083102, doi:10.1063/1.3095763.

crystals

Article

Polaron Trapping and Migration in Iron-Doped Lithium Niobate

Laura Vittadello [1], Laurent Guilbert [2], Stanislav Fedorenko [3] and Marco Bazzan [4,*]

1 School of Physics, Osnabrueck University, Barbarastrasse 7, 49076 Osnabrueck, Germany; laura.vittadello@uni-osnabrueck.de
2 Laboratoire Matériaux Optiques, Photonique et Systèmes, Université de Lorraine et CentraleSupélec, 2 rue E. Belin, F-57070 Metz, France; lorenzo.ghiberti@yahoo.fr
3 Voevodsky Institute of Chemical Kinetics and Combustion, Russian Academy of Sciences, 630090 Novosibirsk, Russia; fedorenk@kinetics.nsc.ru
4 Dipartimento di Fisica e Astronomia, Università di Padova, Via Marzolo 8, 35131 Padova, Italy
* Correspondence: marco.bazzan@unipd.it

Abstract: Photoinduced charge transport in lithium niobate for standard illumination, composition and temperature conditions occurs by means of small polaron hopping either on regular or defective lattice sites. Starting from Marcus-Holstein's theory for polaron hopping frequency we draw a quantitative picture illustrating two underlying microscopic mechanisms besides experimental observations, namely direct trapping and migration-accelerated polaron trapping transport. Our observations will be referred to the typical outcomes of transient light induced absorption measurements, where the kinetics of a polaron population generated by a laser pulse then decaying towards deep trap sites is measured. Our results help to rationalize the observations beyond simple phenomenological models and may serve as a guide to design the material according to the desired specifications.

Keywords: lithium niobate; polarons; photorefractivity; Marcus-Holstein's theory; Monte Carlo simulations

Citation: Vittadello, L.; Guilbert, L.; Fedorenko, S.; Bazzan, M. Polaron Trapping and Migration in Iron-Doped Lithium Niobate. *Crystals* **2021**, *11*, 302. http://doi.org/10.3390/cryst11030302

Academic Editor: László Kovács

Received: 26 February 2021
Accepted: 15 March 2021
Published: 17 March 2021

1. Introduction

Lithium Niobate (LN) stands out among other ferroelectric oxides for its large use in the realisation of acousto-optical, electro-optical and non-linear optical devices. Since its development in the '60 s, this material has evidenced several light-induced effects that were recognized as a complex interplay between charge excitation and migration processes, such as photoconductivity, the bulk photovoltaic effect and ultimately photorefractivity. The interest in those phenomena is timely because they bear a high interest for practical applications: in the field of nonlinear and ultra-fast optics the photorefractive effect is a drawback that limits the use of LN for high intensity multiphoton processes [1], while in photorefractive holography this effect is used to record high quality gratings, optical memories and demonstrate low-intensity all-optical interactions [2]. Integrated optics as well needs to control those phenomena due to the high continuous-wave light intensities obtained in waveguiding regions [3]. Moreover, in analogous materials, charge excitation and transport play a key role in a number of important applications such as ferroelectric photovoltaics [4,5] and in oxide catalysis [6–8].

Thanks to many results obtained especially by O. Schirmer and co-workers [9–14], it is nowadays accepted that charge transport in LN and related materials must be understood in terms of small polarons hopping among regular and/or defective sites. In the initial stage of the process, some charge carriers are photo-generated from deep donor centers and emitted with a preferential direction in the conduction band. Subsequently those "hot" carriers lose energy by interaction with the lattice and finally condensate into a new state which is self-localized by a distortion of the local ionic environment. Under certain conditions the carrier, localized at a single lattice site, and the surrounding deformation can be thought as a quasi-particle that moves as a whole: the small polaron. Its motion takes place by thermally assisted hopping transitions among different sites, until a deep trap is encountered; the polaron is then stably trapped and ready to be photo-excited again [15].

Despite this understanding, it is still challenging to relate the macroscopic observations to basic polaron hopping processes. The technique of choice to investigate the transient decay of a polaron population is time-resolved Light Induced Absorption (LIA) spectroscopy ([1] and refs. therein) in which a polaron population is created in the material by a pulsed photo-excitation process and its decay towards deep trap centers is observed by means of time resolved absorption spectroscopy. One of the main problems of this technique is that the experimental decay curves are difficult to interpret from a microscopic point of view, being the result of the parallel relaxation of the different polarons through a variety of microscopic processes. For this reason experimental curves are generally analyzed phenomenologically with the help of suitable fitting functions such as the widely used Kohlrausch-William-Watts (KWW) stretched exponential law, without a true understanding of the microscopic mechanisms determining the decay.

In this paper we apply a Monte Carlo analysis to investigate the different processes responsible for the decay in the technologically important case of Fe-doped LN, with special focus on the role of deep (Fe) and shallow (Nb_{Li}) defect centers where the polarons can be trapped. We will show how transient measurements can reveal different stages of the decay kinetics corresponding to different types of hopping processes separated by largely different time scales. Furthermore the effect of changing experimental constraints such as sample temperature and composition on the different hopping processes will be elucidated and some criteria will be given to establish a priori what is the regime one should expect on the basis of a given sample composition and temperature.

2. Small Polarons in Fe:LiNbO$_3$

Polarons are quasiparticles made up of an electrical charge that, by interaction with the polar crystalline environment, is able to distort the neighbouring lattice creating a local potential well. As a net result, the particle becomes self-localized. If the confinement effect is strong, the charge is concentrated on a single lattice site (*small* strong-coupling polaron) and randomly moves by thermal-assisted hopping among different sites.

Standard LN crystals of congruent composition contain a high density of point defects either intrinsic (such as the substitutional "antisite" defect Nb^{5+}_{Li} or the Li vacancy V^{-}_{Li}), or extrinsic (dopants or impurities) [16]. Those defects constitute preferential sites for the formation of polarons, so that several types of small electron polarons are recognized in congruent LN [13]: the free polaron (*F*) forming on regular Nb^{5+}_{Nb} sites; the bound polaron (*P*) forming on antisite defects Nb^{4+}_{Li}; the bipolaron (*BP*) which is a combination of a free and a bound polaron on neighbouring sites $Nb^{4+}_{Nb} : Nb^{4+}_{Li}$, with the possible existence of some variants depending on the actual site occupied by the Nb interstitial, which in some cases could be either a vacant Li site, either an octahedral void naturally present in the LN structure [17]. LN may also host O^{-} holes, created when energetic photons capable of band-to band excitation are involved. In this case, hole polarons (*H*) are formed. Finally, impurities with two or more valence states, such as the prototypical case of $Fe^{2+/3+}_{Li}$, have the possibility to capture an electron with a potential that is a mixture of Coulomb attraction and polaronic lattice deformation.

In LN, polarons can be seen as localized defect states in the band gap (Figure 1), which can absorb the light releasing the trapped charge to higher energy levels under the Franck-Condon principle [13]. In Figure 1 the electronic energy is referred to the uncoupled Nb_{Nb} level instead of the bottom of the conduction band, being a more appropriate description when the electron-lattice coupling is strong. Values for $Nb^{4+/5+}_{Nb}$ and $Fe^{2+/3+}_{Li}$ are calculated according to ref. [13] while the one for the $Nb^{4+/5+}_{Li}$ according to ref. [18].

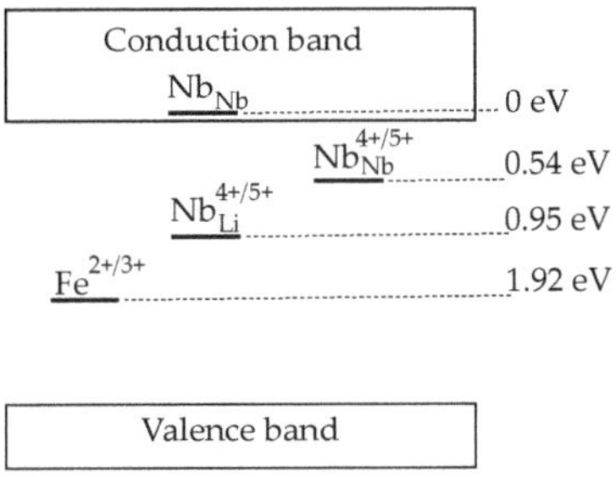

Figure 1. Electronic energy of the different polaron centers in Fe:$LiNbO_3$ with respect to the Nb_{Nb} level in a rigid lattice (i.e., not considering the lowering in energy due to the local lattice distortion associated to the polaronic effect).

The typical experiment dealing with the relaxation of a suddenly-created polaron population is Light Induced Absorption (LIA) [19,20]. The sample is illuminated with a pulsed laser with a pulse duration of few ns and a maximum pulse energy of hundreds of mJ while one or more continuous-wave probe lasers record simultaneously the time evolution of the sample absorption at different wavelengths. The pump laser creates a polaron population by photo-excitation from charged defects (in this case Fe^{2+}) and since polarons can absorb light, the probe beams detect an additional absorption signal that is proportional to the polaron concentration in the sample. In this way the decay of the polaron population as a function of time can be observed. Moreover, since the absorption cross section of the different polaron species is quite different at the different probe wavelengths [20], it is possible to reconstruct the decay kinetic of the different polaron species.

3. Monte Carlo Simulation of Polaron Hopping

According to the Marcus-Holstein polaron hopping model [15,21,22], the non-adiabatic hopping frequency for a $(i \rightarrow f)$ hop is:

$$W_{if}(r,T) = w_{if}(T)\exp\left(-\frac{r}{a_{if}}\right) \tag{1}$$

with:

$$w_{if}(T) = \frac{I_{if}^2}{\hbar}\left(\frac{\pi}{kT\lambda_{if}}\right)^{\frac{1}{2}}\exp\left(-\frac{U_{if}}{kT}\right) \tag{2}$$

In Equation (1), r is the distance between initial and final sites, kT is the absolute temperature (in energy units) and $a_{if} = a_{fi}$ is an orbital parameter describing the overlap between the electronic wavefunctions at sites i and f.

In ref. [18] we determined that it is reasonable to reduce the number of orbital parameters to two. The first one $c = a_{PFe} = a_{FFe} = 1.3$ Å describing "trapping" transitions between F or P and Fe_{Li}. The second one $a = a_{PP} = a_{FF} = a_{PF} = a_{FP} = 1.6$ Å describes electronic transfer of polaron migration, i.e., hopping between Nb ions whatever their positions is regular (Nb_{Nb}) or defective (Nb_{Li}).

In Equation (2) $\lambda_{if} = \lambda_{fi} = (E_i + E_f)$ is the reorganization energy of Marcus' theory corresponding to the energy paid to rearrange the lattice, here equal to the sum of the elastic energies of the two polarons.

The hopping barrier $U_{if}(\neq U_{fi})$ can be computed using the harmonic approximation of small polaron theory [23,24] as:

$$U_{if} = \frac{\left(2E_i + \varepsilon_i - \varepsilon_f\right)^2}{4\left(E_i + E_f\right)} \tag{3}$$

with ε_i and ε_f the binding energies of the electron at zero deformation. When the hop occurs between sites of the same type ($i = f$), $U_{ii} = E_i/2$, recovering the standard result that the hopping activation energy is one half of the polaron stabilization energy [1,13,25]. The pre-exponential factor $I_{if} = I_{fi}$ describes the intrinsic hopping rate between the two sites and is determined by the choice of the $(i,\ f)$ combination. By comparing the results of our simulations to experimental decay curves we estimated that a fair value for all the processes is $I = 0.02\,\text{eV}$. In this situation the non-adiabatic condition is always fulfilled for all polarons species, as discussed in ref. [24].

Those hopping frequencies are used in a dedicated Monte Carlo code based on a classical Gillespie algorithm. The LN structure is generated at the beginning of the code in a $80 \times 80 \times 80$ super-cell with periodic boundary conditions. A given number of empty Nb_{Li} and Fe defects is then placed on a certain number of lattice sites in accordance to a specified sample composition. For sake of simplicity, we will restrict ourselves to the case in which the number of photo-generated polarons is small compared to the amount of antisites and Fe traps. This approximation is valid for antisites, which are present in large amount in standard LN crystals, but may not be so for Fe traps when the concentration of photo-excited donor centers is comparable to the one of traps e.g., in strongly reduced samples with a high $[Fe^{2+}]/[Fe^{3+}]$ ratio. In spite of the fact that this assumption may lead to an inaccurate modelling of the decay shape, our general results remain valid.

Under the assumption of weak concentration, we can perform two important simplifications. (i) The probability that two polarons may interact and/or form bipolaronic states is negligible. This is consistent with the fact that energy level of Fe traps is significantly lower than the bipolaron one (1.92 eV with respect to 1.07 eV, respectively [13]), so that migrating polarons tend to move to Fe traps instead of forming bipolarons. The absence of bipolarons in Fe:LN samples is also corroborated by the observation that in LIA experiments there is no evidence of an increased absorption in the blue spectral region, as it would be the case upon bipolaron formation [26]. (ii) Trap saturation effects are negligible, so that we can assume that the trap concentrations are constant.

Our simulation routine works as it follows: one polaron per time is placed on a randomly chosen starting site, either a regular Nb_{Nb}^{5+} or on an antisite defect according to the user's choice. The program then computes the hopping frequencies towards all possible destination sites comprised in a suitably defined volume and finally chooses the actual destination site by a statistical procedure which weights the possible outcomes according to their respective hopping frequencies using a Gillespie algorithm. The higher the hop frequency towards a given site, the higher the probability that this site will be selected. Note that, according to Equation (1), this probability depends both on distance and on temperature as well as on the type of initial and destination site. This means that (i) long jumps beyond nearest neighbour are always allowed and, if proper conditions are met, may even be more probable than jumps within the first coordination sphere and (ii) the relative weight of the different processes may change according to temperature. The hopping routine is then repeated until a deep Fe trap is reached: in this situation the program records the final polaron position with respect to its original site, the number of different sites encountered during its walk and the time needed to reach the final site. The whole cycle is then repeated until a sufficient statistics is collected. Further details on the code may be found in refs. [18,26].

In the following, we will focus our attention on the decay kinetics of P polarons considering that in some suitably-chosen experimental conditions they are the predominant polaron species present in the sample [18]. In fact, few ns after the pulse, a dynamic equilibrium is established between P and F polarons but since the hopping rate of F polarons is much higher than the one of P polarons, the equilibrium is strongly displaced in favor of the latter ones. In particular we will consider that, in standard congruent samples at room temperature and below, free polarons eventually created by the pulse thermalize quickly to deep traps and to bound polarons, from which they can depart and migrate for short times due to thermal detrapping from charged antisite defects. This process corresponds in our code to a ($P \rightarrow F$) hopping event.

Also, we assume that the excitation wavelength is chosen in such a way that the photo-excitation of electrons from the valence to the conduction band is disfavoured, in order to neglect the presence of hole polarons. Typically the wavelength is in the visible range (532 nm).

Under these conditions, the photo-generation process occurs when a photon is absorbed by a Fe^{2+} donor center. The electronic charge is ejected and subsequently thermalizes to form the polaron in the neighbourhood of its empty donor center. In this situation, all the created polarons would appear close to their respective generating Fe trap and would have a strong chance of being quickly retrapped and disappear from the system soon after the pulse has ended. However the experimentally studied LIA decays, which are the focus of our study, last for long times ranging to milliseconds or even seconds, depending on the experimental conditions. How is this possible? Since the bound polaron formation time (<1 ps, see ref. [27]) is much shorter than the pulse duration while their lifetime is generally much longer, we consider that already during the pulse a given bound polaron concentration is present within the sample. If the pulse energy density D is not too low, there is a fair chance that several polarons absorb additional photons which kicks them far away from the original center, helping to randomize their initial position. In fact, for typical values in the range of $D \sim 200\,\mathrm{mJ/cm^2}$, [1,28,29] the average number of photons absorbed by the polaron is about $s_P \times D/h\nu \approx 3-4$, where $s = 7 \times 10^{-22}\,\mathrm{m^2}$ is the absorption cross section of bound polarons and $h\nu = 2.33\,\mathrm{eV}$ is the photon energy at 532 nm. On the basis of this reasoning, we consider that the polarons remaining after the pulse is ended are those which have absorbed more than one photon and that find themselves in a position completely uncorrelated with respect to their originating site. Accordingly, in our code the initial position of the walking polaron is randomly chosen on a Nb_{Li}^{4+} site.

4. Direct Trapping versus Migration-Accelerated Trapping

In the top part of Figure 2 it is simulated the decay of a bound polaron population for a LN sample doped with 0.02 mol.% Fe^{3+} corresponding to a deep trap concentration of $0.38 \times 10^{25}\,\mathrm{m^{-3}}$. On the bottom figure is reported the occurrence of $P \rightarrow P$ and $F \rightarrow F$ hops (hereby abbreviated as *PP* and *FF*) performed by the different polarons, sorted according to their lifetime. It can be seen that at short times the decay is due to polarons that do not perform any hop among antisites or regular sites and are therefore directly trapped by a Fe trap. At this stage, polarons born near the traps die in one hop, having a high probability of jumping directly to the Fe impurity center. We shall indicate this initial decay mode as *direct trapping regime*. After this stage, we encounter those polarons that live long enough so that the chance to hop on another transport site is not negligible and thus are able to perform a certain number of hops before meeting a Fe trap. This stage is indicated as *migration-accelerated regime*. We can define a characteristic time τ_0 separating the two regimes. τ_0 may be regarded as the average waiting time spent by the polaron on the initial site before hopping on another antisite and/or a regular Nb site. When the trapping processes require a time significantly longer than this, polaron migration has a chance to take place.

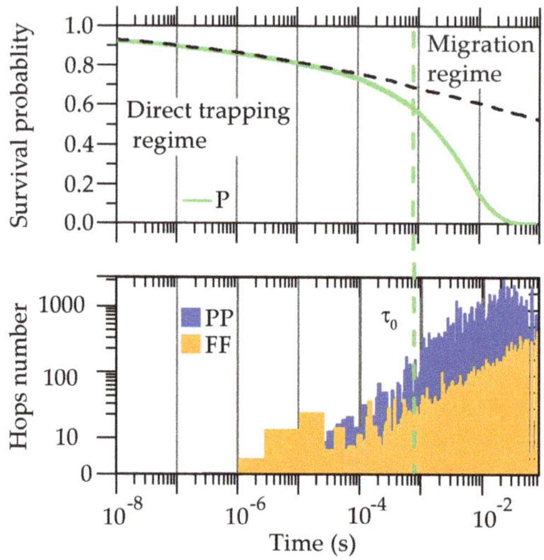

Figure 2. (**Top**) Survival probability of P polarons as measured by a typical LIA experiment. The black dashed line shows the decay shape when polaron migration is prohibited. The green dashed line is placed at $\tau_0 = 8 \times 10^{-4}$ s, as calculated from Equation (12). (**Bottom**) Number of PP and FF hopping processes performed by each polaron during its walk.

5. Estimation of the Escape Time

To perform some quantitative reasoning on τ_0 we need to recall the concept of *trapping radius* already introduced in [18,24,26]. According to Equation (1), for a fixed temperature, fast trapping processes are the ones for which Fe traps are "near" to the starting site. We may therefore define a characteristic distance from a Fe trap at which a trapping process is as fast as the typical hopping rate.

To begin with, we consider a free polaron hopping only among regular Nb_{Nb} sites that comes close to a Fe trap. The trapping radius R_{FFe} for the $F \to Fe$ process is defined as:

$$W_{FFe}(R_{FFe}, T) = \sum_{i}^{\infty} W_{FF}(r_i, T) = w_{FF}(T) S_D \tag{4}$$

where w_{FF} is given by Equations (1) and (2) and $S_D = \sum_i^{\infty} \exp\left(-r_i/a_{FF}\right)$ is a geometric constant factor that depends only on the lattice positions r_i of the Nb_{Nb} sites. For the LN structure, summing up to the $N = 8$ coordination shell of the Nb sublattice and using for the orbital parameter the value $a_{FF} = 1.6$ Å we get $S_D = 1.48$. From Equation (1) we get:

$$R_{FFe} = a_{FFe}\left[\ln \frac{w_{FFe}(T)}{w_{FF}(T)} - \ln S_D\right] \tag{5}$$

R_{FFe} depends on the temperature through the term $\frac{w_{FFe}(T)}{w_{FF}(T)}$. For example, $R_{FFe} = 13$ Å at room temperature, while $R_{FFe} = 20$ Å at 200 K. We may now consider that any Fe trap is surrounded by a sphere of radius R_{FFe} dependent on temperature and that whenever a *F* polaron finds itself inside this sphere it will almost surely be captured by direct transfer to the trap. Polarons spawning outside of these spheres must make a number of jumps before they can get close enough to any trap.

This concept can be readily extended to the case of free polarons captured by antisite shallow traps. In this case the trapping radius is:

$$R_{FP} = a_{FP}\left[\ln \frac{w_{FP}(T)}{w_{FF}(T)} - \ln S_D\right] \tag{6}$$

Let us now consider one polaron starting its walk from a Nb_{Li} antisite, as it is customary to assume in standard LN samples. Let us assume that this starting site is far away from Fe traps. By thermally activated polaron hopping, the charge may leave this site in two possible ways: either hopping on another antisite nearby, either transforming into a free polaron and migrating away on the Nb sublattice. To estimate the first hopping rate (polarons jumping on the disordered antisite lattice) we use a result of percolation theory [30] to take into account the random positions of the Nb_{Li} sites, so that the expression for the waiting time reads:

$$\frac{1}{\tau_{PP}} \approx w_{PP}(T)\exp\left(-1.44\frac{\bar{r}}{a_{PP}}\right) \tag{7}$$

where $\bar{r}$ is a measure of the average distance between two antisites in a sample with antisite concentration N_P given by:

$$\bar{r} = \left(\frac{3}{4\pi N_P}\right)^{\frac{1}{3}} \tag{8}$$

The waiting time for the second process (conversion $P \to F$) is slightly more difficult to compute. A priori, this kind of hop can occur towards a Nb site at any distance from the original Nb antisite. Moreover, since the Nb sites form a nearly-cubic lattice around the starting site, the probability for the polaron to escape is given by the sum of the frequency over all the destination sites, as in Equation (4). However, the Nb_{Nb} sites contained within the antisite trapping sphere of radius R_{FP} (see Equation (6)) correspond to inefficient processes because if the polaron hops in one of them, it will almost surely go back to the origin instead of leaving the initial site. Therefore the (macroscopic) effective waiting time can be computed by summing over all the equivalent destination sites available out of the antisite trapping sphere:

$$\frac{1}{\tau_{PF}} = \sum_{r_i \geq R_{FP}} W_{PF}(r_i, T) \tag{9}$$

By approximating the summation with an integral and substituting Equation (1) we get:

$$\frac{1}{\tau_{PF}} = VN_0 w_{PF}(T)\exp\left(-\frac{R_{FP}}{a_{FP}}\right) \tag{10}$$

with:

$$V = 4\pi a_{PF}(R_{FP})^2\left[1 + \frac{a_{PF}}{R_{FP}} + \left(\frac{a_{PF}}{R_{FP}}\right)^2\right] \tag{11}$$

where $N_0 = 1.87 \times 10^{28}\ \mathrm{m}^{-3}$ is the atomic density of the Nb sublattice and V is the volume of a spherical layer of thickness a_{PF} around a sphere of radius R_{FP}.

The *effective* waiting time on the antisite τ_0 can now be estimated from the parallel between the two above-mentioned processes (Equations (7) and (10)):

$$\frac{1}{\tau_0} = \frac{1}{\tau_{PP}} + \frac{1}{\tau_{PF}} \tag{12}$$

From Equation (12) one can now specify what is the average waiting time on the Nb_{Li} site for a polaron about to leave either by hopping on an antisite nearby or by performing intermediate passages through the Nb_{Nb} sites. In absence of deep traps, this can be understood as the effective hopping rate along the mixed Nb_{Li} and Nb_{Nb} lattices since, in moving from one site to the other, the waiting time due the $F \to F$ processes is negligible. Thus, Equation (12) can be used to compute a trapping radius for the capture process $P \to Fe$ according to the usual equation:

$$W_{PFe}(R_{PFe}, T) = \frac{1}{\tau_0} \tag{13}$$

which gives:

$$R_{PFe} = a_{PFe} \ln (w_{PFe} \tau_0) \tag{14}$$

Note that expression (12) should not be confused with the *emigration* time, i.e., the average time required to leave the starting site without returning to it. A rigorous estimation of the former would require to specify correctly the probability that the moving polaron may be captured back considering all the possible trajectories. In other words, Equation (12) is not verifying the irreversibly of the migration process and must therefore be understood simply as an underestimation of the emigration time. In spite of its limitations, Equation (12) is nevertheless useful as it provides a qualitative understanding of the general behaviour of our system. In the following we shall explore the different properties of τ_0.

From Equations (6)–(8) and (10) it is clear that τ_0 depends both on sample temperature and on the concentration of antisite defects (i.e., on the LN stoichiometry). The temperature dependence is illustrated in Figure 3a. Here it can be noticed that the first part of the decay corresponding to the direct trapping regime is not altered too much by the temperature changes, as the activation energy of the trapping process is very small ($U_{PFe} = 0.04\,\text{eV}$, according to Equation (3) and ref. [26]) so that the effect of lowering the temperature is essentially to move τ_0 to longer times, extending the direct trapping regime to a larger part of the decay curve. In particular, for sufficiently low temperatures, τ_0 is pushed beyond the end of the decay curve so that a pure direct trapping regime is attained.

The composition effect is shown in Figure 3b. Here it can be seen that at room temperature, the first term in Equation (12) dominates only for high antisite concentrations, corresponding to a strongly Li-deficient, sub-congruent LN sample. In this situation the antisite concentration is so high that τ_0 is determined essentially by the direct $P \rightarrow P$ hopping process. By lowering the antisite concentration, the first term in Equation (12) becomes smaller while the second one, which is independent on the sample composition, remains constant. Interestingly, when the sample composition is close to the congruent one i.e., $[\text{Nb}_{\text{Li}}] = 19 \times 10^{25}\,\text{m}^{-3}$, the two terms become comparable at room temperature ($\tau_{PP} = 5.9\,\text{ms}$, $\tau_{PF} = 0.9\,\text{ms}$ hence $\tau_0 = 0.8\,\text{ms}$. From this point on, a further reduction of the antisite concentration has a relatively small effect. In Figure 3b for example by decreasing of one order of magnitude the concentration $[\text{Nb}_{\text{Li}}] = 1.9 \times 10^{25}\,\text{m}^{-3}$ it can be seen that τ_0 barely changes. Note that in those examples we are always assuming the Nb_{Li} as starting site. The situation is however different in the case (experimentally difficult to realize, but not impossible) of null antisite concentration. In this case the starting site of the polaron must be a regular Nb_{Nb} site, which makes a large difference because now the migration occurs solely on the regular pseudo-cubic Niobium sublattice. The new characteristic time in this situation is shifted to much smaller values so that the migration regime sets in very soon, leading to a high mobility and a short lifetime for the polarons (see for comparison the simulated decay of F polarons in Figure 3b). This is the ultimate reason for the well-known threshold dependence of the photorefractive effect upon the sample stoichiometry (see e.g. [16]): the photoconductive properties of LN are only weakly affected by the sample composition until virtually all antisite defects are eliminated.

Finally, in Figure 3c is shown the behaviour of the P polarons decay curve for increasing Fe trap concentrations, corresponding to faster initial decays. For sufficiently high trap concentrations the polaron lifetime eventually becomes shorter than τ_0 and in this case the kinetic is purely determined by direct trapping processes. In this respect, increasing the Fe concentration brings to the same result as lowering the temperature, i.e., it makes the direct trapping processes predominant with respect to diffusive ones. In this respect it should however be stressed that expression (12) is a meaningful expression of the waiting time on the antisites only assuming that the Fe trap concentration is not too high. In this latter case, direct trapping dominates with its own specific kinetic mechanism and hopping on other sublattices becomes not relevant. In the following section we will explore in more details under which conditions this trapping regime is attained.

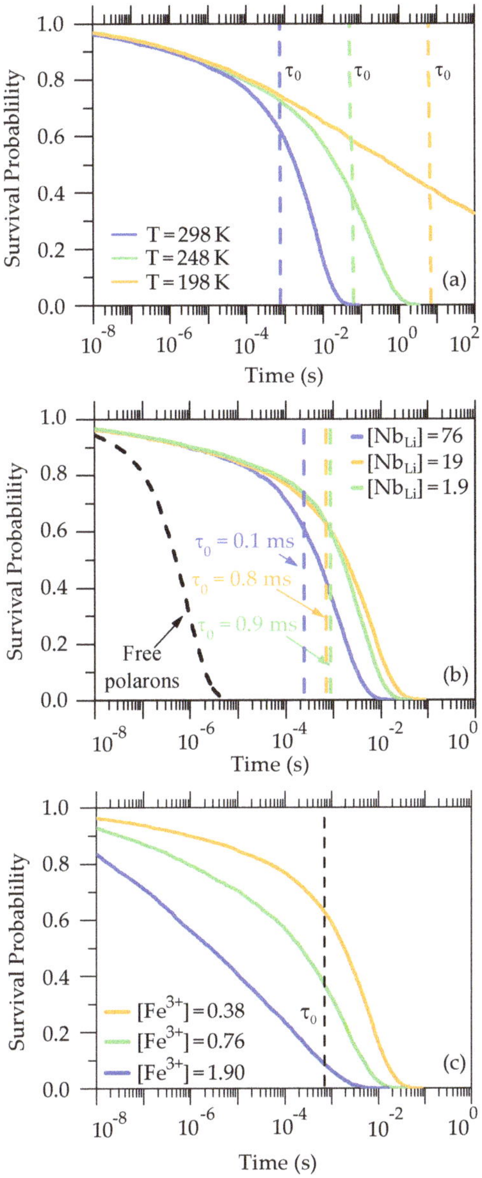

Figure 3. (**a**) Survival probability of P polarons at 298 K, at 248 K and at 198 K in a congruent sample with the same composition as the one in Figure 2. The dashed lines indicate the value of τ_0 as approximately determined by (12). (**b**) Survival probability for P polarons at room temperature in a sub-congruent (**blue**), congruent (**green**) and near-stoichiometric sample (**yellow**). The blackdashed line shows for comparison a decay of a population of F polarons in a fully stoichiometric sample. The numbers in the legend indicate the Nb_{Li} concentration in $\times 10^{25}\,m^{-3}$. (**c**) Survival probability for P polarons at room temperature in a congruent sample for increasing Fe concentrations. The dashed line indicates the hopping characteristic time τ_0. The numbers in the legend indicate the Fe^{3+} concentration in $\times 10^{25}\,m^{-3}$.

6. Pure Trapping Regimes

For practical applications it is useful to provide a criterion to determine at which temperature/composition combination one can consider that a sample is in a trapping or migration regime. As shown in the previous section, there is a characteristic time roughly given by τ_0 separating the two situations. The direct trapping regime is defined by the condition that the trapping frequency is much larger than the inverse of the characteristic hopping time. In other words, for direct trapping processes, *all* the polarons leave their starting site hopping directly towards a deep Fe trap.

An equivalent way to visualize this condition is that, for a given temperature, the trapping spheres around the traps are so large that they start to overlap. Let us consider for instance a condition of pure trapping regime for free polarons in an undoped LN sample with varying concentration of antisite defects. Free polarons would be directly trapped if the antisite concentration satisfies:

$$\chi_{FP} = \frac{4\pi}{3} N_P (R_{FP})^3 \gg 1 \tag{15}$$

where the strong inequality is required since the theoretical expression on the left hand side underestimates the concentration needed to achieve direct trapping, as a consequence of the fact that all the processes are considered irreversible.

The other simple situation involves an iron-doped sample containing no antisites (stoichiometric Fe:LN). In this case the free polarons will disappear by direct Fe trapping for trap concentrations satisfying the following criterion:

$$\chi_{FFe} = \frac{4\pi}{3} N_{Fe} (R_{FFe})^3 \gg 1 \tag{16}$$

In the case of a mixed trap situation such as Fe:LN of varying composition (both deep Fe_{Li} and shallow Nb_{Li} traps), a criterion for the full trapping regime of free polarons is easily obtained by combining together Equations (5)–(15):

$$\chi_{FP} + \chi_{FFe} \gg 1 \tag{17}$$

For reference, in the case of a congruent sample at room temperature, we have $\chi_{FP} = 0.25$ and $\chi_{FFe} = 0.05$.

Using Equation (14) we can also define a criterion on the Fe concentration for which bound polarons are directly trapped:

$$\chi_{PFe} = \frac{4\pi}{3} N_{Fe} (R_{PFe})^3 \gg 1 \tag{18}$$

To check the above criteria, it is possible to use Monte Carlo simulations to quantify, for a given experimental situation, how much hops of a given type the polarons perform on average [26].

In Figure 4a it is reported the percentage of FF hops performed during the polaron walk as a function of the temperature and composition in an undoped LN crystal with varying antisite concentration, where the vertical dashed line at $Nb_{Li} = 19 \times 10^{25}\,\mathrm{m}^{-3}$ indicates the standard congruent composition. Along this line it can be seen that, as the temperature increases, the polaron walk is constituted by a larger fraction of FF hops. At temperatures around 400 K about one half of the hops occurs between regular Nb sites. Conversely, by lowering the temperature and/or increasing the antisite concentration, the diffusion process becomes less and less dependent on free polaron transport. The temperature - composition line corresponding to $\chi_{FP} = 1$ is visible in the bottom right part of the figure. Note that for the typical case of congruent LN, free polaron contribution is not negligible if not at very low temperatures (<200 K).

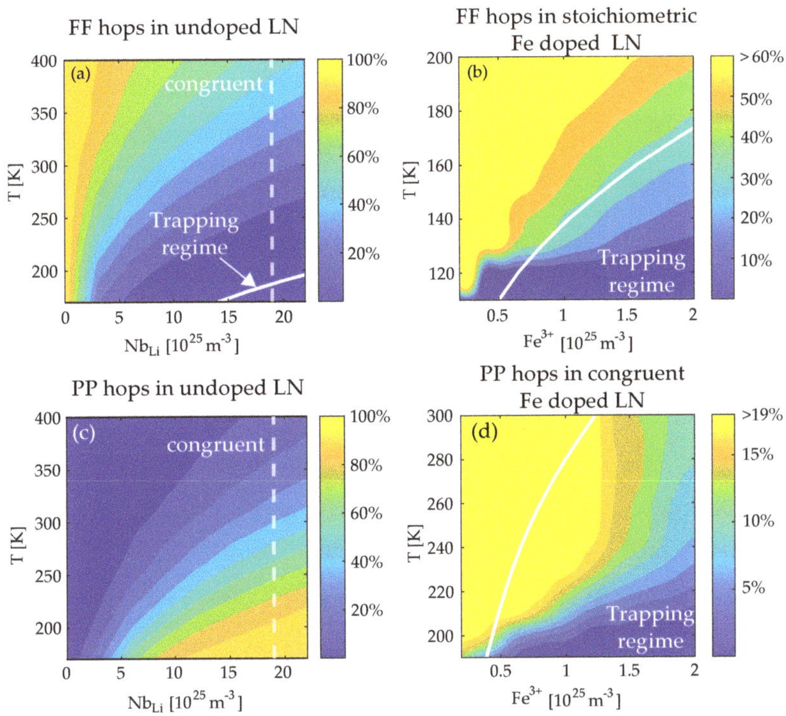

Figure 4. (**a**) Fraction of hops as a free polaron N_{FF}/N_{tot} as a function of temperature and $[Nb_{Li}]$ concentration for an undoped sample ([Fe] = 0). (**b**) Fraction of hops as a free polaron N_{FF}/N_{tot} as a function of the temperature and [Fe] concentration for a stoichiometric sample ($[Nb_{Li}] = 0$). The solid lines represent the theoretical boundary between the pure trapping regime calculated according to Equations (15) and (16). (**c**) N_{PP}/N_{tot} as a function of temperature and $[Nb_{Li}]$ concentration for an undoped sample ([Fe] = 0). (**d**) N_{PP}/N_{tot} as a function of the temperature and [Fe] concentration for a congruent sample ($[Nb_{Li}] = 19 \times 10^{25}\,\mathrm{m}^{-3}$). Note the different temperature range in the two plots.

Figure 4b shows, for the case of a stoichiometric sample (i.e., $[Nb_{Li}] = 0$) doped with Fe, the average value of the fraction of FF hops over the total that a free polaron performs before being trapped by a Fe trap. As expected, by cooling the sample and/or increasing the Fe concentration percentage of hops on the Nb sublattice gets smaller and smaller. Below a certain boundary the polaron enters in a direct trapping regime so that all the created polarons are likely to be immediately captured by a Fe trap. Again, the solid line shows the boundary $\chi_{FFe} = 1$ below which it can be expected the disappearence of $F \rightarrow F$ hops, according to the criterion (16). As it can be seen the simulation is in line with the theoretical estimate and in particular confirms that the pure trapping regime is attained for $\chi_{FFe} \gg 1$.

In Figure 4c are reported the simulation results for the case of P polarons. Here the behaviour is roughly complementary to the one of Figure 4a, so that by lowering the temperature and/or increasing the antisite concentration the fraction of bound-bound hops becomes predominant. In Figure 4d is shown the situation of a congruent sample ($[Nb_{Li}] = 19 \times 10^{25}\mathrm{m}^{-3}$) with varying Fe concentration. Again the direct trapping regime is attained for low temperatures and high Fe concentration, in agreement with the criterion (18). As expected, the condition $\chi_{PFe} = 1$ indicated by a solid line in the figure provides an underestimation, i.e., at a given temperature, more Fe traps than expected are necessary to capture all the polarons. This is due to the fact that our calculations do not verify strictly the irreversibility condition, so that the time needed to truly leave the antisite is longer than τ_0. As a consequence the trapping radius in Equation (14) is an underestimation and a larger amount of Fe traps is necessary to achieve the trapping regime.

7. Conclusions

The outcome of a typical light-induced absorption experiment performed on iron-doped lithium niobate (Fe:LN) has been studied numerically with the help of a Monte Carlo simulation code based on the Marcus-Holstein model for polaron hopping.

Our analysis shows that in general, within the decay process of a polaron population induced by an impulsive illumination, we can distinguish at least two regimes. At short times the decay describes the direct trapping processes of those bound polarons that, at the beginning of the experimental time window, are located close to deep Fe traps and therefore are likely to disappear in a single hopping event due to their high trapping probability. At longer times, we can find those polarons that are located far away from the Fe traps, so that the waiting time on the initial site is comparable with the typical time τ_0 required for the polarons to migrate away, hopping on other transport sites. The two regimes have a quite different behaviour. For example the direct trapping regime is practically independent on temperature due to the low activation energy of the trapping barrier, in agreement with the experimental data [26]. On the other hand the migration-accelerated regime is especially important when the photoconductive properties of the sample are considered, as this is the regime were significant charge transport takes place.

Under appropriate experimental conditions, it may happen that the transient decay of a polaron population (such as those produced in a time-resolved LIA experiment) is predominantly due to one of the two processes, resulting in a direct trapping or migration-accelerated regime. We determined a criterion on the Nb_{Li} antisite and/or on Fe concentration to consider the free polaron contribution negligible in the charge transport processes at a given temperature. In these specific cases the criterion reads:

$$\frac{4\pi}{3}\left[N_{Fe}(R_{FFe})^3 + N_P(R_{FP})^3\right] \gg 1 \quad \text{Negligible free polaron contribution}$$

where R_{FFe} and R_{FP} are the trapping radiuses for the $F \rightarrow Fe$ and $F \rightarrow P$ trapping processes respectively, given by Equations (5)–(15).

On the other hand, for a bound polaron population in a Fe doped sample, the criterion to distinguish between one of the two regimes reads as follows:

$$N_{Fe} \leq \frac{3}{4\pi(R_{PFe})^3} \quad \text{Migration-accelerated trapping regime}$$

$$N_{Fe} \gg \frac{3}{4\pi(R_{PFe})^3} \quad \text{Direct trapping regime}$$

where $R_{PFe} = a_{PFe}\ln(w_{PFe}\tau_0)$ is a characteristic radius of a sphere of strong interaction surrounding the Fe traps. The quantities w_{PFe} and τ_0 are the hopping frequency pre-factor and the typical waiting time for the bound polarons hopping, described by Equations (2) and (12) respectively.

Our results are meaningful in the interpretation of experimental results obtained by time-resolved LIA as they allow a more physical reading of the decay curves. Besides this, they are of strong interest to crystal growers and experimenters to tailor the composition and/or the experimental conditions of a given sample in order to obtain a desired response. For example, a strong photorefractive effect is to be expected in a pure trapping regime, while a strong optical damage resistance is attained in conditions of free polaron diffusion. Further studies are now in progress to investigate, in the case of dominating diffusion conditions, the mobility of small and bound polarons, in order to determine an explicit expression for the sample photoconductivity according to the different regimes.

Author Contributions: Conceptualization, L.G., M.B. and S.F.; methodology, M.B.; software, L.V.; validation, M.B. and L.V.; formal analysis, L.G. and S.F.; investigation, L.V. and M.B.; resources, M.B.; data curation, L.V. and M.B.; writing—original draft preparation, M.B.; writing—review and editing, L.V., S.F. and L.G.; visualization, L.V.; supervision, M.B.; project administration, M.B.; funding acquisition, M.B. All authors have read and agreed to the published version of the manuscript.

Funding: This research was funded by Università degli Studi di Padova, grant number BIRD209412.

Institutional Review Board Statement: Not applicable.

Informed Consent Statement: Not applicable.

Conflicts of Interest: The authors declare no conflict of interest.

Abbreviations

The following abbreviations are used in this manuscript:

LN	Lithium Niobate
KWW	Kohlrausch Williams and Watts stretched exponential function
F	Nb_{Nb} free polarons
P	Nb_{Li} bound polarons
LIA	Light Induced Absorption spectroscopy

References

1. Imlau, M.; Badorreck, H.; Merschjann, C. Optical nonlinearities of small polarons in lithium niobate. *Appl. Phys. Rev.* **2015**, *2*, 040606. [CrossRef]
2. Günter, P.; Huignard, J.P. *Photorefractive Materials and Their Applications 1*; 0342-4111; Springer: New York, NY, USA, 2006; Volume 113.
3. Bazzan, M.; Sada, C. Optical waveguides in lithium niobate: Recent developments and applications. *Appl. Phys. Rev.* **2015**, *2*, 040603, [CrossRef]
4. He, J.; Franchini, C.; Rondinelli, J.M. Lithium Niobate-Type Oxides as Visible Light Photovoltaic Materials. *Chem. Mater.* **2016**, *28*, 25–29. [CrossRef]
5. Tisdale, W.A.; Williams, K.J.; Timp, B.A.; Norris, D.J.; Aydil, E.S.; Zhu, X.Y. Hot-electron transfer from semiconductor nanocrystals. *Science* **2010**, *328*, 1543–1547. [CrossRef]
6. Pelaez, M.; Nolan, N.T.; Pillai, S.C.; Seery, M.K.; Falaras, P.; Kontos, A.G.; Dunlop, P.S.; Hamilton, J.W.; Byrne, J.A.; O'shea, K.; et al. A review on the visible light active titanium dioxide photocatalysts for environmental applications. *Appl. Catal. B Environ.* **2012**, *125*, 331–349. [CrossRef]
7. Migani, A.; Blancafort, L. Excitonic interfacial proton-coupled electron transfer mechanism in the photocatalytic oxidation of methanol to formaldehyde on TiO2 (110). *J. Am. Chem. Soc.* **2016**, *138*, 16165–16173. [CrossRef]
8. Zhong, Y.; Trinh, M.T.; Chen, R.; Purdum, G.E.; Khlyabich, P.P.; Sezen, M.; Oh, S.; Zhu, H.; Fowler, B.; Zhang, B.; et al. Molecular helices as electron acceptors in high-performance bulk heterojunction solar cells. *Nat. Commun.* **2015**, *6*, 1–8. [CrossRef] [PubMed]
9. Schirmer, O.F.; von der Linde, D. Two-photon- and x-ray-induced Nb^{4+} and O^- small polarons in $LiNbO_3$. *Appl. Phys. Lett.* **1978**, *33*, 35–38. [CrossRef]
10. Schirmer, O.; Juppe, S.; Koppitz, J. Electron-spin-resonance, optical and photovoltaic studies of reduced undoped Fe : $LiNbO_3$. *Cryst. Lattice Defects Amorph. Mater.* **1987**, *16*, 353–357.
11. Schirmer, O.; Thiemann, O.; Wöhlecke, M. Defects in $LiNbO_3$. experimental aspects. *J. Phys. Chem. Solids* **1991**, *52*, 185–200. [CrossRef]
12. Schirmer, O.F. O^- bound small polarons in oxide materials. *J. Phys. Condens. Matter* **2006**, *18*, R667. [CrossRef]
13. Schirmer, O.F.; Imlau, M.; Merschjann, C.; Schoke, B. Electron small polarons and bipolarons in $LiNbO_3$. *J. Phys. Condens. Matter* **2009**, *21*, 123201. [CrossRef]
14. Schirmer, O.F.; Imlau, M.; Merschjann, C. Bulk photovoltaic effect of $LiNbO_3$: Fe and its small-polaron-based microscopic interpretation. *Phys. Rev. B* **2011**, *83*, 165106. [CrossRef]
15. Emin, D. *Polarons*; Cambridge University Press: Cambridge, UK, 2013.
16. Tatyana Volk, M.W. *Lithium Niobate*; Springer: Berlin/Heidelberg, Germany, 2009.
17. Schmidt, F.; Kozub, A.L.; Biktagirov, T.; Eigner, C.; Silberhorn, C.; Schindlmayr, A.; Schmidt, W.G.; Gerstmann, U. Free and defect-bound (bi)polarons in $LiNbO_3$: Atomic structure and spectroscopic signatures from ab initio calculations. *Phys. Rev. Res.* **2020**, *2*, 043002. [CrossRef]
18. Guilbert, L.; Vittadello, L.; Bazzan, M.; Mhaouech, I.; Messerschmidt, S.; Imlau, M. The elusive role of NbLi bound polaron energy in hopping charge transport in Fe: LiNbO3. *J. Phys. Condens. Matter* **2018**, *30*, 125701. [CrossRef] [PubMed]

19. Merschjann, C. Optically Generated Small Polarons: Time-Resolved Pump-Multiprobe Experiments in Lithium Niobate vs. Random-Walk Charge-Transport Simulations in Oxide Crystals. Ph.D. Thesis, Fachbereich Physik, Univesitat Osnabrueck, Osnabrück, Germany, 2007.
20. Merschjann, C.; Schoke, B.; Conradi, D.; Imlau, M.; Corradi, G.; Polgár, K. Absorption cross sections and number densities of electron and hole polarons in congruently melting $LiNbO_3$. *J. Phys. Condens. Matter* **2009**, *21*, 015906. [CrossRef] [PubMed]
21. Holstein, T. Studies of polaron motion. *Ann. Phys.* **1959**, *8*, 343–389. [CrossRef]
22. Marcus, R.A. On the Theory of Oxidation-Reduction Reactions Involving Electron Transfer. I. *J. Chem. Phys.* **1956**, *24*, 966–978. [CrossRef]
23. Zylbersztejn, A. Thermally activated trapping in Fe-doped LiNbO3. *Appl. Phys. Lett.* **1976**, *29*, 778–780. [CrossRef]
24. Mhaouech, I.; Guilbert, L. Temperature dependence of small polaron population decays in iron-doped lithium niobate by Monte Carlo simulations. *Solid State Sci.* **2016**, *60*, 28–36. [CrossRef]
25. Austin, I.G.; Mott, N.F. Polarons in crystalline and non-crystalline materials. *Adv. Phys.* **2001**, *50*, 757–812. doi:10.1080/00018730110103249. [CrossRef]
26. Vittadello, L.; Bazzan, M.; Messerschmidt, S.; Imlau, M. Small polaron hopping in Fe: LiNbO3 as a function of temperature and composition. *Crystals* **2018**, *8*, 294. [CrossRef]
27. Beyer, O.; Maxein, D.; Woike, T.; Buse, K. Generation of small bound polarons in lithium niobate crystals on the subpicosecond time scale. *Appl. Phys. B* **2006**, *83*, 527–530. [CrossRef]
28. Berben, D.; Buse, K.; Wevering, S.; Herth, P.; Imlau, M.; Woike, T. Lifetime of small polarons in iron-doped lithium niobate crystals. *J. Appl. Phys.* **2000**, *87*, 1034–1041. [CrossRef]
29. Herth, P.; Granzow, T.; Schaniel, D.; Woike, T.; Imlau, M.; Krätzig, E. Evidence for Light-Induced Hole Polarons in $LiNbO_3$. *Phys. Rev. Lett.* **2005**, *95*, 067404. [CrossRef] [PubMed]
30. Shklovskii, B.I.; Efros, A.L. *Electronic Properties of Doped Semiconductors*; Springer Science & Business Media: New York, NY, USA, 2013; Volume 45.

crystals

MDPI

Article

Electron Polarons in Lithium Niobate: Charge Localization, Lattice Deformation, and Optical Response

Falko Schmidt, Agnieszka L. Kozub, Uwe Gerstmann, Wolf Gero Schmidt and Arno Schindlmayr *

Department Physik, Universität Paderborn, 33095 Paderborn, Germany; schmidtf@campus.uni-paderborn.de (F.S.); agnieszka.kozub@uni-paderborn.de (A.L.K.); uwe.gerstmann@uni-paderborn.de (U.G.); w.g.schmidt@uni-paderborn.de (W.G.S.)

* Correspondence: arno.schindlmayr@uni-paderborn.de

Abstract: Lithium niobate ($LiNbO_3$), a material frequently used in optical applications, hosts different kinds of polarons that significantly affect many of its physical properties. In this study, a variety of electron polarons, namely free, bound, and bipolarons, are analyzed using first-principles calculations. We perform a full structural optimization based on density-functional theory for selected intrinsic defects with special attention to the role of symmetry-breaking distortions that lower the total energy. The cations hosting the various polarons relax to a different degree, with a larger relaxation corresponding to a larger gap between the defect level and the conduction-band edge. The projected density of states reveals that the polaron states are formerly empty Nb 4*d* states lowered into the band gap. Optical absorption spectra are derived within the independent-particle approximation, corrected by the *GW* approximation that yields a wider band gap and by including excitonic effects within the Bethe–Salpeter equation. Comparing the calculated spectra with the density of states, we find that the defect peak observed in the optical absorption stems from transitions between the defect level and a continuum of empty Nb 4*d* states. Signatures of polarons are further analyzed in the reflectivity and other experimentally measurable optical coefficients.

Keywords: lithium niobate; polarons; charge localization; lattice deformation; optical response; density-functional theory; Bethe-Salpeter equation

check for updates

Citation: Schmidt, F.; Kozub, A.L.; Gerstmann, U.; Schmidt, W.G.; Schindlmayr, A. Electron Polarons in Lithium Niobate: Charge Localization, Lattice Deformation, and Optical Response. *Crystals* **2021**, *11*, 542. https://doi.org/10.3390/cryst11050542

Academic Editors: Gábor Corradi and László Kovács

Received: 30 March 2021
Accepted: 2 May 2021
Published: 13 May 2021

1. Introduction

Lithium niobate ($LiNbO_3$, LN) is a transparent ferroelectric solid that is extensively used in optical technologies due to its advantageous combination of functional properties and commercial availability. In particular, its large piezoelectric, electro-optical, acousto-optical, and nonlinear optical coefficients make it an ideal material for a wide range of optical devices, such as optical modulators [1], waveguides [2], optical sensors [3], holographic storage [4], and integrated photonics [5]. By doping with other elements or by manipulating the sample geometry or crystal structure, for example by periodic poling of the ferroelectric domains, the properties of LN can be tailored for specific purposes [6]. To achieve optimal results, it is important to understand how structural details at the atomic scale influence the observable electronic and optical properties of the material.

The nominal composition of $LiNbO_3$ involves an equal number of lithium and niobium atoms, and although it is possible to produce high-quality stoichiometric LN samples by various methods [7], these are not viable for large-scale fabrication. Instead, the Czochralski technique that is typically used to grow LN single crystals for commercial applications from a congruent melt leads to a Li:Nb ratio of about 48.5:51.5 [8]. Even without external doping, this relative surplus of Nb in congruent LN crystals implies a high concentration of intrinsic defects, whose nature has long been a matter of debate, as different conceivable defect models are compatible with the observed nonstoichiometric composition [9]. Nowadays, it is widely accepted that Nb_{Li} antisite defects, where extra niobium atoms are inserted on regular lithium sites, play a central role in explaining this imbalance. As the two atomic species have different charge states, the resulting excess charge must be compensated by other defect types, such as cationic vacancies, however. In the prevalent Li-vacancy model, originally proposed by Lerner et al. [10], each Nb_{Li} antisite defect is compensated by four Li vacancies, denoted as V_{Li}. However, interstitial Nb_V atoms placed at empty cationic sites in the crystal lattice could also contribute to the surplus of niobium [11]. As we showed recently, these models may be partially reconciled, because a Nb_V–V_{Li} defect pair consisting of an interstitial niobium atom and a lithium vacancy can be regarded as a metastable variant of the Nb_{Li} antisite defect, where the antisite niobium atom overcomes a modest energy barrier and migrates to a neighboring empty cationic site [12].

An important consequence of the high defect concentration in LN is the ubiquitous occurrence of different types of polarons [13]. These arise when mobile electrons or holes interact with the ionic lattice to create a local distortion, leading to an attractive effective potential. If the resulting potential well is sufficiently deep, then the charge carrier is trapped and immobilized. The quasiparticle consisting of the self-trapped electron or hole together with the surrounding spatially localized lattice distortion is called polaron or, more specifically, small polaron if the coupling is so strong that the charge carriers are essentially confined to a single lattice site. In LN, three polaron-related optical absorption bands are experimentally observed at 0.9 eV [14], 1.6 eV [15], and 2.5 eV [16]. Based on circumstantial evidence, these are usually attributed to the free electron polaron trapped at a regular Nb lattice site (Nb_{Nb}), the bound polaron formed by a single localized electron at a Nb_{Li} antisite defect, and the bipolaron formed by a pair of bound electrons, one at a Nb_{Li} antisite defect and the other at the neighboring regular Nb_{Nb} atom, respectively, [13]. In particular, the temperature dependence of the relative strengths of the absorption bands at 1.6 eV and 2.5 eV are consistent with the thermal [16] or optical [17] dissociation of bipolarons into single bound polarons, and the band at 0.9 eV could only be observed in reduced samples with a high dopant concentration of Mg ions, which inhibit the formation of Nb_{Li} antisite defects, leaving free polarons as a natural explanation [14]. On the other hand, the assignment to particular defect types rests chiefly on the correlation with electron-paramagnetic-resonance (EPR) signals believed to originate from Nb_{Li} antisite defects [13], but as the actual spatial distribution of the polaron cannot be imaged directly in experiments, the precise relation between the antisite defect and the localized charge accumulation ultimately remains open. In this situation, first-principles simulations of the defect structures and their EPR parameters [12] as well as their optical spectroscopic properties provide valuable further insight.

In an early theoretical study, Donnerberg et al. [18] employed a semiempirical shell model to investigate intrinsic defects in LN. Their results, as well as other subsequent atomistic simulations with improved interaction potentials [19], lent support to the Li-vacancy model and indicated the stability of small electron polarons and bipolarons, but neither the electronic energy levels nor the corresponding orbitals appear explicitly in this approach. Later, first-principles calculations based on density-functional theory yielded accurate quantitative values for the formation energies of several possible intrinsic defects and confirmed the Li-vacancy scenario as the dominant mechanism for the deviation from stoichiometry [20,21]. Furthermore, the localized wavefunctions of the bound polaron and the bipolaron at the Nb_{Li} antisite defect are accessible in this framework [22]. As density-functional theory systematically underestimates the band gap of insulators like LN, the position of the defect levels inside the band gap cannot be obtained reliably in this way, however, precluding a quantitative comparison with the experimentally observed absorption bands. In a first step, the band-gap problem can be solved by including quasiparticle corrections within the *GW* approximation for the electronic self-energy [23,24], the zero-point renormalization due to lattice vibrations [25], and spin-orbit coupling, which altogether raise the band gap of LN by about 2 eV compared to standard density-functional theory [26]. Second, the optical absorption spectrum is strongly influenced by electron-hole attraction effects, which give rise to excitonic resonances. In principle, these can be treated within time-dependent density-functional theory. Numerical results reported by Friedrich et al. for intrinsic [27] and extrinsic [28] defects in LN were in good quantitative agreement with the experimental data, but the uncertainty due to empirical corrections, which are necessary to countervail limitations of standard approximations in time-dependent density-functional theory [29], still precluded a reliable discrimination between competing defect configurations involving antisite or interstitial niobium atoms. As an alternative, we adopt the Bethe–Salpeter equation (BSE), which requires no such empirical corrections, to determine the optical response in this work. Owing to the much higher computational expense, a full solution of the BSE was initially limited to stoichiometric LN [23,30], but we recently succeeded in extending this approach to congruent LN [12], which requires a larger supercell to accommodate the defect.

The aim of this paper is twofold. First, we present a detailed description of the geometric and electronic structure of electron polarons in LN, with special focus on the quasi-Jahn-Teller distortion that breaks the threefold rotational symmetry in the case of free and bound polarons. The resulting tilted configurations were only very recently revealed as the true ground states [12] and are hence not yet widely discussed in the literature, as previous theoretical studies [21,22] were based on simpler, axially symmetric structure models instead. Second, we analyze the contribution of polarons to the dielectric function, which is related to the optical absorption spectrum. Going beyond our previous work [12], we also investigate polaron signatures in other optical coefficients like the reflectivity or the electron-energy-loss function, which can be measured directly in experiments.

This paper is organized as follows. In Section 2, we present the different structure models for free polarons, bound polarons, and bipolarons, and we explain our computational methods. Subsequently, we discuss our results for the polaron-induced lattice deformation in Section 3.1, for the electronic structure in Section 3.2, and for the dielectric function and optical coefficients in Section 3.3. Finally, Section 4 summarizes our conclusions.

2. Models and Methods

Lithium niobate belongs to the trigonal crystal system with the space group *R*3*c* [8]. Its crystal structure consists of octahedral cages formed by oxygen atoms. The cationic sites inside the oxygen cages can either be occupied by a lithium or niobium ion, or remain empty. For defect-free stoichiometric lithium niobate (SLN), the stacking order along the threefold symmetry axis is a periodic repetition of the sequence Li–Nb–vacancy as illustrated in Figure 1a. Due to the ferroelectric distortion, which breaks the inversion symmetry, the cations do not reside in the geometric center of the oxygen cages but are slightly displaced in the vertical direction. To highlight point defects in LN, which can be characterized by a modified occupancy of the cationic sites, we use a schematic representation where each square symbolizes a filled or empty oxygen octahedron.

In the absence of defects, an additional electron introduced into the material will localize at one of the regular Nb_{Nb} atoms, as indicated by the red square in Figure 1b. Together with the resulting lattice relaxation, this gives rise to the free polaron. Congruent LN, on the other hand, features a high concentration of Nb_{Li} antisite defects. In this scenario, an additional electron will localize at the defect site instead, creating a bound polaron as in Figure 1c. The Nb_{Li} antisite atom can also migrate into the next empty oxygen octahedron, producing a Nb_V–V_{Li} defect pair. This configuration can likewise host a bound polaron, depicted in Figure 1d. If not one but two electrons are added to the system, then a bipolaron forms either at the antisite, Figure 1e, or the defect pair, Figure 1f. All defect structures illustrated in Figure 1 are considered in this work.

The primitive rhombohedral unit cell of stoichiometric LN contains two formula units of $LiNbO_3$, amounting to 10 atoms. For the investigation of point defects, we choose a periodically repeated $2 \times 2 \times 2$ supercell with 80 atoms, in which one defect is embedded. The same supercell geometry was used in [21,22]. An explicit comparison of different supercells containing between 80 and 270 atoms found that the formation energies of the defect types considered here remain stable for all supercell sizes, with merely slight numerical variations [20]. This is in agreement with our own test calculations for a larger

$3 \times 3 \times 3$ supercell containing 270 atoms, reported previously in [27], which also confirmed that the relaxed interatomic distances in the immediate vicinity of the defect are almost identical and that the resonances in the imaginary part of the dielectric function align nicely for both supercell sizes, except for a variation in intensity that reflects the different defect concentrations. The present choice is hence justified and serves to limit the computational cost associated with the solution of the BSE, but it will also be validated a posteriori in this work. Incidentally, we note that it is possible to avoid supercells altogether by using perturbation theory and describing the properties of polarons in leading order in terms of the electronic structure of the pristine stoichiometric crystal [31], which is advantageous for materials featuring large polarons, but as electron polarons in LN are known to be small [13], the nonperturbative supercell approach seems appropriate here.

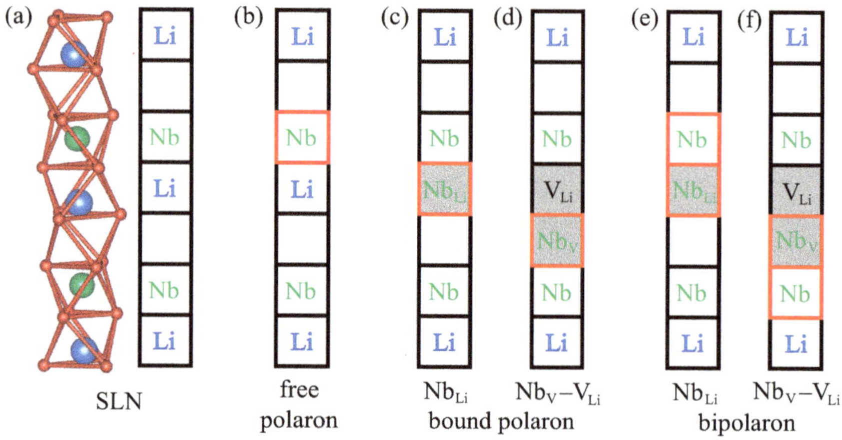

Figure 1. Stacking order of lithium niobate: Stoichiometric crystal (**a**), free polaron (**b**), bound polaron at a Nb_{Li} antisite (**c**) or a $Nb_V{-}V_{Li}$ defect pair (**d**), bipolaron at a Nb_{Li} antisite (**e**) or a $Nb_V{-}V_{Li}$ defect pair (**f**). Black squares represent oxygen octahedra, while red squares indicate the formation of a (bi)polaron at this site. Gray fillings highlight deviations from the stoichiometric solid.

All external and internal degrees of freedom are relaxed within density-functional theory (DFT) using the Quantum ESPRESSO [32] package. We use norm-conserving pseudopotentials and the PBEsol functional [33] to describe electronic exchange-correlation effects. Compared to other common parametrizations, the PBEsol functional yields more reliable lattice parameters for LN [25] and closely related materials [34–36]. The oxygen $2s$ and $2p$ orbitals as well as the lithium $2s$ orbitals are treated explicitly as valence states. The pseudopotential of niobium is optimized for the Nb^{5+} cation configuration, which emerges on regular lattice sites in LN, in order to increase the numerical stability and to allow for smaller cutoff radii; in this case, the $4s$, $4p$, $4d$, and $5s$ orbitals are treated as valence states. This leads to 256 valence bands for the free-polaron system in stoichiometric LN and 260 valence bands for all models of bound polarons and bipolarons, where one Li^+ cation is substituted by a Nb^{5+} ion. We select a kinetic-energy cutoff of 85 Ry for the plane-wave basis set and a shifted Monkhorst-Pack mesh with $2 \times 2 \times 2$ **k** points for the Brillouin-zone integration during the self-consistency cycle, equivalent to $4 \times 4 \times 4$ **k** points in the larger Brillouin zone corresponding to the primitive unit cell. The convergence thresholds for energies and forces during the relaxation are set to 10^{-4} Ry and 10^{-8} Ry/Bohr, respectively. For an accurate description of the Nb $4d$ orbitals, we employ the DFT + U scheme [37]. The values for the Hubbard U parameter are determined self-consistently as 5.2 eV for the Nb_{Li} antisite and Nb_V interstitial atoms and 4.7 eV for regular Nb_{Nb} atoms, including the one hosting the free polaron [12]. For free and bound polarons, which feature a single unpaired electron, we perform spin-polarized calculations; for bipolarons, this is not necessary, because they comprise two electrons with opposite spin orientations.

From the one-particle energies $\epsilon_{n\mathbf{k}}$ and the associated wavefunctions $|n\mathbf{k}\rangle$, we obtain the projected density of states

$$\mathrm{DOS}(\omega) = \sum_n \sum_{\mathbf{k}} \langle n\mathbf{k}|P|n\mathbf{k}\rangle g(\omega - \epsilon_{n\mathbf{k}}) \,, \tag{1}$$

which provides important information about the electronic structure of the material. The summation over n includes 800 bands, sufficient to cover an energy interval up to 10 eV above the valence-band maximum, and the wavevector $\mathbf{k}$ runs over all points of a shifted $4 \times 4 \times 4$ mesh. A Gaussian function g with a broadening of 0.05 eV is applied to compensate for the discrete $\mathbf{k}$-point mesh. The operator P projects the one-particle wavefunctions onto orthogonalized atomic orbitals corresponding to the selected pseudopotentials; the total density of states (DOS) is obtained by setting $P = 1$.

The frequency-dependent dielectric function, from which the absorption spectrum and other optical coefficients may be derived, is constructed at different levels of theory in this work. Within the independent-particle approximation (IPA), the tensor elements in the long-wavelength limit are

$$\varepsilon_{\alpha\alpha}(\omega) = 1 + \frac{16\pi}{\Omega} \sum_{v,c} \sum_{\mathbf{k}} \frac{1}{\epsilon_{c\mathbf{k}} - \epsilon_{v\mathbf{k}}} \frac{|\langle v\mathbf{k}|p_\alpha + i[V^{\mathrm{nl}}, r_\alpha]|c\mathbf{k}\rangle|^2}{(\epsilon_{c\mathbf{k}} - \epsilon_{v\mathbf{k}})^2 - (\omega + i\gamma)^2} \,, \tag{2}$$

where α denotes the spatial direction. We follow the usual convention where the threefold symmetry axis is identified with the z direction, while the x direction is perpendicular to the z direction and lies in the plane spanned by the threefold symmetry axis and one of the three equivalent basis vectors of the primitive rhombohedral unit cell. For the numerical evaluation of the dielectric function, we employ the Yambo package [38]. The first sum in Equation (2) runs over all combinations of valence (v) and conduction (c) bands, the second over the set of wavevectors inside the Brillouin zone. The transition dipole moments include the commutator $[V^{\mathrm{nl}}, r_\alpha]$ with the nonlocal part of the pseudopotentials, Ω is the volume of the supercell, and the broadening γ again compensates for the finite $\mathbf{k}$-point mesh.

For a better quantitative description, we replace the Kohn–Sham eigenvalues $\epsilon_{n\mathbf{k}}$ in a first step by the proper quasiparticle energies $\epsilon^{\mathrm{qp}}_{n\mathbf{k}}$ calculated within the GW approximation, which not only opens the band gap by about 2 eV, in agreement with experimental measurements, but also modifies the dispersion [26]. Within this independent-quasiparticle approximation (IQA), the transition energies between valence and conduction bands are larger than in the IPA, reflecting the widened band gap, but there is no interaction between the created electrons and holes. For the numerical evaluation of the exchange-correlation self-energy, we employ a plasmon-pole approximation for the dynamical screening function as implemented in the Yambo package.

In a second step, we then incorporate the electron-hole interaction by solving the BSE, again with the Yambo package. In this scheme, the macroscopic dielectric function is constructed as

$$\varepsilon_{\alpha\alpha}(\omega) = 1 - \lim_{\mathbf{q}\to 0} \frac{8\pi}{\Omega|\mathbf{q}|^2} \sum_{v,c,\mathbf{k}} \sum_{v',c',\mathbf{k}'} \langle v\mathbf{k}-\mathbf{q}|e^{-i\mathbf{q}\cdot\mathbf{r}}|c\mathbf{k}\rangle \langle c'\mathbf{k}'|e^{i\mathbf{q}\cdot\mathbf{r}}|v'\mathbf{k}'-\mathbf{q}\rangle \sum_\lambda \frac{A^\lambda_{vc\mathbf{k}} \left(A^\lambda_{v'c'\mathbf{k}'}\right)^*}{\omega - E_\lambda + i\gamma} \,, \tag{3}$$

where $\mathbf{q}$ approaches zero in the direction α. The energies E_λ of the interacting electron-hole pairs and the expansion coefficients $A^\lambda_{vc\mathbf{k}}$ of the resulting exciton states in the basis of one-particle wavefunctions are given by the eigenvalues and eigenvectors of the two-particle Hamiltonian

$$H_{vc\mathbf{k},v'c'\mathbf{k}'} = \left(\epsilon^{\mathrm{qp}}_{c\mathbf{k}} - \epsilon^{\mathrm{qp}}_{v\mathbf{k}}\right)\delta_{vv'}\delta_{cc'}\delta_{\mathbf{k}\mathbf{k}'} + \left(2\bar{V}_{vc\mathbf{k},v'c'\mathbf{k}'} - W_{vc\mathbf{k},v'c'\mathbf{k}'}\right) . \tag{4}$$

The first, diagonal term, which equals the transition energies between quasiparticle states in the valence and conduction bands, corresponds to the IQA. The second, nondiagonal term, the kernel of the BSE, is composed of the electron-hole exchange part $\bar{V}$ and the electron-hole attraction W. The latter describes the formation of excitons, i.e., bound electron-hole pairs, which strongly influence the shape of the dielectric function.

From the real and imaginary parts of the complex dielectric function, obtained by solving the BSE, various optical coefficients with a direct relation to experimental measurements may subsequently be derived. In particular, the refractive index $n(\omega)$ and the extinction coefficient $\kappa(\omega)$ are given by

$$n(\omega) = \sqrt{\frac{|\varepsilon(\omega)| + \mathrm{Re}\,\varepsilon(\omega)}{2}} \quad \text{and} \quad \kappa(\omega) = \sqrt{\frac{|\varepsilon(\omega)| - \mathrm{Re}\,\varepsilon(\omega)}{2}}\,. \tag{5}$$

From these, we may obtain the reflectivity $R(\omega)$ and the absorption coefficient $\alpha(\omega)$ according to

$$R(\omega) = \frac{[n(\omega) - 1]^2 + \kappa(\omega)^2}{[n(\omega) + 1]^2 + \kappa(\omega)^2} \quad \text{and} \quad \alpha(\omega) = \frac{2\omega\kappa(\omega)}{c}\,, \tag{6}$$

where c denotes the speed of light. Finally, the electron-energy-loss function equals

$$L(\omega) = -\,\mathrm{Im}\,\frac{1}{\varepsilon(\omega)}\,. \tag{7}$$

3. Results and Discussion

3.1. Structure Optimization

The crystal structure of LN is characterized by a threefold axis of rotational symmetry, which is still preserved even when the spatial inversion symmetry is broken in the ferroelectric phase. As polarons form at regular $\mathrm{Nb_{Nb}}$ atoms or point defects located on this axis, it is natural to assume that the local lattice deformation associated with the polaron formation also preserves the threefold rotational symmetry. Indeed, earlier theoretical studies of polarons in LN incorporated this explicitly as a constraint on the atomic movements in order to simplify the structural optimization [22], and the assumed rotational symmetry was also used to analyze experimental measurements, such as data from electron paramagnetic resonance (EPR) [39]. Recently, it became clear that lower-energy configurations can be reached if the rotational symmetry is broken, however. As an example, the total energy of the optimized tilted geometry obtained from DFT for a bound polaron at the $\mathrm{Nb_{Li}}$ antisite defect is 43 meV lower than for axial symmetry; the corresponding value for a bound polaron at the $\mathrm{Nb_V}$–$\mathrm{V_{Li}}$ defect pair is 38 meV [12]. Further corroboration comes from the fact that EPR parameters calculated within DFT for the tilted configurations are in closer quantitative agreement with high-resolution experimental measurements reported in [40] than those predicted for axial symmetry [12]. From this perspective, the axially symmetric structures can be regarded as the average of three equivalent tilted configurations along the trigonal axes. At high temperatures, the system may then be described in terms of an effective axial symmetry resulting from rapid transitions between these equivalent degenerate configurations. For bipolarons at both defect types, in contrast, the system always relaxes to an axial symmetry, even without explicit constraints.

To explore the spatial extent of the polaronic lattice deformation in relation to the supercell size, we first analyze the displacements of the niobium and the oxygen atoms relative to their positions in stoichiometric LN as a function of the distance to the $\mathrm{Nb_{Li}}$ or $\mathrm{Nb_V}$ defect atom in Figure 2. Only results for the two bipolaron models are displayed, because these exhibit the largest distortion owing to the two localized electronic charges. Overall, the size of the atomic displacements decreases with growing distance and falls below a threshold of 0.05 Å within the displayed range. Earlier test calculations with a larger $3 \times 3 \times 3$ supercell containing 270 atoms [27] yielded very similar shifts for these atoms. We note that some atoms near the boundary of the supercell are affected by finite-size effects and hence excluded from the graph. Furthermore, lithium atoms, which are not directly involved in the formation of the defects and do not contribute to the density of states near the band edges, as shown in the following section, are also omitted in order to avoid cluttering.

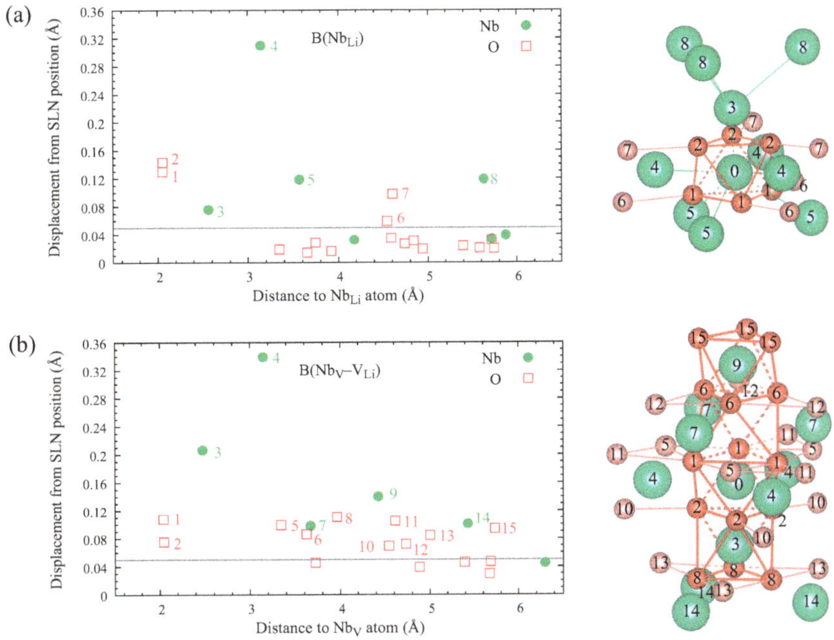

Figure 2. Atomic displacements in the vicinity of a bipolaron at the Nb_{Li} antisite defect (**a**) or the Nb_V–V_{Li} defect pair (**b**) with respect to the positions in stoichiometric LN, plotted as a function of the distance from the defect niobium atom hosting the bipolaron. Green circles and red boxes refer to niobium and oxygen atoms, respectively. The numbering corresponds to the pictured section of the crystal structure, the label "0" marks the defect Nb_{Li} or Nb_V atom hosting the bipolaron. Large green and small red balls represent niobium and oxygen atoms, respectively. Oxygen atoms outside the central pillar of oxygen octahedra are indicated in a lighter color shade.

For bipolarons at the Nb_{Li} antisite, shown in Figure 2a, the entire shell of niobium atoms around the defect (green labels "4", "5", and "8") must be relaxed in order to treat all displacements above 0.05 Å accurately. The oxygen sublattice appears more rigid, in contrast, as only the cage enclosing the Nb_{Li} antisite atom (red labels "1" and "2") plus a few horizontal neighbors (red labels "6" and "7") exhibit significant shifts. An even larger number of atoms must be considered for bipolarons at the Nb_V–V_{Li} defect pair in Figure 2b, including the Nb_{Nb} atom above the V_{Li} vacancy (green label "9"). The oxygen sublattice in particular shows shifts above 0.05 Å over a larger distance in this case, comprising several cages as well as a large number of oxygen atoms outside the central pillar.

These results demonstrate that the lattice deformation is not confined to the oxygen cages directly enclosing the defect niobium atoms but instead extends over several unit cells. Nevertheless, the atomic displacements in the immediate vicinity of the point defects, where the excess electrons are localized, are already accurately obtained with the 2 × 2 × 2 supercell, which was also used in previous studies [21,22]. Although it provides a tight fit, especially for the defect pair, it thus allows an accurate description of the electronic and optical properties if the polarons are sufficiently small. Furthermore, it enables us to apply sophisticated, numerically demanding many-body techniques, such as the BSE, that would be prohibitive for a larger supercell.

The relaxed crystal structures in the vicinity of the polaron sites for all defect models considered in this work are illustrated in Figure 3. In each case, the atomic positions after relaxation, indicated in color, are contrasted with those in the defect-free stoichiometric material, marked by black balls. The illustrations are to scale and indicate the direction and the magnitude of the atomic displacements. For the free polaron and the bound polaron at a Nb_{Li} antisite defect and a Nb_V–V_{Li} defect pair, we show both the axially symmetric configuration, obtained by applying an appropriate constraint, and the corresponding lower-energy tilted configuration. For the bipolaron at both defect types, we only consider axially symmetric configurations, which materialize even without explicit constraints. While the oxygen atoms move by a similar amount in all models, there are noticeable differences regarding the niobium atoms: The free polaron and the bound polaron at a Nb_{Li} antisite defect feature only small displacements, whereas the bound polaron at a Nb_V–V_{Li} defect pair as well as both bipolaron models exhibit much larger shifts.

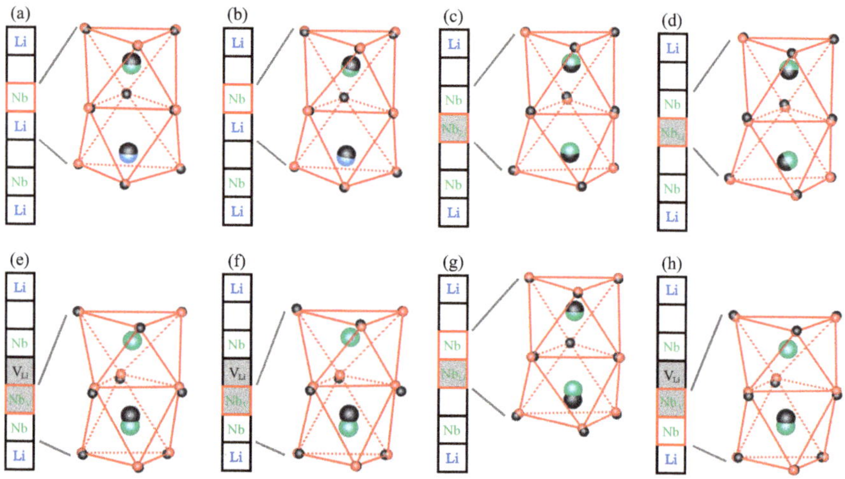

Figure 3. Crystal structure in the vicinity of the free polaron (**a**,**b**), the bound polaron at a Nb_{Li} antisite defect (**c**,**d**) or a Nb_V–V_{Li} defect pair (**e**,**f**), and the bipolaron at Nb_{Li} (**g**) or Nb_V–V_{Li} (**h**). Subfigures (**b**,**d**,**f**) refer to tilted configurations with broken trigonal symmetry, all other structures exhibit axial symmetry. Black balls indicate atomic positions in stoichiometric LN without polarons, while blue, green, and red balls denote lithium, niobium, and oxygen atoms in the relaxed structures, respectively. Note that (**e**,**f**,**h**) show a different segment of the crystal than the other subfigures.

For free polarons, the excess electron is trapped at the regular Nb_{Nb} atom in the top part of the crystal segment illustrated in Figure 3a for the axially symmetric and in Figure 3b for the tilted configuration. Originally positioned near the upper face of its oxygen cage to enlarge the distance to the neighboring Li cation, the Nb_{Nb} ion increases in size due to the electron capture and consequently moves closer to the center of the oxygen octahedron, in turn pushing the neighboring Li atom (bottom) further in the same direction. While both cations shift parallel to the vertical direction in the case of axial symmetry, a small sideways movement of the Nb_{Nb} atom hosting the free polaron as well as the neighboring Li atom are clearly visible if the symmetry constraint is removed, leading to the tilted configuration. The oxygen cage on the other side of the Nb_{Nb} atom is empty and exhibits only little distortion.

If a lithium atom is substituted by niobium, the capture of an excess electron results in a bound polaron at the Nb_{Li} antisite defect, illustrated in Figure 3c for the axially symmetric and in Figure 3d for the tilted configuration. Similar to free polarons, the defect atom carrying the extra electronic charge (bottom) moves towards the center of its oxygen cage, while the neighboring cation, a regular Nb_{Nb} atom (top), is repelled and moves in the same direction. In both configurations, the magnitude of the displacements is moderate, but the tilting is much more pronounced than for free polarons. However, it is essentially limited to the defect Nb_{Li} atom and does not noticeably affect the neighboring cation, in contrast to free polarons. This is most likely due to the larger mass of niobium compared to lithium.

With a second trapped electron at the Nb_{Li} antisite defect, the bound polaron turns into a bipolaron, Figure 3g. The displacement of the Nb_{Li} antisite atom (bottom) is much larger in this case, and the neighboring regular Nb_{Nb} atom (top) moves towards the defect instead of evading it. This indicates a bond similar to a hydrogen molecule, where the two niobium atoms attract each other due to a shared electron pair.

The second defect type examined in this work is the Nb_V–V_{Li} defect pair. Analogous to the Nb_{Li} antisite defect, the capture of an excess electron leads to the formation of a bound polaron, which can be modeled either in an axially symmetric configuration, Figure 3e, or, without geometric constraints, in a fully optimized tilted configuration, Figure 3f. As the interstitial Nb_V atom hosting the bound polaron (top) occupies a previously empty oxygen octahedron, the neighboring regular Nb_{Nb} atom (bottom), originally located near the upper face of its oxygen cage, undergoes a large shift in the direction away from the defect towards the center of the cage. The tilting is weaker than for the Nb_{Li} antisite defect, in accordance with the smaller energy difference between the axially symmetric and the tilted configuration.

For a bipolaron at the Nb_V–V_{Li} defect pair, displayed in Figure 3h, the interstitial Nb_V atom (top) moves further towards the center of its oxygen octahedron than in the case of the single bound polaron, and although the neighboring regular Nb_{Nb} atom (bottom) is still pushed in the same direction, the displacement is manifestly smaller. As a consequence, the distance between the two niobium atoms is smaller for bipolarons than for single polarons, which can again be attributed to the bonding effect arising from the shared electron pair.

3.2. Electronic Properties

Figure 4 shows a larger segment of the central pillars of the optimized lattice structures together with the charge densities of the polaron states for all configurations considered in this work. For single trapped electrons, our results confirm that most of the charge accumulation is indeed localized at one niobium atom, a regular Nb_{Nb} atom in the case of free polarons and a Nb_{Li} or Nb_V defect atom in the case of bound polarons. For the systems with axial symmetry displayed in Figure 4a,c,e, the polaron orbital has a dumbbell shape oriented along the central axis, which reflects the rotational symmetry of the underlying lattice structure, but if the systems are relaxed to the lower-energy tilted configurations, then we find a less elongated clover-leaf shape as shown in Figure 4b,d,f instead. Not all the charge of the polaron is localized at the niobium atom, however: The oxygen atoms that form the octahedral cage around the niobium atom hosting the polaron also attract a significant portion of the charge. For the bound polaron at the Nb_V–V_{Li} defect pair with axial symmetry, there is additionally a significant hybridization with the orbitals of the neighboring regular Nb_{Nb} atom, as seen in Figure 4e. This is facilitated by the fact that the axial symmetry enforces a parallel orientation of the orbitals of the two atoms. The symmetry-breaking tilt diminishes this effect and thereby effectuates a stronger localization at the Nb_V atom, as illustrated in Figure 4f. Although the hybridization also serves to lower the total energy [22], the structural deformation has a larger influence, so that the tilted configuration is in fact more stable for single-electron bound polarons.

The charge density of the bipolaron at a Nb_{Li} antisite defect in Figure 4g appears almost identical to the overlap of the charge densities of a free and a bound polaron, shown separately in Figure 4a,c. This interpretation is in accordance with earlier studies that also characterized the bipolaron as a building-block-like combination of the two single-polaron types [13]. Furthermore, it explains why the structure relaxes to axial symmetry in the case of bipolarons: Although the tilting slightly lowers the energy of single free and bound polarons, the resulting reorientation and shape deformation of the orbitals noted above prevents an effective hybridization. As the energy gain due to the hybridization outweighs the energy difference resulting from the atomic displacements, the system reverts back to an axially symmetric configuration.

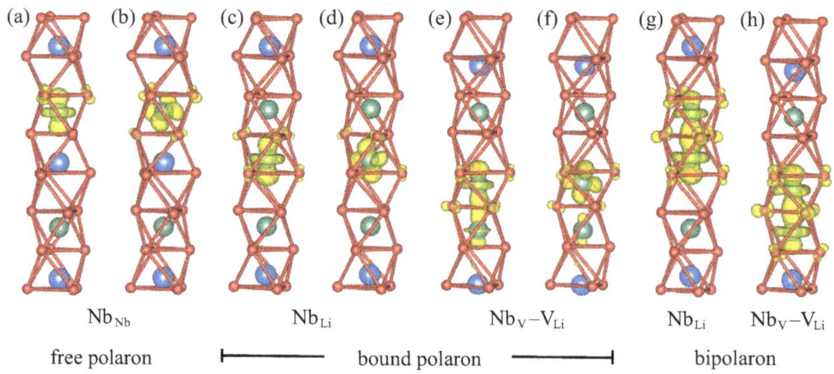

Figure 4. Charge densities of the free polaron (**a**,**b**), the bound polaron at a Nb_{Li} antisite defect (**c**,**d**) or a Nb_V−V_{Li} defect pair (**e**,**f**), and the bipolaron at Nb_{Li} (**g**) or Nb_V−V_{Li} (**h**). Subfigures (**b**,**d**,**f**) refer to tilted configurations, all other structures exhibit axial symmetry. Blue, green, and red balls represent lithium, niobium, and oxygen atoms, respectively. Reproduced from [12].

Lastly, the Nb_V−V_{Li} defect pair may be regarded as a modification of the Nb_{Li} antisite defect where the antisite Nb_{Li} atom migrates into the neighboring empty oxygen octahedron [12]. Concomitantly, as evidenced in Figure 4, the uppermost displayed Li atom also shifts into the neighboring empty octahedron for all models involving the defect pair. The pair of the Nb_V and Nb_{Nb} atoms in the two adjacent oxygen cages that host the bipolaron ultimately bears a strong resemblance to the pair of neighboring Nb_{Li} and Nb_{Nb} atoms in the case of the antisite defect if turned upside down, which can clearly be seen by comparing Figure 4g,h. Due to the very similar local environment, it is not surprising that the charge densities of the bipolarons at the two defects are also almost identical. In particular, the hybridization, which has a twice as large effect on the bipolaron than on the one-electron bound polaron, again explains the relaxation to an axially symmetric configuration.

In Figure 5, we show the projected density of states for all defect configurations as obtained from DFT without quasiparticle corrections. The zero of the energy axis corresponds to the maximum of the bulk valence bands. Only the Nb 4*d*, O 2*s*, and O 2*p* states are displayed, because all other states have negligible contributions in the energy region around the band gap. In agreement with earlier studies [22], we find that the top of the valence bands is dominated by O 2*p* states with minor contributions of Nb 4*d*, while the lowest conduction bands between 3 eV and 5.5 eV are predominantly composed of Nb 4*d* states with an admixture of O 2*p*. The next set of conduction bands above 6.5 eV also exhibits a contribution of O 2*s* states. These features are identical for all configurations, because the valence and conduction bands are bulk properties and independent of the type or symmetry of embedded defects.

The polaron peak is located inside the bulk band gap between 1 eV and 3 eV. As expected, it is dominated by Nb_{Nb} 4*d* states for free polarons, by Nb_{Li} 4*d* states for bound polarons or bipolarons at the antisite defect, and by Nb_V 4*d* states for bound polarons or bipolarons at the Nb_V−V_{Li} defect pair. However, in all cases, there is a also substantial admixture of O 2*p* states from the oxygen atoms surrounding the defect, which accounts for about one third of the density of states. The proportion of Nb_{Nb} 4*d* states from neighboring regular niobium atoms is negligible for all types of bound polarons, while for bipolarons, which extend over two cation sites, the contribution of Nb_{Nb} 4*d* states equals that of the Nb 4*d* states associated with the antisite or interstitial niobium atom. All of these findings are consistent with the charge densities displayed in Figure 4. Our results thus put the common notion of electron polarons in LN as small polarons [13] into perspective: Although the polarons are clearly centered at one or, in the case of bipolarons, two niobium atoms, a significant portion amounting to about one third of the trapped charge is in fact distributed over the surrounding oxygen atoms. On the other hand, the fact that the charge density of the polaron is essentially confined to one oxygen octahedron, or two in the case of bipolarons, justifies a posteriori our use of a $2 \times 2 \times 2$ supercell, which is large enough to reliably isolate each polaron from its periodic images.

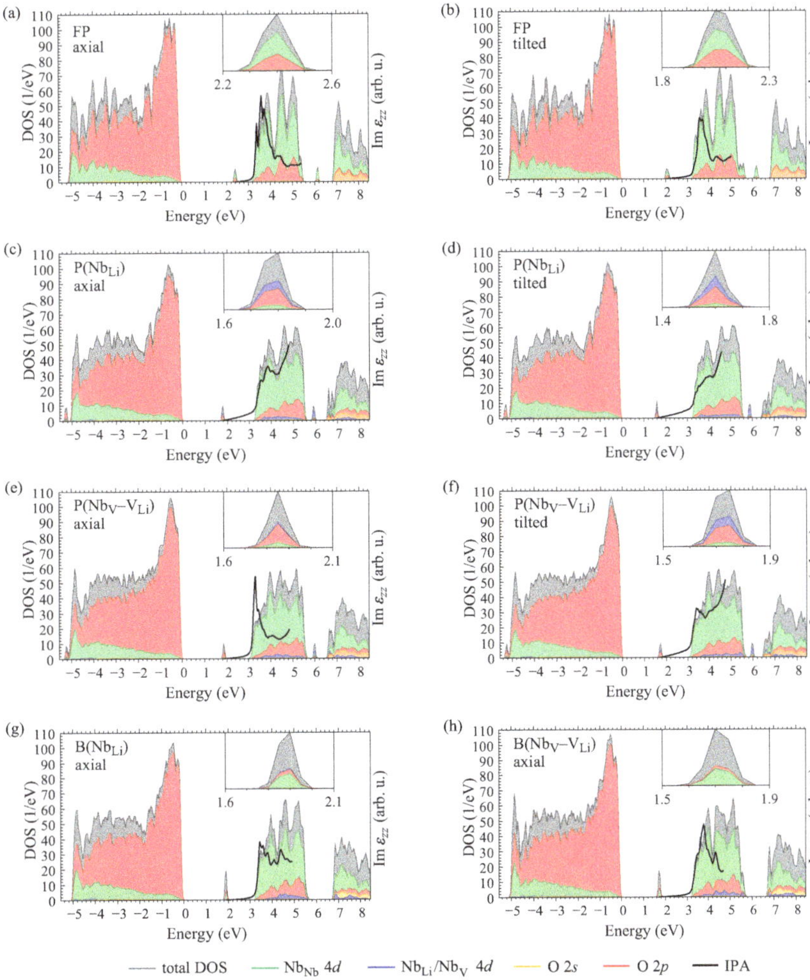

Figure 5. Electron densities of states for the free polaron (**a**,**b**), the bound polaron at a Nb$_{Li}$ antisite defect (**c**,**d**) or a Nb$_V$–V$_{Li}$ defect pair (**e**,**f**), and the bipolaron at Nb$_{Li}$ (**g**) or Nb$_V$–V$_{Li}$ (**h**). Subfigures (**b**,**d**,**f**) refer to tilted configurations, all other structures exhibit axial symmetry. The thick black line shows the imaginary part of $\varepsilon_{zz}(\omega)$ within the independent-particle approximation (IPA), whose low-energy resonances correspond to transitions from the defect state into the conduction band.

As free and bound polarons involve a single trapped excess electron, the system is spin polarized, and the defect level splits. Therefore, the peak inside the band gap stems entirely from the majority spin channel, while another unoccupied defect level with the same composition appears at a higher energy in the minority spin channel, visible at around 6 eV in the density of states. In contrast, bipolarons feature a pair of electrons with opposite spin. As a consequence, the spectral weight of the defect level inside the band gap is doubled, and there is no unoccupied level at higher energy.

A comparison between the axially symmetric and tilted configurations for free and bound polarons reveals that the deformation of the lattice structure has no visible effect on the composition of the polaron peak in the density of states, indicating that no rehybridization takes place. However, the symmetry-breaking distortion lowers the energy of the polaron state by about 0.2 eV, as can clearly be seen in the insets in Figure 5. This energy gain is the driving mechanism for the transition from an axially symmetric to a tilted ground-state configuration for polarons with unpaired electrons in LN.

To relate the observed absorption bands to specific polaron types in congruent LN, we further compare the density of states with the imaginary part of the dielectric function (2) in Figure 5. For the numerical evaluation, we use the same shifted $4 \times 4 \times 4$ **k**-point mesh as for the density of states. The summation includes all valence bands, the defect level inside the band gap, and 267 conduction bands. The broadening is set to 0.001 eV at a photon energy of 0 eV and increases linearly to 0.15 eV at an energy of 5 eV.

Within the IPA, the resonances in the optical spectrum correspond directly to transitions between occupied and unoccupied one-particle states. For congruent LN, the lowest-energy transitions are expected to occur between occupied defect levels inside the band gap and the conduction-band edge. As each of our simulations features only one polaron type and hence a single defect level that acts as an initial state, the low-energy region of $\mathrm{Im}\,\varepsilon_{zz}(\omega)$ is akin to the density of final states in the conduction band, modified by the transition dipole moments. In order to align the curves, we shift the IPA spectrum on the energy axis in Figure 5 so that the defect level, whose dispersion is negligible for the $2 \times 2 \times 2$ supercell used here [27], is taken as the origin. Besides, the spectra are cut at 3 eV, so that transitions at higher photon energies from bulk valence to conduction bands across the band gap are left out. Indeed, for all polaron types, the onset of the absorption, where the imaginary part of the dielectric function assumes nonzero values, coincides precisely with the energy separation between the defect level and the conduction bands in the density of states.

3.3. Optical Properties

The congruency between the density of states and the absorption spectrum in the IPA in Figure 5 affirms that the low-energy absorption peaks arise from transitions between the occupied polaron state inside the band gap and the continuum of Nb 4*d* states at the bottom of the conduction bands, but the IPA ignores quasiparticle corrections to the band structure as well as electron-hole attraction effects, which are both essential for a quantitative comparison with experimental data. In order to successively incorporate these effects, we determine the dielectric function at the three levels of theory described in Section 2, namely the IPA, the IQA, and the BSE. We concentrate on bipolarons in this part, which constitute the ground state of excess electrons in congruent LN and dominate at room temperature, where only a small number of bipolarons are thermally dissociated into free or bound polarons [16]. For the IPA and IQA, we include all 260 valence bands, the defect level, and 939 conduction bands. For the BSE, we reduce these numbers to 65 valence bands, the defect level, and 89 conduction bands to counter the high computational cost associated with the large supercell; we confirmed that these settings are sufficient to describe the optical response accurately up to photon energies around 6 eV. In all calculations, we employ the same shifted $2 \times 2 \times 2$ **k**-point set as for the structure optimization to sample the Brillouin zone, and a constant broadening of 0.1 eV is applied in the entire energy range to smoothen the dielectric function.

In Figure 6, we display the imaginary (top panels) and real (bottom panels) parts of the dielectric function $\varepsilon_{zz}(\omega)$ for bipolarons localized either at the Nb_{Li} antisite defect (left) or at the Nb_V–V_{Li} defect pair (right) for the different approximations. The IPA results are the same as in Figure 5g,h but are here displayed with no shift on the energy axis and calculated with a different **k**-point mesh for consistency with the other schemes. Equivalent results for $\mathrm{Im}\,\varepsilon_{xx}(\omega)$ in relation to Nb_V–V_{Li} can be found in [12]. The resonances at low photon energies, where the dielectric function of stoichiometric LN has no structure [30], arise from the defect-related optical transitions and are characteristic of the specific polaron type. In contrast, bulk transitions dominate at energies larger than the band gap, for example above 3 eV in the case of the IPA. The most pronounced defect-related peak, indicated by the arrow, appears at a lower energy for bipolarons at the Nb_{Li} antisite defect than at the Nb_V–V_{Li} defect pair. If quasiparticle corrections from the *GW* approximation are included, then the band gap increases significantly. Consequently, the dielectric function in the IQA (blue lines) is blueshifted by about 2 eV with respect to the IPA, whereas the shape of the spectrum remains essentially unchanged. In contrast, including electron-hole attraction effects within the BSE (red lines) entails a significant redistribution of spectral weight that enhances the oscillator strengths near the absorption thresholds due to the formation of excitons. In addition, there is a redshift that positions the final spectrum roughly halfway between the IPA and IQA curves. Because of the small number of bands, the BSE results are not fully converged above 6 eV and hence depicted by dashed lines in this region.

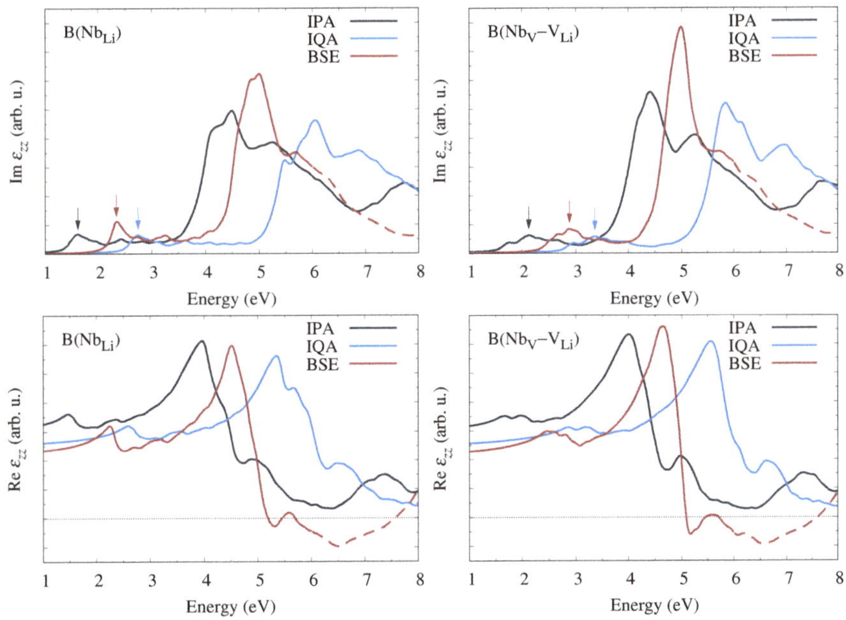

Figure 6. Imaginary (**top**) and real (**bottom**) parts of the ε_{zz} component of the dielectric function for bipolarons localized at a Nb_{Li} antisite defect (**left**) or a Nb_V–V_{Li} defect pair (**right**). We compare results from IPA (black), IQA (blue), and BSE (red). The arrows indicate the positions of the dominant defect-related peaks in the absorption spectra. The dashed horizontal line for $\mathrm{Re}\,\varepsilon_{zz}$ marks zero.

In Figure 7, we compare the $\varepsilon_{zz}(\omega)$ (red) and $\varepsilon_{xx}(\omega)$ (blue) components of the imaginary (solid lines) and real (dashed lines) parts of the dielectric function obtained from the BSE. Our results show that there are clear differences in the calculated optical spectra for the two bipolaron types, especially with respect to the lineshape of the defect peak. In particular, the defect peak is much less pronounced for x-polarized light, which will make an experimental detection more difficult. This strong directional dependence relates to the anisotropy of the bipolaron orbitals displayed in Figure 4g,h, which are oriented along the threefold symmetry axis that corresponds to the z direction. We also note that the positions of the maxima of the defect peaks differ for $\varepsilon_{zz}(\omega)$ and $\varepsilon_{xx}(\omega)$. Although the underlying resonance energies E_λ of the electron-hole pairs in Equation (3) are in fact identical, direction-dependent variations in the oscillator strengths give rise to an apparent shift. Therefore, care must be taken when the calculated spectra are compared with experimental data or analyzed to deduce absorption bands.

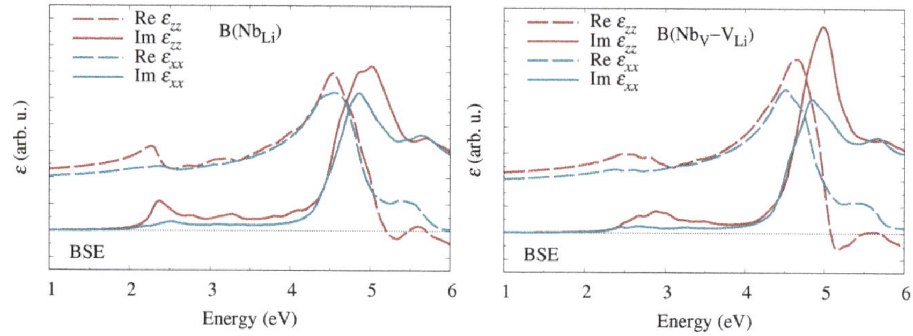

Figure 7. Imaginary (solid lines) and real (dashed lines) parts of the dielectric function ε calculated within the BSE for z-polarized (red) and x-polarized (blue) light. The left and right panel correspond to bipolarons at a Nb_{Li} antisite defect and a Nb_V–V_{Li} defect pair, respectively.

From the complex dielectric function calculated within the BSE, we can finally derive other optical coefficients that are typically measured in experimental spectroscopies, such as the reflectivity $R(\omega)$, the absorption (attenuation) coefficient $\alpha(\omega)$, and the electron-energy-loss function $L(\omega)$. Our results for the z (red) and x (blue) components are shown in Figure 8. The left panels again refer to bipolarons at the antisite defect, the right panels to bipolarons at the defect pair. As for the dielectric function itself, the polaron signatures are much more pronounced in the z components but are clearly present in all optical coefficients. The calculated absorption coefficients are further compared to the experimental value of 2.5 eV that is frequently cited in the literature, for example in [13], as well as a recent measurement by our co-workers [12] that yielded 2.7 eV. Both values were obtained from the peak position in the absorption spectrum of congruent LN, measured in the z direction. Although the two experimental values deviate slightly, indicating variations due to sample preparation and/or measurement technique, both lie between the calculated defect-related absorption maxima for the Nb_{Li} antisite defect and the Nb_V–V_{Li} defect pair. This may indicate that at least some of the antisite niobium atoms indeed migrate to neighboring oxygen octahedra, creating interstitial-vacancy pairs that modify the lineshape and apparent position of the bipolaron absorption band.

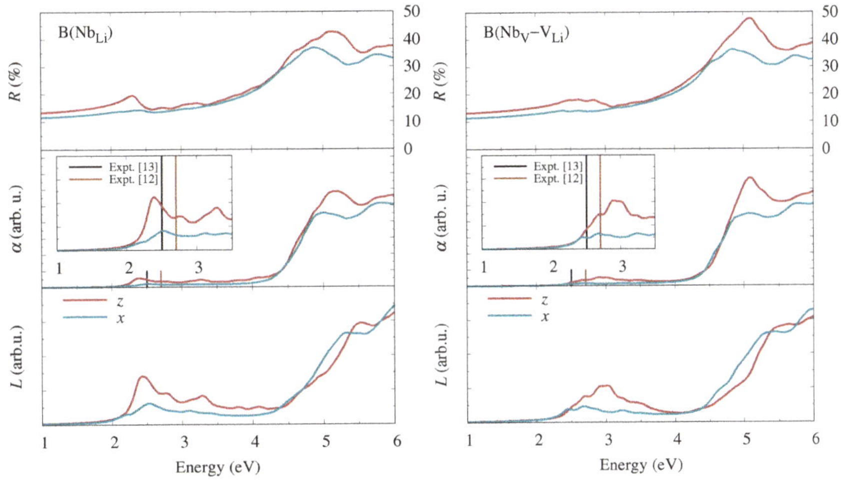

Figure 8. Optical properties derived from the dielectric function within the BSE: reflectivity R (**top**), absorption coefficient α (**middle**), and electron-energy-loss function L (**bottom**) for z-polarized (red) and x-polarized (blue) light. The left panels correspond to bipolarons at a Nb_{Li} antisite defect, the right panels to bipolarons at a Nb_V–V_{Li} defect pair. In the plots of the absorption coefficients, vertical lines indicate the position of the absorption band assigned to bipolarons in different experiments [12,13].

4. Conclusions

Using first-principles calculations based on density-functional theory, we performed a detailed analysis of electron polarons in LN, taking free polarons, bound polarons, and bipolarons into account. In contrast to earlier studies [21,22], we not only considered electron polarons bound to Nb_{Li} antisite atoms but also to another defect compatible with the Li-vacancy model, namely the Nb_V–V_{Li} defect pair, which arises when an antisite niobium atom migrates to a neighboring empty oxygen octahedron. Furthermore, we did not restrict ourselves to axially symmetric structure models but performed a full unconstrained relaxation, which leads to lower-energy tilted configurations where the threefold rotational symmetry is broken. Our results show that the distortion lowers the energy of the occupied defect level inside the band gap and thereby stabilizes the tilted configurations for free and bound polarons, which feature a single unpaired electron. On the other hand, it reduces the hybridization between the orbitals of neighboring niobium atoms along the threefold symmetry axis, which is essential for the formation of bipolarons. Therefore, bipolarons at both defect types relax to an axially symmetric geometry instead. An examination of the charge densities and the projected densities of states reveals that for all configurations, about one third of the trapped electronic charge is not actually localized at the niobium atoms hosting the polaron but distributed over the oxygen atoms that form the octahedral cage enclosing the defect site, well within the limits of the supercell used in this work. The lattice distortion associated with the polaron formation extends farther, especially for bipolarons at Nb_V–V_{Li} defect pairs, but it is not necessary to include its full spatial extent outside the localized charge distribution in calculations of the electronic and optical properties of polarons in LN. This justifies the use of a smaller supercell, which in turn allows us to apply sophisticated many-body techniques that would otherwise be prohibitively expensive.

Hereupon we performed state-of-the-art first-principles calculations of the frequency-dependent complex dielectric function and related optical coefficients. Starting from the independent-particle approximation, we additionally included quasiparticle corrections within the GW approximation for the electronic self-energy as well as excitonic contributions obtained from the Bethe–Salpeter equation. Our results for bipolarons at Nb_{Li} antisite defects and Nb_V–V_{Li} defect pairs show clear signatures at low photon energies in all optical coefficients that are characteristic of the specific defect type. A quantitative comparison of the calculated absorption spectra with experimental measurements of the absorption band attributed to bipolarons [12,13] suggests that at least some of the antisite niobium atoms indeed migrate to neighboring oxygen octahedra and form interstitial-vacancy pairs.

Finally, the structure models established in this work constitute a basis for further theoretical studies. Besides the linear optical properties already addressed here and in previous works [12,27,28,30], the nonlinear response is of particular interest, because it is central to the performance of LN as a key material in many optical technologies. Due to the enormous computational cost [41], significant challenges must still be overcome in order to facilitate accurate quantitative calculations of the polaron contribution to the nonlinear optical coefficients of congruent LN, however.

Author Contributions: Conceptualization, W.G.S. and A.S.; methodology, U.G. and W.G.S.; software, A.L.K. and U.G.; validation, U.G., A.S. and W.G.S.; formal analysis, F.S. and A.L.K.; investigation, F.S. and A.L.K.; resources, W.G.S.; data curation, U.G.; writing—original draft preparation, F.S., A.L.K. and A.S.; writing—review and editing, A.S., U.G. and W.G.S.; visualization, F.S. and A.L.K.; supervision, A.S.; project administration, A.S., U.G. and W.G.S.; funding acquisition, A.S., U.G. and W.G.S. All authors have read and agreed to the published version of the manuscript.

Funding: This research and the APC were funded by the Deutsche Forschungsgemeinschaft (DFG) via Sonderforschungsbereich TRR 142, Project No. 231447078.

Acknowledgments: Computational resources were provided by the Paderborn Center for Parallel Computing (PC^2) and the High Performance Computing Center Stuttgart (HLRS).

Conflicts of Interest: The authors declare no conflict of interest. The funders had no role in the design of the study; in the collection, analyses, or interpretation of data; in the writing of the manuscript, or in the decision to publish the results.

Abbreviations

The following abbreviations are used in this manuscript:

$B(Nb_{Li})$	bipolaron at a Nb_{Li} antisite defect
$B(Nb_V$–$V_{Li})$	bipolaron at a Nb_V–V_{Li} defect pair
BSE	Bethe–Salpeter equation
DFT	density-functional theory
DOS	density of states
EPR	electron paramagnetic resonance
FP	free polaron
IPA	independent-particle approximation
IQA	independent-quasiparticle approximation
LN	lithium niobate
$P(Nb_{Li})$	bound polaron at a Nb_{Li} antisite defect
$P(Nb_V$–$V_{Li})$	bound polaron at a Nb_V–V_{Li} defect pair
PBEsol	Perdew-Burke-Ernzerhof for solids
SLN	stoichiometric lithium niobate

References

1. Binh, L. Lithium niobate optical modulators: Devices and applications. *J. Cryst. Growth* **2006**, *288*, 180–187. [CrossRef]
2. Bazzan, M.; Sada, C. Optical waveguides in lithium niobate: Recent developments and applications. *Appl. Phys. Rev.* **2015**, *2*, 040603. [CrossRef]
3. Zaltron, A.; Bettella, G.; Pozza, G.; Zamboni, R.; Ciampolillo, M.; Argiolas, N.; Sada, C.; Kroesen, S.; Esseling, M.; Denz, C. Integrated optics on lithium niobate for sensing applications. *Proc. SPIE* **2015**, *9506*, 950608.
4. Buse, K.; Adibi, A.; Psaltis, D. Non-volatile holographic storage in doubly doped lithium niobate crystals. *Nature* **1998**, *393*, 665–668. [CrossRef]
5. Qi, Y.; Li, Y. Integrated lithium niobate photonics. *Nanophotonics* **2020**, *9*, 1287–1320. [CrossRef]
6. Arizmendi, L. Photonic applications of lithium niobate crystals. *Phys. Status Solidi A* **2004**, *201*, 253–283. [CrossRef]

7. Lengyel, K.; Péter, Á.; Kovács, L.; Corradi, G.; Pálfalvi, L.; Hebling, J.; Unferdorben, M.; Dravecz, G.; Hajdara, I.; Szaller, Z.; et al. Growth, defect structure, and THz application of stoichiometric lithium niobate. *Appl. Phys. Rev.* **2015**, *2*, 040601. [CrossRef]
8. Volk, T.; Wöhlecke, M. *Lithium Niobate*; Springer: Berlin/Heidelberg, Germany, 2009.
9. Sánchez-Dena, O.; Villalobos-Mendoza, S.D.; Farías, R.; Fierro-Ruiz, C.D. Lithium niobate single crystals and powders reviewed—Part II. *Crystals* **2020**, *10*, 990. [CrossRef]
10. Lerner, P.; Legras, D.; Dumas, J.P. Stoechiométrie des monocristaux de métaniobate de lithium. *J. Cryst. Growth* **1968**, *3–4*, 231–235. [CrossRef]
11. Zotov, N.; Boysen, H.; Frey, F.; Metzger, T.; Born, E. Cation substitution models of congruent $LiNbO_3$ investigated by X-ray and neutron powder diffraction. *J. Phys. Chem. Solids* **1994**, *55*, 145–152. [CrossRef]
12. Schmidt, F.; Kozub, A.L.; Biktagirov, T.; Eigner, C.; Silberhorn, C.; Schindlmayr, A.; Schmidt, W.G.; Gerstmann, U. Free and defect-bound (bi)polarons in $LiNbO_3$: Atomic structure and spectroscopic signatures from ab initio calculations. *Phys. Rev. Res.* **2020**, *2*, 043002. [CrossRef]
13. Schirmer, O.F.; Imlau, M.; Merschjann, C.; Schoke, B. Electron small polarons and bipolarons in $LiNbO_3$. *J. Phys. Condens. Matter* **2009**, *21*, 123201. [CrossRef] [PubMed]
14. Faust, B.; Müller, H.; Schirmer, O.F. Free small polarons in $LiNbO_3$. *Ferroelectrics* **1994**, *153*, 297–302. [CrossRef]
15. Sweeney, K.L.; Halliburton, L.E. Oxygen vacancies in lithium niobate. *Appl. Phys. Lett.* **1983**, *43*, 336–338. [CrossRef]
16. Koppitz, J.; Schirmer, O.F.; Kuznetsov, A.I. Thermal dissociation of bipolarons in reduced undoped $LiNbO_3$. *EPL Europhys. Lett.* **1987**, *4*, 1055–1059. [CrossRef]
17. Jermann, F.; Simon, M.; Bower, R.; Kratzig, E.; Schirmer, O.F. Light-induced absorption changes in reduced lithium niobate. *Ferroelectrics* **1995**, *165*, 319–327. [CrossRef]
18. Donnerberg, H.; Tomlinson, S.M.; Catlow, C.R.A.; Schirmer, O.F. Computer-simulation studies of intrinsic defects in $LiNbO_3$ crystals. *Phys. Rev. B* **1989**, *40*, 11909–11916. [CrossRef]
19. Araujo, R.M.; Lengyel, K.; Jackson, R.A.; Kovács, L.; Valerio, M.E.G. A computational study of intrinsic and extrinsic defects in $LiNbO_3$. *J. Phys. Condens. Matter* **2007**, *19*, 046211. [CrossRef]
20. Li, Q.; Wang, B.; Woo, C.H.; Wang, H.; Wang, R. First-principles study on the formation energies of intrinsic defects in $LiNbO_3$. *J. Phys. Chem. Solids* **2007**, *68*, 1336–1340. [CrossRef]
21. Xu, H.; Lee, D.; He, J.; Sinnott, S.B.; Gopalan, V.; Dierolf, V.; Phillpot, S.R. Stability of intrinsic defects and defect clusters in $LiNbO_3$ from density functional theory calculations. *Phys. Rev. B* **2008**, *78*, 174103. [CrossRef]
22. Nahm, H.H.; Park, C.H. First-principles study of microscopic properties of the Nb antisite in $LiNbO_3$: Comparison to phenomenological polaron theory. *Phys. Rev. B* **2008**, *78*, 184108. [CrossRef]
23. Schmidt, W.G.; Albrecht, A.; Wippermann, S.; Blankenburg, S.; Rauls, E.; Fuchs, F.; Rödl, C.; Furthmüller, J.; Hermann, A. $LiNbO_3$ ground- and excited-state properties from first-principles calculations. *Phys. Rev. B* **2008**, *77*, 035106. [CrossRef]
24. Thierfelder, C.; Sanna, S.; Schindlmayr, A.; Schmidt, W.G. Do we know the band gap of lithium niobate? *Phys. Status Solidi C* **2010**, *7*, 460–463. [CrossRef]
25. Friedrich, M.; Riefer, A.; Sanna, S.; Schmidt, W.G.; Schindlmayr, A. Phonon dispersion and zero-point renormalization of $LiNbO_3$ from density-functional perturbation theory. *J. Phys. Condens. Matter* **2015**, **27**, 385402. [CrossRef] [PubMed]
26. Riefer, A.; Friedrich, M.; Sanna, S.; Gerstmann, U.; Schindlmayr, A.; Schmidt, W.G. $LiNbO_3$ electronic structure: Many-body interactions, spin-orbit coupling, and thermal effects. *Phys. Rev. B* **2016**, *93*, 075205. [CrossRef]
27. Friedrich, M.; Schmidt, W.G.; Schindlmayr, A.; Sanna, S. Polaron optical absorption in congruent lithium niobate from time-dependent density-functional theory. *Phys. Rev. Mater.* **2017**, *1*, 054406. [CrossRef]
28. Friedrich, M.; Schmidt, W.G.; Schindlmayr, A.; Sanna, S. Optical properties of titanium-doped lithium niobate from time-dependent density-functional theory. *Phys. Rev. Mater.* **2017**, *1*, 034401. [CrossRef]
29. Botti, S.; Schindlmayr, A.; Del Sole, R.; Reining, L. Time-dependent density-functional theory for extended systems. *Rep. Prog. Phys.* **2007**, *70*, 357–407. [CrossRef]
30. Riefer, A.; Sanna, S.; Schindlmayr, A.; Schmidt, W.G. Optical response of stoichiometric and congruent lithium niobate from first-principles calculations. *Phys. Rev. B* **2013**, *87*, 195208. [CrossRef]
31. Sio, W.H.; Verdi, C.; Poncé, S.; Giustino, F. Polarons from first principles, without supercells. *Phys. Rev. Lett.* **2019**, *122*, 246403. [CrossRef] [PubMed]
32. Giannozzi, P.; Baroni, S.; Bonini, N.; Calandra, M.; Car, R.; Cavazzoni, C.; Ceresoli, D.; Chiarotti, G.L.; Cococcioni, M.; Dabo, I.; et al. QUANTUM ESPRESSO: A modular and open-source software project for quantum simulations of materials. *J. Phys. Condens. Matter* **2009**, *21*, 395502. [CrossRef] [PubMed]
33. Perdew, J.P.; Ruzsinszky, A.; Csonka, G.I.; Vydrov, O.A.; Scuseria, G.E.; Constantin, L.A.; Zhou, X.; Burke, K. Restoring the density-gradient expansion for exchange in solids and surfaces. *Phys. Rev. Lett.* **2008**, *100*, 136406. [CrossRef]
34. Schmidt, F.; Landmann, M.; Rauls, E.; Argiolas, N.; Sanna, S.; Schmidt, W.G.; Schindlmayr, A. Consistent atomic geometries and electronic structure of five phases of potassium niobate from density-functional theory. *Adv. Mater. Sci. Eng.* **2017**, *2017*, 3981317. [CrossRef]
35. Schmidt, F.; Riefer, A.; Schmidt, W.G.; Schindlmayr, A.; Imlau, M.; Dobener, F.; Mengel, N.; Chatterjee, S.; Sanna, S. Quasiparticle and excitonic effects in the optical response of $KNbO_3$. *Phys. Rev. Mater.* **2019**, *3*, 054401. [CrossRef]

36. Friedrich, M.; Schindlmayr, A.; Schmidt, W.G.; Sanna, S. $LiTaO_3$ phonon dispersion and ferroelectric transition calculated from first principles. *Phys. Status Solidi B* **2016**, *253*, 683–689. [CrossRef]
37. Cococcioni, M.; de Gironcoli, S. Linear response approach to the calculation of the effective interaction parameters in the LDA+*U* method. *Phys. Rev. B* **2005**, *71*, 035105. [CrossRef]
38. Marini, A.; Hogan, C.; Grüning, M.; Varsano, D. Yambo: An ab initio tool for excited state calculations. *Comput. Phys. Commun.* **2009**, *180*, 1392–1403. [CrossRef]
39. Schirmer, O.F.; von der Linde, D. Two-photon- and x-ray-induced Nb^{4+} and O^{-} small polarons in $LiNbO_3$. *Appl. Phys. Lett.* **1978**, *33*, 35–38. [CrossRef]
40. Müller, H.; Schirmer, O.F. Microscopic structure of Nb_{Li} related defects in reduced undoped $LiNbO_3$. *Ferroelectrics* **1992**, *125*, 319–324. [CrossRef]
41. Riefer, A.; Rohrmüller, M.; Landmann, M.; Sanna, S.; Rauls, E.; Vollmers, N.J.; Hölscher, R.; Witte, M.; Li, Y.; Gerstmann, U.; et al. Lithium niobate dielectric function and second-order polarizability tensor from massively parallel ab initio calculations. In *High Performance Computing in Science and Engineering '13*; Nagel, W., Kröner, D., Resch, M., Eds.; Springer: Cham, Switzerland, 2013; pp. 93–104.

crystals

MDPI

Article

Effect of Temperature on Luminescence of $LiNbO_3$ Crystals Single-Doped with Sm^{3+}, Tb^{3+}, or Dy^{3+} Ions

Radosław Lisiecki *, Bogusław Macalik, Robert Kowalski, Jarosław Komar and Witold Ryba-Romanowski

Division of Optical Spectroscopy, Institute of Low Temperature and Structure Research, Polish Academy of Sciences, Okólna 2, 50-422 Wrocław, Poland; b.macalik@intibs.pl (B.M.); r.kowalski@intibs.pl (R.K.); j.komar@intibs.pl (J.K.); w.ryba-romanowski@intibs.pl (W.R.-R.)

* Correspondence: r.lisiecki@intibs.pl

Received: 8 October 2020; Accepted: 11 November 2020; Published: 13 November 2020

check for updates

Abstract: Crystals of $LiNbO_3$ single-doped with Sm^{3+}, Tb^{3+}, or Dy^{3+} and crystal of $LiTaO_3$ single-doped with Tb^{3+} were grown by the Czochralski method. Luminescence spectra and decay curves for $LiNbO_3$ samples containing Sm^{3+} or Dy^{3+} ions were recorded at different temperatures between 295 and 775 K, whereas those for samples containing Tb^{3+} ions were recorded at different temperatures between 10 and 300 K. Optical absorption spectra at different temperatures were recorded within the UV-blue region relevant to optical pumping of the samples. It was found that the effect of temperature on experimental luminescence lifetimes consists of the initial temperature-independent stage followed by a steep decrease with the onset at about 700, 600, and 150 K for Sm^{3+}, Dy^{3+}, and Tb^{3+} ions, respectively. Additionally, comparison of temperature impact on luminescence properties of $LiNbO_3:Tb^{3+}$ and $LiTaO_3:Tb^{3+}$ crystals has been adequately described. Experimental results were interpreted in terms of temperature-dependent charge transfer (CT) transitions within the modified Temperature—Dependent Charge Transfer phenomenological model (TDCT). Disparity of the onset temperatures and their sequence were explained based on the location of familiar zigzag curves connecting the ground state levels of rare earth ions with respect to the band-gap of the host. It was concluded also that $LiNbO_3:Sm^{3+}$ is suitable as an optical sensor within the 500–750 K temperature region whereas $LiNbO_3:Dy^{3+}$ offers the highest sensitivity at lower temperatures between 300 and 400 K.

Keywords: $LiNbO_3$; $LiTaO_3$; oxide crystals; lanthanides; luminescence

1. Introduction

Originally proposed in 1964, $LiNbO_3$ was soon applied in second harmonic generation, electro-optic modulation, acousto-optic, and piezoelectric device application [1]. Subsequent investigation has revealed that it is promising for holographic recording [2,3], optical parametric oscillation [4,5], waveguide lasers [6,7], or self-frequency doubled lasers [8].

However, despite the long-lasting technological success, some fundamental properties of $LiNbO_3$ remain rather puzzling. One of encountered difficulties refers to the reliability of its band gap. Numerous reported band gap values, all inferred from optical absorption spectra, range from 3.28 [9] to 4.3 eV [10] with the most frequently cited value of 3.78 eV. This is due mainly to the fact that the $LiNbO_3$ crystals, suitable for optical application, are fabricated commonly by the Czochralski method from a congruent, lithium deficient melt (Li/Nb = 0.94). Hence, they have a high density of point defects which affect the reliability of the measurement. Even with special care near where stoichiometric crystals have grown, recorded optical absorption spectra still suffered from remaining structural defects. Numerous attempts have been made to determine the band gap from ab initio calculations. The calculated band gap values depended markedly on the method applied, ranging from 3.47 to 6.53 eV [11]. Authors of a subsequent paper [12] entitled: "Do we know the band gap of lithium niobate?" have reported the calculated band gap value of 4.7 eV and more recently, the calculated band gap value of 4.9 eV has been reported in [13].

Much attention has been paid in the past to the nature of defects in crystal structures of isostructural $LiNbO_3$ and $LiTaO_3$ compounds. In the ideal, stoichiometric composition of the cations Nb^{5+}, Li^+, and free sites are located sequentially in distorted octahedra of oxygen ions. The distribution of cations among available sites in congruent $LiNbO_3$ crystals is not easy, however. In the past it has been interpreted in the framework of the Nb-site vacancy model [14]. Later on, it has been concluded that the Nb split models in which the excess of Nb atoms are distributed among the Li and the vacant cation sites cannot be excluded [15]. The same difficulty relates to the location of incorporated luminescent rare earth ions. Investigation of Eu^{3+} sites in $LiNbO_3$:Eu^{3+} and $LiNbO_3$:MgO:Eu^{3+} crystals for different Li/Nb molar ratios has led to the conclusion that Eu^{3+} ions substitute the Li^+ and Nb^{5+} ions [16]. The RBS/channeling spectra of Pr^{3+}, Ho^{3+}, Yb^{3+} ions in $LiNbO_3$ revealed that they substitute Li^+ ions but are slightly shifted from the regular Li^+ position [17]. In a more recent paper dealing with structural and optical properties of powders of $LiTaO_3$ doped with Eu^{3+} ions the authors have observed that the Eu^{3+} ions substitute preferentially the Ta^{5+} ions [18]. Moreover, it was found that this preference increases with increasing Eu^{3+} concentration (an occupation fraction above 60% for Eu^{3+} concentration of 1 at %). In the present work we deal with $LiNbO_3$ crystal single-doped with Sm^{3+}, Tb^{3+}, or Dy^{3+} ions characteristic in that the relaxation of their metastable levels consists essentially of radiative transitions in virtually all inorganic hosts.

In fact, spectroscopic features of Sm^{3+}-doped crystals and glasses stem from the energy level scheme of Sm^{3+} ions that consist of low-energy multiplets from the ^7F and ^7H terms and a higher-energy group of multiplets from quartet terms. The energy difference between the quartet and septet states is large as compared to the phonon energy encountered in inorganic matrices. Therefore, the relaxation of the $^4G_{5/2}$ metastable level of Sm^{3+} consists essentially of radiative transitions providing an efficient visible emission. Its spectral distribution within the blue, yellow, and red regions is weakly affected by the change of a host [19–24]. Spectroscopic properties of Tb^{3+}-doped hosts stem from the energy level scheme of Tb^{3+} ions involving low-energy multiplets from the ^7F term and higher-energy multiplets from quintet terms. The metastable $5D_4$ level of Tb^{3+} is located higher than the next lower energy $^7F_0$ level by about 14,000 cm^{-1}. Therefore, its relaxation consists essentially of radiative transitions providing an efficient visible emission. Literature concerning spectroscopic features of Tb^{3+} doped luminescent materials is rather rich. The mechanism of the $^5D_3$–$^5D_4$ cross-relaxation in $Y_3Al_5O_{12}$:Tb^{3+} was considered in [25]. Several papers were devoted to preparation and spectroscopic features of terbium-doped garnet crystals [26–29]. Other visible phosphors containing Tb^{3+} ions that have been studied include $Y_2LuCaAl_2SiO_{12}$:Ln (Ln = Ce^{3+}, Eu^{3+}, and Tb^{3+}) [30], Tb^{3+} doped lithium lead alumino borate glasses [31], Tb^{3+}-doped $KLu(WO_4)_2$ crystal [32], Tb^{3+}-doped $LiYF_4$ [33], Tb^{3+}-doped K_3YF_6 [34]. Laser potential and laser performance of Tb^{3+}-doped hosts have been considered, too [35]. Spectroscopic properties of Dy^{3+}-doped materials stem from the energy level scheme of Dy^{3+} ions that consists of high-energy multiplets from quartet terms well separated from low-energy multiplets belonging to the ^6H and ^6F terms. The visible emission of Dy^{3+} results from radiative transitions that originate on the metastable $4F_{9/2}$ level and terminate on the lower-energy levels ^6H_J (J = 9/2, 11/2, 13/2, and 15/2). Numerous recent papers have been devoted to Phosphor materials single-doped with Dy^{3+} [36], co-doped with Dy+Eu [37] or triple-doped with Dy+Eu+Tb [38] have been investigated and found promising for novel lighting devices. The potential of Dy^{3+}-doped crystals as luminescent temperature sensors has been considered in numerous papers [39–41]. It has been demonstrated that YAG:Dy^{3+} crystal pumped at 447 nm with a GaN laser diode is able to show yellow laser operation with a slope efficiency of 12% [42]. Optical amplification in Dy^{3+}-doped Gd_2SiO_5, Lu_2SiO_5, and $YAl_3(BO_3)_4$ single crystals has been observed [43] as well. Intention of the present work is to get a closer insight into temperature-dependent processes which are relevant for excited state relaxation of incorporated rare earth ions and to assess the impact of these processes on the potential of system studied for luminescence temperature sensing.

2. Materials and Methods

Single-doped crystals of $LiNbO_3$ containing nominally 0.65 wt% (1.2×10^{20} ions/cm^3) of Sm^{3+}, 2.80 wt% (4.8×10^{20} ions/cm^3) of Tb^{3+}, or 1.94 wt% (3.3×10^{20} ions/cm^3) of Dy $^{3+}$ were grown by the Czochralski method from the congruent melt (Li/Nb = 0.945). By the same method a 3 wt% Tb^3-doped $LiTaO_3$ crystal was grown creating thereby a reference material to help understand the $LiNbO_3$:Tb^{3+} luminescence. The Czochralski growth of $LiNbO_3$ and $LiTaO_3$ crystals has been well known for decades and is described in detail elsewhere, e.g., in [44]. Concentrations given above stem from the trade-off between the intention to achieve required excitation efficiency, on one hand, and to prevent the excessive distortion of the host structure, on the other hand. A Varian 5E UV-VIS-NIR spectrophotometer with instrumental spectral bandwidths of 0.5 nm was employed to record optical absorption spectra in the UV-blue spectral region. To record excitation spectra and luminescence spectra a FLS980 fluorescence spectrophotometer from Edinburg Instruments equipped with a 450 W xenon lamp as an excitation source and a Hamamatsu 928 PMT detector was used. For these measurements, the samples were excited with unpolarized light propagating parallel to the optical axis of the crystal and the luminescence was observed perpendicular to the optical axis of the crystal. Recorded spectra were corrected for the sensitivity and wavelength of the experimental set-up. To record decay curves of luminescence an experimental set-up consisting of a tunable optical parametric oscillator (OPO) pumped by a third harmonic of a Nd:YAG laser, a double grating monochromator with a 1000 mm focal length, a photomultiplier, and a Tektronix MDO 4054B-3 Mixed Domain Oscilloscope was used. For spectroscopic measurement at low temperature the samples were placed in an Oxford Model CF 1204 continuous flow liquid helium cryostat equipped with a temperature controller. A chamber furnace was used for measurements at higher temperature within 295–800 K. The temperature of samples was detected by a copper-constant thermocouple with measurement error less than 1.5 K and controlled by a proportional-integral-derivative (PID) Omron E5CK controller.

3. Results

When interpreting our experimental data, we refer to energy level schemes for Sm^{3+}, Tb^{3+}, and Dy^{3+} ions gathered in Figure 1. The energy levels within respective $4f^5$, $4f^8$, and $4f^9$ configurations are labeled by symbols $^{2S+1}L_J$ of corresponding multiplets.

It can be seen that there is a correspondence between term structures for Sm^{3+} and Dy^{3+} ions in agreement with the principle of electrons and holes in the $4f^n$ configurations. Actually, each multiplet of rare earth ions incorporated in the crystal host is split by the crystal field into crystal field components. In depth analysis of the crystal field splitting of multiplets for Sm^{3+} and Dy^{3+} ions in $LiNbO_3$ has been performed based on low temperature absorption and emission spectra and reported in [45,46], respectively. These results were used to construct corresponding energy level schemes shown in Figure 1. To our knowledge the information concerning the crystal field splitting of Tb^{3+} in $LiNbO_3$ is not available. Therefore, we include in Figure 1 the data for Tb^{3+} in GAGG garnet reported in [29]. It should be noticed here that the agreement between theoretical and experimental crystal field splitting inferred from low temperature spectra of rare earth ions in $LiNbO_3$ is poor owing to multi-site location of incorporated ions combined with a strong inhomogeneous line broadening of their optical spectra. Other consequences result from a strongly defective congruent composition of the hosts crystal. The $LiNbO_3$ forms crystals that belong to trigonal/rhombohedral system, *R3c* space group, hence they are optically uniaxial. Incorporated rare earth ions enter sites with the C_3 local symmetry in the stoichiometric composition but in congruent $LiNbO_3$ they are located in several nonequivalent sites with strongly dissimilar local symmetries. As a consequence, the anisotropy of their transition intensities is no longer consistent with theoretical predictions preventing thereby the reliable interpretation of polarized optical spectra, as observed in [45,46]. To avoid the difficulties mentioned above and remembering that effect of temperature on integrated luminescence intensity is independent of polarization of optical spectra we infer our results from unpolarized luminescence spectra.

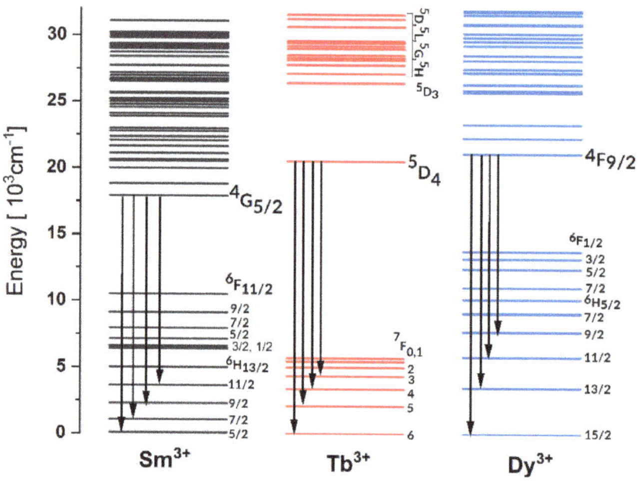

Figure 1. Energy level schemes for Sm^{3+}, Tb^{3+}, and Dy^{3+}. Vertical downward arrows indicate transitions involved.

In the limit of low doping level, the population of an excited multiplet decays by competing radiative transitions and nonradiative multiphonon relaxation consisting of simultaneous emission of several phonons. The contribution of the latter process is commonly assessed from the phenomenological energy gap law which relates the rate of multiphonon relaxation to the order of the process defined as the ratio of energy separation between the excited level in question and a next lower level to the cut-off phonon energy available in the host. It follows from the Raman spectra analysis of $LiNbO_3$ and $LiTaO_3$ crystals that the highest energy stretching vibrations Nb-O are around 610 cm^{-1} (597 cm^{-1} for Ta-O) [47]. Thus, Sm^{3+}, Tb^{3+}, and Dy^{3+} ions in $LiNbO_3$ have single metastable levels able to relax radiatively without multiphonon contribution providing thereby intense luminescence. Downward arrows in Figure 1 indicate radiative transitions observed in the visible region.

3.1. *Effect of Temperature on Luminescence of Sm^{3+} and Dy^{3+} in $LiNbO_3$*

On the right side of Figure 2 we compare survey room temperature luminescence spectra of Sm^{3+} and Dy^{3+} ions in $LiNbO_3$ recorded within the 450–800 nm spectral region. Sm^{3+} luminescence was excited at 402 nm. Its luminescence spectrum consists of bands located at about 567, 600, 646, and 708 nm related to transitions between the metastable $4G_{5/2}$ level and ^6H_J (J = 5/2, 7/2, 9/2, 11/2) terminal levels, respectively. Very weak contribution of the $^4G_{5/2} \rightarrow {^6H_{13/2}}$ transition can be discerned around 800 nm. It should be noticed here that there are other transitions from the $^4G_{5/2}$ level, namely to the multiplets of the ^6F term and the $^4G_{5/2} \rightarrow {^6H_{15/2}}$ transition, all of them located in the near IR. The contribution of these transitions to luminescence spectra of samarium ions in crystals and glasses is marginal and therefore is commonly neglected. Dy^{3+} luminescence shown in Figure 2 was excited at 356 nm. Its luminescence spectrum consists of bands located at about 486, 580, 665, and 754 nm related to transition between the metastable $4F_{9/2}$ level and ^6H_J (J = 15/2, 13/2, 11/2, and 9/2) terminal levels, respectively. Remaining transitions from the metastable $^4F_{9/2}$ level are located in the IR and terminate on multiplets of the ^6F term and on the $^6H_{7/2}$ and $^6H_{5/2}$ multiplets. Their contribution to dysprosium luminescence is very weak, as for samarium luminescence.

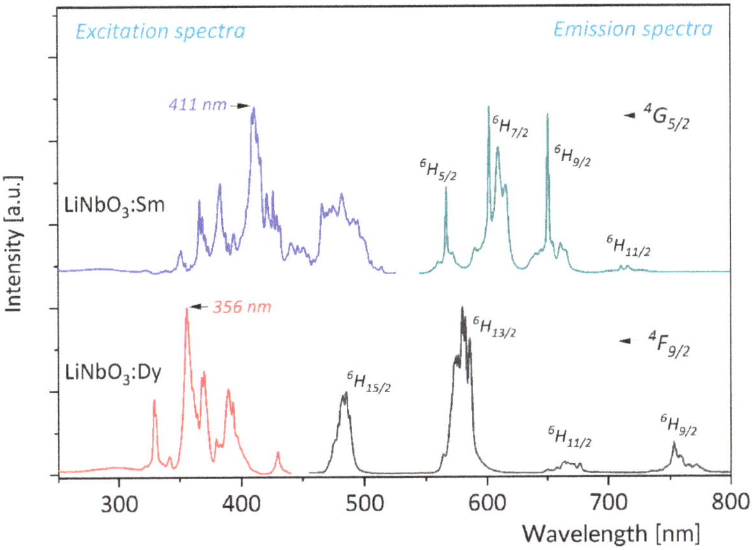

Figure 2. Survey, room temperature luminescence spectra of Sm^{3+} and Dy^{3+} ions in $LiNbO_3$ recorded within the 450–800 nm spectral region (**right**) and excitation spectra of Sm^{3+} and Dy^{3+} luminescence in $LiNbO_3$, monitored at 611 and 573 nm, respectively (**left**).

On the left side of Figure 2 we compare excitation spectra of Sm^{3+} luminescence and Dy^{3+} luminescence in $LiNbO_3$, monitored at 611 and at 573 nm, respectively. Excitation spectra for the two ions show very complex structure of band components corresponding to closely spaced transitions from the ground states to high energy multiplets derived from the ^4F, ^4G, ^4H, ^4I, ^4K, ^4L, ^4M quartet and the ^6P sextet terms. It should be noticed here that despite a rich spectrum, the intensity of absorption in this spectral region is very low because spin forbidden sextet-quartet transitions are involved. It can be seen that the highest energy bands contributing to excitation spectra in Figure 2 are located around 411 and 356 nm although the energy level schemes of two ions contain levels at higher energy. Rather curiously, the wavelength 335 nm correspond to 3.7 eV, a value of the most frequently cited energy gap for $LiNbO_3$. In view of theoretical calculations mentioned above in Section 1 the short wavelength limit of the excitation spectra may not be due to the fundamental UV absorption edge of $LiNbO_3$. It may result also from an absorption of color centers, likely to exist in the matrix and/or transitions between the ground states of incorporated rare earth ions and trapped exciton states, as proposed for $LiTaO_3$:Pr system [47,48].

Figure 3 compares luminescence spectra of Sm^{3+} in $LiNbO_3$ recorded at different temperatures between 295 K and 775 nm. Excitation was at 402 nm. It can be seen that with increasing temperature the contribution of narrow band components to the spectrum vanishes gradually, the bands become smoother and their intensities diminish. To assess the effect of temperature on the overall Sm^{3+} luminescence the spectra recorded at different temperatures were integrated numerically within the 450–750 nm region. Results of this assessment presented in the inset on the right side of Figure 3 reveal the adverse thermally induced quenching of Sm^{3+}emission. The inset on the left side of Figure 3 shows spectra for the high energy part of the spectrum between 460 and 550 nm, where the luminescence grows monotonously.

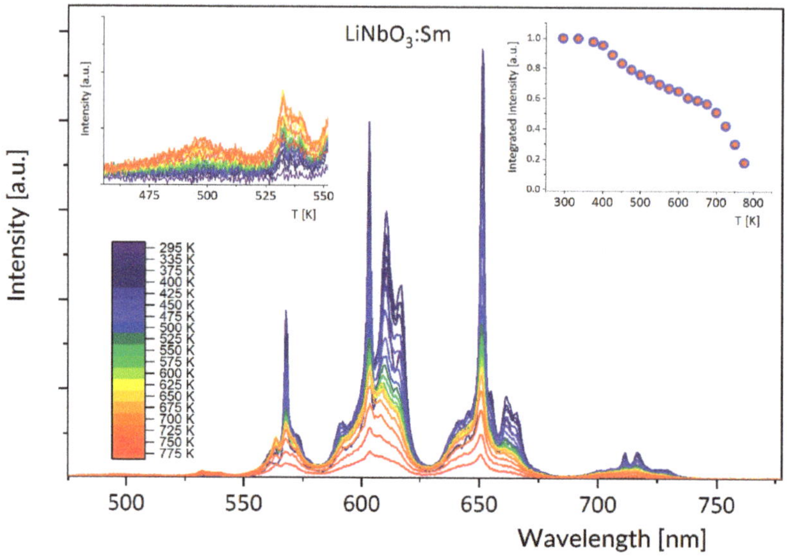

Figure 3. Luminescence spectra of Sm^{3+} in $LiNbO_3$ recorded at different temperatures between 295 and 775 K. Excitation was at 402 nm. Left inset: Magnified spectra of the high energy part of the spectrum between 460 and 550 nm. Right inset: Integrated luminescence intensity plotted versus temperature.

Examination of luminescence spectra of Dy^{3+} in $LiNbO_3$ excited at 356 nm and recorded at different temperatures between 295 and 723 K reveal similar thermal effect.

Namely, the contribution of narrow band components to the spectrum vanishes gradually, the bands become more smooth and their intensities diminish, except for the high energy part of the spectrum between 440 and 500 nm, where the luminescence grows steadily. For the sake of brevity, the spectra shown in Figure 4 are restricted to the short wavelength parts encompassing bands within the 440–550 nm region. The Dy^{3+} luminescence intensities determined by a numerical integration of spectra within the 440–750 nm region are plotted versus temperature in the inset on the right. The inset on the left side shows the onset of steep rise of the sample absorption in the UV region observed at several different temperatures. Examination of plots in insets on the right side of Figures 3 and 4 reveals a thermally enhanced quenching of both the Sm^{3+} and Dy^{3+} luminescence in $LiNbO_3$. This phenomenon may be due to a thermally induced decrease of absorption efficiency in the optical pump region and/or to thermally enhanced increase of nonradiative relaxation rates of the metastable $4G_{5/2}$ and $^4F_{9/2}$ levels of Sm^{3+} and Dy^{3+}, respectively. The contribution of these quenching factors can be assessed based on examination of Figure 5.

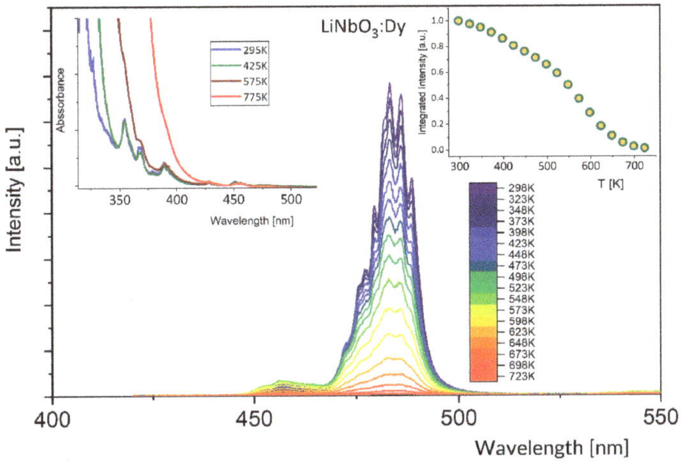

Figure 4. Luminescence spectra between 440 and 500 nm for Dy^{3+} in $LiNbO_3$ recorded at different temperatures between 295 and 723 K. Excitation was at 356 nm. Left inset: The onset of steep rise of the sample absorption in the UV region observed at several different temperatures. Right inset: Integrated luminescence intensity plotted versus temperature.

Table 1. Spectral parameters governing the temperature dependence of $^4G_{5/2}$ (Sm^{3+}) and $^4F_{9/2}$ (Dy^{3+}) lifetimes in $LiNbO_3$ within the framework of the TDCTmodel.

	$LiNbO_3$:Sm	**$LiNbO_3$:Dy**
W_{nr} (1/s)	855	508
E_a (cm^{-1})	16,159	14,795
α (cm^{-1}/K)	23.3	26.1
$\hbar\omega$ (cm^{-1})	672	680
W_r (1/s)	1540	5164

Points in this figure show the temperature dependence of the $^4G_{5/2}$ and $^4F_{9/2}$ experimental lifetimes determined experimentally from respective exponential luminescence decay curves. It can be seen that initially, i.e., up to about 700 K for Sm^{3+} and up to about 600 K for Dy^{3+}, the luminescence lifetimes do not depend on temperature implying that the contribution of the latter factor can be neglected. Therefore, we attribute the decrease of integrated luminescence intensity observed in this initial temperature region to thermally induced increase of a broad-band absorption that affects adversely the efficiency of optical pumping. In fact, the onset of this absorption shifts towards longer wavelengths when the temperature increases, as shown in the inset on the left side of Figure 4. Further increase of the temperature brings about a very steep decrease of the lifetime values. At 775 K, the highest temperature provided by our experimental setup, the Dy^{3+} lifetime is close to zero whereas the Sm^{3+} lifetime attains about 70% of its value at room temperature.

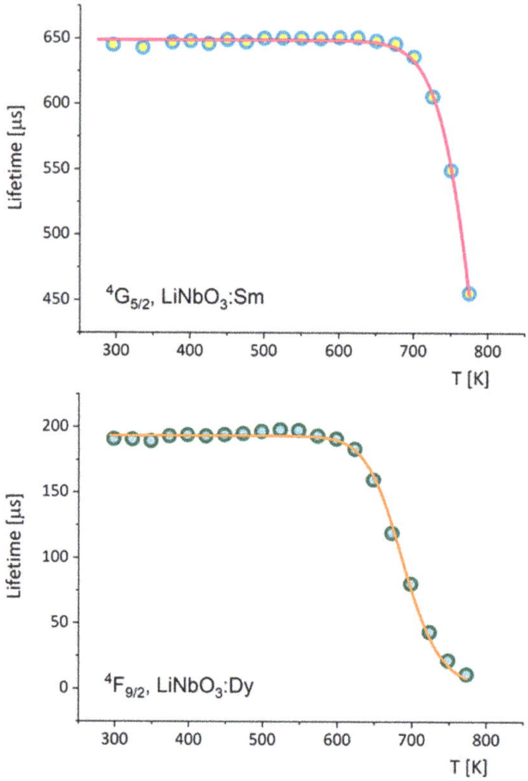

Figure 5. Temperature dependence of $^4G_{5/2}$ (Sm^{3+}) and $^4F_{9/2}$ (Dy^{3+}) lifetimes in $LiNbO_3$. Points indicate experimental data. Solid lines represent theoretical temperature dependence determined from Equation (2) with parameters gathered in Table 1.

To interpret the effect of temperature on the $^4G_{5/2}$ and $^4F_{9/2}$ lifetimes, shown in Figure 5 one should remember that the measured luminescence lifetime τ_{exp} of an excited level of rare earth ion.

$$\tau_{exp} = 1/[W_r + W_{nr}] \tag{1}$$

where W_r is the temperature independent radiative decay rate and W_{nr} denotes the rate of nonradiative decay affected by the temperature. Following the interpretation of data acquired for $LiTaO_3$:Pr^{3+} [47] and recent generalization on spectroscopy of Pr^{3+} in niobate-titanates [48] we interpret the steep decrease of Sm^{3+} and Dy^{3+} lifetimes in terms of temperature-dependent charge transfer (CT) transitions within the modified model (TDCT) proposed by Nikolic et al. [40]. According to the TDCT model the rate $W_{nr}(T)$ of nonradiative charge transfer transitions

$$W_{nr}(T) = W_{nr}(0) \times \left(\frac{1}{T^{*\frac{1}{2}}}\right) \times \exp[(-E_a + \alpha T)/k_B\, T^*] \tag{2}$$

where T denotes the temperature, $W_{nr}(0)$ denotes the rate of the process at 0 K, E_a is the energy barrier to be crossed by phonons, k_B is the Boltzmann constant. The factor αT represents the thermally induced change of the energy E_a assumed to be linear with the slope α.

The factor

$$T^* = \left(\frac{\omega}{2k_B}\right) \times \coth(\omega/2k_B \mathrm{T}) \tag{3}$$

included to the above relation accounts for the non-Arrhenius nature of the excitation energy flow through CT states [49]. The $\hbar\omega$ denotes the energy of the host phonons involved. Solid lines in Figure 5 represent theoretical temperature dependence of Sm^{3+} and Dy^{3+} lifetimes determined from Equation (2) with parameters gathered in Table 1. It can be seen that their agreement with experimental data points is reasonable.

In the following we focus our attention to temperature dependent changes of luminescence bands related to transitions that terminate on the $^6H_{5/2}$ ground state of Sm^{3+} and the $^6H_{15/2}$ ground state of Dy^{3+}. It follows from Figures 3 and 4 that there are contributions of luminescence that grow monotonously with increasing temperature at the expense of the $^4G_{5/2} \rightarrow {}^6H_{5/2}$ and $^4F_{9/2} \rightarrow {}^6H_{15/2}$ bands located at slightly longer wavelengths. Examination of low temperature absorption spectra reported in the past [45,46] reveals that these growing contributions in the Sm^{3+} spectrum is due to transitions from thermally populated $^4F_{3/2}$ level and that in the Dy^{3+} spectrum from thermally populated $^4I_{15/2}$ level. These levels are located, respectively, above the $^4G_{5/2}$ and $^4F_{9/2}$ metastable levels by about 1000 cm^{-1} only and therefore their populations are governed by the Boltzmann statistics. Accordingly, a thermally induced change of fluorescence intensity ratio (FIR) between the $^4F_{3/2} \rightarrow {}^6H_{5/2}$ and $^4G_{5/2} \rightarrow {}^6H_{5/2}$ emission bands for Sm^{3+} and between the $^4I_{15/2} \rightarrow {}^6H_{15/2}$ and $^4F_{9/2} \rightarrow {}^6H_{15/2}$ emission bands for Dy^{3+} can be applied as the temperature sensing parameter. It is known that the luminescence intensities are proportional to the population of involved energy levels and FIR of two thermally coupled levels can be defined by following equation [39]:

$$FIR = \frac{I_{^4I_{15/2}}}{I_{^4F_{9/2}}} = B\exp\left(-\frac{\Delta E}{kT}\right) \tag{4}$$

where B is temperature independent constant, ΔE is the energy gap between the two thermally coupled levels, and k is the Boltzmann constant. Plots of FIR versus temperature for $LiNbO_3$:Sm^{3+} and $LiNbO_3$:Dy^{3+} are compared in Figure 6.

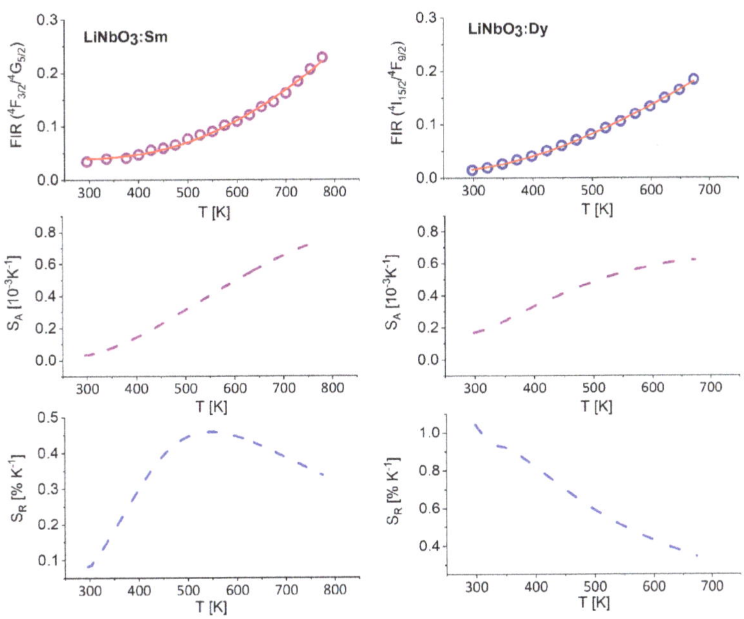

Figure 6. Plots of FIR versus temperature for $LiNbO_3$:Sm^{3+} and $LiNbO_3$:Dy^{3+} (upper graphs) and plots of S_A and S_R parameters versus temperature for $LiNbO_3$:Sm^{3+} (**left**) and $LiNbO_3$:Dy^{3+} (**right**).

Points indicate experimental data determined from spectra in Figures 3 and 4 and the solid lines represent fits of Equation (4) with $\Delta E = 1139\ cm^{-1}$ (for Sm^{3+}) and $\Delta E = 1146\ cm^{-1}$ (for Dy^{3+}).

Optical thermometer may be quantitatively characterized with absolute and relative thermal sensitivity. The former parameter reveals the absolute FIR change with temperature variation and is expressed as:

$$S_A = \frac{dFIR}{dT} = FIR\frac{\Delta E}{kT^2} \tag{5}$$

To reliably compare the thermometers quality, relative sensitivity is usually used because this parameter determines normalized change of *FIR* with temperature variation and is defined as [39]:

$$S_R = \frac{1}{FIR}\frac{dFIR}{dT} \cdot 100\% = \frac{\Delta E}{kT^2} \cdot 100\% \tag{6}$$

Figure 6 compares also plots of S_A and S_R parameters versus temperature for $LiNbO_3$:Sm^{3+} (left) and $LiNbO_3$:Dy^{3+} (right). Examination of the S_R plots indicates that the $LiNbO_3$:Sm^{3+} is mostly suitable for the optical sensor within the 500–750 K temperature region whereas $LiNbO_3$:Dy^{3+} offers the highest sensitivity at lower temperatures between 300 and 400 K. However, the most significant shortcoming of these optical sensors resides in that their luminescence is quenched at temperatures markedly lower as compared to other hosts doped with Sm^{3+} and Dy^{3+}.

3.2. Effect of Temperature on Luminescence of Tb^{3+} in $LiNbO_3$ and $LiTaO_3$

Points in Figure 7 show the temperature dependence of the $^5D_4$ lifetime determined experimentally from luminescence decay curves. Unlike $LiNbO_3$:Sm^{3+} and $LiNbO_3$:Dy^{3+} systems the $LiNbO_3$:Tb^{3+} shows luminescence below ambient temperature only but still the dependence of its lifetime on the temperature consists of the initial part weakly affected by the temperature up to about 130 K and of subsequent steep decrease. For a comparison the experimental $^5D_4$ lifetime for $LiTaO_3$:Tb^{3+} plotted versus temperature also in Figure 7 is nearly constant at temperatures up to about 480 K and next decreases steeply. Solid lines in Figure 7 represent theoretical temperature dependence of the $^5D_4$ lifetime for $LiNbO_3$:Tb^{3+} and $LiTaO_3$:Tb^{3+} determined from Equation (2) with parameters gathered in Table 2. It is worth noticing here that $LiNbO_3$ and $LiTaO_3$ compounds form isostructural crystals characterized by very similar physicochemical properties. Incorporated luminescent rare earth ions are located in sites with the same symmetry in these hosts. They interact with lattice phonons having similar energy distribution with nearly the same cut-off energy corresponding to Nb-O or Ta-O stretching vibrations [47]. Thus, rather small disparity between spectral features of Tb^{3+} in the two hosts can be supposed. Indeed, the survey luminescence spectra of $LiNbO_3$:Tb^{3+} and of $LiTaO_3$:Tb^{3+} compared in Figure 8 corroborate this supposition.

3.3. Discussion

Results presented above imply that the location of the metastable $5D_4$ level of Tb^{3+} with respect to the bottom of conduction band of the host is a factor that governs the luminescence quenching. Obviously, for a given band-gap the energy difference between the bottom of the conduction band and the metastable $5D_4$ level results from the energy difference between the Tb^{3+} ground state and the top of the valence band. The latter difference can be assessed based on generalizations proposed by Dorenbos et al. and by Dorenbos in a series of published papers, e.g., [50,51]. In particular, the ground state of Tb^{3+} located previously 0.7–1.0 eV higher above the valence band than that of Pr^{3+} has been adjusted based on new experimental data showing that these locations are at about the same energy [51]. The band-gap of $LiTaO_3$ has been determined in the past to be 4.59 eV, a value corresponding to the UV absorption edge located at 271 nm [52]. The available information on band-gap of $LiNbO_3$ is not reliable in view of the controversy mentioned above in Section 1 but the disparity of band-gap values for the two hosts is likely be small. In any case, we suppose that peculiarities of luminescence quenching in $LiNbO_3$:Tb^{3+} and $LiTaO_3$:Tb^{3+} are not governed by the band-gaps but stem from the fact that in $LiTaO_3$ the ground state of Tb^{3+} is located lower, i.e., closer to the valence band that in $LiNbO_3$. It follows from the zigzag curves connecting the ground state levels of rare earth ions [50,51] that among rare earth ions considered here the ground state of Tb^{3+} is located at the highest energy, that of Dy^{3+} is located markedly lower and that of Sm^{3+} at the lowest energy. In principle, the energy difference between the top of the valence band and the ground state level of the rare earth ion in question can be determined provided the energy of the charge transfer (denoted also IVCT = intervalence charge transfer) transition and the band-gap of the host are known. Unfortunately, we were not able to obtain these data from our measurement. Figure 2 shows that the IVCT transitions do not contribute to excitation spectra of Sm^{3+} and Dy^{3+} luminescence. Additionally, the IVCT transition does not contribute to the excitation spectrum of Tb^{3+} luminescence in $LiNbO_3$. In [53] the IVCT transition energy of about 3.4 eV for Pr^{3+} in $LiNbO_3$ has been mentioned. Assuming that the band-gap of $LiNbO_3$ equals to 3.8 eV we locate the ground state of Pr^{3+} at about 0.4 eV above the top of the valence band. With this assumption and supposing that the ground states of Pr^{3+} and Tb^{3+} are located at the same energy we locate the $^5D_4$ metastable level of Tb^{3+} at around 0.86 eV below the conduction band. This calculation is rather speculative and the reliability of the $^5D_4$ location is not certain. However, it is worth noticing that it locates the ground states of Sm^{3+} and Dy^{3+} below the top of the valence band accounting for the absence of IVCT band in excitation spectra of their luminescence. The onset of the steep decrease of luminescence lifetime, related with the IVCT, occurs at about 700, 600, and 150 K for Sm^{3+}, Dy^{3+}, and Tb^{3+} ions in $LiNbO_3$, respectively. The $^4G_{5/2}$ metastable level of Sm^{3+} is located at about 17,600 cm^{-1}, i.e., about 2.18 eV above the ground state level. The $^4F_{9/2}$ metastable level of Dy^{3+} is located at about 21,100 cm^{-1} i.e., about 2.61 eV above the ground state level. These values, combined with those inferred from the zigzag curve predict that the energy barrier to be crossed by phonons would be higher, hence the onset temperature would be also higher for Sm^{3+}, in agreement with experimental data. The $^5D_4$ metastable level of Tb^{3+} is located at about 2.55 eV above the ground state level, a value slightly smaller than that for Dy^{3+}. However, this location combined with a relatively large energy inferred from the zigzag curve results in markedly smaller energy barrier hence the lower onset temperature.

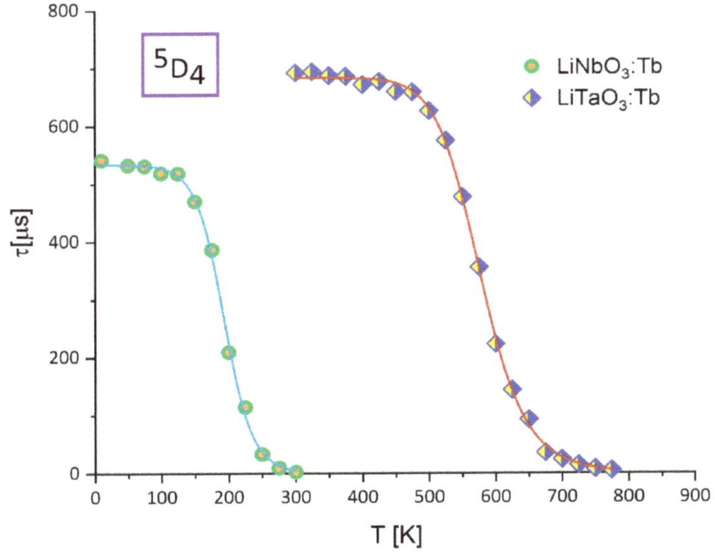

Figure 7. Temperature dependence of $^5D_4$ (Tb^{3+}) lifetimes in $LiNbO_3$ and $LiTaO_3$. Points indicate experimental data. Solid lines represent theoretical temperature dependence determined from Equation (1) with parameters gathered in Table 2.

Table 2. Spectral parameters governing the temperature dependence of $^5D_4$ (Tb^{3+}) lifetimes in $LiNbO_3$ and $LiTaO_3$ within the framework of the TDCT model.

	$LiNbO_3$:Tb	$LiTaO_3$:Tb
W_{nr} (1/s)	1064	807
E_a (cm^{-1})	1984	8128
α (cm^{-1}/K)	17.1	17.6
$\hbar\omega$ (cm^{-1})	693	778
W_r (1/s)	1874	1464

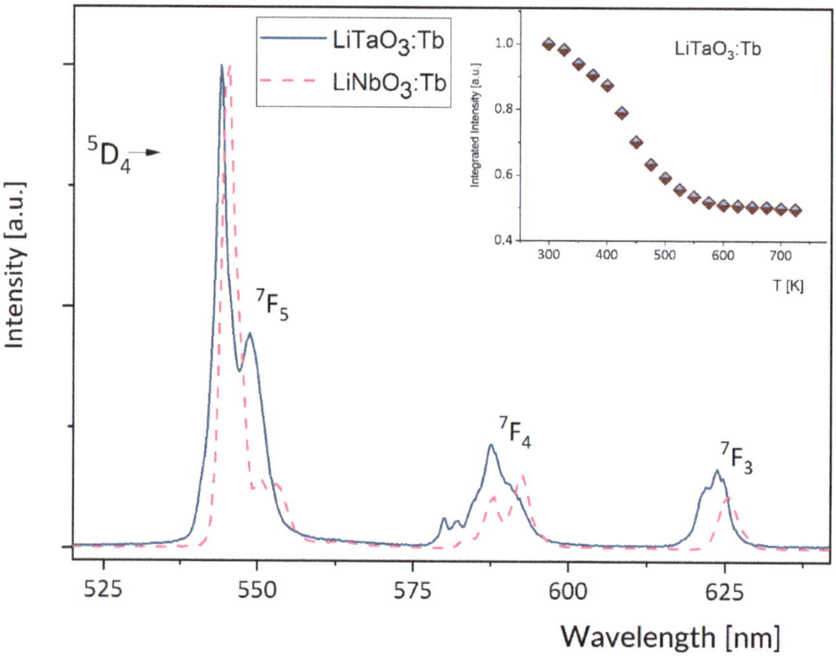

Figure 8. Comparison of the survey luminescence spectrum of $LiNbO_3$:Tb^{3+} recorded at 10 K to survey luminescence spectrum of $LiTaO_3$:Tb^{3+} recorded at 300 K. Inset shows effect of temperature on integrated emission intensity of $LiTaO_3$:Tb^{3+}.

4. Conclusions

The effect of temperature on experimental luminescence lifetimes consists of the initial temperature-independent stage followed by a steep decrease with the onset at about 700, 600, and 150 K for Sm^{3+}, Dy^{3+}, and Tb^{3+} ions, respectively. Experimental results were interpreted in terms of temperature-dependent charge transfer (CT) transitions within the modified phenomenological model TDCT. Disparity of the onset temperatures and their sequence were explained based on the location of familiar zigzag curves connecting the ground state levels of rare earth ions with respect to the band-gap of the host. It was concluded also that $LiNbO_3$:Sm^{3+} is suitable as an optical sensor within the 500–750 K temperature region whereas $LiNbO_3$:Dy^{3+} offers the highest sensitivity at lower temperatures between 300 and 400 K.

Author Contributions: Conceptualization, R.L. and W.R.-R.; methodology, R.L., B.M., R.K. and J.K.; software, J.K.; validation, R.L. and W.R.-R.; formal analysis, R.L., B.M., R.K. and J.K.; investigation, R.L., B.M., R.K. and J.K.; resources, W.R.-R.; data curation, R.L., B.M., R.K. and J.K.; writing—original draft preparation, R.L. and W.R.-R.; writing—review and editing, R.L. and W.R.-R.; visualization, R.L., B.M., R.K. and J.K.; supervision, R.L. and W.R.-R.; project administration, R.L. and W.R.-R.; funding acquisition, W.R.-R.; All authors have read and agreed to the published version of the manuscript.

Funding: The work was financially supported within the statutory funds of the Institute of Low Temperature and Structure Research, Polish Academy of Sciences in Wroclaw.

Conflicts of Interest: The authors declare no conflict of interest.

References

1. Boyd, G.D.; Miller, R.C.; Nassau, K.; Bond, W.L.; Savage, A. $LiNbO_3$: An efficient phase matchable nonlinear optical material. *Appl. Phys. Lett.* **1964**, 5, 234. [CrossRef]
2. Lee, M.; Takedawa, S.; Furukawa, Y.; Kitamura, K.; Hatano, H.; Tanaka, S. Nonvolatile two-color holographic recoding in Tb-doped $LiNbO_3$. *Appl. Phys. Lett.* **2000**, *76*, 1653–1655. [CrossRef]
3. Adibi, A.; Buse, K.; Psaltis, D. Two-center holographic recording. *J. Opt. Soc. Am. B* **2001**, *18*, 584–601. [CrossRef]
4. Myers, L.E.; Eckhardt, R.C.; Fejer, M.M.; Byer, R.L.; Bosenberg, W.R.; Pierce, J.W. Quasi-phase-matched optical parametric oscillations in bulk periodically poled $LiNbO_3$. *J. Opt. Soc. Am.* **1995**, *B12*, 21022116. [CrossRef]
5. Ross, G.W.; Pollnau, M.; Smith, P.G.R.; Clarkson, W.A.; Britton, P.E.; Hanna, D.C. Generation of high-power blue light in periodically poled $LiNbO_3$. *Opt. Lett.* **1998**, *23*, 171–173. [CrossRef]
6. Lallier, E.; Pachole, J.P.; Papuchon, M.; Micheli, M.; Li, M.J.; He, Q.; Ostrovsky, D.B.; Grezes-Besset, C.; Pelletier, E. Nd:Mgo:$LiNbO_3$ waveguide laser and amplifier. *Opt. Lett.* **1990**, *15*, 682–684. [CrossRef]
7. Amin, J.; Aust, J.A.; Veasey, D.L.; Sanfard, N.A. Dual wavelength 980 nm pumped Er/Yb-codoped waveguide laser in Ti:$LiNbO_3$. *Electron. Lett.* **1998**, *34*, 456–458. [CrossRef]
8. Li, R.; Wang, C.J.; Liang, X.; Peng, K.; Xu, G. CW Nd:Mg:$LiNbO_3$ self-frequency-doubling laser at room temperature. *IEEE J. Quantum Electron.* **1993**, *29*, 2419–2420. [CrossRef]
9. Jiangou, Z.; Shipin, Z.; Dingquan, X.; Xiu, W.; Guanfeng, X. Optical absorption properties of doped lithium niobate crystals. *J. Phys. Condens. Matter.* **1992**, *4*, 2977. [CrossRef]
10. Kase, S.; Ohi, K. Optical absorption and interband Faraday rotation in $LiTaO_3$ and $LiNbO_3$. *Ferroelectrics* **1974**, *8*, 419. [CrossRef]
11. Schmidt, W.G.; Albrecht, M.; Wippermann, S.; Blankenburg, S.; Rauls, E.; Fuchs, F.; Rodl, C.; Furthmuller, J.; Hermann, A. $LiNbO_3$ ground- and excited-state properties from first-principles calculations. *Phys. Rev. B* **2006**, *77*, 035106. [CrossRef]
12. Thierfelder, C.; Sanna, S.; Schindlmayr, A.; Schmidt, W.G. Do we know the band gap of lithium niobate? *Phys. Stat. Solidi C* **2010**, *7*, 362–365. [CrossRef]
13. Mamoun, S.; Merad, A.E.; Guilbert, L. Energy band gap and optical properties of lithium niobate from ab initio calculations. *Comp. Mater. Sci.* **2013**, *79*, 125–131. [CrossRef]
14. Abrahams, S.C.; Marsh, P. Defect structure dependence on composition in lithium niobate. *Acta Cryst. B* **1986**, *42*, 61. [CrossRef]
15. Iyi, N.; Kitamura, K.; Izumi, F.; Yamamoto, J.K.; Hayashi, K.; Asano, H.; Kimura, J. Comparative study of defect structures in lithium niobate with different compositions. *J. Solid State Chem.* **1992**, *101*, 340. [CrossRef]
16. Zotov, N.; Boysen, H.; Frey, F.; Metzger, T.; Born, E. Cation substitution models of congruent $LiNbO_3$ investigated by X-ray and neutron powder diffraction. *J. Phys. Chem. Solids* **1994**, *55*, 145–152. [CrossRef]
17. Lorenzo, A.; Jaffrezic, H.; Roux, B.; Boulon, G.; Garcia-Sole, J. Lattice location of rare earth ions in $LiNbO_3$. *Appl. Phys. Lett.* **1995**, *67*, 3735. [CrossRef]
18. Gasparotto, G.; Cebim, M.A.; Goes, M.S.; Lima, S.A.M.; Davolos, M.R.; Varela, J.A.; Paiva-Santos, C.O.; Zaghete, M.A. Correlation between the spectroscopic and structural properties with the occupation of the Eu^{3+} sites in powdered Eu^{3+}-doped $LiTaO_3$ prepared by the Pechini method. *J. Appl. Phys.* **2009**, *106*, 063509. [CrossRef]
19. Shu, S.; Wang, Y.; Ke, Y.; Deng, B.; Liu, R.; Song, Q.; Wang, J.; Yu, R. $NaCaTiTaO_6$:Sm^{3+}: A novel orange-red-emitting tantalate phosphor with excellent thermal stability and high color purity for white LEDs. *J. Alloy. Compd.* **2020**, *848*, 156359. [CrossRef]

20. Su, K.; Zhang, Q.; Yang, X.; Ma, B. Crystal structure and luminescence properties of thermally stable Sm^{3+}-doped $Sr_9In(PO_4)_7$ orange-red phosphor. *J. Phys. D Appl. Phys.* **2020**, *53*, 385101. [CrossRef]
21. Zhang, L.; Che, J.; Ma, Y.; Wang, J.; Kang, R.; Deng, B.; Yu, R.; Geng, H. Luminescent and thermal properties of novel orange-red emitting $Ca_2MgTeO_6{:}Sm^{3+}$ phosphors for white LED's. *J. Lumin.* **2020**, *225*, 117374. [CrossRef]
22. Kovács, L.; Kocsor, L.; Tichy-Rács, É.; Lengyel, K.; Bencs, L.; Corradi, G. Hydroxyl ion probing transition metal dopants occupying Nb sites in stoichiometric $LiNbO_3$. *Opt. Mater. Express* **2019**, *9*, 4506–4516. [CrossRef]
23. Mandula, G.; Kis, Z.; Kovacs, L.; Szaller, Z.; Krampf, A. Site-selective measurement of relaxation properties at 980 nmin Er^{3+}-doped congruent and stoichiometric lithium niobite crystals. *Appl. Phys. B* **2016**, *122*, 72. [CrossRef]
24. Kis, Z.; Mandula, G.; Lengyel, K.; Hajdara, I.; Kovacs, L.; Imlau, M. Homogeneous linewidth measurements of Yb^{3+} ions in congruent and stoichiometric lithium niobate crystals. *Opt. Mater.* **2014**, *37*, 845–853. [CrossRef]
25. Robbins, D.J.; Cockayne, B.; Lent, B.; Glasper, J.L. The mechanism of $^5D_3$–$^5D_4$ cross-relaxation in $Y_3Al_5O_{12}{:}Tb^{3+}$. *Solid State Commun.* **1976**, *20*, 673–676. [CrossRef]
26. Park, J.Y.; Jung, H.C.; Raju, S.R.; Moon, B.K.; Jeong, J.H.; Kim, J.H. Solvothermal synthesis and luminescence properties of Tb^{3+}-doped gadolinium aluminum garnet. *J. Lumin.* **2010**, *130*, 478–482. [CrossRef]
27. Praveena, R.; Shim, J.J.; Cai, P.; Seo, H.J.; Chung, W.; Kwon, T.H.; Jayasankar, C.K.; Haritha, P.; Venkatramu, V. Luminescence properties of $Lu_3Al_5O_{12}{:}Tb^{3+}$ nano-garnet. *J. Korean Phys. Soc.* **2014**, *64*, 1859–1865. [CrossRef]
28. Teng, X.; Wang, W.; Cao, Z.; Li, J.; Duan, G.; Liu, Z. The development of new phosphors 3f Tb^{3+}/Eu^{3+} co-doped $Gd_3Al_5O_{12}$ with tunable emission. *Opt. Mater. (Amst.)* **2017**, *69*, 175–180. [CrossRef]
29. Solarz, P.; Głowacki, M.; Lisiecki, R.; Sobczyk, M.; Komar, J.; Macalik, B.; Ryba-Romanowski, W. Impact of temperature on excitation, emission and cross-relaxation processes of terbium ions in GGAG single crystal. *J. Alloy. Compd.* **2019**, *789*, 409–415. [CrossRef]
30. Hakeem, D.A.; Pi, J.W.; Kim, S.W.; Park, K. New $Y_2LuCaAl_2SiO_{12}{:}Ln$ (Ln = Ce^{3+}, Eu^{3+}, and Tb^{3+}) phosphors for white LED applications. *Inorg. Chem. Front.* **2018**, *5*, 1336–1345. [CrossRef]
31. Deopa, N.; Rao, A.S. Spectroscopic studies of single near ultraviolet pumped Tb^{3+} doped Lithium Lead Alumino Borate glasses for green lasers and tricolour w-LEDs. *J. Lumin.* **2018**, *194*, 56–63. [CrossRef]
32. Loiko, P.; Volokitina, A.; Mateos, X.; Dunina, E.; Kornienko, A.; Vilejshikova, E.; Aguiló, M.; Díaz, F. Spectroscopy of Tb^{3+} ions in monoclinic $KLu(WO_4)_2$ crystal application of an intermediate configuration interaction theory. *Opt. Mater.* **2018**, *78*, 495–501. [CrossRef]
33. Liu, G.K.; Carnall, W.T.; Jones, R.P.; Cone, R.L.; Huang, J. Electronic energy level structure of Tb^{3+} in $LiYF_4$. *J. Alloy. Compd.* **1994**, *207–208*, 69–73. [CrossRef]
34. Gusowski, M.A.; Ryba-Romanowski, W. Unusual behavior of Tb^{3+} in K_3YF_6 green-emitting phosphor. *Opt. Lett.* **2008**, *33*, 1786–1788. [CrossRef] [PubMed]
35. Metz, P.W.; Marzahl, D.T.; Huber, G.; Kränkel, C. Performance and wavelength tuning of green emitting terbium lasers. *Opt. Express* **2017**, *25*, 5716. [CrossRef]
36. Rajendra, H.J.; Pandurangappa, C.; Monika, D.L. Luminescence properties of dysprosium doped YVO_4 phosphor. *J. Rare Earths* **2018**, *36*, 1245–1249. [CrossRef]
37. Liu, Y.; Liu, G.; Wang, J.; Dong, X.; Yu, W. Multicolor photoluminescence and energy transfer properties of dysprosium and europium-doped Gd_2O_3 phosphors. *J. Alloy. Compd.* **2015**, *649*, 96–103. [CrossRef]
38. Rimbach, A.C.; Steudel, F.; Ahrens, B.; Schweizer, S. Tb^{3+}, Eu^{3+}, and Dy^{3+} doped lithium borate and lithium aluminoborate glass: Glass properties and photoluminescence quantum efficiency. *J. Non Cryst. Solids* **2018**, *499*, 380–386. [CrossRef]
39. Perera, S.S.; Rabuffetti, F.A. Dysprosium-activated scheelite-type oxides as thermosensitive phosphors. *J. Mater. Chem. C* **2019**, *7*, 7601. [CrossRef]
40. Nikolić, M.G.; Jovanović, D.J.; Dramićanin, M.D. Temperature dependence of emission and lifetime in Eu^{3+}- and Dy^{3+}-doped $GdVO_4$. *Appl. Opt.* **2013**, *52*, 1716–1724. [CrossRef]
41. Hertle, E.; Chepyga, L.; Batentschuk, M.; Will, S.; Zigan, L. Temperature-dependent luminescence characteristics of Dy^{3+} doped in various crystalline hosts. *J. Lumin.* **2018**, *204*, 64–74. [CrossRef]
42. Bowman, S.R.; O'Connor, S.; Condon, N.J. Diode pumped yellow dysprosium lasers. *Opt. Express* **2012**, *20*, 12906–12911. [CrossRef] [PubMed]

43. Haro-González, P.; Martín, L.L.; Martín, I.R.; Berkowski, M.; Ryba-Romanowski, W. Optical amplification properties of Dy^{3+}-doped Gd_2SiO_5, Lu_2SiO_5 and $YAl_3(BO_3)_4$ single crystals. *Appl. Phys. B Lasers Opt.* **2011**, *103*, 597–602. [CrossRef]
44. Repelin, Y.; Husson, E.; Proust, F.B.C. Raman spectroscopy of lithium niobate and lithium tantalite. Force field calculations. *J. Phys. Chem. Solids* **1999**, *60*, 819–825. [CrossRef]
45. Dominiak-Dzik, G. Sm^{3+}-doped $LiNbO_3$ crystal, optical properties and emission cross-sections. *J. Alloy. Compd.* **2005**, *391*, 26–32. [CrossRef]
46. Dominiak-Dzik, G.; Ryba-Romanowski, W.; Palatnikov, M.N.; Sidorov, N.V.; Kalinnikov, V.T. Dysprosium-doped $LiNbO_3$ crystal. Optical properties and effect of temperature on fluorescence dynamics. *J. Mol. Struct.* **2004**, *704*, 139–144. [CrossRef]
47. Gryk, W.; Dyl, D.; Ryba-Romanowski, W.; Grinberg, M. Spectral properties of $LiTaO_3:Pr^{3+}$ under high hydrostatic pressure. *J. Phys. Condens. Matter* **2005**, *17*, 5381–5395. [CrossRef]
48. Boutinaud, P. Rationalization of the Pr^{3+}-to-transition metal charge transfer model: Application to the luminescence of Pr^{3+} in titano-nobates. *J. Lumin.* **2019**, *214*, 116557. [CrossRef]
49. Englman, R.; Jortner, J. The energy gap law for radiationless transitions in large molecules. *J. Mol. Phys.* **1970**, *18*, 145. [CrossRef]
50. Dorenbos, P.; Krumpel, A.H.; Boutinaud, E.v.P.; Bettinelli, M.; Cavalli, E. Lanthanide level location in transition metal complex compounds. *Opt. Mater.* **2010**, *32*, 1681–1685. [CrossRef]
51. Dorenbos, P. Charge transfer bands in optical materials and related defect level location. *Opt. Mater.* **2017**, *69*, 8–22. [CrossRef]
52. Nikl, M.; Morlotti, R.; Magro, C.; Bracco, R. Emission and storage properties of $LiTaO_3:Tb^{3+}$ phosphor. *J. Appl. Phys.* **1996**, *79*, 2853–2856. [CrossRef]
53. Boutinaud, P.; Bettinelli, M.; Diaz, F. Intervalence charge transfer in Pr^{3+}-and Tb^{3+}-doped double tungstate crystals $KRE(WO_4)_2$ (RE = Y, Gd, Yb, Lu). *Opt. Mater.* **2010**, *32*, 1659–1663. [CrossRef]

Publisher's Note: MDPI stays neutral with regard to jurisdictional claims in published maps and institutional affiliations.

crystals

MDPI

Article

Correlation of Electrical Properties and Acoustic Loss in Single Crystalline Lithium Niobate-Tantalate Solid Solutions at Elevated Temperatures

Yuriy Suhak [1,*], Dmitry Roshchupkin [2], Boris Redkin [3], Ahsanul Kabir [1], Bujar Jerliu [1], Steffen Ganschow [4] and Holger Fritze [1]

1 Institute for Energy Research and Physical Technologies, Clausthal University of Technology, Am Stollen 19B, 38640 Goslar, Germany; ahsanul.kabir@tu-clausthal.de (A.K.); bujar.jerliu@tu-clausthal.de (B.J.)
2 Institute of Microelectronics Technology and High Purity Materials, Russian Academy of Sciences, Academician Ossipyan Str. 6, 142432 Chernogolovka, Russia; rochtch@iptm.ru
3 Institute of Solid State Physics, Russian Academy of Sciences, Academician Ossipyan Str. 2, 142432 Chernogolovka, Russia; redkin@issp.ac.ru
4 Leibniz-Institut für Kristallzüchtung, Max-Born-Str. 2, 12489 Berlin, Germany; steffen.ganschow@ikz-berlin.de
* Correspondence: sensorik@tu-clausthal.de

Citation: Suhak, Y.; Roshchupkin, D.; Redkin, B.; Kabir, A.; Jerliu, B.; Ganschow, S.; Fritze, H. Correlation of Electrical Properties and Acoustic Loss in Single Crystalline Lithium Niobate-Tantalate Solid Solutions at Elevated Temperatures. *Crystals* **2021**, *11*, 398. https://doi.org/10.3390/cryst11040398

Academic Editors: Gábor Corradi and László Kovács

Received: 3 March 2021
Accepted: 7 April 2021
Published: 9 April 2021

Abstract: Electrical conductivity and acoustic loss Q^{-1} of single crystalline Li(Nb,Ta)O_3 solid solutions (LNT) are studied as a function of temperature by means of impedance spectroscopy and resonant piezoelectric spectroscopy, respectively. For this purpose, bulk acoustic wave resonators with two different Nb/Ta ratios are investigated. The obtained results are compared to those previously reported for congruent $LiNbO_3$. The temperature dependent electrical conductivity of LNT and $LiNbO_3$ show similar behavior in air at high temperatures from 400 to 700 °C. Therefore, it is concluded that the dominant transport mechanism in LNT is the same as in LN, which is the Li transport via Li vacancies. Further, it is shown that losses in LNT strongly increase above about 500 °C, which is interpreted to originate from conductivity-related relaxation mechanism. Finally, it is shown that LNT bulk acoustic resonators exhibit significantly lower loss, comparing to that of $LiNbO_3$.

Keywords: lithium niobate-tantalate; piezoelectric; acoustic; high-temperature; sensor; *Q*-factor; BAW resonator

1. Introduction

Piezoelectric resonant sensors are attracting considerable interest since they offer numerous advantages for multiparameter in-situ monitoring and control of industrial processes. In particular, there is an increasing need in sensitive, robust and cost-effective sensors for gas composition, temperature and pressure for the application at high-temperatures or generally in harsh environments [1–3]. The working principles of such sensors are based on frequency shifts that arise from external factors such as temperature or mass load. Therefore, it is important to maximize their resonance quality factor (*Q*-factor), which allows, e.g., improved accuracy and better stability in frequency-control applications [3].

Further, piezoelectric actuators, which can generate movements in the micrometer range upon application of voltage, are demanded in automotive, aerospace and related industrial applications. For example, automotive applications such as fuel injection nozzles require operation at elevated temperatures [3,4].

However, the application temperature of common piezoelectric materials is limited. For example, usage of lead zirconium titanate (PZT) ceramics actuators at high-temperatures is limited by thermal instability at around 200–250 °C for most PZT compositions [3–5]. On the other hand, crystals from the langasite ($La_3Ga_5SiO_{14}$) family show relatively high stability at high temperatures, however their piezoelectric coefficients are too small to use them as actuating devices [2,6,7].

Lithium niobate ($LiNbO_3$, LN) and lithium tantalate ($LiTaO_3$, LT) attracted substantial scientific and industrial interest in the last decades because of its excellent electro-optical, piezoelectric and acoustic properties [8–10]. Both compounds are isostructural (point symmetry group *3m*, space group *R3c*) with only slight differences in the lattice parameters [8,10,11]. Both LN and LT have a congruently melting composition with an approximate lithium content of about 48.4 mol.% Li_2O. These materials possess high piezoelectric coefficients (10 times higher comparing to langasite), which allows generating appreciable and useful strain under an applied electric field, enabling their usage not only as surface or bulk acoustic wave sensors, but also as actuators [12,13]. $LiTaO_3$ shows good thermal stability of sensing-relevant parameters (e.g., resonance frequency or quality factor), however, its application is limited by a relatively low Curie temperature of about 630 °C [14]. The Curie temperature of LN is about 1200 °C, but its thermal stability at elevated temperatures is relatively poor. Li_2O out-diffusion and evaporation and, finally, chemical decomposition [15] impact the performance of LN bulk acoustic resonators, such as quality-factor or resonance frequency [15,16]. Moreover, an increasing material dependent loss for different bulk vibration modes is observed above 700 °C [16]. In this respect, single crystalline lithium niobate-tantalate solid solutions ($Li(Nb,Ta)O_3$, LNT) are expected to combine the best properties, i.e., good thermal stability and high Curie temperature, of both end components of the system.

Several defect models have been proposed in the literature to describe Li-deficiency in congruent $LiNbO_3$, which is commonly formed during the crystal growth. These models consider formation of Li-, Nb- and O-vacancies [17–19]:

$$LiNbO_3 \rightarrow 3Li_2O + 4V_{Li}^{/} + Nb_{Li}^{4\bullet} \quad (1)$$

$$LiNbO_3 \rightarrow 3Li_2O + 4V_{Nb}^{5/} + 5Nb_{Li}^{4\bullet} \quad (2)$$

$$LiNbO_3 \rightarrow Li_2O + 2V_{Li}^{/} + V_{O}^{2\bullet} \quad (3)$$

However, the structural studies and the calculations performed in [17,18] show a low probability for the formation of oxygen vacancies and higher charged ${V_{Nb}}^{5/}$ vacancies during crystal growth. Thus, mechanism (1) is considered to be energetically the most favorable. Here, compensation of Nb antisites occurs by Li-vacancies. In an analogy to the defect model for nonstoichiometric $LiNbO_3$, the congruent $LiTaO_3$ single crystals also contain a large amount of the tantalum antisites and cation vacancies [20–22]. The defect models for LNT are still to be established. However basing on LN and LT studies it is plausible to assume, that the Li-vacancy model could also be considered for congruent LNT crystals.

The high-temperature properties of LN and LT crystals are in general relatively well studied. In particular, the temperature-dependent electrical properties of LT were investigated in [23–26] and of LN in [16,27–32]. For both end components of the LNT system it is established that at elevated temperatures the conductivity is predominantly ionic and attributed to the motion of lithium ions via Li-vacancies [23,24,26–32]. Further, the elastic, dielectric and piezoelectric constants were determined for both end components of the $Li(Nb,Ta)O_3$ system as a function of temperature in [33–35]. In contrast, the electrical, structural and optical properties of LNT are scarcely studied so far [36–39]. Beyond these publications, to the best of our knowledge, no systematic studies on LNT were published. The piezoelectric properties of LN resonators were studied up to 500 °C in [40] and up to 900 °C in [16,41]. Finally, the temperature stability of LN-based piezoelectric transducers was examined in [42,43].

The synthesis of single crystalline LNT solid solutions is challenging, largely because of great difference in melting temperatures T_m of both end components of the system ($T_m(LiNbO_3)$ = 1240 °C; $T_m(LiTaO_3)$ = 1650 °C) [44,45]. The growth of high-quality LNT single crystals appears to have been achieved only recently in the United Kingdom [39,46,47], China [36] and by the groups, involved in this study. Therefore, the high-temperature properties of LNT compounds remain practically unstudied. In particular, the material constants of LNT-solid solutions are not available. To the best of our knowledge, only the electrical conductivity is investigated up to 850 °C. The results are published in [48,49].

Based on the previous studies of acoustic loss in $LiNbO_3$ [16,41] and in other piezoelectric materials [50–52], the following dissipation mechanisms can potentially contribute to the losses:

(1) Intrinsic phonon-phonon interactions,
(2) Anelastic point defect relaxations,
(3) Piezoelectric/carrier relaxation (conductivity-related losses),
(4) Non-material contribution (cables, mounting, etc.).

Phonon–phonon scattering can be the dominant loss mechanism at low temperatures (near and below room temperature) in the case of high-quality piezoelectric crystals [53,54]. This contribution exhibits, however, only weak temperature dependence above room temperature [50,53]. Therefore its contribution is minor at elevated temperatures. Consequently, it is neglected in the current study.

Anelastic relaxations can arise from point defects, which produce local anisotropic distortions in the crystal lattice [55]. The application of acoustic stress leads to time-delayed thermally activated reorientation of these defects. This process depends on frequency and temperature and can be described by a Debye function:

$$Q^{-1}(\omega, T) \approx \frac{\Delta}{T}\frac{\omega\tau}{1+\omega^2\tau^2} \tag{4}$$

where Δ is a temperature-independent constant proportional to the concentration of the defect species, T is the absolute temperature, ω is the angular acoustic frequency (equal to 2πf), and τ is the reorientation relaxation time of the defect. As will be shown in the subsequent sections, no evidence of dominating anelastic point defect relaxation was observed on $Q^{-1}(T)$ dependencies of the samples studied here so far.

Of particular relevance for $Li(Nb,Ta)O_3$ resonators is the dissipation that arises from piezoelectric/carrier relaxation. These conductivity-related losses are caused by the motion of charge carriers in an oscillating piezoelectric field. According to the theory of Hutson and White [56], this contribution has a Debye form, is frequency and temperature dependent and can be approximated by the following equation:

$$Q_c^{-1}(\omega, T) \approx K^2\frac{\omega\tau_c}{1+\omega^2\tau_c^2} \tag{5}$$

where ω is the angular frequency (equal to $2\pi f$); T and K^2 are the absolute temperature and the electromechanical coupling coefficient, respectively and τ_c is the relaxation time, with

$$\tau_c = \frac{\varepsilon_{ij}}{\sigma} \tag{6}$$

Here, σ and ε_{ij} denote the electrical conductivity and the dielectric permittivity, respectively. For the thickness–shear mode of Y-cut crystals from point group *3m*, the electromechanical coupling coefficient is equal to $e_{15}^2/(C_{44}\varepsilon_{11})$ in the reduced-index notation, where e_{15}, C_{44} and ε_{11} are the piezoelectric coefficient, the elastic stiffness and the dielectric permittivity, respectively [57]. For the thickness mode of Z-cut $Li(Nb,Ta)O_3$ crystals the electromechanical coupling factor is defined as $e_{33}^2/(C_{33}\varepsilon_{33})$ [57]

In the current paper the high-temperature electrical and electromechanical properties of single crystalline LNT with different Nb/Ta ratios, grown by the Czochralski technique [58] are studied and compared to the data, reported previously for LN [16,41]. The work presented here is the first step towards systematic studies of high-temperature electrical and acoustic properties of LNT crystals. It seeks to provide a better understanding of physical mechanisms that govern loss in Li(Nb,Ta)O_3 by combining determination of electrical conductivity and loss-related Q^{-1} with the aim to obtain high-temperature stable material with large piezoelectric coefficients for possible actuating applications.

2. Materials and Methods

2.1. Crystals

$LiNb_{0.88}Ta_{0.12}O_3$ and $LiNb_{0.5}Ta_{0.5}O_3$ crystals, grown by the Czochralski technique were used in this study. $LiNb_{0.88}Ta_{0.12}O_3$ was grown at the Institute of Microelectronics Technology and High Purity Materials, Russian Academy of Sciences, Chernogolovka, Russia (IMT). This crystal was grown along the polar axis Z in a Pt crucible (diameter of 60 mm and height of 60 mm). Congruent $LiNbO_3$ and $LiTaO_3$ crystals were used as initial charges. The pulling rate of crystals from the melt was about 0.5 mm per hour. The obtained crystal boules have a diameter and a length of about 20 mm each as shown in Figure 1. The composition of the grown crystals was determined using mass spectrometry with inductively coupled plasma (XSeries II Thermo Scientific spectrometer, Waltham, MA, USA). The full report about the growth process and structural parameters of the obtained crystal is given in [58]. Additionally, the composition of $LiNb_{0.88}Ta_{0.12}O_3$ was studied by energy-dispersive X-ray (EDX) spectroscopy (CamScan 44, Waterbeach, UK). These investigations were performed on two specimens, used for electrical and electromechanical measurements, which are described in Section 2.2. The Nb/Ta ratio was determined along the diameter of the specimens as a line scan with 0.8 mm step. In total 10 measurements per specimen were performed and the uncertainty of the measurements did not exceed 5%. The results are summarized in Table 1. Most important, an insignificant inhomogeneity and small differences in Nb/Ta distribution between two specimens are observed.

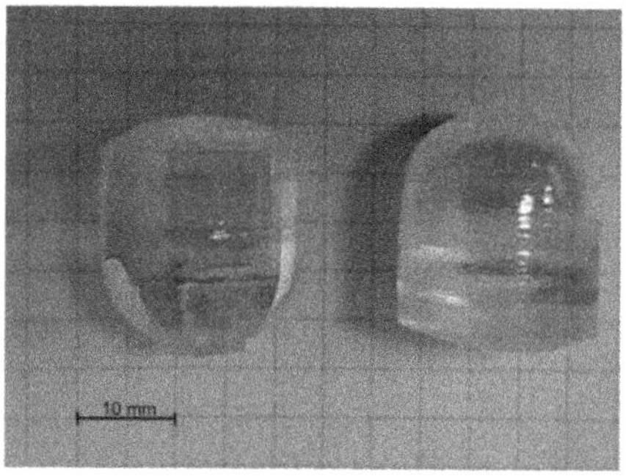

Figure 1. $LiNb_{0.88}Ta_{0.12}O_3$ boules, grown by the Czochralski technique.

Table 1. Nb/Ta ratio in $LiNb_{0.88}Ta_{0.12}O_3$ samples determined by an energy-dispersive X-ray (EDX).

Measurement	LNT88-01			LNT88-02		
	Nb at.%	Ta at.%	Nb/(Nb + Ta)	Nb at.%	Ta at.%	Nb/(Nb + Ta)
1	19.73	2.12	0.90	21.26	2.49	0.90
2	19.75	2.61	0.88	21.45	2.44	0.90
3	20.53	2.52	0.89	21.51	2.54	0.90
4	20.72	2.42	0.90	21.30	2.49	0.90
5	20.91	2.87	0.88	21.85	2.11	0.91
6	21.00	2.73	0.88	21.59	2.06	0.91
7	21.56	2.89	0.88	21.89	2.11	0.91
8	21.38	2.32	0.90	22.36	2.07	0.91
9	21.40	2.8	0.88	21.88	2.00	0.92
10	21.31	2.86	0.88	22.46	1.76	0.93

A single crystal of nominal composition $LiNb_{0.5}Ta_{0.5}O_3$ was grown at the Leibniz-Institut für Kristallzüchtung, Berlin, Germany (IKZ). Initial melt was a mixture of the congruently melting compositions of $LiNbO_3$ and $LiTaO_3$. Considering the strong segregation upon crystallization [46], the melt contained a $LiTaO_3$ mole fraction of only 0.18. The starting materials, previously dried and thoroughly mixed Li_2CO_3 (99.999%, Alfa Aesar, Karlsruhe, Germany), Nb_2O_5 (99.995%, H.C. Starck, Goslar, Germany) and Ta_2O_5 (99.99%, Fox Chemicals, Pfinztal, Germany), were pressed isostatically and sintered at 1100 °C. The obtained blank was melted in an inductively heated 40 mL iridium crucible covered by an active afterheater [59] and thermally isolated by a set of alumina ceramics. Growth atmosphere was high purity Ar with an admixture of approximately 1.3 vol% O_2. The crystal was grown at a rate of 0.5 mm/h on a [00.1]-oriented $LiTaO_3$ seed, provided by CrysTec GmbH, Berlin. To minimize the impact of segregation, only 6% of the melt was crystallized. After growth the crystal was withdrawn from the melt and cooled down to room temperature within 15 h. The obtained crystal was largely transparent of slight yellowish coloration, with a cylindrical part slightly more than 10 mm in both diameter and length. The crystal was not intentionally poled during or after the growth process, i.e., the domain structure after the growth was not defined. However, the existence of piezoelectrically exited resonances confirms that there are obviously sufficiently large areas with identical orientation of the polarization in the sample (see the Results section). Chemical homogeneity was tested on a lengthwise cut (nearly parallel to the [00.1] direction) through the whole crystal using an X-ray fluorescence (XRF) spectrometer with highly focused beam (Tornado M4, by Bruker Nano, Berlin, Germany). Distribution of niobium and tantalum was mapped over the entire sample with a spatial sampling distance of approximately 20 μm. The fluorescence maps were quantified using standard fundamental parameter analysis.

Figure 2a shows the distribution of tantalum (atom fraction = [Ta]/([Ta] + [Nb])) over the whole measured sample area. Although integral compositional variation along the growth direction is small, there exist longitudinal structures with drastically reduced tantalum contents. These structures were the result of cellular growth owing to constitutional supercooling and occur mainly in regions of higher local growth velocity, i.e., the conical part with increasing crystal diameter and in regions with poor solute exchange in the melt, i.e., near the rotation axes. Figure 2b shows the distribution over a (0001) sample cut perpendicular to the growth direction from the approximate axial position indicated by the arrow in Figure 2a. The mean $LiTaO_3$ fraction measured over the full diameter of this sample was 48% with an RMS of 1.4%. Figure 3 presents the distribution of LT fraction along the diameter of the sample, presented in Figure 2b. The measurement was performed along the vertical right edge.

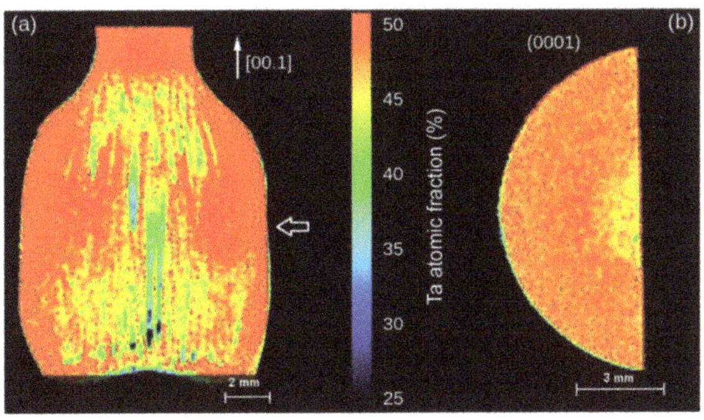

Figure 2. Element concentration mappings of a longitudinal (**a**) and transversal cut (**b**) through the LNT-50 crystal obtained by the XRF analysis. Axial position of the sample in (**b**) is indicated by the arrow in (**a**).

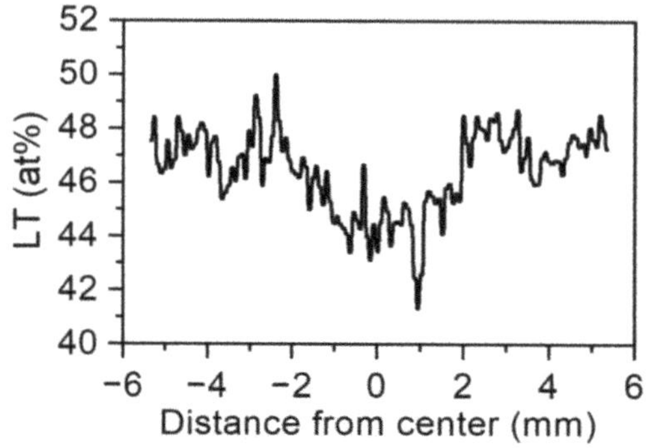

Figure 3. LT distribution in $LiNb_{0.5}Ta_{0.5}O_3$ specimen along the full crystal diameter (right edge) of the sample in Figure 2b.

Additionally, nominally undoped $LiNbO_3$ and $LiTaO_3$ crystals of congruent composition, grown by the Czochralski technique at IMT were used in the current study for the electrical measurement.

2.2. Specimens

Two $LiNb_{0.88}Ta_{0.12}O_3$ specimens, denoted as LNT88-01 and LNT88-02, were used in this study. They were prepared from a Y-cut plate with the approximate dimensions of $18 \times 18 \times 1$ mm^3 that was cut from the middle region of the related crystal boule, described above. Subsequently, the specimens were milled from the plate using a high-precision ultrasonic milling machine (UST-300, Dama Technologies, Häggenschwil, Switzerland) in the form of Y-cut plano-plano discs of 1 mm thickness and 8 mm diameter. Afterwards, they were polished (roughness 1 µm) and coated with keyhole-shaped Pt/Rh electrodes with 300 nm thickness and 4 mm diameter by pulsed laser deposition (PLD). Further, the $LiNb_{0.5}Ta_{0.5}O_3$ sample was prepared as Z-cut half-disc with 10 mm diameter and 0.5 mm thickness, cut from the middle part of the crystal boule, see the right part of Figure 2. This sample was designated as LNT50. Here, the electrodes were prepared by screen printing (print ink: Ferro Corporation, No. 6412 0410, Mayfield Heights, OH, USA) and subsequently annealed at 800 °C for about 30 min. This short thermal treatment was expected not to cause a significant change of the sample properties. Such screen-printed Pt electrodes were excellent for the cases when robust temperature-stable electrodes were required. When annealed in the presence of oxygen platinum demonstrates high thermal and chemical stability, without any transition layers formed on its surface, which provides excellent electrical contact with other materials [60,61]. According to our experience, application of Pt screen-printed electrodes does not impact the high-temperature *Q*-factor of piezoelectric crystals of the langasite family, which show low losses even at elevated temperatures. Consequently, such electrodes are expected to have virtually no influence on the *Q*-factor of LNT at elevated temperatures where materials-related losses dominate.

Additionally Y-cut $LiNbO_3$ and $LiTaO_3$ plates of congruent composition with the dimensions of $10 \times 18 \times 2$ mm^3 and screen-printed electrodes were used in this work for conductivity studies with the aim to compare the electrical properties of LN, LNT and LT.

2.3. Measurements and Analysis

2.3.1. Electrical Conductivity

The studies of electrical conductivity as a function of temperature were performed by means of impedance spectroscopy using an impedance/gain-phase analyzer (Solartron 1260, Ametek Scientific Instruments, Hampshire, UK) in the frequency range from 1 Hz to 1 MHz and in the temperature range from 400 to 700 °C. The excitation voltage was 50 mV. The duration of sweep in in the frequency range that determines the data evaluation from

500 Hz to 1 MHz was about 60 s. The measurements were performed in air at atmospheric pressure while heating at a rate of 1 K/min. Below 400 °C the impedance of the studied specimens was too high to acquire reliable data. For both types of the applied electrodes (PLD and screen-printed) we varied the excitation voltage up to 0.3 V for the comparison purposes. In all the cases the obtained results were the same.

The obtained impedance spectra were visualized in the complex plane, which involves plotting the imaginary part of the impedance against its real part (Nyquist plot). Subsequently, an electrical equivalent-circuit model consisting of a constant phase element (CPE) connected in parallel with a bulk resistance R_B was fitted to the measured data. The low frequency intercepts of R_B-CPE semicircles in such a complex impedance plane were interpreted as bulk resistance and subsequently converted in the bulk conductivity, using the relation $\sigma = t(A \times R_B)^{-1}$, where t and A are the thickness of the sample and the electrode area, respectively.

A Nyquist plot is exemplary given in Figure 4 for the studied specimens at 600 °C. The representation of the resistivity is chosen to eliminate the geometrical factors of the samples (thickness and area). As seen from the figure, slightly depressed semicircles with almost similar resistance were obtained for LN and LT specimens. Such depression results from non-ideal capacitances and corresponds to exponents of the constant phase elements of 0.98 and 0.87 for $LiNbO_3$ and $LiTaO_3$ respectively. The corresponding exponents for LNT50 and LNT88-01 and LNT88-02 samples were equal to 0.98, 0.99 and 1.00 respectively. Further, as seen from Figure 4, the LNT88-01 and LNT88-02 specimens, which were manufactured from the same crystal boule are exhibiting different resistivity. This peculiarity will be discussed in the subsequent section.

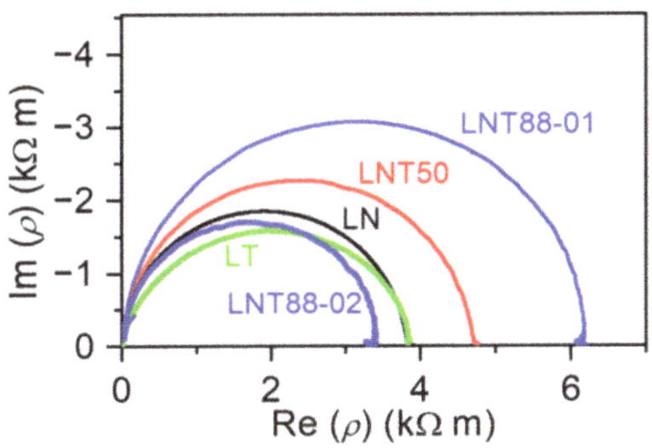

Figure 4. Complex resistivity of different $Li(Nb,Ta)O_3$ samples at 600 °C, presented as the Nyquist plot.

2.3.2. Acoustic Loss

The investigations of acoustic losses were carried out by means of resonant piezoelectric spectroscopy (RPS) on LNT resonators, operated in the thickness–shear mode (TSM) and in the thickness mode (TM). Similarly to conductivity studies, these measurements were performed in air during heating with a rate of 1 K per min from RT to 700 °C. The samples were electrically contacted using platinum foils on each side of the sample. With the objective to minimize damping from mechanical contact, only small areas near the edges of the samples were electrically contacted and mechanically clamped (Figure 5). The electrodes overlapped only in areas that were not mechanically clamped.

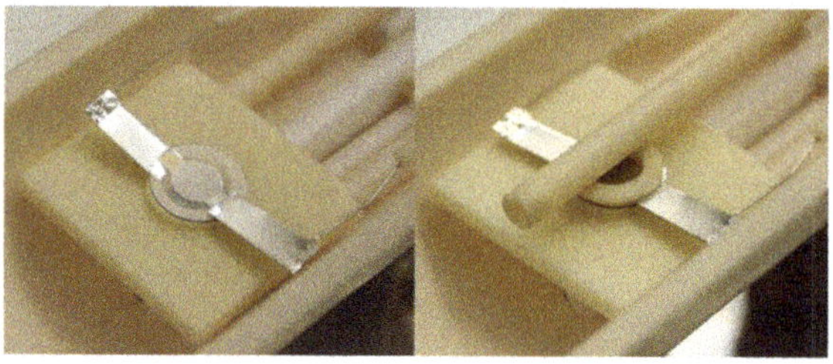

Figure 5. Arrangement of the studied sample for the high-temperature electrical and acoustic measurements. The pictures show the ring-type sample support without (**left**) and with (**right**) the upper Al_2O_3 ring and the rod that mechanically clamps the samples and the Pt foil together.

The real and imaginary parts of the impedance spectra were measured in the vicinity of the resonance frequency using a high-speed network analyzer (Agilent E5100A, Hewlett-Packard, Santa Clara, CA, USA). Subsequently, the impedance was transformed into admittance $Y = Z^{-1}$ and a Lorenz function was fitted to its real part (conductance G). This enabled determination of the resonance frequencies of the fundamental mode and of the subsequent harmonics and of the Q-factor for a given mode. Detailed description of data acquisition is given in [2]. Figure 6 exemplarily shows the real part of admittance, acquired for the LNT88-01 specimen at different temperatures. The frequency corresponds to the 1st harmonic of the TSM resonator. As seen from the figure, the resonance frequency (i.e., the conductance maximum) shifts with the temperature towards lower frequencies. Further, a broadening of the peak is observed at elevated temperatures. This implies an increase of the losses, which will be discussed in detail in Section 3. At 550 °C on the right shoulder of the conductance peak another small peak is observed at about 1.935 MHz, which is interpreted as a spurious mode and understood to originate from the "activity dips" phenomenon [60] (see discussion in Section 3).

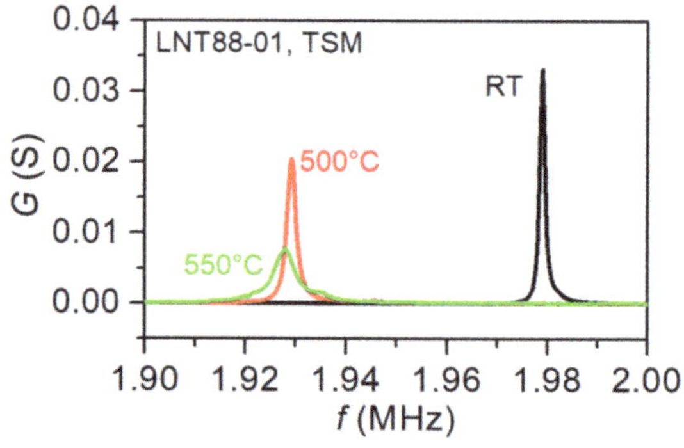

Figure 6. The real part of admittance as a function of frequency of LNT88-01 specimen, operated in the thickness–shear mode (TSM), acquired at different temperatures.

3. Results

3.1. Electrical Conductivity

The temperature dependent electrical conductivity of all studied specimens is shown in Figure 7 in the form of an Arrhenius plot. The measurements were performed in air in the temperature range from about 400 to 700 °C. As seen from the figure, both materials exhibited similar electrical conductivities, reaching at 700 °C values of 1.52×10^{-3} S/m and 1.56×10^{-3} S/m for LN and LT, respectively. Further, the LNT88-01 and LNT88-02 samples are showing different conductivity values in the whole measured temperature range. At 700 °C the σ of LNT88-01 and LNT88-02 equaled 0.9×10^{-3} S/m and

1.9×10^{-3} S/m, respectively. Such differences could not be solely attributed to the uncertainties in conductivity measurements (about 6% at 700 °C). We assumed, that these divergences could be caused by the varying properties within a crystal boule such as an inhomogeneous distribution of Nb and Ta in grown LNT crystals (see also Table 1). We also note that the conductivity of both LN and LT is sensitive to even slight deviations in lithium stoichiometry [16,23,41], which could also be the case of LNT specimens. The conductivity of sample LNT50 equaled 1.2×10^{-3} S/m at 700 °C. The value was 1.3 times lower than that of LN and about 1.3 higher than that of LNT88-01.

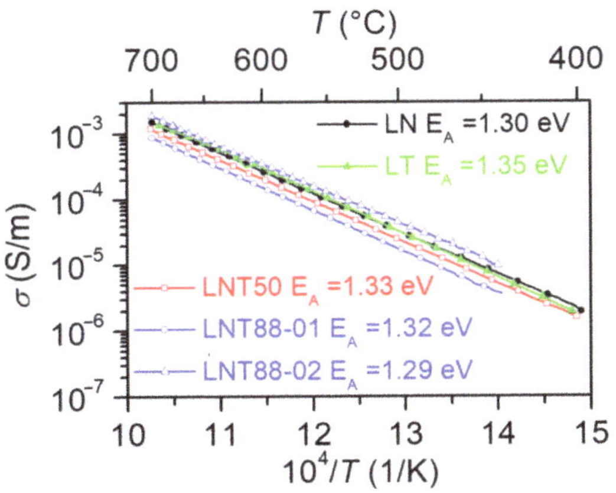

Figure 7. Conductivity of different Li(Nb,Ta)O_3 samples as a function of inverse temperature.

The conductivity of all the studied specimens increased linearly in the Arrhenius presentation (see Figure 7), indicating that it was governed by a single thermally activated process. Previously, the electrical conductivity of LT was investigated up to 800 °C in [23] and the authors assumed that the ionic conduction mechanism with lithium vacancies as a main charge carrier determines the conductivity at elevated temperatures as applied here. Similar conclusions were drawn in [24,26]. Further, the studies, performed in [16,29–32,41] show that the lithium ion migration via lithium vacancies was also the dominating transport mechanism in $LiNbO_3$ at high temperatures. Therefore, it is plausible to assume that in LNT the conduction of Li-ions via Li-vacancies played a major role in the studied temperature range, similarly to both end components of Li(Nb,Ta)O_3-system. Considering that at 700 °C the conductivity is ionic, we claimed that it remains predominantly ionic down to 400 °C since we did not observe here a deviating slope with respect to 700 °C. Consequently, the conductivity σ is written as [23]:

$$\sigma = \frac{\sigma_0}{T} \exp(E_A/kT) \tag{7}$$

where σ_0, T, E_A and k represent a pre-exponential constant, the absolute temperature, an activation energy and the Boltzmann constant, respectively. Following Equations (6) and (7), the relaxation time could be written as:

$$\tau_c = \frac{\varepsilon_{ij} T}{\sigma_0} \exp(E_A/kT) \tag{8}$$

Activation energies and pre-exponential factors, obtained by fitting of Equation (7) to the measured conductivity data are summarized in Table 2. Previously, close values of E_A were obtained in [16,30,31,41] for congruent LN and in [23,26] for congruent LT.

Table 2. Parameters from fits of conductivity of $Li(Nb,Ta)O_3$ specimens.

Sample	Activation Energy (eV)	Pre-Exponential Factor (SK/m)
LN	1.30 ± 0.06	7.31×10^6
LNT88-01	1.32 ± 0.05	6.72×10^6
LNT88-02	1.29 ± 0.04	6.20×10^6
LNT50	1.33 ± 0.07	7.07×10^6
LT	1.35 ± 0.05	1.49×10^7

3.2. Acoustic Loss

The measured loss Q^{-1} of the specimen LNT88-01 is shown in Figure 8 as a function of inverse temperature. The frequencies of the measurements correspond to the first, and fifth harmonics of the resonator that is operated in the TSM mode.

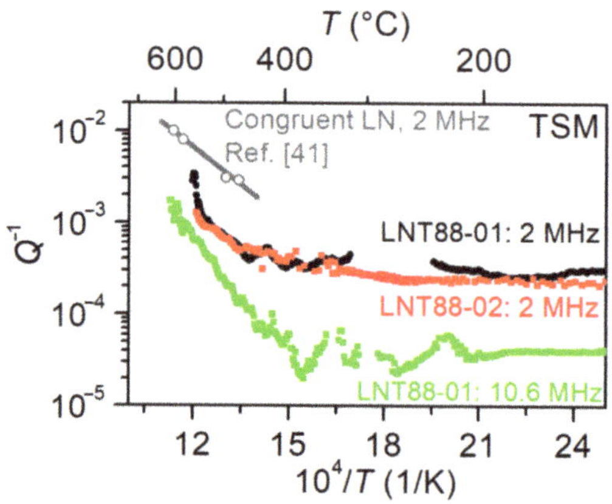

Figure 8. Q^{-1} of LNT88 as a function of inverse temperature, compared to congruent LN, studied in [41].

A substantial number of sharp peaks were observed in the raw data for the first harmonic of LNT88-01 between 250 and 350 °C. These peaks are understood to originate from the "activity dips" phenomenon, caused by the coupling of the main mode with other spurious (parasitic) modes [62]. Such coupling does not allow for correct determination of the Q-factor. Therefore, the Q^{-1} data in temperature ranges where activity dips occurred were removed from the results presented in Figure 8 to provide a more reliable and clear presentation of the measured data.

Further, the loss of higher harmonics was lower than that of the fundamental mode. This peculiarity could be attributed to the non-ideal diameter-to-thickness ratio of the studied resonator. As evaluated in [63] for resonators with a diameter-to-thickness ratio lower than 30, the non-material losses that arise from mounting increase with the frequency decrease must be considered. As it is widely known, the thickness–shear oscillations in any given partially electroded TSM resonator are confined to the area under and close to the electrodes. This phenomenon is known as energy trapping [64] and ensures high Q-factors. Strong energy trapping could be achieved by optimization of the electrode thickness and by choosing the optimal geometry of the resonator and of the electrodes [64,65]. Further, increasing the operational frequencies to higher harmonics could also increase the energy trapping, provided that the confinement of vibration energy is insufficient on the fundamental mode [64–66]. In our study, the geometry of LNT88 specimens was not ideal, which potentially explains relatively high losses at the fundamental mode.

Additionally, the measured loss Q^{-1} of the sample LNT88-02, determined for the first harmonic of a TSM resonator, is also shown in Figure 8 for validation of the results. As seen from the figure, the behavior of both resonators is nearly similar when operated in the fundamental mode. For LNT88-02 the losses between 250 and 350 °C could also be determined, as for this particular resonator no spurious modes were observed in this temperature range.

Figure 8 also compares the Q^{-1} of LNT88 resonators to that of congruent LN specimen, studied previously in [41]. The LN resonator in [41] was prepared as X-cut, operated in TSM and the frequency corresponds to the fundamental mode. In the current study the LNT-88 resonators were prepared as Y-cut discs, however due to the crystal symmetry in both X- and Y-cuts the resonances could be excited in TSM and the material constants, which determine the resonance frequency and electromechanical coupling factor of TSM in X- and Y-cuts are equal ($C_{44} = C_{55}$; $e_{15} = e_{24}$ and $\varepsilon_{11} = \varepsilon_{22}$) [8,34,35,67]. Remarkably, the losses in LNT88 resonators were substantially lower. We note however, that the results overlapped only partially, since the measurements of the LNT88 fundamental mode did not extend to the range of LN in [41].

As also seen from Figure 8, Q^{-1} remained nearly constant below about 450 °C. Above that temperature, a rapid increase is observed for all measured frequencies. In the case of some other materials systems the high-temperature loss is attributed to the piezoelectric/carrier relaxation mechanism [50–52]. The same conclusion follows for LNT if the loss contribution is estimated using the materials constants for LN, which is only doable this way due to missing data for LNT. Inserting materials constants for 600 °C (e_{15} = 4.32 C/m^2; C_{44} = 53.5 GPa; $\varepsilon_{11} = 6.02 \times 10^{-10}$ F/m [34]), and the conductivity values (Table 2) into Equation (5), the calculated contribution of piezoelectric/carrier relaxation matched the experimentally determined loss for the resonance frequencies of resonator LNT88-01, see Figure 9. The differences in absolute values of Q^{-1} could be attributed to the material constants used for these calculations. Again, it has to be emphasized that we used e_{15}, ε_{11} and C_{44} for LN [34], while these values are different for LT and, obviously, also somewhat different for LNT.

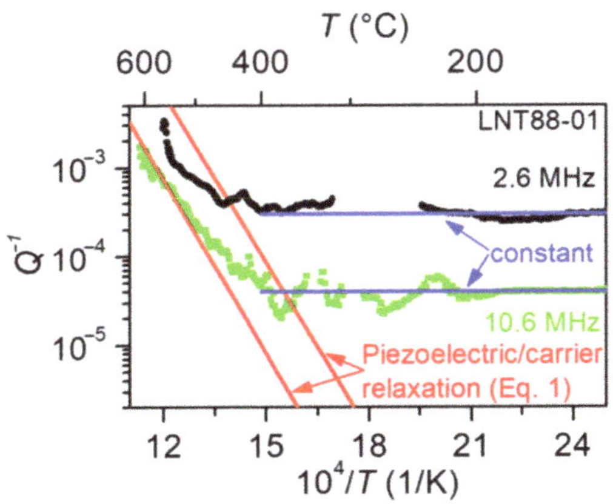

Figure 9. Calculated contribution of piezoelectric/carrier relaxation, compared to the measured Q^{-1} of the LNT88-01 resonator.

Further, in order to minimize the influence of piezoelectric/carrier relaxation at a given anticipated application temperature, the operating frequencies of LNT resonators can be chosen accordingly. For example, a 2 MHz resonator could be operated up to about 450 °C without significant loss increase (Figure 9). Equation (5) enables one to estimate favorable frequencies for the high-temperature applications of LNT TSM resonators. Figure 10 shows the resulting piezoelectric/carrier relaxation losses for different operating frequencies.

As seen from Figure 10, a decrease of the operating frequency to 100 kHz prevented the application of such resonators between about 450 and 1000 °C but enabled its use above that temperature provided that the material was thermally stable. Increase of the frequency to 30 MHz allowed application temperature up to about 750 °C. For both cases it was assumed that $Q^{-1} = 2 \times 10^{-3}$ was the maximum Q^{-1}, required for accurate frequency determination.

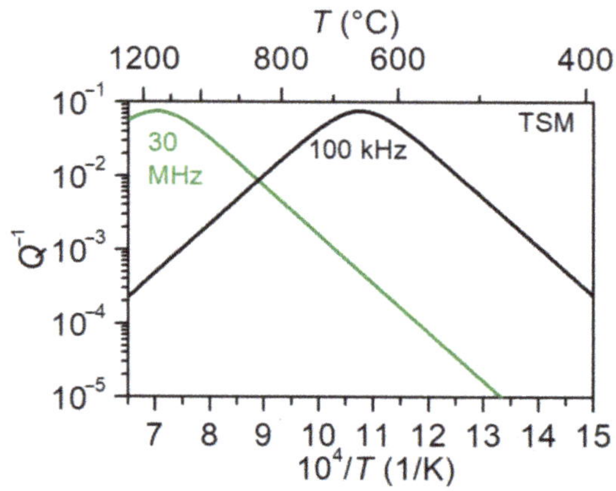

Figure 10. Estimated contribution of the piezoelectric/carrier relaxation to the overall loss of LN TSM resonators, calculated for different operation frequencies.

Figure 11 presents the measured loss Q^{-1} of LNT50 specimen as a function of inverse temperature, compared to that of congruent LN presented in [16]. Both samples are operated in TM. Further, resonator LNT50 is operated at its third harmonic. Reliable measurements for sample LNT50 were obtained only above 400 °C. Below that temperature a substantial number of spurious modes in the vicinity of the resonant mode was observed in the spectra and did not allow for determination of the Q-factor.

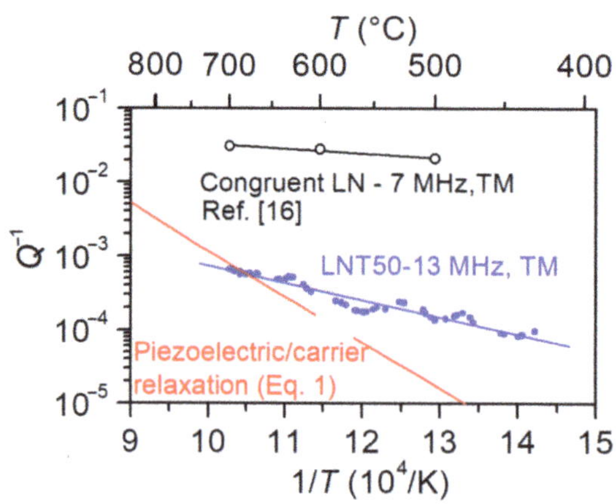

Figure 11. Measured Q^{-1} of LNT50 compared to that of LN, determined in [16].

Similarly to TSM LNT resonators, the piezoelectric/carrier contribution to Q^{-1} could be calculated for LNT50 using Equation (5) and the data, obtained from conductivity measurements (Table 2). For the same reasons as for samples LNT88, these calculations are done using the material constants for $LiNbO_3$ (e_{33} = 1.72 C/m^2; C_{33} = 227.9 GPa; $\varepsilon_{33} = 4.32 \times 10^{-10}$ F/m [35,67]). The result of the calculation is plotted in Figure 11. As seen from figure, the piezoelectric/carrier relaxation might significantly contribute to the overall loss of the 13 MHz TM resonator above about 700 °C, only. Therefore, the exponential increase of Q^{-1} in the range 400–700 °C has presumably a different nature. One possible explanation of this increase is the undefined domain structure of the LNT50. As briefly mentioned in Section 2.1, the crystal was not poled intentionally. The existence of resonances during electrical excitation shows however, that the sample has sufficiently large areas with identical orientation of the polarization. On the other side, some regions of the sample could have a different polarization, which would influence the overall losses. As a consequence, the loss determined here represents the upper limit and would reach lower values for a single-domain structure if not yet present. This issue must be studied in more detail and will be a subject of subsequent investigations.

As seen from Figure 11, the overall loss of LNT50 was substantially lower, comparing to congruent LN in the entire measured temperature range, which enabled improved accuracy in frequency determination of such resonators. It is however worth noting that the operating frequency of the LN resonator, studied in [16] is lower, therefore the maximum of piezoelectric/carrier relaxation contribution is shifted towards lower temperatures in this specimen. This shift could be calculated with the help of Equation (5) and equals only about 50 K, comparing to the LNT50. In other words, the difference of overall Q^{-1} in LN and LNT50 could not be attributed to the different operation frequencies and is rather explained by the intrinsic properties of both materials.

It should be emphasized that the loss occurring at high temperatures is determined by intrinsic materials properties and, therefore, reflected in a correct manner. At lower temperatures, however, the shape and size of the samples impacts the resonator quality factor. In this respect, the dimensions sample LNT50 was not optimal, which primarily results from the relatively small crystal boules of grown LNT (see Section 3). Preparing the specimen with favorable geometry and diameter-to-thickness ratio should result in improved piezoelectric response and, consequently, lower Q^{-1} in the lower part of the measured temperature range. This issue will be the subject of subsequent studies if larger samples are available. Further work is intended to uncover the interplay between microscopic and macroscopic features by regarding a larger temperature range and by extending studies about charge carrier/defect relaxation as given, e.g., in [68] within the framework of collaborative research.

4. Conclusions

In summary, the electrical and electromechanical properties of lithium niobate-tantalate solid solutions with different Nb/Ta ratios were studied at elevated temperatures and compared to those of $LiNbO_3$ and $LiTaO_3$. The measured temperature dependence of electrical conductivity in air up to 700 °C shows similar behavior for all studied samples. Consequently, the dominating transport mechanism in LNT in the range 400–700 °C was attributed to the lithium ion migration via lithium vacancies as found in LN.

Further, the study of the acoustic loss in LNT resonators, operated in the thickness-shear mode, revealed rapid, frequency dependent loss increase at elevated temperatures. This loss increase was correlated with conductivity measurements and attributed to the piezoelectric/carrier relaxation mechanism. Further, it is shown that minimization of the influence of conductivity-related losses at a given anticipated application temperature could be achieved by an appropriate choice of operating frequencies.

Finally, the study revealed that the temperature-dependent loss in LNT specimens was generally lower than that of congruent LN for both thickness and thickness–shear mode resonators. The domain structure of the LNT50 specimen, operated in the thickness mode, was unknown. However, clear resonances were observed, which indicates largely uniform polarization of the domains or even a single domain. Since different orientations of polarization cannot be excluded, the loss was potentially affected. As a consequence, the Q^{-1} presented here for the LNT50 specimen is the upper limit and would reach lower for a single-domain structure if not yet present.

Author Contributions: Conceptualization, Y.S. and H.F.; Crystal growth and structural analysis D.R., B.R. and S.G.; Sample preparation, S.G., B.J. and Y.S.; Electrical and acoustic measurements, B.J., A.K. and Y.S.; Writing—original draft preparation, Y.S.; Writing—review and editing, H.F. All authors have read and agreed to the published version of the manuscript.

Funding: Research grants from the Deutsche Forschungsgemeinschaft (DFG, German Research Foundation, FR1301/32-1, SU1261/1-1, GA2403/7-1, FR1301/42-1) and Russian Foundation for Basic Research (grant 19-52-12044) supported this work.

Institutional Review Board Statement: Not applicable.

Informed Consent Statement: Not applicable.

Data Availability Statement: All relevant data presented in the article are stored according to institutional requirements and as such are not available online. However, all data used in this manuscript can be made available upon request to the authors.

Acknowledgments: The authors from Clausthal University of Technology acknowledge the support of the Energie-Forschungszentrum Niedersachsen, Goslar, Germany. M. Brützam und M. Stypa from Leibniz-Institut für Kristallzüchtung, Berlin, Germany are thankfully acknowledged for their technical assistance in crystal growth. The authors are indebted to S. Lenk (Clausthal University of Technology) for the elemental analysis by EDX. We acknowledge the support by Open Access Publishing Fund of Clausthal University of Technology.

Conflicts of Interest: The authors declare no conflict of interest.

References

1. Tuloup, C.; Harizi, W.; Aboura, Z.; Meyer, Y.; Khellil, K.; Lachat, R. On the use of in-situ piezoelectric sensors for the manufacturing and structural health monitoring of polymer-matrix composites: A literature review. *Compos. Struct.* **2019**, *215*, 127–149. [CrossRef]
2. Fritze, H. High-temperature bulk acoustic wave sensors. *Meas. Sci. Technol.* **2011**, *22*, 12002. [CrossRef]
3. Stevenson, T.; Martin, D.G.; Cowin, P.I.; Blumfield, A.; Bell, A.J.; Comyn, T.P.; Weaver, P.M. Piezoelectric materials for high temperature transducers and actuators. *J. Mater. Sci. Mater. Electron.* **2015**, *26*, 9256–9267. [CrossRef]
4. Khesro, A.; Wang, D.; Hussain, F.; Sinclair, D.C.; Feteira, A.; Reaney, I.M. Temperature stable and fatigue resistant lead-free ceramics for actuators. *Appl. Phys. Lett.* **2016**, *109*, 142907. [CrossRef]
5. Zhang, S.; Xia, R.; Lebrun, L.; Anderson, D.; Shrout, T.R. Piezoelectric materials for high power, high temperature applications. *Mater. Lett.* **2005**, *59*, 3471–3475. [CrossRef]
6. Ohsato, H.; Iwataki, T.; Morikoshi, H. Crystal structure and piezoelectric properties of four component langasite $A_3BGa_3Si_2O_{14}$ (A = Ca or Sr, B = Ta or Nb). *Trans. Electr. Electron. Mater.* **2012**, *13*, 171–176. [CrossRef]
7. Suhak, Y.; Schulz, M.; Sotnikov, A.; Schmidt, H.; Ganschow, S.; Sakharov, S.; Fritze, H. Dielectric, piezoelectric and elastic constants of $Ca_3TaGa_3Si_2O_{14}$ single crystals at elevated temperatures. *Integr. Ferroelectr.* **2018**, *537*, 255–263. [CrossRef]
8. Volk, T.; Wöhlecke, M. *Lithium Niobate: Defects, Photorefraction and Ferroelectric Switching*; Springer: Berlin/Heidelberg, Germany, 2008.
9. Sugak, D.; Zhydachevskii, Y.; Suhak, Y.; Buryy, O.; Ubizskii, S.; Solskii, I.; Schrader, M.; Becker, K.D. In-situ investigation of optical absorption changes in $LiNbO_3$ during reducing/oxidizing high temperature treatments. *J. Phys. Cond. Matter* **2007**, *19*, 086211. [CrossRef]
10. Solskii, I.M.; Sugak, D.Y.; Vakiv, M.M. Growing large size complex oxide single crystals by Czochralski technique for electronic devices. *Acta Physica. Polonica A* **2013**, *124*, 314–320. [CrossRef]
11. Bordui, P.F.; Notwood, R.G.; Bird, C.D.; Carella, J.T. Stoichiometry issues in single-crystal lithium tantalate. *J. Appl. Phys.* **1995**, *78*, 4647–4650. [CrossRef]
12. Kawamata, A.; Hosaka, H.; Morita, T. Non-hysteresis and perfect linear piezoelectric performance of a multilayered lithium niobate actuator. *Sens. Actuators A* **2007**, *135*, 782–786. [CrossRef]
13. Nakamura, K.; Shimizu, H. Hysteresis-free piezoelectric actuators using $LiNbO_3$ plates with a ferroelectric inversion layer. *Ferroelectrics* **1989**, *93*, 211–216. [CrossRef]
14. Samuelsen, E.J.; Grande, A.P. The ferroelectric phase transition in $LiTaO_3$ studied by neutron scattering. *Z. Physik B* **1976**, *24*, 207–210. [CrossRef]
15. Damjanovic, D. Materials for high-temperature piezoelectric transducers. *Curr. Opin. Solid St. M.* **1998**, *3*, 469–473. [CrossRef]
16. Weidenfelder, A.; Shi, J.; Fielitz, P.; Borchardt, G.; Becker, K.D.; Fritze, H. Electrical and electromechanical properties of stoichiometric lithium niobate at high temperatures. *Solid State Ion.* **2012**, *225*, 26–29. [CrossRef]
17. Schirmer, O.F.; Thiemann, O.; Wöhlecke, M. Defects in $LiNbO_3$—1. Experimental Aspects. *J. Phys. Chem. Solids* **1991**, *52*, 185–200. [CrossRef]
18. Donnerberg, H.; Tomlinson, S.M.; Catlow, C.R.A.; Schirmer, O.F. Computer-simulation studies of intrinsic defects in $LiNbO_3$ crystals. *Phys. Rev. B* **1989**, *40*, 11909–11916. [CrossRef]

19. Abrahams, S.C.; Marsh, P. Defect Structure Dependence on Composition in Lithium Niobate. *Acta Crystallogr. B* **1986**, *42*, 61–68. [CrossRef]
20. Furukawa, Y.; Kitamura, K.; Suzuki, E.; Niwa, K. Stoichiometric $LiTaO_3$ single crystal growth by double crucible Czochralski method using automatic powder supply system. *J. Cryst. Growth* **1999**, *197*, 889–895. [CrossRef]
21. Kitamura, K.; Furukawa, Y.; Niwa, K.; Goplan, V.; Mitchell, T.E. Crystal growth and low coercive field 180° domain switching characteristics of stoichiometric $LiTaO_3$. *Appl. Phys. Lett.* **1998**, *73*, 3073–3076. [CrossRef]
22. Krampf, A.; Imlau, M.; Suhak, Y.; Fritze, H.; Sanna, S. Evaluation of similarities and differences of LiTaO3 and $LiNbO_3$ based on high-T-conductivity, nonlinear optical fs-spectroscopy and ab initio modeling of polaronic structures. *New J. Phys.* **2021**, in press. [CrossRef]
23. Huanosta, A.; West, A.R. The electrical properties of ferroelectric $LiTaO_3$ and its solid solutions. *J. Appl. Phys.* **1987**, *61*, 5386–5391. [CrossRef]
24. Sinclair, D.C.; West, A.R. Electrical properties of a $LiTaO_3$ single crystal. *Phys. Rev. B* **1989**, *39*, 13486–13492. [CrossRef]
25. Ming, D.; Reau, J.M.; Ravez, J.; Gitae, J.; Hagenmuller, P. Impedance-spectroscopy analysis of a $LiTaO_3$-type single crystal. *J. Solid State Chem.* **1995**, *116*, 185–192. [CrossRef]
26. Palatnikov, M.N.; Sandler, V.A.; Yatsenko, A.V.; Sidorov, N.V.; Evdokimov, S.V.; Makarova, O.V. Anisotropic electrical conductivity and dielectric properties of $LiTaO_3$ crystals in the temperature range 290–900 K. *Inorg. Mater.* **2015**, *51*, 685–695. [CrossRef]
27. Smyth, D.M. Defects and transport in $LiNbO_3$. *Ferroelectrics* **1983**, *50*, 93–102. [CrossRef]
28. Lanfredi, S.; Rodrigues, A.C.M. Impedance spectroscopy study of the electrical conductivity and dielectric constant of polycrystalline $LiNbO_3$. *J. Appl. Phys.* **1999**, *86*, 2215–2219. [CrossRef]
29. Chen, R.H.; Chen, L.; Chia, C. Impedance spectroscopic studies on congruent $LiNbO_3$ single crystal. *J. Phys. Condens. Matter* **2007**, *19*, 086225. [CrossRef]
30. Ruprecht, B.; Rahn, J.; Schmidt, H.; Heitjans, P. Low-Temperature DC Conductivity of $LiNbO_3$ Single Crystals. *Z. Phys. Chem.* **2012**, *226*, 431–437. [CrossRef]
31. El-Bachiri, A.; Bennani, F.; Bousselamti, M. Ionic and polaronic conductivity of lithium niobate. *Spectrosc. Lett.* **2014**, *47*, 374–380. [CrossRef]
32. Efremov, V.V.; Shcherbina, O.B.; Palatnikov, M.N.; Masloboeva, S.M. Comparative investigation of electrophysical characteristics of ceramic and single crystal $LiNbO_3$. *J. Phys. Conf. Ser.* **2020**, *1658*, 012010. [CrossRef]
33. Yamada, T.; Iwasaki, H.; Niizeki, N. Piezoelectric and elastic properties of $LiTaO_3$: Temperature characteristics. *Jpn. J. Appl. Phys.* **1969**, *8*, 1127–1132. [CrossRef]
34. De Castilla, H.; Belanger, P.; Zednik, R.J. High temperature characterization of piezoelectric lithium niobate using electrochemical impedance spectroscopy resonance method. *J. Appl. Phys.* **2017**, *122*, 244103. [CrossRef]
35. Chen, F.; Kong, L.; Song, W.; Jiang, C.; Tian, S.; Yu, F.; Qin, L.; Wang, C.; Zhao, X. The electromechanical features of $LiNbO_3$ crystal for potential high temperature piezoelectric applications. *J. Mater.* **2019**, *5*, 73–80. [CrossRef]
36. Rüsing, M.; Sanna, S.; Neufeld, S.; Berth, G.; Schmidt, W.G.; Zrenner, A.; Yu, H.; Wang, Y.; Zhang, H. Vibrational properties of $LiNb_{1-x}Ta_xO_3$ mixed crystals. *Phys. Rev. B* **2016**, *93*, 184305. [CrossRef]
37. Riefer, A.; Sanna, S.; Schmidt, W.G. $LiNb_{1-x}Ta_xO_3$ electronic structure and optical response from first principles. *Ferroelectrics* **2013**, *447*, 78. [CrossRef]
38. Huband, S.; Keeble, D.S.; Zhang, N.; Glazer, A.M.; Bartasyte, A.; Thomas, P.A. Crystallographic and optical study of $LiNb_{1-x}Ta_xO_3$. *Acta Cryst. B* **2017**, *73*, 498. [CrossRef]
39. Glazer, A.M.; Zhang, N.; Bartasyte, A.; Keeble, D.S.; Huband, S.; Thomas, P.A. Observation of unusual temperature-dependent stripes in $LiTaO_3$ and $LiNb_{1-x}Ta_xO_3$ crystals with near-zero birefringence. *J. Appl. Cryst.* **2010**, *43*, 1305. [CrossRef]
40. Palatnikov, M.N.; Sandler, V.A.; Sidorov, N.V.; Makarova, O.V.; Manukovskaya, D.V. Conditions of application of $LiNbO_3$ based piezoelectric resonators at high temperatures. *Phys. Lett. A* **2020**, *384*, 126289. [CrossRef]
41. Weidenfelder, A.; Schulz, M.; Fielitz, P.; Shi, J.; Borchardt, G.; Becker, K.D.; Fritze, H. Electronic and ionic transport mechanisms of stoichiometric lithium niobate at high temperatures. *MRS Proc.* **2013**, *1519*, 1–7. [CrossRef]
42. Parks, D.A.; Zhang, S.; Tittmann, B.R. High-Temperature (>500 °C) Ultrasonic transducers: An experimental comparison among three candidate piezoelectric materials. *IEEE Trans. Ultrason.* **2013**, *60*, 1010–1015. [CrossRef]
43. Hauser, R.; Reindl, L.; Biniasch, J. High-temperature stability of $LiNbO_3$-based SAW devices. *IEEE Ultrason. Symp.* **2003**, *1*, 192–195.
44. Garavito, R.; Villafuerte-Castrejón, M.E.; Valenzuela, R.; West, A.R. Phase equilibria in the system Li_2O-Nb_2O_5-Ta_2O_5. *Br. Ceram. Trans. J.* **1985**, *84*, 104–108.
45. Fernandez-Ruiz, R.; Bermudez, V. Determination of the Ta and Nb ratio in $LiNb_{1-x}Ta_xO_3$ by total reflection X-ray fluorescence spectrometry. *Spectrochim. Acta B* **2005**, *60*, 231. [CrossRef]
46. Wood, I.G.; Daniel, P.; Brown, R.H.; Glazer, A.M. Optical birefringence study of the ferroelectric phase transition in lithium niobate tantalate mixed crystals: $LiNb_{1-x}Ta_xO_3$. *J. Phys. Cond. Matter* **2008**, *20*, 235237. [CrossRef]
47. Bartasyte, A.; Glazer, A.M.; Wondre, F.; Prabhakaran, D.; Thomas, P.A.; Huband, S.; Keeble, D.S.; Margueron, S. Growth of $LiNb_{1-x}Ta_xO_3$ solid solution crystals. *Mater. Chem. Phys.* **2012**, *134*, 728. [CrossRef]
48. Bak, K.Y.; Tan, K.B.; Khaw, C.C.; Zainal, Z.; Tan, P.Y.; Chon, M.P. Structural and electrical properties of Nb-substituted $LiTa_{1-x}Nb_xO_3$. *Sains. Malays.* **2014**, *43*, 1573–1582.

49. Xue, D.; Betzler, K.; Hesse, H. Dielectric properties of lithium niobate–tantalate crystals. *Solid State Commun.* **2000**, *115*, 581–585. [CrossRef]
50. Johnson, W.L.; Kim, S.A.; Uda, S.; Rivenbark, C.F. Contributions to anelasticity in langasite and langatate. *J. Appl. Phys.* **2011**, *110*, 123528. [CrossRef]
51. Hirschle, C.; Schreuer, J. High-temperature ultrasound attenuation in langasite and langatate. *IEEE Trans. Ultrason. Ferroelect. Freq. Control* **2018**, *65*, 1250–1257. [CrossRef] [PubMed]
52. Suhak, Y.; Schulz, M.; Johnson, W.L.; Sotnikov, A.; Schmidt, H.; Fritze, H. Electromechanical properties and charge transport of $Ca_3TaGa_3Si_2O_{14}$ (CTGS) single crystals at elevated temperatures. *Solid State Ion.* **2018**, *317*, 221–228. [CrossRef]
53. Mason, W.P. *Physical Acoustics, Principles and Methods*; Academic Press: New York, NY, USA, 1965; Volume 3B.
54. Tabrizian, R.; Rais-Zadeh, M.; Ayazi, F. Effect of Phonon Interactions on Limiting the fQ Product of Micromechanical Resonators. *IEEE Transducers* **2009**, 2131–2134. [CrossRef]
55. Nowick, A.S.; Berry, B.S. *Anelastic Relaxation in Crystalline Solid*; Academic Press: New York, NY, USA, 1972.
56. Hutson, A.R.; White, D.L. Elastic wave propagation in piezoelectric semiconductors. *J. Appl. Phys.* **1962**, *33*, 40–47. [CrossRef]
57. Ikeda, T. *Fundamentals of Piezoelectricity*; Oxford University Press: Oxford, UK, 1990.
58. Roshchupkin, D.; Emelin, E.; Plotitcyna, O.; Rashid, F.; Irzhak, D.; Karandashev, V.; Orlova, T.; Targonskaya, N.; Sakharov, S.; Mololkin, A.; et al. Single crystals of ferroelectric lithium niobate–tantalate $LiNb_{1-x}Ta_xO_3$ solid solutions for high-temperature sensor and actuator applications. *Acta Cryst. B* **2020**, *76*, 1071–1076. [CrossRef]
59. Reiche, P.; Hermoneit, B.; Schultze, D. A modified heater system for rf-Czochralski equipments. *Cryst. Res. Technol.* **1985**, *20*, 845–849. [CrossRef]
60. Vechembre, J.B.; Fox, G.R. Sintering of screen-printed platinum thick films for electrode applications. *J. Mater. Res.* **2001**, *16*, 922–931. [CrossRef]
61. Golosov, D.A.; Okojie, J.E.; Zavadski, S.M.; Rudenkov, A.S.; Melnikov, S.N.; Kolos, V.V. Stability of the platinum electrode during high temperature annealing. *Thin Solid Films* **2018**, *661*, 53–59. [CrossRef]
62. Ballato, A. Electronic Activity Dip Measurement. *IEEE Trans. Instr. Meas.* **1978**, *27*, 59–65. [CrossRef]
63. Ballato, A.; Gualtieri, J.G. Advances in high-Q piezoelectric resonator materials and devices. *IEEE Trans. Ultrason. Ferroelect. Freq. Control* **1994**, *41*, 834–844. [CrossRef]
64. Shockley, W.; Curran, D.R.; Koneval, D.J. Trapped energy modes in quartz filter crystals. *J. Acoust. Soc. Am.* **1967**, *41*, 981. [CrossRef]
65. Wang, J.; Shen, L.; Yang, J. Effects of electrodes with continuously varying thickness on energy trapping in thickness-shear mode quartz resonators. *Ultrasonics* **2008**, *48*, 150–154. [CrossRef] [PubMed]
66. Cassiede, M.; Paillol, J.H.; Pauly, J.; Daridon, J.-L. Electrical behaviour of AT-cut quartz crystal resonators as a function of overtone number. *Sens. Actuators A* **2010**, *159*, 174–183. [CrossRef]
67. Ledbetter, H.; Ogi, H.; Nakamura, N. Elastic, anelastic, piezoelectric coefficients of monocrystal lithium niobate. *Mech. Mater.* **2004**, *36*, 941–947. [CrossRef]
68. Vittadello, L.; Bazzan, M.; Messerschmidt, S.; Imlau, M. Small polaron hopping in Fe:$LiNbO_3$ as a function of temperature and composition. *Crystals* **2019**, *8*, 294. [CrossRef]

crystals

Article

Investigations of $LiNb_{1-x}Ta_xO_3$ Nanopowders Obtained with Mechanochemical Method

Leonid Vasylechko [1], Volodymyr Sydorchuk [2], Andrey Lakhnik [3], Yuriy Suhak [4], Damian Wlodarczyk [5], Stepan Hurskyy [1,4], Uliana Yakhnevych [1,*], Yaroslav Zhydachevskyy [1,5], Dmytro Sugak [1,6], Ihor I. Syvorotka [6], Ivan Solskii [6], Oleh Buryy [1], Andrzej Suchocki [5] and Holger Fritze [4]

1 Department of Semiconductor Electronics, Lviv Polytechnic National University, 12, Bandery Str., 79013 Lviv, Ukraine; crystal-lov@polynet.lviv.ua (L.V.); stepan.hurskyy@tu-clausthal.de (S.H.); zhydach@ifpan.edu.pl (Y.Z.); dm_sugak@yahoo.com (D.S.); oburyy@yahoo.com (O.B.)
2 Institute for Sorption and Problems of Endoecology, NASU, 13 Gen. Naumov St., 03164 Kyiv, Ukraine; bilychi@ukr.net
3 G.V. Kurdyumov Institute of Metal Physics NASU, 36 Acad. Vernadsky Bvd., 03142 Kyiv, Ukraine; lakhnik@imp.kiev.ua
4 Institute for Energy Research and Physical Technologies, Clausthal University of Technology, Am Stollen 19B, 38640 Goslar, Germany; sensorik@tu-clausthal.de (Y.S.); sensorikK@tu-clausthal.de (H.F.)
5 Division of physics and technology of wide-band-gap semiconductor nanostructures, Institute of Physics PAS, 32/46 Al. Lotników, 02668 Warsaw, Poland; wlodar@ifpan.edu.pl (D.W.); suchy@ifpan.edu.pl (A.S.)
6 Scientific Research Company 'Electron-Carat', 202 Stryjska St., 79031 Lviv, Ukraine; ii.syvorotka@carat.electron.ua (I.I.S.); solskii@carat.lviv.ua (I.S.)
* Correspondence: uliana.v.yakhnevych@lpnu.ua

Citation: Vasylechko, L.; Sydorchuk, V.; Lakhnik, A.; Suhak, Y.; Wlodarczyk, D.; Hurskyy, S.; Yakhnevych, U.; Zhydachevskyy, Y.; Sugak, D.; Syvorotka, I.I.; et al. Investigations of $LiNb_{1-x}Ta_xO_3$ Nanopowders Obtained with Mechanochemical Method. *Crystals* **2021**, *11*, 755. https://doi.org/10.3390/cryst11070755

Academic Editors: László Kovács and Gábor Corradi

Received: 21 May 2021
Accepted: 25 June 2021
Published: 28 June 2021

Abstract: Nanocrystalline compounds $LiNb_{1-x}Ta_xO_3$ of various compositions (x = 0, 0.25, 0.5, 0.75, 1) were synthesized by high-energy ball milling of the initial materials (Li_2CO_3, Nb_2O_5, Ta_2O_5) and subsequent high-temperature annealing of the resulting powders. Data on the phase composition of the nanopowders were obtained by X-ray diffraction methods, and the dependence of the structural parameters of $LiNb_{1-x}Ta_xO_3$ compounds on the value of x was established. As a result of the experiments, the optimal parameters of the milling and annealing runs were determined, which made it possible to obtain single-phase compounds. The Raman scattering spectra of $LiNb_{1-x}Ta_xO_3$ compounds (x = 0, 0.25, 0.5, 0.75, 1) have been investigated. Preliminary experiments have been carried out to study the temperature dependences of their electrical conductivity.

Keywords: nanoparticles; nanopowders; lithium niobate-tantalate; X-ray diffraction; Raman spectroscopy; temperature dependence of electroconductivity

1. Introduction

Lithium niobate ($LiNbO_3$, LN) and tantalate ($LiTaO_3$, LT) are among the most studied oxide compounds in modern materials science. This interest is due to the widest application of these materials in functional electronics. Analytical studies of the global market for LN and LT sales performed by various marketing agencies [1–4] show that as of the end of 2020, their consumption in monetary terms exceeded $40 billion per year, and by 2027, it may reach more than $75 billion. Such impressive sales are due to the variety of electronics industry branches and devices that use LN and LT. Accordingly, the forms in which these materials are used in practice are also different—single crystals, thin films, micro- and nanopowders, ceramics. Moreover, they can have a different chemical composition—congruent, stoichiometric, and contain various dopants of metal ions. It is obvious that such a variety of forms and applications again requires research to modify and optimize material properties.

$LiNb_{1-x}Ta_xO_3$ (LN-LT) solid solutions have recently been studied (see, for example, [5–8]), while they open up prospects for combining the advantages of both materials.

In particular, the LN-LT can be expected to exhibit high piezoelectric coefficients (close to LN) and the temperature stability of the LT properties. However, the growth of LN-LT single crystals is challenging because of the deviation from stoichiometry inherent in LN and LT crystals, the uneven distribution of cations caused by fluctuations in the temperature fields in the crystal growth zone, and differences in the melting temperatures T_m of both end components of the system ($T_m(LiNbO_3)$ = 1240 °C; $T_m(LiTaO_3)$ = 1650 °C).

Meanwhile, LN-LT nanoparticles can be attractive both for practical application and from the point of view of fundamental research, for example, to determine the dependence of the $LiNb_{1-x}Ta_xO_3$ structural characteristics depending on the composition.

Today, wet chemistry methods are most commonly used to obtain LN and LT nanoparticles. Among them, the sol–gel method predominates, when Nb or Ta pentaetoxides and Li ethoxide or salts—acetates, citrates, nitrates [1–5,9–14]—are chosen as starting materials for obtaining xerogels. Since alcoxides are extremely sensitive to moisture and susceptible to hydrolysis, the entire synthesis procedure must be carried out in a box filled with pure dry N_2 or Ar. In [7,15], LN nanoparticles were synthesized by the sol–gel method using a solution of $NbCl_5$ in toluol and $LiNO_3$ as starting materials. However, even in this case, part of the technological operations must be performed in a dry N_2 or Ar environment. In addition, in the sol–gel method, NbF_5 or TaF_5, obtained by the action of HF on niobium and tantalum oxides, can be used as starting materials (see, e.g., [16,17]).

Another method to obtain LN nanoparticles is the solution combustion method. For example, the authors of [18] have successfully synthesized $LiNbO_3$ and $LiNbO_3$:Fe nanopowders. The starting materials were $LiNO_3$, aqueous iron nitrate $Fe(NO_3)_3 \cdot 9H_2O$, ammonium niobium (V) oxalate hydrate $C_4H_4NNbO_9 \cdot nH_2O$, and glycine $C_2H_5NO_2$. After mixing and long stirring of the starting materials, a precursor was formed, which was calcinated at a temperature of 600 °C for 1 h in an O_2 atmosphere.

Another wet method is solvothermal synthesis, which was used to obtain $LiTaO_3$ nanopowders [19]. The authors used an autoclave, 80% of the volume of which was filled with a Li_2CO_3 solution in a mixture of distilled water and glycol with the addition of an appropriate amount of Ta_2O_5. The autoclave was kept at 240 °C, and the contents were constantly stirred for 12 h under autogenous pressure. The resulting LT nanopowders were washed, centrifuged, and calcinated.

The preparation of LN and LT nanoparticles by wet chemistry shows that there are certain difficulties in their application. First of all, this is the multistage nature of technological processes (dissolution, mixing, stirring, centrifugation, high-temperature treatment, calcination, etc.): the use of a large number of organic and inorganic reagents, including hazardous; performing synthesis in a special environment; use of relatively high temperatures. These factors make the process of obtaining nanoparticles time-consuming and expensive. In addition, an increase in the number of technological operations and the number of reagents used increases the probability of errors in the synthesis. Finally, the production of nanopowders in large quantities is complicated using the methods described above.

In contrast to wet chemistry methods, high-energy ball milling with subsequent annealing of powders looks promising from a technological point of view with relatively simple production organization. The synthesis process consists of only two stages—milling and subsequent annealing, which could be performed in air. Only Li_2CO_3, Nb_2O_5, and Ta_2O_5 can be used as starting materials. Today, we are aware of only three works that describe the successful synthesis of LN nanopowders by the mechanochemical method [20–22].

In addition, in [23,24], in order to obtain nanocrystals of lithium niobate, the preliminary synthesized $LiNbO_3$ phase was milled. In [23], LN obtained by the solid-phase reaction between lithium carbonate and niobium pentoxide was subjected to milling. In [24], the remnants of the melt solidified in the crucible after the completion of the procedure for growing the LN single crystal by the Czochralski method were used for milling. To the best of our knowledge, there are no works devoted to the mechanochemical synthesis of LT nanopowders. Furthermore, the synthesis and properties of nanopowders of $LiNb_{1-x}Ta_xO_3$ solid solutions, with $0 < x < 1$, remain completely unexplored.

This work focuses on the preparation of $LiNb_{1-x}Ta_xO_3$ nanopowders of different compositions (x = 0, 0.25, 0.5, 0.75, 1), using the mechanochemical synthesis method (high-energy ball milling of the starting materials Li_2CO_3, Nb_2O_5, Ta_2O_5 and subsequent annealing). Furthermore, the crystal structure, Raman spectra, and electrophysical properties of obtained LN-LT compounds are studied.

2. Materials and Methods

The mixed lithium niobate–tantalate nanopowders with nominal compositions $LiNb_{1-x}Ta_xO_3$ (x = 0, 0.25, 0.5, 0.75 and 1) were obtained by high-energy ball milling mixtures of Li_2CO_3, Nb_2O_5, and Ta_2O_5 powders (manufactured by Alfa Aesar, purity 4N) taken in molar ratios corresponding to stoichiometric compositions. The masses of the components for obtaining compounds with a certain x value are given in Table 1.

Table 1. Weights of reagents Li_2CO_3, Nb_2O_5, and Ta_2O_5, calculated to obtain 10 g of stoichiometric compounds $LiNb_{1-x}Ta_xO_3$.

Sample No.	Composition	Weight, g		
		Li_2CO_3	Nb_2O_5	Ta_2O_5
S01 (x = 0)	$LiNbO_3$	2499	8989	–
S02 (x = 0.25)	$LiNb_{0.75}Ta_{0.25}O_3$	2175	5868	3252
S03 (x = 0.5)	$LiNb_{0.5}Ta_{0.5}O_3$	1926	3463	5758
S04 (x = 0.75)	$LiNb_{0.25}Ta_{0.75}O_3$	1727	1554	7748
S05 (x = 1)	$LiTaO_3$	1566	–	9367

The synthesis was performed with the planetary ball mill machine Pulverisette-7. The rotation speed was equal to 600 rpm, and the duration of milling was about 10–15 h. A total of 134 balls of zirconium dioxide with a diameter of 5 mm and total weight of 91.5 g were used as working bodies. The mass ratio ball/sample was about 10. Milling was performed in 15 min cycles; subsequently, a reverse was carried out after each cycle. Based on the results of previous investigations (see, e.g., [21] where $LiNbO_3$ nanoparticles were synthesized by a mechanochemical technique), one can assume that the surfaces of particles are activated, but the synthesis exactly of an LN-LT compound was not fully completed after milling. This is confirmed by X-ray analysis of the powders performed after milling. As an example, Figure 1 shows a diffraction pattern of a powder with x = 0.5 compared to the reference pattern of a lithium niobate powder from the PDF database. The results of X-ray phase analysis indicate the formation of a predominantly amorphous precursor and partially a perovskite-like phase of lithium niobate.

To obtain nanocrystalline LN-LT particles, subsequent annealing of powders was performed. To determine the range for selecting the optimal annealing temperature, thermal analysis curves of the powders after milling were recorded. These dependences for all samples had a similar character. As an example, Figure 2 shows the thermogravimetric curves for milled mixture corresponding to $LiNb_{0.5}Ta_{0.5}O_3$ powder. As can be seen, sharp mass loss is observed within 400–550 °C (TG and DTG curves). This event corresponds to the decomposition of lithium carbonate according to (1):

$$Li_2CO_3 = Li_2O + CO_2 \tag{1}$$

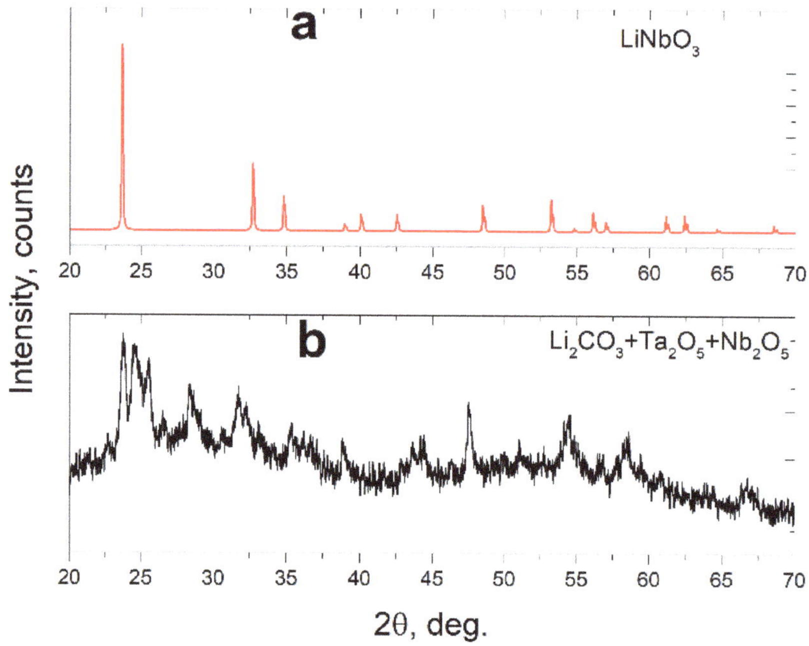

Figure 1. Reference diffraction patterns of lithium niobate (**a**) and diffraction patterns of the precursor obtained after milling a mixture of powders of lithium carbonate, niobium pentoxide, and tantalum pentoxide (**b**).

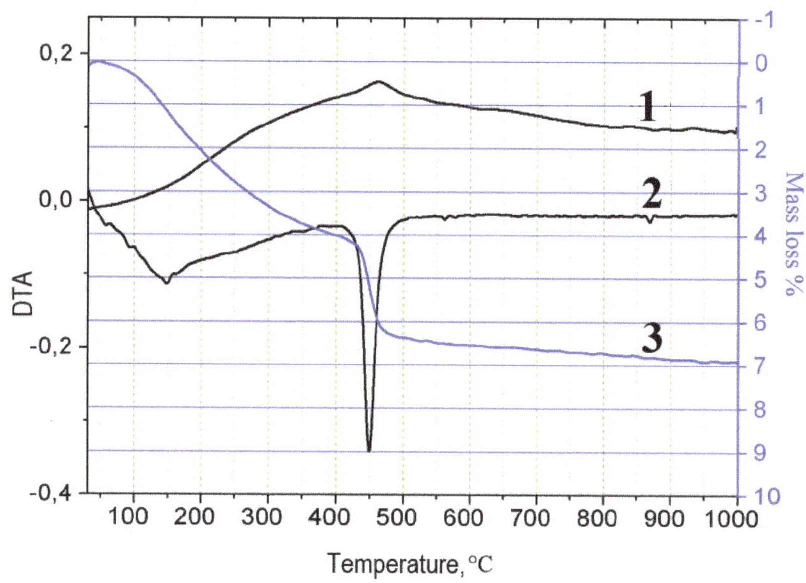

Figure 2. DTA–DTG curves obtained for $LiNb_{0.5}Ta_{0.5}O_3$ powders, which underwent a high-energy ball milling process for 15 h at a rotation speed of 600 rpm (1—DTA; 2—DTG; 3—TG).

At the same time, it is known that lithium carbonate decomposes above 800 °C [25]. Therefore, significant lowering of the temperature of this process is observed after milling the reaction mixture. In addition, the experimental value of mass loss in the temperature range 400–550 °C is about 20% of the theoretical value calculated by (1). The latter indicates that about 80% of lithium carbonate is decomposed at the stage of milling. Above 550 °C, the interaction of the formed lithium oxide with niobium (tantalum) oxides according to Reaction (2) occurs:

$$Li_2O + Nb_2O_5(Ta_2O_5) = 2LiNb(Ta)O_3. \tag{2}$$

Based on the obtained results, three annealing temperatures in air were chosen for the experiments—550, 700, and 800 °C.

Phase compositions of obtained nanoparticles were studied by X-ray phase analysis using the modernized DRON-3M diffractometer. Crystal structure parameters (unit cell dimensions, positional and displacement parameters of atoms) of both series of the materials were derived by full profile Rietveld refinement by using the WinCSD program package for structural analysis [26].

The micro-Raman spectra of LN-LT nanopowders were registered by confocal Raman microscope spectrometer MonoVista CRS+. The laser beam (λ = 532 nm) was focused in a 1 mm spot on the surface of pressured nanopowder.

Temperature dependencies of electrical conductivity in the range from 300 to 820 °C were obtained via impedance measurements in the frequency range from 1 Hz to 1 MHz using impedance gain-phase analyzer (Solartron 1260, Ametek Scientific Instruments, Hampshire, UK). For this experiment, pressed pellets with thickness varying from 1.25 to 1.7 mm and diameter of 10 mm were formed from $LiNb_xTa_{1-x}O_3$ nanopowders with $x = 0$, 0.25, 0.5, 0.75, and 1. Throughout the preparation process, the samples were heated up to 210 °C at the rate of 2 °C/min, while the pressure applied was 190 MPa at all times. Additionally, a constant voltage of 1 kV has been applied to the samples with an intention to electrically polarize them. The so-obtained pellets were subsequently annealed in air at 600 °C for 6 h. Platinum electrodes (5 mm in diameter) were deposited on both sides of each sample via screen printing (print ink: Ferro Corporation, No. 6412 0410). The samples were subsequently thermally treated for 1 h at 800 °C to ensure electrode adhesion.

Scanning electron microscopy (SEM) imaging of $LiNb_{0.5}Ta_{0.5}O_3$ sample has been performed after initial annealing at 600 °C (Figure 3) and after impedance measurements (Figure 4). The comparison of these two images reveals an increase in average grain size from approximately 100 to 200 nm due to temperature treatment during electrode preparation and the impedance studies. The SEM analysis after the impedance spectroscopy experiment, performed on a larger fragment of the sample, also demonstrates a high homogeneity of grain size distribution along the area.

Obtained impedance spectra of $LiNb_xTa_{1-x}O_3$ pressed samples are represented in form of Nyquist diagrams. Subsequently, an electrical equivalent-circuit model consisting of a constant phase element (CPE) connected in parallel with a bulk resistance R_B is fitted to the measured data. The intercepts of semicircles in the range of low frequencies are interpreted as samples resistance and subsequently converted to conductivity using the relation $\sigma = t(A \times R_B)^{-1}$, where t and A are the thickness of the sample and the electrode area, respectively.

Figure 3. Scanning electron microscopy image of pressed $LiNb_{0.5}Ta_{0.5}O_3$ after initial annealing at 600 °C.

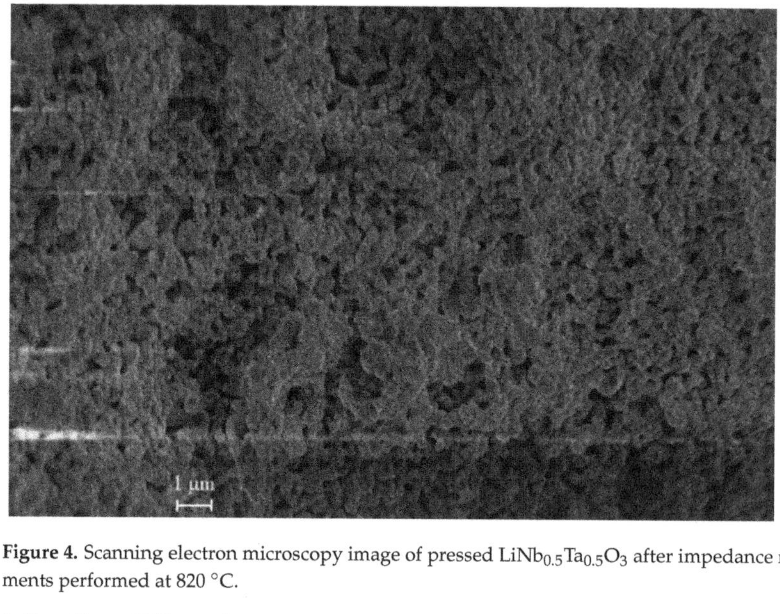

Figure 4. Scanning electron microscopy image of pressed $LiNb_{0.5}Ta_{0.5}O_3$ after impedance measurements performed at 820 °C.

3. Results and Discussion

3.1. X-ray Diffraction

X-ray diffraction (XRD) study of all the $LiNb_{1-x}Ta_xO_3$ samples annealed at 550 °C revealed a rhombohedral $LiNbO_3$-type structure as the main phase with some amount of the parasitic phase $Li(Nb,Ta)_3O_8$, and unreacted Ta_2O_5 and Nb_2O_5 (Table 2). Only the sample with $x = 0$ that corresponds to the 'pure' $LiNbO_3$ shows single phase composition as in work [21].

Table 2. Lattice parameters (*a*, *c*), average grain size, and microstrains values (<ε>) of two $LiNb_{1-x}Ta_xO_3$ series heat treated at 550 and 800 °C.

Sample	x	T, °C	Parasitic Phases	a, Å	c, Å	D_{ave}, nm	<ε>, %
S01	0	550	–	5.1483(3)	13.8484(9)	41	0.088
		800	–	5.1517(2)	13.8335(7)	206	0.108
S02	0.25	550	$Li(Nb,Ta)_3O_8$	5.1477(3)	13.822(1)	50	0.143
		800	$Li(Nb,Ta)_3O_8$	5.1521(3)	13.8148(8)	171	0.105
S03	0.5	550	Ta_2O_5 + $LiNb_3O_8$	5.1534(4)	13.808(1)	63	0.107
		800	$Li(Nb,Ta)_3O_8$	5.153(1)	13.778(3)	97	0.128
S04	0.75	550	Ta_2O_5	5.149(2)	13.788(5)	31	0.093
		800	$Li(Nb,Ta)_3O_8$	5.1558(7)	13.753(2)	92	0.139
S05	1	550	Ta_2O_5	5.1529(6)	13.767(2)	66	0.114
		800	$Li(Nb,Ta)_3O_8$	5.1593(4)	13.745(2)	80	0.135

Observable broadening of the Bragg's peaks on the nanocrystalline character for the powders was revealed. Additional heat treatment of the materials at 800 °C led to the narrowing of the diffraction peaks and to the considerable change of the phase composition of the $LiNb_{1-x}Ta_xO_3$ samples with x from 0.5 to 1, in which the increase of monoclinic $LiNb_3O_8$ or $Li(Nb,Ta)_3O_8$ phases and disappearing of individual Ta_2O_5 and Nb_2O_5 oxides were detected (see, e.g., Figure 5). No significant changes of the phase composition in the samples with x = 0 and 0.25 were observed after such annealing of the samples.

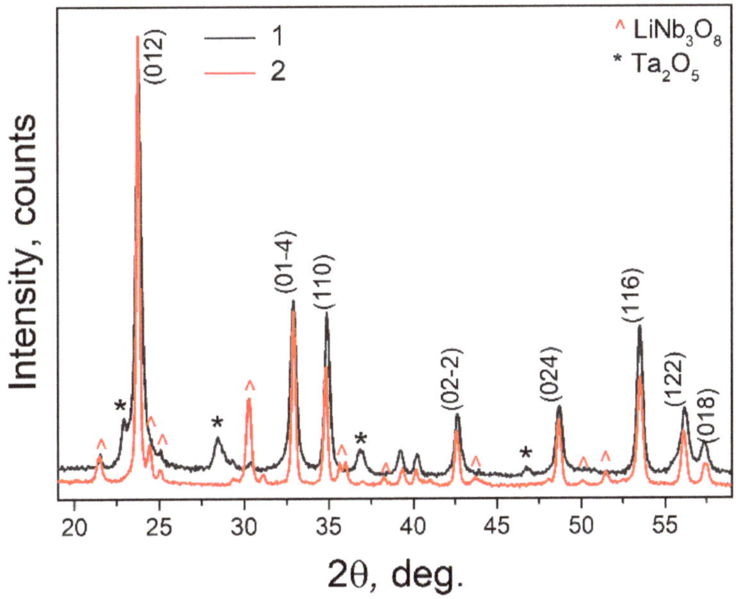

Figure 5. XRD patterns of the "S04" sample with a nominal composition $LiNb_{0.25}Ta_{0.75}O_3$ heat treated at 550 °C and 800 °C. For the main rhombohedral LN-LT phase, the Miller's indices are given (**1**—S04 at 550 °C; **2**—S04 at 800 °C).

As an example, Figure 6 demonstrates graphical results of Rietveld refinement of $LiNb_{0.75}Ta_{0.25}O_3$ material heat treated at 800 °C.

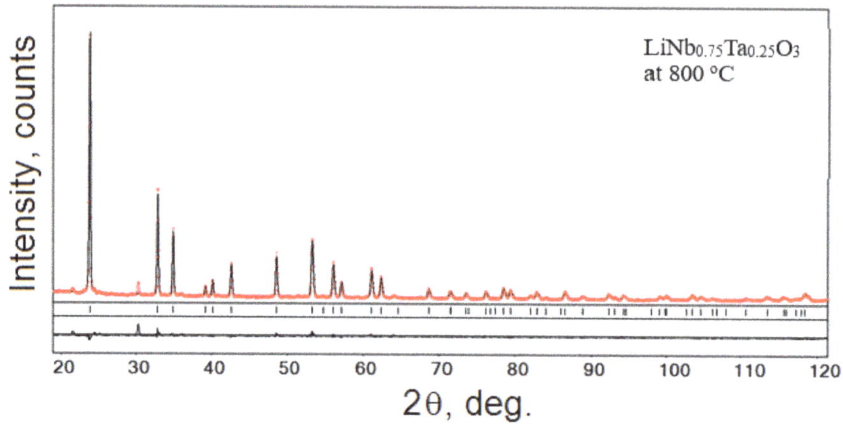

Figure 6. Results of Rietveld refinement of the $LiNb_{0.75}Ta_{0.25}O_3$ material heat treated at 800 °C. An experimental XRD pattern (dots) is shown in comparison with the calculated pattern. The difference between measured and calculated profiles is shown as a curve below the diagram. Short vertical bars indicate the positions of diffraction maxima in space group *R*3*c*.

The WinCSD programme package was also used for the evaluation of microstructural parameters of the powders, as presented in Table 1. The average grain size and microstrains $<\varepsilon> = <\Delta d>/d$ associated with the dispersion of interplanar distances *d* were derived from the analysis of angular dependence of the Bragg's peaks profiles. For the correction of instrumental broadening, the LaB_6 external standard was used. It was revealed that the average grain size of the $LiNb_{1-x}Ta_xO_3$ powders annealed at 550 °C is between 31 and 66 nm. Increase of thermal annealing temperature led to essential growth of grain size, being especially pronounced for the nominally pure $LiNbO_3$ and $LiNb_{0.75}Ta_{0.25}O_3$ sample. Corresponding changes for tantalum-reach materials are much less pronounced: the average grain size of $LiNb_{1-x}Ta_xO_3$ at 800 °C powders with *x* from 0.5 to 1 lies between 80 and 97 nm (see Table 2).

Analysis of the obtained structural parameters revealed that an increase of Ta content in both of the $LiNb_{1-x}Ta_xO_3$ specimens treated at 550 and 800 °C leads to the increase of the *a*-parameter and simultaneous decrease of the *c*-parameter (Figure 7a). As a result, a significant decrease of the *c*/*a* ratio and minor decrease of the unit cell volume in both $LiNb_{1-x}Ta_xO_3$ series is observed. It is observed that similar to the compositional effect on the unit cell dimensions of $LiNb_{1-x}Ta_xO_3$ materials, there is an increase of the heat treatment temperature from 550 to 800 °C, which led to increase of the *a*-parameter and simultaneous reduction of the *c*-parameter (see Figure 7).

The comparison of the obtained structural parameters of LN-LT samples with the corresponding structural data for nominally pure LN and LT [27–33] points to the formation of the continuous $LiNb_{1-x}Ta_xO_3$ solid solution.

As can be seen from the results presented in Table 2 and Figure 7, after annealing at T = 550 °C, the phase analysis indicates that niobium and/or tantalum oxides prevail among the traces of parasitic phases in the resulting nanopowders with the lithium niobate structure. At the same time, after annealing at T = 800 °C, the predominant parasitic phase is the compound $Li(Nb,Ta)_3O_8$, which, as compared to $Li(Nb,Ta)O_3$, contains less lithium. The latter is due to the sublimation of lithium in the form of oxide [34]. This result indicates that the optimal annealing temperature of nanopowders lies in the range of 550–800 °C. In order to optimize the heat treatment conditions for obtaining single-phase nanocrystalline powders, a number of experiments were carried out on the milling and annealing of equiatomic lithium niobate–tantalate $LiNb_{0.5}Ta_{0.5}O_3$ at different milling times and different annealing temperatures.

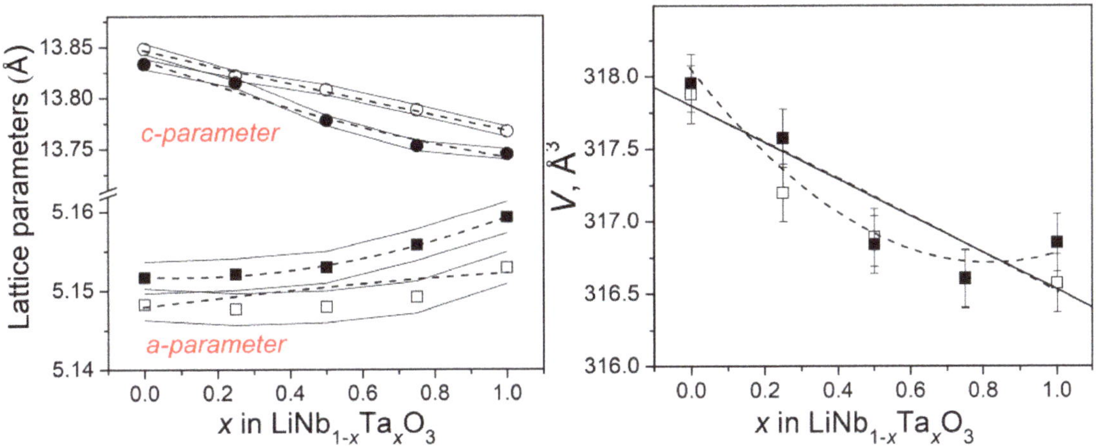

(a) **(b)**

Figure 7. Concentration dependence of the lattice parameters (**a**) and unit cell volume (**b**) in $LiNb_{1-x}Ta_xO_3$ at 550 °C and $LiNb_{1-x}Ta_xO_3$ at 800 °C series. The dashed lines are a guide for eyes (empty symbols—samples annealed at 550 °C; solid symbols—samples annealed at 800 °C).

Based on the data of X-ray phase analysis of the synthesized compounds, it was found that the best results were achieved with a milling time of 12–15 h and a heat treatment temperature of 650–700 °C for 5 h. Such modes ensure the absence of parasitic phases (within the accuracy of the measurement method) and the absence of violation of stoichiometric ratio Li/Nb. Diffraction patterns of powders obtained in optimal conditions are shown in Figure 8.

3.2. Raman Spectra

The micro-Raman spectra of LN-LT nanopowders with different Nb and Ta content, annealed at 550 °C, are shown in Figure 9, and the positions of the observed bands are indicated in Table 3. As it is seen from Figure 9, the Raman spectra of LN-LT with different x are generally similar; however, some peculiarities are observed. Particularly, in the spectrum of pure LN, 15 Raman bands can be distinguished, 11 of which can be attributed to A1 and E vibrational modes. The A1 modes are polarized along the Z-axis, while the doubly degenerate E modes correspond to ionic motions along the X or Y-axis [35,36]. In the spectrum of LN-LT (x = 0.25), 25 Raman bands can be distinguished; for x = 0.5, the number of bands is equal to 20, for x = 0.75 to 15 and for pure LT to 17 (see Table 3). The differences in the number of distinguished bands could probably be caused by the overlapping of some bands with those with higher intensity. The observed differences are associated with different compositions of the nanopowders as well as with probably non-optimal technology regimes of nanopowders synthesis. Particularly, the following main specific features of the LN-LT nanopowders Raman spectra were revealed.

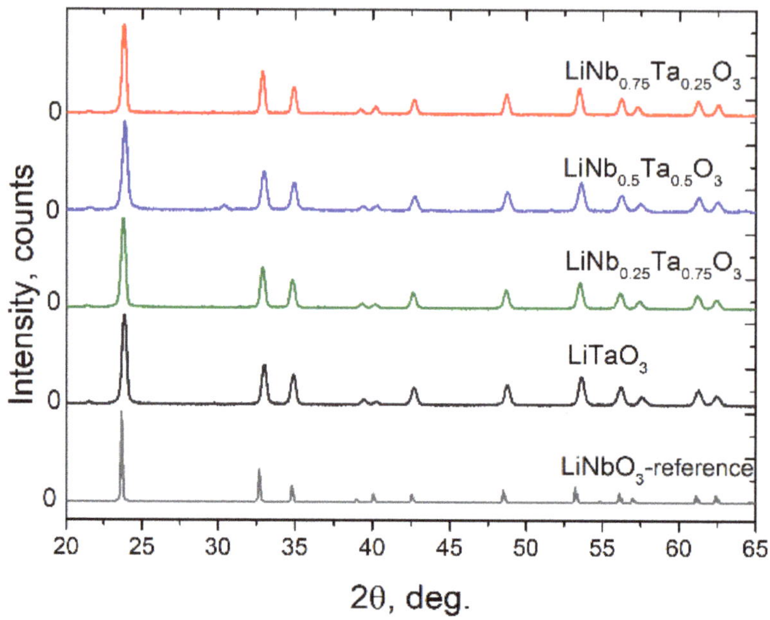

Figure 8. Diffraction patterns of $LiNb_{1-x}Ta_xO_3$ nanopowders (x = 0.25; 0.5; 0.75; 1) obtained by mechanosynthesis in optimal modes (milling speed—600 rpm, milling time—15 h, annealing at 700 °C for 5 h), as well as the reference diffractogram of $LiNbO_3$.

1. Few low-frequency bands near 73–75 and 90–97 cm^{-1} are observed for the nanopowders with $x \neq 0$. These bands cannot be linked with known Raman bands of LT [35–37]. In accordance with the results of [38], these bands could be induced by the presence of $Li(Nb,Ta)_3O_8$ phase, as identified by X-ray diffraction technique (see Table 2).
2. A more intensive band near 117–125 cm^{-1} is observed only for the LN-LT sample with x = 0.5. The presence of this band could be attributed to the contribution of $Li(Nb,Ta)_3O_8$ and Ta_2O_5 additional phases in this sample. Note that the authors in [38] observed the close bands at 116 and 136 cm^{-1} and attributed them to $LiNb_3O_8$. The bands near 100 cm^{-1} were attributed to Ta_2O_5 by the authors of [39]. As it is followed from the XRD data (see Table 1), the simultaneous presence of $Li(Nb,Ta)_3O_8$ and Ta_2O_5 phases occurs only in the sample with x = 0.5, so the overlay of corresponding bands can result in a peculiar form of its spectrum. Furthermore, the sample with x = 0.5, i.e., with the composition intermediate between pure LN and pure LT, ought to essentially reveal the bands of both crystals, so it is no wonder that the spectrum of this sample has the most complex character. Furthermore, the sample with x = 0.5 reveals a significant increase of the bands' intensities near 260 and 630–670 cm^{-1} that visually looks like a widening of intensive neighboring peaks. This result is in good agreement with [40], where two intensive neighboring bands in the region of 600 cm^{-1} were also observed for the LN-LT sample with x = 0.553. Finally, it should be noted that the bands near 600 cm^{-1} are considerably broad for all investigated samples in comparison with the other observed bands. This is consistent with the results in [41] where it is concluded that the band at 600 cm^{-1} is broader for non-poled LN samples (particularly, nanopowders) than for polarized.
3. The Raman spectra of LN and LT nano- and micropowders are shown in Figure 10 for comparison purposes. The latter were obtained by the crushing of LN and LT single crystals grown at SRC 'Electron-Carat'. As seen from Figure 10, the band observed at about 1008–1009 cm^{-1} for LT nanopowder is not pronounced for LT

micropowder as well as for LN compounds. The similar band can be observed in Figure 9 for nanopowders with $x \neq 0$. As it is seen from Figure 9, the intensity of this band increases with increasing of x. Moreover, for $x = 0.5$, this band splits into two with the frequencies of 994 and 1008 cm^{-1}. As it is shown in [39], this band is absent in Ta_2O_5 Raman spectrum. Since the data about Raman scattering in $LiNb_3O_8$ are not available in this spectral range, we cannot exclude that this band is linked with the $LiNb_3O_8$ (or $LiTa_3O_8$) phase. However, as it is seen from Table 2, the $LiTa_3O_8$ phase is absent in the pure LT sample annealed at 550 °C as well as in the sample with $x = 0.75$ (within the limits of accuracy). Thus, we have to conclude that the nature of this band cannot be clearly determined from current experiments and requires additional studies.

4. Contrary to the results in [42], we did not observe any remarkable effect of grain size reduction, i.e., decreasing of the intensities of all Raman bands caused by grain size decrease (see Figure 10).
5. Increasing the spectral range up to 4000 cm^{-1} allows revealing the weak vibrations at 1600 and 3400 cm^{-1} (looking as low-intensive wide bands) that can be caused by the traces of OH– groups, which are always present in LN and LT as well as the traces of HCO_3^- groups (near 1750 and 2900 cm^{-1}) present in synthesized compounds, which is probably due to the use of lithium carbonate as a component of the initial mixture.

Similarly to X-ray studies, we have measured the Raman spectra of nanopowder samples obtained under optimal conditions. These spectra are shown in Figure 11.

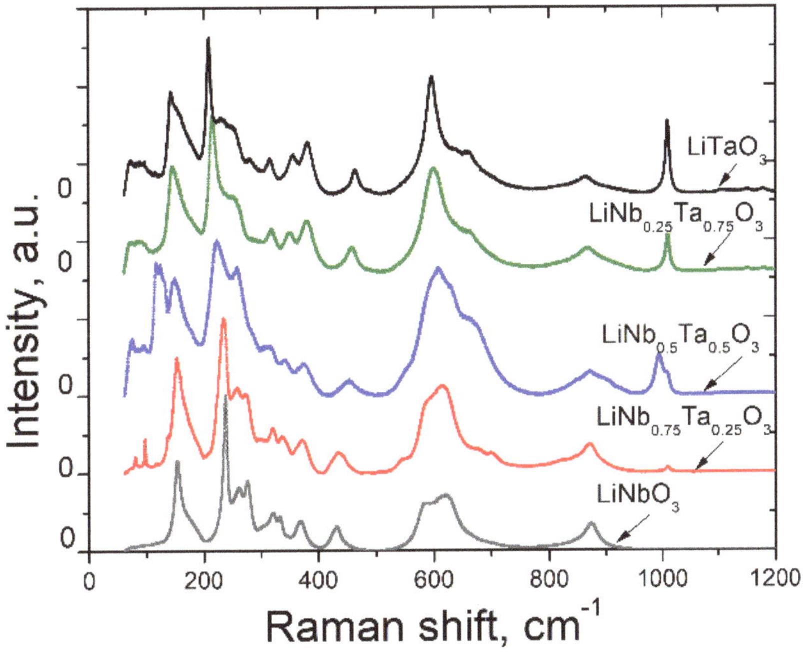

Figure 9. Raman spectra of $LiNb_{1-x}Ta_xO_3$ nanopowders obtained in non-optimal synthesis conditions.

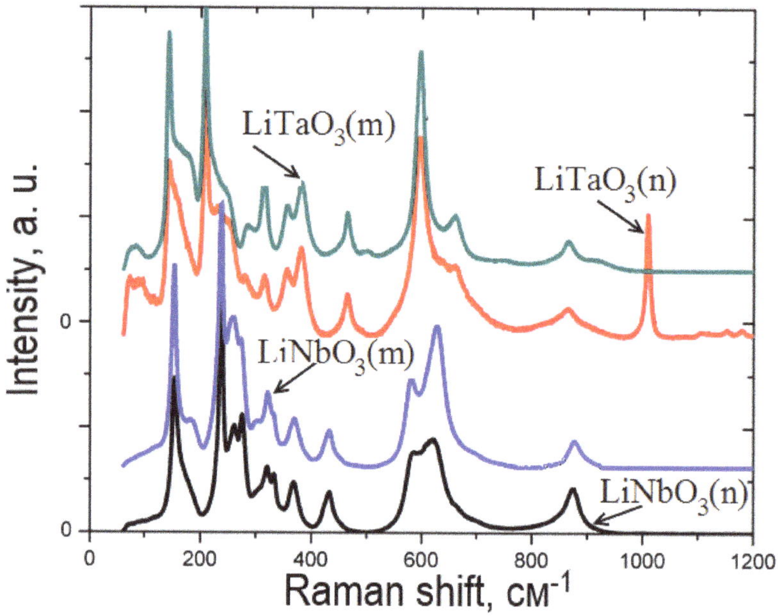

Figure 10. Raman spectra of nanopowders (n) of $LiNbO_3$ and $LiTaO_3$ synthesized by mechanochemical treatment and the ones of micropowders (m) of $LiNbO_3$ and $LiTaO_3$ obtained by crushing.

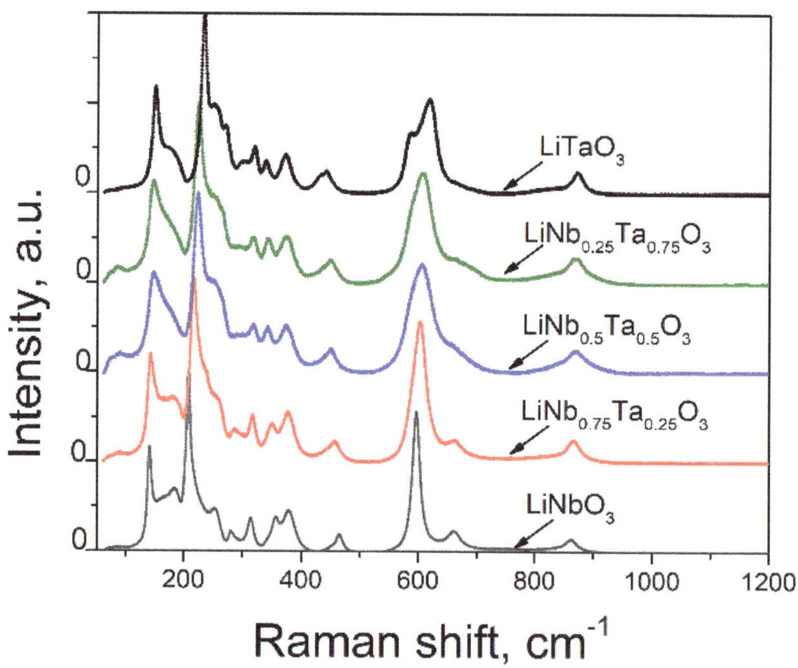

Figure 11. Raman spectra of $LiNb_{1-x}Ta_xO_3$ nanopowders obtained in optimal synthesis conditions.

Table 3. The observed bands in Raman spectra of $LiNb_{1-x}Ta_xO_3$ nanopowders.

$x = 0$		$x = 0.25$		$x = 0.5$		$x = 0.75$		$x = 1$	
Raman Shift, cm^{-1}	Intensity, a. u.	Raman Shift, cm^{-1}	Intensity, a. u.	Raman Shift, cm^{-1}	Intensity, a. u.	Raman Shift, cm^{-1}	Intensity, a. u.	Raman Shift, cm^{-1}	Intensity, a. u.
117.3	556	63.5	151	75.5	3779	73.8	2037	73.1	2275
153.8	5873	73.2	391	95.8	3327	90.2	2046	97	2194
180	1089	80.8	1117	117.1	8657	146.9	6845	143.2	6681
238.8	10,002	97.5	2255	124.2	8534	168.4	4115	162	4985
262.6	4068	117.7	715	131	7601	215.3	10,093	209.1	10,133
276.7	4481	138	2445	150.2	7676	240.2	5079	232.5	4917
303	1582	153.8	7500	224.3	10,117	252.7	4891	250.5	4384
321	2491	168.8	4176	258.3	8375	319	2778	280.8	2294
333	2220	208.6	1390	312.4	3266	351.1	2550	316	2327
369	1904	235.9	9978	342.3	2388	380.5	3293	356.2	2579
432.8	1550	259	5508	375.7	2129	458.3	1570	381.5	3387
584.6	3074	272.7	5113	453.1	997	600.1	6663	465.1	1618
621.3	3557	298.6	2242	608.4	8264	660.7	2598	597	7575
700.8	464	320.8	2914	629.7	7113	868.2	1535	637	2886
875	1681	336.4	2391	664.1	4978	1009.7	2344	660.2	2741
		371.7	2150	849.6	1035			865.8	1122
		436	1339	872.9	1571			1009.2	4684
		547.3	945	906	954				
		597.7	5018	994.1	2675				
		615.5	5608	1008.1	1536				
		676	1572						
		700	1300						
		821.8	808						
		872	1841						
		1009.7	341						

As seen from Figure 11, the spectra for the $LiNb_{1-x}Ta_xO_3$ are generally consistent with those recorded for samples, which were obtained in non-optimal conditions (see also Figure 9). The only difference is the band observed near 1000 cm^{-1} for the $LiNb_{1-x}Ta_xO_3$ specimens with $x = 0.25$, 0.5, 0.75, and 1, obtained in a non-optimal regime (Figure 9). This band vanishes from the spectra of specimens, which are obtained in optimal regimes. The nature of this band remains not clear and requires additional studies.

3.3. Electrical Conductivity

Impedance data of the pressed $LiNb_xTa_{1-x}O_3$ samples are exemplarily shown for 400 °C (Figure 12) and 600 °C (Figure 13) in the form of Nyquist diagrams. The representation of the resistivity is chosen to eliminate the geometrical factors of the samples (thickness and area). At 400 °C, the impedance data of all samples show a slightly depressed single arc semicircle. Such depression could be attributed to the non-ideal capacitance of the samples, which corresponds to low values of the CPE exponents obtained from fitting of an R_b-CPE equivalent circuit model to experimental data. In particular, at 400 °C, exponent values vary between 0.74 ($LiTaO_3$) and 0.85 ($LiNb_{0.75}Ta_{0.25}O_3$) and slightly decrease with the temperature increase. Another possible interpretation implies that the obtained impedance spectra may consist of two overlapping semicircles, with similar relaxation times, representing grain interior and grain boundary conduction mechanisms, respectively [43]. We note that in case of nanostructured materials, the grain boundary contributions may significantly overlap with that of the grain interiors, making it hard to distinguish between them [43,44].

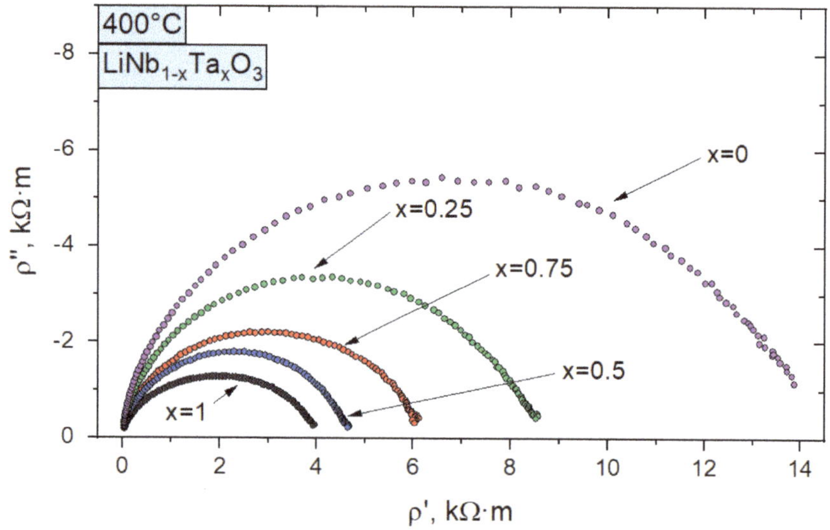

Figure 12. Complex impedance of pressed $LiNb_{1-x}Ta_xO_3$ samples measured at 400 °C.

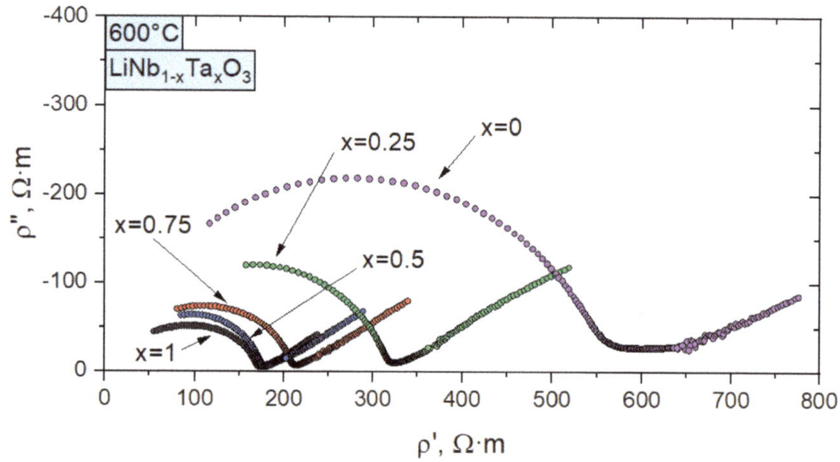

Figure 13. Complex impedance of the pressed $LiNb_{1-x}Ta_xO_3$ samples measured at 600 °C.

Impedance spectra of all samples at temperatures above 500 °C reveal the existence of linear region of $\rho''(\rho')$ dependence, which follows the semicircle intercept at lower frequencies. This is exemplarily shown in Nyquist diagram at 600 °C (Figure 13). At this temperature, the impedance of the samples has been additionally measured down to 0.01 Hz in order to examine the low-frequency region of the $\rho''(\rho')$ dependence in more detail; however, no changes in the line slope were observed. According to [45], such peculiarity at low frequencies can be attributed to the electrode effect, which is typical for an ionic conductor: mobile charge carriers in form of ions are blocking the metal–sample interface. On the other hand, a similar behavior of Z″(Z′) dependence at low frequencies, which was observed in [46] for polycrystalline lithium niobate samples at temperatures above 550 °C, was associated with a grain boundaries conduction mechanism.

The temperature dependencies of $LiNb_xTa_{1-x}O_3$ conductivity are shown in Figure 14 in form of Arrhenius plots. Generally, the behavior of conductivity is similar to that of single crystalline $LiNb_xTa_{1-x}O_3$ reported in [47]. Furthermore, the conductivity tends to correlate with Nb/Ta ratio, increasing with an increase of Ta content and reaching the values at 820 °C of 0.022 S/m and 0.076 S/m for x = 0 and x = 1, respectively. However, the conductivity of $LiNb_xTa_{1-x}O_3$ specimen with x = 0.5 deviates from this correlation, which could not be solely attributed to experiment uncertainty. The latter, according to our estimations, is approximately 6% at 800 °C. This issue requires additional studies and will be a subject of subsequent investigations.

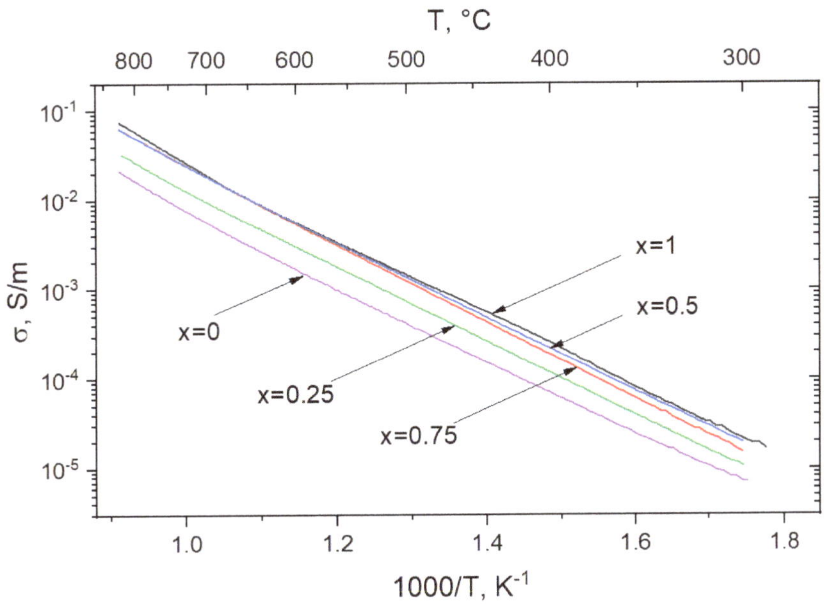

Figure 14. Temperature dependence of conductivity of $LiNb_{1-x}Ta_xO_3$ pressed pellets.

The absolute values of conductivity (Figure 14) are generally higher, comparing to the obtained previously for single crystalline and polycrystalline $LiNbO_3$ and $LiTaO_3$ [45–52]. As reported in [51], such a difference is associated with a greater number of free ions in polycrystalline $LiNbO_3$ and $LiTaO_3$ due to a developed system of grain boundaries and much more defective structure in general [51]. Consequently, the variation of conductivity, reported for different polycrystalline samples, can be attributed to differences in the average grain size in particular.

The conductivity of all samples increases linearly in Arrhenius presentation up to around 620 °C. After a transition temperature range of 620–670 °C, we observed the conductivity increase as linear again; however, the slope changes (see mark in Figure 14). This indicates that the conductivity is governed by a different process than that at lower temperatures. The obtained results enable the determination of activation energy, E_A using the relation:

$$\sigma = \frac{\sigma_0}{T} e^{-\frac{E_A}{kT}} \quad (3)$$

where σ_0, T, E_A, and k represent the pre-exponential constant, absolute temperature, activation energy, and the Boltzmann constant, respectively. Activation energies and pre-exponential factors, obtained by fitting the Arrhenius equation to the measured conductivity data, as well as the temperature ranges for fitting are summarized in Table 4. The corresponding fits for low and high-temperature regions are exemplarily shown for the $LiTaO_3$ specimen in Figure 15.

Table 4. Activation energies E_A and pre-exponential coefficients σ_0 for conductivity of $LiNb_{1-x}Ta_xO_3$ pressed pellets at different temperatures.

Composition	300 ÷ 620 °C		670 ÷ 820 °C	
	E_A, eV	σ_0, $\cdot 10^6$ S/m	E_A, eV	σ_0, $\cdot 10^6$ S/m
$LiNbO_3$	0.88 ± 0.031	0.17	1.05 ± 0.071	1.54
$LiNb_{0.75}Ta_{0.25}O_3$	0.88 ± 0.011	0.32	1.01 ± 0.073	1.59
$LiNb_{0.5}Ta_{0.5}O_3$	0.88 ± 0.015	0.58	1 ± 0.07	2.61
$LiNb_{0.25}Ta_{0.75}O_3$	0.91 ± 0.015	0.83	0.99 ± 0.069	2.58
$LiTaO_3$	0.86 ± 0.015	0.48	1.09 ± 0.072	8.33

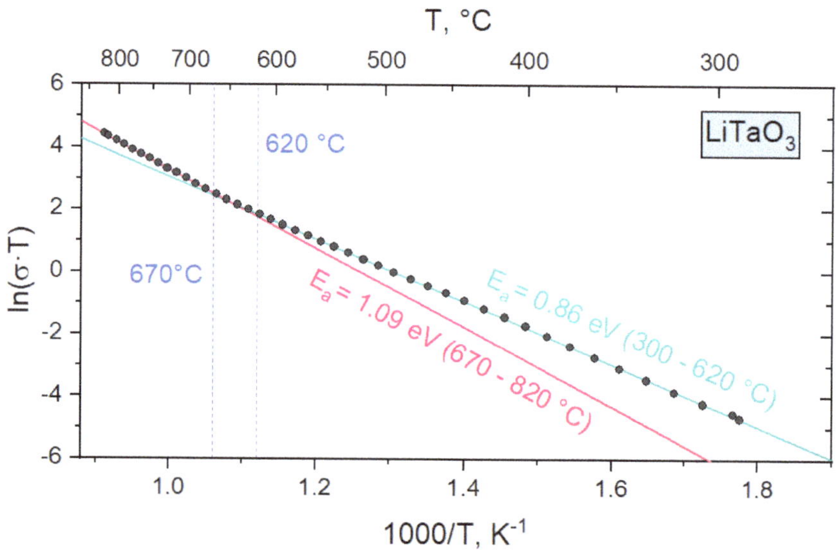

Figure 15. ln(σ·T) vs. 1000/T dependence of $LiTaO_3$ samples (dots) and linear fits of its separate regions (lines).

Generally, the activation energies obtained in our study (Table 4) are consistent with the values obtained previously for polycrystalline lithium niobate and/or lithium tantalate [43,46,50–53], varying from 0.63 to 1.25 eV. It should be noted that the wide range of activation energies reported [43,46,50–53] is associated with their strong dependence on samples grain size, which is in particular shown in a comparative impedance spectroscopy study of single-, micro-, nanocrystalline, and amorphous lithium niobate [43]. Similar to our study, the authors in [43] observed slightly depressed semicircles on $\rho''(\rho')$ dependence of nanocrystalline $LiNbO_3$ and attributed such depression to the superposition of two conduction mechanisms, arising from the grain interior and grain boundary, respectively. The nanopowders studied in [43] were obtained by high-energy ball milling with the grains size of about 20 nm.

Qualitatively similar dependencies were observed for polycrystalline samples of non-stoichiometric $LiNbO_3$ in [53]. Here, the authors have obtained the $LiNbO_3$ micropowders by the ball milling technique with the grain sizes varying from 2 to 3 m. Obtained powders were subsequently annealed and isostatically pressed in pellets at 2500 bars [53]. It is assumed in [53] that the conductivity of $LiNbO_3$ in low and high-temperature regions is governed by the polaronic and ionic conduction mechanisms, respectively. The activation energies obtained in [53] vary from 0.78 to 0.88 eV for the low-temperature region and from 1.07 to 1.20 eV for the high-temperature region, which is in good correlation with the values obtained in our study.

Furthermore, the authors in [51] studied the electrical properties of polycrystalline $LiNbO_3$ up to 800 K and obtained the activation energy of 0.88 eV, which is very close to the corresponding value in our work. The sol–gel method was applied for the synthesis of nanopowders. Subsequently, the powders were annealed at 1200 °C for 3 h and pressed in pellets [51]. The average grain size was between 0.2 and 2 m. Since the measurements in [51] did not extend to higher temperatures, no evidence of a second conducting mechanism was observed.

4. Conclusions

In summary, it is shown in the current study that the mechanosynthesis is the most simple and most accessible method of lithium niobate and tantalate nanoparticles production. This method allows obtaining $LiNb_{1-x}Ta_xO_3$ nanopowders in only two stages: high-energy ball milling of the raw powders (Li_2CO_3, Nb_2O_5, Ta_2O_5) and subsequent annealing of the obtained precursors. Furthermore, this method allows completely excluding usually undesirable wet chemistry procedures from the technological process.

A number of $LiNb_{1-x}Ta_xO_3$ nanopowders (x = 0, 0.25, 0.5, 0.75, 1) is obtained by means of the mechanosynthesis technique for the first time, to the best of our knowledge.

The X-ray analysis of $LiNb_{1-x}Ta_xO_3$ nanopowders obtained at different conditions allowed for determination of the optimal parameters for milling and annealing runs: milling at 600 rpm for 12–15 h and subsequent annealing of powders in air at temperatures 650–700 °C for 5 h. This allowed obtaining single-phase $LiNb_{1-x}Ta_xO_3$ nanopowders for all the studied values of x.

The crystal structure of the samples was determined by the X-ray diffraction technique. The full-profile Ritveld refinement was used for determination of the crystal structure parameters and micro-structure parameters of $LiNb_{1-x}Ta_xO_3$ nanopowders. It is shown that the replacement of niobium by tantalum in the $LiNb_{1-x}Ta_xO_3$ structure asymmetrically influences the parameters of the unit cell: increasing the Ta content x leads to increasing the lattice parameter a and the simultaneous decreasing of parameter c accompanied by a slight decreasing of the unit cell volume. It is shown that the average size of crystallites varies from 31 nm (treatment only at 550 °C) to 206 nm (additional treatment at 800 °C) for different samples.

Raman scattering in $LiNb_{1-x}Ta_xO_3$ nanopowders (x = 0, 0.25, 0.5, 0.75, 1) obtained by mechanosynthesis was studied, to the best of our knowledge, for the first time. It is shown that the obtained Raman spectra are generally similar. Some features (band shifts, changes in their intensity, the formation of new bands) are attributed to the different values of x. For samples obtained in non-optimal milling and annealing runs, bands typical for the parasitic phases $LiNb(Ta)_3O_8$, Nb_2O_5, and/or Ta_2O_5 are observed in the spectra.

The measured temperature dependence of electrical conductivity in air up to 820 °C shows similar behavior for all studied $LiNb_{1-x}Ta_xO_3$ samples and tends to correlate with the Nb/Ta ratio. Two linear regions are observed in the Arrhenius presentation, which is attributed to different conductivity mechanisms. Activation energies vary from 0.86 $\pm$ 0.015 eV to 0.91 $\pm$ 0.015 eV for different compositions in the low-temperature region and from 0.99 $\pm$ 0.069 eV to 1.09 $\pm$ 0.072 eV at high temperatures.

Author Contributions: Conceptualization, D.S.; Data curation, S.H. and U.Y.; Investigation, L.V., Y.S., D.W., S.H. and Y.Z.; Methodology, V.S. and A.L.; Resources, I.I.S. and I.S.; Writing—original draft, O.B.; Writing—review and editing, A.S. and H.F. All authors have read and agreed to the published version of the manuscript.

Funding: Research grants from the Bundesministerium für Bildung und Forschung (BMBF, Federal Ministry of Education and research, FKZ: 01DK20009) and Ministry of Education and Science of Ukraine (M/48-2020) in frames of the joint German-Ukrainian project 'Nanocrystalline piezoelectric compounds $LiNb_{1-x}Ta_xO_3$ for high-temperature applications' supported this work. Furthermore, the work was partially supported by the Polish National Science Center (grant No. 2019/33/B/ST8/02142) and by the Deutsche Forschungsgemeinschaft (DFG, German Research Foundation, www.for5044.de; SU1261/1-1, FR1301/42-1).

Data Availability Statement: All relevant data presented in the article are stored according to institutional requirements and as such are not available online. However, all data used in this manuscript can be made available upon request to the authors.

Acknowledgments: Stepan Hurskyj and Uliana Yakhnevych acknowledge the partial support of the Ukrainian Ministry of Education and Science through the project DB/KINETYKA (No 0119U002249). The authors from Clausthal University of Technology acknowledge the support of the Energie-Forschungszentrum Niedersachsen, Goslar, Germany. The authors acknowledge support by Open Access Publishing Fund of Clausthal University of Technology.

Conflicts of Interest: The authors declare no conflict of interest.

References

1. $LiNbO_3$ Crystal Market by Type (Acoustic Grade, Optical Grade, Others), by Application (Electro-Optical, Surface Acoustic Wave, Piezoelectric Sensors, Non-Linear Optical, Other) and Region—Global Forecasts to 2027. Available online: https://www.reportsanddata.com/report-detail/global-linbo3-crystal-market-by-manufacturers-countries-type-and-application-forecast-to-2022 (accessed on 21 May 2021).
2. Lithium Niobate Modulator Market—Global Industry Analysis, Size, Share, Growth, Trends, and Forecast. 2019–2027. Available online: https://www.transparencymarketresearch.com/lithium-niobate-modulator-market.html (accessed on 21 May 2021).
3. Global Lithium Tantalate Crystal Market 2019–2027—Industry Analysis, Size, Share, Growth, Trends and Forecast. Available online: https://www.prnewswire.com/news-releases/global-lithium-tantalate-crystal-market-2019--2027--industry-analysis-size-share-growth-trends-and-forecast-300980802.html (accessed on 21 May 2021).
4. Lithium Tantalate Crystal Market (Grade: Optical Grade, SAW Grade, and Others; and Application: Electro-Optical Modulators, Pyroelectric Detectors, Piezoelectric Transducers, Acoustic Wave Devices, and Others)—Global Industry Analysis, Size, Share, Growth, Trends, and Forecast. 2019–2027. Available online: https://www.transparencymarketresearch.com/lithium-tantalate-crystal-market.html (accessed on 21 May 2021).
5. Xue, D.; Betzler, K.; Hesse, H. Dielectric properties of lithium niobate-tantalate crystals. *Solid State Commun.* **2000**, *115*, 581–585. [CrossRef]
6. Wood, I.G.; Daniels, P.; Brown, R.H.; Glazer, A.M. Optical birefringence study of the ferroelectric phase transition in lithium niobate tantalate mixed crystals: $LiNb_{1-x}Ta_xO_3$. *J. Phys. Condens. Matter* **2008**, *20*, 235237. [CrossRef] [PubMed]
7. Bartasyte, A.; Glazer, A.M.; Wondre, F.; Prabhakaran, D.; Thomas, P.A.; Huband, S.; Keebleand, D.S.; Margueron, S. Growth of $LiNb_{1-x}Ta_xO_3$ solid solution crystals. *Mater. Chem. Phys.* **2012**, *134*, 728–735. [CrossRef]
8. Roshchupkin, D.; Emelin, E.; Plotitcyna, O.; Fahrtdinov, R.; Irzhak, D.; Karandashev, V.; Orlova, T.; Targonskaya, N.; Sakharov, S.; Mololkin, A.; et al. Single crystals of ferroelectric lithium niobate–tantalate $LiNb_{1-x}Ta_xO_3$ solid solutions for high-temperature sensor and actuator applications. *Acta Crystallogr. B* **2020**, *76*, 1071–1076. [CrossRef]
9. Nashimoto, K.; Cima, M.J. Epitaxial $LiNbO_3$ thin films prepared by a sol-gel process. *Mat. Lett.* **1991**, *10*, 348–354. [CrossRef]
10. Hirano, S.; Yogo, T.; Kikuta, K.; Morishita, T.; Ito, Y.; Hirano, S.; Yogo, T.; Kikuta, K.; Morishita, T. Preparation of potassium tantalate niobate by sol-gel method. *J. Amer. Cer. Soc.* **1992**, *75*, 1701–1704. [CrossRef]
11. Ono, S.; Hirano, S. Processing of highly oriented lithium tantalate films by chemical solution deposition. *J. Mater. Res.* **2002**, *17*, 2532–2539. [CrossRef]
12. Takahashi, M.; Yamauchi, K.; Yagi, T.; Nishiwaki, A.; Wakita, K.; Ohnishi, N.; Hotta, K.; Sahashi, I. Preparation and characterization of high-quality stoichiometric $LiNbO_3$ thick films prepared by the sol–gel method. *Thin Solid Film.* **2004**, *458*, 108–113. [CrossRef]
13. Satapathy, S.; Gupta, P.K.; Srivastava, H.; Srivastava, A.K.; Wadhawan, V.K.; Varma, K.B.R.; Sathe, V.G. Effect of capping ligands on the synthesis and on the physical properties of the nanoparticles of $LiTaO_3$. *J. Cryst. Growth* **2007**, *307*, 185–191. [CrossRef]
14. Gou, J.; Wang, J.; Huang, Z.H.; Jiang, Y.D. Preparation of $LiTaO_3$ nano-crystalline films by sol-gel process. *Key Engin. Mat.* **2013**, *531–532*, 446–449. [CrossRef]
15. Johann, F.; Jungk, T.; Lisinski, S.; Hoffmann, Á.; Ratke, L.; Soergel, E. Sol-gel derived ferroelectric nanoparticles investigated by piezoresponse force microscopy. *Appl. Phys. Lett.* **2009**, *95*, 202901. [CrossRef]

16. Liu, Y.G.; Hu, J.H.; Huang, Z.H.; Fang, M.H. Preparation of $LiTaO_3$ nanoparticles by a sol–gel route. *J. Sol. Gel. Sci. Technol.* **2011**, *58*, 664–668. [CrossRef]
17. Mathew, T.V.; Kuriakose, S. Synthesis and characterization of sodium–lithium niobate ceramic structures and their composites with biopolymers. *J. Adv. Cer.* **2013**, *2*, 11–20. [CrossRef]
18. Li, Y.-Y.; Chen, H.-L.; Chen, G.-J.; Kuo, C.-L.; Hsieh, P.-H.; Hwang, W.-S. Investigation of the defect structure of congruent and Fe-doped $LiNbO_3$ powders synthesized by the combustion method. *Materials* **2017**, *10*, 380. [CrossRef]
19. Gao, L.; Chen, H.; Wang, Z.; Zou, X. Solvothermal synthesis and characterization of lithium tantalate nanoparticles. *Key Eng. Mater.* **2014**, *602–603*, 19–22. [CrossRef]
20. de Figueiredo, R.S.; Messai, A.; Hernandes, A.C.; Sombra, A.S.B. Piezoelectric lithium niobate obtained by mechanical alloying. *J. Mater. Sci. Lett.* **1998**, *17*, 449–451. [CrossRef]
21. Khalameida, S.; Sydorchuk, V.; Leboda, R.; Skubiszewska-Zięba, J.; Zazhigalov, V. Preparation of nano-dispersed lithium niobate by mechanochemical route. *J. Therm. Anal. Calorim.* **2014**, *115*, 579–586. [CrossRef]
22. Fierro-Ruiz, C.D.; Sánchez-Dena, O.; Cabral-Larquier, E.M.; Elizalde-Galindo, J.T.; Farías, R. Structural and magnetic behavior of oxidized and reduced Fe doped $LiNbO_3$ powders. *Crystals* **2018**, *8*, 108. [CrossRef]
23. Kar, S.; Logad, S.; Choudhary, O.P.; Debnath, C.; Verma, S.; Bartwal, K.S. Preparation of lithium niobate nanoparticles by high energy ball milling and their characterization. *Univers. J. Mater. Sci.* **2013**, *1*, 18–24. [CrossRef]
24. Kocsor, L.; Péter, L.; Corradi, G.; Kis, Z.; Gubicza, J.; Kovács, L. Mechanochemical reactions of lithium niobate induced by high-energy ball-milling. *Crystals* **2019**, *9*, 334. [CrossRef]
25. Timoshevskii, A.N.; Ktalkherman, M.G.; Emel'kin, V.A.; Pozdnyakov, B.A.; Zamyatin, A.P. High-temperature decomposition of lithium carbonate at atmospheric pressure. *High. Temperature* **2008**, *46*, 414–421. [CrossRef]
26. Akselrud, L.; Grin, Y. WinCSD: Software package for crystallographic calculations (Version 4). *J. Appl. Crystallogr.* **2014**, *47*, 803–805. [CrossRef]
27. Abrahams, S.C.; Marsh, P. Defect structure dependence on composition in lithium niobate. *Acta Crystallogr. Sect. B Struct. Sci.* **1986**, *42*, 61–68. [CrossRef]
28. Ohgaki, M.; Tanaka, K.; Marumo, F. Structure refinement of lithium (I) niobium (V) trioxide, $LiNbO_3$, withanharmonic thermal vibration model. *Mineral. J.* **1992**, *16*, 150–160. [CrossRef]
29. Zotov, N.; Boysen, H.; Schneider, J.; Frey, F. Application of combined neutron and X-ray powder diffraction refinements to the structure of congruent lithium niobate. *Mater. Sci. Forum* **1994**, *166*, 631–636. [CrossRef]
30. Abrahams, S.C.; Reddy, J.M.; Bernstein, J.L. Ferroelectric lithium niobate. 3. Single crystal X-ray diffraction study at 24 °C. *J. Phys. Chem. Solids* **1966**, *27*, 997–1012. [CrossRef]
31. Iyi, N.; Kitamura, K.; Izumi, F.; Yamamoto, J.K.; Hayashi, T.; Asano, H.; Kimura, S. Comparative study of defect structures in lithium niobate with different compositions. *J. Solid State Chem.* **1992**, *101*, 340–352. [CrossRef]
32. Abrahams, S.C.; Bernstein, J.L. Ferroelectric lithium tantalate. 1. single crystal X-ray diffraction study at 24 °C. *J. Phys. Chem. Solids* **1967**, *28*, 1685–1692. [CrossRef]
33. Abrahams, S.C.; Hamilton, W.C.; Sequeira, A. Ferroelectric lithium tantalate. 2. Single crystal neutron diffraction study at 24 °C. *J. Phys. Chem. Solids* **1967**, *28*, 1693–1698. [CrossRef]
34. Wang, S.Y.; Jiang, H. Synthesis and Characterization of $LiNbO_3$ Powders by Thermal Decomposition Method at Low Temperature. *Ferroelectrics* **2011**, *413*, 212–219. [CrossRef]
35. Gorelik, V.S.; Abdurakhmonov, S.D.; Sidorov, N.V.; Palatnikov, M.N. Raman scattering in lithium niobate and lithium tantalate single crystals and ceramics. *Inorg. Mater.* **2019**, *55*, 524–532. [CrossRef]
36. Ridah, A.; Bourson, P.; Fontana, M.D.; Malovichko, G. The composition dependence of the Raman spectrum and new assignment of the phonons in $LiNbO_3$. *J. Phys. Cond. Matter* **1997**, *9*, 9687–9693. [CrossRef]
37. Repelin, Y.; Husson, E.; Bennani, F.; Proust, C. Raman spectroscopy of lithium niobate and lithium tantalate. Force field calculations. *J. Phys. Chem. Solids* **1999**, *60*, 819–825. [CrossRef]
38. Bartasyte, A.; Plausinaitiene, V.; Abrutis, A.; Stanionyte, S.; Margueron, S.; Boulet, P.; Kobata, T.; Uesu, Y.; Gleize, J. Identification of $LiNbO_3$, $LiNb_3O_8$ and Li_3NbO_4 phases in thin films synthesized with different deposition techniques by means of XRD and Raman spectroscopy. *J. Phys.: Condens. Matter.* **2013**, *25*, 205901. [CrossRef]
39. Verma, N.; Mari, B.; Singh, K.C.; Jindal, J.; Mollar, M.; Rana, R.; Pereira, A.L.J.; Manjón, F.J. Structural and optical properties of $Ta_2O_5:Eu^{3+}$: Mg^{2+} or Ca^{2+} phosphor prepared by molten salt method. *AIP Conf. Proc.* **2016**, *1724*, 020082. [CrossRef]
40. Rüsing, M.; Sanna, S.; Neufeld, S.; Berth, G.; Schmidt, W.G.; Zrenner, A.; Yu, H.; Wang, Y.; Zhang, H. Vibrational properties of $LiNb_{1-x}Ta_xO_3$ mixed crystals. *Phys. Rev. B.* **2016**, *93*, 284305. [CrossRef]
41. Golubović, A.; Gajić, R.; Hinić, I.; Šćepanović, M. IR and Raman spectra of poled and unpoled $LiNbO_3$ single crystals. *J. Alloy. Comp.* **2008**, *460*, 74–78. [CrossRef]
42. Heitjans, P.; Masoud, M.; Feldhoff, M.; Wilkening, M. NMR and impedance studies of nanocrystalline and amorphous ion conductors: Lithium niobate as a model system. *Faraday Discuss.* **2007**, *134*, 67–82. [CrossRef]
43. Chen, R.H.; Chen, L.F.; Chia, C.T. Impedance spectroscopic studies on congruent $LiNbO_3$ single crystal. *J. Phys. Condens. Matter* **2007**, *19*, 086225. [CrossRef]
44. Huanosta, A.; West, A.R. The electrical properties of ferroelectric $LiTaO_3$ and its solid solutions. *J. Appl. Phys.* **1987**, *61*, 5386–5391. [CrossRef]

45. Lanfredi, S.M.; Rodrigues, A.C. Impedance spectroscopy study of the electrical conductivity and dielectric constant of polycrystalline. *J. Appl. Phys.* **1999**, *86*, 2215–2219. [CrossRef]
46. Suhak, Y.; Roshchupkin, D.; Redkin, B.; Kabir, A.; Jerliu, B.; Ganschow, S.; Fritze, H. Correlation of Electrical Properties and Acoustic Loss in Single Crystalline Lithium Niobate-Tantalate Solid Solutions at Elevated Temperatures. *Crystals* **2021**, *11*, 398. [CrossRef]
47. Rahn, J.; Huger, E.; Dorrer, L.; Ruprecht, B.; Heitjans, P.; Schmidt, H. Li self-diffusion in lithium niobate single crystals at low temperatures. *Phys. Chem. Chem. Phys.* **2012**, *14*, 2427–2433. [CrossRef] [PubMed]
48. Weidenfelder, A.; Schulz, M.; Fielitz, P.; Shi, J.; Borchardt, G.; Becker, K.-D.; Fritze, H. Electronic and Ionic Transport Mechanisms of Stoichiometric Lithium Niobate at High-Temperatures. *MRS Online Proc. Libr.* **2013**, *1519*, 330. [CrossRef]
49. Barik, R.; Satpathy, S.K.; Behera, B.; Biswal, S.K.; Mohapatra, R.K. Synthesis and Spectral Characterizations of Nano-Sized Lithium Niobate ($LiNbO_3$) Ceramic. *Micro Nanosyst.* **2020**, *12*, 1–6. [CrossRef]
50. Efremov, V.V.; Shcherbina, O.B.; Palatnikov, M.N.; Masloboeva, S.M. Comparative investigation of electrophysical characteristics of ceramic and single crystal $LiNbO_3$. *J. Phys. Conf. Ser.* **2020**, *1658*, 012010. [CrossRef]
51. El Bachiri, A.; Bennani, F.; Bousselamti, M. Dielectric and electrical properties of $LiNbO_3$ ceramics. *J. Asian Ceram. Soc.* **2016**, *4*, 46–54. [CrossRef]
52. Masoud, M.; Heitjans, P. Impedance Spectroscopy Study of Li Ion Dynamics in Single Crystal, Microcrystalline, Nanocrystalline and Amorphous $LiNbO_3$. *Defect Diffus. Forum* **2005**, *237–240*, 1016–1021. [CrossRef]
53. El Bachiri A.; Bennani F.; Bousselamti M. Dielectric and electrical properties of $LiNbO_3$ ceramics. *J. Asian Ceram. Soc.* **2016**, *4*, 46–54. [CrossRef]

crystals

MDPI

Article

Wavelength-Tunable Nonlinear Mirror Mode-Locked Laser Based on MgO-Doped Lithium Niobate

Lina Zhao *, Luyang Tong, Fangxin Cai, Ye Yuan and Yangjian Cai *

College of Physics and Electronics, Center of Light Manipulations and Applications & Shandong Provincial Key Laboratory of Optics and Photonic Device, Shandong Normal University, Jinan 250014, China; 2020010065@stu.sdnu.edu.cn (L.T.); 2018020558@stu.sdnu.edu.cn (F.C.); 2019020530@stu.sdnu.edu.cn (Y.Y.)

* Correspondence: lnzhao@sdnu.edu.cn (L.Z.); yangjiancai@suda.edu.cn (Y.C.)

Received: 1 September 2020; Accepted: 23 September 2020; Published: 24 September 2020

check for updates

Abstract: We present a high-power, wavelength-tunable picosecond Yb^{3+}: $CaGdAlO_4$ (Yb:CALGO) laser based on MgO-doped lithium niobate (MgO:LN) nonlinear mirror mode locking. The output wavelength in the continuous wave (CW) regime is tunable over a 45 nm broad range. Mode locking with a MgO:LN nonlinear mirror, the picosecond laser is tunable over 23 nm from 1039 to 1062 nm. The maximum output power of the mode-locked laser reaches 1.46 W, and the slope efficiency is 18.6%. The output pulse duration at 1049 nm is 8 ps. The laser repetition rate and bandwidth are 115.5 MHz and 1.7 nm, respectively.

Keywords: mode-locked laser; nonlinear mirror mode locking; lithium niobate

1. Introduction

Diode-pumped all-solid-state ultrafast lasers have received wide attention due to their broad applications in nonlinear frequency conversion, micro-machining, and medicine, since they have the advantages of simplicity, compactness, and variable wavelength [1–5]. The commonly used technology is passive mode locking based on a semiconductor saturable absorber mirror (SESAM), graphene, and a carbon nanotube (CNT) saturable absorber [6–8]. However, a SESAM requires complex fabrication and works under a lower damage threshold. For graphene and CNT saturable absorber mode locking, the power is constricted at the milliwatt level [7,8]. Nonlinear mirror mode locking (NLM) is an alternate passive mode locking technique to obtain ultrafast lasers with higher than average power. A nonlinear mirror consists of a frequency doubling crystal and a dichroic mirror [9]. The dichroic mirror partially reflects the fundamental wavelength (FW) and highly reflects a second harmonic wavelength (SH). In the intracavity, the partial FW is converted into an SH when it first passes the nonlinear crystal, and the SH is totally reflected by the dichroic mirror (DM) while the unconverted FW is partially reflected by the DM. For the backward passing, the SH is back-converted into the FW, due to the different phases of the FW and SH introduced by the air between the nonlinear crystal and the dichroic mirror. For the whole process, the different reflectivity between the FW and the SH leads to a nonlinear positive feedback. The intensity-dependent nonlinear reflectivity results in passive mode locking, thus it acts like a fast saturable absorber with a negative equivalent Im(χ^3). NLM is advantageous with its high damage threshold, large nonlinear loss modulation, and the capability of being able to work at any wavelength for which phase matching is satisfied in the nonlinear crystal's transparent range, thus ensuring a wavelength-tunable laser with high power. The nonlinear crystals for NLM are generally divided into two types: birefringence phase-matching (BPM) crystals (BBO, BiBO, LBO, KTP) [10–16] and quasi-phase-matching (QPM) crystals (periodically poled LN, LT, KTP) [17–22]. The peak power of the output laser is commonly at the kilowatt level while a bulk laser crystal is utilized, as shown in Figure 1. Combined with a thin-disk (TD) laser crystal, NLM could generate ultrafast lasers within several hundreds of femtoseconds and with peak powers up to 10 MW [23,24].

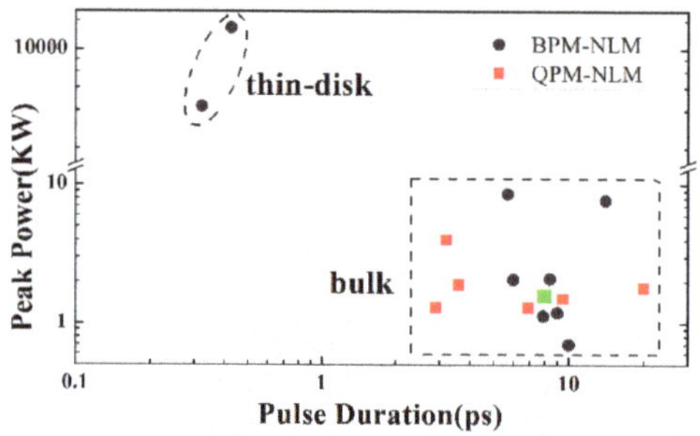

Figure 1. Peak power and pulse duration of NLM lasers around 1 μm. Black dots are NLM lasers based on BPM, red squares are NLM lasers based on QPM, green dot is our work. Dashed oval is thin-disk laser, dashed rectangle is bulk laser.

To date, most of the studies on NLM have focused on a particular wavelength, such as laser sources around 1, 1.3, and 2 μm [10–24]. However, tunable picosecond lasers with output powers at the watt level are also of interest. In this study, we demonstrate a picosecond-tunable Yb:CALGO laser which is achieved by MgO:LN nonlinear mirror mode locking. The wavelength tuning range is 23 nm from 1039 to 1062 nm. The pulse duration is 8 ps at 1049 nm. The average output power reaches 1.46 W, with a slope efficiency of 18.6%. The laser repetition rate and bandwidth are 115.5 MHz and 1.7 nm, respectively.

2. Materials and Methods

The schematic experimental setup of the tunable nonlinear mirror mode-locked Yb:CALGO laser is shown in Figure 2. The gain medium is an a-cut, 3 at.% Yb:CALGO crystal with dimensions of 4 mm × 4 mm × 8 mm, and both end faces are antireflection coated (R < 1%) at the wavelengths of 976 nm and 1030–1080 nm. The pump light is a laser diode (LD) array operating at 976 nm. The maximum output power of the pump light is 30 W. The LD has a fiber with a core diameter of 105 μm and a numerical aperture of 0.22. A 1:1 optical coupling system is used to project the pump light onto Yb:CALGO. M1 is an input mirror with a transmissivity of 98% at 976 nm and a reflectivity of 99.5% at 1030–1080 nm on the left side. Two concave mirrors, M2 and M3, are folding mirrors, with radii of 500 and 200 mm, respectively. M4 is highly reflected at the SH and partially reflected at the FW. M4 and MgO:LN form a nonlinear mirror. A birefringent plate (BF) is used to constrain the laser wavelength within the phase-matching bandwidth of the second harmonic generation process. M1, M2, M3, and M4 form a z-type cavity and the length of the resonant cavity is 1.3 m. The size of MgO:LN is 5 mm × 5 mm × 5 mm (w × h × l), and the phase-matching angle of MgO:LN is cut to be $\theta = 79.2°$, $\varphi = 90°$.

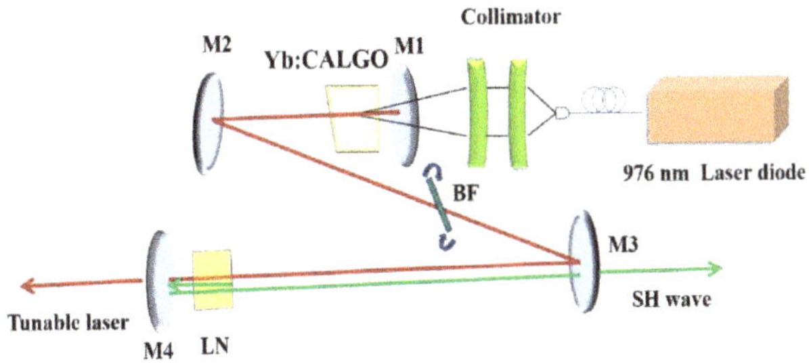

Figure 2. Schematic diagram of the experimental device of Yb:CALGO tunable mode-locked laser. M1 is plane input mirror, M2 and M3 are folding mirrors, M4 is output coupler, LN and M4 form a nonlinear mirror. BF is the birefringent filter.

3. Results

In the experiment, we tested four output couplers (OCs) at the FW (T= 1%, 2%, 5%, 9%). The maximum output power was achieved with the T = 9% OC, thus, subsequent experiments were measured based on the T = 9% OC. The birefringent filter was placed at the Brewster angle and its orientation was then tuned to select the FW. The wavelength-tunable operation of the CW laser was firstly tested under a pump power of 5.8 W. The wavelength was tunable over 45 nm from 1022 nm to 1067 nm, as shown in Figure 3a. The output power achieved its maximum when the wavelength was around 1049 nm, and it decreased when the wavelength was away from 1049 nm on both sides. The wavelength was then fixed at 1049 nm. The oscillation started at a threshold of about 1.8 W, and the maximum output power exceeded 1.7 W under a pump power of 9.8 W, with a slope efficiency of 22.2%. The output power increased linearly as the pumping power increased, as shown in Figure 3b.

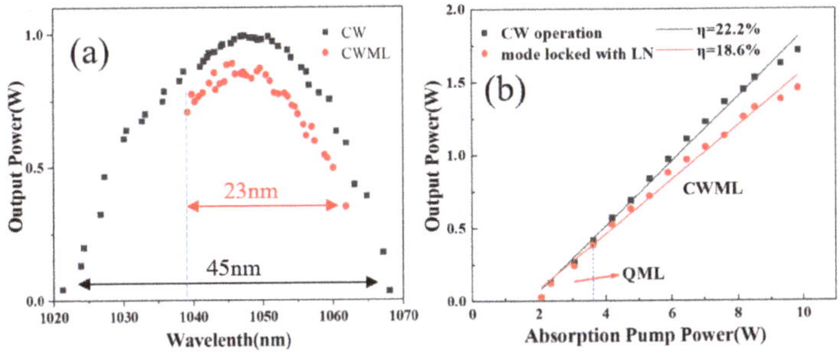

Figure 3. (**a**) Tuning wavelength under CW and CWML operation. (**b**) Output power dependence on absorption pump power during CW and mode-locking operation.

The MgO:LN crystal was then inserted into the cavity and the position of MgO:LN was located near to M4. When MgO:LN was adjusted to the phase-matching angle, a green light transmitted through M3 could be observed. Besides, the relative phase shift between the FW and SH could be adjusted by moving the MgO:LN crystal along the transverse direction to ensure that the SH had the correct phase and efficient back conversion. The lasing threshold was 1.7 W and Q-switch mode locking was observed by the InGaAs detector (EOT, ET-3000, Traverse City, MI, USA) and digital oscilloscope (LeCroy, HDO4104A, New York, NY, USA). When the distance between MgO:LN and the OC was less than 5 mm and the pump power reached 3.6 W, CW mode locking (CWML) was realized. The pulse sequence was stable when the pump power was under 10 W. However, if the pump power exceeded 10 W, the thermal lens effect caused an unstable CWML output. We measured the

CWML output power as the pump power increased from 3.6 to 9.8 W, and the highest output power at 1049 nm was 1.46 W, with a slope efficiency of 18.6%, as shown in Figure 3b.

The tunable wavelength of the mode-locked laser was achieved according to the following procedures: (i) tuning orientation of the birefringent filter (selecting the FW), (ii) adjusting the phase-matching angle of the nonlinear crystal (satisfying the phase-matching condition of SH generation) and transverse position of the nonlinear crystal in millimeters (relative backward phase shift). The mode-locked laser wavelength was tunable over 23 nm from 1039 to 1062 nm (Figure 3a). For a wavelength that was shorter than 1039 nm or longer than 1062 nm, the obvious work-off effect made the mode locking extremely unstable. To obtain a mode-locked laser covering the entire CW range (45 nm), using MgO:LN with another appropriate phase-matching angle is recommended.

The pulse train at the timescale of 40 ns is shown in Figure 4a. The nearby pulse interval was 8.6 ns, which was consistent with the roundtrip time of the 1.3 m cavity length. The pulse train was also traced over 10 ms (Figure 4b), which showed that a pulse-to-pulse amplitude fluctuation was less than 2% at this timescale. The radio frequency (RF) waveform was detected by the spectrum analyzing function of the oscilloscope. As shown in Figure 4c, the fundamental frequency was 115.5 MHz when a resolution bandwidth (RBW) was set to be 5 kHz. According to the resonant cavity length, the frequency was calculated to be 115.38 MHz. The signal-to-noise ratio was 60 dB. Figure 4d shows an RF waveform in the range of 750 MHz with an RBW of 1MHz.

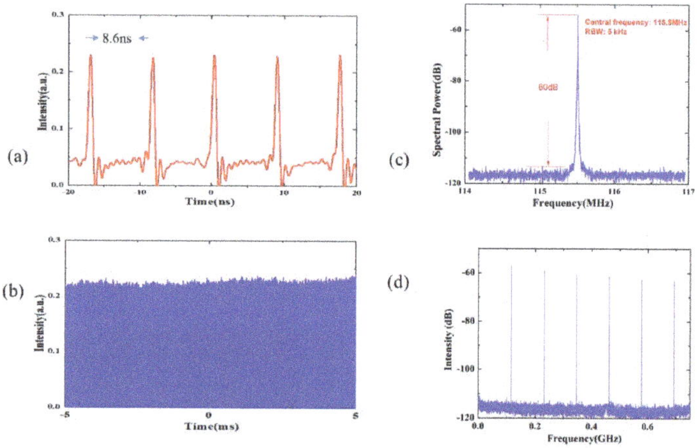

Figure 4. (**a**) The pulse train at a timescale of 40 ns. (**b**) The pulse train at a timescale of 10 ms. (**c**) The RF spectrum of a CWML laser with an RBW of 5 kHz. (**d**) RF spectrum with a span of 750 MHz and an RBW of 1 MHz.

The spectrum property of the CWML operation was measured by a spectrum analyzer (AVASPEC-3648, Apeldoorn, The Netherlands). Figure 5a shows that the output tunable laser wavelength was from 1039 to 1062 nm. The full width at half maximum (FWHM) spectral bandwidth at a representative output wavelength (1049 nm) was ~1.7 nm. The autocorrelation trace was measured using an autocorrelator (APE pulseCheck USB IR, Berlin, Germany), as shown in Figure 5b. Assuming a Gaussian pulse shape, the pulse duration was about 8 ps. The peak power was about 1.58 kW at the maximum output power and the pulse energy was 12.6 nJ, accordingly. As the FW and SH have different group velocities in MgO:LN, the group velocity mismatch (GVM) is the major limiting effect for pulse shortening. Considering the time delay from the GVM between the FW and the SH in MgO:LN, the SH pulse after double passing through MgO:LN (length L_c) is delayed by approximately $\tau_c = 2\delta_c L_c$, where $\delta_c = 1/v_{g2\omega} - 1/v_{g\omega} = 0.6$ ps/mm is the GVM parameter of MgO:LN. For the 5 mm long crystal used, the SH was relatively delayed by 6 ps to the FW. According to the Fourier transform limit, the pulse duration should be 0.95 ps if the spectral bandwidth is 1.7 nm. Taking GVM into account, the final pulse duration was calculated to be 6.95 ps, which was consistent with our experiment result.

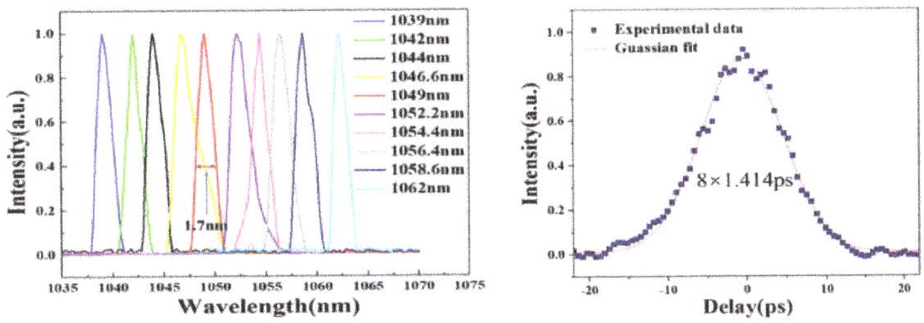

(a) (b)

Figure 5. (**a**) The wavelength-tunable mode-locked pulse spectrum. (**b**) Autocorrelation trace of the mode-locked pulses.

4. Discussion

It is worth noting that, firstly, benefiting from its high damage threshold, the output power of NLM could reach the watt level. It is higher than mode locking using graphene and a CNT absorber, whose output power is commonly limited to the milliwatt level. Secondly, previous works around 1 μm on the NLM technique mostly focused on a particular wavelength (Table 1). However, the capability of being able to work at any wavelength in the nonlinear crystal's transparent range was not fully utilized. Here, Yb:CALGO is utilized as a gain medium, since compared with other laser crystals in Table 1, it has a broad emission spectrum from 990 to 1050 nm [25]. In our experiment, the wavelength for CW operation could be extended to 1067 nm. Moreover, the BF is tuned to select the FW. MgO:LN is used as a nonlinear crystal. The phase-matching angle of MgO:LN could be adjusted to realize high nonlinear conversion efficiency. Therefore, Yb:CALGO combined with a MgO:LN-based NLM technique is an efficient method to obtain a high-power, wavelength-tunable picosecond laser.

Table 1. Passively mode-locked solid-state 1 μm lasers based on NLM.

Nonlinear Crystal	Phase Matching Method	Laser Crystal	Laser Output Characteristics					Reference	Year
			Wavelength (nm)	Pulse Width (ps)	Repetition Rate (MHz)	Power	Peak Power (kW)		
LBO	BPM	Nd:YAG	1064	10	100	0.7 W	0.7 kW	[10]	1994
KTP	BPM	Nd:YVO_4	1064	7.9	150	1.35 W	1.14 kW	[11]	1997
LBO	BPM	Nd:YVO_4	1064	6	200	2.5 W	2.08 kW	[12]	2001
LBO	BPM	Yb:YAG	1031	9	82	0.9 W	1.2 kW	[13]	2003
KTP	BPM	Nd:$GdVO_4$	1063	57	137	11.3 W	1.4 kW	[14]	2010
BiBO	BPM	Nd:$GdVO_4$	1063	5.7	144	7.1 W	8.6 kW	[14]	2010
BiBO	BPM	Nd:YVO_4	1064	14	110	12 W	7.7 kW	[15]	2010
BiBO	BPM	Nd:$GdVO_4$	1063	12.7	100	16.8 W	13 kW	[16]	2012
BBO	BPM	TD-Yb:YAG	1030	0.323	17.8	21 W	3.6 MW	[22]	2017
BBO	BPM	TD-Yb:YAG	1030	0.426	9.3	66 W	16.7MW	[23]	2019
PPKTP	QPM	Nd:YVO_4	1064	2.9	117	0.5 W	1.3 kW	[17]	2010
PPMgSLT	QPM	Nd:$GdVO_4$	1064	3.2	107	1.4 W	4 kW	[18]	2010
PPMgSLT	QPM	Nd:YVO_4	1342	3.6	120	0.8 W	1.9 kW	[19]	2011
PPLN	QPM	Nd:YVO_4	1342	9.5	101	1.52W	1.5 kW	[20]	2011
PPLN	QPM	Nd:YVO_4	1064	2.8	186	1.3 W	2.49 kW	[21]	
MgO:LN	BPM	Yb:CALGO	1039–1062	8	115.5	1.46 W	1.58 kW	Our work	2019

As mentioned above, GVM is the major factor that limits pulse shortening. In MgO:LN, the nonlinear process is o + o→e, where the FW is ordinary polarization and the SH is extraordinary polarization. After first passing MgO:LN, the FW goes faster than the SH due to GVM. If another birefringence crystal is inserted between MgO:LN and M4, and the polarization of the FW is parallel to the slow axis and SH polarization is parallel to the fast axis, the FW will goes slower than the SH. Therefore, the pulses of the FW and SH will overlap again and GVM is compensated [26].

5. Conclusions

In conclusion, we report a MgO:LN nonlinear mirror mode-locked picosecond Yb:CALGO laser. The mode-locked laser wavelength is tunable from 1039 to 1062 nm. The maximum output power of the mode-locked laser reaches 1.46 W at 1049 nm, and the slope efficiency is 18.6%. The output pulse duration is ~8 ps. The repetition rate is 115.5 MHz and the bandwidth of the mode-locked laser is 1.7 nm.

Author Contributions: Writing—review and editing, L.Z., L.T., and Y.C., investigation, Y.Y. and F.C., data collection, L.Z., L.T., and F.C.; All authors have read and agreed to the published version of the manuscript.

Funding: This work was supported by National Natural Science Foundation of China (No. 91950106, 11404196, 11525418, 91750201, 11974218), and National key Research and Development Project of China (2019YFA0705000).

Acknowledgments: Innovation Group of Jinan (2018GXRC010).

Conflicts of Interest: The authors declare no conflict of interest.

References

1. Baxter, J. Optical coherence tomography: Tunable for medicine. *Nat. Photonics* **2012**, *6*, 70.
2. Malinauskas, M.; Žukauskas, A.; Hasegawa, S.; Hayasaki, Y.; Mizeikis, V.; Buividas, R.; Juodkazis, S. Ultrafast laser processing of materials: From science to industry. *Light Sci. Appl.* **2016**, *5*, e16133. [CrossRef] [PubMed]
3. Zhang, F.; Wu, Y.J.; Liu JPang, S.Y.; Ma, F.K.; Jiang, D.P.; Wu, Q.H.; Su, L.B. Mode locked Nd^{3+} and Gd^{3+} co-doped calcium fluoride crystal laser at dual gain lines. *Opt. Laser Technol.* **2018**, *100*, 294–297. [CrossRef]
4. Nie, W.; Jiang, S.; Li, R.; Ren, F.; Clayton, A.H.A.; Juodkazis, S.; Chen, F. Plasmon-induced photoluminescence and Raman enhancement in Pr:CaF_2 crystal by embedded silver nanoparticles. *Appl. Surf. Sci.* **2020**, *530*, 147018. [CrossRef]
5. Zhang, F.; Zhu, H.; Liu, J.; He, Y.; Jiang, D.; Tang, F.; Su, L. Tunable Yb: CaF_2–SrF_2 laser and femtosecond mode-locked performance based on semiconductor saturable absorber mirrors. *Appl. Opt.* **2016**, *55*, 8359–8362. [CrossRef] [PubMed]
6. Lin, H.; Zhang, G.; Zhang, L.; Lin, Z.; Pirzio, F.; Agnesi, A.; Petrov, V.; Chen, W. Continuous-wave and SESAM mode-locked femtosecond operation of a Yb:$MgWO_4$ laser. *Opt. Express* **2017**, *25*, 11827–11832. [CrossRef] [PubMed]
7. Kivisto, S.; Hakulinen, T.; Kaskela, A.; Aitchison, B.; Brown, D.P.; Nasibulin, A.G.; Kauppinen, E.I.; Harkoen, A.; Okhotnikov, O.G. Carbon nanotube films for ultrafast broadband technology. *Opt. Express* **2009**, *17*, 2358–2363. [CrossRef] [PubMed]
8. Sun, Z.; Hasan, T.; Torrisi, F.; Popa, D.; Privitera, G.; Wang, F.; Bonaccorso, F.; Basko, D.M.; Ferrari, A.C. Graphene Mode-Locked Ultrafast Laser. *ACS Nano* **2010**, *4*, 803–810. [CrossRef]
9. Stankov, K.A.; Jethwa, J. A new mode-locking technique using a nonlinear mirror. *Opt. Commun.* **1988**, *66*, 41–46. [CrossRef]
10. Cerullo, G.; Danailov, M.B.; De Silvestri, S.; Laporta, P.; Magni, V.; Segala, D.; Taccheo, S. A diode-pumped nonlinear mirror mode-locked Nd:YAG laser. *Appl. Phys. Lett.* **1994**, *65*, 2392–2394. [CrossRef]
11. Agnesi, A.; Pennacchio, C.; Reali, G.C.; Kubecek, V. High-power diode-pumped picosecond Nd^{3+}:YVO_4 laser. *Opt. Lett.* **1997**, *22*, 1645–1647. [CrossRef]
12. Agnesi, A.; Lucca, A.; Reali, G.; Tomaselli, A. All-solid-state high-repetition-rate optical source tunable in wavelength and in pulse duration. *J. Opt. Soc. Am. B* **2001**, *18*, 286–290. [CrossRef]
13. Saikawa, J.; Taira, T. Second-harmonic nonlinear mirror CW mode locking in Yb:YAG microchip lasers. *Jpn. J. Appl. Phys.* **2003**, *42*, L649–L651. [CrossRef]
14. Thomas, G.M.; Bauerle, A.; Farrell, D.J.; Damzen, M.J. Nonlinear mirror modelocking of a bounce geometry laser. *Opt. Express* **2010**, *18*, 12663–12668. [CrossRef] [PubMed]

15. Thomas, G.M.; Chard, S.P.; Damzen, M.J. High power modelocking of a stigmatic bounce geometry laser using a nonlinear mirror. *Appl. Phys. B* **2010**, *101*, 553–557. [CrossRef]
16. Thomas, G.M.; Omatsu, T.; Damzen, M.J. High-power neodymium-doped mixed vanadate bounce geometry laser, mode locked with nonlinear mirror. *Appl. Phys. B* **2012**, *108*, 125–128. [CrossRef]
17. Iliev, H.; Chuchumishev, D.; Buchvarov, I.; Petrov, V. Passive mode-locking of a diode-pumped Nd:YVO_4 laser by intracavity SHG in PPKTP. *Opt. Express* **2010**, *18*, 5754–5762. [CrossRef]
18. Iliev, H.; Buchvarov, I.; Kurimura, S.; Petrov, V. High-power picosecond Nd:$GdVO_4$ laser mode locked by SHG in periodically poled stoichiometric lithium tantalite. *Opt. Lett.* **2010**, *35*, 1016–1018. [CrossRef]
19. Iliev, H.; Buchvarov, I.; Kurimura, S.; Petrov, V. 1.34-mum Nd:YVO_4 laser mode-locked by SHG-lens formation in periodically-poled stoichiometric lithium tantalite. *Opt. Express* **2011**, *19*, 21754–21759. [CrossRef]
20. Liu, Y.H.; Xie, Z.D.; Pan, S.D.; Lv, X.J.; Yuan, Y.; Hu, X.P.; Lu, J.; Zhao, L.N.; Chen, C.D.; Zhao, G.; et al. Diode-pumped passively mode-locked Nd:YVO_4 laser at 1342 nm with periodically poled $LiNbO_3$. *Opt. Lett.* **2011**, *36*, 698–700. [CrossRef]
21. Lin, S.; Huang, C.H. Effects of nonlinear phase in cascaded mode-locked Nd:YVO_4 laser. *Opt. Express* **2019**, *27*, 504–511. [CrossRef] [PubMed]
22. Cheng, H.; Jiang, X.D.; Hu, X.P.; Zhong, M.L.; Lv, X.J.; Zhu, S.N. Diode-pumped 1988-nm Tm:YAP laser mode-locked by intracavity second-harmonic generation in periodically poled $LiNbO_3$. *Opt. Lett.* **2014**, *39*, 2187–2190. [CrossRef] [PubMed]
23. Saltarelli, F.; Diebold, A.; Graumann, I.J.; Phillips, C.R.; Keller, U. Modelocking of a thin-disk laser with the frequency-doubling nonlinear-mirror technique. *Opt. Express* **2017**, *25*, 23254–23266. [CrossRef] [PubMed]
24. Graumann, I.; Saltarelli, F.; Lang, L.; Wittwer, V.J.; Sudmeyer, T.; Phillips, C.R.; Keller, U. Power-scaling of nonlinear-mirror modelocked thin-disk lasers. *Opt. Express* **2019**, *27*, 37349–37363. [CrossRef] [PubMed]
25. Petit, J.; Goldner, P.; Viana, B. Laser emission with low quantum defect in Yb:$CaGdALO_4$. *Opt. Lett.* **2005**, *30*, 1345–1347. [CrossRef]
26. Cerullo, G.; Magni, V.; Monguzzi, A. Group-velocity mismatch compensation in continuous-wave lasers mode locked by second-order nonlinearities. *Opt. Lett.* **1995**, *20*, 1785–1787. [CrossRef]

crystals

Article

Electro-Optic Control of Lithium Niobate Bulk Whispering Gallery Resonators: Analysis of the Distribution of Externally Applied Electric Fields

Yannick Minet [1,2], Hans Zappe [2], Ingo Breunig [1,3,*] and Karsten Buse [1,3]

1 Laboratory for Optical Systems, Department for Microsystems Engineering—IMTEK, University of Freiburg, Georges-Köhler-Allee 102, 79110 Freiburg, Germany; yannick.minet@imtek.uni-freiburg.de (Y.M.); karsten.buse@ipm.fraunhofer.de (K.B.)
2 Gisela and Erwin Sick Chair of Micro-Optics, Department for Microsystems Engineering—IMTEK, University of Freiburg, Georges-Köhler-Allee 102, 79110 Freiburg, Germany; zappe@imtek.uni-freiburg.de
3 Fraunhofer Institute for Physical Measurement Techniques IPM, Georges-Köhler-Allee 301, 79110 Freiburg, Germany
* Correspondence: optsys@ipm.fraunhofer.de

Abstract: Whispering gallery resonators made out of lithium niobate allow for optical parametric oscillation and frequency comb generation employing the outstanding second-order nonlinear-optical properties of this material. An important knob to tune and control these processes is, e.g., the linear electro-optic effect, the Pockels effect via externally applied electric fields. Due to the shape of the resonators a precise prediction of the electric field strength that affects the optical mode is non-trivial. Here, we study the average strength of the electric field in z-direction in the region of the optical mode for different configurations and geometries of lithium niobate whispering gallery resonators with the help of the finite element method. We find that in some configurations almost 100% is present in the cavity compared to the ideal case of a cylindrical resonator. Even in the case of a few-mode resonator with a very thin rim we find a strength of 90%. Our results give useful design considerations for future arrangements that may benefit from the strong electro-optic effect in bulk whispering gallery resonators made out of lithium niobate.

Keywords: electro-optics; whispering gallery resonators; lithium niobate

Citation: Minet, Y.; Zappe, H.; Breunig, I.; Buse, K. Electro-Optic Control of Lithium Niobate Bulk Whispering Gallery Resonators: Analysis of the Distribution of Externally Applied Electric Fields. *Crystals* **2021**, *11*, 298. https://doi.org/10.3390/cryst11030298

Academic Editor: Gábor Corradi

Received: 26 February 2021
Accepted: 14 March 2021
Published: 17 March 2021

1. Introduction

Non-centrosymmetric materials offer second-order non-linearities that allow for, e.g., second harmonic generation, optical parametric oscillation and difference frequency generation. Furthermore, such materials bear the linear electro-optic effect, the so-called Pockels effect and the piezo-electric effect [1]. Due to this, non-centrosymmetric materials are the centerpiece of many setups for optical frequency conversion, e.g., for optical parametric oscillators for mid-infrared spectroscopy and frequency comb conversion into the molecular footprint region [2,3], as a single photon source for quantum imaging [4] and infrared spectroscopy [5], to generate terahertz radiation [6] and to generate tunable continuous wave light in the visible spectrum [7]. One of the most prominent non-centrosymmetric materials is lithium niobate. Due to high second-order nonlinear-optical coefficients, high transparency in the visible and infrared optical wavelengths and large Pockels coefficients lithium niobate is an excellently suited optical material.

In the last two decades whispering gallery resonators (WGRs) have developed to a well established and powerful platform employing non-centrosymmetric materials for frequency conversion. The WGRs are monolithic resonators with spheroidal geometry manufactured from the bulk material or chip-integrated on wafers. They provide a high quality factor making them particularly interesting for non-linear optical applications. Moreover the mirror-less, monolithic devices reduce the footprint drastically compared with other resonator design schemes. The remarkable features of this platform are demonstrated in many experiments such as frequency comb generation based on third-order [8,9], and second-order optical non-linearity [10,11], electro-optically generated combs [12,13], optical parametric oscillation, second harmonic generation [14] and as a platform for quantum optics [15]. All applications can benefit from the Pockels effect. In the simplest case, the resonance frequency of the WGR can be stabilized to the pump laser frequency. In Figure 1a we illustrate a typical setup used in experiments recently that involve the linear electro-optic effect in bulk WGRs, e.g., electro-optic combs [13], second-order-non-linearity-based frequency combs [10], efficient microwave-to-optical photon conversion [16] and adiabatic frequency conversion [17]. The main principle can be summarized as follows: By generating an electric field in the resonator, the refractive index is changed and hence the resonance frequency due to the Pockels effect according to [18]:

$$\Delta\nu_0 = \frac{n^2}{2}\nu_0 r_i E_z \tag{1}$$

there n is the refractive index, ν_0 the WGR resonance frequency, E_z the electric field and r_i the electro-optic coefficient. This way, for example, the second harmonic generation light can be maximized [19]. Furthermore, the frequency comb threshold via second harmonic generation can be minimized [20]. For the Pockels-effect-based adiabatic frequency conversion the resonance frequency is changed with a voltage pulse faster than the photon life time. This allows to generate frequency shifts of several GHz within nanoseconds [17]. Also an electro-optically tunable higher-order optical filter made out of lithium niobate WGRs has made use of the linear electro-optic effect [18]. Knowledge of the electric field strength at the position of the light field and its homogeneity are needed to be able to compare the experimental data and theoretical predictions [15]. In the case of a cylindrical WGR, as illustrated in Figure 1b, estimating the electric field strength E_z would be simple. According to the plate capacitor formula, we obtain

$$E_0 = \frac{U}{\Delta z} \tag{2}$$

with the thickness Δz and the electric potential difference U. A straight sidewall would lead to an optical mode extended in z-direction up to the electrodes of the resonators, which would cause a damping of the light field. A WGR possesses a spheroidal geometry as shown in Figure 1. This makes the prediction of the electric field at the location of the optical mode non-trivial. The optical mode is not located between the electrodes of the resonator, the light traveling around the rim experiences in general a weaker externally generated electric field $E_z < E_0$. Furthermore, the electric field can become inhomogeneous through the cross section of the mode.

The aim of our work is to analyze the electric field strength E_z in the area of the optical mode and its homogeneity. The strength of E_z addresses the two commonly used Pockels-coefficients of LN, r_{31} and r_{33} [17,19,20]. They change the refractive index for light polarized perpendicularly to and parallel to the symmetry axis for z-cut lithium niobate. Furthermore, due to symmetry properties of lithium niobate this orientation is the best suited for most applications in bulk WGRs. We analyze the electric field distribution in two common mm-sized WGR configurations, for conventional geometries as well as the so called few- or single mode geometry [21,22] made out of z-cut bulk lithium niobate. By introducing a small bulge, single mode operation in bulk WGRs is possible [21,22]. Furthermore we analyze the effect of introducing a distance between the rim of the resonator and the beginning of the electrode to prevent electric breakdowns in air when applying high voltages to the electrodes of the resonator. Additionally, we consider two different dielectric constants, for the unclamped ϵ^{UC} and for the clamped ϵ^{C} case. In the experiments mentioned before different frequencies of the external electric fields were used. For external electric fields with a frequency well below 1 MHz we assume the unclamped case with ϵ^{UC}. But for example in the Pockels-based adiabatic frequency conversion frequencies of the external electric field are significantly higher, and the resonator is effectively clamped, partially or fully making ϵ^C applicable in the latter case. Due to this we simulate with ϵ^C. Although the dielectric constant for the unclamped and clamped case in z-direction is $\epsilon_{33} \approx 30$, the radial component ϵ_{11} changes from 84 to 43. This will influence the electric field strength at the optical mode.

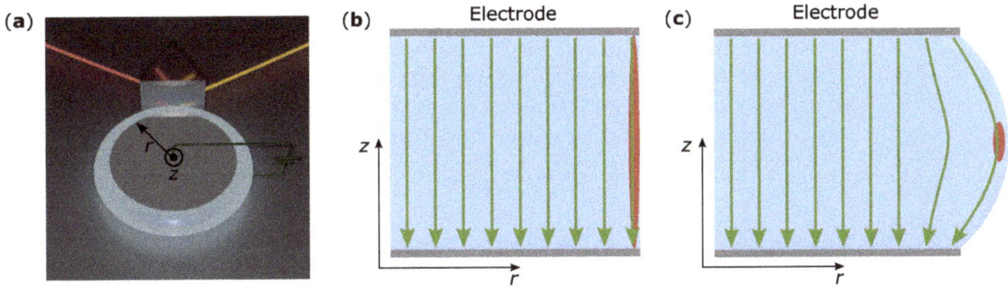

Figure 1. (**a**) Artist drawing of a setup used in recent experiments of frequency conversion schemes involving bulk lithium niobate whispering gallery resonators involving the electro-optic effect. The laser light is coupled into the WGR via a coupling prism. The resonator is coated with a metal electrode. (**b**,**c**) The electric field distribution indicated by the green lines for a cylindrical resonator with an idealized homogeneous field and for a commonly used resonator geometry. The red area indicates the size and position of the optical mode, not to scale.

2. Materials and Methods

The simulations in this paper were performed using the finite element method [23]. We used the commercially available software COMSOL Multiphysics to calculate the external electric field and the spatial distribution of the optical mode. Since the WGRs show rotation symmetry with respect to the z-axis we can reduce the computational effort drastically by solving the problem in a 2D section. As illustrated in Figure 2 both geometries considered in this work include a lithium niobate WGR coated at the +z-side and −z-side with metal electrodes to allow the application of an external electrical field. The WGR is surrounded by air. For the permittivity of lithium niobate we consider two different cases. The unclamped case with ϵ^{UC} with a radial component $\epsilon_{11}^{UC} = 84$ and for the z-axis direction $\epsilon_{33}^{UC} = 30$. For the clamped case $\epsilon_{11}^{C} = 43$ and $\epsilon_{33}^{C} = 29.6$ [24]. We furthermore assume that no space charge fields are present in the crystals. One needs to be aware that all lithium niobate crystals exhibit some photo-refractive sensitivity that stems from light induced space charge fields. Thus for intense and blue/ultraviolet light space charges may modify the electric fields in the regions of the optical modes, most likely reducing the fields because of the generated photoconductivity.

We perform calculations using two different types of geometries. The first one is shown in Figure 2a and very typical [11,16,17,20]. Here most resonators are mm-sized and have a thickness of a few hundred micrometers. For a lithium niobate WGR which should couple extraordinarily polarized light of a wavelength of $\lambda = 1$ µm via a rutile prism, a ratio of the major radius R_1 and the minor radius R_2 of $R_2/R_1 \approx 0.4$ is a good choice for a sufficient coupling efficiency [25]. For the simulations shown in Section 3.1 using this ratio, we vary the thickness between a minimum of 50 µm and a maximum of $\Delta z = 2 \times R_2$ for major radii of $R_1 = 100$, 500 and 1000 µm. In case of $\Delta z = 2 \times R_2$ we obtain a semicircular bulge. Furthermore we investigate the effect of a gap between the resonator rim and the beginning of the electrode. To this end we introduce a distance Δd between the rim of the resonator and the beginning of the electrode for a WGR with $\Delta z = 500$ µm and $R_1 = 1000$ µm as is illustrated in Figure 2a. We define the rim as the spatial position where the curvature begins. This investigation we consider to be important for two different reasons: (1) Lithium niobate can withstand very high electric field strengths of up to 65 kV mm^{-1} [26]. However, air has a breakdown voltage of 3 kV mm^{-1} only, which limits the maximum voltage that can be applied to the resonator [27]. The shortest path between the electrodes through the air can be enlarged and hence the maximum voltage until an electric breakdown occurs. Furthermore, insulators can be applied to further increase the breakdown voltage. However, smaller electrodes reduce the field at the rim for a given voltage and it is open which effect will win. (2) This gap can also stem from the fabrication process of the WGR. During the shaping and polishing of the resonator the electrode can be partially removed. By introducing the gap Δd we want to account for this, by either applying a gap just to the top electrode , in the following referred to as asymmetric case or to both electrodes, referred to as symmetric case in the following. Since the bulk WGRs are typically mounted on a post and therefore one side is protected, we consider the asymmetric case for a damage of the electrode during the fabrication process. The symmetric case is considered when applying an isolation layer on both electrodes.

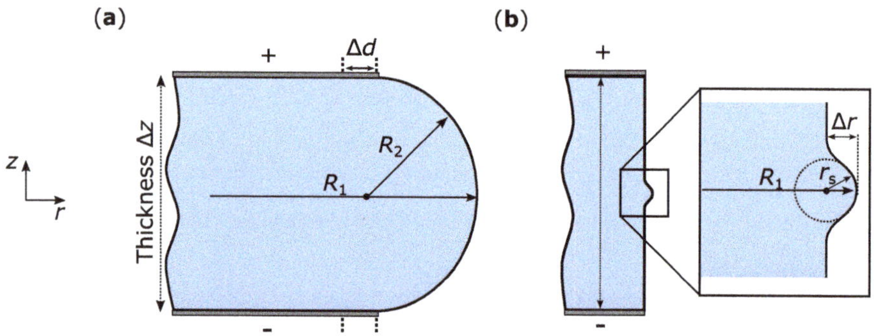

Figure 2. Geometries used for the simulation. We assume an axially symmetrical 2D model. The resonator has at the bottom and top a metal electrode. In (**a**) the geometry with a constant ratio is illustrated. Here R_1 is the major radius, R_2 the minor radius and Δz the thickness. For this shape we also investigated a variation of the distance Δd between the rim of the resonator and the electrode. (**b**) The few-mode geometry. The small bulge at the equator of the WGR is used to guide the light. Here r_s is the radius of the small circle and the distance Δr between the rim of the resonator and R_1.

Another investigated geometry is that of few-mode WGRs. As shown in Figure 2b, this is characterized by a small convex bulge at the equator of the resonator. The major radius R_1 of the resonator is always kept at 1 mm. The shape of the small bulge is characterized by two parameters. The first, r_s, describes the radius of the circle, and Δr, which characterizes the distance of the bulge from the straight part of the resonator. We keep the radius $r_s = 5\,\mu m$ constant, while the distance Δr is varied. We simulate for thicknesses of $\Delta z = 500\,\mu m$ and $\Delta z = 50\,\mu m$. In our model we approximate the shape of the bulge with a semicircle. In order to provide a smooth transition between semicircular bulge and the bulk material, the shape of the bulge is approximated by a cubic Bézier curve. The formula with weights can be found in the Appendix A.

The following simulation and analysis procedure is performed: For each geometrical configuration we perform an eigenfrequency analysis to compute the intensity distribution of the fundamental mode. In a second step we compute the electric field distribution for a voltage of $U = 1\,V$ applied to the electrodes. This is done by solving the electrostatic Maxwell equations. Finally, we determine for each geometry the average strength of the external electric field and the distribution in the area of the optical mode where the intensity is larger than 10% of the maximum intensity of the mode. We always use a wavelength $\lambda = 1\,\mu m$ and the refractive index n of 2.16 [28]. In the simulation we use a free triangular mesh with a locally dense meshing. Inside the WGR the maximum element size is set to $4\,\mu m$. This value decreases towards the position of the optical mode. Here, the maximum size is set to $\lambda/10$, i.e., to 100 nm.

3. Results

The results section is divided in two parts: The first part is dealing with the simulation in the conventional resonator geometry with different ratios of the thicknesses Δz and the major radii R_1. Additionally we discuss the influence of a gap Δd between the rim of the resonator and the electrode. In the second part we discuss few-mode resonators for different Δr and Δz.

3.1. Conventional Resonator Geometry

In Figure 3 the electric field distribution in z-direction for two exemplary thicknesses $\Delta z = 500\,\mu\text{m}$ and $\Delta z = 50\,\mu\text{m}$ for $R_1 = 1000\,\mu\text{m}$ is shown. In both cases the permittivity is assumed to be ϵ^C. This results in a ratio of $\Delta z/R_1 = 0.5$ in the first and $\Delta z/R_1 = 0.05$ in the second example.

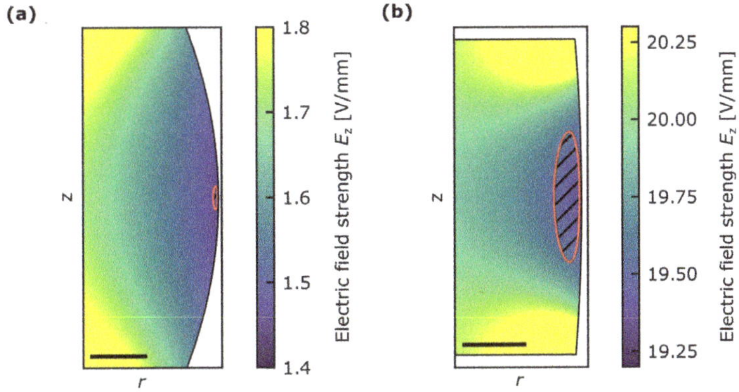

Figure 3. Two examples of electric field distributions E_z for different thicknesses. The distribution is plotted based on ϵ^C. The red area indicates where the intensity of the fundamental optical mode is larger than 10% of its maximum. The black bar has in (**a**) a size of 50 µm and in (**b**) of 10 µm. For better visibility the range of field values is limited such that the region of interest provides enough contrast.

Comparing the field strengths in Figure 3a,b we obtain the expected behavior, that with decreasing thickness and a constant voltage applied to the electrodes, the electric field increases. Since the ratio R_2/R_1 is constant, the horizontal distance between the electrode and the position of the optical mode decreases, while the size of the mode is constant for a given R_1. This gives an additional gain in the effective electric field strength. In Figure 4a the average value of E_z is shown for the investigated thicknesses between $\Delta z = 800\,\mu\text{m}$ and $\Delta z = 50\,\mu\text{m}$ of a WGR with $R_1 = 1000\,\mu\text{m}$ in the unclamped case. Furthermore, in Figure 4b we compare our electric field strength determined for the unclamped and clamped cases with the field strength of a respective plate capacitor for different ratios $\Delta z/R_1$ and for radii of $R_1 = 100$, 500 and 1000 µm.

For the resonator with $\Delta z/R_1 = 0.5$ we obtain an electric field strength of 79.5% (72.5%) of the strength a plate capacitor of similar thickness would have. With decreasing thickness the mismatch becomes less, ending up with a strength of 98% (97%) of the case of a respective plate capacitor for $\Delta z/R_1 = 0.05$. The values within brackets represent the clamped case. To quantify the homogeneity of the external applied electric field we calculate the maximum values of the external electric field $E_{z,\text{max}}$ and the minimum values $E_{z,\text{min}}$ in the considered area of the optical mode. Then we derive $\Delta E_z/E_z = (E_{z,\text{max}} - E_{z,\text{min}})/E_z$. It is as small as 2% for all investigated thicknesses in case of the fundamental optical mode.

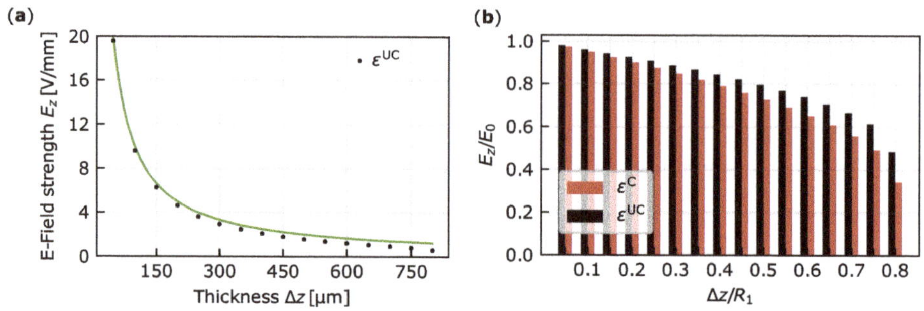

Figure 4. (**a**) The average electric field strength in the area where the intensity is larger than 10 % of its maximal intensity of the fundamental optical mode for $R_1 = 1000$ μm and different thicknesses Δz. The green line indicates the field strength E_z according to $E_z = E_0 = U/\Delta z$. (**b**) Comparison between the electric field E_z and the respective field in a plate capacitor E_0 for different $\Delta z/R_1$. The error bars have point size.

In the next set of simulations we determine the average strength of E_z for the variation of the distance Δd as indicated in Figure 2 for a standard 500-μm-thick resonator. The results are shown in Figure 5a. For each Δd we calculate E_z for both dielectric tensor values for lithium niobate.

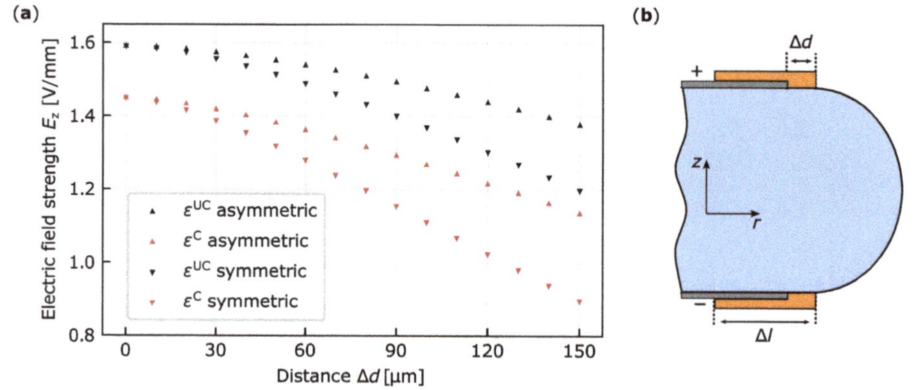

Figure 5. (**a**) The electric field strength in the area of the optical mode for different distances Δd of the electrode from the rim of the resonator between 0 and 150 μm. The major radius of the WGR is $R_1 = 1000$ μm and the thickness $\Delta z = 500$ μm. (**b**) Schematic of a possible configuration with an additional isolation layer with length Δl covering the top and the bottom electrode.

We investigate the symmetric case, where the electrodes on the top and bottom move both the same way, as well as the asymmetric case, where only the top electrode changes its distance from the rim. In all cases the strength of E_z decreases nearly linearly after $\Delta d = 30$ μm. Our measure for the field inhomogeneities $\Delta E_z/E_z$ stays below 2% across the area of the fundamental mode even for increasing values of distance Δd. To estimate the possible gain in the breakdown voltage we calculate the shortest distance of the electrodes for this configuration through the air, assuming this to be the relevant distance. For an electrode reaching the rim we obtain an arc length of 0.54 mm. A straightforward way of increasing this distance and thus reducing the risk of a breakdown in air would be to apply isolating layers with a very high dielectric strength in the gaps Δd and on top of both electrodes as shown in Figure 5b. As an example, we assume a $\Delta l = 300$ μm long layer on both sides. Hence the relevant distance between the electrodes doubles. Assuming that $\Delta d = 20$ μm is needed to apply the isolation layer sufficiently, we obtain a decrease in the electric field strength assuming ϵ^{UC} of 1% only. In order to be able to contact the resonator the isolation layer cannot cover the whole electrode.

3.2. Few-Mode Resonator

Two examples for the obtained strength distributions of the electric field E_z for the few-mode geometry are displayed in Figure 6 for a resonator with $R_1 = 1000\,\mu m$ and $\Delta z = 500\,\mu m$. For $\Delta r = 1\,\mu m$ the optical mode is almost completely located in the part of the resonator which is between the electrodes, while for $\Delta r = 5\,\mu m$ the optical mode is located in the bulge outside the electrodes. The change of the mean electric field strength E_z in the area of the optical mode with increasing Δr from 1 to 5 µm is shown in Figure 7. Additionally, the number of the obtained transversal modes for each configuration is displayed. The size of the error bar is given as $\Delta E_z = E_{z,max} - E_{z,min}$, with the minimum values of the external applied electric field $E_{z,min}$ and the maximum value $E_{z,max}$ in the considered area of the optical mode. For $\Delta r = 0$ and 0.5 µm no solution for the optical mode could be found in our simulation. For each Δr we calculated E_z for both dielectric constants. In Figure 7b we compare both E_z to the respective field strength E_0 of a plate capacitor with the same thickness. The highest values for 91.2% (90%) are obtained for $\Delta r = 1\,\mu m$. The value in the brackets represents the determined strength for ϵ^{C}. At $\Delta r = 5\,\mu m$ we determine for ϵ^{UC} a relative strength of $E_z/E_0 = 60\%$ which is 9.5% larger than that for ϵ^{C}. The relative electric field inhomogeneity $\Delta E_z/E_z$ across the area of the fundamental mode for $\Delta r = 5\,\mu m$ assuming ϵ^{UC} is determined to be 24% and for ϵ^{C} 41% of average value. When the thickness of the resonator is reduced to $\Delta z = 50\,\mu m$, the relative strength E_z/E_0 changes in the sub-percent level compared to the WGR with $\Delta z = 500\,\mu m$. Also the field homogeneity does not change significantly.

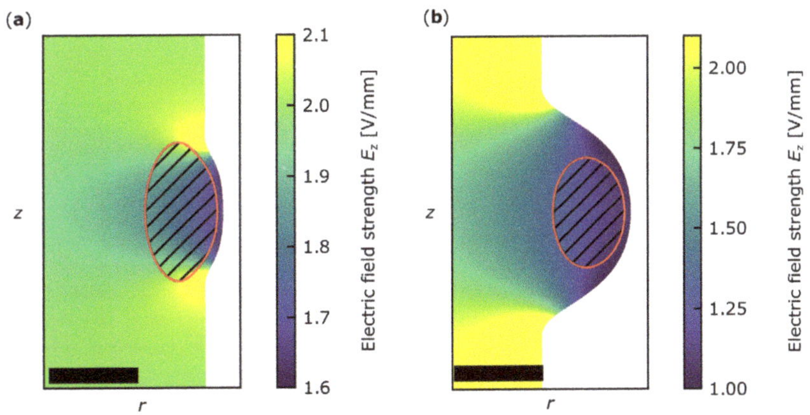

Figure 6. Two examples for different Δr in the case of a few-mode WGR with a semicircular bulge with radius $r_s = 5\,\mu m$, $R_1 = 1000\,\mu m$ and $\Delta z = 500\,\mu m$. The electric field strength is plotted using the unclamped dielectric constant. We simulate in (**a**) $\Delta r = 1\,\mu m$ and in (**b**) $\Delta r = 5\,\mu m$. The color code represents the electric field strength E_z. The red area indicates the area where the intensity of the fundamental optical mode is larger than 10% of its the maximum value. The black bar has a size of 5 µm. For better visibility the range of E_z field values is limited such that the region of interest provides enough contrast.

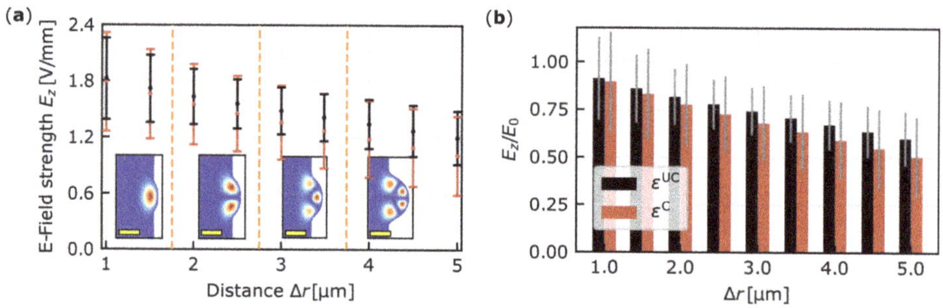

Figure 7. (**a**) The average electric field strength E_z in the few-mode resonator with small radius $r_s = 5\,\mu m$ in the area where the normalized intensity of the fundamental optical mode is larger than 10%. The red data points represent the strength E_z with ϵ^C and black the one with ϵ^{UC}. The error bars represent ΔE_z. The inset show the highest possible optical mode. The yellow bar has a size of 5 µm. (**b**) Comparison between the electric field E_z and the respective field in a plate capacitor E_0. The gray bars as a measure for the field inhomogeneity are given by $\Delta E_z / E_0$.

4. Discussion

For the conventional WGR geometry the performed simulations validate that with decreasing thickness the electric field strength is increasing. This is so far the expected behavior considering a simple plate capacitor model. Depending on the dielectric constant used, E_z develops differently. For the unclamped case, ϵ^{UC} is twice as large in the radial component as ϵ^C and the E-field lines are thus more strongly drawn into the bulge providing a larger field strength and higher homogeneity of the field. For example for a WGR with $\Delta z/R_1 = 0.5$ the deviation from the respective electric field strength of a plate capacitor between both WGRs is just 7%. This difference decreases to 0.7% at $\Delta z/R_1 = 0.05$. The homogeneity of E_z considering the difference between the minimum and the maximum value is less than 2% for each $\Delta z/R_1$ constant. We investigated for three different radii different ratios of $\Delta z/R_1$ as long as the minimal dimensions of $\Delta z = 50\,\mu m$ and $R_1 = 100\,\mu m$ are not violated. Due to the characteristics of the electric field we can use this ratio to determine the behavior of the electric field E_z for typical dimensions of bulk WGRs made out of lithium niobate.

For a WGR with $R_1 = 1000\,\mu m$ and $\Delta z = 500\,\mu m$ and with a gap Δd we determine a reduction of the average electric field strength E_z of less than 2%, if Δd is less than 30 µm in the asymmetric case and 20 µm in the symmetric case. Also here for the unclamped case the reduction is lower compared to the clamped case. The relative inhomogeneity of the electric field $\Delta E_z / E_z$ is for all investigated cases less than 2%. These findings confirm that small damages of the electrode during the fabrication process have only minor impact on E_z. We find that it is possible to increase the effective distance between the electrodes and thus increase the voltage at which the electrical breakdown in air takes place significantly by applying an isolation layer with only a minor decrease of the effective electric field E_z at the cross section of the optical mode.

For the few-mode resonator the difference between the electric field strength E_z for the clamped and unclamped dielectric tensor is larger compared with that for the conventional geometry. For this micrometer-sized bulge the effect of a twice as large radial component of ϵ^{UC} compared with ϵ^C comes obviously more into play as our results show. The larger Δr, the greater the difference between both values is. For $\Delta r = 5\,\mu m$ the difference between both strengths of the dielectric tensors is 9.4%. Also the relative field inhomogeneity $\Delta E_z / E_z$ is larger compared to that of the conventional WGR geometry and varies between 17% and 23% for ϵ^{UC} and from 28% up to 42% for ϵ^C. The present study has only investigated the dielectric constants for the unclamped and clamped cases. Depending on the frequency that is used for the external electric field, one might excite resonances. For lithium niobate we expect these resonances for frequencies of some MHz [29]. Here the values for the dielectric constant might differ drastically. So far we discussed only the fundamental optical mode. Higher-order modes in conventional resonators will have a reduced homogeneity since they are spatially more expanded compared to the fundamental optical mode. However, the next higher order of transversal modes does not obtain significant deviation from the respective values of the fundamental mode with respect to strength and homogeneity. Applying an external electric field to a lithium niobate WGR due to the piezoelectric effect the resonance frequency is changed as well. However, this contribution is more than one order of magnitude lower than the frequency shift induced by the Pockels effect and can be neglected in most cases [17,19,20].

5. Conclusions

To conclude: For all geometries used most of the expected electrical field is indeed present in the region where the light circulates. However, for small rims and overlaying rims just half of the field expected from the plate capacitor formula reaches the relevant region. The inhomogeneity of the electric field distribution is in most cases excellent amounting to just 2% or less, again except for small and overlaying rims where several 10% of inhomogeneity might be present. The cases of the clamped versus the free resonators for quickly/slowly varying external applied electric fields yield differences of the order of 10% or more for the fields achieved. The simulations provide relevant information how to optimally design the resonators for Pockels tuning and to compare more precisely experimental data with theoretical expectations.

Author Contributions: Formal analysis, Y.M.; writing—review and editing, Y.M., I.B., H.Z. and K.B.; supervision, I.B., H.Z. and K.B. All authors have read and agreed to the published version of the manuscript.

Funding: This research received no external funding.

Institutional Review Board Statement: Not applicable.

Informed Consent Statement: Not applicable.

Data Availability Statement: The data presented in this study are available on request from the corresponding author.

Acknowledgments: Y.M. gratefully acknowledges financial support from the Gisela und Erwin Sick Stiftung. The authors thank Jan Szabados for helpful discussions.

Conflicts of Interest: The authors declare no conflict of interest.

Abbreviations

The following abbreviation are used in this manuscript:

Whispering gallery resonator	WGR

Appendix A. Bézier Curve

We used the following cubic Bézier curve to model the small bulge symmetrically around the center $z = 0$:

$$\mathrm{b}(t) = \frac{\sum_{i=0}^{3} \mathrm{b}_i w_i B_i^3(t)}{\sum_{i=0}^{3} w_i B_i^3(t)}, \quad t \in [0,1]. \tag{A1}$$

$B_i^p(t)$ are Bernstein basis function, with control points in the 2D euclidean space $\begin{pmatrix} r \\ z \end{pmatrix}$,
$b_0 = \begin{pmatrix} R_1 - \Delta r \\ \sqrt{r_s^2 - (r_s - \Delta r)^2}\sqrt{2} \end{pmatrix}$, $b_1 = \begin{pmatrix} R_1 - \Delta r \\ \sqrt{r_s^2 - (r_s - \Delta r)^2} \end{pmatrix}$, $b_2 = \begin{pmatrix} R_1 \\ \sqrt{r_s^2 - (r_s - \Delta r)^2} \end{pmatrix}$,
$b_3 = \begin{pmatrix} R_1 \\ 0 \end{pmatrix}$, all weight $w_i = 1$.

References

1. Boyd, R.W. *Nonlinear Optics*, 3rd ed.; Academic Press: Burlington, MA, USA, 2008.
2. Schliesser, A.; Picqué, N.; Hänsch, T.W. Mid-infrared frequency combs. *Nat. Photonics* **2012**, *6*, 440–449. [CrossRef]
3. Vainio, M.; Halonen, L. Mid-infrared optical parametric oscillators and frequency combs for molecular spectroscopy. *Phys. Chem. Chem. Phys.* **2016**, *18*, 4266–4294. [CrossRef] [PubMed]
4. Lemos, G.B.; Borish, V.; Cole, G.D.; Ramelow, S.; Lapkiewicz, R.; Zeilinger, A. Quantum imaging with undetected photons. *Nature* **2014**, *512*, 409–412. [CrossRef] [PubMed]
5. Kalashnikov, D.A.; Paterova, A.V.; Kulik, S.P.; Krivitsky, L.A. Infrared spectroscopy with visible light. *Nat. Photonics* **2016**, *10*, 98–101. [CrossRef]
6. Sowade, R.; Breunig, I.; Camara Mayorga, I.; Kiessling, J.; Tulea, C.; Dierolf, V.; Buse, K. Continuous-wave optical parametric terahertz source. *Opt. Express* **2009**, *17*, 22303. [CrossRef]
7. Sowade, R.; Kießling, J.; Breunig, I. One light source—All colors. *Opt. Photonik* **2013**, *8*, 46–49. [CrossRef]
8. He, Y.; Yang, Q.F.; Ling, J.; Luo, R.; Liang, H.; Li, M.; Shen, B.; Wang, H.; Vahala, K.; Lin, Q. Self-starting bi-chromatic $LiNbO_3$ soliton microcomb. *Optica* **2019**, *6*, 1138. [CrossRef]
9. Wang, C.; Zhang, M.; Yu, M.; Zhu, R.; Hu, H.; Loncar, M. Monolithic lithium niobate photonic circuits for Kerr frequency comb generation and modulation. *Nat. Commun.* **2019**, *10*, 978. [CrossRef]
10. Szabados, J.; Puzyrev, D.N.; Minet, Y.; Reis, L.; Buse, K.; Villois, A.; Skryabin, D.V.; Breunig, I. Frequency comb generation via cascaded second-order nonlinearities in microresonators. *Phys. Rev. Lett.* **2020**, *124*, 203902. [CrossRef]
11. Hendry, I.; Trainor, L.S.; Xu, Y.; Coen, S.; Murdoch, S.G.; Schwefel, H.G.L.; Erkintalo, M. Experimental observation of internally pumped parametric oscillation and quadratic comb generation in a $\chi(2)$ whispering-gallery-mode microresonator. *Opt. Lett.* **2020**, *45*, 1204–1207. [CrossRef]
12. Zhang, M.; Buscaino, B.; Wang, C.; Shams-Ansari, A.; Reimer, C.; Zhu, R.; Kahn, J.M.; Lončar, M. Broadband electro-optic frequency comb generation in a lithium niobate microring resonator. *Nature* **2019**, *568*, 373–377. [CrossRef]
13. Rueda, A.; Sedlmeir, F.; Kumari, M.; Leuchs, G.; Schwefel, H.G.L. Resonant electro-optic frequency comb. *Nature* **2019**, *568*, 378–381. [CrossRef]
14. Breunig, I. Three-wave mixing in whispering gallery resonators. *Laser Photonics Rev.* **2016**, *10*, 569–587. [CrossRef]
15. Strekalov, D.V.; Marquardt, C.; Matsko, A.B.; Schwefel, H.G.L.; Leuchs, G. Nonlinear and quantum optics with whispering gallery resonators. *J. Opt.* **2016**, *18*, 123002. [CrossRef]
16. Rueda, A.; Sedlmeir, F.; Collodo, M.C.; Vogl, U.; Stiller, B.; Schunk, G.; Strekalov, D.V.; Marquardt, C.; Fink, J.M.; Painter, O.; Leuchs, G.; Schwefel, H.G.L. Efficient microwave to optical photon conversion: An electro-optical realization. *Optica* **2016**, *3*, 597. [CrossRef]
17. Minet, Y.; Reis, L.; Szabados, J.; Werner, C.S.; Zappe, H.; Buse, K.; Breunig, I. Pockels-effect-based adiabatic frequency conversion in ultrahigh-Q microresonators. *Opt. Express* **2020**, *28*, 2939–2947. [CrossRef] [PubMed]
18. Savchenkov, A.A.; Ilchenko, V.S.; Matsko, A.B.; Maleki, L. High-order tunable filters based on a chain of coupled crystalline whispering gallery-mode resonators. *IEEE Photonics Technol. Lett.* **2005**, *17*, 136–138. [CrossRef]
19. Fürst, J.U.; Strekalov, D.V.; Elser, D.; Lassen, M.; Andersen, U.L.; Marquardt, C.; Leuchs, G. Naturally phase-matched second-harmonic generation in a whispering-gallery-mode resonator. *Phys. Rev. Lett.* **2010**, *104*, 153901. [CrossRef]
20. Szabados, J.; Sturman, B.; Breunig, I. Frequency comb generation threshold via second-harmonic excitation in $\chi(2)$ optical microresonators. *APL Photonics* **2020**, *5*, 116102. [CrossRef]
21. Ferdous, F.; Demchenko, A.A.; Vyatchanin, S.P.; Matsko, A.B.; Maleki, L. Microcavity morphology optimization. *Phys. Rev. A* **2014**, *90*. [CrossRef]
22. Grudinin, I.S.; Yu, N. Dispersion engineering of crystalline resonators via microstructuring. *Optica* **2015**, *2*, 221. [CrossRef]
23. Archambeault, B.; Brench, C.; Ramahi, O.M. (Eds.) *EMI/EMC Computational Modeling Handbook*, 2nd ed.; The Springer International Series in Engineering and Computer Science; Springer: Boston, MA, USA, 2001; Volume 630. [CrossRef]
24. Weis, R.S.; Gaylord, T.K. Lithium niobate: Summary of physical properties and crystal structure. *Appl. Phys. A* **1985**, *37*, 191–203. [CrossRef]
25. Strekalov, D.V.; Savchenkov, A.A.; Matsko, A.B.; Yu, N. Efficient upconversion of subterahertz radiation in a high-Q whispering gallery resonator. *Opt. Lett.* **2009**, *34*, 713–715. [CrossRef]
26. Luennemann, M.; Hartwig, U.; Panotopoulos, G.; Buse, K. Electrooptic properties of lithium niobate crystals for extremely high external electric fields. *Appl. Phys. B* **2003**, *76*, 403–406. [CrossRef]
27. Tipler, P.A. *College Physics*; Worth: New York, NY, USA, 1987; p. 467.
28. Umemura, N.; Matsuda, D.; Mizuno, T.; Kato, K. Sellmeier and thermo-optic dispersion formulas for the extraordinary ray of 5 mol.% MgO-doped congruent $LiNbO_3$ in the visible, infrared, and terahertz regions. *Appl. Opt.* **2014**, *53*, 5726–5732. [CrossRef]
29. Abarkan, M.; Aillerie, M.; Kokanyan, N.; Teyssandier, C.; Kokanyan, E. Electro-optic and dielectric properties of zirconium-doped congruent lithium–niobate crystals. *Opt. Mater. Express* **2014**, *4*, 179. [CrossRef]

crystals

MDPI

Article

Study of Type II SPDC in Lithium Niobate for High Spectral Purity Photon Pair Generation

Ilhwan Kim, Donghwa Lee and Kwang Jo Lee *

Department of Applied Physics, Institute of Natural Science, Kyung Hee University, Yongin-si 17104, Korea; dlfghks383@gmail.com (I.K.); fairytale095@gmail.com (D.L.)
* Correspondence: kjlee88@khu.ac.kr

Abstract: Recent advances of high-quality lithium niobate (LN) on insulator technology have revitalized the progress of novel chip-integrated LN-based photonic devices and accelerated application research. One of the promising technologies of interest is the generation of entangled photon pairs based on spontaneous parametric down-conversion (SPDC) in LNs. In this paper, we investigated, theoretically and numerically, Type II SPDC in two kinds of LNs—undoped and 5-mol% MgO doped LNs. In each case, both non-poled and periodically poled crystals were considered. The technique is based on the SPDC under Type II extended phase matching, where the phase matching and the group velocity matching are simultaneously achieved between interacting photons. The proposed approach has not yet been reported for LNs. We discussed all factors required to generate photon pairs in LNs, in terms of the beam propagation direction, the spectral position of photons, and the corresponding effective nonlinearities and walk-offs. We showed that the spectral positions of the generated photon pairs fall into the mid-infrared region with high potential for free-space quantum communication, spectroscopy, and high-sensitivity metrology. The joint spectral analyses showed that photon pairs can be generated with high purities of 0.995–0.999 with proper pump filtering.

Keywords: lithium niobate; parametric down-conversion; photon-pair generation; extended phase matching

Citation: Kim, I.; Lee, D.; Lee, K.J. Study of Type II SPDC in Lithium Niobate for High Spectral Purity Photon Pair Generation. *Crystals* **2021**, *11*, 406. https://doi.org/10.3390/cryst11040406

Academic Editors: Gábor Corradi and László Kovács

Received: 3 March 2021
Accepted: 9 April 2021
Published: 10 April 2021

1. Introduction

Lithium niobate ($LiNbO_3$, LN) has been the most preferred nonlinear optic platform for decades due to its high material quality, mature manufacturing technology, large second-order nonlinearity and electro-optic (EO) modulation efficiency [1]. In particular, the ion-cut technology for manufacturing wafer-size, low propagation loss, sub-micron-thick crystalline LN thin films has made significant advances over the past decade [2,3]. This technique allows the successful integration of undoped or MgO/Fe/Er/Tm doped LN on insulators such as SiO_2/Si, Si, sapphire, and quartz [4]. This rapid development of LN-on-insulator (LNOI) technology has revitalized the progress of novel chip-integrated LN-based photonic devices and accelerated application research [5–17]. The monolithically integrated LN modulator with on-chip loss of less than 0.5 dB and the high modulation rate of 210 Gb/s has been successfully demonstrated, which was followed by many promising applications including on-chip photonic integration, Kerr frequency comb or supercontinuum generation, and nonlinear-optic quasi-phase-matching (QPM) frequency conversion on an integrated platform [5–7]. Chip-based LNOI light sources including frequency combs, on-chip self-referencing systems, and a two-octave spanning supercontinuum have been demonstrated [8–10]. On-chip integration of spectrometers, modulators, resonators, frequency shifters, and periodically poled LN (PPLN) waveguides for integrated photonic circuit construction has also been reported recently [11–17].

One of the important applications of second-order nonlinearity is the generation of entangled photon pairs based on spontaneous parametric down-conversion (SPDC) in $\chi^{(2)}$-crystals [18]. In particular, two photons generated simultaneously via Type II SPDC have orthogonal polarizations in the basis that are distinguished by crystallographic axes. In this case, the entangled photon pairs are obtained from the spatial region where two beams (or cones) with each polarization overlap, or from the Sagnac-loop interferometer configuration [19,20]. In addition, the extended phase matching (EPM) is defined as when group velocity (GV) matching is additionally achieved between photons interacting under Type II SPDC, where high-purity photon pairs can be generated in a wide bandwidth [21–25]. Several studies on the photon-pair generation based on Type II SPDC in PPLN waveguides have been reported so far, but in all cases the EPM scheme was not used or the frequencies of the generated photon pairs were not identical to each other [26–30]. The experimentally reported brightness values are 6×10^5/s/GHz and 3×10^5/s/GHz for non-diffused and Ti-indiffused PPLN waveguides, respectively [26,27]. The rates of photon-pair generation are 1.2×10^3 pairs/s/mW and 2.87×10^7 pairs/s/mW for a bulk PPLN and hybrid PPLN waveguide, respectively [28,29]. Experimental results for the generation of photon pairs in thin-film PPLN have only been reported recently, where Type 0 SPDC were used [31,32]. The brightness is 69×10^6/s/mW/nm, and the photon-pair generation rates are 91 pairs/(s GHz Mw) for Type 0 and 44,360 pairs/(s GHz mW) for Type I, respectively [31,32]. In addition, Type II EPM approach has not yet been reported for non-poled LNs. However, in order to generate high-purity photon pairs that are indistinguishable except for the state of polarization, SPDC based on frequency-degenerate Type II EPM is required. Therefore, it is timely to investigate the usefulness of LN as another suitable candidate for the generation of frequency-degenerate, entangled photon pairs while meeting all requirements.

In this paper, we investigate, both theoretically and numerically, Type II SPDC in two kinds of LNs (i.e., undoped and 5-mol% MgO doped LNs), which are expected to be useful as material platforms for the generation of entangled photon pairs of high spectral purity. In each case, both non-poled and PP crystals will be considered. The 5-mol% MgO-doping prevents photorefractive damage to LN (optically induced refractive index change when the crystal is exposed to blue or green CW light) [33,34]. The technique is based on the SPDC under Type II EPM, where the phase matching (QPM or birefringent PM) and the GV matching are simultaneously achieved between the interacting photons in both non-poled and PP crystals. First, we investigate the Type II EPM characteristics for non-poled and PPLN in two types of LN, respectively, in terms of the beam propagation direction, the spectral position of photon pairs, and the corresponding effective nonlinearities and beam walk-offs. We will show that the spectral positions of the generated photon pairs fall into the mid-infrared (mid-IR) region with high potential for free-space quantum communication, spectroscopy, and high-sensitivity metrology [35–38]. The joint spectral analyses showed that photon pairs can be generated with high purities of 0.995–0.999 with proper pump filtering.

2. Materials and Theories

LN belongs to the trigonal point group 3 m at room temperature, and exhibits negative uniaxial birefringence [1]. In this case, the order of refractive-index (RI) magnitudes is given by $n_e < n_o$, where n_e and n_o represent the RIs of the extraordinary and ordinary waves in the principal axes of the index ellipsoid [39,40]. Figure 1 illustrates the frequency and polarization relationships of pump, signal, and idler photons interacting via Type II SPDC. In a non-poled LN shown in Figure 1a, an input pump photon with a frequency 2ω produces a pair of photons (signal and idler) with the same frequency ω and polarization states perpendicular to each other. In this case, the pump wave vector that satisfies Type II EPM generally does not propagate along the crystallographic axis, resulting in a spatial walk-off between the interacting beams as illustrated in Figure 1a. For a PPLN case, the pump beam propagates along the y-direction perpendicular to the ferroelectric domain wall, as shown in Figure 1b. Under the first-order QPM, an x-polarized pump photon with a frequency 2ω produces a pair of signal-idler photons with the same frequency ω, and the pair has x- and z-polarization states (or vice versa), respectively. The arrow in each ferroelectric domain represents the direction of spontaneous polarization (P_S), and Λ is the poling period for the first-order QPM.

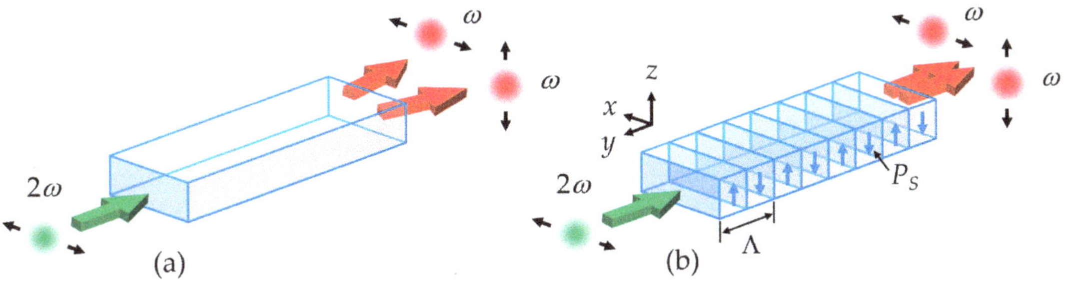

Figure 1. Schematic diagrams showing the frequency and polarization relationships of pump, signal, and idler photons interacting via Type II spontaneous parametric down-conversion (SPDC) in (**a**) a non-poled lithium niobate (LN) and (**b**) a periodically poled LN (PPLN).

For a non-poled LN shown in Figure 1a, the Type II PM condition can be expressed as,

$$\Delta k = |k_e(2\omega,\theta) - k_o(\omega) - k_e(\omega,\theta)| = 0, \tag{1}$$

where each k represents the wave number of the interacting wave, and is defined as $k_o(\omega) = (\omega/c)n_o(\omega)$ or $k_e(l\omega,\theta) = (l\omega/c)n_e(l\omega,\theta)$. Here, l can be 1 or 2, and c denotes the speed of light in vacuum. For light propagating at the angle θ to the optic axis of the LN, the RI of extraordinary wave, $n_e(\theta)$, can be derived as follows, using the definition in [18]:

$$n_e(l\omega,\theta) = \frac{n_o(l\omega)n_e(l\omega)}{\sqrt{n_o^2(l\omega)\sin^2\theta + n_e^2(l\omega)\cos^2\theta}}. \tag{2}$$

The RIs (n_e and n_o) of two kinds of LNs considered in this work (i.e., undoped and 5-mol% MgO-doped LNs) are found in [39,40].

For the PPLN configuration shown in Figure 1b, each of the three interacting photons is polarized along one of the principal axes of the index ellipsoid, respectively. The first order QPM condition under the Type II interaction (when using the nonlinear optic coefficient d_{15}) is then expressed as,

$$\Delta k_Q = |k_o(2\omega) - k_o(\omega) - k_e(\omega)| - \frac{2\pi}{\Lambda} = 0, \tag{3}$$

where $k_o(l\omega) = (l\omega/c)n_o(l\omega)$ and $k_e(\omega) = (\omega/c)n_e(\omega)$. Here, l can be 1 or 2.

The temporal walk-off between the interacting photons due to the difference in GV is expressed as the time delay (ΔT) per unit crystal length as follows:

$$\frac{\Delta T}{L} = \frac{\Delta n_g}{c}, \tag{4}$$

where L and Δn_g represent the crystal length and the group index difference between interacting photons, respectively [41]. Then, when $\Delta n_g = 0$ in Equation (4), GV matching is achieved, which can be simplified as:

$$2n_e^{(g)}(2\omega,\theta) = n_o^{(g)}(\omega) + n_e^{(g)}(\omega,\theta), \tag{5}$$

$$2n_o^{(g)}(2\omega) = n_o^{(g)}(\omega) + n_e^{(g)}(\omega), \tag{6}$$

where Equations (5) and (6) correspond to the case of non-poled LN and PPLN, respectively. Each superscript g in Equations (5) and (6) means the group index. Now Type II EPM for each case shown in Figure 1 is defined as when Equations (1) and (5) (for a non-poled LN in Figure 1a), or Equations (3) and (6) (for a PPLN in Figure 1b) are satisfied simultaneously. For a non-poled LN, Equations (1) and (5) can both be expressed as two-variable functions for θ and the pump wavelength (λ_p). Therefore, by solving the system of Equations (1) and (5), a solution set of λ_p and θ that satisfies Type II EPM can be obtained, where pure biphoton-states are produced. For a PPLN, Equations (3) and (6) are both given as a function of λ_p. Therefore, once a λ_p that satisfies the GV matching condition of Equation (6) is obtained, the poling period (Λ) for the first QPM is determined by Equation (3). PP structures with specific periods (>a few μm, typically) can be easily fabricated with current ferroelectric domain engineering techniques. Therefore, an SPDC-based photon-pair source that satisfies Type II EPM can be implemented in a PPLN.

In negative uniaxial crystals with the point symmetry 3 m, such as LN, the effective nonlinear-optic coefficient for Type II PM in a non-poled LN is given by [42],

$$d_{eff} = d_{22}\cos^2(\theta+\rho)\sin 3\phi, \tag{7}$$

where ϕ and θ are the azimuthal and polar angles in the spherical coordinate, respectively. The nonlinear optic coefficient, d_{22}, is known to be 2.1 ± 0.21 pm/V at 1064 nm [43,44]. For a negative uniaxial crystal with the point symmetry 3 m, the spatial walk-off angle (ρ) between the wave vector and the Poynting vector within the crystal can be simplified to:

$$\tan\rho = \frac{\sin 2\theta}{2}\left(\frac{1}{n_e^2(\omega)} - \frac{1}{n_o^2(\omega)}\right)n_e^2(\omega,\theta), \tag{8}$$

where $n_e(\omega,\theta)$ is given in Equation (2). To derive this expression, we used the definition in [45]. The azimuth angle has no effect on the PM condition as shown in Equations (1) and (2), so the maximum d_{eff} can be obtained by setting ϕ to $\pi/6$. As can be appreciated from Equations (7) and (8), d_{eff} and ρ are both functions of λ_p and θ. Therefore, for a non-poled LN, d_{eff} can be obtained by substituting the solution sets of λ_p and θ that satisfy Type II EPM (i.e., Equations (1) and (5)). For a PPLN, since Type II PM uses d_{15}, d_{eff} is given as $(2/\pi)d_{15}$ for the first-order QPM. The SPDC efficiency is proportional to the square of the d_{eff} for a given direction of beam propagation [4]. The nonlinear optic coefficients d_{15} of two kinds of LNs (i.e., undoped and 5 mol% MgO-doped LNs) are found in [43,44,46].

The spatial walk-off between the interacting beams does not occur in the QPM situation shown in Figure 1b. However, when the beams generally do not propagate along the crystallographic axis, as shown in Figure 1a, the spatial walk-off occurs due to the difference in direction of the wave vector and the Poynting vector within the crystal. For the Type II interaction in a negative uniaxial crystal defined in Equation (1), the relationships of Poynting vectors (**s**) of the interacting waves are given as illustrated in Figure 2b, while the PM is collinear as shown in Figure 2a. For the ordinary wave, the wave vector and the Poynting vector are parallel to each other (i.e., $\mathbf{k}_o(\omega)//\mathbf{s}_o(\omega)$), but this is not the case for the extraordinary waves. ρ_ω and $\rho_{2\omega}$ in Figure 2b are the walk-off angles of the extraordinary waves with frequencies ω and 2ω (Equation (8)), respectively, and in this study the relationship of $\rho_{2\omega} > \rho_\omega$ is valid. Then, for an SPDC in a non-poled LN, the largest walk-off angle (w) between the Poynting vectors of the interacting waves is equal to $\rho_{2\omega}$. As expected in Equation (8), w is also the two-variable function for λ_p and θ, which can be obtained by substituting the solution set of λ_p and θ that satisfies Type II PM. Then the maximum deviation between the interacting beams after passing through the crystal of length L can be expressed as:

$$\Delta = L\tan w. \tag{9}$$

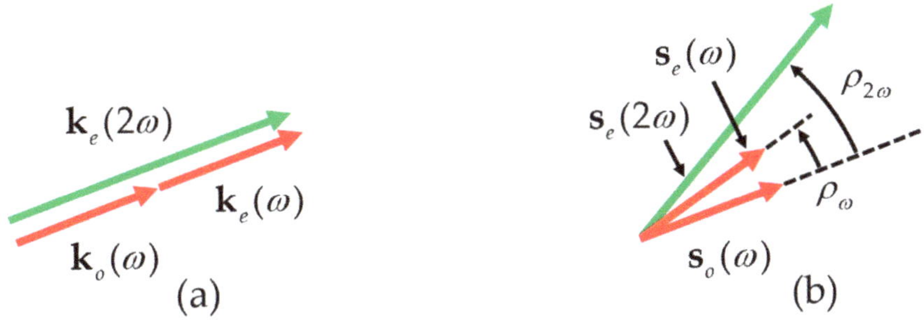

Figure 2. Example schematic diagrams showing the relationships of (**a**) wave vectors ($\mathbf{k}_e$, $\mathbf{k}_o$) and (**b**) Poynting vectors ($\mathbf{s}_e$, $\mathbf{s}_o$) of the interacting waves under Type II PM. Each ρ represents the walk-off angle between **k** and **s**. The angular scale is exaggerated.

The joint spectral analysis—the construction of the signal-idler joint spectral amplitude (JSA) and the calculation of the purity of the biphoton state via Schmidt decomposition—is commonly used to quantify the heralded-state spectral purity of the SPDC output [47–49]. Here, the spectral purity is a parameter that describes the degree of spectral uncorrelation between the signal and the idler photons. The signal-idler biphoton state $|\psi\rangle$ generated from SPDC can be expressed as,

$$|\psi\rangle = \int_0^\infty \int_0^\infty d\omega_s d\omega_i f(\omega_s, \omega_i) \hat{a}_s^\dagger(\omega_s) \hat{a}_i^\dagger(\omega_i) |0\rangle |0\rangle, \tag{10}$$

where $\hat{a}_s^\dagger$ and $\hat{a}_i^\dagger$ are the creation operators of the signal and idler photons, respectively, and ω_s and ω_i are the corresponding frequencies [49]. The normalized correlation function, $f(\omega_s, \omega_i)$, is expressed as the product of the pump envelope function and the PM function as follows:

$$f(\omega_s, \omega_i) = \alpha(\omega_s, \omega_i) \varphi(\omega_s, \omega_i), \tag{11}$$

which represents the biphoton JSA [47]. Assuming a pump with Gaussian spectral shape, the pump envelope (PE) function can be written as,

$$\alpha(\omega_s, \omega_i) \propto \exp\left[-\frac{(\omega_s + \omega_i - \omega_p)^2}{\sigma_p^2} \right], \tag{12}$$

where ω_p and σ_p represent the center frequency and bandwidth of the pump, respectively. The PM function is given in the form of a sinc function as shown below,

$$\varphi(\omega_s, \omega_i) \propto \mathrm{sinc}\left(\frac{\Delta k' L}{2} \right), \tag{13}$$

where $\Delta k'$ represents the phase-mismatch defined in Equation (1) (Δk for a non-poled LN) or Equation (3) (Δk_Q for a PPLN). For the given k_p-direction (i.e., θ) that satisfies the EPM, Equation (11) is given as a function of the signal and idler wavelengths (or λ_s and λ_i), thus the JSA can be plotted on a two-dimensional plane as a function of λ_s and λ_i. The purity can be calculated via the following Schmidt decomposition,

$$f(\omega_s, \omega_i) = \sum_j \sqrt{c_j} |\zeta_{s,j}\rangle |\zeta_{i,j}\rangle, \tag{14}$$

where Schmidt coefficients, c_j, are a set of non-negative real numbers satisfying the normalization condition, $\sum_j c_j = 1$. $|\zeta_{s,j}\rangle$ and $|\zeta_{i,j}\rangle$ represent the orthonormal basis states called

Schmidt modes [48]. Then the spectral purity, P, is defined as the sum of squares of Schmidt coefficients, as in [49]:

$$P = \sum_j c_j^2. \tag{15}$$

Therefore, once we plot the JSA as functions of λ_s and λ_i in a two-dimensional plane, P can be calculated via the Schmidt decomposition in Equation (15).

3. Simulations and Discussion

Figure 3 shows the numerical simulation results of Type II EPM in a PPLN (Figure 3a) and a 5-mol% MgO-doped PPLN (MgO:PPLN) (Figure 3b) at temperature t = 50 °C. Λ satisfying the first-order QPM (Equation (3)) and the temporal walk-off per unit crystal length ($\Delta T/L$) defined in Equation (4) are plotted as a function of λ_p. The refractive indices used for the simulations are found in [39,40]. As indicated by the dashed circles and arrows in Figure 3, the λ_p values for GV matching (i.e., $\Delta T/L$ = 0) are 1817.03 nm (for a PPLN) and 1757.35 nm (for a MgO:PPLN), and the corresponding Λs are 18.04 μm and 18.35 μm, respectively. Examples of pump light sources at these wavelengths include Fabry-Pérot lasers, distributed feedback lasers, and super-luminescent diodes [50–52]. PP crystals with such long periods can be readily fabricated using current electric poling technology with the sub-μm resolution. Type II EPM properties calculated for two kinds of PPLNs are summarized in Table 1. The signal/idler wavelength in each case falls into the mid-IR range (i.e., 3634.06 nm and 3514.70 nm). LN-based photon pair sources in this spectral range can be utilized in absorption spectroscopy to control the combustion and leakage of methane (CH_4) and to detect formaldehyde (CH_2O) exposure [53,54]. In addition, the generation and detection of nonclassical light in the mid-IR range has good potential in an emerging field of high-sensitivity metrology, gravitational wave detection, and free-space quantum key distribution [35–38]. As indicated in Figure 3, Λ has an extremum at each λ_p for GV matching because the derivative of Λ by λ_p is proportional to Δn_g [41]. As described in Section 2, the Type II QPM uses the nonlinear optic coefficients d_{15}, and the values are listed in Table 1 together with references.

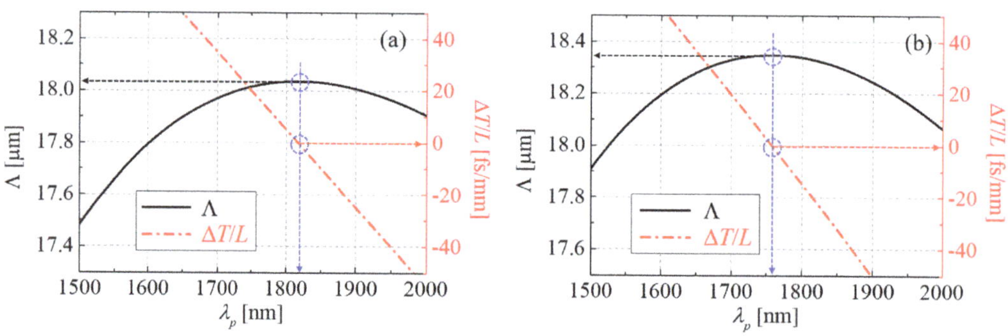

Figure 3. Numerical simulation results of Type II extended phase matching (EPM) in (**a**) a PPLN and (**b**) a 5-mol% MgO:LN. The poling period (Λ) satisfying the first-order quasi phase matching (QPM) and the temporal walk-off per unit crystal length ($\Delta T/L$) are plotted as a function of λ_p.

Table 1. Type II EPM properties of two kinds of PPLNs: the poling periods (Λ), the resonant pump wavelength (λ_p), and the corresponding signal/idler wavelengths.

Crystal	$\lvert d_{15} \rvert$ [pm/V] [1]	Λ [μm]	λ_p [nm]	$\lambda_{s,i}$ [nm]
PPLN	4.35 ± 0.44 [43,44]	18.04	1817.03	3634.06
5-mol% MgO:PPLN	4.4 [46]	18.35	1757.35	3514.70

[1] Assuming that Kleinman symmetry is valid, $d_{15} = d_{31}$.

Figure 4 shows the numerical simulation results of Type II EPM in a non-poled LN (Figure 4a) and a non-poled MgO:LN (Figure 4b) at t = 50 °C. In each graph, the solid black and dash-dot red lines represent the PM and GV matching properties calculated using Equations (1) and (5), respectively. The intersection point of two curves represents the k_p-direction (i.e., θ) and the corresponding λ_p satisfying Type II EPM. The pump wavelengths for GV matching (i.e., $\Delta T/L = 0$) are 1007.99 nm (for a LN) and 1018.15 nm (for a MgO:LN), where the k_p-directions are θ = 70.47° and θ = 74.58°, respectively. Type II EPM properties calculated for two non-poled LNs are summarized in Table 2. The calculated wavelengths of photon pairs are 2015.99 nm and 2036.30 nm, which are within the silica fiber transparency (<2300 nm) [55]. It is widely known that communication C-band is the best option for low-loss optical communication. However, it is necessary to use a wider bandwidth (e.g., S + C + L bands) to accommodate the increasing communication traffic [56,57]. Here, the S- and L-bands are used as many passive optical networks (PONs) and dense wavelength division multiplexing (DWDM) systems. Therefore, LN-based quantum light sources have good potential for demonstrating fiber-based broadband quantum communication. The generation and detection of nonclassical light about 2 μm also has good potential, especially in gravitational wave detection [35].

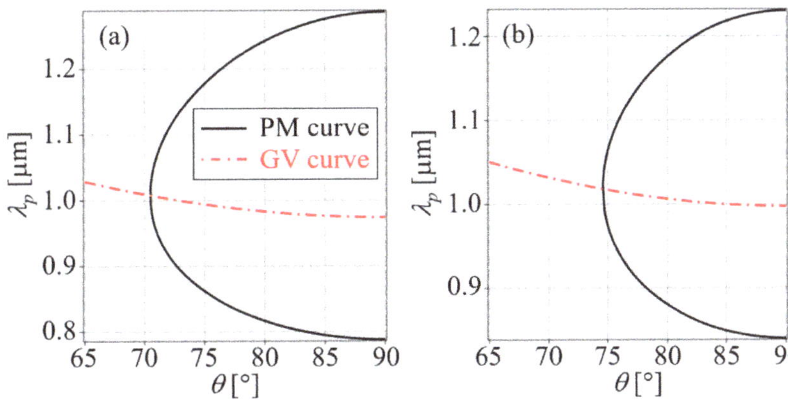

Figure 4. Numerical simulation results of Type II EPM in (**a**) a LN and (**b**) a 5-mol% MgO:LN. In each graph, the intersection point of the PM and GV curves represent the k_p-direction (i.e., θ) and the corresponding λ_p satisfying Type II EPM.

Table 2. Type II EPM properties of two kinds of non-poled LNs: the PM direction (θ), the resonant pump wavelength (λ_p) and the corresponding signal/idler wavelengths, ($\lambda_{s,i}$), the effective nonlinear optic coefficients (d_{eff}), the angle of walk-off between the interacting photons (w), and the beam deviation (Δ).

Crystal	θ [°]	λ_p [nm]	$\lambda_{s,i}$ [nm]	d_{eff} [pm/V]	w [°]	Δ [μm/mm]
LN	70.47	1007.99	2015.99	0.21	1.30	22.73
5-mol% MgO:LN	74.58	1018.15	2036.30	0.13	1.01	17.60

Since the SPDC efficiency is proportional to the square of d_{eff}, it is critical to estimate the effective nonlinearity for the given k_p-direction that satisfies Type II EPM. Figure 5 shows d_{eff} and $\rho_{2\omega}$ plotted as a function of λ_p, which are calculated numerically using Equations (7) and (8). The λ_p values in the horizontal axes in Figure 5 correspond to the λ_p range satisfying Type II PM shown in Figure 4. Here, λ_p is the solution to Equation (1), which is a solution pair with the PM direction, θ. The d_{eff} values calculated for Type II EPM are 0.21 pm/V for a LN and 0.13 pm/V for a MgO:LN as shown in Figure 5a. Over the entire

PM wavelength range, d_{eff} is larger for LN than for MgO:LN. In the PPLN configuration shown in Figure 1b, d_{eff} is simply given as $(2/\pi)d_{15}$ for the first-order QPM. Figure 5b shows the walk-off angle $\rho_{2\omega}$, calculated for the given PM solution set of λ_p and θ. Overall, also in the case of LN, the walk-off is larger than that of MgO:LN. The calculated value of $\rho_{2\omega}$ is used to calculate d_{eff} with Equation (8), and it also directly means the maximum walk-off angle w between the interacting photons, as discussed in Section 2. The w values are calculated as 1.30° for a LN and 1.01° for a MgO:LN, which correspond to the beam deviation (Δ) 22.73 μm/mm and 17.60 μm/mm, respectively. Here, each Δ was calculated using Equation (9), which can be sufficiently overcome by a large beam window in thick crystals. The calculated values of d_{eff}, w, and Δ are also summarized in Table 2.

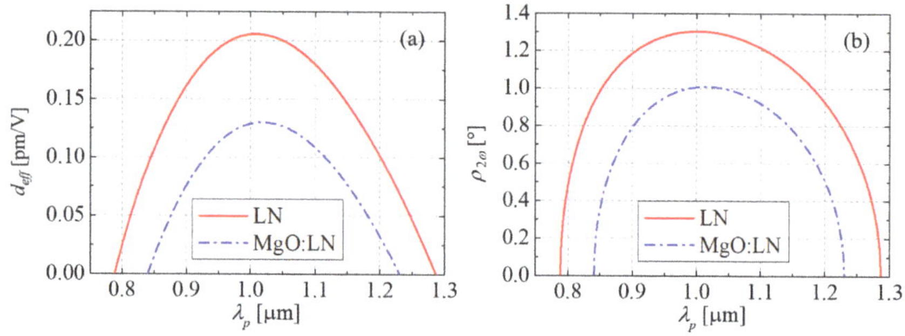

Figure 5. (**a**) Effective nonlinear optic coefficients (d_{eff}) under Type II EPM, numerically calculated for the two kinds of non-poled LNs: LN and 5-mol% MgO:LN; (**b**) the walk-off angle $\rho_{2\omega}$, calculated for the PM solution set of λ_p and θ.

Now we investigate the temperature behaviors of the Type II EPM properties of LNs. Figure 6 shows the change in λ_p for Type II EPM (corresponding to the intersection points in Figure 4) and the PM direction θ at that point, which are plotted as a function of t in two kinds of non-poled LNs. Here, $\Delta\lambda_p$ means the difference in λ_p between the temperature t and t = 50 °C (i.e., $\Delta\lambda_p \equiv \lambda_p(t) - \lambda_p(50\ ^\circ C)$). As the temperature increases, λ_p shifts to longer wavelengths ($\Delta\lambda_p > 0$) for LN (Figure 6a) and shorter wavelengths ($\Delta\lambda_p < 0$) for 5-mol% MgO:LN (Figure 6b). In the case of LN, $\Delta\lambda_p$ is only 1.87 nm (corresponding to 3.74 nm spectral change in $\Delta\lambda_{s,i}$) for large temperature changes from 50 °C to 150 °C, which is sufficiently small compared to the spectral bandwidth of the photon pair (~7.6 nm) to be described in a later paragraph. Since the spectral change for a temperature change of 100 °C is within the signal/idler spectral bandwidth, the LN-based photon-pair source operates stably against temperature change. For 5-mol% MgO:LN, the Type II EPM point does not exist at temperatures above 80 °C as plotted in Figure 6b. $\Delta\lambda_p$ for temperature changes from 50 °C to 80 °C is −0.16 nm (corresponding to a 0.32 nm spectral change in $\Delta\lambda_{s,i}$), which is also much smaller than the spectral bandwidth of the photon pair (~7.8 nm) to be described in a later paragraph. Therefore, both LN and 5-mol% MgO:LN can be suitable platforms for quantum light sources that are insensitive to temperature changes.

Figure 7a,b shows the temperature dependence of the first-order QPM (Equation (3)) and the temporal walk-off (Equation (4)) between the interacting photons. The solid black line in each graph represents the change in λ_p as a function of t of a PP crystal with the fixed poling period designed for Type II EPM at t = 50 °C ($\equiv\Lambda_{50^\circ C}$). The dash-dot red line in each graph represents the change in λ_p for no temporal walk-off as a function of t. In both cases of PPLN (Figure 7a) and 5 mol%-MgO:PPLN (Figure 7b), λ_p for the first-order QPM appears even when the temperature of PP crystals change. However, the GV matching is not achieved except for t = 50 °C. Figure 7c shows the PM function (Equation (13)) plotted as a function of t, which means the spectral responses of the PP crystals to temperature. The bandwidths (full-width-half-maximum) are calculated as >52 °C for PPLN and 1.90 °C for 5 mol%-MgO:PPLN. Therefore, in application as a quantum light source, PPLN is not sensitive to temperature changes, while 5 mol%-MgO:PPLN needs to be maintained at an appropriate temperature.

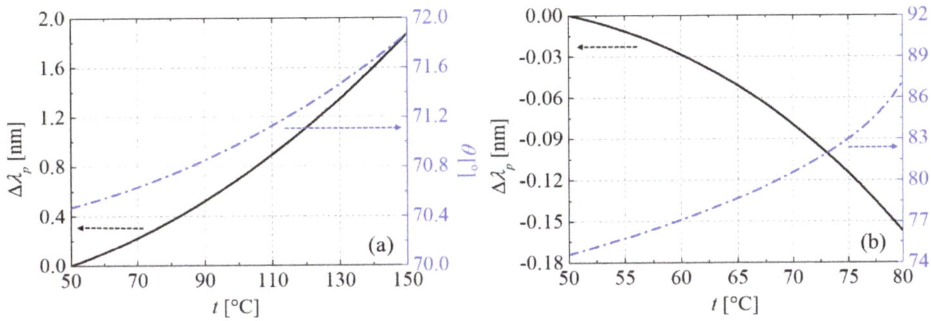

Figure 6. Temperature behaviors of the resonant λ_p and the corresponding PM direction (θ) in two kinds of non-poled LNs: (**a**) LN and (**b**) 5 mol%-MgO:LN.

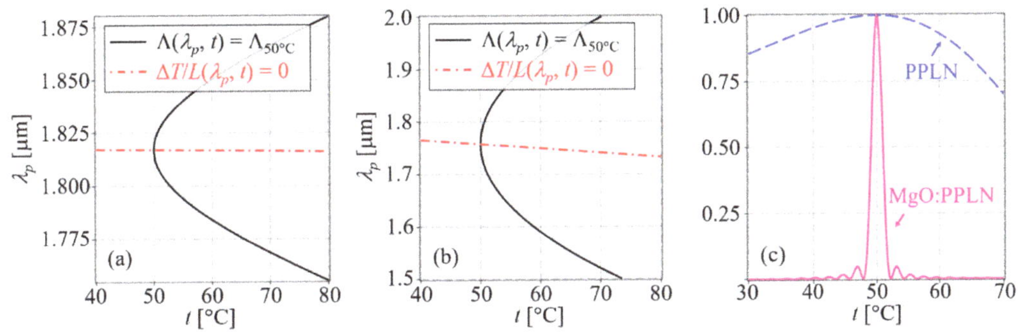

Figure 7. Temperature behaviors of the resonant λ_p for the first order QPM and the temporal walk-offs in (**a**) PPLN and (**b**) 5 mol%-MgO:PPLN; (**c**) the spectral responses of the PP crystals to temperature.

In order to quantify the heralded-state spectral purity of SPDC output, we now discuss the analysis of the signal-idler JSA properties including the calculation of the purity of the biphoton state. Figure 8 shows the joint spectral properties of Type II SPDC in the cases listed in Tables 1 and 2. The density plots of the JSAs (Equation (11)) for LN and 5 mol% MgO:LN are shown in Figure 8a,b, respectively. JSAs of PPLN and 5 mol% MgO:PPLN are plotted in Figure 8c,d, respectively. The crystal length used for the calculations is 10 mm for all four cases. The bandwidths of PE functions (Equation (12)) chosen for higher purities are 2.27 nm, 2.33 nm, 5.33 nm, and 5.17 nm, respectively, in the order of Figure 8a–d. As can be seen in Figure 8, the JSAs show typical circular shapes as in the case of PPKTP under Type II EPM [24]. The contour lines for the joint spectral intensity (JSI) = 0.5 are plotted in Figure 9, where Figure 9a–d, in turn, correspond to Figure 8a–d. Here, the JSI is defined as $|\mathrm{JSA}|^2$. For clear comparison of the spectral bandwidths of the photon pairs, all contour plots in Figure 9 are displayed in a 20 nm × 20 nm window. The contour lines show almost perfect circular shapes with the diameters of 7.6 nm, 7.8 nm, 17.3 nm, and 17.8 nm, in the order of Figure 9a–d. The purities calculated using Schmidt decomposition (Equation (15)) are as high as 0.995, 0.995, 0.999, and 0.998, respectively, in the order of Figure 8a–d. The results show that very high-purity photon pairs can be generated in both PP and non-poled LNs.

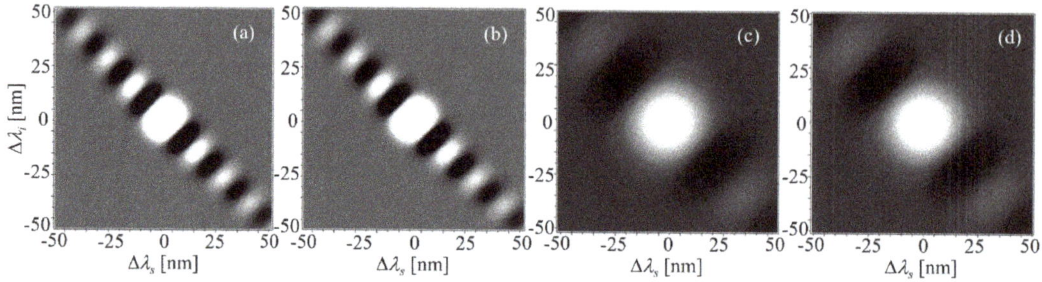

Figure 8. Joint spectral amplitudes (JSAs) of Type II SPDC in (**a**) LN, (**b**) 5 mol%-MgO:LN (**c**) PPLN, and (**d**) 5 mol%-MgO:PPLN. The calculated purities are 0.995, 0.995, 0.999, and 0.998, respectively.

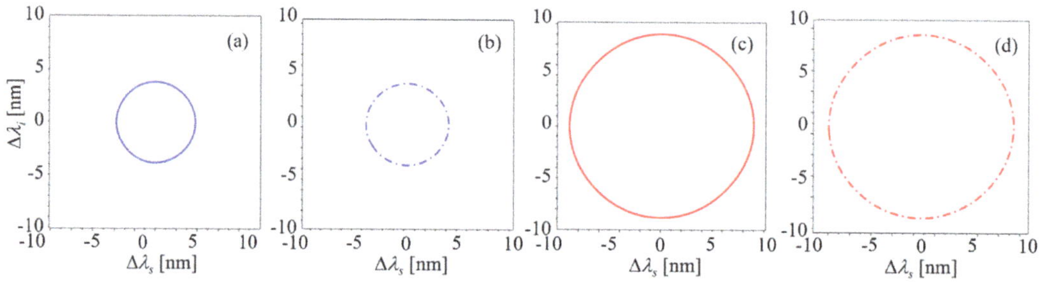

Figure 9. Contour lines for joint spectral intensity (JSI) = 0.5 for the cases shown in Figure 8: (**a**) LN, (**b**) 5 mol%-MgO:LN (**c**) PPLN, and (**d**) 5 mol%-MgO:PPLN. The diameters of the circles are 7.6 nm, 7.8 nm, 17.3 nm, and 17.8 nm, in the order of Figure 9a–d. Each result was calculated for a crystal length of 10 mm.

4. Conclusions

We have theoretically and numerically investigated Type II SPDC properties of two kinds of LNs—undoped LN and 5-mol% MgO:LN. In both non-poled and PP crystals, SPDC properties under Type II EPM were theoretically investigated, in terms of PM, GV matchings, effective nonlinearities, and spatial walk-offs between the interacting photons. The spectral position of photon pairs can be chosen in the mid-IR region including 2015.99 nm, 2036.30 nm, 3634.06 nm, and 3514.70 nm, which has high potential for free-space quantum communication, spectroscopy, and high-sensitivity metrology. The heralded-state spectral purities of SPDC output were quantified via JSA analyses, and the results show that the purities calculated with proper pump filtering are as high as 0.995–0.999. While the acceptable temperature bandwidth of a MgO:PPLN-based SPDC source operating with a stable spectrum is less than 1.9 °C, other counterparts based on PPLN or two kinds of non-poled LNs are not sensitive to temperature changes. Nevertheless, practically all four cases do not require temperature precision equipment, which is a huge advantage for practical use of LN-based quantum light sources outside the laboratory. LN-based quantum light sources are expected to provide a variety of functions for SPDC-based quantum communication, information processing, metrology, and spectroscopy.

Author Contributions: Conceptualization, K.J.L.; Simulation, D.L.; writing—original draft preparation, I.K.; writing—review and editing, K.J.L.; supervision, K.J.L. All authors have read and agreed to the published version of the manuscript.

Funding: This research was funded by the National Research Foundation of Korea (Grant NRF-2019R1F1A1063937); Korea Institute of Science and Technology (Grant 2E29580-19-147); and an Institute for Information & Communications Technology Promotion (IITP) grant funded by the Korea government (MSIT) (No. 2020-0-00947).

Institutional Review Board Statement: Not applicable.

Informed Consent Statement: Not applicable.

Data Availability Statement: The data presented in this study are available in this article.

Conflicts of Interest: The authors declare no conflict of interest.

References

1. Volk, T.; Wöhlecke, M. *Lithium Niobate*; Springer: Berlin, Germany, 2008; pp. 1–7.
2. Jia, Y.; Wang, L.; Chen, F. Ion-cut lithium niobate on insulator technology: Recent advances and perspectives featured. *Appl. Phys. Rev.* **2021**, *8*, 011307. [CrossRef]
3. Poberaj, G.; Hu, H.; Sohler, W.; Günter, P. Lithium niobate on insulator (LNOI) for micro-photonic devices. *Laser Photon. Rev.* **2012**, *6*, 488–503. [CrossRef]
4. Jinan Jingzheng Electronics Co.,Ltd. (nanoln). Available online: https://www.nanoln.com (accessed on 25 February 2021).
5. Wang, C.; Zhang, M.; Chen, X.; Bertrand, M.; Shams-Ansari, A.; Chandrasekhar, S.; Winzer, P.; Lončar, M. Integrated lithium niobate electro-optic modulators operating at CMOS-compatible voltages. *Nature* **2018**, *562*, 101–104. [CrossRef] [PubMed]
6. Wang, C.; Langrock, C.; Marandi, A.; Jankowski, M.; Zhang, M.; Desiatov, B.; Fejer, M.M.; Lončar, M. Ultrahigh-efficiency wavelength conversion in nanophotonic periodically poled lithium niobate waveguides. *Optica* **2018**, *5*, 1438–1441. [CrossRef]
7. Wang, C.; Zhang, M.; Yu, M.; Zhu, R.; Hu, H.; Lončar, M. Monolithic lithium niobate photonic circuits for Kerr frequency comb generation and modulation. *Nat. Commun.* **2019**, *10*, 978. [CrossRef]
8. Yu, M.; Wang, C.; Zhang, M.; Lončar, M. Chip-based lithium-niobate frequency combs. *IEEE Photon. Technol. Lett.* **2019**, *31*, 1894–1897. [CrossRef]
9. Yu, M.; Desiatov, B.; Okawachi, Y.; Gaeta, A.L.; Lončar, M. Coherent two-octave spanning supercontinuum generation in lithium-niobate waveguides. *Opt. Lett.* **2019**, *44*, 1222–1225. [CrossRef]
10. Okawachi, Y.; Yu, M.; Desiatov, B.; Kim, B.Y.; Hansson, T.; Lončar, M.; Gaeta, A.L. Chip-based self-referencing using integrated lithium niobate waveguides. *Optica* **2020**, *7*, 702–707. [CrossRef]
11. Pohl, D.; Reig Escalé, M.; Madi, M.; Kaufmann, F.; Brotzer, P.; Sergeyev, A.; Guldimann, B.; Giaccari, P.; Alberti, E.; Meier, U.; et al. An integrated broadband spectrometer on thin-film lithium niobate. *Nat. Photonics* **2020**, *14*, 24–29. [CrossRef]
12. Xu, M.; He, M.; Zhang, H.; Jian, J.; Pan, Y.; Liu, X.; Chen, L.; Meng, X.; Chen, H.; Li, Z.; et al. High-performance coherent optical modulators based on thin-film lithium niobate platform. *Nat. Commun.* **2020**, *11*, 3911. [CrossRef]
13. Desiatov, B.; Shams-Ansari, A.; Zhang, M.; Wang, C.; Lončar, M. Ultra-low loss integrated visible photonics using thin-film lithium niobite. *Optica* **2019**, *6*, 380–384. [CrossRef]
14. Holzgrafe, J.; Sinclair, N.; Zhu, D.; Shams-Ansari, A.; Colangelo, M.; Hu, Y.; Zhang, M.; Berggren, K.K.; Lončar, M. Cavity electro-optics in thin-film lithium niobate for efficient microwave-to-optical transduction. *Optica* **2020**, *7*, 1714–1720. [CrossRef]
15. Luke, K.; Kharel, P.; Reimer, C.; He, L.; Loncar, M.; Zhang, M. Wafer-scale low-loss lithium niobate photonic integrated circuits. *Opt. Express* **2020**, *28*, 24452–24458. [CrossRef] [PubMed]
16. Shao, L.; Sinclair, N.; Leatham, J.; Hu, Y.; Yu, M.; Turpin, T.; Crowe, D.; Lončar, M. Integrated microwave acousto-optic frequency shifter on thin-film lithium niobate. *Opt. Express* **2020**, *28*, 23728–23738. [CrossRef]
17. Jankowski, M.; Langrock, C.; Desiatov, B.; Marandi, A.; Wang, C.; Zhang, M.; Phillips, C.R.; Lončar, M.; Fejer, M.M. Ultrabroadband nonlinear optics in nanophotonic periodically poled lithium niobate waveguides. *Optica* **2020**, *7*, 40–46. [CrossRef]
18. Boyd, R.W. *Nonlinear Optics*, 4th ed.; Academic Press: San Diego, CA, USA, 2020; pp. 77–79.
19. Kwiat, P.; Mattle, K.; Weinfurter, H.; Zeilinger, A.; Sergienko, A.V.; Shih, Y. New high-intensity source of polarization-entangled photon pairs. *Phys. Rev. Lett.* **1995**, *75*, 4337–4341. [CrossRef] [PubMed]
20. Steinlechner, F.; Gilaberte, M.; Jofre, M.; Scheidl, T.; Torres, J.P.; Pruneri, V.; Ursin, R. Efficient heralding of polarization-entangled photons from type-0 and type-II spontaneous parametric downconversion in periodically poled $KTiOPO_4$. *J. Opt. Soc. Am. B* **2014**, *31*, 2068–2076. [CrossRef]
21. Giovannetti, V.; Maccone, L.; Shapiro, J.H.; Wong, F.N.C. Extended phase-matching conditions for improved entanglement generation. *Phys. Rev. A* **2002**, *66*, 043813. [CrossRef]
22. Jin, R.-B.; Shimizu, R.; Wakui, K.; Fujiwara, M.; Yamashita, T.; Miki, S.; Terai, H.; Wang, Z.; Sasaki, M. Pulsed Sagnac polarization-entangled photon source with a PPKTP crystal at telecom wavelength. *Opt. Express* **2014**, *22*, 11498–11507. [CrossRef]
23. Li, Y.; Zhou, Z.-Y.; Ding, D.-S.; Shi, B.-S. CW-pumped telecom band polarization entangled photon pair generation in a Sagnac interferometer. *Opt. Express* **2015**, *23*, 28792–28800. [CrossRef]
24. Weston, M.M.; Chrzanowski, H.M.; Wollmann, S.; Boston, A.; Ho, J.; Shalm, L.K.; Verma, V.B.; Allman, M.S.; Nam, S.W.; Patel, R.B.; et al. Efficient and pure femtosecond-pulse-length source of polarization-entangled photons. *Opt. Express* **2016**, *24*, 10869–10879. [CrossRef] [PubMed]
25. Kim, I.; Lee, D.; Lee, K.J. Numerical investigation of high-purity polarization-entangled photon-pair generation in non-poled KTP Isomorphs. *Appl. Sci.* **2021**, *11*, 565. [CrossRef]
26. Fujii, G.; Namekata, N.; Motoya, M.; Kurimura, S.; Inoue, S. Bright narrowband source of photon pairs at optical telecommunication wavelengths using a type-II periodically poled lithium niobate waveguide. *Opt. Express* **2007**, *15*, 12769–12776. [CrossRef] [PubMed]

27. Martin, A.; Issautier, A.; Herrmann, H.; Sohler, W.; Ostrowsky, D.B.; Alibart, O.; Tanzilli, S. A polarization entangled photon-pair source based on a type-II PPLN waveguide emitting at a telecom wavelength. *New. J. Phys.* **2010**, *12*, 103005. [CrossRef]
28. Cheng, X.; Sarihan, M.C.; Chang, K.-C.; Lee, Y.S.; Laudenbach, F.; Ye, H.; Yu, Z.; Wong, C.W. Design of spontaneous parametric down-conversion in integrated hybrid Si_xN_y-PPLN waveguides. *Opt. Express* **2019**, *27*, 30773–30787. [CrossRef]
29. Kuo, P.S.; Verma, V.B.; Nam, S.W. Demonstration of a polarization-entangled photon-pair source based on phase-modulated PPLN. *OSA Continuum* **2020**, *3*, 295–304. [CrossRef]
30. Kuo, P.S.; Gerrits, T.; Verma, V.; Nam, S.W.; Slattery, O.; Ma, L.; Tang, X. Characterization of type-II spontaneous parametric down-conversion in domain-engineered PPLN. In Proceedings of the SPIE Advances in Photonics of Quantum Computing Memory, and Communication IX, San Francisco, CA, USA, 13–18 February 2016; SPIE: Bellingham, WA, USA, 2016.
31. Zhao, J.; Ma, C.; Rüsing, M.; Mookherjea, S. High quality entangled photon pair generation in periodically poled thin-film lithium niobate waveguides. *Phys. Rev. Lett.* **2020**, *124*, 163603. [CrossRef]
32. Chen, J.-Y.; Sua, Y.M.; Ma, Z.-H.; Tang, C.; Li, Z.; Huang, Y.-P. Efficient parametric frequency conversion in lithium niobate nanophotonic chips. *OSA Continuum* **2019**, *2*, 2914–2924. [CrossRef]
33. Schlarb, U.; Betzler, K. 'Influence of the defect structure on the refractive indices of undoped and Mg-doped lithium niobate. *Phys. Rev. B* **1994**, *50*, 751. [CrossRef]
34. Zelmon, D.E.; Small, D.L.; Jundt, D. Infrared corrected Sellmeier coefficients for congruently grown lithium niobate and 5 mol. % magnesium oxide–doped lithium niobate. *J. Opt. Soc. Am. B* **1997**, *14*, 3319–3322. [CrossRef]
35. Prabhakar, S.; Shields, T.; Dada, A.C.; Ebrahim, M.; Taylor, G.G.; Morozov, D.; Erotokritou, K.; Miki, S.; Yabuno, M.; Terai, H.; et al. Two-photon quantum interference and entanglement at 2.1 μm. *Sci. Adv.* **2020**, *6*, eaay5195. [CrossRef]
36. Rosenfeld, L.M.; Sulway, D.A.; Sinclair, G.F.; Anant, V.; Thompson, M.G.; Rarity, J.G.; Silverstone, J.W. Mid-infrared quantum optics in silicon. *Opt. Express* **2020**, *28*, 37092–37102. [CrossRef] [PubMed]
37. Mancinelli, M.; Trenti, A.; Piccione, S.; Fontana, G.; Dam, J.S.; Tidemand-Lichtenberg, P.; Pedersen, C.; Pavesi, L. Mid-infrared coincidence measurements on twin photons at room temperature. *Nat. Commun.* **2017**, *8*, 15184. [CrossRef]
38. Sua, Y.M.; Fan, H.; Shahverdi, A.; Chen, J.Y.; Huang, Y.P. Direct generation and detection of quantum correlated photons with 3.2 μm wavelength spacing. *Sci. Rep.* **2017**, *7*, 17494. [CrossRef]
39. Edwards, G.J.; Lawrence, M. A temperature-dependent dispersion equation for congruently grown lithium niobate. *Opt. Quant. Electron.* **1984**, *16*, 373–375. [CrossRef]
40. Gayer, O.; Sacks, Z.; Galun, E.; Arie, A. Temperature and wavelength dependent refractive index equations for MgO-doped congruent and stoichiometric $LiNbO_3$. *Appl. Phys. B* **2008**, *91*, 343–348. [CrossRef]
41. Lee, K.J.; Lee, S.; Shin, H. Extended phase matching properties of periodically poled potassium niobate crystals for mid-infrared polarization-entangled photon-pair generation. *Appl. Opt.* **2016**, *55*, 9791–9796. [CrossRef] [PubMed]
42. Shoji, I.; Nakamura, H.; Ohdaira, K.; Kondo, T.; Ito, R.; Okamoto, T.; Tatsuki, K.; Kubota, S. Absolute measurement of second-order nonlinear-optical coefficients of β-BaB_2O_4 for visible to ultraviolet second-harmonic wavelengths. *J. Opt. Soc. Am. B* **1999**, *16*, 620–624. [CrossRef]
43. Miller, R.C.; Nordland, W.A.; Bridenbaugh, P.M. Dependence of second-harmonic-generation coefficients of $LiNbO_3$ on melt composition. *J. Appl. Phys.* **1971**, *42*, 4145–4147. [CrossRef]
44. Roberts, D.A. Simplified characterization of uniaxial and biaxial nonlinear optical crystals: A plea for standardization of nomenclature and conventions. *IEEE J. Quant. Electron.* **1992**, *28*, 2057–2074. [CrossRef]
45. Gehr, R.J.; Kimmel, M.W.; Smith, A.V. Simultaneous spatial and temporal walk-off compensation in frequency-doubling femtosecond pulses in β-BaB_2O_4. *Opt. Lett.* **1998**, *23*, 1298–1300. [CrossRef]
46. Shoji, I.; Kondo, T.; Kitamoto, A.; Shirane, M.; Ito, R. Absolute scale of second-order nonlinear-optical coefficients. *J. Opt. Soc. Am. B* **1997**, *14*, 2268–2294. [CrossRef]
47. Grice, W.P.; Walmsley, I.A. Spectral information and distinguishability in type-II down-conversion with a broadband pump. *Phys. Rev. A* **1997**, *56*, 1627. [CrossRef]
48. Law, C.K.; Walmsley, I.A.; Eberly, J.H. Continuous frequency entanglement: Effective finite Hilbert space and entropy control. *Phys. Rev. Lett.* **2000**, *84*, 5304. [CrossRef] [PubMed]
49. Mosley, P.J.; Lundeen, J.S.; Smith, B.J.; Walmsley, I.A. Conditional preparation of single photons using parametric downconversion: A recipe for purity. *New. J. Phys.* **2008**, *10*, 093011. [CrossRef]
50. Nanoplus GmbH. Available online: https://nanoplus.com/en/products/fabry-perot-lasers-fp/1700-nm-2400-nm (accessed on 31 March 2021).
51. Nanoplus GmbH. Available online: https://nanoplus.com/en/products/distributed-feedback-lasers-dfb/1650-nm-1850-nm (accessed on 31 March 2021).
52. Nanoplus GmbH. Available online: https://nanoplus.com/en/products/superluminescent-diodes-sld/1700-nm-2300-nm (accessed on 31 March 2021).
53. Nanoplus GmbH. Available online: https://nanoplus.com/en/applications/applications-by-gas (accessed on 15 February 2021).
54. Whittaker, R.; Erven, C.; Neville, A.; Berry, M.; O'Brien, J.L.; Cable, H.; Matthews, J.C.F. Absorption spectroscopy at the ultimate quantum limit from single-photon states. *New J. Phys.* **2017**, *19*, 023013. [CrossRef]
55. Thorlabs Inc. Available online: https://www.thorlabs.com/thorproduct.cfm?partnumber=SM2000 (accessed on 25 February 2021).

56. Iqbal, M.A.; Krzeczunowicz, L.; Phillips, I.; Harper, P.; Forysiak, W. Evaluation of performance penalty from pump-signal overlap in S+C+L band discrete Raman amplifiers. In Proceedings of the Optical Fiber Communication Conference (OFC), San Diego, CA, USA, 8–12 March 2020.
57. Hamaoka, F.; Nakamura, M.; Okamoto, S.; Minoguchi, K.; Sasai, T.; Matsushita, A.; Yamazaki, E.; Kisaka, Y. Ultra-wideband WDM transmission in S-, C-, and L-bands using signal power optimization scheme. *J. Lightwave Technol.* **2019**, *37*, 1764. [CrossRef]

crystals

Article

Effects of the Domain Wall Conductivity on the Domain Formation under AFM-Tip Voltages in Ion-Sliced $LiNbO_3$ Films

Radmir Gainutdinov * and Tatyana Volk

Shubnikov Institute of Crystallography of FSRC "Crystallography and Photonics" RAS, 119333 Moscow, Russia; volk@crys.ras.ru

* Correspondence: radmir@crys.ras.ru

Received: 1 December 2020; Accepted: 16 December 2020; Published: 19 December 2020

Abstract: The specified domain patterns were written by AFM-tip voltages in $LiNbO_3$ films composing LNOI ($LiNbO_3$-on-insulator). The domain wall conductivity (DWC) was estimated in the written patterns. This estimate was based on the effects of load resistors R_L inserted between DWs and the ground, on the features of occurring domains. In this case, the domain formation is controlled by the ratio between R_L and the DWs' resistance R_{DW}. Starting from the comparison of patterns appearing at different R_L, the value of R_{DW} in a specified pattern was estimated. The corresponding DWC is of $\sigma_{DW} \approx 10^{-3}$ (Ohm cm)$^{-1}$ which exceeds the tabular bulk conductivity of $LiNbO_3$ by no less than twelve orders of magnitude. A small DW inclination angle of $(10^{-4})^0$ responsible for this DWC is not caused by any external action and characterizes the domain frontal growth under an AFM-tip voltage.

Keywords: lithium niobate; LNOI; ferroelectric domains; domain-wall conduction; AFM

1. Introduction

Ferroelectric materials have undergone a renaissance in the last two decades due to the discovery of many novel applications based on micro- and nanosized ferroelectric domains. The main examples of these applications are the nonlinear optical-frequency conversion [1] and an ultrahigh-density data storage [2]. More recently, interest has centered on the conductivity of charged domain walls (CDWs) which is a fundamental property of ferroelectric domains. The authors of [3] were the first to discuss the screening effects at the boundary of "encountering" ("head-to-head") 180°-domains. The specificity of ferroelastic (twin) domain walls (DWs) was discussed in the pioneering works of Salje and co-authors, e.g., [4,5]. Particularly, the superconductivity at DWs in weakly reduced WO_3 and trapping of O-vacancies by DWs in $CaTiO_3$ was reported in [4] and [5], respectively. Later on, an enhanced domain-wall conductivity (DWC) up to the superconductivity was observed in a large amount of ferroelectric films and crystals; the detailed bibliography can be found in several reviews [6–8]. Taking into account a small DW thickness and a spatial mobility of ferroelectric domains, CDWs can be regarded as the field controlled nanosized wires. In the context of various applications, a concept of "DW nanoelectronics" was proposed by Catalan and coauthors [7].

DWC is caused by a DW inclination with respect to the direction of spontaneous polarization P_s and subsequent accumulation of a screening charge with the density of $\sigma = 2P_s \sin\theta$ (where θ is the DWs' inclination angle). The authors of [9] for the first time proved experimentally the relation of DWC to the DW inclination angle in $LiNbO_3$ crystals. In the recent TEM experiments [10], an essential meandering of 180° DWs has been found in PPLN, which predicts the appearance of DWC even at non-inclined DWs; for this case, the 2D resistor network approach was proposed [11].

The largest part of DWC studies were performed in $LiNbO_3$ [9–15], which is the most convenient object for these investigations not only because of its practical potentials, but due to the fact that domain patterns fabricated in this material by any technique, are stable.

The pioneering work [9] concerning the specified DWC in $LiNbO_3$ was mentioned above. The authors of [12] succeeded in enhancing DWC in $LiNbO_3$ by 3–4 orders of magnitude compared to [9], by means of the DWs' field shaping in combination with a photoactive illumination. In [14] DWC was measured in the domain patterns with a specified DW inclination angle, written by AFM-tip voltages in $LiNbO_3$: 5% Mg plates.

The results described below were obtained in the domain patterns fabricated by AFM-tip voltages in $LiNbO_3$ films forming LNOI ($LiNbO_3$-on-insulator). For the further discussion, we dwell shortly on this device. The comprehensive bibliography can be found, e.g., in the reviews [8,16–18]. LNOI (schematically presented in Figure 1a) consists of an ion-sliced single-crystal single-domain $LiNbO_3$ film bonded on an insulating substrate (thin SiO_2 layer in our case); this sandwich is fixed onto a $LiNbO_3$ plate. In the LNOI samples under investigations, a thin metal layer serving as an electrode is inserted between $LiNbO_3$ film and SiO_2. Recently, we presented the AFM -tip domain writing in $LiNbO_3$ films forming LNOI [19,20] and DWC measurements in the written patterns [13]. Later on, AFM domain writing in LNOI and DWC observations in the written patterns were reported in [15,21,22].

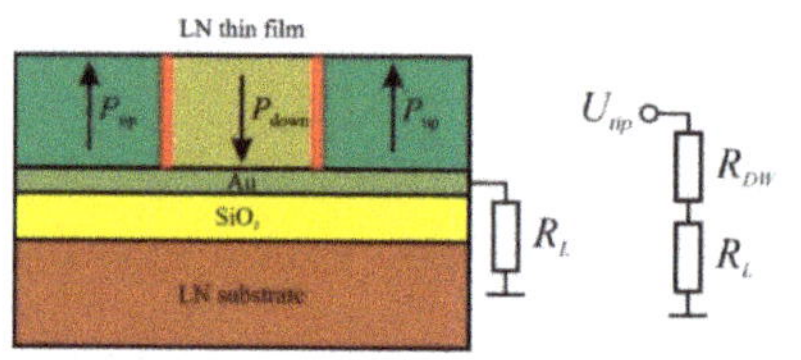

(a) (b)

Figure 1. The schematic presentation of a domain written by AFM-tip voltage in LNOI (**a**) and the equivalent circuit of the domain writing (**b**). In (**a**) the red vertical lines represent DWs, the horizontal green and yellow layers show the metal and SiO_2 ones, respectively; in (**b**) R_{DW} is the DWs' total resistance. In (**a**,**b**) R_L is the load resistor.

According to the general concept, DWs represent regions of charge accumulation. As a result, an electrostatic repulsion between the closely spaced domains can be expected. Actually, some experimental data provide indirect evidence for this effect. When AFM domain writing in thin $LiNbO_3$ crystals, the neighboring domains became unstable [23] as the inter-domain distance Λ was reduced. Similarly, when AFM domain writing in He-$LiNbO_3$ optical waveguides, the domain sizes tended to decrease as Λ was decreased [24]. An instability and chaotic behavior observed in the domain patterns written by AFM-tip voltages in relatively thick $LiNbO_3$ films [25], can be attributed to an inter-domain electrostatic repulsion.

One can expect that an electrostatic repulsion should be suppressed by the DWs' grounding leading to the elimination of accumulated charge. This effect was actually observed when AFM domain writing in LNOI [20]. If the metal layer was grounded when writing, the written patterns were regular and stable. Oppositely, if the metal layer was disconnected from the ground, the formed patterns were chaotic and unstable.

For clarity, the further reasoning underlying the problem setting is illustrated by Figure 1b. A domain pattern can be regarded as a system of conducting nanowires with a resistance R_{DW}, embedded into an insulating matrix. Two abovementioned extreme scenarios, namely, the regular and chaotic patterns occurring at $R_L = 0$ and $R_L = \infty$, can be interpreted as the results of grounding DWs and their disconnection from the ground, respectively.

The aim of the present research was to examine an "intermediate" case, namely, the domain formation in the conditions of R_{DW} grounding through a load resistor $R_L \neq 0$ (Figure 1b). It was expected that the characteristics of the occurring domain patterns will be affected by R_L insertion, the effects being controlled by the ratio between R_{DW} and R_L.

Experiments were performed in LNOI samples provided by NANOLN Electronics (Jinan, China).

2. Experimental Results

2.1. The Effects of R_L Insertion on the Domain Formation

The panels shown in Figure 2a–c, present the phase PFM images of the patterns written at different conditions. One can detect three types of scenarios, depending on the writing conditions.

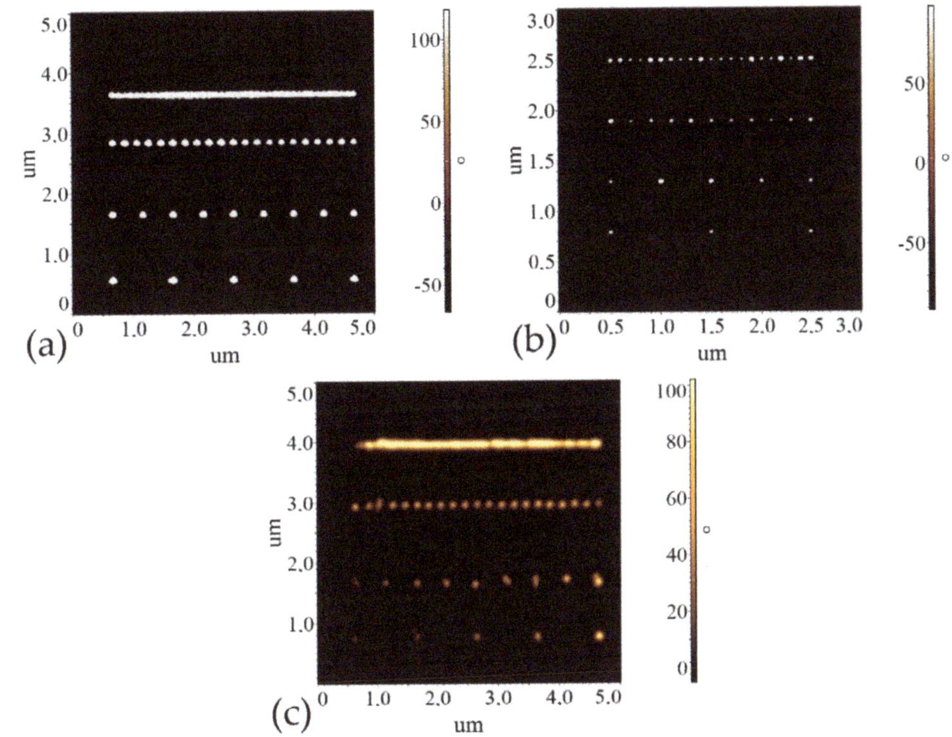

Figure 2. PFM images of the domain panels observed under different conditions of AFM-tip writing. (**a**) U_{tip} = 35 V, t_p = 0.1 ms, R_L = 0; almost the same images are observed up to $R_L = 10^9$ Ohm (see below); (**b**) U_{tip} = 21 V, t_p = 500 ms, R_L = 0; (**c**) Utip = 50 V, tp = 5 s, $R_L = 10^{10}$ Ohm.

Let us start from the writing at R_L = 0 (Figure 2a). In accordance with our recent results [19,20], domains appear on applying certain threshold U_{tip}. At U_{tip} lower than the threshold, no domains are detected obviously because of the backswitching.

In 500 nm thick films the threshold U_{tip} of the domain appearance is of 18 V; for thicker films it is somewhat larger. Figure 2a presents the results of writing at t_p = 0.1 ms, U_{tip} = 35 V. Qualitatively similar panels are visualized on writing in the ranges of U_{tip} > 22 V and $t_p \geq 0.1$ ms (the shortest experimental t_p). The exposure characteristics of the domain diameter $D(U_{tip})$ and $D(t_p)$ were presented in [19]. In agreement with [19,20], the written domains are fully stable. The dependences of the domain sizes $D(U_{tip})$ and $D(t_p)$ are not affected by the inter-domain spacing. A specific feature of these patterns is the domain merging as the distance between the neighboring DWs is decreased (the upper rows in Figure 2a). This merging ignores the domain sizes and occurs when the distance between the neighboring DWs becomes lesser than 20–30 nm. The domain merging phenomenon was described in detail in our preceding publications [19,20].

We dwell shortly on the "pre-threshold" U_{tip} range from 18 to 22 V (Figure 2b) ($R_L = 0$). Figure 2b gives an example of panel written at $U_{tip} = 20$ V, $t_p = 500$ ms. The appearing domains are unstable since on turning-off U_{tip}, the domain diameter D is slowly shrinking during hours. At the smallest $\Lambda = 100$ nm (the upper row in Figure 2b) the domain sizes become somewhat irregular.

The insertion of load resistors R_L leads to the following results. Up to certain R_L value denoted below as R_{crit}, the scenario is very similar to that observed at $R_L = 0$, the writing conditions being the same (Figure 2a). At $R_L > R_{crit}$ the writing characteristics change abruptly (Figure 2c). Now, to create a stable pattern with the domain sizes close to those observed at $R_L = 0$ under the same U_{tip}, the exposure time should be increased up to seconds, which exceeds t_p required for writing at $0 \leq R_L \leq R_{crit}$ more than by order of magnitude.

Let us discuss in order the phenomena described. In further consideration an ohmic contact between DWs and the metal layer was assumed.

A written pattern shorted to the metal layer, in its turn grounded through a resistor R_L can be represented jointly by an equivalent circuit consisting of two resistors, R_{DW} and R_L, connected in series (Figure 1b); here R_{DW} is the total DWs resistance. As the resistance of the surrounding crystal bulk is by orders of magnitude larger than R_{DW}, the voltage drop on it is neglected.

The phenomena occurring at $R_L = 0$ can be qualitatively accounted for by the elimination of an inter-domain repulsion due to the DW's grounding. As a result, the times required for the formation of an equilibrium domain structure are rather short. Due to the disappearance of an inter-domain electrostatic repulsion, nothing prevents the domains from merging when they come together. The physical meaning of the critical distance of 20–30 nm between the neighboring DWs at which the domains start to merge, is unclear yet.

As follows from Figure 1b, at $R_L \neq 0$ the domain formation is governed by the ratio between R_L and R_{DW}. Provided that $R_{DW} \geq R_L$, the voltage drop occurs mainly on R_{DW}; oppositely, in the case that $R_L \geq R_{DW}$, the voltage drops dominantly on R_L. Therefore, in the latter case to write a pattern with the same characteristics as at $R_L = 0$, the exposure times should be increased essentially. This is actually observed in the case illustrated by Figure 2c. At $U_{tip} = 35$ V, $R_L = 0$ the exposure times of writing are of tenth of seconds, whereas when inserting $R_L = 10^{10}$ Ohm, t_p becomes of the order of seconds.

In the framework of our consideration, a threshold change in the domain formation takes place at $R_{crit} \approx R_{DW}$. In other words, at $R_L = R_{crit}$ the voltage is divided approximately equally between R_L and R_{DW}. It should be taken into account (see the experimental procedure) that when writing any next panel, the value of R_L was changed by an order of magnitude. Due to this, the switch from the scenario illustrated by Figure 2a to that shown in Figure 2c occurs in a step-like manner. Therefore, for a given domain pattern, R_{crit} can be evaluated within an order of magnitude.

We dwell briefly on the domain formation in the "pre-threshold" voltage range Utip = 18–20 V (Figure 2b). The observed instability of domain patterns can be qualitatively accounted for by the fact that these voltages are insufficient to provide a complete grounding, so an inter-domain repulsion still persists. This qualitative explanation is supported by the fact that the threshold U_{tip} at which the writing characteristics change abruptly, grows with the film thickness being of 22 and 25 V for the 500 and 700 nm thick films, respectively. Note, an instability of AFM written domains in LNOI mentioned in [21], might be related to a non-complete grounding.

Summing up this section, a load resistor R_{crit} at which the results of writing change radically (from the panel 2a to 2c), is of the order of magnitude of DWs' resistance R_{DW}. This approach paths the way to estimate R_{DW} in a given domain pattern by investigating the dependence of the occurring domain structures on the inserted resistors R_L.

The following remark is relevant here. Our consideration was illustrated by Figure 2a–c presenting the static ultimate structures. These structures result from the process, involving various stages, the slowest of which being obviously the screening of the bound charge by the free carriers. It occurs within a time of $\tau_M = \varepsilon\varepsilon_0/\sigma$ (where $\varepsilon\varepsilon_0$ and σ are the dielectric permittivity and conductivity, respectively). Taking the obtained below DW conductivity $\sigma \approx 2 \times 10^{-7}$ (Ohm cm)$^{-1}$ and $\varepsilon_{33} \approx 30$ for $LiNbO_3$, we come to $\tau_M \approx 10^{-7}$ s. Therefore, the final DWs' equilibrium (static) state visualized by the PFM scanning is attained within the times by three orders shorter than the shortest exposure times t_p.

2.2. *DWC Estimation Based on the Examination of Domain Patterns*

A simplified reasoning presented above, paves the way to evaluate DWC in a given domain pattern. It can be done by investigating the effects of specified resistors R_L inserted between the metal layer and the ground, on the domain formation. In this section we present an example of this evaluation.

These experiments were performed by writing a row of circled domains by means of step-by-step AFM-tip movement. Figure 3a,b present, respectively, the amplitude PFM image of a fragment of this row and the scanline of an individual domain. The domains were written by U_{tip} = 27 V, t_p = 500 ms; the specified distance between the writing points was of Λ = 100 nm.

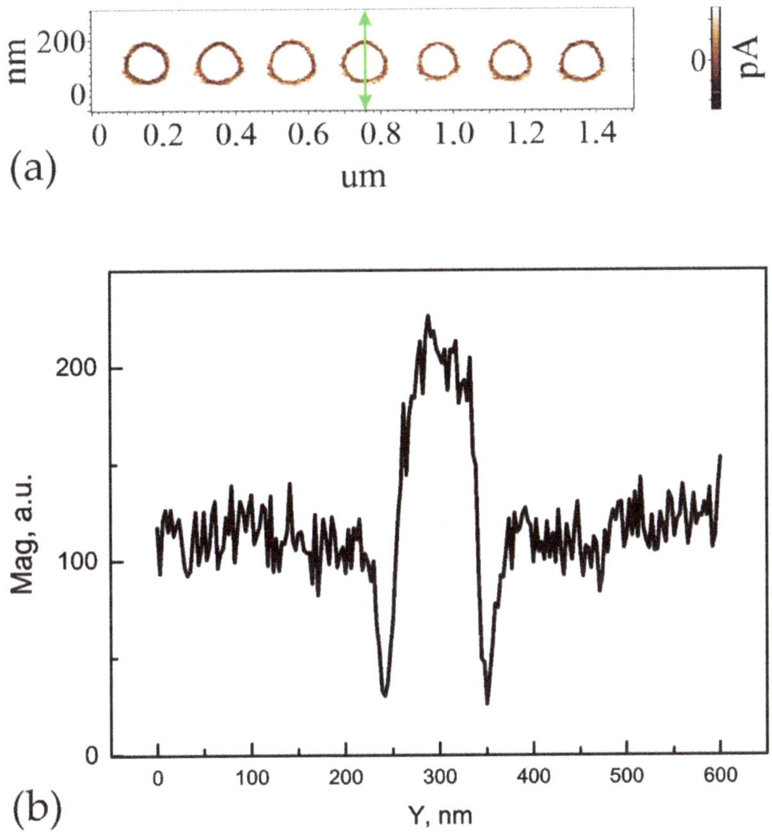

Figure 3. The amplitude PFM image of an AFM-written domain chain underlying the below calculations of DWC (**a**) and the scanline of an individual domain (marked with a green arrow in PFM image) composing this chain (**b**); Utip = 27 V, tp = 500 ms; the AFM-tip radius r ≤ 10 nm.

The sequence of these measurements was the same as when writing the domain panels discussed above. First, the domain row was written at $R_L = 0$ (the reference case). Then, the writing procedure was repeated with the inserted load resistors $R_L \neq 0$, at each subsequent writing step R_L being raised by an order of magnitude. Up to $R_L = 10^9$ Ohm the shape of the PFM image (illustrated by Figure 3a) remained almost unchanged. At $R_L = 10^{10}$ Ohm a drastic change of the writing conditions was observed, namely to write a row identical to that shown in Figure 3a, tp should be increased up to several seconds, thus more than at ten times. In the framework of our consideration this jump takes place at R_L exceeding R_{DW}. Summing up, in the given pattern $R_{DW} \approx R_L = 10^9$ Ohm (correct to first order).

3. Discussion

This section consists of two parts. First, for the pattern shown in Figure 3, the DW conductivity σ_{DW} was calculated based on the above experimental estimate $R_{DW} \approx 10^9$ Ohm. Second, the DW inclination angle in the pattern shown in Figure 3 was evaluated in the framework of the approach proposed in [14].

3.1. DWC Calculations

Two independent approaches were used to calculate DWC. The first approach is as follows. A single domain can be represented by a cylinder with the diameter D; a DW is represented by a thin-walled tube with the wall thickness of w. The domains written by the U_{tip} amplitudes large enough, penetrate through the whole film thickness, thus the tube length is taken equal to the film thickness L. The resistance of a cylindrical domain with the DW thickness of w can be presented as:

$$R_{DW} = \rho_{DW}\,(L/\pi Dw) \equiv 1/\sigma_{DW}\,(L/\pi Dw) \tag{1}$$

As shown above, the conditions of the domain appearance change radically at certain load resistor R_{crit}. In the framework of our consideration it is equal to R_{DW} (correct to first order). From this it follows:

$$\sigma_{DW} = L/(\pi Dw R_{crit}) \tag{2}$$

(for the notations see above).

All terms in this equation are specified except for the value of DW thickness w, which requires to be discussed in more detail. For different materials, e.g., BFO, BTO, and PZT, the value of $w \approx 10$ nm was reported repeatedly (for references see [8]). In our measurements, to find this value, we used a tip with $r \leq 10$ nm for the PFM scanning. This small tip radius permitted us to improve the accuracy of the DW's thickness estimation. The repeated measurements in varied domain patterns analogous to that shown in Figure 3a, gave an average value of $w \approx 20$ nm with the error of 10%. This value is in line with the literature data mentioned above [8].

Substituting $R_{crit} = 10^9$ Ohm, L = 500 nm, D = 100 nm, $w = 20$ nm to Equation (2) we obtain $\sigma_{DW} = 8 \times 10^{-4}$ (Ohm cm)$^{-1}$. This estimate of DWC is conservative, since for DW thicknesses lesser than 20 nm, σ_{DW} would be larger.

We now turn to the alternative estimation of σ_{DW}, which is based on the approach proposed in [14]. The tip contact is considered as a point contact width d. The lines of the current density are curving and spreading out of the DW plane when moving away from the upper electrode. Then:

$$R_{DW} = \rho_{DW}\,(F/w) \equiv (1/\sigma_{DW})(F/w) \tag{3}$$

where F is a dimensionless factor governed by the contact geometry; other notations see above. According to [14], the dependence of F on the contact geometry is satisfactorily approximated by a logarithmic function:

$$F = 0.72 \cdot \lg(5L/a) \approx 0.72 \cdot \lg(5L/R_{tip}) \tag{4}$$

where a is the point contact width.

As was shown above, $R_{crit} = R_{DW}$. Taking into account Equation (4) we obtain:

$$\sigma_{DW} = F/(wR_{crit}) \approx 0.72 \cdot lg(5L/R_{tip})/(wR_{crit}) \quad (5)$$

Substituting into Equation (5) the values used above for calculations by Equation (2) ($R_{crit} = 10^9$ Ohm, L = 500 nm, R_{tip}= 35 nm, w = 20 nm) we obtain $\sigma_{DW} = 7 \times 10^{-4}$ (Ohm cm)$^{-1}$. This is very close to the above alternative estimate $\sigma_{DW} = 8 \times 10^{-4}$ (Ohm cm)$^{-1}$. The value of σ_{DW} obtained by two methods exceeds the tabular bulk conductivity of $LiNbO_3$ not less than by twelve orders of magnitude.

3.2. The DW Inclination Angle

We now evaluate the DW inclination angle θ responsible for the σ_{DW} value estimated above. In these simplified calculations we follow [14]. The conductivity is expressed as $\sigma = ne\mu$, where n is the screening charge concentration, e is the elementary charge and μ is the charge mobility; so $n = \sigma/e\mu$. Taking $\mu \leq 10^{-2}$ cm^2/Vs and substituting it together with $\sigma_{DW} \approx 8 \times 10^{-4}$ (Ohm cm)$^{-1}$ into the expression for the concentration of compensating electrons, we obtain $n \geq 0.6 \times 10^{16}$ cm^{-3}. The inclination angle θ required to provide this compensating charge is calculated from the expression $new = 2P_s \sin\theta$ [14]. Substituting $n \geq 0.6 \times 10^{16}$ cm^{-3}, w = 20 nm and P_s = 70 μC cm^{-2} to this expression, we obtain an inclination angle of $\theta \approx (7 \times 10^{-3})°$. The validity of this estimate is supported by the results of [14], underlying our consideration. In that work, the calculations were performed for the angle $\theta \approx 1°$ specified experimentally. As seen, the inclination of DWs in our case is by orders of magnitude lesser than the angles achieved with the aid of various experimental tricks (e.g., [9,12,14]). A negligible smallness of θ finds the following obvious explanation. As opposed to the aforementioned works aimed at rising DWC by increasing θ, in our case a domain was growing without any external action, so to say, "on its own". Assuming the frontal DW motion to be the dominant mechanism of the domain growth, the DWs' inclination angle is governed by the axial field distribution. So, the value of θ specified by the domain frontal growth itself, should be very small. Certain analogy can be drawn with the electron-beam EB domain writing in $LiNbO_3$ [26]. In that case the domains were growing axially under a local field created by EB-irradiation of the polar plane. The DWs in the formed domains appeared to be inclined by a very small angle.

4. Materials and Methods

The samples under study were LNOI wafers composed of a +Z-cut ion-sliced single-domain $LiNbO_3$ film bonded onto SiO_2 coated $LiNbO_3$ plate. The film thicknesses in the samples under study were of 300, 500 and 700 nm. The thickness of SiO_2 layers was of 1.4 μm. A 100 nm thick Au/Cr layer inserted between the bottom film surface and SiO_2 layer, served as an electrode. The total size of the samples was of X × Y × Z = (11 × 9 × 0.5) mm^3.

The methods of the domain examination with the aid of scanning probe microscopy are commonly known, so needless to describe them in detail. The only detail should be mentioned. Two types of AFM tips were utilized for different aims. The domain writing and subsequent PFM scanning were performed using Si tip with Pt coating HA_FM/Pt (Tipsnano, Tallinn, Estonia) with the tip curvature radius R ≤ 35 nm. To determine the DWs' thickness w in the written patterns, we used the Si tips with boron doped diamond coating HA_HR_DCP (Tipsnano) having a tip curvature radius R ≤ 10 nm. All AFM experiments were carried out with a NTEGRA PRIMA SPM system (NT–MDT, Moscow, Russia).

To search for the effects expected, the panels (illustrated by Figure 2) consisting of domain rows with the specified distances Λ between the writing points were written by the stepwise tip movement; in a given row Λ = const. On writing, each next panel was PFM–scanned. Any panel consisting of the domain rows with Λ decreasing from 1000 to 100 nm (bottom-up) was written under constant exposure conditions (U_{tip} = const, t_p = const).

When studying the effects of load resistors $R_L \neq 0$ inserted between the metal layer and the ground on the domain occurrence, the sequence of operations was as follows. First, a reference panel was written under the given writing conditions (U_{tip} = const, t_p = const) at $R_L = 0$. Then, the following panels were written with the same U_{tip} and t_p, a load resistor R_L being inserted between LNOI and the ground. When writing each subsequent panel, the R_L value was increased (or decreased) by an order of magnitude. The domain panels written with $R_L \neq 0$ were compared to the reference one.

The DWC measurements were performed in a row of circled domains written by step-by-step AFM-tip movement (Figure 3a,b). The sequence of these measurements was the same as when writing the domain panels (Figure 2). First, the domain row was written at $R_L = 0$ (the reference case). Then, the writing procedure was repeated with the inserted load resistors $R_L \neq 0$. At each subsequent writing step, R_L was raised by an order of magnitude

5. Conclusions

The main results obtained here can be summarized as follows: The AFM-tip domain writing in $LiNbO_3$ films was investigated under various conditions of the DW's grounding through the specified load resistors R_L inserted between DWs and the ground. The features of arising patterns are determined by the ratio between the domain-wall resistance R_{DW} and R_L. The formed patterns depend critically on R_L. This permitted us to estimate DWC in a specified pattern. Importantly, this estimate was based on the observations of ferroelectric phenomena. It was performed by comparing the patterns occurring at given Utip and tp with varied R_L. The calculations of DWC on the basis of two independent approaches gave the very close values of σ_{DW} = (7–8) $\times 10^{-4}$ (Ohm cm)$^{-1}$. In the framework of the approach developed in [14] this DWC corresponds to a small inclination angle of $\theta \approx (7 \times 10^{-3})°$. The observed DWC not caused by any external manipulation, is related to a DW inclination accompanying the domain frontal growth under an AFM-tip field.

In our opinion, an important general conclusion is as follows: When creating a small-scale domain pattern, no matter by which method, an electrostatic repulsion between the neighboring DWs and, thus, the conditions of DWs' grounding affect substantially the features of the occurring terminal domain structure.

Author Contributions: R.G. performed the experiments and analyzed the data, T.V. analyzed the data and wrote the paper. All authors have read and agreed to the published version of the manuscript.

Funding: This research received no external funding.

Acknowledgments: The authors are grateful to Boris Sturman for the fair criticism and useful remarks. We are indebted to NANOLN Electronics (Jinan, China) for providing us the LNOI samples. This work was supported by the Ministry of Science and Higher Education of the Russian Federation. The equipment of the Shared Research Center (Project No. RFMEFI62114x0005) was used in experiments.

Conflicts of Interest: The authors declare no conflict of interest.

References

1. *Ferroelectric Crystals for Photonic Applications*; Springer Science and Business Media LLC: Berlin/Heidelberg, Germany, 2009; pp. 1–132.
2. Cho, Y.; Fujimoto, K.; Hiranaga, Y.; Wagatsuma, Y.; Onoe, A.; Terabe, K.; Kitamura, K. Tbit/inch2 ferroelectric data storage based on scanning nonlinear dielectric microscopy. *Appl. Phys. Lett.* **2002**, *81*, 4401–4403. [CrossRef]
3. Vul, B.M.; Guro, G.M.; Ivanchik, I.I. Encountering domains in ferroelectrics. *Ferroelectrics* **1973**, *6*, 29–31. [CrossRef]
4. Calleja, M.; Dove, M.T.; Salje, E.K.H. Trapping of oxygen vacancies on twin walls of $CaTiO_3$: A computer simulation study. *J. Physics Condens. Matter* **2003**, *15*, 2301–2307. [CrossRef]

5. Aird, A.; Salje, E.K.H. Sheet superconductivity in twin walls: Experimental evidence of. *J. Physics Condens. Matter* **1998**, *10*, L377–L380. [CrossRef]
6. Seidel, J.; Martin, L.W.; He, Q.; Zhan, Q.; Chu, Y.-H.; Rother, A.; Hawkridge, M.E.; Maksymovych, P.; Yu, P.; Gajek, M.; et al. Conduction at domain walls in oxide multiferroics. *Nat. Mater.* **2009**, *8*, 229–234. [CrossRef]
7. Catalán, G.; Seidel, J.; Ramesh, R.; Scott, J.F. Domain wall nanoelectronics. *Rev. Mod. Phys.* **2012**, *84*, 119–156. [CrossRef]
8. Bednyakov, P.S.; Sturman, B.I.; Sluka, T.; Tagantsev, A.K.; Yudin, P.V. Physics and applications of charged domain walls. *NPJ Comput. Mater.* **2018**, *4*. [CrossRef]
9. Schröder, M.; Haußmann, A.; Thiessen, A.; Soergel, E.; Woike, T.; Eng, L.M. Conducting Domain Walls in Lithium Niobate Single Crystals. *Adv. Funct. Mater.* **2012**, *22*, 3936–3944. [CrossRef]
10. Gonnissen, J.; Batuk, D.; Nataf, G.F.; Jones, L.; Abakumov, A.M.; Van Aert, S.; Schryvers, D.; Salje, E.K.H. Direct Observation of Ferroelectric Domain Walls in LiNbO3: Wall-Meanders, Kinks, and Local Electric Charges. *Adv. Funct. Mater.* **2016**, *26*, 7599–7604. [CrossRef]
11. Wolba, B.; Seidel, J.; Cazorla, C.; Godau, C.; Haußmann, A.; Eng, L.M. Resistor Network Modeling of Conductive Domain Walls in Lithium Niobate. *Adv. Electron. Mater.* **2018**, *4*, 1700242. [CrossRef]
12. Godau, C.; Kampfe, T.; Thiessen, A.; Eng, L.M.; Haußmann, A. Enhancing the Domain Wall Conductivity in Lithium Niobate Single Crystals. *ACS Nano* **2017**, *11*, 4816–4824. [CrossRef] [PubMed]
13. Volk, T.; Gainutdinov, R.V.; Zhang, H.H. Domain-wall conduction in AFM-written domain patterns in ion-sliced $LiNbO_3$ films. *Appl. Phys. Lett.* **2017**, *110*, 132905. [CrossRef]
14. Werner, C.S.; Herr, S.J.; Buse, K.; Sturman, B.; Soergel, E.; Razzaghi, C.; Breunig, I. Large and accessible conductivity of charged domain walls in lithium niobate. *Sci. Rep.* **2017**, *7*, 9862. [CrossRef] [PubMed]
15. Lu, H.; Tan, Y.; McConville, J.P.V.; Ahmadi, Z.; Wang, B.; Conroy, M.; Moore, K.; Bangert, U.; Shield, J.E.; Chen, L.; et al. Electrical Tunability of Domain Wall Conductivity in $LiNbO_3$ Thin Films. *Adv. Mater.* **2019**, *31*, e1902890. [CrossRef] [PubMed]
16. Poberaj, G.; Hu, H.; Sohler, W.; Gunter, P. Lithium niobate on insulator (LNOI) for micro-photonic devices. *Laser Photon- Rev.* **2012**, *6*, 488–503. [CrossRef]
17. Boes, A.; Corcoran, B.; Chang, L.; Bowers, J.; Mitchell, A. Status and Potential of Lithium Niobate on Insulator (LNOI) for Photonic Integrated Circuits. *Laser Photon- Rev.* **2018**, *12*, 1700256. [CrossRef]
18. Rusing, M.; Weigel, P.O.; Zhao, J.; Mookherjea, S. Toward 3D Integrated Photonics Including Lithium Niobate Thin Films: A Bridge Between Electronics, Radio Frequency, and Optical Technology. *IEEE Nanotechnol. Mag.* **2019**, *13*, 18–33. [CrossRef]
19. Gaĭnutdinov, R.V.; Volk, T.; Zhang, H.H. Domain formation and polarization reversal under atomic force microscopy-tip voltages in ion-sliced $LiNbO_3$ films on $SiO_2/LiNbO_3$ substrates. *Appl. Phys. Lett.* **2015**, *107*, 162903. [CrossRef]
20. Volk, T.; Gainutdinov, R.; Zhang, H. Domain Patterning in Ion-Sliced LiNbO3 Films by Atomic Force Microscopy. *Crystals* **2017**, *7*, 137. [CrossRef]
21. Shao, G.-H.; Bai, Y.-H.; Cui, G.-X.; Li, C.; Qiu, X.-B.; Geng, D.-Q.; Wu, D.; Lu, Y.-Q. Ferroelectric domain inversion and its stability in lithium niobate thin film on insulator with different thicknesses. *AIP Adv.* **2016**, *6*, 075011. [CrossRef]
22. Slautin, B.N.; Turygin, A.; Greshnyakov, E.D.; Akhmatkhanov, A.R.; Zhu, H.; Shur, V.Y. Domain structure formation by local switching in the ion sliced lithium niobate thin films. *Appl. Phys. Lett.* **2020**, *116*, 152904. [CrossRef]
23. Kan, Y.; Bo, H.; Lu, X.; Xu, T.; Jin, Y.; Wu, X.; Huang, F.; Zhu, J. Decay properties of artificial two-domain structures in LiNbO3 crystals studied by scanning probe microscope. *Appl. Phys. Lett.* **2010**, *97*, 202903. [CrossRef]
24. Ofan, A.; Lilienblum, M.; Gaathon, O.; Sehrbrock, A.; Hoffmann, A.; Bakhru, S.; Bakhru, H.; Irsen, S.; Jr, R.M.O.; Soergel, E. Large-area regular nanodomain patterning in He-irradiated lithium niobate crystals. *Nanotechnology* **2011**, *22*, 285309. [CrossRef] [PubMed]
25. Ievlev, A.V.; Jesse, S.; Morozovska, A.N.; Strelcov, E.; Eliseev, E.A.; Pershin, Y.V.; Kumar, A.; Shur, V.Y.; Kalinin, S. Intermittency, quasiperiodicity and chaos in probe-induced ferroelectric domain switching. *Nat. Phys.* **2013**, *10*, 59–66. [CrossRef]

26. Dierolf, V.; Sandmann, C. Direct-write method for domain inversion patterns in $LiNbO_3$. *Appl. Phys. Lett.* **2004**, *84*, 3987–3989. [CrossRef]

crystals

Article

"Seeing Is Believing"—In-Depth Analysis by Co-Imaging of Periodically-Poled X-Cut Lithium Niobate Thin Films

Sven Reitzig [1,*], Michael Rüsing [1], Jie Zhao [2,†], Benjamin Kirbus [1], Shayan Mookherjea [2] and Lukas M. Eng [1,3]

1 Institut für Angewandte Physik, Technische Universität Dresden, 01062 Dresden, Germany; Michael.Ruesing@tu-dresden.de (M.R.); benjamin.kirbus@tu-dresden.de (B.K.); lukas.eng@tu-dresden.de (L.M.E.)
2 Department of Electrical and Computer Engineering, University of California, San Diego, CA 92161, USA; jie.2.zhao@nokia-bell-labs.com (J.Z.); smookherjea@ucsd.edu (S.M.)
3 ct.qmat: Dresden-Würzburg Cluster of Excellence—EXC 2147, TU Dresden, 01062 Dresden, Germany
* Correspondence: sven.reitzig@tu-dresden.de; Tel.: +49-351-463-43354
† Since February 2021 affiliated with Nokia Bell Labs, Murray Hill, NJ 07974, USA.

Abstract: Nonlinear and quantum optical devices based on periodically-poled thin film lithium niobate (PP-TFLN) have gained considerable interest lately, due to their significantly improved performance as compared to their bulk counterparts. Nevertheless, performance parameters such as conversion efficiency, minimum pump power, and spectral bandwidth strongly depend on the quality of the domain structure in these PP-TFLN samples, e.g., their homogeneity and duty cycle, as well as on the overlap and penetration depth of domains with the waveguide mode. Hence, in order to propose improved fabrication protocols, a profound quality control of domain structures is needed that allows quantifying and thoroughly analyzing these parameters. In this paper, we propose to combine a set of nanometer-to-micrometer-scale imaging techniques, i.e., piezoresponse force microscopy (PFM), second-harmonic generation (SHG), and Raman spectroscopy (RS), to access the relevant and crucial sample properties through cross-correlating these methods. Based on our findings, we designate SHG to be the best-suited standard imaging technique for this purpose, in particular when investigating the domain poling process in x-cut TFLNs. While PFM is excellently recommended for near-surface high-resolution imaging, RS provides thorough insights into stress and/or defect distributions, as associated with these domain structures. In this context, our work here indicates unexpectedly large signs for internal fields occurring in x-cut PP-TFLNs that are substantially larger as compared to previous observations in bulk LN.

Keywords: thin film lithium niobate; TFLN; LNOI; x-cut LN; ferroelectric domains; domain walls; piezoresponse force microscopy; second-harmonic generation; Raman scattering

Citation: Reitzig, S.; Rüsing, M.; Zhao, J.; Kirbus, B.; Mookherjea, S.; Eng, L.M. "Seeing Is Believing"—In-Depth Analysis by Co-Imaging of Periodically-Poled X-Cut Lithium Niobate Thin Films. *Crystals* **2021**, *11*, 288. https://doi.org/10.3390/cryst11030288

Academic Editor: Gábor Corradi

Received: 26 February 2021
Accepted: 12 March 2021
Published: 15 March 2021

1. Introduction

In recent years, periodically-poled thin film lithium niobate (PP-TFLN) has emerged as a promising platform for realizing modern-type integrated nonlinear and quantum optical devices. When compared to the standard bulk periodically-poled lithium niobate (PPLN) platform, PP-TFLN offers higher conversion efficiencies, a significantly reduced footprint size, as well as an excellent integrability into nanooptical systems [1–7]. Unlike bulk PPLN, the application of the TFLN platform is not limited by the restrictions that weak optical confinement imposes to footprint and device integration [8–10]. Subwavelength optical confinement is provided by a large change of the refractive index between TFLN and the substrate material. This is mandatory and fundamental in order to achieve an ultra-high frequency conversion for next-generation applications.

Similar to other ferroelectrics, efficient nonlinear conversion may be achieved by applying the straightforward concept of quasi phase matching (QPM) to periodically-poled ferroelectric (FE) domain structures. For efficient and narrow-band QPM devices, it is thus required that the domain grid matches closely with its design parameters [11,12]. In particular, early studies on nonlinear optical devices in TFLN reported on strong deviations between simulated and effectively measured conversion efficiencies and spectra, a fact that was attributed to inhomogeneities in the poling procedure. Nevertheless, these assertions were off ground, and could not be verified to date, since they simply lack the appropriate analysis methods [1]. Hence, adequate inspection techniques are needed to quantify the necessary parameters, to the benefit of proposing dedicated process protocols for the optimized and highly accurate domain fabrication in TFLN. Key indicators for high-grade domain grids in TFLN are: (a) The appropriate poling period; (b) the duty cycle and homogeneity; (c) the domain grid length; and (d) the depth of poled domains. Therefore, inspection methods need to provide access to these parameters.

(a) Poling period inspection (resolution): Due to the strong confinement and the associated large dispersion of optical modes, TFLN requires poling periods that are significantly shorter as compared to bulk LN. For a typical application at telecom frequencies, poling periods Λ between 2 and 5 μm must be realizable in TFLN, while for similar devices in bulk LN 10–30 μm periods usually would be sufficient [2]. This makes the fabrication of domain grids in TFLN much more demanding. Furthermore, recent works have demonstrated that even sub-μm periodicities in PP-TFLN waveguides can be manufactured [13,14], an achievement that has not yet been possible with bulk devices. Therefore, any investigation method must provide an optical resolution in the lower μm and even sub-μm range, as well as be able for imaging over the full length of the poled areas.

(b) Duty cycle and homogeneity control (imaging contrast): For an optimal conversion efficiency, a duty cycle of the periodic domain grid close to 50% is desired. Furthermore, irregular variations in the duty cycle will significantly broaden the conversion spectrum and hence reduce that efficiency [11,12]. Therefore, methods with a clear imaging contrast are required to analyze the duty cycle along its full poling length.

(c) Domain grid length (speed): The nonlinear conversion efficiency scales quadratically with the interaction length, while the spectral width decreases. Hence, for narrow band, low pump power, and high efficiencies, device lengths in the range of mm to cm or even longer are requested [2,9,15]. Ideally, an appropriate inspection method then must be able to investigate these large areas in a reasonable time, i.e., within less than some hours, solely to facilitate fabrication and structure optimization.

(d) In-depth domain growth control (depth sensitivity): To achieve optimal efficiencies, all inverted domains along the waveguide structure need to penetrate across the full depth of the TFLN. In most designs, x-cut TFLN is used for fabrication, facilitating domain poling via surface electrodes that are easily fabricated and removed by standard lithography, deposition, and etching techniques. Hence, the PP domain structure is arranged in-plane, requiring that domain growth fully proceeds along both the in-plane polar axis and into the TFLN (non-polar) depth, as shown in Figure 1a. To optimize the fabrication of PP domain grids as well as for delivering decent conversion estimates, these depths of inverted domains need to be known and measured with high accuracy. Typical optical applications involve TFLN feature thicknesses of 300–800 nm. Hence, imaging methods that offer a depth-sensitivity of a few nm are highly desirable.

Furthermore, it is highly beneficial that an imaging method is not only sensitive to the domain structure, but may be able to simultaneously detect parameters that potentially might influence the domain growth dynamics and/or the linear and nonlinear optical properties [16]. Examples into this context are defects, dopant concentration variations (e.g., from waveguide diffusion [17]), built-in electric fields or inhomogeneous stress distributions [18]. Here, an appropriate imaging method might allow analyzing the interaction between the domain structures and its growth, as well as other influencing factors. It is known from many previous works that domain walls (DWs) are accompanied by defects, electric fields, and significant mechanical stress, that might reach out by as much as some μm [19]. For TFLN, this is particularly relevant, since, on the one hand, typical film thicknesses are on the order of 500 nm, resulting in extremely strained domain structures. On the other hand, poling periods in TFLN are significantly shorter than in bulk LN, which might impact the material properties or lead to DW-DW-interactions [20]. Additionally in x- or y-cut TFLN, other sources such as strong built-in fields stemming from charged head-to-head (h2h) and tail-to-tail (t2t) DWs need to be mentioned. Hence, a fundamental study of this broad palette of influencing parameters especially in TFLN is highly desired, since having been mostly neglected so far.

In total, it is critical to apply imaging techniques to the domain structures that enable monitoring and quantifying the aforementioned key performance indicators, in order to optimize the conversion efficiencies of PP-TFLN based devices. Several analysis methods have been applied so far, with limited success:

In situ transmission monitoring methods [4,21] are very fast and non-destructive, for example, as electrooptical [21] or SH-conversion-efficiency [4] monitoring of that amount of light which is transmitted through the waveguide while poling the surrounding. The poling procedure is then stopped when reaching a preset indicator value, e.g., a certain peak efficiency. While these methods provide almost instant feedback, they only deliver averaged and indirect information on the duty cycle and poling quality. In addition, they provide no insights into the microscopic domain distribution, the shape, depth or additional influence factors as mechanical stress.

For z-cut PPLN bulk wafers, selective etching can be considered as "the" standard technique. Selective etching [1] relies on the fact that oppositely-poled crystalline facets along the FE axis of LN show different etching rates in an hydrofluoric acid (HF) etch solution, typically exhibiting a deeper etch on the negatively-poled face [22]. This allows directly accessing the domain shape via topography on the z-surface, which is then analyzed by standard techniques such as light microscopy, AFM techniques, profilometers or electron microscopy. While being widely applied for z-cut bulk devices due to its large-scale imaging potential, it might be readily applicable to z-cut TFLN structures, as well [6]. Nevertheless, its destructive nature prohibits an application for standard process control in thin films: A conservative estimate of the etch-depth resolution certainly yields values on the order of 10–100 nm. While for a bulk crystal with a waveguide depth in the order of several μm, a surface grid of a few tens of nanometer can be polished without impacting the design properties of a waveguide structure. Nevertheless, this is not possible in TFLN. Here, even a thickness change on the order of tens of nanometers will significantly impact the propagating modes [2]. Furthermore, in TFLN structures, HF may impact the buried oxide layer as well and, hence, endanger the bonding. In addition, x-cut LN which is often used due to the above-mentioned advantages in domain fabrication, cannot be readily analyzed via selective etching. As HF etching affects the FE z-axis of the LN crystal only, it is necessary to expose the respective crystal surface by an additional step, e.g., via focused ion beam milling (see Figure 1b), a process step that dramatically limits fast fabrication. Finally, these adaptions lead to a low information content of the analysis: Rather than a large-scale domain pattern, selective TFLN etching only reveals the one-dimensional duty cycle and poling period information along the exposed z-face of the thin film. Hence, only small areas (<40 μm) of poled domain structures can be investigated in this way in x-cut TFLNs [1,5].

Hence, for the investigation of domain structures in TFLN, the non-destructive scanning probe or optical methods are preferred, which in addition, do not require any special sample preparation. As shown from previous investigations, SHG microscopy and piezoresponse force microscopy (PFM) can be successfully applied to inspect and analyze TFLN [1–3], analogously to their use for poled bulk LN [2,3,14,21,23]. Both methods directly detect a change in a material property which is connected to the FE domain orientation, namely the piezoelectric and nonlinear optical tensor, respectively. Hence, no special sample preparation for imaging is required. While to date, the specifics of the SHG contrast mechanism in TFLN have been thoroughly analyzed both by simulation and experiments, PFM has only been used as a mere imaging technique on TFLN, but no attempts have been undertaken to further analyze the obtained signals, for instance, by explaining apparent differences between bulk LN and TFLN imaging.

Raman spectroscopy (RS) constitutes another (nonlinear) optical method which is well suited for imaging FE domain structures. While commonly applied to visualize bulk LN domain structures [24–28], no other works, to the best of our knowledge, so far have ever reported on the RS analysis and imaging of domain structures in TFLN. RS not only allows giving clear pictures of the domain distribution, but is also sensitive to other effects, such as defects, electric fields or stress distributions [29–31] which are often related to DWs. The potential to gain further insights into the TFLN system makes RS a promising and complementary tool for TFLN imaging and spectroscopy.

To date, most studies on PP-TFLN always rely on interpreting the results obtained from one single technique used for visualizing and analyzing the fabricated domain structures; comparing the different responses by complementary methods from one and the same sample spot is not reported yet. Here, a systematic comparison would allow not only to validate the potential and strength of every such method involved, but, furthermore, also to reveal the true domain shape and characteristics through cross-correlating the different contrast mechanisms of each technique. Such a comparison is more than relevant and needed, as contrast mechanisms and imaging results might always be influenced by the specific conditions of both the thin-film system and setup, as shown, for example, by SHG microscopy [23].

Therefore, we will combine here PFM, SHG, and RS for the analysis of x-cut TFLN down to the nanometer resolution. We will start by briefly reviewing the capabilities of every method and identify their strengths and weaknesses in order to determine the key performance factors in TFLN. Furthermore, we will identify the challenges for each method that may be addressed in future work.

2. Materials and Imaging Methods

2.1. Sample

The sample for the study here was fabricated from commercial ion-sliced TFLN (congruent 5% MgO-doped LN; NanoLN, Jinan Jingzheng Electronics Co., Ltd., Jinan, Shandong, China). The sample consists of a 300-nm thin single-crystalline lamella of a single FE domain oriented along the x-axes (x-cut TFLN), chip-bonded to a SiO_2/Si substrate (1800 nm SiO_2 on 500 μm Si {100}) to yield a lithium niobate on the insulator (LNOI) platform. In x-cut LN, the spontaneous polarization points parallel to the sample surface (in-plane polarization; z-axis). Therefore, electric field poling is readily achieved through electrodes manufactured solely onto the sample's top surface, as depicted in Figure 1a. The poling electrodes were structured with standard photolithography and lift-off, and consist of a 10-nm Cr adhesion layer and a 100-nm thin gold layer. Poling electrodes on the sub-μm sample in Section 3.2 were patterned via electron beam lithography. For poling, electrodes were contacted via electrical probes and a single, high voltage poling pulse surpassing the coercive field was applied. Details on the fabrication process and protocol are reported elsewhere [13,21].

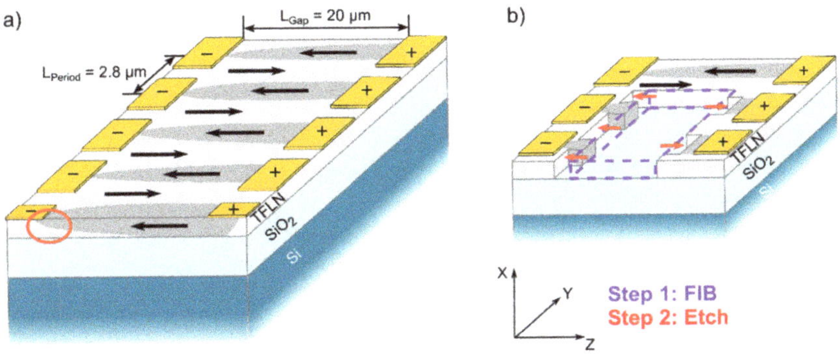

Figure 1. (**a**) Sketch of the periodically-poled x-cut thin film lithium niobate (TFLN) sample, consisting of a 300 nm lithium niobate (LN) thin film on a silicon substrate with silicon oxide (1800 nm). TFLN polarization is inverted by surface Au electrodes (period: 2.8 µm, gap distance: 20 µm). The inverted domains (dark grey) might not reach across the full depth of the LN thin film, as indicated in the foremost domain (red circle), since inversion along the *z*-axis proceeds faster than along x. (**b**) Selective etching scheme on x-cut TFLN. After exposure of the LN z-face by focused ion beam (FIB) (purple dashed box), subsequent etching with a higher etch rate on the negative polar face reveals the domain pattern, which is then ready to be visualized by standard imaging techniques.

2.2. Imaging Techniques

2.2.1. Piezoresponse Force Microscopy—PFM

Piezoresponse force microscopy (PFM) is a scanning probe technique based on atomic force microscopy (AFM). PFM is considered one of the standard techniques for imaging and analyzing FE materials and their domain structures [32–34]. PFM makes use of the fact that the orientation of FE domains directly determines the orientation of the piezoelectric tensor. Hence, by mapping the piezoelectric response, i.e., its phase and amplitude, FE domains can be visualized. Moreover, when quantifying the whole piezoelectric tensor in crystalline ferroelectrics, PFM may deliver the crystallographic sample orientation at the 1-nm length scale [35]. A typical PFM setup is sketched in Figure 2a. Here, an AC-voltage is applied to the conductive AFM cantilever tip, which is in contact with the grounded sample. The high electric field at the tip apex leads to a piezoelectric response of the sample, e.g., micromechanical expansion or contraction, of the piezoelectric material. This leads to an accompanied motion of the AFM tip, e.g., normal (deflection), lateral (torsion) [36] or buckling motion, which is then monitored by the AFM setup. The resolution of a PFM setup directly scales with the AFM tip diameter [37], but may be influenced by other parameters as well, such as the applied peak voltage or sample properties, e.g., depolarization charges [38]. In principle, a lateral resolution down to the 1-nm range is possible. More in-depth reviews on PFM, its capabilities, and different operation modes can be found in various reviews [32–34].

The PFM images in this study were recorded using a commercial Cypher AFM (Asylum Research, Oxford Instruments) and its built-in Vector PFM mode that allows analyzing lateral, as well as normal piezoelectric motions. For imaging, standard Pt-coated silicon tips were used (tip radius < 50 nm). For high-resolution imaging down to the 10-nm length scale, we used full-metal tips (Rocky Mountain tips; Pt-Ir; tip radius < 20 nm).

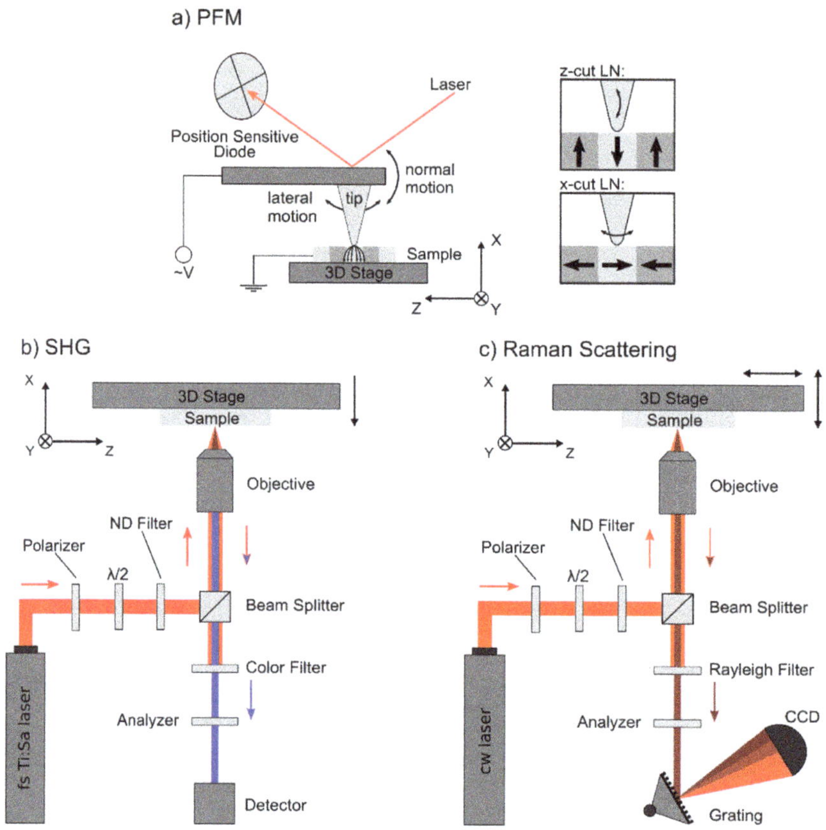

Figure 2. Schematic drawings of the different tools and setups as applied for imaging x-cut TFLN in this work. (**a**) In piezoresponse force microscopy (PFM), the normal or lateral motion of the cantilever tip is induced by the piezoelectric response to an electric field applied between the conductive tip and (insulating) sample. Insets to the right: Polarization directions and respective main tip motions for t-cut and x-cut LN samples. (**b**) In second-harmonic generation (SHG), the fundamental laser light is focused onto the sample through an objective, while the backscattered SH signal is collected via the same objective. Light with the fundamental wavelength is blocked via color filters, allowing only the second-harmonic signal to be detected. (**c**) The cw laser in the Raman backscattering setup is polarized, focused, and recollected analogously to (**b**). The Rayleigh filter blocks off most incident light (Rayleigh scattering), while the scattered "Raman" light is spectrally analyzed using a grating and appropriate charge-coupled device (CCD) camera.

2.2.2. Second-Harmonic Generation Microscopy—SHG

Second-harmonic generation (SHG) microscopy is an optical microscopy technique that relies on the second-order nonlinear optical effect of second-harmonic generation as its contrast mechanism. In the SHG process, two photons at the pump wavelength are annihilated to create a single photon at half the wavelength (twice the energy of a single photon). Analogously to PFM, SHG and hence the SH tensor are dependent on the crystal symmetry, i.e., FE domains and their orientation. Therefore, it can be ideally used to visualize domain structures. Due to its nature as a nonlinear, optical process, it requires high pump intensities usually provided from ultra-fast, fs-pulsed laser light sources.

For this work, we used two different SHG setups, one self-built SHG microscope, as well as a commercial microscope (SP5 MP by Leica Microsystems, Wetzlar, Hessen, Germany). Figure 2b schematically depicts their optical layout. Both SHG systems use a tunable fs-Ti: Sapphire laser as the pump source (self-built: Griffin by KMLabs, Boulder,

Colorado, USA, tuning range: 760–840 nm; Leica: MaiTai by Spectra-Physics, Stahnsdorf, Brandenburg, Germany, tuning range 720–900 nm). The pump light is directed to the sample via a dichroic mirror and a high numerical aperture (NA) air objective (self-built: NA = 0.9, Leica: NA = 0.8). The locally generated SH light is collected in back-reflection through the same objective. To block off any pump light from the detector (self-built: Silicon single photon avalanche diode; Leica: Photo-multiplier tubes) appropriate short-pass filters are inserted into the detection path. Since SH light is only generated point-wise at the focal location, a scanning procedure is needed in order to generate SHG images; hereto galvo mirror scanners steering the pump beam are installed in the Leica microscope, whereas the sample is mounted on a 3D-scanning piezo stage with a fixed focus in the home-built setup.

2.2.3. Raman Micro-Spectroscopy—RS

The Raman effect describes the inelastic scattering of photons with low energy excitations. In crystalline media, Raman scattering (RS) is primarily linked to the excitation of phonons. Scattering properties such as the phonon frequency or the scattering cross section are very sensitive to changes in the crystal structure, i.e., the geometry of the crystal lattice and its atomic composition. As vibrational motions and involved atoms significantly vary between different phonons, the specific influence of structural changes on certain phonon properties can give valuable insights into the nanoscopic impacts of microscopic effects, such as the position of DWs, which represent a specific change in the crystal structure. Hence, by spatially mapping changes in the phonon spectrum, DWs and other influences can be readily distinguished.

RS experiments are carried out here using a Horiba LabRAM HR Evolution commercial Raman spectroscope (HORIBA Jobin Yvon GmbH, Bensheim, Hessen, Germany), (λ = 633 nm, P = 17 mW cw, 1800 L/mm grating). A scheme of the used backscattering setup is shown in Figure 2c. The high NA = 0.9 of the microscope objective provides a good signal strength and a small focal spot (lateral Rayleigh criterion: 429 nm; axial Rayleigh criterion: 1.56 µm) to detect signals mainly from TFLN. Spectra were recorded in x(zz)-x (A_1(TO) modes) and x(zy)-x (E(TO) modes) Porto geometries [20]. To ensure a constant focus for the whole measurement, an automated focusing routine is applied to every measuring spot, consisting of a depth sweep with an automated fitting procedure, in order to allocate that focus depth at which the largest signal intensity for the $A_1(TO_1)$ peak (~250 cm^{-1}) or the $E(TO_1)$ mode (~155 cm^{-1}), respectively, is recorded. Images are generated by analyzing phonon peak properties (peak intensity, frequency shifts) of hyperspectral Raman maps. In this work, a lateral point-to-point separation of 100 nm in y-direction and 500 nm in z-direction was chosen with a higher sampling rate along the *y*-axis, i.e., perpendicular to the created domain walls.

3. Results and Discussion

To compare the potential and restrictions of each imaging technique listed in Section 2, the PP-TFLN sample with a domain period of Λ = 2.8 µm is analyzed with PFM, SHG, and RS. A selection of the results is displayed in Figure 3. Here, for each method, we selected a signal property that provides a DW contrast. All images show a similar representation of the domain structure, demonstrating that every method in principle is able to visualize the same structure, despite the fact that each method probes different material properties, i.e., the piezoelectric tensor for PFM, the nonlinear tensor in SHG, and the phonon spectrum for RS. However, the colormaps differ in properties such as the signal-to-noise ratio, number of pixels, and most importantly the perceived DW width. Partly, this is due to the different spatial resolution, but also since different material properties change at different length scales close to DWs [19,27]. The respective imaging details will be discussed in the following sections.

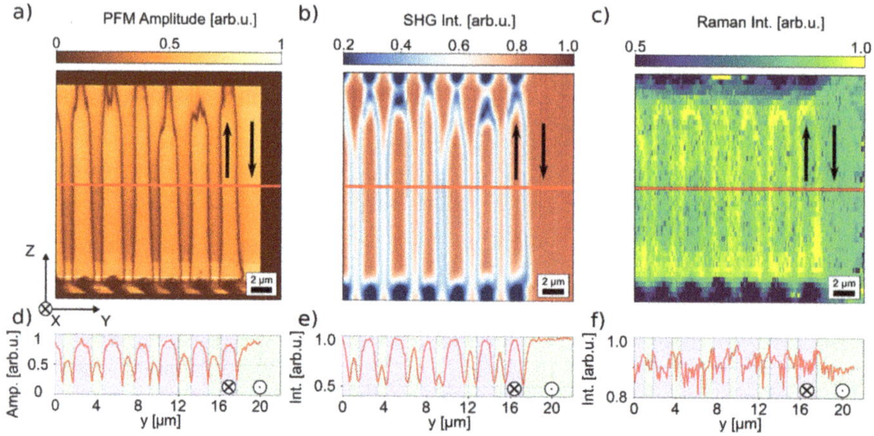

Figure 3. Periodically-poled thin film lithium niobate (PP-TFLN) in-plane ferroelectric domains imaged consecutively by (**a**) piezoresponse force microscopy (PFM), (**b**) second-harmonic generation (SHG), and (**c**) Raman spectroscopy (RS) over exactly the same sample area. Note the pronounced domain wall (DW) contrast by all techniques. The colormaps in (**d–f**) show cross-sections taken along the indicated red lines of the respective method (**a–c**). As a guide to the eye, green (pristine) and blue (inverted) domain areas mark the differently oriented domains. All measurements show the same domain pattern despite their different contrast mechanisms.

3.1. Piezoresponse Force Microscopy

PFM is a widespread method for the analysis and imaging of FE thin films or bulk crystals [32–34]. The contrast in PFM relies on directly probing the piezoelectric tensor, which is connected to the FE domain orientation. In PFM, the tip is in contact with the sample, and an AC-voltage is applied between the conductive tip and (insulating) sample. Depending on the sample orientation, i.e., the piezoelectric tensor orientation, the sample will react with a piezoelectric deformation, e.g., contraction or expansion, to the electric field. In this simple picture, the domain contrast can be explained as follows: Lithium niobate is a uniaxial FE crystal, where only two domain orientations are allowed along its crystallographic z-direction. Here, the piezoelectric tensor is inverted in domains of antiparallel order parameter orientation. Hence, the piezoelectric response will be 180° phase-shifted between two such adjacent domains. In a simple Ising-type picture, the polarization on DWs in ferroelectrics shows a tanh shape with a width of a few unit cells [19]. The piezoelectric tensor is changed accordingly, with a magnitude of 0 in the center and maximum values within domains. Hence, a 180° phase shift is observed in the PFM phase, while the PFM amplitude will show a minimum at the DW center. In principle, the limiting factor in the resolution in PFM is the tip radius. Hence, resolutions down to single digit nm have been reported [19]. In real experiments, the observed resolution is often limited by additional factors, such as the applied electric field, sample geometry, the dielectric properties, screening charges in the bulk or at the sample surface [38]. In depth reviews on PFM can be found elsewhere [32–34,38].

Figure 4 shows a typical PFM result on our x-cut TFLN sample, with a-c displaying the topography, the PFM phase and PFM amplitude, recorded simultaneously. As expected, the topography image exhibits only weak features with heights <0.5 nm that are not correlated to the domain structure. In contrast, the electrodes are visible at the lower sample edge with heights of >50 nm. The phase signal provides a net domain polarity contrast of 180° between antiparallel domains, as expected. Conversely, the amplitude signal shows a dominant DW contrast with a signal width (FWHM) of <200 nm. To create images such as these, typical scan rates around ~1 Hz per line are recommended, leading to a total acquisition time per frame of minutes when using scan increments of 100 nm.

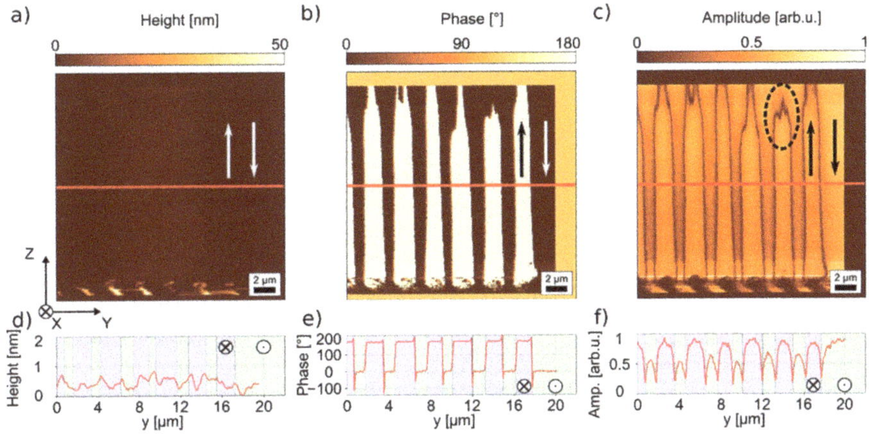

Figure 4. PFM colormaps of (**a**) topography, (**b**) phase, and (**c**) amplitude signal for an x-cut PP-TFLN sample, and (**d**–**f**) respective line scan graphs taken along the red line in (**a**–**c**). While the height signal shows no topographical features in poled regions, the phase signal provides domain inversion, while the amplitude signal allows imaging with a superb DW contrast. The circled area in the amplitude map (**c**) might be indicative for an incomplete in-depth domain inversion close to the (−) electrode (see text).

The high lateral resolution makes PFM the method of choice when it comes to imaging and detecting sub-micrometer polar features, as shown in Figure 5 for a TFLN sample with a periodicity of $\Lambda = 0.8$ µm. Such samples are not accessible with diffraction-limited (optical) methods, as these periodicities lie in the range of the Abbe limit. Similar to Figure 4, the results on the submicron structure in Figure 5 show no domain related contrast in the topography. While the PFM phase shows a domain polarity contrast, the PFM amplitude allows distinguishing between individual domains with DW-DW distances of less than 200 nm and individually resolved DWs (see Figure 5b,d).

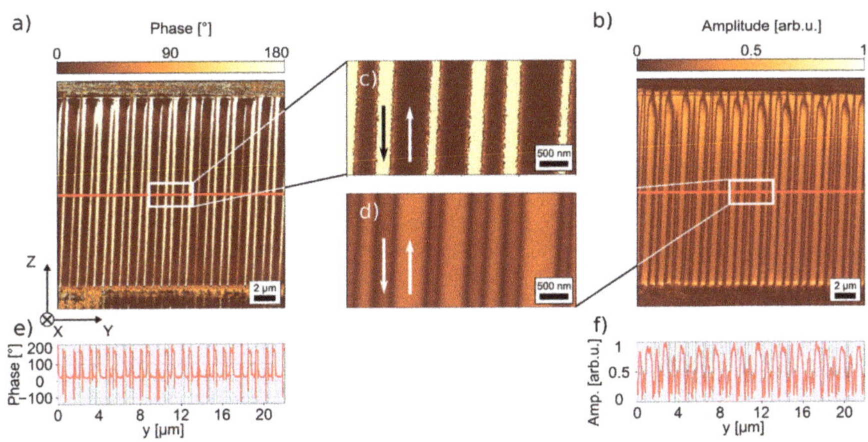

Figure 5. PFM colormaps of an x-cut PP-TFLN sample with sub-micrometer periodicity. (**a**) Phase map providing the domain polarity contrast. (**b**) Amplitude map with the domain wall contrast. (**c**) and (**d**) Zoomed-in map segments of the phase (**c**) and amplitude (**d**) maps that demonstrate high-resolution imaging of sub-µm domains. (**e**) and (**f**) Line scan graphs of (**e**) phase and (**f**) amplitude maps along the red lines indicated in the respective colormaps.

In principle, PFM is also sensitive to vertically stacked domains in the depth of the sample. If domains of different orientations are stacked upon each other, their piezoelectric response will partly cancel. Hence, a decrease in PFM amplitude may be expected. However, as the electric field rapidly decays with the increasing distance away from the tip, i.e., typically with $1/r$, where r is the tip radius, PFM provides only a limited and nonlinear depth resolution. Experiments performed on vertically stacked domains in z-cut bulk lithium niobate indicated that domains up to a depth of 1.7 μm could be detected with a normal (deflection) mode (90% criterion) [39]. This suggests that in principle PFM allows imaging the domain constitutions into the full depth of these TFLN samples, as film thicknesses typically measure 300–700 nm. Indeed, a comparison of the SHG and PFM images in Figure 3 suggests that some of these domains might not be fully inverted in the full TFLN sample depth, as deduced from the reduced PFM amplitude of the encircled domain in Figure 4c. As seen close to the top sample edge where the negative electrode for poling is located, the color contrast is slightly shaded as compared to the fully poled domains. This information is corroborated between both the PFM and SHG information. Nevertheless, further experiments and theoretical analysis are necessary to calibrate the depth resolution in PFM for this type of sample, since the amplitude does not decrease linearly with the domain depth. Furthermore, in clear contrast to previous depth-resolved PFM experiments on z-cut bulk LN, x-cut TFLN features a different crystal orientation. Hence, smaller piezoelectric tensor elements and/or different cantilever motions are addressed. Moreover, the used substrate material stack will certainly influence the electric field distribution, as well.

In conclusion, PFM unambiguously allows imaging the FE domain structures due to the well-defined signal origin and behavior. The advantage of PFM is its capability of high-resolution imaging since it is not limited by diffraction. This makes PFM the best-suited method for imaging micrometer- and sub-micrometer-sized structures. However, due to its limited scan speed, visualization by PFM is limited to smaller areas. Hence, for analyzing mm-sized PP-TFLN structures as requested for narrow-band and highly efficient optoelectronic applications, PFM needs to be complemented by additional techniques, e.g., SHG imaging. In z-cut TFLN, PFM has another particular advantage, as it also allows writing and subsequently analyzing FE domain structures. Recently, this has seen widespread use for studies of conductive domain walls [40,41] in TFLN.

3.2. Second-Harmonic Generation Microscopy

Second-harmonic generation microscopy sees widespread use for imaging and analyzing FE domain structures. It is frequently applied to bulk samples, since it allows fast, large-scale, and non-destructive 3D imaging with a diffraction-limited optical resolution [25,42–45]. The high speed of this technique facilitates in situ imaging of FE domains and domain walls in bulk samples under external stimuli, such as electric fields or temperature [46–49]. In SHG microscopy, the contrast mechanism relies on probing the second-order susceptibility tensor $\chi^{(2)}$, whose properties are directly connected to the FE domain orientation. Lithium niobate is a uniaxial ferroelectric. Here, only two domain orientations are allowed. The tensor $\chi^{(2)}$ is inverted, i.e., switched from $+\chi^{(2)}$ to $-\chi^{(2)}$ between antiparallel domains. The SHG process is a coherent, optical process. Whenever the sign of the nonlinear tensor $\chi^{(2)}$ is reversed, the generated SH light will see an additional 180° phase shift compared to the SH light generated in a region with a non-inverted nonlinear tensor. According to the experimental and theoretical analysis, FE DWs are only a few unit cells wide [19], i.e., significantly smaller than the diffraction limited optical focus spot (<500 nm diameter). Neglecting any other effects accompanying a DW (which may further influence the SHG response, e.g., strain fields, defects or electric fields) this allows explaining the appearance of DW in a SHG image in the simple picture. Assuming a focus is placed symmetrically on top of a domain wall, the light generated within one side of the DW will destructively interfere with the SHG signal on the other side of the DW, due to the inverted tensor. Hence, a DW will appear as a dark line within an SHG image. This simple picture holds true, if a tensor element can be directly addressed with the pump light polarization, and SHG imaging is performed near the surface (correct beam focusing, normal dispersion, i.e., $n_{SHG} > n_{fund}$). It should be noted that the contrast mechanism may be modified for large numerical apertures [50] or that other contrast mechanisms, e.g., the so called Čerenkov contrast or modified SH tensors at DWs, can play a significant role in bulk systems. However, as theoretical and experimental works suggest [23], these effects can be neglected for x-cut thin films, since the largest tensor element d_{33} can be directly addressed, and imaging is always near the surface (i.e., film thicknesses and focus depth are similar). Therefore, SHG microscopy is well suited for non-destructive, large-scale imaging of periodically-poled structures.

Due to its nature as a coherent optical process, SHG microscopy is further sensitive to the inversion depth well below the optical depth of focus. If domains of opposing directions are stacked vertically, the signal generated from the opposing domains will interfere destructively resulting in a reduced signal as compared to the fully inverted or non-inverted single domain. As SHG scales quadratically with the interaction, this allows sensitively detecting the stacked domains with an accuracy of tens of nanometers [23]. This is possible for TFLN, since the coherent interaction length of SHG in the forward direction ("co-propagating" phase matching) measures about 1.28 μm for 800 nm fundamental light, which is larger than the typical film thickness used in integrated optics (<1 μm). The light generated in the forward direction can be measured even in the backward detection due to the high reflectivity of the substrate. In SHG, the light is also directly generated in the backward direction. However, due to the short coherent interaction length (~45 nm) and the quadratic scaling with the interaction length, this counter-propagating light is two to three orders of magnitude weaker in intensity. It should be noted that using a nonlinear substrate, e.g., LN rather than a silicon wafer, may significantly influence the detected SH light, since the SH light is not only generated in the film, but also in the substrate. Further insights into the SHG imaging process can be found in our previous paper [23].

To demonstrate the sensitivity for the inversion depth of SHG microscopy, we have performed numerical simulations of the imaging process for a typical x-cut film of 300 nm thickness on a 1.8 μm oxide layer, similar to our structure in Figure 1a. For simulation, we used the code that was previously developed [19,23]. Figure 6a depicts the simulation setup. Here, we assume an inverted domain of 2 μm width with a hexagonal cross-section, as it can be expected for x-cut films. For the simulation, the domain extends infinitely along the z-axis. Figure 6b shows the calculated focus field for a NA of 0.8 and a wavelength of 800 nm, which is used for all simulations reported in this paper. Figure 6b shows that reflections play a crucial role in the fundamental field. As shown previously, similar conclusions can be drawn for the reflected SHG signal, which is the main SHG signal that is detected. In our simulations, the focus is moved in increments of 50 nm and the reflected SHG intensity collected by the objective lens is calculated. This simulation was performed for different depths of inverted domains in 50 nm increments, and the result is plotted in Figure 6c. These simulations show that only for a fully inverted domain (h = 300 nm), the SHG signal level within the domain recovers and reaches up to the bulk value (h = 0 nm). For this case, only vertical DWs lead to a signal decrease, i.e., a domain wall contrast is observed. The simulation demonstrates that due to the coherent nature of the process, domain inversion depths of much less than the optical resolution can be detected by SHG.

To show that the contrast and imaging mechanism is independent of the setup, e.g., NA or wavelength, we have imaged the same area with our two setups and at two different wavelengths. These results are displayed in Figure 7. Overall, a similar domain image is obtained independent of the setup. In detail, the resolution decreases from (a) to (c), which is a result of the reduced NA (self-built: NA = 0.9, Leica: NA = 0.8) and the increased wavelength (c). To realize an image as shown in Figure 7a, the self-built system requires an integration time typically on the order of 10 to 50 ms. Hence, assuming a step width of 200 nm along each direction, typical measurement times in the order of 10–30 min are required with the self-built setup. Due to the high dynamical range of the photo-multiplier tube and the beam scanning rather than sample scanning, the commercial system only requires integration times of much less than 1 ms per pixel. Complete frames, as shown in Figure 7b,c, can be visualized within a second or less, allowing for fast imaging. Via moving the sample on a piezo stage and stitching these frames, large images (>1 cm^2) can be created in a reasonable time.

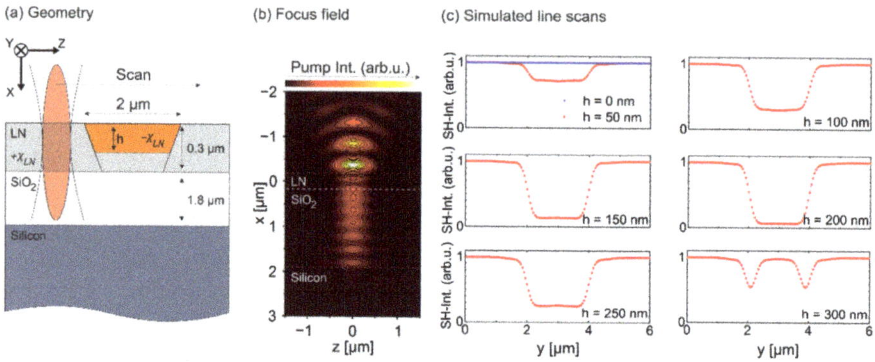

Figure 6. SHG simulations for x-cut TFLN with NA = 0.8 and λ_0 = 800 nm. (**a**) Sample geometry and simulation layout. (**b**) Focus field simulation for all sample layers. (**c**) Simulated SHG intensities for different poling depths of inverted domains in increments of 50 nm domain depth. Only a completely inverted domain allows imaging with a clean DW contrast.

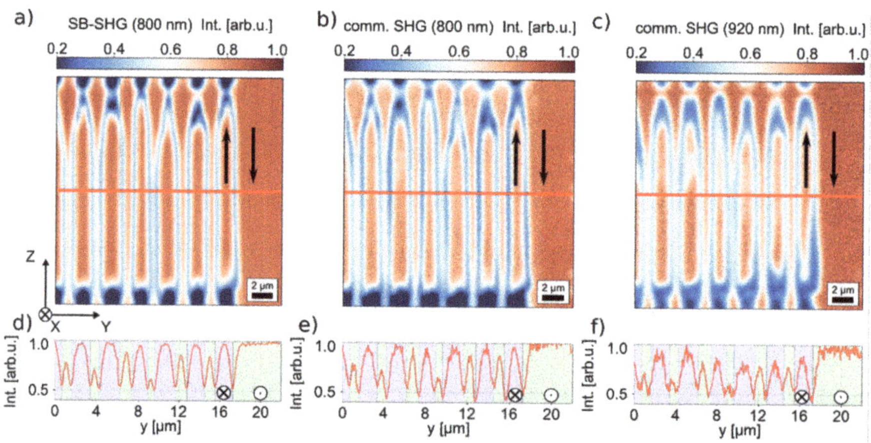

Figure 7. SHG micrographs obtained with (**a**) the self-built setup (NA = 0.9)typical measurement times in the order o, and (**b**) the commercial Leica SP5 MP (NA = 0.8), both recorded at an incident wavelength of 800 nm. (**c**) SHG image with 920 nm incident light measured with the Leica SP5 MP. The DW contrasts of all images are comparable, indicating no dependencies on the wavelength or setup over the analyzed wavelength range. (**d–f**) Line scan graphs of the maps (**a–c**) along the red lines indicated in the respective colormaps.

To demonstrate the capability of SHG for large-scale imaging of complete chips, we have visualized the domain structures along a full 5 mm electrode with the Leica microscope. The result is displayed in Figure 8a. In Subfigure b–d, detailed zoomed-in images taken from the large scan are shown. The zoomed-in frames exhibit similar features as the images in Figure 7. This means that even at this large scale, no loss of information is observed and the poling quality, e.g., poling period, homogeneity, duty cycle or poling depth, can be analyzed for a full wafer. The large image in (a) is cut from a stitched image of 20 frames, each with a size of approximately 260 × 260 µm^2. To account for the non-perfectly planar mounting of the TFLN sample (approximately 50 µm height difference along 5 mm), a depth scan over 50 μm with a step size of 5 μm was conducted for each frame. Due to the slightly different focus in each frame, slightly different overall intensities are observed, leading to the large-scale intensity oscillations observed along the chip. As the previous theoretical analysis indicates, the focus position does not influence the domain wall contrast [23]. Taking the images at all different planes into account, a total of 200 frames were scanned for this image. Assuming an imaging time of ~1 s per frame and accounting for (automatic) sample repositioning, a total imaging time in the range of only a few minutes is required for the example shown here. The scan speed may be further increased though at the cost of signal-to-noise ratio. Whenever a low magnification objective lens is used with a larger field of view, even higher imaging speeds are possible at the cost of lateral resolution. This example demonstrates that complete chips (~cm^2) can be imaged within an hour or even less.

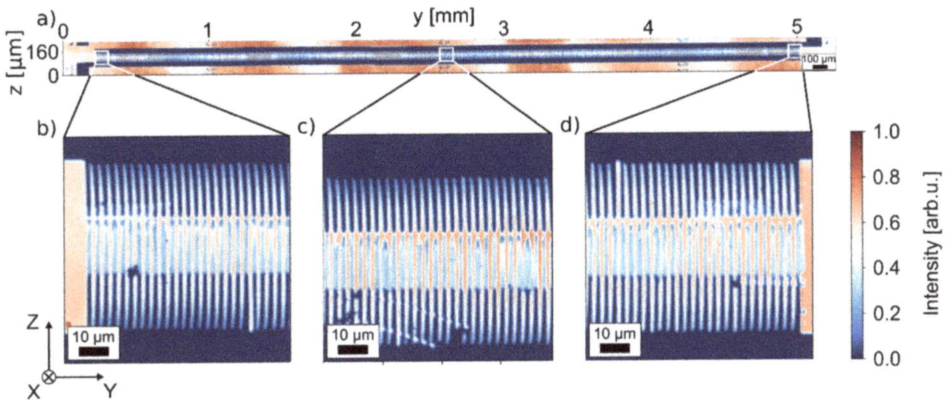

Figure 8. Large-scale SHG imaging of the complete, 5-mm-long PP-TFLN sample. (**a**) Overview micrograph obtained by stitching 20 individual measurements together. (**b–d**) Detailed high-resolution scans at different locations along the waveguide structure.

As demonstrated, the imaging process and contrast mechanism of SHG microscopy in TFLN can be well understood based on simple assumptions. Nevertheless, there are several open questions that need to be addressed in future works. A key aspect of our model is that FE DWs are assumed to constitute a Heaviside function flipping instantaneously from $+\chi^{(2)}$ to $-\chi^{(2)}$. Based on this model, the contrast can be solely explained by an interference process of SH light generated in opposing domains. As we have shown, this model may well explain our observations made in x-cut TFLN, where also the largest tensor element is addressed. However, various works on FE DWs demonstrated that DWs exhibit an internal structure and are accompanied by other effects, such as strain or electric fields [19], which have been demonstrated to span over several μm around the DW and are likely to have an influence on the local second-harmonic response in the vicinity of the DW [51]. In this context, techniques such as SHG polarimetry allow proving the existence of deviations from the ideal Ising wall in ferroelectrics [42]. Therefore, SHG polarimetry is indicated for TFLN in future work. This might help better understand the fundamental physics of domains in TFLN, e.g., the nature of the buried h2h and t2t DWs close to the electrodes. An additional substructure of $\chi^{(2)}$ can be readily included in the numerical model to provide further insight into DW properties in TFLN. In the context of electric or strain fields, correlative imaging with multiple methods, e.g., RS or PFM, might give a further understanding of the influence of these fields, as well as on the SHG properties, which will more accurately allow for the interpretation of measurements.

In conclusion, SHG provides fast imaging and analysis of domain structures in TFLN devices. In principle, high resolution images of complete TFLN chips (~cm^2) containing hundreds of poled areas can be imaged within less than an hour when using modern microscopes, while detailed images of smaller areas (<500 × 500 μm^2) can be taken in a matter of seconds. For z-cut TFLN or bulk samples, this imaging speed can only be matched by selective etching and subsequent microscopy imaging, yet at the cost of sample destruction. Even more, the key advantage of SHG imaging on TFLN is that it can unambiguously distinguish between fully or only partly penetrated domains with sub-diffraction-limited depth resolution, providing valuable information in the performance analysis. Due to the non-destructive nature and high speed, SHG microscopy can provide valuable information to design and fabrication, e.g., by selecting only suitable areas within the electrodes for device fabrication.

3.3. *Hyperspectral Raman Imaging*

Raman scattering microscopy is a common technique for the investigation of crystal structures and has been widely used for imaging LN DWs [24–28]. The RS contrast of DWs in poled LN has been ascribed to two different mechanisms [27]:

(1) LN DWs that are described as large two-dimensional defects by Stone and Dierolf, introduce a quasi-momentum to the RS process, which is directed perpendicularly to the plane of the DW. If measured on the crystal z-face, the detected signal originating from the DW does not stem from phonons propagating along the beam direction, but from obliquely propagating phonons. Therefore, the selection rules are lifted at DWs, and A_1(TO) phonons or mixed longitudinal optical-transversal optical (LO-TO) phonons can be detected, which are usually not accessible in that geometry [28].

(2) In the vicinity of DWs, a large strain [52] and electric fields [53] have been observed. Based on these findings, Fontana et al. concluded that these fields change the RS efficiencies due to elastooptic or electrooptic coupling [54]. Capek et al. conclude that the observed phonon frequency shifts at DWs are also caused by local fields [26].

For Raman scattering on the incident x-face of bulk LN, no directional dispersion as in case 1 has been observed in the used scattering configurations [27,55]. Instead, the DW signature here presumably results only from phonon frequency shifts due to strain and electric fields, and intensity changes resulting from elastooptic and electrooptic coupling, as explained in case 2. In RS experiments on poled TFLN structures, we expect a similar behavior as for these investigations on non-polar surfaces of LN bulk samples, although possible large-area stress fields resulting from the bonding process of the LN thin film and the confinement along one dimension, can additionally affect the detected scattering signal and lead to different results.

Although a common analysis method for poled bulk LN, Raman micro-spectroscopy, to the best of our knowledge, has not yet been applied for TFLN imaging beyond our own earlier work [56]. This might be due to the increased challenges faced when inspecting thin films in general: Due to the low interaction volume of the thin film with the incident laser photons, the generated scattering signal is weak in absolute terms and also compared to the substrate signal. This fact calls for long acquisition times (in our measurement, we applied acquisition times per pixel of 2 × 5 s/px in x(zz)-x and 2 × 150 s/px in x(zy)-x), precise focusing and careful peak analysis.

The Raman spectra in a virgin domain and on a domain wall are compared in Figure 9. Only the four indicated phonon peaks are generated in the LN thin film, whereas the SiO_2/Si substrate causes a strong background with the characteristic triply degenerate Si peak at 521 cm^{-1}. For this reason, it is essential to keep the laser focus in the thin film and maximize the LN scattering signal, which is achieved by the autofocusing routine detailed in Section 2. This special treatment massively prolongs the measure time per pixel, which makes Raman investigations the slowest of the techniques shown in this work (the total acquisition time for the shown Raman colormaps including autofocus is ~15 days; the pure acquisition time without autofocus amounts to 27.6 h).

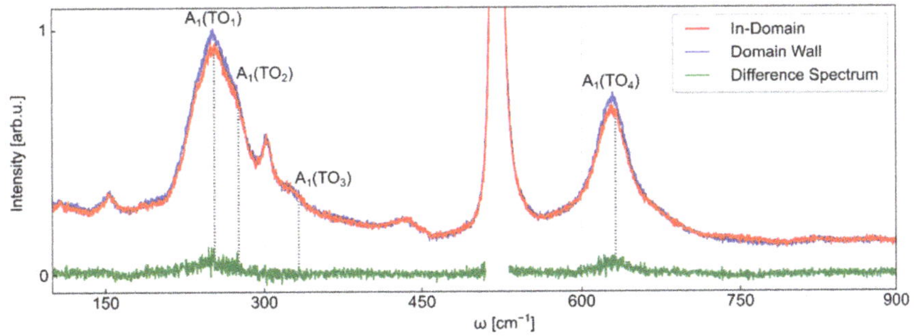

Figure 9. Comparison of Raman spectra taken inside a domain (red) and directly at the DW (blue) in x-cut TFLN on SiO_2/Si in x(zz)-x geometry. The expected LN phonon peaks are marked in the figure. As seen in the different spectrum (green), the detected LN phonon modes $A_1(TO_1)$, $A_1(TO_2)$, and $A_1(TO_4)$ experience spectral changes between the domain walls and poled domains. Due to the small scattering volume of the thin film, the phonon peaks of the Si substrate form a strong background.

For Raman imaging of the x-cut TFLN sample, we analyzed the intensities, frequencies, and widths (FWHM) of A_1(TO) and E(TO) peaks. A complete collection of A_1(TO) spectral features in TFLN is provided in the Supplementary Materials. The TFLN phonon features partly deviate from the bulk behavior, as a comparison of the $A_1(TO_4)$ features shows. Whereas only the peak frequency analysis yields a notable DW contrast in bulk LN [20], TFLN DWs are accessible via the frequency, scattering intensity, and FWHM of $A_1(TO_4)$. This different behavior might be attributed to the influence of the substrate bonding on the acoustooptic coupling and a stronger inhomogeneity of the local stress distribution due to the low film thickness.

Figure 10 shows a selection of hyperspectral colormaps generated from A_1(TO) phonon features with different contrasts. Imaging with the domain wall (Figure 10a) as well as the domain polarity contrast (Figure 10b) can be achieved by addressing different phonon properties from the same measurement. The different phonon responses can be explained by the atom motions associated with the respective phonons [57,58]. Whereas the dominant motion in $A_1(TO_4)$ is a vibration of the oxygen hexagon, the $A_1(TO_2)$ phonon is dominated by a vibration of lithium atoms along the z-axis. As the polarity change in inverted domains is caused by a dislocation of niobium atoms within their respective oxygen cages as well as hopping of lithium atoms into adjacent oxygen cages, the change of polarization directly affects the vibrational motion of the $A_1(TO_2)$ phonon. On the other hand, the atom movement in $A_1(TO_4)$ is not affected by the specific polarization direction, but only in the transitional region, i.e., in the vicinity of the DW. Therefore, $A_1(TO_2)$ phonon properties provide a domain polarity contrast, whereas $A_1(TO_4)$ phonon properties enable a DW contrast.

Paying closer attention to the DW contrast mapping in Figure 10a (e.g., in the area marked by the dashed circle), the image shows similar features such as the SHG map in Figure 7c, which is subject to poling depth investigations. However, different from SHG, we assume that this signal does not allow accessibility to the poling depth, but results from the presence of buried domain walls. The Raman signal is generated in full depth of the illumination spot and hence is sensitive to structural changes in full depth of the LN thin film. Whenever a poled domain does not extend to the full depth of the thin film, a DW in the y-z-plane is generated. This DW is accessible via Raman imaging mechanisms that are sensitive to DWs, such as $A_1(TO_4)$ phonon properties. Therefore, the Raman scattering investigation may be facilitated to monitor the completeness of domain in-depth poling, however, does not yield quantitative information on the poling depth.

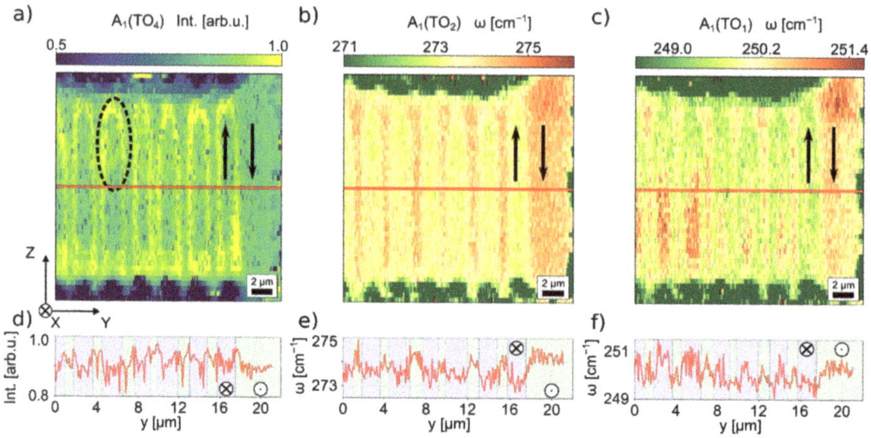

Figure 10. Hyperspectral Raman colormaps of an x-cut PP-TFLN sample (**a–c**), and (**d–f**) the respective cross-sections taken along the indicated red lines in (**a–c**). The step size of the measure points is 0.5 µm along the TFLN *z*-axis and 0.1 µm along the *y*-axis. (**a**) The $A_1(TO_4)$ intensity provides a net DW contrast with a suspected sensitivity to check for poling completeness, for instance, as in the circled area. (**b**) The $A_1(TO_2)$ frequency reveals a domain polarity contrast. (**c**) The frequency of $A_1(TO_1)$ also shows the DW contrast with a suspected sensitivity to check for poling completeness, and is assumed to be sensitive to stress fields, as well.

A peculiar behavior can be seen for the $A_1(TO_1)$ peak frequency mapping in Figure 10c. Similar to the $A_1(TO_4)$ intensity, this feature shows a domain wall contrast with the assumed sensitivity for poling completeness, but in the unpoled region of the sample (upper right area of the colormap), inhomogeneities of the signal, i.e., a phonon frequency change of ~1.5 cm^{-1}, can be seen although there is no evident structuring. The cross correlation with PFM and SHG measurements shows that no structural features are to be expected in this area. A possible explanation for this observation is a sensitivity to mechanical stress and/or electric fields. Previous works on the pressure dependence of LN phonon features [30] suggest that the isotropic pressure dependence coefficient of $A_1(TO_1)$ is ~0 cm^{-1}/kbar, which would directly contradict this assumption. However, that work gives no information on the influence of uni- or biaxial stress fields. Stone et al. [29] ascertained the scattering dependence on apparent electric fields, but only listed data for $A_1(LO)$ and E(TO) phonons. Assuming that the coefficient of $A_1(TO_4)$ is in a comparable order of magnitude such as $A_1(LO_4)$, the electric field is estimated to be in the order of 45 kV/mm. In both cases of strain and electric field dependence of the phonon frequency, future work is indicated to extend the fundamental knowledge on these dependencies and evaluate our assumptions.

An example of Raman imaging with E(TO) modes is shown in Figure 11. Here, we have analyzed a different electrode of the same sample that provides larger domains, which allows a more clear visualization of the phonon frequency shift between oppositely-poled domains. The resulting larger domain sizes allow a clearer distinction of the phonon properties. The phonon frequency of the $E(TO_2)$ mode provides a clear domain polarity contrast. The frequency change between the oppositely-poled domains has been applied to the pressure and electric field correlation studies. With a pressure dependency coefficient of 0.05 cm^{-1}/kbar [30], the $E(TO_2)$ frequency change of -0.27 cm^{-1} from pristine to poled domain would correspond to -5.4 kbar, i.e., a tensile force acting on the LN thin film due to the polarization change. With an electric field coefficient of -0.01 cm^{-1}/(kV/mm) [29], a change of electric field by 27 kV/mm between oppositely-poled domains can be deduced. While this number is larger than the coercive field in undoped, bulk LN [59], several works reported that for poling doped TFLN (which even has a lower coercive field) high electric fields need to be applied reaching values of up to 40–60 kV/mm in order to achieve successful poling. While some authors suspected this to be an effect of the poling electrode design and to facilitate nucleation, our results indicate that these high fields maybe necessary to overcome internal limits, e.g., imposed fields and charges required to stabilize h2h and t2t DWs. Further research is necessary.

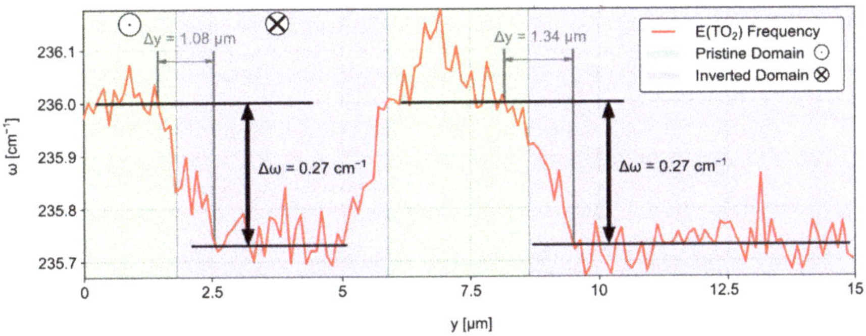

Figure 11. Line scan of the $E(TO_2)$ phonon frequency over poled TFLN domains. A sample section with large domains has been chosen for a clear visualization of the phonon frequency shift between pristine and inverted domains, which are highlighted in blue and green as a guide to the eye. The analyzed DW interaction length Δy is larger than the resolution limit, suggesting long-range stress fields in the vicinity of the DW.

The presented analysis shows the high versatility of Raman microscopy on TFLN. This method not only provides the domain polarity and domain wall contrast for the inspection of poling period and duty cycle, but is capable of detecting buried DWs and thus probing the in-depth poling completeness. Additionally, stress and electric fields are also potentially accessible. Regarding the multitude of possible influence factors on the detected signals, the challenge is to unequivocally assign the observed features to certain material properties. The cross-correlation with other methods is a viable way to achieve this goal, combined with extended fundamental investigations of the specific property sensitivities. The resolution, similar to SHG imaging, is diffraction limited. However, this resolution already allows experimental statements, as we observe signatures for strain fields that spread significantly further than the optical resolution. Similar observations have been made in bulk in the past as well, e.g., fields of shear strain reaching 10 μm from the DW have been observed [52].

The biggest challenge for the application of RS to TFLN is the low scattering volume of the thin film. The necessary long acquisition times limit the pixel number and applications in the process control. Potentials to improve the signal and hence application times, can be a higher pump power, lower spectral resolution or improved focusing. However, even in bulk, typical acquisition times in the order of 1 s are typical and may be the limit. Raman spectroscopy is thus mainly suited for fundamental investigations of sample properties and the validation of other methods via cross-correlation.

Alongside fundamental investigations that facilitate the quantification of electric and stress fields, angular dispersion experiments on TFLN can give insights on whether the phonon properties are further affected by the boundaries of the thin film. DFT calculations can also help theoretically understand the differences to bulk LN systems [58,60].

In summary, the application of Raman spectroscopy provides imaging with DW and polarity contrast, a sensitivity towards the completeness of in-depth poling, and access to apparent strain and electrical fields. Due to its long acquisition time of several seconds per pixel, the number of data points is considerably lower compared to PFM and SHG, and the application is limited to fundamental investigations. For a better understanding and quantification of the obtained results, further experiments are suggested.

4. Conclusions

In this work, we have performed imaging of PP-TFLN devices with PFM, SHG microscopy, and Raman spectroscopy. Moreover, we have discussed the underlying contrast mechanisms and the differences to imaging in related bulk structures. While each method addresses a different physical property to reveal the domain structure, i.e., piezoelectric tensor, the SH tensor, and the phonon-properties, the resulting images are comparable. Hence, each method is capable of visualizing and analyzing the domain structures. Furthermore, each method gives access to additional properties of the domain structure.

Table 1 summarizes the imaging contrasts, resolution, and speed of all the discussed methods. Similar to its use in bulk PPLN, selective etching allows fast imaging of z-cut TFLN. However, due to the additional need to expose z-faces, its speed and information content is limited for x-cut TFLN. If a waveguide is already fabricated, in situ transmission investigations provide almost instant feedback during fabrication. However, since only integral information is retrieved, little quantitative information on the domain structure can be gained. SHG is a promising candidate for standard imaging on TFLN. It offers access to all the determined key performance indicators, i.e., poling period, duty cycle, poling grid length, and poling depth. In the case of submicron poled TFLN structures, however, the high resolution of PFM scans is vital for the acquisition. PFM potentially offers access to most key performance indicators, as well. For fundamental optimization, the sensitivity of Raman scattering for defects, mechanical stress, and/or electric fields gives additional insights into the material system which are not approachable by other techniques. Due to its simultaneous sensitivity to domain structures, it allows studying the interactions and dependencies of domain structures and defects or fields.

Table 1. Comparison of the techniques applied to the analysis of x-cut TFLN, with respect to the following parameters: Detectable imaging contrasts, contrast mechanisms, spatial resolution, measurement speed, acquisition time, destructive nature of the technique. PFM: Piezoresponse force microscopy; SHG: Second-harmonic-generation microscopy: RS: Raman micro-spectroscopy.

	In Situ	Etching	PFM	SHG	Raman
Contrast mechanism	device performance	etch rate	electrooptic tensor	$\chi^{(2)}$	phonon properties
DW contrast	-	x	x	-	x
Polarity contrast	-	-	x	x	x
In-depth poling	-	-	(x)	x	(x)
Strain & el. fields	-	-	-	-	x
Resolution	-	50–100 nm	1–100 nm	0.5–1 μm	0.5–1 μm
Typical frame size	no imaging	μm^2 to mm^2 (z-cut) 1–20 μm^2 (x-cut)	100–1000 μm^2	mm^2	100 μm^2
Time per frame	In situ	-	~1–10 min	1 s–10 min	1 d–1 week
Destructivity	no	yes	no	no	no

For future developments, the combination of these imaging methods paves the way for highly efficient TFLN platforms. However, a complete understanding of the underlying imaging mechanisms and quantification of signals is necessary for further TFLN optimization. For PFM, the sensitivity for in-depth poling needs to be validated. In SHG measurements, the depth sensitivity of the signal must be further quantified. For RS, the quantification of the stress sensitivity and further understanding of the different imaging mechanisms will provide deeper insights into the nature of poled TFLN samples.

With these future investigations, the imaging and in-depth signal analysis of TFLN structures provide a valuable toolkit for the development of nonlinear optical devices with maximum conversion efficiency.

Supplementary Materials: The following are available online at https://www.mdpi.com/2073-4352/11/3/288/s1. Figure S1: Colormaps of A_1(TO) frequencies, intensities, and peak widths.

Author Contributions: Conceptualization, S.R., M.R., and L.M.E.; methodology, S.R. and M.R.; formal analysis, S.R. and M.R.; investigation, S.R., B.K., and M.R.; resources, J.Z., S.M., and L.M.E.; writing—original draft preparation, S.R. and M.R.; writing—review and editing, M.R. and L.M.E.; visualization, S.R.; supervision, L.M.E.; project administration, L.M.E.; funding acquisition, S.M. and L.M.E. All authors have read and agreed to the published version of the manuscript.

Funding: This research was funded by Deutsche Forschungsgemeinschaft (DFG) via FOR5044, EN 434/41-1 and INST 269/656-1 FUGG, Sandia National Laboratories (SigmaNONlin), and the National Science Foundation (EFMA-1640968).

Institutional Review Board Statement: Not applicable.

Informed Consent Statement: Not applicable.

Acknowledgments: The authors would like to thank the Light Microscopy Facility of the CMCB Technology Platform at TU Dresden for the use of the SHG microscope Leica SP5 MP. Open Access Funding by the Publication Fund of the TU Dresden.

Conflicts of Interest: The authors declare no conflict of interest.

References

1. Chang, L.; Li, Y.; Volet, N.; Wang, L.; Peters, J.; Bowers, J.E. Thin film wavelength converters for photonic integrated circuits. *Optica* **2016**, *3*, 531–535. [CrossRef]
2. Zhao, J.; Rüsing, M.; Javid, U.A.; Ling, J.; Li, M.; Lin, Q.; Mookherjea, S. Shallow-etched thin-film lithium niobate waveguides for highly-efficient second-harmonic generation. *Opt. Express* **2020**, *28*, 19669. [CrossRef] [PubMed]
3. Zhao, J.; Ma, C.; Rüsing, M.; Mookherjea, S. High Quality Entangled Photon Pair Generation in Periodically Poled Thin-Film Lithium Niobate Waveguides. *Phys. Rev. Lett.* **2020**, *124*, 163603. [CrossRef]
4. Rao, A.; Abdelsalam, K.; Sjaardema, T.; Honardoost, A.; Camacho-Gonzalez, G.F.; Fathpour, S. Actively-monitored periodic-poling in thin-film lithium niobate photonic waveguides with ultrahigh nonlinear conversion efficiency of 4600%W^{-1} cm^{-2}. *Opt. Express* **2019**, *27*, 25920–25930. [CrossRef]
5. Wang, C.; Langrock, C.; Marandi, A.; Jankowski, M.; Zhang, M.; Desiatov, B.; Fejer, M.M.; Lončar, M. Ultrahigh-efficiency wavelength conversion in nanophotonic periodically poled lithium niobate waveguides. *Optica* **2018**, *5*, 1438–1441. [CrossRef]
6. Lu, J.; Surya, J.B.; Liu, X.; Bruch, A.W.; Gong, Z.; Xu, Y.; Tang, H.X. Periodically poled thin-film lithium niobate microring resonators with a second-harmonic generation efficiency of 250,000%/W. *Optica* **2019**, *6*, 1455–1460. [CrossRef]
7. Honardoost, A.; Abdelsalam, K.; Fathpour, S. Rejuvenating a Versatile Photonic Material: Thin-Film Lithium Niobate. *Laser Photon. Rev.* **2020**, *14*, 1–20. [CrossRef]
8. Rüsing, M.; Eigner, C.; Mackwitz, P.; Berth, G.; Silberhorn, C.; Zrenner, A. Identification of ferroelectric domain structure sensitive phonon modes in potassium titanyl phosphate: A fundamental study. *J. Appl. Phys.* **2016**, *119*, 044103. [CrossRef]
9. Rusing, M.; Weigel, P.O.; Zhao, J.; Mookherjea, S. Toward 3D Integrated Photonics Including Lithium Niobate Thin Films: A Bridge Between Electronics, Radio Frequency, and Optical Technology. *IEEE Nanotechnol. Mag.* **2019**, *13*, 18–33. [CrossRef]
10. He, M.; Xu, M.; Ren, Y.; Jian, J.; Ruan, Z.; Xu, Y.; Gao, S.; Sun, S.; Wen, X.; Zhou, L.; et al. High-performance hybrid silicon and lithium niobate Mach–Zehnder modulators for 100 Gbit s^{-1} and beyond. *Nat. Photon.* **2019**, *13*, 359–364. [CrossRef]
11. Santandrea, M.; Stefszky, M.; Ansari, V.; Silberhorn, C. Fabrication limits of waveguides in nonlinear crystals and their impact on quantum optics applications. *New J. Phys.* **2019**, *21*, 033038. [CrossRef]
12. Fejer, M.M.; Magel, G.A.; Jundt, D.H.; Byer, R.L. Quasi-Phase-Matched 2nd Harmonic-Generation—Tuning and Tolerances. *IEEE J. Quantum Electron.* **1992**, *28*, 2631–2654. [CrossRef]
13. Zhao, J.; Rüsing, M.; Roeper, M.; Eng, L.M.; Mookherjea, S. Poling thin-film x-cut lithium niobate for quasi-phase matching with sub-micrometer periodicity. *J. Appl. Phys.* **2020**, *127*, 193104. [CrossRef]
14. Nagy, J.T.; Reano, R.M. Submicrometer periodic poling of lithium niobate thin films with bipolar preconditioning pulses. *Opt. Mater. Express* **2020**, *10*, 1911–1920. [CrossRef]
15. Weigel, P.O.; Mookherjea, S. Design of folded hybrid silicon carbide-lithium niobate waveguides for efficient second-harmonic generation. *J. Opt. Soc. Am. B* **2018**, *35*, 593–600. [CrossRef]
16. Zisis, G.; Ying, C.Y.J.; Ganguly, P.; Sones, C.L.; Soergel, E.; Eason, R.W.; Mailis, S. Enhanced electro-optic response in domain-engineered $LiNbO_3$ channel waveguides. *Appl. Phys. Lett.* **2016**, *109*, 021101. [CrossRef]
17. Cai, L.; Wang, Y.; Hu, H. Low-loss waveguides in a single-crystal lithium niobate thin film. *Opt. Lett.* **2015**, *40*, 3013–3016. [CrossRef] [PubMed]
18. Tejerina, M.R.; Torchia, G.A. Computation of the expansion parameters of femto-waveguides using a two dimensional μ-Raman map and guided modes. *J. Appl. Phys.* **2013**, *114*, 153106. [CrossRef]

19. Gopalan, V.; Dierolf, V.; Scrymgeour, D.A. Defect–Domain Wall Interactions in Trigonal Ferroelectrics. *Annu. Rev. Mater. Res.* **2007**, *37*, 449–489. [CrossRef]
20. Turygin, A.P.; Alikin, D.O.; Kosobokov, M.S.; Ievlev, A.V.; Shur, V.Y. Self-Organized Formation of Quasi-Regular Ferroelectric Nanodomain Structure on the Nonpolar Cuts by Grounded SPM Tip. *ACS Appl. Mater. Interfaces* **2018**, *10*, 36211–36217. [CrossRef]
21. Zhao, J.; Rüsing, M.; Mookherjea, S. Optical diagnostic methods for monitoring the poling of thin-film lithium niobate waveguides. *Opt. Express* **2019**, *27*, 12025–12038. [CrossRef] [PubMed]
22. Sones, C.L.; Mailis, S.; Brocklesby, W.S.; Eason, R.W.; Owen, J.R. Differential etch rates in z-cut $LiNbO_3$ for variable HF/HNO_3 concentrations. *J. Mater. Chem.* **2001**, *12*, 295–298. [CrossRef]
23. Rüsing, M.; Zhao, J.; Mookherjea, S. Second harmonic microscopy of poled x-cut thin film lithium niobate: Understanding the contrast mechanism. *J. Appl. Phys.* **2019**, *126*, 114105. [CrossRef]
24. Nataf, G.F.; Guennou, M.; Haußmann, A.; Barrett, N.; Kreisel, J. Evolution of defect signatures at ferroelectric domain walls in Mg-doped $LiNbO_3$. *Phys. Status Solidi (RRL)—Rapid Res. Lett.* **2015**, *10*, 222–226. [CrossRef]
25. Berth, G.; Quiring, V.; Sohler, W.; Zrenner, A. Depth-Resolved Analysis of Ferroelectric Domain Structures in Ti:PPLN Waveguides by Nonlinear Confocal Laser Scanning Microscopy. *Ferroelectrics* **2007**, *352*, 78–85. [CrossRef]
26. Capek, P.; Stone, G.; Dierolf, V.; Althouse, C.; Gopolan, V. Raman studies of ferroelectric domain walls in lithium tantalate and niobate. *Phys. Status Solidi (c)* **2007**, *4*, 830–833. [CrossRef]
27. Rüsing, M.; Neufeld, S.; Brockmeier, J.; Eigner, C.; Mackwitz, P.; Spychala, K.; Silberhorn, C.; Schmidt, W.G.; Berth, G.; Zrenner, A.; et al. Imaging of 180o ferroelectric domain walls in uniaxial ferroelectrics by confocal Raman spectroscopy: Unraveling the contrast mechanism. *Phys. Rev. Mater.* **2018**, *2*, 103801. [CrossRef]
28. Stone, G.; Dierolf, V. Influence of ferroelectric domain walls on the Raman scattering process in lithium tantalate and niobate. *Opt. Lett.* **2012**, *37*, 1032–1034. [CrossRef]
29. Stone, G.; Knorr, B.; Gopalan, V.; Dierolf, V. Frequency shift of Raman modes due to an applied electric field and domain inversion in $LiNbO_3$. *Phys. Rev. B* **2011**, *84*, 134303. [CrossRef]
30. Mendes-Filho, J.; Lemos, V.; Cerdelra, F. Pressure dependence of the Raman spectra of $LiNbO_3$ and $LiTaO_3$. *J. Raman Spectrosc.* **1984**, *15*, 367–369. [CrossRef]
31. Fontana, M.D.; Bourson, P. Microstructure and defects probed by Raman spectroscopy in lithium niobate crystals and devices. *Appl. Phys. Rev.* **2015**, *2*, 040602. [CrossRef]
32. Soergel, E. Visualization of ferroelectric domains in bulk single crystals. *Appl. Phys. A* **2005**, *81*, 729–751. [CrossRef]
33. Gruverman, A.; Alexe, M.; Meier, D. Piezoresponse force microscopy and nanoferroic phenomena. *Nat. Commun.* **2019**, *10*, 1–9. [CrossRef]
34. Uršič, H.; Prah, U. Investigations of ferroelectric polycrystalline bulks and thick films using piezoresponse force microscopy. *Proc. R. Soc. A Math. Phys. Eng. Sci.* **2019**, *475*, 20180782. [CrossRef]
35. Eng, L.M.; Güntherodt, H.-J.; Schneider, G.A.; Köpke, U.; Saldaña, J.M. Nanoscale reconstruction of surface crystallography from three-dimensional polarization distribution in ferroelectric barium–titanate ceramics. *Appl. Phys. Lett.* **1999**, *74*, 233–235. [CrossRef]
36. Abplanalp, M.; Eng, L.; Günter, P. Mapping the domain distribution at ferroelectric surfaces by scanning force microscopy. *Appl. Phys. A* **1998**, *66*, S231–S234. [CrossRef]
37. Soergel, E. Piezoresponse force microscopy (PFM). *J. Phys. D Appl. Phys.* **2011**, *44*. [CrossRef]
38. Scrymgeour, D.A.; Gopalan, V. Nanoscale piezoelectric response across a single antiparallel ferroelectric domain wall. *Phys. Rev. B* **2005**, *72*, 1–16. [CrossRef]
39. Johann, F.; Ying, Y.J.; Jungk, T.; Hoffmann, A.; Sones, C.L.; Eason, R.W.; Mailis, S.; Soergel, E. Depth resolution of piezoresponse force microscopy. *Appl. Phys. Lett.* **2009**, *94*, 172904. [CrossRef]
40. Volk, T.R.; Gainutdinov, R.V.; Zhang, H.H. Domain-wall conduction in AFM-written domain patterns in ion-sliced $LiNbO_3$ films. *Appl. Phys. Lett.* **2017**, *110*, 132905. [CrossRef]
41. Kämpfe, T.; Wang, B.; Haußmann, A.; Chen, L.-Q.; Eng, L.M. Tunable Non-Volatile Memory by Conductive Ferroelectric Domain Walls in Lithium Niobate Thin Films. *Crystals* **2020**, *10*, 804. [CrossRef]
42. Cherifi-Hertel, S.; Bulou, H.; Hertel, R.; Taupier, G.; Dorkenoo, K.D.; Andreas, C.; Guyonnet, J.; Gaponenko, I.; Gallo, K.; Paruch, P. Non-Ising and chiral ferroelectric domain walls revealed by nonlinear optical microscopy. *Nat. Commun.* **2017**, *8*, 15768. [CrossRef] [PubMed]
43. Bozhevolnyi, S.I.; Hvam, J.M.; Pedersen, K.; Laurell, F.; Karlsson, H.; Skettrup, T.; Belmonte, M. Second-harmonic imaging of ferroelectric domain walls. *Appl. Phys. Lett.* **1998**, *73*, 1814–1816. [CrossRef]
44. Flörsheimer, M.; Paschotta, R.; Kubitscheck, U.; Brillert, C.; Hofmann, D.; Heuer, L.; Schreiber, G.; Verbeek, C.; Sohler, W.; Fuchs, H. Second-harmonic imaging of ferroelectric domains in $LiNbO_3$ with micron resolution in lateral and axial directions. *Appl. Phys. A* **1998**, *67*, 593–599. [CrossRef]
45. Uesu, Y.; Yokota, H.; Kawado, S.; Kaneshiro, J.; Kurimura, S.; Kato, N. Three-dimensional observations of periodically poled domains in a $LiTaO_3$ quasiphase matching crystal by second harmonic generation tomography. *Appl. Phys. Lett.* **2007**, *91*, 182904. [CrossRef]
46. Kämpfe, T.; Reichenbach, P.; Haußmann, A.; Woike, T.; Soergel, E.; Eng, L.M. Real-time three-dimensional profiling of ferroelectric domain walls. *Appl. Phys. Lett.* **2015**, *107*, 152905. [CrossRef]

47. Godau, C.; Kämpfe, T.; Thiessen, A.; Eng, L.M.; Haußmann, A. Enhancing the Domain Wall Conductivity in Lithium Niobate Single Crystals. *ACS Nano* **2017**, *11*, 4816–4824. [CrossRef]
48. Wehmeier, L.; Kämpfe, T.; Haußmann, A.; Eng, L.M. In Situ 3D Observation of the Domain Wall Dynamics in a Triglycine Sulfate Single Crystal upon Ferroelectric Phase Transition. *Phys. Status Solidi (RRL) Rapid Res. Lett.* **2017**, *11*, 1–6. [CrossRef]
49. Kirbus, B.; Godau, C.; Wehmeier, L.; Beccard, H.; Beyreuther, E.; Haußmann, A.; Eng, L.M. Real-Time 3D Imaging of Nanoscale Ferroelectric Domain Wall Dynamics in Lithium Niobate Single Crystals under Electric Stimuli: Implications for Domain-Wall-Based Nanoelectronic Devices. *ACS Appl. Nano Mater.* **2019**, *2*, 5787–5794. [CrossRef]
50. Spychala, K.J.; Mackwitz, P.; Rüsing, M.; Widhalm, A.; Berth, G.; Silberhorn, C.; Zrenner, A. Nonlinear focal mapping of ferroelectric domain walls in $LiNbO_3$: Analysis of the SHG microscopy contrast mechanism. *J. Appl. Phys.* **2020**, *128*, 234102. [CrossRef]
51. Cherifi-Hertel, S.; Voulot, C.; Acevedo-Salas, U.; Zhang, Y.; Crégut, O.; Dorkenoo, K.D.; Hertel, R. Shedding light on non-Ising polar domain walls: Insight from second harmonic generation microscopy and polarimetry analysis. *J. Appl. Phys.* **2021**, *129*, 081101. [CrossRef]
52. Jach, T.; Kim, S.; Gopalan, V.; Durbin, S.; Bright, D. Long-range strains and the effects of applied field at 180° ferroelectric domain walls in lithium niobate. *Phys. Rev. B* **2004**, *69*, 1–9. [CrossRef]
53. Zhang, Y.; Guilbert, L.; Bourson, P. Characterization of Ti: $LiNbO_3$ waveguides by micro-raman and luminescence spectroscopy. *Appl. Phys. A* **2004**, *78*, 355–361. [CrossRef]
54. Fontana, M.D.; Hammoum, R.; Bourson, P.; Margueron, S.; Shur, V. Raman Probe on PPLN Microstructures. *Ferroelectrics* **2008**, *373*, 26–31. [CrossRef]
55. Yang, X.; Lan, G.; Li, B.; Wang, H. Raman Spectra and Directional Dispersion in $LiNbO_3$ and $LiTaO_3$. *Phys. Status Solidi (b)* **1987**, *142*, 287–300. [CrossRef]
56. Reitzig, S.; Rüsing, M.; Kirbus, B.; Gössel, J.; Singh, E.; Eng, L.M.; Zhao, J.; Mookherjea, S. μ-Raman Investigations of Periodically-Poled X-Cut Thin-Film Lithium Niobate for Integrated Optics. In Proceedings of the 2020 Joint Conference of the IEEE International Frequency Control Symposium and International Symposium on Applications of Ferroelectrics (IFCS-ISAF), Keystone, CO, USA, 19–23 July 2020; pp. 1–4. [CrossRef]
57. Caciuc, V.; Postnikov, A.V.; Borstel, G. Ab initio structure and zone-center phonons in $LiNbO_3$. *Phys. Rev. B* **2000**, *61*, 8806. [CrossRef]
58. Rüsing, M.; Sanna, S.; Neufeld, S.; Berth, G.; Schmidt, W.G.; Zrenner, A.; Yu, H.; Wang, Y.; Zhang, H. Vibrational properties of$LiNb_{1-x}Ta_xO_3$ mixed crystals. *Phys. Rev. B* **2016**, *93*, 184305. [CrossRef]
59. Gopalan, V.; Mitchell, T.E.; Furukawa, Y.; Kitamura, K. The role of nonstoichiometry in 180° domain switching of $LiNbO_3$ crystals. *Appl. Phys. Lett.* **1998**, *72*, 1981–1983. [CrossRef]
60. Sanna, S.; Neufeld, S.; Rüsing, M.; Berth, G.; Zrenner, A.; Schmidt, W.G. Raman scattering efficiency in$LiTaO_3$and$LiNbO_3$ crystals. *Phys. Rev. B* **2015**, *91*, 8806. [CrossRef]

crystals

MDPI

Article

Design, Simulation, and Analysis of Optical Microring Resonators in Lithium Tantalate on Insulator

Siyang Yao [1], Huangpu Han [2,3], Shangen Jiang [1], Bingxi Xiang [1,*], Guangyue Chai [1] and Shuangchen Ruan [1]

1 College of New Materials and New Energies, Shenzhen Technology University, Shenzhen 518118, China; 2017323004@stumail.sztu.edu.cn (S.Y.); 2017323002@stumail.sztu.edu.cn (S.J.); chaiguangyue@sztu.edu.cn (G.C.); scruan@sztu.edu.cn (S.R.)
2 School of Mechanical Engineering, Zibo Vocational Institute, Zibo 255314, China; pupuhan@126.com
3 School of Physics, Shandong University, Jinan 250100, China
* Correspondence: xiangbingxi@sztu.edu.cn; Tel.: +86-135-3085-0236

Abstract: In this paper we design, simulate, and analyze single-mode microring resonators in thin films of z-cut lithium tantalate. They operate at wavelengths that are approximately equal to 1.55 μm. The single-mode conditions and transmission losses of lithium tantalate waveguides are simulated for different geometric parameters and silica thicknesses. An analysis is presented on the quality factor and free spectral range of the microring resonators in lithium tantalate at contrasting radii and gap sizes. The electro-optical modulation performance is analyzed for microring resonators with a radius of 20 μm. Since they have important practical applications, the filtering characteristics of the microring resonators that contain two straight waveguides are analyzed. This work enhances the knowledge of lithium tantalate microring structures and offers guidance on the salient parameters for the fabrication of highly efficient multifunctional photonic integrated devices, such as tunable filters and modulators.

Keywords: microring resonator; varFDTD; lithium tantalate thin film; electro-optical devices

Citation: Yao, S.; Han, H.; Jiang, S.; Xiang, B.; Chai, G.; Ruan, S. Design, Simulation, and Analysis of Optical Microring Resonators in Lithium Tantalate on Insulator. *Crystals* **2021**, *11*, 480. https://doi.org/10.3390/cryst11050480

Academic Editors: Gábor Corradi and Alessandro Chiasera

Received: 18 March 2021
Accepted: 24 April 2021
Published: 25 April 2021

1. Introduction

An essential component in the construction of high-density photonic integrated circuits is an optical microring resonator. This is due to its compact size, simple structure and outstanding wavelength-selective properties [1–3]. Such a component consists of a bus waveguide that couples to a micrometer-size ring resonator via an evanescent field. The characteristic frequency spectrum of the microring, which is dependent upon its size, is a defining factor in the transmission of a selected wavelength of light into another waveguide. The microring resonators are widely employed in the design and manufacture of optical filters [4–6], modulators [7,8], optical switches [9,10], optical delay lines [11], and Kerr frequency combs [12–15], etc. Multiple material systems have been used to fabricate microring resonators, such as silicon on insulator (SOI) [1,16–18], silicon nitride (SiN) [5,19], and lithium niobate on insulator (LNOI) [20–23].

As a material in microring resonators, lithium tantalate ($LiTaO_3$ or simply LT) crystals have much promise because of their superior electro-optical (EO, γ_{33} = 27.4 pm/V), nonlinear optical, ferroelectric, and piezoelectric properties [24–28]. Moreover, compared with $LiNbO_3$, LT crystals have a higher optical damage threshold (with a laser radiation induced damage of 240 MW/cm^2) [29]. This enables the crystal to be used in integrated photonic chips, especially in high input power fields, such as broadband electro-optic frequency comb generation. In fact, LT crystals are already widely employed in integrated photonics and surface acoustic wave (SAW) devices [26–29]. In recent years, lithium tantalate on insulator (LTOI) is increasingly favored as the material of choice for integrated electro-optic and SAW devices [30–32]. This is because of its high-index contrast—which leads to a robust light guidance and a high-performance integrated device with a small footprint—that is particularly suitable for microring resonators. Due to formation of a high refractive index contrast between the ring core and the surrounding materials, a small radius microring and a large free spectral range (FSR) are possible. LTOI also offers the opportunity of electronically controlling the transmission spectrum, via the efficient EO effects of LT crystals that can enable extreme compactness and ultra-fast switching and modulation. Moreover, the integration of lithium tantalate thin films onto silicon substrates enables LTOI to be compatible with a complementary metal-oxide-semiconductor (CMOS); this could lead to significantly decreased research costs and an extensive increase in production [30]. However, to our knowledge, microring resonators in LTOI have not been reported elsewhere in the literature.

In this paper, we present the design, simulation and analysis of a single-mode microring resonator that is based on LTOI. By use of a full-vectorial finite difference method, we offer a study of the single-mode conditions and transmission losses of LT waveguides that contains different geometric parameters and SiO_2 cladding layer thicknesses. In order to obtain improved wavelength filtering effects, both the radius of the microring (R) and the size of the gap between the ring and linear waveguides have optimized values in the 2.5-dimension variational finite-difference time-domain (varFDTD) using Lumerical's MODE solutions software. Such parameters have direct effects on the quality factor (known as the Q factor) and FSR. The electro-optic modulated microring resonators, which are significant in practical applications, are also described.

2. Device Description

From top to bottom, the LTOI presented in this work contains a z-cut LT thin film, a SiO_2 cladding layer, and a Si substrate. The research center of NANOLN Corporation, for example, could easily perform such a fabrication. The schematic for this microring resonator is shown in Figure 1. The optical device consists of a ring resonator coupled to a straight waveguide channel, both of which are made from the z-cut LT thin film. If the optical signal is resonant within the ring, a coupling occurs from the channel into the cavity. To calculate the single-mode conditions, and the propagation losses of the LT waveguides, a full-vectorial finite-difference method is employed. Computation of the Q factor and FSR of the simulation is achieved with a varFDTD method that uses perfectly matched layers (PML) boundary conditions [33]. The varFDTD is the direct space and the time solution for solving Maxwell's equations in complex geometrical shapes. It is based on collapsing a 3D geometry into a 2D effective indices set, which allows 2D FDTD to be used to solve this problem. By performing Fourier transforms, results such as normalized transmission, far field projection, and Poynting vector can be obtained [23].

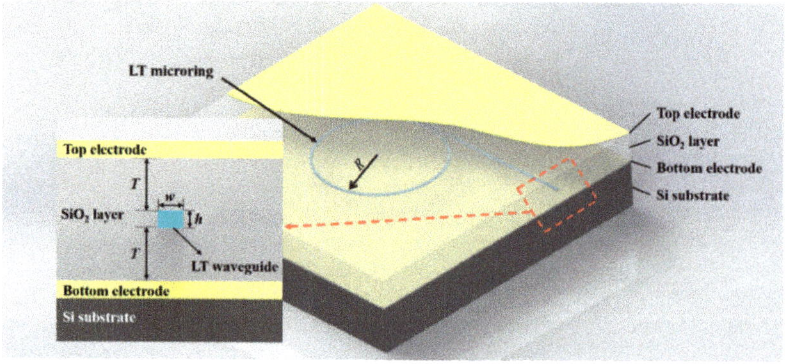

Figure 1. Schematic showing the waveguide-coupled microring resonator on LTOI. Inset: cross-section of the LT waveguide structure.

3. Results and Discussion

To prevent signal distortion during the transmission, single-mode conditions within the waveguide are required. To achieve this necessity, a suitable waveguide thickness (h) and width (w) are needed. The refractive indexes for the LiTaO3, the SiO2 cladding layer, and the Si substrate at the simulation wavelength of 1.55 μm are shown in Table 1. As shown in Figure 2a, via a simulation we determine the curve of the effective refractive index on changing the waveguide thickness (using a waveguide width of 0.7 μm and operating at the wavelength λ = 1.55 μm). The single-mode conditions of the transverse electric (TE) and transverse magnetic (TM) modes are 0.77 μm and 0.78 μm, respectively. The thickness of the LT film is set to 0.5 μm to ensure the single-mode condition in the simulation.

Table 1. The refractive indexes for $LiTaO_3$, SiO_2, and Si at λ = 1.55 μm.

Material	$LiTaO_3$	SiO_2	Si
Refractive index	$n_o = 2.1189$ $n_e = 2.1228$	1.46	3.47

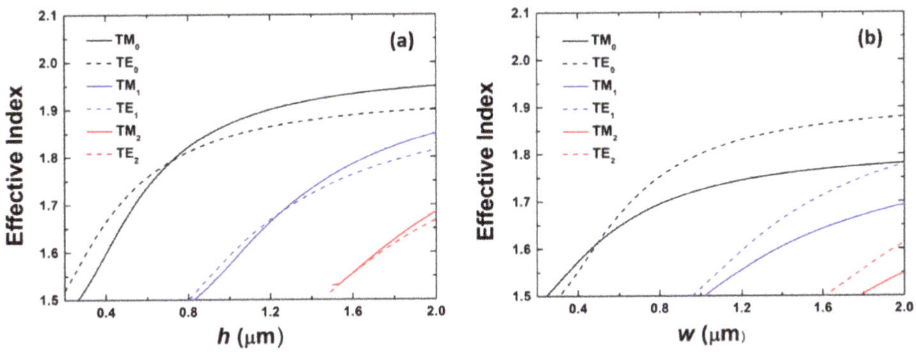

Figure 2. Graphs showing the effective index of the TE and TM light modes as a function of (**a**) the film thickness with a waveguide width of 0.7 μm and (**b**) the waveguide width with a film thickness of 0.5 μm in the LT waveguide. Both are calculated at λ = 1.55 μm.

In addition, the waveguide width also needs to be optimized. On setting the waveguide thickness to its optimal value of 0.5 μm, we calculate the effective refractive index as a function of the LT ridge waveguide width—as shown in Figure 2b. For the LT ridge waveguide, the first order TE and TM modes first appear for the LT thicknesses at 0.92 μm and 0.97 μm, respectively. To guarantee the single-mode condition, a waveguide width of 0.7 μm is chosen. This is because only the TM modes can employ the coefficient γ_{33} in a z-cut LT crystal, the TM modes are calculated in the following calculation.

Since the thickness of the SiO_2 layer (T) affects the transmission losses of the LT waveguides, we simulate these losses for different layer thicknesses. The results are illustrated in Figure 3. They show that the transmission losses decrease with increasing thickness of the SiO_2 layer. The Au films in the LT-SiO_2-Au structure are used as the electrodes in the electro-optic modulator. At the same thickness of the SiO_2 layer, the losses for a LT-SiO_2-Au structure (when the Au film thickness is 0.3 μm) are less than those of LT-SiO_2-Si. This is because electromagnetic fields are more likely to leak into the Si substrate (during the light propagation) since the refractive index of silicon (n_{Si}) is the highest among the relevant materials. Therefore, inclusion of the gold film enables greater isolation of the fields from the surroundings and, thus, the leakage of light is more preventable. In addition, the waveguide light mode has contact with the Au films during its propagation, which enables the excitation of the surface plasmons, and increases the absorption losses. But in comparison to the leakage loss, the absorption loss is much lower.

When the thickness of the SiO_2 layer is greater than 2 μm, the losses from the LT waveguide are negligible. Therefore, the optimal value for the SiO_2 layer thickness is set at 2 μm.

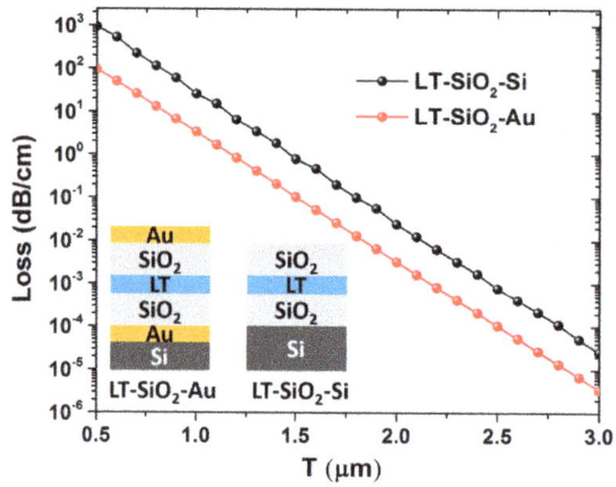

Figure 3. Graph showing the transmission loss of the LT waveguides with different thicknesses of the SiO_2 layer for TM light mode.

For our purposes, the key parameters of the optical device are the radius of the microring and the gap between the ring and linear waveguides. These factors are used to determine the Q factor and FSR of the microring resonators. We calculate these parameters for differing microring radii and gap sizes, while operating at around $\lambda = 1.55$ μm. As shown in Figure 4a, when the microring radius is lower than 10 μm, the Q factor becomes rapidly enhanced with increasing radius size. However, the Q factor value is essentially unchanged when the radius is greater than 20 μm. The Q factor is increased with larger gaps. The propagation losses of the microrings are caused by the following factors: (1) the electromagnetic fields leak into the Si substrate when the light propagates through the resonators (but lower than 2.5×10^{-2} dB/cm with a SiO_2 layer thickness of 2 μm), and (2) the radiative losses within the curved waveguide (which is roughly 8×10^{-2} dB/cm with a bending radius of 20 μm). Additionally, in practice, the structures will contain a residual roughness for the etching surface of the waveguide which causes scattering losses. When the radius is greater than 20 μm, the Q-factor may decrease as the radius increases due to the existence of scattering loss. As shown in Figure 4b, the FSR for TM mode reach a maximum at a radius of 5 μm and then it decreases with increasing ring radius. Figure 5 shows the transmission spectrum of a microring with a radius of 20 μm for the TM mode.

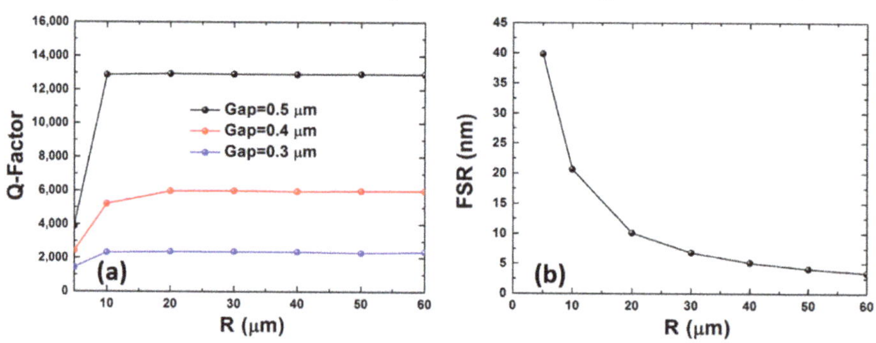

Figure 4. Graphs for (**a**) the Q-factor of the microring resonators at different ring radii and gap sizes and (**b**) the FSR at different microring radii for a TM light mode at around $\lambda = 1.55$ μm.

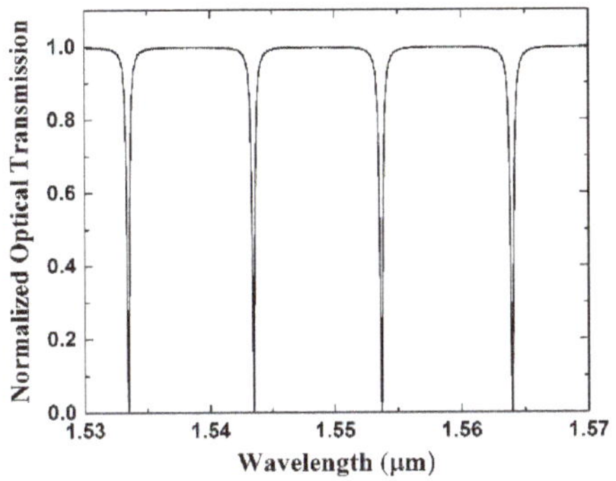

Figure 5. Transmission spectrum of the microring with radius 20 μm for TM light mode.

The characterization of the electronic tuning of the optical resonances in the microring resonators is determined using optical transmission simulations. As we have seen, the schematic of the electrode structure for the microring resonator is shown in Figure 1. This depicts an LT microring resonator that is embedded within a SiO_2 layer, with the electrodes positioned either side (above and below) the SiO_2 layer. The tuning range for the microring resonators is highly dependent upon the strength of the EO effects. The result of a simulation for the optical field is shown in Figure 6. Here, the LT waveguide has a thickness of 0.5 μm, a width of 0.7 μm, and a SiO_2 layer thickness of 2 μm. As seen in Figure 6, most of the optical power (TM mode) is confined within the electro-optic active material, i.e., the LT core. Electrodes can be designed, and positioned close to the waveguides, that circumvent substantially larger optical transmission losses.

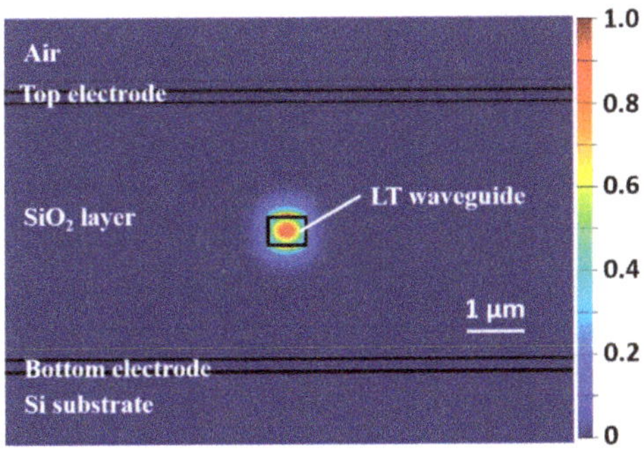

Figure 6. Diagram showing the optical field inside the waveguide.

The variation of extraordinary refractive index of LT (Δn_e) after application of an electrostatic field is expressed as:

$$\Delta n_e = -\frac{1}{2} n_e^3 \gamma_{33} E_z \tag{1}$$

where n_e is the extraordinary refractive index of LT. When the electrostatic field intensity, E_z, is 1 V/μm, the variation in the refractive index at λ = 1.55 μm is $\Delta n_e = 1.3 \times 10^{-4}$.

Observation of the EO effect requires an electric field to be applied between the electrodes. Figure 7 shows the EO modulation characteristic spectra at different DC voltages (using a TM mode) for such a situation. A greater resonance wavelength shift is exhibited for increasing electric field strengths. In detail, the resonance wavelength displacement is 80 pm, 160 pm, 240 pm, and 320 pm when the electric field intensity is set at 1 V/μm, 2 V/μm, 3 V/μm, and 4 V/μm, respectively.

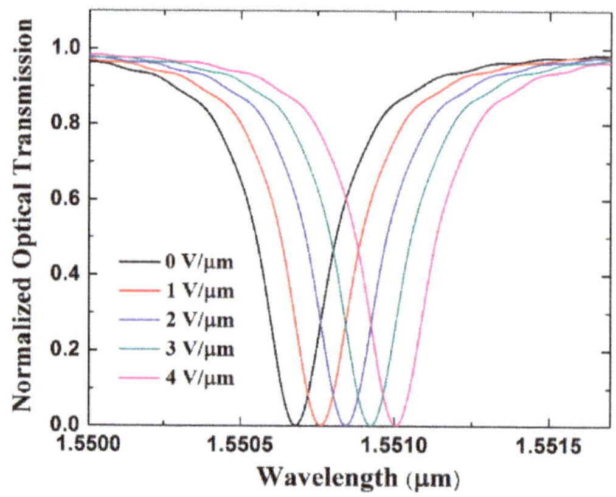

Figure 7. Graph showing the wavelength shift of the transmission spectrum at different electric field intensities. In this case, a TM mode and a microring with a 20 μm radius are used.

Tunable filtering is an important application for the LT microring because it is an attractive candidate for a narrower bandwidth integrated optical filter for the wavelength division multiplexing (WDM) systems. Figure 8 shows the transmission spectra for the drop port of the LT microring filters, at different electric field intensities, with a double straight waveguide structure. The structure of the microring filter is shown in the inset of Figure 8. The resonance wavelength shift is 320 pm in an electric field intensity of 4 V/μm.

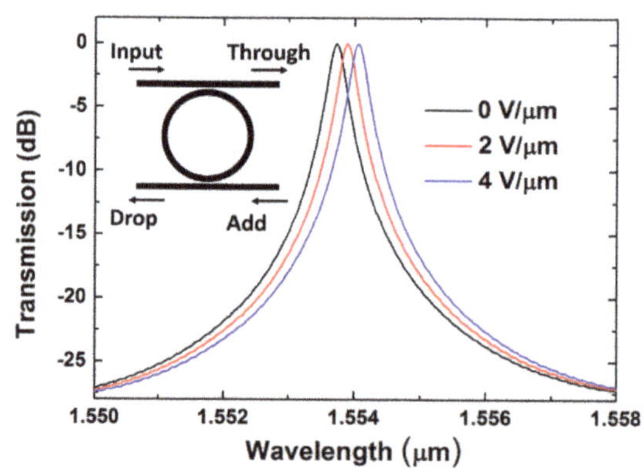

Figure 8. Graph showing the transmission spectra for the drop port of the LT microring at different electric field intensities.

Figure 9 shows the electrostatic field strength when a voltage of 1V is applied to the electrodes. As we can see, the attainable electric field in the LT film is relatively weak; this is because the adjacent cladding materials exhibit a dielectric constant, ε, that is one order of magnitude smaller, i.e., ε_{SiO2} = 3.9 and ε_{LT} = 42.8 [34]. Hence, the electric field strength in the LT thin film is significantly smaller than the SiO_2 layer. The electrostatic field intensity in the *z*-direction (E_z) is 0.033 V/μm when close to the center of the waveguide, for cases when a voltage of 1V is applied to the electrodes (for which h = 0.5 μm, w = 0.7 μm, and T = 2 μm). The problem of the relatively small E_z field in the LT waveguides can be improved by changing the geometry of the electrodes or by using a x- or y-cut configuration. For example, using an incompletely etched waveguide structure (leaving a certain thickness of slab across the chip) on the x- or y-cut LTOI enables a relatively strong electric field strength in the LT waveguide [35,36]. However, compared with the microrings on the z-cut LTOI, the microrings on the x- or y-cut LTOI can apply an electric field to only part of the microring, so the length of the effective electro-optic modulation is shorter [15,35,36]. Assuming that there is a uniform electric field in the waveguide, and the electric field strength is equal to that at the center of the waveguide. The EO tunability of LTOI microring is about 2.6 pm/V in this work. Table 2 reports the comparison of the results of different types for $LiNbO_3$ and $LiTaO_3$ tunable microrings. As we can infer from this table, LTOI photonics provides a promising approach for tunable microring resonators. The geometry of the electrodes, and the cavity-photon lifetime of the resonator, will affect the modulation rate of the microring [35,37,38]. Electrode design is especially crucial for modulators, especially for high bit rate modulation formats. For example, a top electrode with a ring structure can achieve a relatively high modulation rate [37]. A further study of the influence of the electrode geometry on the modulation rate will be provided in future work.

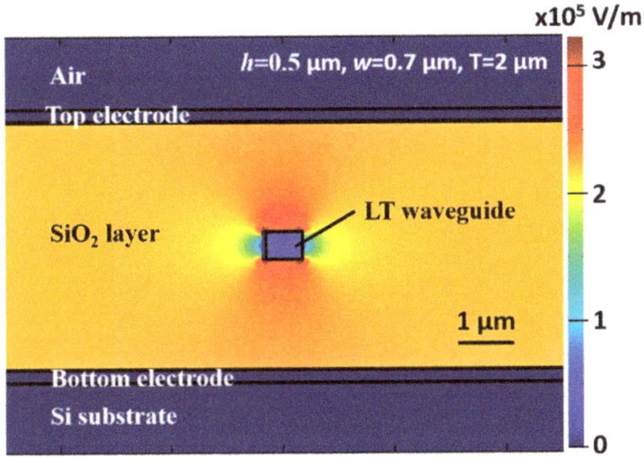

Figure 9. Diagram showing the electrostatic field strength when a voltage is applied to the electrodes.

Table 2. Comparison between the different types of $LiNbO_3$ and $LiTaO_3$ tunable microrings.

Material	Mode	EO Tuning (pm/V)	Ref
x-cut LNOI	TE	2.4	[15]
x-cut LNOI	TE	7	[35]
y-cut LNOI	TE	0.32	[36]
z-cut LNOI	TM	3	[39]
z-cut LNOI	TM	1.05	[20]
z-cut LNOI	TM	2.15	[40]
z-cut LTOI	TM	2.6	This work

4. Conclusions

In this paper we design, simulate, and analyze a single-mode microring resonator that is based on LTOI. The waveguide width and the LT film thickness are optimized to ensure single-mode conditions. The propagation losses of the LT waveguides at different SiO_2 layer thickness are analyzed and discussed. The effects on the Q factor and the FSR that arise due a change in the ring radius and the gap size between the ring and linear waveguides are quantified and then discussed. It is also determined that the Q factor increases when the ring radius increases, the Q factor stabilizes when the radius is greater than 20 μm, and FSR decreases with increasing radius. We simulate and discuss the electro-optic tunable microring resonator, which is a highly significant topic because of its practical applications. For example, an EO modulation spectrum of the microring (with a gap = 0.4 μm and a radius = 20 μm) is produced that is displaced by 80 pm when an electric field intensity of 1 V/μm is applied. We expect that our results and analysis will provide useful guidance to laboratory works on microring resonator in LTOI.

Author Contributions: Conceptualization, B.X.; methodology, B.X.; validation, S.Y. and S.J.; investigation, S.Y., H.H. and B.X.; data curation, S.Y.; writing—original draft preparation, S.Y.; writing—review and editing, B.X., G.C., and S.R.; supervision, B.X.; funding acquisition, B.X. All authors have read and agreed to the published version of the manuscript.

Funding: This work was supported by the Shenzhen Science and Technology Planning (NO. JCYJ20190813103207106), the Project of Youth Innovative Talents in Higher Education Institutions of Guangdong (NO. 2018KQNCX399), and the National Natural Science Foundation of China (NO. 61935014).

Conflicts of Interest: The authors declare no conflict of interest.

References

1. Bogaerts, W.; De Heyn, P.; Van Vaerenbergh, T.; De Vos, K.; Selvaraja, S.K.; Claes, T.; Dumon, P.; Bienstman, P.; Van Thourhout, D.; Baets, R. Silicon microring resonators. *Laser Photonics Rev.* **2012**, *6*, 47–73. [CrossRef]
2. Vahala, K.J. Optical microcavities. *Nature* **2003**, *424*, 839–846. [CrossRef] [PubMed]
3. Han, H.; Xiang, B. Simulation and analysis of electro-optic tunable microring resonators in silicon thin film on lithium niobate. *Sic. Rep.* **2019**, *9*, 6302. [CrossRef] [PubMed]
4. Mak, J.C.C.; Sacher, W.D.; Xue, T.Y.; Mikkelsen, J.C.; Yong, Z.; Poon, J.K.S. Automatic resonance alignment of high-order microring filters. *IEEE J. Quantum. Elect.* **2015**, *51*, 1–11. [CrossRef]
5. Barwicz, T.; Popovic, M.A.; Rakich, P.T.; Watts, M.R.; Haus, H.A.; Ippen, E.P.; Smith, H.I. Microring-resonator-based add-drop filters in SiN: Fabrication and analysis. *Opt. Express* **2004**, *12*, 1437–1442. [CrossRef]
6. Chen, G.P.; Jiang, C. Reverse design of microring resonator channel dropping filters. *Results Phys.* **2020**, *19*, 103380. [CrossRef]
7. Xu, Q.F.; Manipatruni, S.; Schmidt, B.; Shakya, J.; Lipson, M. 12.5 Gbit/s carrier-injection-based silicon micro-ring silicon modulators. *Opt. Express* **2007**, *15*, 430–436. [CrossRef]
8. Cao, W.; Hagan, D.; Thomson, D.J.; Nedeljkovic, M.; Littlejohns, C.G.; Knights, A.; Alam, S.U.; Wang, J.J.; Gardes, F.; Zhang, W.W.; et al. High-speed silicon modulators for the 2 μm wavelength band. *Optica* **2018**, *5*, 1055–1062. [CrossRef]
9. Wang, Y.H.; Lv, P.; Zhang, Y.L.; Song, M.X.; Liu, C.L.; Wang, G.F.; Wang, C.X.; Qin, Z.K. Analysis of characteristics of a parallel channel microring resonator electro-optic switch array. *Optik* **2018**, *165*, 332–340. [CrossRef]
10. Emelett, S.J.; Soref, R. Design and simulation of silicon microring optical routing switches. *J. Lightwave Technol.* **2005**, *23*, 1800–1807. [CrossRef]
11. Cardenas, J.; Foster, M.A.; Sherwood-Droz, N.; Poitras, C.B.; Lira, H.L.R.; Zhang, B.B.; Gaeta, A.L.; Khurgin, J.B.; Morton, P.; Lipson, M. Wide-bandwidth continuously tunable optical delay line using silicon microring resonators. *Opt. Express* **2010**, *18*, 26525–26534. [CrossRef] [PubMed]
12. Zhang, M.; Buscaino, B.; Wang, C.; Shams-Ansari, A.; Reimer, C.; Zhu, R.R.; Kahn, J.M.; Loncar, M. Broadband electro-optic frequency comb generation in a lithium niobate microring resonator. *Nature* **2019**, *568*, 373–377. [CrossRef] [PubMed]
13. Kovach, A.; Chen, D.Y.; He, J.H.; Choi, H.; Dogan, A.H.; Ghasemkhani, M.; Taheri, H.; Armani, A.M. Emerging material systems for integrated optical Kerr frequency combs. *Adv. Opt. Photonics* **2020**, *12*, 135–222. [CrossRef]
14. Fujii, S.; Tanabe, T. Dispersion engineering and measurement of whispering gallery mode microresonator for Kerr frequency comb generation. *Nanophotonics* **2020**, *9*, 1087–1104. [CrossRef]
15. Wang, C.; Zhang, M.; Yu, M.J.; Zhu, R.R.; Hu, H.; Loncar, M. Monolithic lithium niobate photonic circuits for Kerr frequency comb generation and modulation. *Nat. Commun.* **2019**, *10*, 978. [CrossRef]
16. Reed, G.T.; Mashanovich, G.; Gardes, F.Y.; Thomson, D.J. Silicon optical modulators. *Nat. Photonics* **2010**, *4*, 518–526. [CrossRef]
17. Xu, Q.F.; Fattal, D.; Beausoleil, R.G. Silicon microring resonators with 1.5-μm radius. *Opt. Express* **2008**, *16*, 4309–4315. [CrossRef]
18. Mi, G.C.; Horvath, C.; Aktary, M.; Van, V. Silicon microring refractometric sensor for atmospheric CO_2 gas monitoring. *Opt. Express* **2016**, *24*, 1773–1780. [CrossRef] [PubMed]
19. Zamora, V.; Lutzow, P.; Weiland, M.; Pergande, D. Investigation of cascaded SiN microring resonators at 1.3 μm and 1.5 μm. *Opt. Express* **2013**, *21*, 27550–27557. [CrossRef]
20. Guarino, A.; Poberaj, G.; Rezzonico, D.; Degl'Innocenti, R.; Gunter, P. Electro-optically tunable microring resonators in lithium niobate. *Nat. Photonics* **2007**, *1*, 407–410. [CrossRef]
21. Lu, J.J.; Surya, J.B.; Liu, X.W.; Bruch, A.W.; Gong, Z.; Xu, Y.T.; Tang, H.X. Periodically poled thin-film lithium niobate microring resonators with a second-harmonic generation efficiency of 250,000%/W. *Optica* **2019**, *6*, 1455–1460. [CrossRef]
22. Yu, M.J.; Okawachi, Y.; Cheng, R.; Wang, C.; Zhang, M.; Gaeta, A.L.; Loncar, M. Raman lasing and soliton mode-locking in lithium niobate microresonators. *Light. Sci. Appl.* **2020**, *9*, 9. [CrossRef]

23. Han, H.P.; Xiang, B.X.; Zhang, J.L. Simulation and analysis of single-mode microring resonators in lithium niobate thin films. *Crystals* **2018**, *8*, 342. [CrossRef]
24. Casson, J.L.; Gahagan, K.T.; Scrymgeour, D.A.; Jain, R.K.; Robinson, J.M.; Gopalan, V.; Sander, R.K. Electro-optic coefficients of lithium tantalate at near-infrared wavelengths. *J. Opt. Soc. Am. B* **2004**, *21*, 1948–1952. [CrossRef]
25. Barie, N.; Wessa, T.; Bruns, M.; Rapp, M. Love waves in SiO_2 layers on STW-resonators based on $LiTaO_3$. *Talanta* **2004**, *62*, 71–79. [CrossRef]
26. Lobino, M.; Marshall, G.D.; Xiong, C.; Clark, A.S.; Bonneau, D.; Natarajan, C.M.; Tanner, M.G.; Hadfield, R.H.; Dorenbos, S.N.; Zijlstra, T.; et al. Correlated photon-pair generation in a periodically poled MgO doped stoichiometric lithium tantalate reverse proton exchanged waveguide. *Appl. Phys. Lett.* **2011**, *99*, 081110. [CrossRef]
27. Kawachi, O.; Mineyoshi, S.; Endoh, G.; Ueda, M.; Ikata, O.; Hashimoto, K.; Yamaguchi, M. Optimal cut for leaky SAW on $LiTaO_3$ for high performance resonators and filters. *IEEE Trans. Ultrason. Ferroelectr. Freq. Control* **2001**, *48*, 1442–1448. [CrossRef] [PubMed]
28. Takai, T.; Iwamoto, H.; Takamine, Y.; Yamazaki, H.; Fuyutsume, T.; Kyoya, H.; Nakao, T.; Kando, H.; Hiramoto, M.; Toi, T.; et al. High performance SAW resonator on new mutilayered substrate using $LiTaO_3$ crystal. *IEEE Trans. Ultrason. Ferroelectr. Freq. Control* **2017**, *64*, 1382–1389. [CrossRef]
29. Yan, X.S.; Liu, Y.A.; Ge, L.C.; Zhu, B.; Wu, J.W.; Chen, Y.P.; Chen, X.F. High optical damage threshold on-chip lithium tantalate microdisk resonator. *Opt. Lett.* **2020**, *45*, 4100–4103. [CrossRef]
30. Yan, Y.Q.; Huang, K.; Zhou, H.Y.; Zhao, X.M.; Li, W.Q.; Li, Z.X.; Yi, A.L.; Huang, H.; Lin, J.J.; Zhang, S.B.; et al. Wafer-scale fabrication of 42° rotated Y-cut $LiTaO_3$-on-insulator (LTOI) substrate for a SAW resonator. *ACS Appl. Electron. Mater.* **2019**, *1*, 1660–1666. [CrossRef]
31. Tauzin, A.; Dechamp, J.; Madeira, F.; Mazen, F.; Zussy, M.; Deguet, C.; Clavelier, L.; Moulet, J.S.; Richtarch, C.; Akatsu, T.; et al. 3-inch single-crystal $LiTaO_3$ films onto metallic electrode using Smart Cut^{TM} technology. *Electron. Lett.* **2008**, *44*, 822–824. [CrossRef]
32. Ma, C.D.; Lu, F.; Xu, B.; Fan, R.R. Visualized strain profile in the process of crystal ion slicing of $LiTaO_3$. *J. Phys. D Appl. Phys.* **2016**, *49*, 205301. [CrossRef]
33. Berenger, J.P. A perfectly matched layer for the absorption of electromagnetic waves. *J. Comput. Phys.* **1994**, *114*, 185–200. [CrossRef]
34. Smith, R.T.; Welsh, F.S. Temperature dependence of the elastic, piezoelectric, and dielectric constants of lithium tantalate and lithium niobate. *J. Appl. Phys.* **1971**, *42*, 2219–2230. [CrossRef]
35. Wang, C.; Zhang, M.; Stern, B.; Lipson, M.; Lončar, M. Nanophotonic lithium niobate electro-optic modulators. *Opt. Express* **2018**, *26*, 1547–1555. [CrossRef] [PubMed]
36. Mahmoud, M.; Cai, L.; Bottenfield, C.; Piazza, G. Lithium niobate electro-optic racetrack modulator etched in Y-cut LNOI platform. *IEEE Photonics J.* **2018**, *10*, 6600410. [CrossRef]
37. Chen, L.; Xu, Q.; Wood, M.G.; Reano, R.M. Hybrid silicon and lithium niobate electro-optical ring modulator. *Optica* **2014**, *1*, 112–118. [CrossRef]
38. Chen, L.; Chen, J.; Nagy, J.; Reano, R.M. Highly linear ring modulator from hybrid silicon and lithium niobate. *Opt. Express* **2015**, *23*, 13255–13264. [CrossRef] [PubMed]
39. Krasnokutska, I.; Tambasco, J.L.J.; Peruzzo, A. Tunable large free spectral range microring resonators in lithium niobate on insulator. *Sci. Rep.* **2019**, *9*, 11086. [CrossRef]
40. Siew, S.Y.; Saha, S.S.; Tsang, M.; Danner, A.J. Rib Microring Resonators in Lithium Niobate on Insulator. *IEEE Photonics Technol. Lett.* **2016**, *28*, 573–576. [CrossRef]

MDPI
St. Alban-Anlage 66
4052 Basel
Switzerland
Tel. +41 61 683 77 34
Fax +41 61 302 89 18
www.mdpi.com

Crystals Editorial Office
E-mail: crystals@mdpi.com
www.mdpi.com/journal/crystals

www.ingramcontent.com/pod-product-compliance
Lightning Source LLC
Chambersburg PA
CBHW061554220926
51850CB00005B/334